SELECTED SCIENTIFIC

WORKS OF

Hans Christian Ørsted

SELECTED SCIENTIFIC

WORKS OF

Hans Christian Ørsted

TRANSLATED AND EDITED BY

KAREN JELVED, ANDREW D. JACKSON,

AND

OLE KNUDSEN

WITH AN INTRODUCTION BY

ANDREW D. WILSON

PRINCETON UNIVERSITY PRESS

PRINCETON, NEW JERSEY

Ørsted, Hans Christian, 1777–1851.
[Selections. English. 1997]
Selected scientific works of Hans Christian Ørsted / translated and edited by
Karen Jelved, Andrew D. Jackson, and Ole Knudsen ;
with an introduction by Andrew D. Wilson.
p.   cm.
Includes bibliographical references and indexes.
ISBN 0-691-04334-5 (cloth : alk. paper)
1. Physics.   I. Jelved, Karen, 1944–   .   II. Jackson, Andrew D., 1941–   .
III. Knudsen, Ole, 1939–   .   IV. Title.
QC21.2.O7713   1997
530—dc21   97-6179

This book has been composed in Times Roman
by Wilsted & Taylor Publishing Services

Princeton University Press books are printed on
acid-free paper and meet the guidelines for permanence
and durability of the Committee on
Production Guidelines for Book Longevity of the
Council on Library Resources

http://pup.princeton.edu

Printed in the United States of America

2 4 6 8 10 9 7 5 3 1

TABLE OF CONTENTS

PREFACE

THE SEEDS of this project are to be found in a lecture entitled "The Ørsted Invisibility," which Gerald Holton delivered to the American Physical Society some years ago.[1] In the summer of 1993, Holton visited Gjorslev Castle in the county of Stevns south of Copenhagen, where he chanced upon a slim volume by Ørsted in the library. In a subsequent discussion with Yoyo Tesdorpf Jones, who grew up at Gjorslev, he again noted the particular neglect of Ørsted by historians and expressed his conviction that it would be useful if an English edition of Ørsted's scientific works were available. Yoyo Jones has boundless energy and a passion for introducing others to the very best of Denmark. The Ørsted Project was born.

Yoyo consulted with another old friend, Abraham Pais, for suggestions as to who might undertake such a translation. Pais suggested that we might be interested. We had the necessary language skills and scientific knowledge between us, and Bram knew that we were looking for a collaborative project. Lacking expertise in the scientific literature of the nineteenth century, we were aware that we could not carry this project alone. A card-carrying historian of science was required. We contacted Professor Olaf Pedersen at the University of Aarhus, who gave the project his strong support though he did not feel that he had the time to be an active part of it. Instead, he led us to Ole Knudsen, and discussions with Ole convinced us that he was precisely the historian we needed. He became an instant old friend with his thoroughly professional approach and sly wit. We agreed on the distribution of tasks. Karen Jelved would be responsible for actually producing the translation with constant (incessant) advice from Andrew Jackson, who would also serve as an advisor on matters of science. Ole Knudsen would perform the selection of texts, provide advice on matters of the history of science, and cast a critical eye on the translation as it was produced. Yoyo Jones would oversee the financial side of the project once she found funds to administer.

We also agreed that our starting point should be the three-volume edition of Ørsted's collected scientific works from 1920. This edition, published under the auspices of the Royal Danish Academy of Sciences and Letters, was prepared almost single-handedly by Dr. Kirstine Meyer to mark the centenary of Ørsted's discovery of electromagnetism. While her efforts received the nominal support of Niels Bohr, the magnitude of her task and the scholarly value of her edition have been largely unrecognized and underappreciated. In addition to presenting extremely clean versions of Ørsted's scientific writings in their original languages, Meyer provided a long biographical essay (in English), which remains a useful and almost unique primer to the scientific Ørsted.

---

[1] G. Holton, *The Advancement of Science, and Its Burdens*. Cambridge and New York: Cambridge University Press, 1986, pp. 197–201.

Ørsted, like many present-day scientists, was given to publishing the same idea in more than one place. Thus, we decided upon a selection of approximately eighty of Ørsted's scientific writings of greatly varying length, ranging from 6 lines to 122 pages. The selection was made by Ole Knudsen. The works presented here, which comprise roughly two-thirds of Ørsted's total scientific production, are intended to be representative without being too repetitive. Some topics, such as his long treatise on the geology of the island of Bornholm, have been omitted as being of limited interest to a non-Danish audience. Other, more central topics have been treated somewhat more thoroughly. For example, we have chosen to present both the Danish-language version of "Fundamentals of the Metaphysics of Nature" (Chapter 6) and the closely related Latin text "On the Structure of the Elementary Metaphysics of External Nature" (Chapter 7). These two works, inspired by Kant's *Metaphysische Anfangsgründe der Naturwissenschaft*, reveal subtly different aspects of Ørsted's philosophical thinking which would seem to be essential for understanding the context of his discovery of electromagnetism two decades later. Similarly, we have included both a translation of the Latin original of his most significant paper entitled "Experiments on the Effect of the Electric Conflict on the Magnetic Needle" (Chapter 39) and the contemporary English version (Chapter 39a, presumably prepared by Ørsted), which appeared in Thomson's *Annals of Philosophy*.

Once the texts had been selected, all that was missing was a publisher and funding. A proposal was soon produced, and letters in support of the proposal were provided by Holton, Pais, and Chen Ning Yang. Gerald Holton's long-standing interest in Ørsted has already been noted. Abraham Pais turned to the history of science and scientific biography after a long and distinguished career as a theoretical physicist. His life of Einstein, *Subtle Is the Lord*, is one of the finest scientific biographies ever written. Chen Ning Yang's contributions to gauge theories in our century have defined modern thinking in elementary particle physics. Consequently, he has a deep interest in the very first gauge theory (electromagnetism), its discoveries, and its discoverers. The enthusiastic initial endorsements and the continuing support of these three distinguished scholars are gratefully acknowledged.

Armed with proposal and endorsements, we began to look for a publisher. We quickly settled on Princeton University Press. Yoyo Jones went in search of funding, and she met with success. This translation has been funded completely by the Villum Kann Rasmussen Fund. The enthusiasm of Aino Kann Rasmussen and the generosity of the Villum Kann Rasmussen Fund literally made this edition possible.

With a publisher in tow and support arranged, there was nothing left but the translation itself. That required a copy of Kirstine Meyer's edition. Fortunately, the Royal Danish Academy still had copies, and Erik Dal and Thor A. Bak, the president and the secretary of the Academy, readily supplied us with two of them. They emerged from the archives still wrapped in proof sheets for the 1920 Danish Railway timetable. The unbound edition proved fragile; our working copy was quickly reduced to a single-page system.

The original texts included in our selection of Ørsted's scientific works were written in one of five different languages: Danish, German, French, English, or Latin.[2] They reflect the importance for a scholar speaking an "unknown" language like Danish to master other tongues if he wanted to participate in the scientific dialogue. Ørsted came to all of these languages early in life. As children, Hans Christian and his brother, Anders Sandøe, were cared for by a German wig-maker and his wife, who taught them to read German at an early age in order to keep them occupied. Rudkøbing was not a large town, and people soon recognized that the boys had talent. Many participated in their early education. The burgomaster gave one brother lessons in French and the other lessons in English. They both received what Ørsted later described as "indifferent" lessons in Latin.

Ørsted was proud of his linguistic skills. He was primarily responsible for constructing the modern Danish terminology in chemistry and even made suggestions for similar vocabularies in Swedish, German, and Dutch. Indeed, the translation of Ørsted's chemical vocabulary can be particularly difficult. While all of his chemical researches were performed after the publication of Lavoisier's *Traité élémentaire de Chimie* in 1789, Ørsted moved freely between the old and new chemical terminologies. The most useful English-language sources for contemporary chemical terminology include Robert Kerr's 1790 translation of Lavoisier, John Dalton's *A New System of Chemical Philosophy* from 1803, and Sir Humphry Davy's collected works. A useful reference for the older Danish chemical terminology can be found in Kurt Bærentsen's *De uorganiske kemiske monografier i Pharmacopoea Danica 1772*.[3] In his work entitled *The Incompleat Chymist*, Jon Eklund has provided a marvelously useful "essay on the eighteenth-century chemist in his laboratory, with a dictionary of obsolete chemical terms of the period."[4] This essay also contains a number of useful references.

Ørsted's linguistic innovations are easy to keep straight in Danish. The most successful of his inventions—from the Danish words for *oxygen* and *hydrogen* to the words for *butterfly* and *whale*—are either part of the language today or well documented in *Ordbog over det Danske Sprog*, the complete dictionary of the Danish language in twenty-seven volumes.[5] Ørsted's inventive and sometimes idiosyncratic use of language was not limited to Danish and can provide challenges for the translator. While his mastery of foreign languages was impressive, his skills were not perfect, as will be seen from the articles that he wrote in English. Wherever possible, we have chosen to reproduce those English-language versions of Ørsted's articles that were "communicated by the author." These are given in their original form with no attempt at linguistic improvements. It should be noted that there are frequent differences between versions of the same article in different languages.

The articles that Ørsted wrote in English also serve as a stylistic model. They

[2] A footnote may be in order here, concerning the name of the author, which appears in slightly different versions (mostly *Oe*) in languages which do not have the Danish letter Ø. We have generally chosen to use the Danish spelling, Ørsted, regardless of the spelling of the original.

[3] We thank Henning Aldershvile for discovering Bærentsen and Kurt Bærentsen for kindly providing us with a copy of this dissertation.

[4] J. Eklund, Smithsonian Studies in History and Technology, No. 33, pp. 19–49, 1975.

[5] Our thanks to Birgit Christiansen for finding a copy of this work and having it delivered to our door.

provide an accurate sense of his voice and a fair representation of his often romantic and occasionally ponderous style. It has been our intention throughout to stay as
close as possible to Ørsted's style, language, and vocabulary. It is remarkable how
consistent his tone was and how independent of the language in which he chose to
write. Fortunately, his language also transmits a strong sense of the man, which we
have attempted to preserve.[6] Previously, it has been difficult to get a real taste of
Ørsted's language. This edition is the first to present him in a single language. It has
been our desire to translate, not to interpret, and with the exception of a few obvious writing or typographical errors, which we have emended in footnotes, we have
not seen it as our task to make Ørsted more easily understandable. We have also
taken some pains to avoid verbal anachronisms. If the text is occasionally not completely clear, we suggest that this is not necessarily the fault of the translator. It is
interesting to note that the more uncertain Ørsted is about his subject, the more
opaque his language becomes. When his mind is clear, so is his language.

Karen Jelved has translated all Danish and French texts. A preliminary translation of the German articles was performed by Martina Sonntag (who dealt with
Ørsted's two long chemical texts—*Materials for a Chemistry of the Nineteenth
Century* [Chapter 12] and *View of the Chemical Laws of Nature* [Chapter 30]) and
by Susan Adami (who handled the rest). The final versions of these German
texts—and the sole responsibility for their accuracy—are ours. The Latin evaluation of Ørsted's thesis at the beginning of "On the Origin and Use of Amniotic
Fluid" (Chapter 1) was kindly translated by Inger Grete Jacobsen. Ørsted's dissertation (Chapter 7) has been translated from the Latin by Fritz Saaby Pedersen,
Steen Ebbesen, and Russell L. Friedman, who also translated two brief Latin quo-
tations. In the case of the dissertation, we have made only minor suggestions aimed
at maintaining a homogeneous Ørsted voice. Of course, all the texts have been read
for language and content by Ole Knudsen.

This edition contains three kinds of footnotes. Some pertain to the text itself.
These are typically Ørsted's own and are left unmarked. Others were provided by
the editor(s) of the journal in which the paper was published and are marked as
such. Finally, there are the footnotes of the editors of this edition. They include our
comments and those of Kirstine Meyer from the 1920 edition of Ørsted's scientific
works. It should be noted that Meyer's footnotes have been included in this edition
only when we have been able to verify them. These (given in square brackets) include summaries of the publication history of Ørsted's scientific articles. The accuracy of such citations was one of the very few weak points of the 1920 edition.
We emphasize that all references to other Ørsted publications have been checked
by us. For the convenience of readers wishing to consult the Meyer edition, we have
included volume and page references marked "KM." In passing we note that,
while the entries in the Index of Names are also those of that edition, we have verified all information and made corrections as required.

---

[6] It is perhaps easy for modern readers to find irony or mockery in the linguistic excesses of romanticism. Ørsted was a man of extraordinary decency; there were no thorns hidden in his roses. When he describes a scholar as *skarpsindig*, he means "astute" or "perspicacious" with none of the edge of the dictionary alternative, "shrewd."

All the practical work of verifying information and checking potential errors involved considerable time spent in libraries in various parts of the world. We take this opportunity to thank the staffs of the Royal Library (Det Kongelige Bibliotek) and the Science Library (Danmarks Natur- og Lægevidenskabelige Bibliotek) in Copenhagen; the Frank Melville Jr. Memorial Library at the State University of New York at Stony Brook, whose interlibrary loan division was helpful in getting a number of copies for us; the Department of the History of Science at the University of Aarhus; and the Dibner Institute for the History of Science and Technology at the Massachusetts Institute of Technology. The original editions of most of the articles containing figures were found at the Science Library in Copenhagen, and we left the reproduction of these in the capable hands of their photographer, Bent Grøndahl. One of the more interesting tasks was associated with Ørsted's reports to the Royal Danish Academy. We note that none of these contributions from *Videnskabernes Selskabs Oversigter* were originally given titles. We have chosen to adopt the titles used by Kirstine Meyer. In order to provide the dates of the meetings (only some of which are given in Kirstine Meyer's edition), it was necessary to go to the original, handwritten protocols of the meetings themselves in the archives of the Royal Danish Academy. Pia Grüner's generous assistance and encouragement and her lively sense of humour made this an enjoyable undertaking.

While we do not wish to dwell on our decisions regarding the translation of individual words, two exceptions seem appropriate. Ørsted had a considerable number of terms at his disposal for the designation of physicists and their profession. Since honest people can disagree about the rendering of these terms in English, we have tried to maintain a consistent, and therefore reversible, translation. The issue is complicated by the fact that "physicists" did not exist in English until the Rev. William Whewell introduced the term in 1840. While some of Ørsted's English-speaking contemporaries did not appreciate this word, in which "four sibilant consonants fizz like a squib,"[7] Ørsted would surely have approved. Although he almost always used the term *natural philosopher* when writing in English, there is a hint of desperation in his 1831 letter to Whewell,[8] where Ørsted was reduced to writing "distinguished scientific characters." Thus, the Danish terms *Naturphilosoph, Naturforsker* (or the equivalent *Naturgransker*), and *Fysiker* will be rendered as "natural philosopher," "student of nature," and "physicist," respectively. In the interest of reversibility, we have accepted the last in spite of the anachronism. The Danish *Videnskabsmand* is rendered as "man of science" when it appears before 1840 and as "scientist" after the introduction of this term. Similarly, *Naturphilosophie* is rendered as "natural philosophy," *Naturvidenskab* as "natural science." The only difficulty here lies in the rendering of *Fysik* and *Naturlære*.[9] There appears to be no clear-cut distinction between these terms either in Danish or German. Gehler's 1798 dictionary of scientific terms (in German) suggests that some regarded *Physik* as the more general term, others *Naturlehre*,

---

[7] *Blackwood's Magazine*, Vol. 54, p. 524. 1843. Cited in the *Oxford English Dictionary*, 1971 compact edition.

[8] "On the Compressibility of Water" (Chapter 65, this volume).

[9] In German, *Physik* and *Naturlehre*.

while others again considered them indistinguishable. This remains the case. We have chosen to translate both terms as "physics" except where Ørsted suggests a distinction.

Late in the spring of 1995, we started to look for a historian of science who would be willing to provide us with an introduction to this edition. While there are a number of Danish experts, including Ole Knudsen himself, we felt that the perspective of an English-language scholar would be more appropriate. Therefore, it seemed natural to contact L. Pearce Williams, whose expertise on Ørsted, Faraday, and the history of electromagnetism is widely recognized and highly regarded. When asked, he declared himself willing to write an introduction but suggested that it might be a better idea to give the opportunity to a member of the next generation of scholars, a suggestion that we endorsed completely. He recommended Andrew Wilson, who accepted and produced an introduction that is particularly well suited to present Ørsted to a new audience. His contribution is yet another illustration of the collaborative spirit that has made this project so satisfying.

Karen Jelved and Andrew D. Jackson
Setauket, New York
December 1995

INTRODUCTION

WITHOUT question, Hans Christian Ørsted's discovery of electromagnetism ranks among the greatest experimental discoveries in the history of science. It opened a new era not only in the development of physics, leading to the seminal theories and discoveries of Ampère, Faraday, Maxwell, Hertz, Einstein, et al., but also in the technological development of world civilization, where its economic, social, environmental, and military consequences have fundamentally changed our lives. Ørsted's scientific work and achievements, however, were not restricted to this single discovery. He also conducted pioneering research on the compressibility of gases and fluids, especially water, and invented the necessary instruments for his experiments. The Ørsted piezometer continued to be used in compression research well into this century. Ørsted also conducted important experimental research in the production of chlorine compounds and succeeded in producing silicon chloride and aluminum chloride. Finally, during the 1840s, he conducted significant research on diamagnetism.

In spite of the range of his achievements and the magnitude of his discovery of electromagnetism, Ørsted's life and work have remained relatively unknown outside his native Denmark. Popular works on Ørsted are nonexistent, and there is still no Ørsted biography. Moreover, the scholarly literature on his intellectual and scientific development and his scientific achievements has been sparse and limited mostly to short essays. In fact, Kirstine Meyer's lengthy essays on Ørsted's scientific life and work and his professional life in Denmark, both of which appeared in 1920 as introductory essays to the three-volume centenary edition of Ørsted's scientific writings, have remained the most extensive studies of Ørsted for the past seventy-five years. They are also the only works that have utilized materials from Ørsted's *Nachlaß*, and that have surveyed the entire corpus of his scientific writings. To date, Ørsted's philosophical and poetic works have remained virtually untouched by scholars.

It is difficult to explain fully why historians of science have not more thoroughly combed Ørsted's works, especially his scientific writings, in their attempts to understand how and why he was led to his discovery of electromagnetism. It is equally difficult to explain why they have not relied on them in their efforts to grasp Ørsted's own understanding and theoretical explanation of his discovery, and how his other scientific research related to his galvanic and magnetic studies. Undoubtedly, one reason has been that access to Ørsted's works has been hampered by linguistic obstacles. He published in several languages (one is reminded of Schrödinger in this regard), writing principally in Danish and German, occasionally in French, sometimes in English, and rarely in Latin. For this reason he has remained a remote figure to readers limited to one language, especially to those restricted to

English. Here, for the first time, his major scientific writings have been made available in a single language, eliminating many of the language barriers, and, also for the first time, introducing Ørsted's scientific writings in a meaningful way to an English-reading audience. The works selected and translated here provide a full picture not only of the scientific background to Ørsted's epochal discovery, but also of his scientific career and other scientific achievements. It is an introduction that has long been needed, and that should spawn new and more extensive Ørsted scholarship.

Linguistic impediments, however, are not the only reason Ørsted and his work have remained relatively unknown. Throughout his career, he was deeply concerned with the philosophical foundations of natural science and was guided in both his experimental and theoretical research by strong metaphysical beliefs. When reading through Ørsted's scientific writings, one finds that philosophical elements are so thoroughly woven into them that, without some understanding of their philosophical underpinnings, many of his most important works are exceedingly difficult to comprehend completely while others remain strange and utterly obscure to the modern reader. Even an attempt to restrict attention to the purely physical foundations of Ørsted's research will fall short as they, too, cannot be fully comprehended or appreciated without some familiarity with their deeper metaphysical origins and foundations. As is well known, Ørsted's philosophical ideas and metaphysical worldview were strongly associated with Kantianism, Romanticism, and *Naturphilosophie*. They have therefore been either summarily denounced or studiously avoided by the majority of historians of science. Consequently, Ørsted and his philosophical brethren have not received their due attention. What follows is an attempt to provide some of the broader context needed for a better understanding of the works presented in this volume by laying out some of the most important elements of the philosophical and conceptual foundations of Ørsted's scientific research, particularly in regard to his discovery of electromagnetism and his theoretical understanding of the physics behind this natural phenomenon that has dramatically changed our world.

DURING the spring of 1820, Ørsted delivered a series of lectures on electricity, galvanism, and magnetism to a group of advanced students at the University of Copenhagen. To meet the challenge of satisfying the expectations of his well-schooled auditors, the 42-year-old professor set about reworking his standard lectures on the subject, delving deeper into an area of research that had occupied him off and on over the previous two decades. The particulars of the research that again attracted him were, by the standards of the day, rather unusual. Around the beginning of the century, a handful of experimental philosophers had demonstrated physical connections between certain chemical phenomena and electricity, galvanisms, and magnetism, respectively, as well as between lightning and magnetism. From around 1805, Ørsted believed it was possible to add to this list of connections by producing a magnetic effect from a galvanic circuit, thereby revealing a physical connection between electricity and magnetism. Such a connection, however, had been officially declared impossible by the mainstream scientific community, especially in England and, most formidably, in France. Sober-minded men of science in those countries and across Europe viewed any attempt to produce an interaction between electricity and magnetism as a complete waste of time since, as André-Marie Ampère had maintained, "electrical and magnetic phenomena are due to two different fluids which act independently of each other."[1] As almost everyone had expected, Ørsted and others had failed in their early attempts to demonstrate the indemonstrable.

In the face of experimental failure, scientific orthodoxy, and French pronouncements, Ørsted remained firm in his conviction regarding the physical connection between electricity and magnetism and the possibility of demonstrating it. His renewed investigations in early 1820 only strengthened his resolve. He decided, in fact, to devote one of his new lectures to the "impossible" electromagnetic connection. Moreover, and much more radically, he decided that, as part of the lecture, he would demonstrate publicly that "an electrical discharge might produce some effect upon the magnetic needle placed out of the galvanic circuit."[2] He also decided that he would perform his demonstration only after first rehearsing it in private, to ensure that Nature was willing to comply with the experimental arrangement.

Unfortunately, affairs of the workday prevented Ørsted from conducting a practice run-through of what certainly would be the highlight, not only of that evening's lecture, but of the entire course. Lacking a rehearsal, he felt that a modicum of discretion was in order and, before entering the lecture hall, decided to forgo the demonstration. Once he began speaking, however, he changed his mind. The lecture was proceeding much better than anticipated, and his other demonstrations had been working wonderfully well. Under such propitious circumstances, he moved on to his untried experiment.

Unlike Ørsted, Nature remained little moved by the uplifted spirit and orchestrated excitement in the lecture hall. She was not yet ready to reveal completely a part of herself that she had hitherto preferred to keep hidden. As Ørsted later recounted, "The magnetical needle, though included in a box, was disturbed; but . . . the effect was very feeble."[3] Elsewhere he explained that "although the effect was unquestionable, it appeared to me nevertheless so confused. . . ."[4] In the end, the physical effect of the demonstration had been lamentably uncertain and rather anticlimactic. Its pedagogical effect, on the other hand, had been much clearer. In Ørsted's words, "the experiment made no strong impression on the audience."[5]

Ørsted's busy schedule prevented him from following up on his demonstration immediately. Indeed, he had to wait a full three months, until July of 1820, before he could resume the experiment. At that time and after five dozen attempts, he was satisfied that he had produced a clear demonstration of the magnetic effect caused

---

[1] As quoted in Robert C. Stauffer, "Speculation and Experiment in the Background of Ørsted's Discovery of Electromagnetism," *ISIS*, Vol. 48, p. 43. 1957.

[2] "Observations on Electro-magnetism" (Chapter 42, this volume), p. 431.

[3] "Thermo-Electricity" (Chapter 62, this volume), p. 547.

[4] See footnote 2.

[5] See footnote 4.

by the galvanic circuit. He then published a brief account of his demonstration in Latin, under the title *Experimenta circa effectum conflictus electrici in acum magneticam* (Experiments on the Effect of the Electric Conflict on the Magnetic Needle [Chapter 39, this volume]), thereby introducing electromagnetism to the scientific world.

Ørsted's announcement, which provided instructions for reproducing his demonstration, immediately forced other natural scientists to rethink their understanding of the physical nature and action of electricity and magnetism. Their task now became one of constructing a physical explanation for Ørsted's effect that would fit into the already existing theoretical framework of imponderable fluids and Newtonian mechanics. The year following Ørsted's notice brought a spate of different electromagnetic theories along these lines. Ørsted, too, had outlined a physical explanation in his initial publication, but it was one that fell outside the usual understanding of electrical and magnetic phenomena and Newtonian natural philosophy. It was based, as the title of the notice indicates, on an effect he called the "electric conflict." He briefly described the physics of the electric conflict and, hence, the cause of the electromagnetic effect as follows: "The electric conflict can act only on the magnetic particles of matter. All non-magnetic bodies seem to be penetrable by the electric conflict, whereas magnetic bodies, or rather their magnetic particles, seem to resist the passage of this conflict; hence, they can be moved by the impetus of the contending powers."[6] He had further argued that "the electric conflict is not confined to the conductor, but is dispersed quite widely in the circumjacent space," and that "this conflict moves in circles."

Nowhere in the notice, however, did Ørsted explain exactly what the electric conflict was. Consequently, his scientific colleagues found the physics of the electric conflict to be obscure at best. In London Michael Faraday confessed that he did not quite understand Ørsted's theory. In his efforts to figure it out, he could only conclude that Ørsted's explanation seemed "to require that there be two electric fluids; that they be not either combined or separate, but in the act of combining so as to produce an electric conflict; that they move nevertheless separate from each other, and in opposite spiral directions, through and round the wire and that they have entirely distinct and different magnetical powers."[7] With his usual insight, Faraday was correct in believing that he did not understand Ørsted's theory.

The situation was hardly better in Paris, the center of the scientific world. There Ampère had been among the first to come to grips with Ørsted's demonstration and to extend it with a series of related experiments. By the end of September 1820, his investigations had already led him to offer a rival "electrodynamic" theory to Ørsted's electromagnetic theory, in which he explained the physical connection and implied identity between electricity and magnetism by reducing magnetism to an epiphenomenon of the motion of the two electrical fluids moving in opposite directions. Like Faraday, Ampère also seems to have met with serious difficulty (at

---

[6] "Experiments on the Effects of the Electric Conflict on the Magnetic Needle" (Chapter 39, this volume), p. 416.

[7] Michael Faraday, "Historical Sketch of Electro-magnetism," in *Annals of Philosophy*, New Series, February 1822, p. 108.

least according to Ørsted) in comprehending the details of Ørsted's theory. Almost two and a half years later, during one of his extended European tours, Ørsted expended considerable time and energy in the French capital trying to explain the physics of the electric conflict and his general electromagnetic theory to Ampère and several other leading men of French science. His efforts met with little success. In a letter to his wife, dated 23 February 1823, he told of his frustration, writing, "On many occasions I have noticed that it is nearly impossible to make my theory understandable to the French without also explaining to them some features of the philosophy of nature. While I am often tempted to declare myself against natural philosophy in Germany when I see the misuses of it, I see myself all the more called upon to defend it in France . . ."[8] He went on to tell her that he recognized "an essential difference in the scientific way of thinking" between himself and the French "which I would not have imagined to be so great if I had not so often felt its living presence."

Ørsted always believed that his electromagnetic theory should be perfectly intelligible strictly as a physical explanation, without having to include the metaphysical foundations from which it had been derived and which, to his way of scientific thinking, guaranteed its rigor and validity. His ability to convey his ideas clearly, though, constantly fell short of his desires. All too often he had to resort to metaphysical arguments, which often made matters worse since the metaphysics of his theory appeared to be too foreign to be fully comprehensible to most other natural scientists, including his French hosts and someone as well versed in and sympathetic to Kantian philosophy as Ampère. Nevertheless, Ørsted persisted in his efforts, and in April 1823 he again had the opportunity to explain his theory. On 25 April he spent nearly three hours expounding his ideas to Ampère, Fourier, Cuvier, Dulong, Chevreul, and two of Ampère's young supporters, Savary and De-Montferrand. At the end of that day, Ampère remained "very displeased" that Ørsted still held to his electromagnetic theory and the physics of the electric conflict, but Ørsted believed he had finally made some headway in communicating the substance of his theory. As he wrote to his wife, "I succeeded completely in demonstrating that my theory accounts for all the phenomena. . . . Even Ampère's two disciples declared that my theory was capable of explaining all the phenomena."[9] Savary and DeMontferrand, however, were unwilling to give Ørsted complete satisfaction. As he went on to tell his wife, "They assert that Ampère's can also do that, and . . . his theory is nothing other than the opposite of mine since he has transferred the rotations I discovered from the conductor to the magnet. . . ." To top everything off, following this assertion by Ampère's disciples, it occurred to Fourier that Ørsted had obviously derived his theory from that of Ampère. Indeed, before he took his leave of the gathering, Ørsted found it "most strange" that he "had to prove to Fourier that my theory was older than Ampère's, which, by the way, was easy since I have already given it in my first notice." It must have been mildly disconcerting to Fourier to learn that not every intelligible physical theory

[8] *Breve fra og til Hans Christian Ørsted*, ed. by Mathilde Ørsted, Vol. 2, pp. 54–55. Copenhagen 1870.
[9] Ibid., pp. 65–66.

had French origins, but when all was said and done, it was undoubtedly reassuring to know that, probably because of its pedigree, the physics of Ørsted's theory was certainly wrong, at least to the French scientific way of thinking.

Even before Faraday had tried to make sense of the electric conflict, and before Ampère challenged his electromagnetic theory during his stay in Paris, Ørsted had been aware of the fact that the initial enunciation of his theory had been incomplete and vague. He had tried to correct this by presenting a somewhat lengthy account of his theoretical understanding of the nature of electricity and electromagnetism in a follow-up article to his original notice, entitled "Observations on Electro-magnetism" (Chapter 42, this volume), which was published in Schweigger's *Journal für Chemie und Physik*, Thomson's *Annals of Philosophy*, and the *Journal de physique* in 1821. In the article he explained that his electrical investigations had, from as early as 1806, led him to believe that "electrical conduction consists in an incessant disruption and restoration of equilibrium and so contains an abundance of activity which is not suspected according to the view of a mere continuous [electrical] flow."[10] His subsequent elucidation, however, was still far from clear:

> I therefore considered conduction to be an electric conflict (*conflictus*), and, especially in my investigations on the heat produced by an electrical discharge, I found myself induced to show that the two opposing electrical forces were indeed united in the wire heated by their activity but were not combined in a state of complete rest so that they are still capable of manifesting considerable activity, only in a form completely different from the one properly called electrical. . . . Since I had long viewed the forces which manifest themselves in electricity as the general forces of nature, I also had to derive magnetic effects from them. I therefore expressed the conjecture that "the electrical forces in one of the states in which they happen to be closely bound could produce some effect on a magnet as magnet."[11]

Continued reference to an "electric conflict," to "opposing electrical forces" that manifest activity under different "forms," and to the "forces which are developed by electricity as the general forces of nature" was indeed cryptic to normal scientific understanding and, as we have seen, had been of no help to either Faraday or Ampère. As Ørsted soon realized, his extensive discussion with Ampère et al. had failed to end the confusion over his theory and its physical foundations. He therefore felt compelled in 1827 to attempt, once again, to explain the foundation of his electromagnetic theory. This time the opportunity to clarify himself came in an article on thermoelectricity, written for Brewster's *Edinburgh Encyclopedia*, which appeared in print in 1830, three years after Ørsted had completed his manuscript.

Ørsted devoted a significant portion of the article to the historical background of the discovery of thermoelectricity by Thomas Seebeck in 1822, including his own

---

[10] *Journal für Chemie und Physik*, ed. by Dr. Schweigger and Dr. Meinecke, Vol. 32, pp. 199–231. Nuremberg 1821. There are differences between this text and the version from Thomson's *Annals of Philosophy* given below, "Observations on Electro-magnetism" (Chapter 42, this volume) p. 430.

[11] See note 10, preceding.

discovery of electromagnetism. The article, in fact, seems to be as much concerned with Ørsted and his discovery as with Seebeck and his. Indeed, rather than end the essay with Seebeck or some aspect of thermoelectric effects, Ørsted concluded with a few theoretical considerations relating to "whether or not magnetism and electricity are identical"; i.e., he once again took up his cause against his French rival, Ampère. In defense of his theoretical views Ørsted argued:

> There has been a good deal of misunderstanding in the discussions on this subject. Mr. Ampère pretends that the discoverer of electromagnetism, though he had earlier admitted the identity of these effects has, in his first paper upon electromagnetism, denied it. We must here remark that the words have two acceptations; in one of these Professor Ørsted is perhaps the most earnest supporter of this identity, in the other he is a no less decided opponent of it. His opinion is, that all effects are produced by one fundamental power, operating in different forms of action. These different forms constitute all the dissimilarities. Thus, for instance, pressure upon the mercury of the barometer, wind and sound, are only different forms of action of the same powers. It is easy to see that this fundamental identity extends to all mechanical effects. All pressures are produced by the same powers as that of air; all communications of motion, and likewise all vibrations, owe their origin to the same expansive and attractive powers, by which each body fills its space, and has its parts confined within this space. This fundamental and universal identity of mechanical powers has for a long time been more or less clearly acknowledged; ... Thus acknowledging the fundamental and universal identity of powers, effects must be considered as different, when their form of action differs, and therefore magnetism, in this acceptation of the term, is far from being identical with electricity. It would likewise be erroneous to pretend that all chemical effects are produced by electricity; but the truth seems to be, that the chemical effects are produced by the same powers which, in another form of action, produce electricity.[12]

Earlier in the article he had already explained that he had been led to a belief in the underlying unity of nature by the "philosophical principle, that all phenomena are produced by the same original power," and that the fundamental powers of expansion and attraction were therefore probably "only two different forms of one primordial power." He also reiterated that "he did not consider the transmission of electricity through a conductor as an [*sic*] uniform stream, but as a succession of interruptions and re-establishments of equilibrium, in such a manner, that the electrical powers in the current were not in quiet equilibrium, but in a state of continual conflict."

As these passages indicate, Ørsted still explained the foundations of his theory in rather broad and vague terms; he still had not explained his theory or the nature of the electric conflict fully or clearly. Moreover, he now even seemed to muddy the waters by referring to a single primordial force behind all physical phenomena, a

---

[12] "Thermo-Electricity" (Chapter 62, this volume), p. 579.

force alien to the orthodox dynamical theories of matter derived from Kant's original theory. On the other hand, the passages above indicate that Ørsted's understanding of electricity, magnetism, and, indeed, of physical effects and nature in general had continued over the years to be based on the apparently contradictory concepts of conflict and identity and on the further concepts of force and form of action. From the time of his epochal discovery to the time of his death, however, he never explained these concepts or his theory satisfactorily. It is only by understanding these concepts, though, that one can begin to comprehend Ørsted's theoretical speculations regarding the nature of electricity and of the physical world as a whole, something neither Faraday nor Ampère was willing, or perhaps even able, to do.

Fortunately, it is possible to make some sense of Ørsted's obscure and fragmented remarks and to understand the foundations of his research by turning to the scientific and philosophical concerns expressed in his scientific writings prior to 1820, beginning at the time when he was a student at the University of Copenhagen in the mid 1790s. The remainder of this introduction will therefore be devoted to the development and substance of the major conceptual foundations of his mature scientific thought, especially as they concern the theoretical entities of the electric conflict, form of action, and the primordial force, or *Urkraft*.

Like so many other students at the close of the eighteenth century, Ørsted had immersed himself in the new philosophy emerging from Germany, and even though his major areas of study were in pharmacology, mathematics, physics, and chemistry, he had begun to study the works of Kant and Fichte by his second year at the University of Copenhagen. As Schelling's works appeared, he also read them. Early on, he had been especially intrigued by Kant's critical philosophy, which had been introduced into the curriculum at the university in 1791 by Christian Hornemann and Børge Riisbrigh. However, he strayed from the mainstream of Danish Kantian studies, which tended to focus on Kant's ethical theories, by reading the *Metaphysical Foundations of Natural Science* (1786). As its title indicates, this work introduced him to Kant's understanding of the metaphysical foundations of natural science, including the necessary metaphysical foundations of scientific knowledge. In the process, Ørsted was introduced to Kant's taxonomic system regarding the division of the sciences and to his dynamical theory of matter based on the attractive and repulsive interactions of two fundamental forces, or *Grundkräfte*. After reading through Kant's work, Ørsted readily rejected atomism in favor of a dynamical view of matter and nature but could not accept other important elements of Kant's metaphysical system. He was particularly troubled by two major elements in Kant's metaphysics of scientific knowledge. The first was Kant's criteria regarding the requirements for a body of knowledge to be accepted as "actual" (*eigentliche*) scientific knowledge. The second was Kant's severely limited application of dynamism, restricting it solely to the science of mechanics.

In regard to Ørsted's first concern, Kant had maintained that a bona fide natural science must be grounded in a priori first principles and laws from which the rest of the science can be systematically derived. Ørsted wholeheartedly agreed with this requirement but was convinced that Kant himself had failed to fulfil it. According to Ørsted, Kant had, in fact, undermined the apodictic certainty of his dynamic *Naturmetaphysik* by mixing a posteriori and a priori concepts in his derivation of the existence of matter and its fundamental qualities. Additionally, and in line with Kantian revisionists such as Reinhold, Fichte, and Schelling, Ørsted believed that a thoroughly systematic, scientific body of knowledge should ultimately be grounded in a single, unconditioned, a priori metaphysical first principle rather than a plurality of first principles, as was characteristic of the foundations of Kant's critical philosophy and metaphysics of nature.

As part of his argument, Kant had further maintained that, in addition to being a priori, the fundamental principles of a natural science must be capable of being presented *in concreto* in our faculty of intuition, since, he maintained, it is only by means of our intuition that a priori concepts and ideas can be directly related to material objects of our sense experience. He consequently insisted that a doctrine of material nature (*Körperlehre*) contains only as much actual science as the mathematics contained in it, having already argued in his first *Critique* that it is only by means of mathematics that something can be presented in intuition a priori.

This mathematical requirement led directly to Ørsted's second major disagreement with Kant's definition of science and his taxonomic system, in that Kant could not consider certain disciplines, such as chemistry and psychology, which did not seem to be capable of mathematization, to be actual natural sciences. In the case of chemistry, which was Ørsted's principal concern, Kant maintained that, since chemical principles are ultimately only empirical, and "the laws from which the given facts are explained through reason are merely laws of experience, they therefore entail by themselves no awareness of their necessity (are not apodictically certain). . . ."[13] Consequently:

> So long as no concept is discovered for the chemical actions of substances on one another, which allows itself to be constructed, i.e., so long as no law of the approach and removal of the parts of matter can be specified, according to which approximately in proportion to their densities etc. a clear idea is given of their motions together with their future motions, and which allows itself to be presented in space a priori (a demand which scarcely will ever be fulfilled), chemistry can never be more than a systematic art or experimental doctrine. Thus at no time can it be an actual science, since its principles are merely empirical and permit no presentation a priori in intuition. Consequently, the fundamental principles do not make the possibility of chemical phenomena in the least conceivable because they are unfit for the application of mathematics.[14]

Ørsted could in no way accept the declaration that chemistry, by its very nature, is barred from being an actual natural science, and that it would most likely continue to be so into the distant future. Already at this early stage of his career, he had

---

[13] Immanuel Kant, *Metaphysische Anfangsgründe der Naturwissenschaft*, in *Gesammelte Schriften*, published by the Royal Prussian Academy of Sciences, Vol. 4, p. 468. Berlin 1903.
[14] Ibid., pp. 470–71.

become deeply committed to a worldview in which nature was understood as a universally connected, continually active, living, divine creation filled with polar forces in relative degrees of conflict and equilibrium. According to this view of nature, he believed that all branches of natural science should be unified into a comprehensive, universal science that would correspond to the integrated organization of nature as a whole. This view of natural science further required that the first principles of every branch of science be the same, just as the first principles of every natural phenomenon are the same. In line with Kant's dynamical theory, this meant for Ørsted that every natural phenomenon, including the existence and qualities of matter in general (which he took as the basis of all physical science) and other so-called physical substances, such as the electrical and magnetic fluids and the effects they produce, should be explicable in terms of the interactions of the two fundamental and generic attractive and repulsive forces of nature.

This view of nature and science espoused by Ørsted was not unique but was one that he shared with the various scientists and men of letters associated with the Romantic movement in Germany, and that had been derived in large measure from the recent philosophical works of Friedrich Schelling, particularly his *Ideas for a Philosophy of Nature* (1797) and *On the World-Soul* (1798). In those works Schelling had responded to the Cartesian separation of mind and matter, which had effectively destroyed any notion of a unified and integrated cosmos. The idea of a fragmented and disconnected cosmos struck Schelling not only as philosophically absurd and aesthetically unpleasing, but, even worse, potentially atheistic. In the *Ideas* he therefore stressed that "the system of nature is at the same time the system of our mind."[15] He further maintained that the two had to be synthesized into a unified philosophical whole. The groundwork for a philosophical system of the mind, he believed, had already been partially completed, more or less adequately, by Kant and Fichte. Their systems, though, had remained too much under the influence of Cartesianism and therefore offered no real link between the essential nature of mind and the essence of external nature. Moreover, as far as Schelling was concerned, there was still no system of physical nature that could be wedded to a legitimate philosophy of mind. The most obvious candidate, the mechanical system of nature derived from Newton's *Principia*, was wholly unsuited to the task since, like its philosophical counterparts, it provided no bridge between the world of dead brute matter and that of living organisms and consciousness. Consequently, Schelling set out to provide a sufficient revision of the philosophies of Kant and Fichte and, more important, to establish the foundation and general outline for a universal theory of nature. The bringing together of his philosophy of nature with a revised and improved version of the critical philosophy would thus be the first major step toward completing the synthesis he envisioned and toward restoring the unity of the cosmos.

Schelling believed he had found the key to a new universal theory of nature in the most recent discoveries and advances in chemistry and general physics. In the wake of Luigi Galvani's discovery of animal electricity, Alessandro Volta's dis-

[15] F.W.J. Schelling, *Ideas for a Philosophy of Nature*, translated by E. E. Harris and P. Heath, p. 30. Cambridge: Cambridge University Press 1988.

covery of contact electricity, and Johann Ritter's early attempts to unify them into a general theory of galvanism that was valid for both organic and inorganic nature, it appeared to Schelling in the *Ideas* that, contrary to Kant's injunction, "the new system of chemistry, the work of a whole era, spreads its influence ever more widely over the other branches of natural science and, employed over its whole range, may very well develop into the universal system of nature."[16] Schelling took the initial steps toward establishing this status for chemistry by offering the outline for a comprehensive chemical science based on a dynamic theory of matter and the dialectical interactions of the two *Grundkräfte*. Unfortunately, his empirical understanding of chemical processes and phenomena was sorely wanting, and only his most devoted disciples adhered to the details of his system of chemistry.

Nevertheless, as was the case for the Romantics, Ørsted's reading of Schelling's *Ideas* was decisive in the development of his general metaphysical view of nature and in determining his understanding of the purpose and goal of natural science. Even though he believed that Schelling had failed in presenting the first principles of a true science of chemistry and consequently had failed to an equal degree in the construction of a system of chemistry itself, he found great value in Schelling's accounts of the correspondence between mind and nature and of nature as a dynamically active, integrated whole. As he later told his friend Adam Oehlenschläger in a letter dated 1 November 1807:

> Schelling's accomplishment, as you know, is that he founded the philosophy of nature and with this affected all sciences. The genius of this was not that he construed nature and, least of all, the way he construed it, but his accomplishment was to see nature as a single organism. These words, "Nature is a productive product," would . . . suffice to mark him as a comprehensive and profound spirit. He has repeated this in a great many forms. The most beautiful and most felicitous is that nature is nothing other than a revelation of the Godhead.[17]

Thus, with the basic elements of Kant's dynamical theory of matter and Schelling's general *Naturmetaphysik* in hand, Ørsted set out toward the end of his university studies to contribute to the Schellingian synthesis by laying an adequate metaphysical foundation for a general dynamic science of nature with chemistry at its center. As part of this effort, though, he needed to revise Kant's views of the division of the sciences and the place of chemistry in the study of nature. He began to do this in 1798 in the first installment of his *Chemical Letters*.[18] Shortly thereafter, he found the opportunity to present more commentary, both direct and indirect, on Kant's definition of natural science etc., with the publication of the first volume of a new edition of Adam Hauch's *Foundations of Physics*, published in two parts in 1798 and 1799. Hauch's *Foundations* was an elementary textbook and the only book of its kind in Danish. Much to Ørsted's disappointment, it presented a system of natural science based on atomistic foundations. In order to prevent Hauch's atomism from being the only system of nature available to Danish students (at least

---

[16]Ibid., p. 59.   [17]*Breve* (see footnote 8 above), Vol. I, p. 230.
[18]"Letters on Chemistry. First Letter" (Chapter 2, this volume).

those capable of reading only Danish), he responded by offering two revised outlines of the dynamical theory of matter presented in the *Metaphysical Foundations*. The first was contained in an essay review of the first part of Hauch's book; the second was a longer essay entitled *Fundamentals of the Metaphysics of Nature, Partly According to a New Plan*.[19] He followed these with his doctoral dissertation, *Dissertation on the Structure of the Elementary Metaphysics of External Nature*.[20] In between the *Fundamentals* and the *Dissertation*, he published a review of the second part of Hauch's book. Three years later, in 1802, M. H. Mendel wrote and published, with Ørsted's approval, a revised and condensed version of these earlier works in German under a title reflecting Schelling's founding work in *Naturphilosophie*, viz., *Ideas toward a New Architectonic for the Metaphysics of Nature*. The *Ideas* marks the end of Ørsted's earliest essays directly concerned with the metaphysical foundations of natural science.

In all of these early essays, Ørsted criticized the acceptance of atomism as the ultimate foundation of science and nature and argued in favor of dynamism as the true foundation of both. As expected, his attack utilized elements he had gleaned from his study of both Kant and Schelling. Drawing on Kant, he attacked the metaphysical foundations of atomism, arguing that they failed to provide the a priori necessary first principles. In particular, he maintained that the basis of the atomistic explanation of the existence of matter was rooted in a logical fallacy and misinterpretation of the Law of Non-Contradiction. Consequently, it could at best only be accepted as a mere hypothesis, rather than an apodictically established truth. As he explained in the *Fundamentals*:

> The only attempt that has been made to prove the atomistic system a priori is the well-known proof of the impenetrability of bodies. It was believed that it was contrary to the principle of contradiction to say that a body could be in the same space as another at the same time. No doubt one would never have thought of this proof if this theorem had not been expressed in the following way: A thing cannot be and not be at the same time. If this is expressed in different words: A thing cannot have contradictory predicates, then it is not possible to see how to complete the proof as there is no contradiction in two bodies occupying one and the same space unless one presupposes the impenetrability that one wants to prove.
>
> Consequently, the foundations of the atomistic system are shaky. Its principal axioms, that there are atoms and empty spaces, are unproved and therefore only hypotheses which we are entitled to reject if they do not fully explain the phenomena, or if we can explain them by means of other, proved theorems. With the foundation follows the collapse of the entire structure. . . .[21]

With Schelling, he criticized atomism for failing to provide a comprehensive and integrated body of knowledge capable of explaining all natural phenomena,

---

[19] *Fundamentals of the Metaphysics of Nature* (Chapter 6, this volume).

[20] *Elementary Metaphysics of External Nature* (Chapter 7, this volume).

[21] *Fundamentals of Metaphysics of Nature* (Chapter 6, this volume), p. 75.

not just phenomena of a restricted range of nature. He insisted on this in his review of the second part of Hauch's textbook by declaring, "Physics is philosophy of nature, so it must explain the whole of nature, and therefore also every single event in nature."[22] Ørsted had to admit, however, that the dynamical system could not account for every event in nature either, but he could argue, and did, that it explained a significantly broader range of natural phenomena, especially chemical phenomena, than did atomism.

In the *Fundamentals of the Metaphysics of Nature*, Ørsted believed he had conclusively demonstrated the superiority of a force-based, dynamical explanation of the existence, qualities, and many physical effects of matter. In the end he believed that the advantage of the dynamic system over atomism was to be found in the a priori connection between mind and nature. Indeed, according to Ørsted, the metaphysical foundations of dynamism present "the laws of nature as founded on human cognition so that we know beforehand that there are no exceptions to these because, in order to imagine that anything happened according to natural laws which were at variance with the ones we had proved in this way, we would have to change our cognition, that is, become other beings."[23] Even with this intrinsic connection, however, he knew that more work needed to be done before dynamism would supplant atomism as the foundation for natural science. Nevertheless, he remained confident that dynamism would eventually supersede its rival since, he believed, "[a]ll that is needed to make the dynamic system that of all astute natural philosophers is a few more successful attempts at uniting the dynamic system with all of natural philosophy. . . ."

The desire to create a comprehensive science of nature continued to guide Ørsted's scientific interests and research for the rest of his career. In many ways, all of his subsequent scientific work can be understood as an extended attempt, both experimentally and theoretically, to unite the dynamic system with the rest of nature. To this end, he was always attracted to natural scientists and scientific theories that promised to help him in this effort. In the years immediately following his university studies, for instance, he took a keen interest in the writings of his friend Henrik Steffens, one of Schelling's most ardent followers, especially his *Contributions to the Internal Natural History of the Earth* (Copenhagen, 1801). In the *Contributions*, Steffens described nature as an integrated, hierarchical, and dynamically evolving chain of being. The evolving interconnectedness of nature was at the core of his understanding not only of the natural history of the earth, and of mineral chemistry in general, but also of plant and animal life.

Ørsted was also drawn to the galvanic and electrochemical experiments of Johann Ritter, having already become interested in the new and exciting field of galvanic studies in the late 1790s and having already begun to carry out his own experiments. In Ritter he encountered an experimental researcher who, like himself, was attempting to understand and interpret electrical, galvanic, magnetic, and chemical phenomena, not in isolation but in their relation to each other and to the

[22]H. C. Ørsted, "Recension over Begyndelsesgrunde til Naturlæren," 1799. In his *Videnskabelige Skrifter*, ed. by Kirstine Meyer, Vol. 3, p. 40. Copenhagen 1920.

[23]*Fundamentals of the Metaphysics of Nature* (Chapter 6, this volume), p. 76.

organization of nature as a whole. In particular, Ritter had come to believe that his synthesis of the then competing galvanic theories offered by Galvani and Volta would demonstrate galvanism to be *the* fundamental force of nature. More importantly, he believed his synthesis would demonstrate that galvanism was the essential, and hitherto missing, link joining the organic and inorganic realms of nature. For Ritter, galvanism and galvanic studies thus provided the key to creating a unified, comprehensive force-based theory of nature. During an extended scientific tour from 1801 to 1803, Ørsted visited Ritter in Oberweimar and spent several weeks with him conducting galvanic experiments and learning how galvanism fit into the general organization of nature. Their interaction left a deep and lasting impression on Ørsted, who came away believing Ritter to be the true father of modern chemistry. Nor surprisingly, Ritter's imprint can be found in the majority of Ørsted's publications in the years following their meeting.

During his stay with Ritter, Ørsted turned his attention away from galvanic studies long enough to read the *Prolusiones ad chemiam saeculi decimi noni*, which had been published in 1800 by Jakob Joseph Winterl, Professor of Chemistry and Botany at the University of Buda. Here, again, he thought he had found a kindred philosophical and scientific spirit, for Winterl's *Prolusiones* was a grand speculative effort to formulate a comprehensive system of chemistry designed to replace Lavoisier's chemical system. Winterl believed, in fact, that he had successfully produced a comprehensive science of chemistry, founded on experimental evidence (an erroneous belief) and based on two fundamental physical principles, Andronia and Thelyke. According to Winterl, these principles, which he also viewed as fundamental substances, corresponded to the principles of acidity and alkalinity respectively, as well as to positive and negative electricity and other analogous polarities in nature. By identifying such analogical connections throughout nature, Winterl argued that the principles of acids and bases, electricity and magnetism, and heat and light are ultimately the same, viz., ultimately reducible to Andronia and Thelyke. According to Winterl, all chemical phenomena could in fact be explained in terms of different combinations and concentrations of these two fundamental elements. By means of Andronia and Thelyke, the underlying unity of nature could thus be revealed.

As had been the case with the empirical details of Schelling's and Steffens's systems, Ørsted also harbored reservations about the experimental foundations and validity of Winterl's chemical philosophy. Nevertheless, he found considerable value in Winterl's approach to nature. In much the same manner as he did with Ritter's work, Ørsted immediately became an enthusiastic supporter of Winterl's system, and during the remainder of his European tour in 1803 he tried to enlighten others about both Ritter's and Winterl's fundamental contributions to science. As part of his efforts, he offered summaries of their work in an essay surveying the most recent progress in physics written for Friedrich Schlegel's journal, *Europa*.[24] He also published a lengthy, and somewhat revised, account of Winterl's system

---

[24] "A Review of the Latest Advances in Physics" (Chapter 11, this volume).

entitled *Materials for a Chemistry of the Nineteenth Century.*[25] At the conclusion of the *Materials,* Ørsted made it clear that the greatest attribute of Winterl's work lay in the promise it held for a comprehensive and systematic science of nature:

> The constituent principles of heat, which are important in alkalis and acids, in electricity, and in light, are also the principles of magnetism, and thus we would have the unity of all the forces which act together to govern the entire universe, and previous physical knowledge would therefore unite into one coherent natural philosophy (even with this, our natural philosophy is not yet complete), for do friction and impact not produce both heat and electricity, and are dynamics and mechanics not thereby perfectly intertwined? (If it should be necessary, this will become even more obvious once we are able to survey at a glance all Ritter's beautiful and relevant discoveries, some of which were made a long time ago.) Our natural philosophy, therefore, will no longer be a collection of fragments on motion, on heat, on air, on light, on electricity, on magnetism, and who knows what else, but with one system we shall embrace the entire world.[26]

Now, it was all well and good for Schelling, Steffens, Ritter, Winterl, and even Ørsted to identify whole series of analogies between physical phenomena and to assert the unity of nature. However, it was much more problematic to demonstrate the inner connectedness of nature physically, i.e., to reveal metaphysical reality in physical appearances. Ørsted was well aware of the problems associated with such a task and was equally cognizant of the imperative of undertaking and completing it. In 1809, six years after writing the *Materials,* he instructed one of his students on just this point, telling him, "It is also my firm conviction, and my lectures bear witness thereof, that a great fundamental unity pervades the whole of nature; but just when one has become convinced of this, it becomes doubly necessary to direct one's whole attention to the world of the manifold, wherein this truth above all finds its confirmation. If one does not do this, unity itself remains an unfruitful and empty idea which leads to no true insight."[27] As early as 1799, he had asserted that proper scientific method should combine theoretical speculation and experimental demonstrations, with speculation providing both the foundation and the framework for empirical investigations. Prior to 1803, however, he had made no significant headway in combining the two or in effectively revealing the hidden unity of nature, for neither his experimental papers nor his theoretical essays indicate that he had any real insight as to how to demonstrate the underlying, common dynamic principles and causes of all natural phenomena.

His situation began to improve, albeit slowly, after 1803 and the publication of the second, revised and enlarged, edition of Schelling's *Ideas toward a Philosophy of Nature.* In this new edition, Schelling added supplemental sections based on his

---

[25] Chapter 12, this volume.   [26] Ibid., p. 164.

[27] Johannes Carsten Hauch, *H. C. Ørsted's Leben. Zwei Denkschriften von Hauch und Forchhammer,* translated by H. Sebald, p. 13. Spandau 1853. As quoted in Stauffer, p. 39 (see footnote 1 above).

more recent *naturphilosophische* researches. He now offered a more specific theoretical outline for examining the formal relationships between the two *Grundkräfte* and the physical effects produced by changing those relationships. More particularly, he now explained the differences found between kinds of matter and between physical effects in terms of formal (rather than substantive) differences that occur as a result of the physical "unfolding" of a fundamental metaphysical unity existing behind the world of appearance. The result was a revised version of his philosophy of nature in which the physical world is understood in terms of the geometrical properties of dynamic (i.e., force-based) processes. As Schelling put it:

> This unfolding of identity is . . . that of the embodiment of unity into multiplicity, the absolute form of which is absolute space, as its relative form is the line. All forms whereby things are separated . . . become, therefore, mere forms of space, and since space, in its identity as image of the absolute in difference, simply includes the three unities again, these forms are the three unities or dimensions of space. Now that, generally speaking, all bodily differences must reduce uniquely to their relation to the three dimensions of space, and that bodies in all qualities are divided into the three classes, that they display the preponderance of the first dimension and of absolute coherence, or that of the other dimensions and of relative connection, or finally the greater or lesser indifference of both in fluidity—this follows already from the general proof but can also be based on complete induction.
>
> With this there is an end to all those qualitative differences of matter which a false physics fixes and makes permanent in the so-called basic substance: All matter is intrinsically one, by nature pure identity; all difference comes solely from the form and is therefore merely ideal and quantitative.
>
> It has been demonstrated in the forgoing that magnetism, as a process, as a form of activity, is the process of length, electricity the process of breadth, just as the chemical process, on the other hand, is that which alone affects cohesion or form in all three dimensions, and hence in the third.
>
> . . . We have ruled out all fixed qualitative oppositions of particular matters, whose action has for long enough been vainly invoked for the comprehension of these phenomena. Their ground and source lies in the form and inner life of the bodies themselves. . . . The difference of their forms rests only on the different relations of the same activity to the three dimensions.[28]

By focusing on what he called forms of action (*Wirkungsweise*) and forms of activity (*Formen der Thätigkeit*), i.e., by understanding physical effects, such as magnetism, electricity, and chemical properties, as processes and epiphenomena of the interactions between the two *Grundkräfte*, rather than as fundamental substances, and by providing causal explanations for physical differences in terms of the geometrical forms of those processes, Schelling believed he had finally secured the necessary a priori metaphysical foundations of a universal science of na-

---

[28] Schelling (see footnote 15 above), p. 137.

ture. He also thought he had fulfilled the one Kantian demand of natural science that chemistry still had not succeeded in meeting, and that had troubled both Schelling and Ørsted, viz., the need to make chemistry a mathematical science. After laying out his new metaphysical foundations, he explained the relationship between geometry and chemistry by indicating how the physical effects of magnetism, electricity, and chemical processes (including galvanism) find their geometrical analogues in the three dimensions of space. Earlier he had expressed these spatial relationships schematically as "Mag : El = | : L . El : Galv = L : △."[29]

Schelling's emphasis on the formal and quantitative differences of physical phenomena provided Ørsted with the theoretical clue he had been searching for that would lead to the demonstration of the underlying unity of nature and, in turn, to the establishment of a universal science of nature. In 1805 he published a pair of essays, "On the Harmony Between Electrical Figures and Organic Forms" (Chapter 18, this volume) and "New Investigations into the Question: What Is Chemistry?" (Chapter 19, this volume) in which he emphasized the causal relations between fundamental natural forces, natural forms, and physical effects. Indeed, his purpose, as he put it in the "Harmony" essay, was "to suggest the connection which we find in all nature between force and form."[30] Starting with the geometrical patterns produced in Lichtenberg figures, which he accepted not only as symbolic manifestations of the fundamental forms of positive and negative electricity, but as *the* fundamental forms of nature, Ørsted traced a plethora of formal and polar analogies found throughout organic and inorganic nature. From these he concluded that "the electro-chemical process is the formative process" found throughout terrestial nature, but that this process is reducible to the conflict and geometrical relationships between the two *Grundkräfte*. He attempted to explain this further in terms of Schelling's outline:

Schelling has shown that three instances must be distinguished in the construction of matter by the attractive and repulsive forces. The first, in which the contrast between these two forces merely assumes the form of the line, the second, in which it is in the form of the surface, the third, in which both these interpenetrate and thus form the final dimension of space and matter, depth. Every time one body produces an internal change in another, whereby matter is really reconstructed, one or more of these actions must reappear. Thus, the longitudinal function manifests itself as magnetism, the latitudinal function as electricity, and the depth function as penetration or a chemical process. Each of these dynamic processes is the interaction of opposite fundamental forces in a different form. The transition to form occurs when a magnetic, electric, or chemical plus and minus are roused in a homogeneous substance, or, in other words, when indifference becomes difference.[31]

[29] F.W.J. Schelling, "Allgemeine Deduktion des dynamischen Processes," 1800. In his *Werke*, Münchner Jubiläumsdruck, Vol. 2, p. 707. Munich 1965.

[30] "On the Harmony Between Electrical Figures and Organic Forms" (Chapter 18, this volume), p. 191.

[31] Ibid., p. 190.

He invoked Schelling again in his next essay, "What Is Chemistry?" (Chapter 19, this volume). Here, in answer to the questions "What are the fundamental chemical forces?" and "What are the forms in which they work?" he wrote, "Using philosophical arguments I could easily demonstrate that two opposite fundamental forces are at work throughout nature, in alternating expansions and contractions; I could show how their effects have as many basic forms as space has dimensions, and finally I could point out that all these forms, to varying degrees, must be discernible in every effect."[32] Later in the essay, after asserting that "the same forces which manifest themselves in electricity also manifest themselves in magnetism, although in another form," he again explained the connections between the two *Grundkräfte*, geometrical form, and physical effects (this time including the production of heat and light in electrical phenomena) in terms of Schelling's theoretical outline:

> A brief outline of what we know about the effects of these forces is sufficient to show us the possibility that all the different forces of nature can be traced back to those two fundamental forces. How could there be three more different effects than heat, electricity, and magnetism! Yet, all of these are due to the effect of the same fundamental forces, only in different forms. Magnetism acts only in a *line* which is determined by the two opposite poles and the intermediate point of equilibrium. Purely electrical effects only follow *surfaces.* Heat works equally freely in *all directions* in a body. It cannot be denied that this difference actually exists. . . . However, it can hardly fail to attract the greatest attention that these three effects assume forms which correspond to the three dimensions of space and their realizations: line, surface, and body. It seems obvious to me at a glance that there can be no additional forms for the effects of the fundamental forces, but philosophy alone can provide complete certainty in this matter.[33]

Even in these essays, however, Ørsted was still only able to offer general descriptions and philosophical arguments regarding the connections between force and form found in nature; empirical demonstrations remained out of reach. Schelling's new theoretical outline seemed to promise further progress in this regard, but its precise application remained rather obscure. The prospect of a universal science of nature thus remained a rather distant hope. By the close of 1805, Ørsted could only lament that "[w]e have fragments of such a science, for example, physical astronomy, geology, and meteorology, but the complete science does not exist yet, and can never be reached by the path of experience. It is well known that Schelling, through speculation, has produced an attempt which, as such, is of incalculable value, but the combined efforts of a great number of blessed geniuses are probably required for the accomplishment of this task."[34]

Even though Ørsted had to be satisfied with scientific fragments for the time being, he remained convinced that geometrical arrangements of the *Grundkräfte*

---

[32] "What Is Chemistry?" p. 195.   [33] Ibid., p. 197.
[34] Ibid., p. 199.

were the key to demonstrating the underlying connection between, and common cause of, different physical effects and to transforming chemistry into a mathematical science. He also continued to rely on theoretical speculation to guide him in his empirical research. His resolve began to produce results the following year, in 1806, when he could offer not only philosophical arguments to demonstrate the alternating expansions and contractions of the *Grundkräfte* but also, in the case of electricity, present experimental confirmation of this fundamental dynamic action.

The line of research enabling him to demonstrate the dynamical cause of electrical effects had begun two years earlier, in 1804. At that time, his interest in the physics behind the production of natural geometrical forms had led him, in addition to the study of Lichtenberg figures, to the acoustical researches of Chladni and the geometrical figures he had produced with sand on vibrating metal plates. He was especially interested in discovering what insight Chladni figures could add to the understanding of the properties of electricity already revealed in Lichtenberg figures. His subsequent experiments on acoustical vibrations, along with important electrical and galvanic experiments, provided him with empirical demonstrations of the alternating expansions and contractions of the *Grundkräfte*. He presented the results of these efforts by way of a physical explanation of electrical induction, polarization, and conduction in "On the Manner in Which Electricity Is Transmitted (A Fragment)" (Chapter 21, this volume).

As with his line of research, Ørsted's physical understanding of electrical propagation did not emerge suddenly in 1806. Elements of it can be found prior to his propagation article, especially in his two papers from 1805 and in an article on the series of acids and bases, written earlier in 1806. In all three articles, he discussed the existence and effects of the *Grundkräfte* in general and with regard to electricity in terms of their omnipresence, dormancy, and conflict. We have already seen this in the passage from the "Harmony" essay where he mentioned without further comment that "[t]he transition to form occurs when a magnetic, electrical, or chemical plus and minus are roused in a homogeneous body, or, in other words, when indifference becomes difference."[35] In "What Is Chemistry?" he explained this more fully but still rather obscurely: "However, if heat is nothing but the phenomenon of the struggle to combine of the same forces which are found separated in electricity and magnetism . . . , we are compelled to assume that these forces lie dormant in every body and in each of its parts so that they must be considered absolutely necessary for their constitution."[36] Finally, in "The Series of Acids and Bases," the concepts and language of the dormancy and conflict of forces were used again by Ørsted in a passing description of the physical mechanism of electrical effects in which he asserted that a weak degree of one kind of electricity "merely arouses its opposite,"[37] while a stronger degree "not only eliminates this opposite but even produces the same electricity."

[35] "On the Harmony Between Electrical Figures and Organic Forms" (Chapter 18, this volume), p. 190.

[36] "What Is Chemistry?" (Chapter 19, this volume), p. 197.

[37] "The Series of Acids and Bases" (Chapter 24, this volume), p. 238.

The concepts of the dormancy and conflict of the *Grundkräfte*, which Ørsted presented in bits and pieces in 1805 and early 1806, became the implicit foundations for the relatively more detailed account of the physics (or "internal mechanism,"[38] as he called it) of the propagation of electricity offered in his 1806 "fragment." His explanation, however, because of its brevity and conceptually condensed nature, is difficult to understand. Even with the information contained in his earlier statements on this subject, his explanation is, in the end, vague and incomplete. He began his paper with an attempt to explain electrical induction and polarization in terms of a temporal and spatial progression of the creation and destruction of infinitely small electrical polarities. More specifically, he explained the polarization that results in an initially nonelectrified conducting cylinder (*BC*) when it is brought into contact with an electrified body (*A*) as follows: Based on the fact that the transmission of electricity from the end of the cylinder closer to the electrified body to the far end is not instantaneous but requires a certain amount of time, and based on his prior theoretical conclusion that the two *Grundkräfte* "lie dormant in every body and in each of its parts,"[39] Ørsted conceived the duration of the electrical transmission as divided into infinitely small units of time and the cylinder as divided into infinitely small units of space. He then provided the following analysis:

> We want to think of such a small part of space as represented by *Bb*, where, in the first small part of time, an infinitesimally small electric polarity is aroused; if, e.g., *A* becomes positive, *Bb* becomes negative at *B* and positive at *b*. However, in the next instant, *A* will seek to enlarge the negative zone, whereby the positive will be enlarged as well, while the positive in *b* will strive to produce a negative one even further towards *C*. And the entire process continues in this manner until the negative extends over the front half, the positive over the back half, and the middle remains indifferent.[40]

As described here, it is difficult to envision this as a continuous process that would necessarily result in a polar distribution of electrical charge. It is possible that Ørsted had not fully thought out the details of this process since he already knew that it was true from the nature of things, and since he was not concerned with trying to demonstrate the validity of his explanation experimentally, having admitted that "we can have no hope of ever finding" these infinitely small alternations of positive and negative principles in experience because electricity in good conductors spreads far too rapidly. Moreover, as the rest of his paper shows, Ørsted was not particularly interested in explaining the infinitesimal physics behind electrical polarity but was far more concerned with demonstrating a general physical property of nature; i.e., he first wanted to demonstrate (macroscopically) that electricity is propagated via a form of undulatory motion characterized by alternating

---

[38] "On the Manner in Which Electricity Is Transmitted (A Fragment)" (Chapter 21, this volume), p. 210.

[39] "What Is Chemistry?" (Chapter 19, this volume), p. 197.

[40] "On the Manner in Which Electricity Is Transmitted" (Chapter 21, this volume), pp. 210–11.

positive (expanding) and negative (contracting) zones. He then wanted to demonstrate that "[t]his mechanism for the transmission of actions through undulation is certainly common in all of nature. . . ."[41] He fulfilled both intentions by presenting the results of various electrical and galvanic experiments, all of which provided observable evidence of various kinds of expansions and contractions, from the alternating expanding and contracting zones along a metal wire melted by a strong electrical discharge, to the graphic patterns of alternating expansions and contractions collected on paper from the clouds produced by a metal wire vaporized by electrical discharges, to the color patterns displayed by electrical sparks. He then argued, by analogy, that undulatory motion also characterizes the propagation of magnetic effects and of sound through air and solid bodies, as demonstrated by the production of Chladni figures.

The dialectical process of expansion and contraction created by the opposition between and intermediate syntheses of the positive and negative forces of electricity, as described in the article on electrical propagation and in passing in his articles from 1805 and early 1806, constituted the physical effect that Ørsted would later call the electric conflict. On the basis of the various fragments in Ørsted's works from 1805 and 1806 describing this dialectical process, the physical mechanism of the conflict of electricity can be described in general terms as follows. Every part of every physical body contains the two fundamental forces of nature. Under normal circumstances, these forces lie dormant in a state of relative equilibrium or indifference. When the *Grundkräfte* outside of the body, but in close proximity to it, have been disturbed from their state of indifference in such a way as to take on the form of electrical action, that form of action disrupts the relative indifference of the initially nonelectrified body or conductor, causing the *Grundkräfte* throughout it to react to the external activity. This reaction occurs initially in the form of opposition and conflict, with the appropriate *Grundkraft* resisting the influence of the predominant force in its near vicinity. The resistance mounted by that *Grundkraft* results in a higher concentration of it in one location, thereby disrupting the equilibrium of the *Grundkräfte* in the body and thus creating the conditions for an internal opposition and resistance to the newly concentrated *Grundkraft*. This progression of conflict and resolution continues until the conflict between opposite principles reaches a new and different state of relative equilibrium, e.g., a polarized (re)distribution of electric charge over the body. The process of conflict and resolution between the *Grundkräfte* also accounts for the transmission of electric effects, or action, through a conductor in general, as manifested by alternating expansions and contractions under the right conditions. Ørsted, in fact, had tried to show that alternating conflict and resolution between the *Grundkräfte* is *the* fundamental mechanism behind the communication, or propagation, of all physical effects in nature. The precise nature of the resultant effect, e.g., as either magnetism or electricity, depends solely on the form in which the *Grundkräfte* are acting, which is directly determined by the intensity and quantity of the *Grundkräfte* either in free space or in

[41] Ibid., pp. 212–13.

physical objects and particles of matter and by the degree to which they are bound together. These factors, in turn, determine the formal (spatial) range of the specific effect in terms of its ability to act in either one, two, or three dimensions.

The principle of progressive dialectical conflict and resolution was accepted by Ørsted as one of the truly fundamental universal laws of nature. It was a process that he saw occurring throughout the scale of nature, and that he believed was responsible for keeping nature alive and active, while also guaranteeing the continuous creation of ever higher and more perfect natural forms. The universality of this principle even extended into the realm of human affairs and history. As he wrote the following year in an essay on the history of chemistry, "[i]t is quite in the nature of the human spirit to work in alternating expansions and contractions."[42] Indeed, for Ørsted, dialectical progress was perhaps the strongest expression of the unity between mind and nature dreamed of by Schelling and the Romantics. What he had to say in this regard is so important for understanding the foundations of his scientific research and worldview in general that it is well worth quoting at length.

[A]ll human knowledge has always, although with varying clarity, intervened in the essence of things. This has always developed through a continually renewed struggle, which, however, has resolved itself in perfect harmony. And it is not just science, not just human nature, it is all of nature that develops according to these laws. To show this to its full extent would be to present a complete natural science and a complete history. Therefore, here as above, I am forced to content myself with a simple account of a single observation. The evolution of the earth seems to me to be the most suitable for this. We are able to penetrate into the darkness which envelops the history of our globe as we penetrate into its bowels and compare the deeper layers with the older and the newer ones. By examining these layers and the fossilized or imprinted creatures which are found in them, we learn that the earth began with enormous creative forces but with little definite direction. Through alternating expansions and contractions, it has gradually killed and buried its earlier creatures in order to make room for the present chain of creations with man at the head. It is clear to every open-eyed observer of nature that the creative and organizing forces have alternated, with an ever increasing predominance of the organizing ones, however, and that it has arrived at the stage of development at which it now stands only after many struggles. In short, the development of the earth is like that of the human spirit. This harmony between nature and spirit is hardly accidental. The further we progress, the more perfect you will find it and all the more readily agree with me in assuming that both natures are shoots from a common root. . . . We have glanced at a higher physics, where the development of science itself, along with all its apparent contradictions, belongs to the laws of nature. It shows us that all and everything has grown from one common root and will develop into a common life. However, where something must be and work and grow, there the forces must have abandoned

---

[42] "Reflections on the History of Chemistry, A Lecture" (Chapter 25, this volume), p. 258.

pure equilibrium, and a struggle must have begun. One force must have won but only for a time. Another must since have achieved dominance, but that, too, must have yielded when it had created its product and threatened to continue and disturb the rest. While thus

> Everything 'twixt love and hate,
> To the last link must alternate,

while the scholar himself must take part in this alternation because his own human passions are swayed by the influence of external nature, he can still, if only he fixes his eye on the steady unity in this confusion, maintain a confidence and a tranquility, indeed, a blessedness that no might in the world can destroy.[43]

Ørsted never strayed from the views presented in this passage. They are a powerful expression of the deep Romantic foundations of his understanding of life and the world, an understanding in which the order of the cosmos resided in the continuous oscillation between polar opposites, between love and hate, between conflict and harmony, between expansion and contraction, between the two *Grundkräfte*.

Ørsted brought together the fruits of all of his theoretical speculation and experimental findings in a single synthetic work in 1812. Its title was *View of the Chemical Laws of Nature* (Chapter 30, this volume) and was his greatest effort at laying the metaphysical and physical foundations for a comprehensive dynamical science of chemistry. It was also his last great attempt to lay the physical foundations for the mathematical treatment of chemical phenomena. He set out to accomplish his first goal through a systematic presentation of his theory of the *Grundkräfte* and their various physical specifications, e.g., as either positive or negative electricity; north or south magnetism; or in the context of chemical action, *Brennkraft* or *Zündkraft*. He intended to accomplish his second goal by analyzing the correlations between the two fundamental forces, natural forms, and physical effects. In both endeavors, the form of action of the *Grundkräfte*, for which he now adopted the term *Wirkungsform*, had become the central concept. Indeed, the idea of *Wirkungsform* had fully become one of the essential concepts of his dynamical theory of nature.

In addition to presenting the most sustained and coherent treatment of his understanding of the two *Grundkräfte* and of their various forms of action, the *Chemical Laws* also contains elements of Ørsted's deepest theoretical and metaphysical beliefs, which he had only hinted at in the history of chemistry essay when he referred to the "common root" from which both mind and nature have sprung, and which exists beneath the apparent conflict between polar opposites. In the *Chemical Laws*, Ørsted revealed, in fact, that the cosmos is not ultimately dualistic, filled with two opposing fundamental forces, but that it is intrinsically One and filled with a single primordial force, or *Urkraft*.

He presented his views on the ultimate unity of nature in the section of the *Chemical Laws* containing his general observations on the two fundamental forces

---

[43] Ibid., pp. 259–60.

and his analysis of the essential difference between atomism and dynamism. Atomism, he argued, "wants to compose the whole from already-finished parts,"[44] by attempting to explain everything "solely on the basis of the size, the form, and the motion of atoms." As one would expect, for Ørsted, this approach to the study of nature had had a "limiting and misleading influence on natural philosophy." Dynamism, on the other hand, offers a clearer perspective on nature through the realization that "there is nothing dead and rigid in nature, but that every thing exists only as a result of an evolution, that this evolution proceeds according to laws, and that, therefore, the essence of every thing is based on the totality of the laws or on the unity of the laws. . . ." In other words, according to Ørsted, dynamism "allows the whole to develop together with its parts as different forms of an elemental force." Earlier he had already proposed that the two *Grundkräfte* be regarded as "two opposite aspects of a single, space-filling activity," such that we "imagine all of space pervaded by an elemental force, of which our two forces would merely be different modes of action."

But what exactly was this omnipresent *Urkraft*? To answer this, we need to return to Ørsted's understanding of natural laws and their unity. According to Ørsted, as we have already seen, every physical phenomenon must be considered and understood not in isolation but in terms of its connection to the rest of nature. Given this, he argued in the *Chemical Laws* that every object must be "regarded as an active agent of a more comprehensive whole, which again belongs to a higher whole so that only the great All sets the limit of this progression. And thus the universe itself would be regarded as the totality of the evolutions, and its law would be the unity of all other laws." Moreover, as we have already seen, Ørsted believed that mind and nature were intrinsically and essentially connected to each other. He therefore continued his line of reasoning with the following:

> . . . what finally gives the study of nature its ultimate meaning is the clear understanding that natural laws are identical with laws of reason, so they are in their application like thoughts; the totality of the laws of an object, regarded as its essence, is therefore an idea of Nature, and the law or the essence of the universe is the quintessence of all ideas, identical with absolute Reason. And so we see all of nature as the *manifestation* of one infinite force and one infinite reason united, as the *revelation of God*.[45]

A year earlier in his *First Introduction to General Physics*, Ørsted had declared in a very similar discussion of the unity of the laws of nature that "the world is merely the revelation of the combined creative power and reason of the Godhead."[46] He further declared in that work that "the investigation of nature must arise from the fundamental truth: That all nature is the revelation of an infinitely rational will, and that it is the task of science to apprehend as much of it as possible with finite

---

[44] *Views of the Chemical Laws of Nature* (Chapter 30, this volume), p. 384.
[45] Ibid.
[46] *First Introduction to General Physics* (Chapter 29, this volume), p. 286.

forces."[47] Four years earlier, in 1808, he had published a philosophical dialogue on mysticism in which his interlocutors reached the same fundamental truth: "What we have called the essence of nature is thus an infinite, omnipresent, eternal essence, an omnipotent creator and preserver of a world of reason, the great mystery of Life. God! And the world is the revelation of the Godhead."[48] We have already seen that, as early as 1807 in his letter to Oehlenschläger, Ørsted had considered Schelling's greatest achievement to be his understanding that "nature is nothing other than the revelation of the Godhead,"[49] an idea that Ørsted had first encountered in his reading of Schelling's *On the World-Soul* in 1798.

In the end, Ørsted's *Urkraft* was nothing other than an aspect of God manifested in nature, i.e., of God's creative Will, which, in turn, is organized by a second aspect of God's being, i.e., of divine reason manifested in nature as the universal laws of the universe. Indeed, in much the same vein as Herder, for whom God was the "primordial force of all forces,"[50] and as Ritter, who shortly before his death wrote that one "should not hesitate for a moment to call the active principle in force God,"[51] Ørsted also believed that all of nature was nothing but the evolutionary and temporal actualization of God. His commitment to dynamism and the view that all natural phenomena can be reduced to being merely modifications of the two *Grundkräfte*, which are only different aspects of an underlying *Urkraft*, was rooted in his deep theological beliefs. It was the unquestionable certainty of the existence and attributes of God that, in the end, served as the ultimate foundation for his metaphysical and physical beliefs, and that guaranteed the validity both of his understanding of the nature and actions of natural forces and of his knowledge that electricity and magnetism are ultimately identical, being only modifications of the universal *Urkraft*, i.e., of God. Likewise, he found the ultimate, unconditioned first principle for all of nature and natural science in God. All natural science could therefore be deduced a priori from the qualities of God himself. In this way, Ørsted believed that knowledge of nature is at the same time knowledge of the Divine. Indeed, as he told an audience at the University of Copenhagen in 1814, "we [. . . seek] to confirm our conviction of the harmony between religion and science, by contemplating how the man of science, if he fully understands his own endeavours, must regard the pursuit of science as an exercise of religion."[52]

In the years following 1830, Ørsted continued to conduct experimental research on galvanic and electromagnetic subjects (e.g., diamagnetism), as well as on the compressibility of water. The time had passed, however, for his greatest scientific work, and during the final two decades of his life, he devoted more attention to

---

[47] Ibid., p. 292.

[48] "Samtale over Mysticismen," in *Samlede og Efterladte Skrifter*, Vol. 5, p. 94. Copenhagen 1851.

[49] *Breve* (see footnote 8 above), Vol. 2, p. 230.

[50] J. G. Herder, *Gott. Einige Gespräche*, 1787 and 1800. In *Sämmtliche Werke*, ed. by Bernhard Suphan, Vol. 16, p. 453. Berlin 1887. See also p. 480.

[51] J. W. Ritter, *Fragmente aus dem Nachlaße eines jungen Physikers*, p. 156. Heidelberg 1810.

[52] H. C. Ørsted, "Videnskabsdyrkningen, betragtet som Religionsuddøvelse," in *Aanden i Naturen*, 3rd ed., Vol. 1, pp. 136–37. Copenhagen 1856.

writing philosophical essays and dialogues and to expounding his general philosophy of nature and its theological foundations. Throughout these years, he continued to be committed to revealing the presence of God (and of the spiritual in general) in the material world. When he died in 1851, he was at work on what he intended to be the greatest and most complete exposition of his scientific, philosophical, and theological beliefs. It was to be given the title *The Way from Nature to God*. It is perhaps most fitting that this, his final word on matters physical and metaphysical, necessarily remained a fragment of the whole.

Andrew D. Wilson
Keene, New Hampshire
October 1995

SELECTED SCIENTIFIC

WORKS OF

Hans Christian Ørsted

1

Response to the Prize Question in Medicine
Set by the University of Copenhagen in the Year 1797:
On the Origin and Use of Amniotic Fluid[1]

Chemia, oculus Medicinæ alter[2]

DEDICATED TO MY FIRST TEACHER OF CHEMISTRY, MY PRECIOUS FATHER,
SØREN KRISTIAN ØRSTED, APOTHECARY IN RUDKJØBING,
AS A SMALL TOKEN OF CONTINUED ASSIDUITY,
BY HANS KRISTIAN ØRSTED, GRADUATE IN PHARMACY.

HEREBY I deliver to the judgment of the public and, in particular, to that of my fellow students the prize paper in medicine, with which I was fortunate enough to win the announced academic prize in the last prize competition. I think that my fellow students would prefer to see it printed unchanged, as it was when it was submitted, and I owe them the courtesy to fulfill this presumed wish, but as I owe to Truth the correction of the errors to which my judges have drawn my attention, this has been done in the added rectifying notes, designated by the Greek letters: α, β, γ, δ.

The evaluation of the prize paper was as follows:[3]

The author of this paper presents a treatise of brilliant clarity and perspicacity, composed with great skill. He has treated the chemical analysis particularly thoroughly and has splendidly solved and completed the proposed task. In addition, he has displayed a modesty which is unusual and singular in our time and in this way endeavoured to reconcile divergent opinions of various

---

[1] [*Bibliothek for Physik, Medicin og Oeconomie*, Vol. 13, No. 3, pp. 217–77. Copenhagen 1798. KM I, pp. 3–31. Originally published in Danish.]

[2] [Chemistry, the second eye of medicine.]

[3] [Auctor hujus commentationis offert tractatum lucido ordine, perspicuitate & solerti opera conscriptum, præsertim in analysi chemica industriam suam monstravit, materiam propositam bene absolvendo & perficiendo, singulari præterea usus est modestia hodie rara, qva contrarias auctorum sententias reconciliare studuit. Illi ergo laudem et Premium promeritum denegari non posse, omnes censemus, solummodo optantes, ut nonnulla momenta qvæ passim in veritates physicas offendunt, prius corrigantur, qvam in lucem publicam aliqvando prodeat tractatus.]

writers. Therefore, we all agree that he cannot be denied praise and is worthy of the prize. However, we shall request that occasional details which are in conflict with physical truths be corrected before the treatise is eventually published.

M. Saxtorph, Tode, Bang, Aasheim.

Of all the branches of the knowledge of nature, none exists in which there is more uncertainty than the one which is concerned with investigating the way in which nature produces the animal body. It seems as if nature has swathed this sacred thing in an impenetrable darkness and will only allow its favourites to sense what is hidden in it. However, human thirst for knowledge will not be limited; it also wishes to discover the laws which rule nature in this its most splendid task. With the bright colours of imagination it paints what reliable experience has not shown it and creates easily overthrown theories where facts do not guide it. However, if we do not want to give up for ever the hope of attaining some certainty in this, we must first investigate more closely the part of it which immediately offers itself to our senses, for the disagreement of the greatest natural philosophers proves that even in this we have not proceeded as far as we could have. No matter how distant the desired goal seems to be, we should not lose courage but risk every attempt our powers will permit if we had the slightest hope that our efforts would not be completely in vain. Urged by this happy hope, I began writing about the origin and use of amniotic fluid, and encouraged by trust in the lenient indulgence of my judges I completed these pages. If they are fortunate enough to win their approval, the first of my wishes will have been fulfilled.

Students of nature have always disagreed about the origin and use of amniotic fluid. One side asserted that it came from the fœtus, while the other tried to prove that it came from the mother. On the one hand, a nourishing power was attributed to it while, on the other, this was completely denied. If an Egyptian darkness shall not forever hide the truth from us, we must, with all possible assiduity, try to discover the nature of amniotic fluid and separate it into its constituents with the aid of chemistry. Though many wise men have already trodden this path, the discord which rules among them in this matter and the conflict which even appears to exist between unquestionable experiences show that not all that the inquisitive student of nature might wish in this issue has been done. Perhaps the torch of chemistry has not yet illuminated this matter sufficiently, perhaps an observation which escaped even the most penetrating eye may some day be discovered by a weaker one and thus place all this in a clearer light. This thought has made me indulge the hope that perhaps I could contribute something to end the chemical controversy about the nature of amniotic fluid. However, before I proceed to the chemical investigation, I shall point out a few things which ought to be known about amniotic fluid, especially with regard to its sensible properties.

Amniotic fluid is contained in the innermost membrane of the ovum, which is called amnion, and thus differs from the false amniotic fluid, which is contained between the two innermost membranes of the ovum, the chorion and the amnion. This false amniotic fluid gradually disappears whereas the true remains until birth.

Amniotic fluid is already found in the ovum before the embryo is formed[4] and grows along with the ovum at the beginning of pregnancy, but later on it decreases. When the embryo has been formed, its weight can be said to be inversely proportional to that of the embryo. When the embryo has the size of an ant, this fluid weighs approximately 6 *lod*.[5] Verheyen found $1^{1}/_{2}$ pounds of amniotic fluid with a calf which weighed only 3 *quintin*.[6] Van den Bosch found over 30 ounces with a calf which weighed only 7 ounces.[7] Van Doevern[8] found 36 ounces in a woman who had died in the 3rd or 4th month of pregnancy.[9] However, as the fœtus grows, its weight and that of the amniotic fluid approach each other more, so that the former always outweighs the latter from the 6th month of pregnancy on, except in extremely rare cases. In one human ovum, brought whole into the world, Wrisberg found 14, and in another 18, ounces of amniotic fluid.[10] When one collects that which runs out during delivery, one usually gets 4, 5 up to 8 ounces, and I know of only one instance when 11 were obtained; the remainder is used to moisten and lubricate the external female genitals.

As far as the specific gravity of amniotic fluid is concerned, we can determine with certainty that it is heavier than water, but it is not possible to determine it more accurately as amniotic fluid has different densities at different times. However, it is possible to determine that the amniotic fluid increases during pregnancy, in weight as well as in density.

At the beginning of pregnancy, the amniotic fluid is limpid or faintly yellow. During delivery it loses its limpidity and becomes somewhat milky. The cause of this is undoubtedly that the amniotic fluid, at the time of birth, comes into such violent motion that its viscous constituents enable it to combine with the oleaginous substance which covers the surface of the fœtus, whereby it must naturally become milky and opaque. As chemical analysis shows that amniotic fluid contains some lixivial salt, it would also be possible for this to combine with the oleaginous substance and thus form a soap,[11] whereby this dissolution would happen all the more readily. In unnatural cases this fluid has also been found to have a red or some other unusual colour. Thus van Doveren[12] found it to be intensely red[13] so that he could hardly distinguish the position of the fœtus.

There is no appreciable smell to fresh amniotic fluid, but when it has been left for some time, and especially when it is boiled, it emits an odour which is very close to that of milk. Its taste is not unlike that of whey. When a little of it is taken between the fingers, it turns out to be sticky and viscous.

---

[4] Blumenbach's *Institutiones Physiologiæ*, a footnote to §571.

[5] Haller's *Elementa physiologiæ*, p. 192. [This weight unit, corresponding to $^{1}/_{32}$ of a Danish pound of 500 grams, was used in Denmark until abolished by a law of February 19, 1861.]

[6] Loc. cit. [4 *quintin* = 1 *lod*. See the previous footnote.]

[7] Van den Bosch, *De natura et utilitate liqvoris amnii*, §9.

[8] [van Doeveren.]　　　　　[9] Ibid.　　　　　[10] Ibid.

[11] Here I have overlooked the fact that lixivial salt is not easily found in a caustic state in the animal body since it has so much opportunity to become saturated with carbonic acid. (α.)

[12] [van Doeveren.]

[13] In a woman who had died of an acute illness. Van den Bosch, loc. cit., §19.

But now to the chemical properties of amniotic fluid. It is truly exceptional to see experiments, made by the most astute men, which are so completely at variance with each other as the ones concerning amniotic fluid. According to the experiments of some, it coagulated both when heated and when mixed with acids, alcohol, or astringent vegetable matter, while others have seen the opposite result in their experiments. Of course, the former had to assume its similarity to the lymph of the blood, which is a somewhat diluted albumen, whereas the latter have compared it now to mucus, now to gluten,[14] and now to urine. I shall mention the experiments and experiences which speak for one view as well as those which speak for the other and also try to show that they can be combined with each other quite easily, and that the conflict between them is only apparent. The most, and the most important, experiments have been conducted by Van den Bosch. Roederer has also performed some with similar results. In order, if possible, to attain more certainty in this matter, I have repeated their experiments myself and added a couple of my own.

Before I proceed, I consider it necessary to point out that the amniotic fluid which Bosch and I have used for our experiments was taken during delivery. It is highly likely that Roederer did the same since he, being an *accoucheur*, had so much opportunity to obtain it. As it is necessary for chemical experiments to have it as free of foreign components as possible, it first has to be filtered in order to separate it from the mucus and the foreign components which it contains. Roederer and Bosch say that they have done the same.

When one mixes blue plant juices with amniotic fluid, they do not immediately change colour, but when it has been left for some time (with syrup of violet for one hour, with tincture of brassica for half an hour), it turns green. When it has first been boiled down by half, this change happens sooner. Van den Bosch wants to prove from this that it contains some alkali,[15] but I do not think that this experiment warrants this conclusion. On the contrary, it seems to me to show that lixivial salt must have been formed after the syrup was added. This also happens with other animal fluids which are inclined to putrefy because they create ammonia from their hydrogen during this process. When it has been evaporated to half, the change of colour of the plant juice happens more quickly because volatile lixivial salt has been produced during evaporation. In these phenomena, then, amniotic fluid does not appear to be different from other watery animal fluids, for instance water distilled from blood and gluten.

According to the experiments of some, amniotic fluid coagulates with the aid of sulphuric acid and nitric acid.[16] However, I have not succeeded in doing this, nor have Bosch[17] and Roederer.[18] When sulphuric acid, in large quantities, is mixed with amniotic fluid, heat is generated, but there is no sediment, which, on the other hand, is obtained if only a little is added. The cause of this must be that a great deal of acid dissolves the sediment again. This explains, then, why Roederer got some sediment, Bosch none. More sediment is obtained with nitric acid and also with

---

[14] [fibrin.]   [15] Loc. cit., §14.   [16] Haller, loc. cit., p. 194.
[17] Loc. cit., §15.   [18] *Opuscula medica*, pp. 99–100.

muriatic acid, though not so much as with nitric acid. Nor have I been able to cause amniotic fluid to coagulate with any of the other acids; nor with alum either, although others say that they have succeeded in that.[19]

Pure lixivial salts produce strange clouds in amniotic fluid, the carbonates, on the other hand, less so. Roederer, however, got them with the so-called *oleum tartari per deliqvium*, rather dense, like the ones which are seen when alcohol is mixed with water in which sugar has been dissolved.[20]

According to the assertion of some,[21] amniotic fluid also coagulates with alcohol, but according to the experiments conducted by Roederer, Bosch, and myself, it only produced rather dense milky clouds, which are best seen when it has been left for some time.

According to Roederer's experiments and my own, astringent vegetable matter, e.g., an infusion of gall-apples, makes amniotic fluid rather cloudy because of intense sedimentation. Haller, on the other hand, reports that amniotic fluid has been made to coagulate by this means.

Nitrate of mercury, mixed with amniotic fluid, first gives a milky sediment, but this later turns faintly rosy. This provides grounds for the assumption that amniotic fluid must contain phosphoric acid as it is possible to make phosphorus from this sediment. Fourcroy has demonstrated that the red sediment which is obtained from blood serum and other animal fluids by means of nitrate of mercury consists of phosphorus and quicklime.[22] It must, however, be pointed out that much more sediment is obtained from blood serum than from amniotic fluid, from which it follows that the latter must contain less phosphorus than the former.

We have an admirable number of experiments on the behaviour of amniotic fluid when heated but, unfortunately, just as contradictory as the ones mentioned above. Rhades made it coagulate at 188° when it was obtained from animals, and even more easily when it came from human beings.[23] Besides him, Haller mentions many who have made this experiment with the same result, but, on the other hand, there are just as many whose experiments gave the opposite result. Mauriceau, Roederer, Bosch, and a great many others, whom Haller enumerates, as well as myself, have not been able to cause it to coagulate when heated. I put the amniotic fluid in a flat-bottomed vessel so that oxygen, which Fourcroy[24] not without reason considers to be the cause of the coagulation of albumen, could all the more easily combine with it from the surrounding air and placed it in a sand bath. At the beginning I gave it low heat and later more intense until it boiled, but no coagulum appeared in it. I continued the boiling until about $^9/_{10}$ had evaporated, but I got nothing except a tough membrane on its surface, which cannot in any way be considered a coagulum. I attempted to dissolve this membrane in a solution of lixivial salt and in water, but I did not succeed. This convinces us that amniotic fluid cannot be gluten as its residue following evaporation can again be dissolved in water. I placed the fluid remaining after the evaporation in a cold place in order to see if it would not

[19] Haller, loc. cit., p. 194.

[20] Loc. cit.   [21] Haller, loc. cit.

[22] Fourcroy, *Elements de Chemie*, 4th edition, Vol. 4, p. 313ff.

[23] Haller, loc. cit.   [24] Crell's *Chem. Anal.*, 1793, Vol. 2, p. 445.

crystallize. After 24 hours some very small crystals did appear, but in such a small quantity that I could not examine them more closely. The distillation of amniotic fluid has the same result as that of other animal fluids with regard to the fluids which are produced by it. First, pure water is distilled, then alkaline spirits, and finally empyreumatic oil. Because of the watery nature of amniotic fluid, it was to be expected that it would yield little of the latter, and indeed the result of the distillation was consistent with this assumption. Only a few drops of this oil were distilled. The residue is particularly worthy of examination in order to determine what salts amniotic fluid contains. When the soluble part of the residue is dissolved in water, this acquires an alkaline taste, and the yellow sediment it gives when mixed with a solution of corrosive sublimate confirms this testimony of our senses. When sulphuric acid is poured on this leached salt, muriatic acid is produced, which clearly manifests itself by its odour. It is very difficult to determine whether this proves the presence of some common salt or some digestive salt when one does not have more than what is obtained from the 8 ounces that I used. That it is only present in small quantities can be seen from the fact that it is very difficult to make it crackle in fire. No iron can be detected in the residue from the distillation by means of a magnet. I have not here determined how much weakly and how much more strongly alkaline water I got as the latter differs from the former only in regard to the quantity of its alkali, and more is obtained of either one or the other according to whether the receiving vessel is removed sooner or later.

As it is beyond doubt that amniotic fluid contains many foreign elements, I left it for 24 hours in a securely corked bottle so that it would not putrefy. After that period of time, I found that a quantity of mucus had sunk to the bottom, and some grease floated on top. The former amounted to about one-twelfth of the fluid put aside, the latter to one-sixteenth. I have never had so much of it that I could make a chemical investigation, and it is impossible to keep it until more is obtained because it putrefies so quickly.

If amniotic fluid is exposed to the influence of air, it undergoes the same changes as other animal fluids under similar circumstances, viz., it putrefies. In this putrefaction it should be noted that amniotic fluid does not begin with a sour fermentation, whereby it differs from gluten. Otherwise, it does not differ from other animal fluids during putrefaction.

It is now possible to deduce the basic constituents of amniotic fluid from these experiments. The water filled with lixivial salt that was first distilled contained, as water, oxygen and hydrogen and, as volatile lixivial salt, hydrogen and nitrogen. The empyreumatic oil, which was distilled partly on its own and partly mixed with the alkaline water imparting a yellowish colour to it (which I forgot to mention before), consists of hydrogen, carbon, and very little oxygen. The leached residue showed that it contained fixed lixivial salt and common or digestive salt. The experiment with nitrate of mercury showed that it contained phosphorated quicklime. Thus, the basic constituents of amniotic fluid are hydrogen, nitrogen, carbon, oxygen, phosphorus, and quicklime.

Now, we have here a number of experiments and experiences of amniotic fluid, but how contradictory! On the one hand, it is said that amniotic fluid coagulates when heated, when mixed with acid, with alcohol, or with astringent vegetable matter, but in all this amniotic fluid resembles the lymph of the blood and albumen, so according to these experiments we must assume its principal constituent to be albumen. On the other hand, there are even more numerous experiments which seem to indicate the opposite. Were we to combine these experiments with each other, we must either assume that the amniotic fluid used in those experiments was not in its natural state, or we must prove that albumen can exist in two different states, in one of which it coagulates, in the other it does not. The answer which has most often been given to this question, that the amniotic fluid which does not coagulate must have putrefied, will hardly satisfy anyone who, like me, has had an opportunity to see that it does not coagulate no matter how recently it has been taken from the mother.

Chemical experience proves that albumen can actually be found in a state in which it cannot be made to coagulate. When one part of the lymph of the blood, which is almost pure albumen diluted with water, is mixed with 6 to 8 parts water, it will no longer coagulate with heat. In order to obtain more certainty in this matter, I mixed 12 to 16 parts of water with one part albumen and subjected this mixture to the same chemical experiments that I had conducted with amniotic fluid, and the experiments gave me the same results. Like amniotic fluid, it could not be made to coagulate by any of the usual means, and when most of it had evaporated over heat, the same membrane appeared on its surface as the one on amniotic fluid. This membrane, like the one on amniotic fluid, also had the property that it could not be dissolved in water or alkali. If we now consider all the various experiments in which amniotic fluid has successfully been made to coagulate, it seems quite certain that it is diluted albumen, which only occasionally has a concentration such that it is able to coagulate.

But why, it might be objected, was it only the anatomists, who took amniotic fluid from dead bodies, who succeeded in coagulating amniotic fluid and not the *accoucheurs*, who took it from the womb at birth? Does that not lead to the suspicion that the amniotic fluid which was used in the experiments of the former was not in a natural state? I believe that I can answer this with an unequivocal "No"! It is true that it has generally been anatomists who succeeded in this, and that they took fluids from dead bodies, but was it always from those who had died of a disease? No, undoubtedly only rarely. The older anatomists, and it is from these that we have the greatest number of experiments on this, did not often have the opportunity to conduct their experiments on humans, as it was less often their lot than that of later anatomists to obtain human bodies for dissection, so they obtained it from animals, and these were likely to be healthy except in rare cases. I shall confirm my assertion by mentioning some of these experiments. By means of heat, Vieussens coagulated the amniotic fluid of a ewe which was four months gone.[25] It is fairly certain that this animal was not sick because this excellent anatomist, who otherwise so carefully describes the illness from which a human being had died, does not mention any disease here. Rhades found the amniotic fluid of animals

[25] Vieussens, *Novum vasorum corporis humani systema*, 1655 [1705], 12, p. 22 & p. 13.

coagulable at 188° and that of human beings even more susceptible.[26] Langlej[27] found the amniotic fluid of the hen coagulable when it had reached some maturity, otherwise not.[28] Hertod[29] also found it coagulable in a dog.[30] Maday found it coagulable in a three-month-old aborted ovum.[31]

In view of all this clear evidence, it cannot be denied that amniotic fluid is sometimes coagulable. Now it is only a matter of answering the question: Why is the fluid that runs out during delivery not similarly so? I must answer this by demonstrating that amniotic fluid loses its coagulability during birth. The nature of things might already make us suspect this. The oleaginous substance which covers the surface of the fœtus, and which is often seen floating in the amniotic fluid itself, the mucus which the fœtus secretes, especially during the final period, in a not inconsiderable quantity, cannot but mix with the amniotic fluid in the course of the motion created both by the contraction of the womb and by the movement of the fœtus itself. Hereby, part of the albumen in the amniotic fluid must combine with the oleaginous elements which partly cover the surface of the fœtus and partly float in it, just as albumen, through shaking and mixing, combines with oil; the mucus must similarly combine with some of the particles in the albumen and thereby form a rather viscous substance. Since the most mucous substance also adheres most easily to what it comes in contact with, these albuminous compounds are mostly exhausted in lubricating the vagina as the fœtus must pass through it. The mucus and oleaginous matter still left must remain on the filtre when the amniotic fluid is strained in order to make it clear. Even if this were allowed to remain, it could still not be made to coagulate because the mucus and the grease keep the albumen so strongly combined with them. This shows that amniotic fluid must lose much of its albumen during delivery, indeed even a little before that time, whereby the small quantity left becomes so dilute that it can no longer be made to coagulate. These inferences, together with the experiments mentioned, already seem to have proved to a high degree of probability that amniotic fluid loses its coagulability when the time of delivery is at hand. However, I have one proof remaining from Malpighi, which raises this my opinion to an absolute certainty. "*I will add this one thing,*" says this great man about the hen's egg, "*that the weak solution of albumen inside the chorion sometimes disappears when put over fire, particularly from the tenth day until the chick is hatched. The same often happens to the juice contained in the amnion; in a very young chick it is seen to coagulate to some extent, in a mature chick, and particularly when it is just about to be hatched, it has turned salt and evaporates over fire.*"[32]

To this may be added that, according to chemical analysis, the constituents

---

[26] Haller, loc. cit., 194.   [27] [Langly.]

[28] Ibid.   [29] [Hertodt.]   [30] Ibid.

[31] Maday [Madai], *Specimen, sistens anatomen ovi foecundati, sed deformis*, §1, No. VII.

[32] Malpighi, *Diss. de formatione pulli*, p. 12, in the version which is found in his *Opera omnia Londini*, 1687, Part 2. [Hoc unum addam tenue albumen, corio contentum, igni appositum interdum evanescere, præcipue a die decima ad pulli exitum. Idem freqventer accidit contento succo in amnio, qvi in tenello tamen pullo secundum aliqvam suam partem concrevit, in adulto vero & mox nascituro præcipue, salsus redditus igne evaporat.]

of amniotic fluid are in no way different from those of albumen, which are, like those of amniotic fluid: hydrogen, carbon, nitrogen, oxygen, phosphorus, and quicklime.[33]

As a consequence of all this, I believe that we may rightly assume that *amniotic fluid consists primarily of a thin, watery solution of albumen*, which, neither at the beginning nor at the end of pregnancy, contains enough particles of this substance to be made to coagulate but often exhibits this property between these two extremities.

Since we have now become reasonably familiar with the nature of amniotic fluid, we may hope to be able to determine something about its origin and use with greater certainty. We shall first deal with its origin.

I considered the determination of the origin of amniotic fluid the most difficult part of my project. This is a problem which neither Haller, the best physiologist of his time, nor Blumenbach, one of the best of ours, thought that they could solve. Therefore, I feel certain of forgiveness should I not be able to advance something satisfactory in this regard.

The less one has been able to say anything definite about the origin of amniotic fluid, the more fertile have been the theories. Most of these have little in their favour and irrefutable objections against them. I shall mention the ones I know and ask indulgence if I should have forgotten some, which is quite possible as they are strewn about in so many books of various contents that it is not possible to read them all.

Theories about the origin of amniotic fluid are divided into two groups. Those which suppose the source of this fluid to be in the mother belong to one, those which suppose it to be in the fœtus to the other. The latter I can refute all at once, but the former I must partly treat at somewhat greater length.[34]

Those who believe that amniotic fluid has its origin in the fœtus assume that it is:

1) The sweat of the fœtus.
2) The urine of the fœtus.
3) A mixture of both.
4) A juice which flows from the breasts of the fœtus.
5) The spit of the fœtus.
6) A mixture of its spit, snot, and urine.
7) A liquid which flows from its anus.

All these theories are refuted together 1) by the observation that the weight of the amniotic fluid has an inverse relation to that of the fœtus, as has been shown previously; 2) by the fact that they are all at variance with the results of the chemical analysis of amniotic fluid; 3) these theories are most completely refuted by the

---

[33] Gren's *Handbuch der gesamten Chemie* §1600.

[34] Most of the theories that I shall mention here are to be found in Haller, *Elem. physiol.*, Part 8, pp. 196–98, and in Temel [Themel], *Commentatio qva nutritionem foetus in utero per vasa umbilicalia solum fieri, occasione monstri ovilli sine ore et faucibus nati, ostenditur*. In the case of those which are not to be found in either of these places, I shall mention the book from which I have taken them.

fact that amniotic fluid exists before the fœtus or at least before it is perceptible to our senses.

Those who suppose the source of amniotic fluid to be in the mother assume that it comes either indirectly or directly from her. The former consider the placenta to be the source of it, the latter consider the uterus to be so. To the former belong: A, those who assume the umbilical cord to be the place where amniotic fluid originates; B, those who assume it to be the membranes of the ovum; C, those who think the placenta is the source; D, the hypothesis which Roederer adopted, that the ovum was supposed to be a *folliculus mucosus*;[35] E, the opinion adopted by Bonfilius[36] that amniotic fluid is in the ovum before the embryo, and that it grows with it.

To A belong those who consider the source of amniotic fluid to be:

α) The small warts on the umbilical cord,
β) its fluid ducts,
γ) the ends of its arteries.

These opinions cannot be maintained in view of the experience that oviparous quadrupeds have amniotic fluid but no umbilical cord. Nor are they compatible with the fact that amniotic fluid is in the ovum before the embryo, thus also before the umbilical cord has been formed, as I have just mentioned.

Under B belong those who consider the source of amniotic fluid to be in:

α) the glands of the chorion,
β) the lymphatic vessels of the chorion,
γ) the glands of the amnion.

These three hypotheses give sources whose existence has never been established anatomically. Without anatomical demonstration, they remain mere figments of the imagination, and the theories built upon them rest on sand and can be overturned by the slightest puff. De la Motte[37] mentions that Mery[38] showed him lymphatic vessels, full of a clear fluid, which wound not only through the membranes which contained this fluid but through all the parts which are used for reproduction and disappeared immediately following the death of the animal. However, this lacks confirmation by other anatomists, and it is also tempting to ask whether it would not be more reasonable to assume that these vessels absorbed their fluid from the amniotic fluid since they disappeared after death than to assume that they are sources of amniotic fluid. This becomes even more reasonable since Brugmans saw similar lymphatic vessels under the skin of the fœtus, which also lost their fluid upon the death of the fœtus.[39]

To C belongs the opinion of those who assume that amniotic fluid comes from small lactiferous ducts, which originate in the mother and terminate in the amnion. It is obvious that this opinion conflicts with the chemical analysis of amniotic fluid.

[35] Roederer, *Opuscula medica*, p. 100.
[36] [This is probably a reference to Silvestro Bonfioli, a seventeenth-century Italian physician and astrologer.]
[37] *Traité des accouchements*, Book 1, Chap. 24.   [38] [Méry.]
[39] Bosch, *De liqvore amnii*, §62.

D. Roederer's hypothesis was based on the chemical analysis which he himself had made of amniotic fluid. As he could not make it coagulate when heated, but, on the contrary, it could be made to evaporate almost completely and finally burn with an unpleasant odour, he held it to be a mucus and considered the ovum to be a *folliculus mucosus*. He believed his opinion to be even further confirmed by the fact that amniotic fluid gradually becomes thicker, like mucus. The chemical analysis which I mentioned above best illustrates how weakly founded this hypothesis is.

E. It is easy to see how we shall judge Bonfilius's opinion. By assuming that amniotic fluid comes into being with the ovum and grows with it, he has not at all answered the question: How does the amniotic fluid originate? In order to have answered this, he should also have explained in what way its growth occurred. Otherwise, however, this opinion deserves some attention as no significant objection can be made against it as far as I know, and as, with its modest form of explanation, it permits us to explain the growth correctly once we acquire greater knowledge. Why should we insist on explaining the increase of the amniotic fluid when we are not able to explain the growth of the membranes which surround it? A little more will be said about this opinion below.

Levret has declared himself in favour of the view that the amniotic fluid comes from the uterus. "The interior of the uterus," he says, "is perforated by a great number of very small holes which are not discovered immediately, except under certain circumstances which make them perceptible: These small openings penetrate a membrane which is so fine that anatomists have denied its existence; this membrane covers a substance which is, as it were, pulpous or cellular."[40] He believes that the clear and lymphatic part of the amniotic fluid comes from these fine openings, but he regards the part which makes it unclear as the excretions of the fœtus.[41] This view cannot possibly be considered proved until it is shown that there is a fluid in the cellular tissue of the womb which is similar to the one which is found in the innermost membrane of the ovum. As long as no such thing has been proved, I see no reason to accept the opinion of this otherwise highly deserving man.

The cause of the incompleteness of all these theories is undoubtedly that there has been a desire to indicate far too accurately the place where this fluid has its origin as it is possible that it comes from several sources at the same time, and that too little thought has been given to the nature of this fluid. If we compare amniotic fluid with other animal fluids, it is obvious at first glance that it is very similar to pericardial fluid[42] and that, after that, it resembles blood serum, which consists of albumen, diluted with water, and some salts. To some extent, this view can help us

[40] Levret, *L'Art des accouchements*, §165.

[41] Ibid., §316. In his often-cited physiology, p. 196, Haller attributes to Levret the opinion that it comes from the capillary tubes of the chorion, but the third edition, which I have used here, does not state anything like that. I am not in possession of the second edition, which Haller used.

[42] Bosch also compares these two fluids to each other, but for quite different reasons than I. He approves of the comparison because pericardial fluid does not coagulate easily; I, on the other hand, because it does coagulate to some extent and therefore contains a little albumen, like amniotic fluid.

formulate a reasonable theory of the origin of amniotic fluid. The fluid in the pericardium is secreted from the blood and is nothing but secreted blood serum. As this fluid has the same relation to blood serum as amniotic fluid, it seems resonable that it also originates in the same way.[43] But where are the arteries from which it should come? Several people, including Bosch in more recent times, have answered this question by indicating the arteries which wind through the amnion as this source. However, his arguments do not seem to me to prove what they were supposed to. He quite adroitly counters the objection that blood vessels cannot be discovered in the amnion by showing that they are found in animals, and therefore it can be inferred, by analogy, that they must also be found in humans. Besides, he also refers to Wrisberg's *Observationes de structura corporis humani*, p. 22, where this excellent anatomist says that he has seen a few blood vessels in the amnion. However, the best experiment he can cite is Monro's, according to which warm water, injected into the umbilical cord, was exuded again in the amnion. This experiment, however, does not prove that all amniotic fluid comes from these arteries; it merely makes it reasonable that some of it comes from them. It is impossible that it should all come from them since amniotic fluid, as I have already mentioned above, exists before the embryo or the umbilical cord, from where these arteries were supposed to originate. Still, in view of the experiment mentioned, it seems reasonable to me that the amnion contributes to the increase of amniotic fluid by means of these arteries, but it has undoubtedly been proved most decisively, by its existence before the embryo, that some of it must come from the mother, and this does not even require support from the experience that amniotic fluid has been found to be yellow from saffron which the mother had consumed. There is no doubt that a large part of the amniotic fluid comes from some lymph in the blood vessels of the placenta being transformed by the heat from the surrounding parts into vapours which penetrate the membranes of the ovum into the cavity of the amnion, where it loses some caloric and thus becomes liquid again. In this way it is easy to explain that amniotic fluid is diluted lymph, for it is mainly the watery parts which are thus transformed into vapour and carry some particles of albumen and the salts of the blood with them. In the discussion on the use of amniotic fluid, I shall try to explain how it gradually becomes more dense in the amnion itself, and also how its weight relative to the fœtus is reduced.

I do not think that we know anything more definite about the origin of amniotic fluid at the present time. Perhaps some fortunate discoveries of the future will be able to shed more light on this issue. We must only try as much as possible to avoid theories about it as long as we do not have sufficient data on which to build them. I have tried particularly to be brief concerning this topic as I do not consider it to be very useful to enumerate and refute at great length hypotheses which few or none accepts any longer.

Enough on the origin of amniotic fluid. In our investigations of it, Experience, this ever-faithful guide, abandoned us while we, surrounded by a chaos of hypotheses, did not know where to turn. We now want to proceed to deal with its use, where, less confused by conjectures, we can dedicate ourselves to our investiga-

---

[43] The analogy, then, is against Bonfilius's view.

tions without being interrupted at every moment by futile disputes. However, we do not find complete certainty in this part either. Therefore, we begin with that which we can determine with certainty and end with that which the disagreement of most scholars makes us doubt.

It is not only the fœtus but also the mother who profits considerably from the amniotic fluid, which also promotes the entire process of birth. As there is general agreement on its usefulness in these two respects, we shall begin with that here.

During pregnancy, the womb experiences two kinds of expansion, one passive and one active. Amniotic fluid plays an important role in both, in that it causes the former and assists the latter. As soon as the ovum, which has entered the womb after conception, has grown to the size that it can no longer float in the fluids which are found in the womb but touches its sides, it unites with the womb, as is well known. But no sooner has this happened than the womb begins to swell and reduce its cavity. It would be impossible for the ovum to expand the womb again if this did not contain some such incompressible fluid, like amniotic fluid, which is capable of resisting the reaction of the womb and of touching it at many points. The fluid causes the ovum to assume the same shape as the interior surface of the womb since, being fluid, it yields to any pressure. Thus it is able to act with even greater force because it comes into contact with the womb at more points, which might be regarded as the end points of levers.[44] However, in addition to the gradual re-expansion of the womb, which thus makes its reaction weaker and weaker every day, amniotic fluid also has the function of assisting the active dilation of the womb. Gradually, as the substance of the womb is distended by the in-flowing fluids, it begins to expand by itself even if the fruit is somewhere other than in the womb. It is natural that the fluid makes this work easier by making the ovum suitable to support the uterus at every point.

It is of even more important and more obvious utility during parturition. When the labour pains begin, and the womb contracts, this contraction takes place both in the neck of the womb and in the fundus of the uterus. Therefore the fœtus would rather be crushed to death than expelled from the mother if it did not in itself possess a motive force,[45] whereby it is primarily driven towards the neck of the womb and thus forces this to open where it is weakest, and this place is the external orifice of the uterus, which has absolutely no power to resist. As this motive force continues to act on it, the external orifice of the uterus must also dilate more and more until finally it is completely open. The motive force which causes all this is the

[44]The theorem which I have put forward here, from the generally accepted theory of the science of midwifery, according to which amniotic fluid is supposed to cause the dilation of the womb through its own expansion, is hardly in agreement with nature because the uterus does not expand equally at every point, nor even most in the places where it is weakest, which, however, should be the case if the water had brought this about solely through its increasing quantity and its thus added volume. It may also be pointed out here that it is not in accordance with the complete concept of the lever, which also requires rigidity, that the lines on the surface of the ovum could be regarded as levers with which they do seem to have some similarity in that one end must rise when the other falls. This similarity has undoubtedly induced Levret, Stein, and other famous *accoucheurs* that I have followed to use this analogy. (β)

[45]It is not to be feared that the fœtus would be crushed by the contraction of the womb as its capacity is increased by the contraction because the uterus changes from an elliptical to a more spherical shape. (γ)

weight of the fœtus.[46] As the amniotic fluid increases this, it will already be seen from this that it has an important function during birth. However, this is far from the most important one; it has two other important functions during birth. At first glance it will be noticed that the opening of the uterus and of the external female genitals is far smaller than the fœtus. Parturition would take place with a violence that would cost the life of both mother and fœtus without the presence of amniotic fluid. No sooner has labour begun to contract the uterus than the water is pushed forward by the contracted fundus of the uterus so that its membranes resemble a distended bladder in front of the head of the fœtus. As a fluid, it easily assumes the shape of any more solid body that presents itself. Therefore, it gradually forces itself into the narrow passage which the fœtus must pass, and evenly and gradually it dilates first the external orifice of the uterus and then the vagina, as the wedge widens the crack. Thus it clears a passage for the fœtus and frees it from the peril, in which it otherwise stood, of having its head crushed by the excessive resistance of the opposing solid parts and reduces the pains and the danger which threaten the mother by making the pressure which the birth canal and the surrounding parts must suffer less violent so that they are not so easily disrupted as otherwise. And when the membranes finally burst, it flows out, lubricates the birth canal, and reduces the volume of the ovum, whereby it creates for the fœtus an easier egress from the prison in which it was confined.

Besides thus promoting birth directly by moistening and somewhat dilating the birth canal, amniotic fluid is also of indirect use by facilitating the task of the *accoucheur*. Without it, the womb would close tightly around the hand of the *accoucheur* when he wanted to make the version, and there would be no room to turn the fœtus. Therefore, it is an important practical rule that one tries to keep as much fluid as possible in the womb when one bursts the membranes until the version has been executed. Mauriceau, in his *Observations sur grossesse, & L'accouchement des femmes, obs. XXVI*, tells a remarkable story about a woman, whose water had broken on the first day of labour so that it was not possible to turn the fœtus even though both he himself and an Englishman, Chamberlen, tried to do so. It also renders the *accoucheur* a not unimportant service by informing him of the onset and the phases of the delivery and of the position of the fœtus. There is no clearer indicator that labour has started, and that it is not false, than that the amniotic fluid is pushed in front of the head of the fœtus because this happens every time the fundus of the uterus contracts, and it is no longer possible to feel the head of the fœtus. At the beginning of labour it assumes a more convex position in its membranes than in the last two phases, and in that way the experienced *accoucheur* can easily see

[46]This sentence must be qualified as the fœtus is not perpendicular to the external orifice of the uterus in every position that the womb has, and this is required if it is to act with its weight. I am now more inclined to believe that the dilation of the external orifice of the uterus happens because of a stimulus as I have been convinced by Professor Aasheim's lectures and instructive conversations that a stimulus dilates a contractile circle. This is not the place to mention all his perceptive arguments, but one, taken from our subject, would be appropriate here, and that is that if the ovum were to dilate the external orifice of the uterus by its mere pressure, its edge would be forced outwards instead of being dilated and disappearing little by little. (δ)

how far the birth has progressed. It can also easily be seen from the flat or pointed position of the water whether the delivery is perfect, or whether the feet are forward. The trained eye also recognizes the other positions of the fœtus fairly well from the positions of the water.

As far as the mother is concerned, she also profits not inconsiderably from this fluid. Without it, every movement of the fœtus in the sensitive uterus would be felt most painfully by the mother, but this rather dense medium prevents it from making its movements with such speed that the resulting bumps would become very painful. The fœtus would also be far more trouble to its mother because of its weight if our fluid, by surrounding it, did not ensure that it made contact with the mother at many more points, whereby the pressure at each individual point is felt less. The irritation and the pressure which the mother must suffer in the absence of the water could also easily induce labour too soon and thereby cause premature birth. I have referred above to the important use of this fluid to the mother during delivery.

The uses of amniotic fluid to the fœtus are extremely important and manifold, and it would not even be able to live without it, of which we are convinced by the fact that the fœtus faces certain death if it breaks prematurely. If there should be exceptions to this, it must be because a little might collect again in the ovum. If the fœtus were inside the womb without this, it would be squeezed particularly at certain protruding points, and these parts would be injured or misshapen as this pressure would not allow it to grow as freely as its formative urges require. On the other hand, now that it has equal pressure everywhere from the surrounding fluid substance, it acquires a solidity which enables it to withstand the impacts which occur at each individual point. To my mind, therefore, amniotic fluid can be regarded as a ligament which holds all the parts of the fœtus together and through its pressure on it serves the same function that air does for the born human. Experience shows that man cannot endure in places where the air is very thin, e.g., on high mountains, but that he is overcome by dangerous attacks, which originate in excessive sublimation, e.g., violent hæmorrhage, and seems to verge on the disintegration of all his parts. Only denser air ends these inconveniences through its uniformly greater pressure. A fluid, like amniotic fluid, which has a density far greater than air, would be the most suitable to replace air for the fœtus as its far softer and less dense parts need far greater pressure to be held together properly than the more solid ones of the born human. By this, however, I do not mean to claim that the density of amniotic fluid is the smallest that would serve the fœtus.

Amniotic fluid can also be regarded as a warm bath which not only makes the skin and the fibres soft and elastic but also promotes the circulation of the blood. It also prevents the parts of the fœtus from growing together with each other and with the membranes of the ovum, just as the pericardium becomes attached to the heart itself when that special fluid which moistens it is absent. And although one sometimes sees the parts of the fœtus, or two fœtuses which do not have separate membranes, grown together, this would happen far more often if this fluid did not penetrate between all the limbs of the fœtus, thus preventing them from getting so close to each other that they could join.

If not surrounded by amniotic fluid, the fœtus would also feel very intensely every bump or other such strong impact which its mother experienced. Every impact is transmitted more quickly through the solid parts of a body than through a medium which possesses such a small degree of elasticity as amniotic fluid. We are convinced of this particularly by the experiments which Wünsch conducted in order to show that a sound wave is transmitted far more quickly and is felt far more intensely when it passes through a solid elastic body than when it passes through air.[47]

Besides the important use which I have attributed above to amniotic fluid during birth as regards the fœtus, it can also be pointed out that it frees the umbilical cord from the pressure which it would otherwise suffer, especially during birth, whereby the circulation of the blood in it, and consequently also that of the fœtus, would be interrupted and the fœtus killed, for it can no more dispense with communication with the placenta than the born human with air.

The question of whether amniotic fluid is nourishing or not has divided students of nature into three groups, one of which claims that the fœtus swallows some of it, the second that it merely absorbs it through the skin, and the third denies it all ability to nourish. All three groups have important arguments and authorities in their favour; we shall hear them. Those who claim nourishment of the fœtus through the mouth state that amniotic fluid has been seen in both the œsophagus and the stomach of the fœtus. However, it is rather uncertain whether it is the same fluid they have seen. Rather it seems to be a secretion of the arteries which has mixed with the mucus which so often covers these parts. The following is very closely connected with this argument. Excrement has been found with the fœtus, which must imply a digestion and something that is digested, and they thought this was amniotic fluid. To this it can be answered that, in that case, the fœtus must eliminate its excrement now and then during the long period it stays in the womb, but we find no trace of this in the amnion, and besides no more excrement than usual has been found in a fœtus with a closed anus.[48] Temel also found excrement with a sheep born without a mouth and throat.[49] Regner de Graaf found meconium in the large intestine of a headless pup.[50] This excrement could not possibly have come from swallowed amniotic fluid, so we must seek its origin elsewhere. I agree with Blumenbach in thinking that it owes its origin to the gall bladder, for once gall begins to be secreted in the fœtus, meconium is also found with it, and it has been impossible to discover any meconium in deformed babies born without a liver and consequently without a gall bladder.[51] Some have thought that the fœtus could also absorb it through the openings of the lachrymal ducts in the event that the mouth was closed. But could the headless animal obtain it in that way? Some have also wanted to prove that it was necessary for amniotic fluid to moisten the internal parts of the fœtus as they would otherwise grow together, but this argument falls when we see

[47] Wünsch, *Kosmologische Unterhaltungen*, Part 2.
[48] Buffon & Daubanton [Daubenton], *Histoire naturelle*. Amsterdam, 1766. Vol. 11, p. 145.
[49] Introduction to his treatise referred to above.
[50] R. de Graaf, *De mulierum organis*, p. 289.   [51] Blumenbach, loc. cit., §625.

that these parts can be moistened in other ways. Of even less consequence is the argument that it would be useful in training the digestive organs and keeping the œsophagus, the stomach, etc., dilated and open if some amniotic fluid were swallowed. In natural science, we must use induction in order to prove that everything is arranged in the best way, and if we now wanted to prove part of this theorem by using it in the induction, our argument becomes circular. Besides, we do not have sufficiently profound insight into nature to determine what is best; rather we must be content with the realization that we find what we are capable of understanding completely. It is also said that children are often seen born with an open mouth, so nothing is more reasonable than that some amniotic fluid must enter there. If this always happened, it would be a not unimportant argument for this view, but this is not the case; this is seen only occasionally and then most often in dead children.

Consequently, if it has not been proved that the mouth of the fœtus is generally open, we must also abandon the argument that it was reasonable that the muscles of the stomach and the diaphragm, through their motion and pressure, must force some amniotic fluid to pass through the œsophagus down to the stomach. More important than all these arguments in favour of nourishment through the mouth seems to be the one mentioned by Heister, who says that he saw a calf which was still confined inside the womb, quite frozen in the amniotic fluid, and that this fluid was connected to a mass found inside the calf itself; he saw this twice.[52] Here it must be noticed that the calf had to be dead before it could freeze in this manner, and therefore it would be possible for it to have had its mouth closed when alive but open after death. Moreover, it is a well-known fact in physics that water increases its volume noticeably upon freezing, whereby it must exert pressure on the calf with far greater intensity and thus penetrate all its orifices. However, the most important of all the arguments usually advanced in favour of nourishment through the mouth seems to be this one: Children have been seen born without an umbilical cord, in which case they could not have got their nourishment directly from the mother, so they must have lived on the nourishment which they have been able to take in through the mouth. If this fact were true, there could not be many objections to it, but Blumenbach[53] rightly considers accounts of this to be fables. I believe that even if we did not have critical historical reasons to doubt the correctness of this information, we would be sufficiently convinced of its incorrectness by the nature of the case. If these accounts were true, why can a fœtus not endure being deprived of its connection with the mother for even a few minutes when the umbilical cord is compressed during birth if, according to these accounts, it can endure it for a long period during pregnancy. Besides, I do not see how this fact could be so advantageous to this theory as it would force it to assume that amniotic fluid contained all the constituents that the complete human being has, but this does not agree with the chemical analysis, which, for example, does not find the iron of the blood in it.

Now, after having thus refuted the arguments with which people have wished to prove the nourishment of the fœtus through the mouth, it will not be difficult to

---

[52] Heister, *Compendium anatomicum*, p. 271.   [53] *Inst. Physiol.*, 574.

refute it completely with a few more remarks. The proof which is usually adduced against oral nourishment from the watery nature of amniotic fluid and the imperfection of the digestive organs of the fœtus appears to me not to be particularly weighty as, on the one hand, the fluid may well be sufficient for the as yet imperfect creature in spite of its wateriness since it does contain some nourishing elements, and, on the other, such a weak fluid is easily digested by the weak digestive organs of the fœtus.

On the other hand, the experience that a fœtus has lived for a long time inside the womb even though the amniotic fluid had run off is a somewhat more important argument though it might be possible to answer to this that amniotic fluid is considered to be a contributory source of nourishment for the fœtus which might, at least for some time, be missing.

I do not consider very important either the objection derived from the fact that the mouth of the fœtus is most often closed and from its bowed head since there are experiences which seem to point towards the opposite.

However, I shall now turn to some objections which I think are not easily refuted.

Children who are brought into the world prematurely can lie for several days without taking the slightest nourishment, during which time they sleep continuously, and only when this time has passed do they take any nourishment. How can this phenomenon be consistent with the oral nourishment of the fœtus in the womb? If the fœtus were already used to feeding itself in this way, why could it not do the same outside the womb? It is true that Haller[54] mentions that he and others have seen chickens and even quadrupeds open and close their mouths while they were still confined in their membranes, but this has hardly been without their having stayed somewhat longer in the amniotic fluid than until the appointed time of birth, and because of this they have almost instinctively opened their mouths to take nourishment.

Nor do I know how this view can be made consistent with the fact that living children are often brought into the world without a head or without a mouth, for if it served to nourish the fœtus in this way, it would not be reasonable to assume that they could live entirely without it. I have already mentioned examples of such deformities above. Nor is it more consistent with this view that so little excrement is found with the fœtus even though it has not been able to deposit any in the amnion, for it is not very likely that no greater quantity of excrement should be produced from all the amniotic fluid which had to be swallowed during pregnancy than that contained in the fœtus itself.

I think that the hypothesis of nourishment of the fœtus through the mouth has been sufficiently refuted by the arguments given here. However, one more phenomenon remains whose explanation seems to demand its acceptance, by which I mean that the quantity of amniotic fluid decreases proportionally as the weight of the fœtus increases, but if we add the phenomenon which accompanies it, viz., that the density of this water also increases, we can easily see that more explana-

[54]Loc. cit., p. 201.

tion is required here. Therefore, I shall attempt to explain them both with other arguments.

It is well known that we inhale air not only through our lungs but also through all the pores of our skin, which again exhale a substance that is quite similar to the exhalation of the lungs. As amniotic fluid performs the function of air in so many respects, by its pressure, by allowing the free movement of the foetus, etc., why should it not take the place of this elastic fluid in this respect also?

However, before I proceed to show this, I must be permitted to put forward something about inhalation through the skin. I shall begin by describing some experiments by Mr. Riegels. He had a box of glass made, two feet high and two feet wide, and in this he put rats and frogs in such a way that their heads were in the air, and then he placed them in the hottest midday sun from 11 AM to 2 PM. Now such a strong heat was created in the box that his hand could not endure it, and in a quarter of an hour two rats and two frogs covered the panes with their exhalation so that they became opaque. The pulmonary respiration of the rats increased noticeably, they yawned and twitched their nostrils intensely; so did the frogs. The frogs emitted drops of water from all over their body, and the hair of the rats became quite wet. After half an hour, the frogs died. When they were opened, their internal membranes were dry and easily breakable and, like the lungs, had lost all elasticity. On another occasion, he put a dog and a rat in the same way in a box which was filled with the most rotten gutter muck so that it would be filled with mephitic vapours, but the animals lived in it for four days.[55]

It is evident from these experiments: 1) that respiration through the skin is necessary to the sustenance of animal life, 2) that the membrane of skin or fat is the instrument of this respiration, 3) that the water in the last experiment must have been decomposed into its constituents, oxygen and hydrogen, and given off the former to the acidification of the blood, in order to combine with the blood just for a few moments and then separate from it again in other compounds. However, it might be asked, would it not be possible for the air, in the last experiment, to penetrate the pores of the wooden box and thus provide the animal with the air necessary for dermal respiration? But to this we can answer that the air must meet with resistance from the water in the box which would not be able to absorb very much air as it would soon fill up with the carbonic acid gas that the animals exhaled.

On the other hand, I think that another objection which might be made to this view of the nature of inhalation ought to be countered more carefully. It might be thought that the hydrogen which the water must give off to the blood along with its oxygen would be as harmful as the oxygen would be useful, but this is seen to be invalid when we consider that the direct benefit of inhalation does not consist in removing carbon and hydrogen from the blood but in oxidizing its iron, which, during the circulation of the blood, has been partly deoxidized by the great quantities of hydrogen and carbon with which it came into contact. Through this oxidation of the iron a quantity of caloric is generated, which combines partly with some of the

---

[55] *Physisk-oeconomisk og medico-chirurgisk Bibliothek*, April 1795.

carbon and oxygen of the blood to form carbonic acid gas, partly with some water to form aqueous vapours, and partly also with some carbon, hydrogen, and oxygen to form fat. It is now easy to see that the hydrogen gas which the blood gets from the water can combine again in even greater amounts with the fat and other compounds with which we are not so familiar.

It might also be objected that, according to the antiphlogistians, water is compounded during breathing from its constituents, hydrogen and oxygen, so it would not be reasonable that it should also be decomposed by the blood. However, this view does not seem to me to be in accordance with experience, for why would we otherwise exhale primarily water in the thinnest air, such as on high mountains, where, after all, we must inhale less oxygen during the same period of time than in places with more air.

Now, if we add to this that it must be assumed that fish decompose water in the same way when breathing, I hope to have shown the possibility that the fœtus could derive the same advantage from amniotic fluid as the born human being with regard to dermal respiration of the surrounding air. I shall now seek to render this reasonable with several arguments.

1) It seems reasonable that the fœtus cannot be content with the oxygen that it can get from blood from the placenta, which must take the place of the lungs, as this blood receives its oxygen from what has passed through a large part of the mother and there deposited a portion of its oxygen. The born animal is not content with the oxygen that it can take from the air with full lungs, so how much less satisfied will the fœtus be with the little which it only gets, so to speak, second-hand?

2) The surface of the fœtus is usually covered by a fatty crust. This proves that the fœtus must deprive the fluid which surrounds it of some of its aqueous parts so that it becomes more solid and more cohesive. Some may well have wanted to regard this crust as a deposit or sediment from the amniotic fluid, but in that case it would be found only on one side of the fœtus and not be as even as it generally is. It is true that this crust seems to be able to prevent the aqueous parts from penetrating the pores of the skin, but this obstacle is not so important as it may appear to some at first glance, partly because it is very soft so that the water can probably penetrate it as easily as it can penetrate the pores of the skin, and partly because the amniotic fluid disintegrates some of it, indeed, entire pieces fall away so that the fœtus is also given some opportunity to absorb the aqueous parts in this manner.

3) The decreasing weight of the fluid can also be explained very easily from this as it is natural that the larger fœtus must decompose more water than the smaller.

4) The increasing consistency of amniotic fluid can also be explained from this, for as the fœtus attracts the oxygen at the expense of the aqueous parts, they must necessarily decrease gradually and the water thereby gain in viscosity.

5) Our fluid fills the same cavities in the fœtus as the air in the born human being. Our deserving divisional surgeon, Dr. Herholdt, has found that it fills both the tympanic cavity and the trachea.[56] This similarity with air further justifies us in attributing the same function to it also in other respects.

We might ask if it was reasonable that the fœtus again expelled some of the fluid that it received through the pores of the skin as the born human expels some of the air inhaled. I think so, for although it is true that the fatty membrane lacks some of the elasticity that it acquires after birth, there is another way in which the fluid can be expelled, viz., by heat. When oxygen combines with blood, heat is generated, partly because the blood loses some of its capacity for heat, and partly because the fluid, which contains some caloric, is decomposed. This liberated caloric transforms the fluid which has not been decomposed into vapour, thus increasing its volume and forcing it to seek a way out through the pores through which it entered. It is quite reasonable that these vapours carry some fat with them from the glands in the fatty membrane of the fœtus. When these vapours arrive at the surface of the fœtus, they are forced to give off some of their caloric to the rest of the amniotic fluid and thus return to the liquid state. As the fat contained in the vapours also loses some of its caloric, it becomes thicker by nature and adheres to the surface of the child. This partly accounts for the fatty crust which covers the fœtus.

It still remains for me to investigate the view of those who believe that amniotic fluid is resorbed by the fœtus without being separated into its constituents. Levret[57] mentions an argument against the resorption of the amniotic fluid by the amnion, which I think can also be used against that by the fœtus. If a venereal pregnant woman is rubbed with mercury in the usual way, the amniotic fluid contains a portion of this metal when the fœtus has been cured, and the more of it that has been given to the mother, the more is found in the fluid. It is true that Rudolf[58] objects to this that the fluids of a person cured of a venereal disease contain mercury for some time after the cure, but in that event it must decrease gradually. The other argument he mentions, that the pores through which the amniotic fluid is resorbed must be regarded as capillary tubes in which the fluid can rise only if it is lighter than these, i.e., not heavy mercurial particles, I consider to be a misunderstanding because the fluid does not have to rise everywhere in these capillaries as there are many into which it must penetrate downwards according to the laws of gravity, and all the more must penetrate into these.

Van der Bosch[59] also mentions some experiments by Brugmanns,[60] which were

---

[56] *Physisk-oekonomisk og medico-chirurgisk Bibliothek*, October 1797, p. 105. Confirmed by Professor Wiborg [Viborg], p. 175.

[57] *L'Art des accouchements*, p. 56.

[58] [Rudolph.] *De partu sicco*. This treatise, as well as those by Bosch, Madai, and Temel, is to be found in Sclegel's [Schlegel's] *Sylloge operum minorum præstantiorum, ad artem obstetriciam spectantium*.

[59] Loc. cit., §62.   [60] [Brugmans.]

supposed to prove that amniotic fluid is absorbed undecomposed by the fœtus. He has seen in several fœtuses of animals, when he raised the skin while they were still warm, that there was a fluid in their lymphatic vessels which disappeared soon after they died. However, I think that we may rightly ask: Was it really amniotic fluid that he saw? He mentions that a rabbit showed no signs of life when he pulled it out of the womb, but when it was placed in amniotic fluid which had artificially been kept at its natural temperature, it evinced signs of life and showed the fluid in its lymphatic vessels more clearly, but this might be due to the fact that the fœtus was not able to endure the cold atmosphere as it was used to the warmer amniotic fluid. Therefore, it was natural that the vital functions should stop and the force which produced these fluids and gave them the appropriate expansion by means of heat should disappear. Consequently, I do not believe that these experiments confirm this view as they may just as well be explained in another way.

CONTENTS

2

Letters on Chemistry

FIRST LETTER[1]

DURING our last conversation I promised to explain to you the systematics of chemistry in letters, both because you like this form best, and also because it is best suited for a clear presentation. I am delighted to fulfill my promise, both for your sake and for that of science, which, as you know, I find so much pleasure in communicating to others. I am obliged to begin with the exposition of several concepts which at first glance do not seem to be of any particular interest, but I hope that their importance in what follows will reconcile you to their present dryness. The first thing we must do in the systematic exposition of any science is to determine what it is, and where its place is among the other sciences. As chemistry is part of natural philosophy, you will most easily get an idea of it when you contemplate the various parts of which that consists and see what place chemistry occupies among these. When we first look at the objects which surround us, we must necessarily take notice of their external qualities: their shape, colour, smell, and taste must attract our attention before their internal qualities. By means of these external qualities, we are also able to distinguish one object from another. All the creations of nature have even been classified according to these characteristics in a scientific manner, and this systematic knowledge of natural objects is called the description of nature or natural history, which latter name, however, should rather be reserved for an account of the various natural phenomena which have occurred. This knowledge of the external features of objects, however, cannot satisfy a thinking individual, who discovers many changes and is not satisfied until he has discovered their causes. The science of those qualities of objects which cause these changes is called *physics* or *natural philosophy*. If we delve still deeper into nature, we will not be content with distinguishing things and knowing their causes, but we will occupy ourselves with the investigation of the constituents of objects, and then we obtain chemical knowledge. I shall illustrate all this with examples: If I give you sufficiently many characteristics of hemlock that you are able to distinguish it from all other plants, this belongs to the description of nature, but if I show you what

[1][*Bibliothek for Physik, Medicin og Oeconomie*, Vol. 14, pp. 152–60. Copenhagen 1798. KM III, pp. 3–6. Originally published in Danish.]

causes this, as well as other plants, to turn pale when they are deprived of light, I have obtained this information from natural philosophy, but if I also decompose the plant into its constituents and show that it contains water, mucilage, earth, etc., I enter the realm of chemistry. We see from this that the description of nature must always precede natural philsosophy, and that chemistry is incomprehensible without both these sciences although, on many occasions, a knowledge of chemistry must also be presupposed in natural philosophy.

The definition of chemistry will cause you to ask a question, the answer to which must cast light on a great many natural phenomena. Chemistry, you see, shall teach us to recognize the constituents of bodies, that is, show us a body which is composed of several others. Therefore, it must readily occur to you to ask: What causes several different bodies to combine into a single coherent one? Before I answer the question, I shall raise another myself: What causes several small parts of one and the same kind to unite into a coherent object? I shall first answer my own question because it will then be all the easier to answer yours later. A famous thinker has already answered my question, and that is the illustrious Kant. This man's perspicacious investigations have taught us that every body has two fundamental forces, the cohesive force and the expansive force. The latter force causes a body to occupy a volume, the former prevents one single body from occupying all space, for if the cohesive force did not hold the parts of a body together, the expansive force would spread them beyond all bounds so that no trace of the body would remain. I cannot, without straying too far from my goal, inform you of the profound investigations on which depends the evidence that bodies really do possess these fundamental forces, but I shall mention some experiences which, to some extent, could serve to convince you of the existence of the cohesive force at any rate. If you take two polished glass plates and put one on top of the other, you observe that they stick together; if you put one drop of mercury next to another, they unite; and if you dip your finger into water, the moisture sticks to it; all this proves that one body displays a cohesive force with another. This cohesive force, however, is not equally strong between all bodies; thus a gold plate adheres more firmly to the surface of mercury than a copper plate, and this again more firmly than an iron plate. Now, if you had placed a small drop of mercury on a copper plate and touched this drop from above with a gold plate, it is clear that the gold plate, having a much stronger cohesive force with mercury than the copper, must attract the drop and remove it from the copper plate. If you now imagine that the mercury, *secundum artem*, had been so thoroughly combined with the copper that you could not distinguish one from the other with the naked eye, you will still discover that the gold attracts the mercury, and the copper is separated from it.[2] In this operation, you notice two things happening: first, that the mercury combines with the copper and, as it were, silvers it, and later, that the gold removes it. Chemists have named this kinship or affinity.

This ability of bodies to combine with each other is divided in such a way that

---

[2]Mercury does not unite with copper except through heat and friction or through even more complicated artifices. Gold, on the other hand, unites so easily with it that the smallest contact between them produces a white spot on the gold.

affinity, in so far as it merely causes the combination of two bodies with each other, is called chemical affinity while, on the other hand, the operation whereby a body separates a compound into its constituents by combining with one of them and thereby separating it from the other is called elective affinity. Thus, in the above-mentioned example, copper and mercury manifested chemical affinity, gold, on the other hand, elective affinity by combining with mercury and separating it from copper. Chemical affinity, then, is nothing but a manifestation of the cohesive force between bodies, and bodies which do not display this force with each other, like oil and water, are said to have no affinity to each other.

The occupation of chemists is then, as we have seen, to separate bodies into their constituents, that is, into parts of different natures, but in addition to this decomposition, there is another into parts of an identical kind. The first decomposition is called chemical, the second mechanical. Consequently, it is a chemical decomposition when saltpetre is decomposed into its two constituents, nitric acid and potash, but it is a mechanical decomposition to grind rhubarb into a powder. In order to reach the goal of discovering the constituents of bodies, chemistry can choose one of two courses: It either decomposes a body into its constituents, which is called analysis, or it compounds it from them, which is called synthesis. It may seem as if analysis would be sufficient to discover the constituents of bodies, but on closer inspection you will soon notice that by choosing this course alone you can easily be deceived. An example will illustrate this: When you collect everything that is produced when a piece of meat rots, you will still not be able to reconstruct the meat from the substances which are produced in its putrefaction. This shows that although you have all the constituents of the meat, they must have combined with each other in a new way so that, by studying them more closely, you do not obtain much clearer knowledge of the composition of meat than you had before. If, on the other hand, you could have reconstructed it from those constituents, you would have had some understanding of them. In order to discover the constituents of bodies, chemistry also uses reagents, which, when added to others, reveal a certain constituent in them. Thus, tincture of gall is used to detect the presence of iron as this makes every solution of iron black. Similarly, it is possible to detect whether water contains lime by adding vitriolic acid, which in this case immediately makes it cloudy, etc.

There is one more thing to observe in chemical operations, and that is whether what is obtained by them has been produced through a combination of several individual bodies, or whether it already existed before the operation and was only revealed through this. Thus, if potash is added to a solution of silver in nitric acid, this will combine with the nitric acid, and the silver will settle (be precipitated). Here, the silver had been formed before the operation and was just developed through this. After the operation, on the other hand, saltpetre is obtained from the nitric acid and the potash, and this was produced during the operation. Therefore, the former is called an educt (extracted), the latter a product (produced). I shall not trouble you with further introduction to chemistry although much more is generally included, and I hope you will be grateful to me for sparing you.

3

Letters on Chemistry

SECOND LETTER, ON HEAT[1]

IN MY last letter, I showed you that bodies possess two fundamental forces, the expansive force and the cohesive force. I now embark on a subject which seems to be connected with these, that is, the theory of heat. Heat expands all bodies which it can act on and thus seems to be causally related to the expansive force, but the investigations which have hitherto been made have not taught us in what way. The feeling which heat produces is all too familiar to you for a description of it to be of any interest; also we describe most poorly that of which our senses inform us most fully. On the other hand, I hope that a narration about the other effects of heat will better satisfy your curiosity. I merely want to point out that this feeling, as a natural phenomenon, must have a cause. This cause has been called caloric fluid, which is considered to be a substance which penetrates all bodies and makes them more or less hot, depending on whether it is present in larger or smaller amounts. That it is literally a substance which causes heat has not been completely established, but in the following I shall call the cause of heat caloric and leave undecided the question of its nature.[2] It is this substance which, by penetrating all bodies, produces the remarkable effect on them that they increase in volume. The experiments which could convince you of the validity of this assertion are innumerable. I shall mention several of the easiest ones. If you hold a limp, tied bladder over the fire, it will become distended when heated because the heat expands the air in it. When a metal ball which fits exactly into a hole is heated, its volume increases in such a way that

---

[1] [*Bibliothek for Physik, Medicin og Oeconomie*, Vol. 14, pp. 313–27. Copenhagen 1798. KM III, pp. 7–14. Originally published in Danish.]

[2] However, I have always considered the cause of heat to be material for the very reason that it can pass from one body to another, and that, you know, is an old rule: *Qualitates non migrant a corpore ad corpus*. The structure of the argument which has since convinced me of the material nature of the cause of heat is approximately the following: Bodies have only two fundamental forces, the expansive force and the cohesive force, and it is possible to derive all their other forces from these. The cause of heat acts in such a way that it cannot be conceived as one of these original or derived forces, so it must be material. I shall try to prove, with all possible stringency, the truth of both these premises in a treatise on *Natural Philosophy*, which I shall shortly publish in *Filosofisk Repertorium*. [Cf. *Fundamentals of the Metaphysics of Nature* (Chapter 6, this volume).]

it cannot be inserted until it has cooled again. When a metal wire is heated, it becomes longer and longer, the more it is heated, and you will even be able to measure
the intensity of a fire by means of such a metal bar. I shall briefly show you the procedure and illustrate it with a figure.

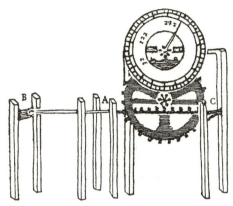

Here you see a rough sketch of this instrument. The part of the bar whose extremities I have designated B and A is exposed to fire. As the heat now expands it,
the cogs at the other end must naturally move forward and thereby turn the gear
(the pinion) which is at the centre of the wheel. This gear is fixed to the wheel, and
therefore this also turns and moves the indicator with its cogs. The number which
the indicator now points to shows the temperature, for instance 32 indicates the
temperature of freezing water, 212 that of boiling water, etc. You will get a clearer
idea of this division into degrees below when I have occasion to speak about it in
connection with thermometers.

How the caloric acts on fluids through its expansive force is best seen in thermometers. These instruments generally consist of a glass tube which terminates in

a ball at the bottom. It contains a fluid on which the heat can act. If you
place such an instrument in melting snow or ice, its fluid will only be at
quite a low level, as at A. If, on the other hand, you place it in a warm
room or in tepid water, it will be a good deal higher, and in boiling water
much higher still. If we put a mark, B, at the point to which it rises in boiling water and divide the space between them into 80 equal parts, we are
able to measure the temperature of water, air, and other bodies provided
it does not exceed that of boiling water. If we also divide the space
below B[3] into parts of the same size as those above this point, these are
called degrees below freezing or degrees below zero, the ones above B
degrees above freezing or degrees above zero, B itself zero. This graduation is
called Reaumur's[4] scale. Somewhat different from this is that of Farenheit,[5] which
puts 32 at Reaumur's zero, 212 at Reaumur's 80th degree, and 0 at the point to
which liquids are forced by an artificial cold, obtained by mixing sal ammoniac

---

[3] [Here and below Ørsted means A.]   [4] [Réaumur.]
[5] [Fahrenheit.]

and snow. The liquid with which thermometers are usually filled is either spirits of wine or mercury. In its water-free state the former cannot tolerate the heat of boiling water and must therefore be mixed with water. Colour is also added to make it easier to observe its rise and fall. The mercury thermometer is the most accurate as it rises and falls more steadily than the ones with spirits of wine.

There exists yet another peculiar thermometer, in which the expansion of air indicates the changes in temperature. In the adjoining sketch, *B* is a ball which contains air, and *A* is another ball which, together with part of the tube, contains spirits of wine. If the air in the upper part of the instrument is expanded, it must force the liquid down so that this thermometer falls lower, the warmer it becomes. This thermometer has the advantage over the others that heat affects it more strongly so that it rises or falls with the smallest change in temperature, but on the other hand, it cannot be used to measure temperatures as high as that of boiling water. The other thermometers are capable of measuring all temperatures up to that of boiling mercury when they are filled with this substance. The pyrometer which I described at the beginning of this letter is used to measure even higher temperatures. However, it is so complicated that it is easily damaged, and as it cannot measure the temperature at which iron melts, it still leaves much to be desired. Wedgwod's[6] pyrometer is much to be preferred in this respect. It consists of clay cubes which shrink with heat and thus indicate the temperature. In order to measure how much smaller they have become, one uses four metal bars which are placed upright in such a way that they slant down towards each other, like these two lines $\bigvee$. The smaller the cubes become, the deeper they obviously sink between these bars, on which numbers are engraved to indicate how high the temperature has been in each case. It seems as if the fact that the clay cubes become smaller with heat is at variance with the general law of nature that heat expands all bodies, but this conflict is only apparent since clay contains a large quantity of moisture, which is driven off by the heat, and this is why the heat reduces its volume, not because it possesses a peculiar force to do so.

I must give you another example of the expansive force of heat which I shall later use to explain one of the most important natural phenomena. If you take an empty bottle and hold it over the heat from a candle until it reaches a fairly high temperature, the heat will expand the air in the bottle, whereby it will occupy a greater volume than before, and some of the air must therefore leave the bottle in order to make room for the rest. If you put this bottle, while it is still hot, with the neck down into water which must be warm so that the bottle does not burst at the transition from heat to cold, the air will return to its former smaller volume as the bottle cools. This, however, will leave a vacuum, so water must flow in to fill it up. You will therefore be delighted to see water rise, little by little, in the bottle and, contrary to the laws of gravity, rise above the level of the rest of the water. Now you easily understand that if the water had not been there, air would have flowed into the space which the water filled here and thus created a wind. The same thing must

[6][Wedgwood's.]

happen when part of the earth's atmosphere has been heated for some time and then is cooled down again. Air must also flow from other atmospheric regions to this place and thus generate that motion in the air which we call wind.

Through all these examples I hope to have convinced you of the expansive force of the caloric, and we now want to consider the other effects of this substance. The more a body is heated, the more it must expand. When the expansive force of a body is increased, the cohesive force must, of course, be weakened, and therefore it becomes possible for the cohesive force of a solid to be so weakened by the caloric that it quite dissolves and becomes a fluid. As the heat increases, so does the elasticity of the fluid so that it either acquires such a large volume that it is called a vapour or even comes into such close contact with the air that it becomes a gas. Gases and vapours differ from each other only in that the material in the former can never be deprived of its caloric without combining with another body, while the material of vapours can easily be deprived of its caloric without combining with another body. Thus you see that the steam rising from boiling water gives off its heat and changes into drops of water merely through contact with a colder body. On the other hand, you cannot deprive the air which surrounds us of its gaseous state, whether by cooling, compression, or the like, but if it is to lose it, it must combine with another body, for instance when it is partly sucked in by a burning body. Consequently, matter can be imagined in four states: solid, liquid, vaporous, and gaseous, all depending on the elasticity which heat gives them. Therefore it is often possible to make a body pass from one state to another by either giving or depriving it of caloric. Thus ice can be made liquid by being heated, and water can be changed into a vaporous state by the same means. Conversely, liquid water can be turned into a solid by being deprived of heat. Distillation is based on this. When water is to be distilled, it must be put into a still, which is heated. This changes the water into steam, which rises and, because of its great volume, everywhere seeks a way out, which it finds through the pipes which open in the top of the still. These pipes are passed through cold water, which cools them and consequently deprives the steam which passes through them of its caloric so that it must again turn into drops and, as such, run from the pipes into the vessels which have been prepared. We explain sublimation in the same way. This differs from distillation only in that the vapours in the former operation are not forced out of the vessel in which it is performed as they are in the latter. In sublimation the vapours of the heated body are merely forced upwards by the heat and cooled down in the upper part of the sublimating vessel into a solid, whereby it becomes what it was before. It might seem to you that these two operations were useless since they do not change the nature of the body, but it only seems so, for the bodies which are subjected to distillation are always compound and of such a nature that some parts of them can more easily be changed into vapours than others. Therefore, when these different constituents are to be separated from each other, they must be subjected to such a temperature that it can drive off one of them without having very much effect on the other. This happens with mash which contains spirit of wine, which is much more easily changed into a vapour than water. Therefore, spirit of wine can be distilled at a suitable temperature without any fear that the rest will go with it. Some bodies are extremely difficult to vaporize, such as gold, silver, and iron, and are therefore called non-volatile, while others, which are more easily vaporized, are called volatile, such as spirit of wine, naphtha, and Hoffmann's anodyne. When these volatile bodies evaporate, a considerable coldness is felt. You will convince yourself of the truth of this by putting a very volatile substance, such as spirit of wine, on your hand, and you will then feel an intense cooling at the spot where it was applied. The same thing will happen with water, which also produces some coldness when you put some drops of it on your skin. This is easy to understand as heat is required to make them evaporate. You feel the same coldness when you hold a piece of ice in your hand, for this also requires heat to become liquid. However, in spite of all the heat of which it deprives your hand, the water thus produced does not become even slightly warmer than the ice. The ice, then, has absorbed heat without becoming perceptibly warmer. Hence we see that a body can receive heat which combines with it in such a way that the heat is not observable but merely makes it more liquid than before. In this case, the heat is called latent or bound, but when it is in a state in which it can be felt, it is called free. Free heat flows from one body to another and thereby increases its temperature and rouses sensations in us. Bound heat, on the other hand, is not noticed by our senses and cannot leave the body to which it is bound without changing it into a more solid body. Each time a body passes from a more to a less solid state, it must bind heat, for example, when it passes from a solid to a liquid, from a liquid to a vaporous, and from a vaporous to a gaseous state. Therefore, the thermometer does not rise in melting ice or snow until it has melted completely, no matter how much it is heated.

In the same way we can also explain the fact that when water which has been heated to 172° Farenheit is mixed with ice, which is 32°, in equal proportions, the water will melt the ice, but a thermometer placed in it will not rise at all but stay at the 32°. The hot water, whose temperature fell from 172° to 32°, thus lost 140° of heat, which the ice or the snow must have absorbed and bound in order to become liquid, for if the hot water had been mixed with cold water that was not warmer than the ice, that is, 32°, the temperature of the mixture would have been 102°, which is half-way between the other two. This is also the reason why crystallized salts produce some coldness when they melt in water, for they always contain some water of crystallization, which is in a solid state in the salt, that is, like ice, and must bind heat when melting. It can also be understood from this that boiling water can never reach more than a certain temperature, for when the water has reached 212°, it begins to evaporate, whereby all the heat which would have made the water exceed this temperature is bound. This is also the reason why the bottom of a kettle, in which something is being boiled, is never particularly hot as the vapour which is created there requires all the caloric for its formation. The only thing which sometimes causes water to need more or less heat to boil is different atmospheric pressure. The atmosphere is not equally dense or equally heavy at all times and, therefore, does not always exert the same pressure on the bodies it surrounds, but as this pressure can only prevent vapours from expanding and rising, they must develop with more or less difficulty, as this obstacle is bigger or smaller, and therefore need more or less heat. I also take this opportunity to explain to you several natural phenomena, which it has not been possible to explain until quite recently. They are the following: Man can only reach a certain temperature, 96° Farenheit, and if someone is placed in air which has a higher temperature, this will not be transmitted to the body. The cause of this is that the animal body evaporates more, the more it is heated so that all the heat exceeding 96° is used to evaporate water. The other phenomenon is very similar to this: In summer, it is much cooler under the trees than under any other cover which can keep out the rays of the sun. Here the generation of vapour plays the same role as in the previous example, for the tree evaporates more during the hottest time, and therefore it must constantly deprive the surrounding air of some of its caloric and thus cool it. Consequently, such countries which have many forests are colder than others at the same latitude. Here I have entered the domain of physics and will do the same in the following as it is necessary to apply the theory to natural phenomena in order to make it come alive.

4

Letters on Chemistry

THIRD LETTER[1]

WHEN you saw in my last letter that bodies bind caloric each time they pass from one state to another less solid one, you probably thought at once that bodies must liberate caloric when, conversely, they pass into a more solid state, and you have not deceived yourself in this. There are very many cases in nature which could be explained by this thesis alone and thereby also serve as proof of its validity. Thus you see that heat is generated even at the transition of water to solidity (to ice). You merely place a thermometer in water which is exposed to the cold, and you will see it rise at the very moment when the ice crust begins to appear on the surface of the water and, on the other hand, fall when the ice has formed because now much more[2] heat is generated, and the air soon absorbs the small amount of caloric which the thermometer had received. Now you also understand why the lid on a kettle in which water is being boiled easily becomes warmer than its bottom since the aqueous vapours, as they hit the upper part, lose some of the caloric which is necessary to keep them in the vaporous state, for which reason they now return to their former liquid state and therefore liberate even more caloric as their new state is closer to the solid state than the one they were in previously. For the same reason, the head of a still is always very hot, as are the pipes, in which an even larger quantity of steam is condensed. By the same token, it is easy to understand that the air must become warmer just before rain falls because on this occasion the aqueous vapours in the air are condensed and therefore liberate caloric. After the rain, on the other hand, a pleasant cooling is felt, partly because the condensation of the vapours, which causes the heating of the air, now ceases, and partly because some of the fallen water again passes into vapour and consequently binds heat. The same principle also explains why many efflorescent salts (salts which have lost their water of crystallization) become warmer when they are dissolved in water, for the water is thereby transformed into a solid state, and consequently heat must be liberated.

---

[1] [*Bibliothek for Physik, Medicin og Oeconomie*, Vol. 16, pp. 18–31. Copenhagen 1799. KM III, pp. 14–21. Originally published in Danish.]

[2] [less.]

This is what happens to burnt lime, which heats intensely when cold water is poured over it because part of the water is transformed into a solid state.

Bodies, then, contain caloric in two states: bound and free. In the latter state, as I have already mentioned, it manifests a tendency to spread uniformly among all bodies with which it can come into contact. When two substances having different temperatures are mixed, the caloric must therefore pass from the warmer to the colder body until both have reached the same temperature. In this way the colder gains as much caloric as the warmer loses, and when the bodies are of the same kind, the temperature of the mixture will be halfway between those of the two bodies before mixing. Thus a mixture of a pound of water at 110° and a pound at 44° reaches a temperature of 72°.[3] If, on the other hand, two different bodies are mixed, the result is often very different from this. If a pound of water at 110° is mixed with a pound of mercury at 44°, a thermometer placed in this mixture will read 107°. The water thus lost 3°, while the mercury gained 63° and therefore needs no more heat to rise 63° than water to be driven 3° higher, or mercury gains 21° with the same amount of caloric with which water gains 1°. This assertion can be further confirmed when the experiment is made with the change that the water is only 44° but the mercury 110°, and then the temperature of the mixture is 47° so that the mercury here has had to lose 63° in order to give the water 3°, that is, the mercury has lost 21° for each 1° the water has gained.

When one body thus requires more heat in order to be raised to a certain temperature than another, it is said to have greater susceptibility to (capacity for) heat than the other. Therefore, as an example, water has a heat susceptibility 21 times greater than that of mercury. You can easily see that when two bodies with different heat susceptibilities are brought to the same temperature according to the thermometer, the one with the greater susceptibility must also contain more caloric, in the same proportion as it has greater susceptibility, and this is called *specific* heat. Thus water contains 21 times more *specific* heat than mercury.

Besides this difference between bodies with regard to the amount of caloric they absorb, they also differ from each other with regard to the speed at which they permit heat to flow through them. Thus heat is transmitted much faster through a piece of iron than through an equally large piece of wood, which you can see, for instance, from the fact that when an iron with a wooden handle is placed in the fire, it does not easily become very hot, whereas a handle of the same metal soon becomes so hot that it cannot be touched. It stands to reason that the faster a body transmits heat, the faster it heats when brought to the fire and cools when taken into the cold, so the rate at which heat is transmitted through a body can be determined by the speed with which it cools down when hot. This ability of bodies to transmit heat is called *heat-conducting force*. When it is found in a high degree in a body, this is called a *good* heat conductor, in the opposite case a *poor* one. A good heat conductor, for example a piece of metal, always feels rather cold when we touch it because, on this occasion, the heat is quickly conducted away from our body so that we feel

[3] [77°.]

cold. If, on the other hand, we touch a poor heat conductor, no such sensation occurs, which is easy to understand; so, in daily life, the *good* heat conductors are called *cold*, the *poor warm* bodies.

The application of this principle to everyday life is very widespread and important. It is very often used unconsciously. It is thus an application of this principle when woollen clothes are worn in winter and linen in summer, for the former is a poorer heat conductor than the latter, for which reason it does not so easily conduct heat away from the body. It is for the same reason that water pumps are covered with straw in winter as this, being a *poor* heat conductor, prevents the water from giving off its caloric to the atmospheric air. Liquids are generally *good* heat conductors, for the reason that when heated there occurs internal motion which greatly promotes the transmission of heat. I shall give a more detailed example of this: If a kettle of water is put on the fire, the particles which are closest to the bottom are naturally heated first, whereby they expand and thus become lighter than the rest of the water and are forced to rise to the surface as the lighter always rises to the top. Through this motion, the colder water particles now occupy the place of the warmer until these, through heating, are also forced to rise to the surface. Thus the water keeps moving up from the middle of the bottom and down along the sides. One water particle, then, receives its heat not only from another but also directly from the fire, whereas the parts of a solid body which do not touch the fire cannot be heated except by the heat which one part gives to the other. That this explanation is correct can be confirmed by a great many experiments, of which I shall only mention one to you, which is that water is not heated so quickly when it is mixed with feathers or another substance which is capable of reducing the speed of motion. The famous Count Rumford has made this and a number of other experiments, from which he draws the conclusion that liquids do not have any heat-conducting power at all, but that all transmission of heat in these must be attributed to motion. I think that this conclusion is hasty, and that the only thing that can be concluded from his experiments is that liquids owe much, indeed maybe most, of the ease with which heat is transmitted in them to motion. It would lead me into far too much detail to enumerate all his experiments in order to prove my assertion, so it must suffice here to point out to you that the experiment which I have mentioned here proves nothing more than this.

From what I have said here about the transmission of heat in fluids, you will easily understand that this must happen more quickly upwards than downwards as the heating, when it begins from below, makes the heated particles, which have now become lighter, rise to the top, but when heat begins to spread from above, it cannot produce any motion as the particles become lighter and therefore have even less tendency to sink to the bottom than before. Therefore, an iron bar, which is red-hot at one end, is also heated much more quickly if it is held downwards than if it is held upwards, not because the heat moves faster upwards than downwards in the iron bar itself, but because the surrounding air is heated and rises when the heating starts from below, whereby it then transmits some of its heat to the upper and colder parts. There are some students of nature who, because of this rapid transmission of heat upwards, think themselves able to prove that caloric has a power to rise, contrary to the general law according to which all bodies must fall to the ground, but the explanation which I have given here, and which most well-informed students of nature have now accepted, is undeniably easier to understand and in agreement with nature itself. Besides, a force which neutralizes gravity is completely contrary to the nature of bodies, for each body must have a force of attraction, which I proved in my first letter, and this is not different from gravity.

I have now given you the fundamental features of the theory of heat, whose applicability in the sciences as well as in everyday life is so widespread that it would be impossible for me to explain it to you in its entirety. I shall just give you one example of its importance to the theory of chemical instruments, of which not a few can be explained by it and, partly through its guidance, could be designed more efficiently. One example of these is the common distillation apparatus, which could be made more perfect by adopting a number of changes in accordance with theory. Common stills have their furnace under them, whereby, it is true, a great deal of the heat reaches the liquid contained in them, but much of it spreads sideways and downwards, where it is of no use. If, on the other hand, the furnace were inside the still so that it would be surrounded on all sides by the liquid to be distilled, it is easy to understand that much fuel would be saved thereby. You can see how this could be arranged most conveniently in the following figure, in which I have also shown the other improvements which I think ought to be made.

In this figure you find the furnace (B in Fig. 1), which is best made of sheet iron, almost at the bottom of the still. The mouth (a) must be just wide enough to permit the insertion of fuel so that less heat will escape. This oven, like any other, must have a grate (b) and a pipe (c) on top, which can serve as a chimney. Such a still can also be made of wood, which has the advantage that the heat does not escape so easily as when it was made of metal since wood is a poorer heat conductor than metal. Moreover, very much of the cost is saved in this way because copper, which is generally used for this purpose, is far more expensive than wood. It is true that some heat escapes unused through the mouth of the furnace with this arrangement, too, but it shares this defect with the old apparatus. The heat which escapes unused with the smoke through the tube is insignificant as a good part of its path is through the matter to be distilled, and it thereby gives up most of its heat. It is also possible to make an important improvement in the head (D) of the still by making it tapered like a cone and not round like a hemisphere as usual, for the old device has the defect that many vapours which are cooled and return to a liquid state fall perpendicularly from the round part of the head back into the still so that they now need new heat to be returned to a vaporous state. Because of this defect, a not insignificant amount of caloric is wasted, which could otherwise have been used to drive the rest upwards. With the new arrangement, however, this defect is not present as the vapours which are liquefied in the head run down along its sides, whereby they could of course fall back into the still, but this is prevented by the channel at the edge of the head, whose cross section you see in the figure. The condensed vapours now run from this channel out into the pipes. The opening in the head of the still is usually fairly small, but this, too, is incorrect because the vapours which strike the

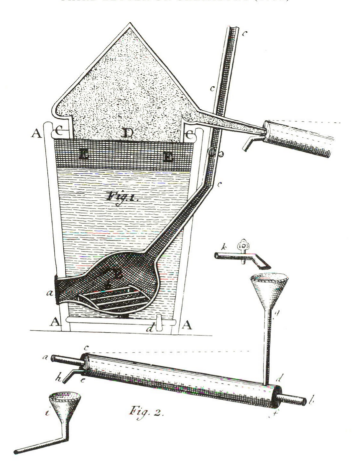

head are liquefied and drop into the still. It is very important that the pipes from the head are well-cooled as many vapours could otherwise escape without being condensed and thus be lost completely. It is easy to understand that this goal is realized by passing the pipes through cold water, and that it is attained all the more perfectly, the longer the distance which the vapours must traverse under water. As a curved line between two points is always longer than a straight line, it would be better to make the pipes wind through the water than to make them completely straight. Generally, it is strongly recommended to keep the head of the still as cold as possible so that the vapours there can be liquefied all the more completely. This attention seems immaterial to me as long as it is certain that the vapours are properly cooled in the pipes. If this does not happen, it is probably useful.

These winding tubes cause great inconvenience in that they are very difficult to keep clean, so other ways have been tried to provide proper cooling. The most successful experiment we have in this respect is by Göttling, who makes the straight tube (*ab* Fig. 2) from the still pass through another much wider tube (*cdef*), which, like the tube of the still itself, slopes a little downwards and is closed at the lower end (*df*) but opens into a tube (*h*) at the higher end. At the lower end a tall funnel is placed, which gradually fills with water which must rise as much in the tube as it does in the funnel, from which it follows that it must, little by little, run out of the tube (*h*) again. As the hottest water is always the lightest and therefore must float on top of the rest, you easily understand that it must always leave the tube at (*h*).

There is no question that if a distillation apparatus were arranged in the manner explained here, one would not only save more than half the fuel, but one would even obtain a better product of the distillation and in greater quantity than usual.

5

Letters on Chemistry

FOURTH LETTER[1]

WE NOW proceed to the contemplation of that remarkable phenomenon in nature whereby the most concentrated and the greatest quantity of heat that we know of is generated; you understand that I can only mean combustion. The explanation of this phenomenon has the most profound effect on the whole of chemical theory; indeed, it serves as the foundation of the entire system which is now accepted in this science, and the older but now rejected system also centred completely on this topic. Therefore it must be very important for us to consider all the circumstances which accompany this phenomenon and, with this end in view, not content ourselves with observing it as chance so often shows it to us but also try to discover its true cause through our own experiments, often made with controlled changes.

With even the slightest attention, one must notice that an influx of air furthers combustion. Therefore a fire which starts in calm weather is not so dangerous as the one which starts when there is a wind, therefore a fire burns more quickly in a stove when a small door is opened than when it is closed, and therefore we blow on a fire when we want it to burn. Indeed, it would even seem that the presence of air is absolutely necessary because it is possible to smother a fire completely by blocking the supply of air, which we often see exemplified in everyday life. However, not satisfied with this testimony from our daily experience, since something foreign and unknown to us might possibly be involved, we confirm this truth further through experiments. When air is pumped out of a glass vessel, especially adapted for the purpose, gunpowder, wool, sulphur, and other combustible substances which are confined in it cannot be ignited even if a burning-mirror or a burning-glass is focused on them, but if air is allowed to re-enter, these bodies will again prove combustible. In this way we have proved the law that *no body can burn without the presence of air.*

However, it is not sufficient that air is present during combustion because many times it happens that a fire goes out even though air is present. Thus it often happens that a candle which is taken into a basement in which there are many fer-

[1] [*Bibliothek for Physik, Medicin og Oeconomie*, Vol. 16, pp. 165–77. Copenhagen 1799. KM III, pp. 21–27. Originally published in Danish.]

menting substances is suddenly extinguished. The same thing happens when a candle is placed very close to a vessel containing a fermenting substance. There are also natural grottoes and caves in which no combustion can take place. From this we conclude that *there must be several kinds of air which are unfit to maintain a fire, but that the air which surrounds the entire globe (atmospheric air) is suitable.* This is also confirmed by the fact that we can artificially produce gases which are unfit to maintain combustion. We shall become better acquainted with these gases below.

Atmospheric air, or at least one of its constituents, is therefore necessary in order to produce and maintain combustion, but how it manifests this effect is a problem which we have yet to solve. Daily experience is not sufficient here, so we must make an experiment which may shed light on the matter. For this purpose we make experiments on combustion in enclosed atmospheric air and notice the changes which the air undergoes during them, with regard to both its quantity and its character. We also observe the changes in the burning body, with regard to both quantity and character. One of the easiest experiments which can be made in this connection is the following: A piece of phosphorus is fastened with a needle to a cork or some other floating body and placed on the surface of the water in a basin, the phosphorus is ignited, and a bell jar, or lacking that a large beer glass, is immediately placed over it. If we now observe this combustion, we find that the air, little by little, decreases in volume, and the water rises and fills the space of the vanished air. When somewhat more than a quarter of the air has thus been replaced by the water, the combustion ceases completely, and no further change takes place except that the water rises a little more because air contracts when cooled. No matter how much phosphorus is left, the combustion can no longer be continued, and any burning body which is placed in the residual air is extinguished at once. If we now examine the burnt phosphorus, we find it transformed into an acid which for the most part has settled on the sides of the glass, and, which is even more remarkable, it has gained in weight.

Several extremely interesting conclusions can be drawn from this. Air decreases in volume with combustion, so it seems likely that *part of it has been absorbed by the burning body.* This is also confirmed by the fact that this weighed more after combustion than before. Some of the air remained, and this could not be consumed by any burning body. From this we conclude that atmospheric air is not a chemically pure substance but is *composed of two constituents, one of which serves to maintain combustion, the other does not.* Finally we were able to observe that the phosphorus was transformed into an acid, which leads us to assume that *the part of the atmospheric air which is consumed during combustion contains the cause of the acidity.*

The results of the experiments remain the same if another combustible substance is substituted for phosphorus. They differ only in that almost all other bodies are themselves transformed into a gas by the combustion, so it is very difficult to weigh them afterwards. I shall try to show you how this can happen by means of an example. When a piece of burning charcoal is put into such an enclosed space as the one we mentioned before, the air in which the combustion takes place will undergo the same changes as in the previous experiment, but the burnt charcoal will appear to us to have vanished completely. However, if we examine the water with which the apparatus is sealed, we find that it has acquired a slightly acidulous taste and all the other properties of a weak acid, whereby we are entitled to conclude that, in this combustion, too, an acid has been produced which, however, has been absorbed by the water. If the bell jar is sealed with mercury instead of water, far more air than usual is found after the combustion. Consequently, the acid produced in the combustion of charcoal manifests itself as a gas when it does not come into contact with a body which absorbs it. This acid, which we shall henceforth call carbonic acid, combines very easily with the lime in lime water, whereby this turns cloudy. If lime water is now used to seal the jar in which the charcoal is burnt, this will absorb all the carbonic acid produced, and its weight will be found to be somewhat greater than that of the charcoal used and the lime water combined. In this way, then, we can also convince ourselves that the charcoal increases in weight and is transformed into an acid by combustion.

As we thus see every burning body increase in weight, we are entitled to assume that *combustion consists in the burning body attracting an element from the air.* This is confirmed in the most perfect manner when the experiments on combustion are made in such a way that the air to be used, as well as the substance to be burnt, is first weighed. It will then be discovered that, after the combustion, the body has gained exactly as much in weight as the air has lost.

During combustion a large quantity of caloric is liberated. The cause of this is easily understood from what I have explained about the caloric in my previous letters, that is, that this is liberated when a body is transformed into a denser state than the one it was in before, which happens here to the part of the air which combines with the burning body. This shows that the part of the air which is consumed during combustion consists of caloric, which has no perceptible weight, and a heavy element. As we have seen before, this element can be regarded as the cause of the acidity and can therefore be called acid-generating matter or, more briefly, acidic matter.[2] We do not intend this designation to express the opinion that this substance itself is a genuine acid but rather the indisputably demonstrated fact that those substances with which it combines in any considerable amount are thereby transformed into acids and all the more perfectly, the more of this substance they combine with. As we chemists now call each individual kind of air a gas, we could call the part of the air which consists of oxygen dissolved in caloric *oxygen gas.* The other part of the air, which has the property that it smothers flames and also animals which are placed in that gas alone, we could call *azote.*

There is another phenomenon associated with combustion which we have not yet explained, and that is the strong light which burning bodies emit. This has been explained in many different ways, none of which is entirely convincing. In this letter I dare not embark on all the arguments for and against, but I shall merely tell you that, in my opinion, it is most easily explained by assuming that light, like oxygen gas, consists of its own element and caloric, and that this element is contained in

[2] [oxygen. The old Danish term for oxygen, *surstof,* literally means "acidic matter."]

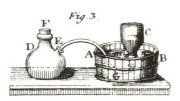

the burning body, where, for want of caloric, it cannot reveal itself as light, but by means of the caloric generated during combustion, it acquires what it lacked before, radiates in all directions, and appears as brilliant light.

Some other time I shall acquaint you with the arguments in favour of this opinion, which I have simply stated here in order to leave no feature of combustion unexplained.

I hope that you will find the evidence that I have already put forward in favour of our theory of combustion as rigorous as can rightly be demanded of a theory which must serve as the basis for so many others. However, you undoubtedly want more light shed on this important matter. I shall try to fulfill this wish, partly by showing you how to produce in isolated form that part of the air which causes combustion and partly by comparing combustion to other phenomena which have some similarity to it.

In order to show you how oxygen gas is produced in pure form, I must begin with one of these comparisons. You know from daily experience that almost all metals, through exposure to air, lose their metallic lustre and elasticity. Generally this is called *to rust* or *to tarnish*. In scientific language they are said to be calcined. This happens more easily when assisted by fire than through the effect of air alone. The air in which a metal is calcined undergoes the same transformation as that in which combustion takes place, that is, it decreases in volume and becomes unfit for the maintenance of fire and animal life. Metal which is calcined also increases in weight, like the bodies that burn, and this increase in weight corresponds exactly to the loss of weight of the air so that we are entitled to make a law which says that *the calcination of a metal consists in its combination with the oxygen in the atmosphere.* Some metals have the property that, when heated until they become glowing, they give off their oxygen, in combination with the caloric, as oxygen gas. Mercury is such a metal. This liquid metal attracts the oxygen in the air as soon as it is brought to the temperature at which it boils. Hereby it is transformed into a red powder, which is called *mercurius præcipitatus per se*. If this metallic calx is exposed to such a strong heat that it starts to glow, it will give off its oxygen in a gaseous state. In order to collect the gas, the calx of mercury is placed in a flask such as the one depicted in Fig. 3. This flask has two necks, one of which is equipped with a cork, while the other is joined to an S-shaped tube. One end of this tube goes down into a box or a vessel with water. Over the opening of the tube which terminates in the water a bottle is placed, which is initially filled completely with water. When the flask with the two necks is heated appropriately, which can be done by means of an Argand burner, the gas will rise in the flask, and as it cannot escape, it must force its way into the tube, driving the water out, and into the bottle covering its mouth. Since the bottle is also filled with water, the gas cannot find a stable place here either until it has risen to the top of the bottle. Hereby the water is forced out of the bottle, little by little, until no water is left, then it is taken away and replaced by another. This gas is the purest oxygen gas and therefore manifests the ability of air to maintain combustion to a much higher degree than the atmosphere itself. If phosphorus is placed in this gas and ignited, it will burn with a flame whose brightness far exceeds that given by the same combustion in ordinary air. It will then be seen that the oxygen gas is consumed so that none at all is left, but the jar under which the combustion takes place is completely filled with water. If the burnt phosphorus is now weighed, it will be found to have gained in weight by exactly as much as the weight of the oxygen gas employed so that this experiment can serve as further confirmation of the theory of combustion which I advanced above. Another no less interesting experiment can be made with this gas by placing in it a watch spring or a thin spiral steel wire with ignited tinder at the end. The tinder, and immediately after that the steel wire, will burst into a violent flame, and the steel, melted and burnt, will fall down in small pieces.

Another phenomenon which also has much similarity to combustion is the breathing of animals. When a small animal has been in an enclosed space for a few minutes, its breathing becomes laboured, and immediately afterwards it convulses and dies. After the experiment, the air is found to be decreased and transformed in the same way as after an experiment on combustion. In an enclosed space filled with pure oxygen gas, such an animal can live much longer and seems to feel much better. Breathing, then, is essentially like combustion, but there are many other circumstances connected with it which I am not in a position to explain to you until later.

6

Fundamentals of the Metaphysics of Nature
Partly According to a New Plan[1]

PROMPTED BY *FOUNDATIONS OF PHYSICS*
BY THE LORD STEWARD HAUCH

IF A body of empirical knowledge is to be able to claim the name of science in the true sense of this word, these experiences must be joined according to certain general and necessary laws which themselves cannot be drawn from experience but must be proved without its help (a priori). If this is not the case with an organized body of experience, it does not at all satisfy the scholar but leaves him standing at a limit which he is not certain is extreme and shows him laws which he dare not assume to be general and necessary because he knows that experience can only teach us what is but not what necessarily must be. However, no matter how important it is to have an a priori foundation for the natural sciences, the efforts of philosophers have rarely been applied to this subject. It is true that Kant's metaphysics of nature met with much approval, but hardly any of his works has been less read and pondered, no doubt because of the prejudice among philosophers that it is not really necessary in order to understand the other parts of philosophy and because of the habit among empiricists who constantly go from experiences back to principles, not from principles forward to experiences. No small contribution to the lack of attention to the metaphysics of nature comes from the fact that those who knew it thoroughly were rarely versed in empirical knowledge so that they were not able to demonstrate its many agreements with experience, whereby it would find favour with the empiricist. In order to achieve completeness in our knowledge of nature, we must start from two extremes, from experience and from the intellect itself. The former method is regressive, beginning with composite facts and resolving these until it arrives at the most simple; the latter is progressive and thus begins with the simplest and progresses towards the most composite. Consequently, the former method must conclude with natural laws which it has abstracted from experience, while the latter must begin with principles, and gradually, as it develops more and more, it also becomes ever more detailed. Of course, I speak here about the method

---

[1] A special reprint of *Philosophisk Repertorium* (Copenhagen 1799, printed by Boas Brünnich). [KM I, pp. 33–78. Originally published in Danish.]

as manifested in the process of the human intellect itself, not as found in textbooks, where the laws of nature which have been abstracted from the consequent experiences are placed first because they are required to explain the experiences. When the empiricist in his regression towards general laws of nature meets the metaphysician in his progression, science will reach its perfection. Only then will the latter see the end of his progression and the former the end of his regression, which he would otherwise not know when to end because he must always assume that there was still another step backwards. The chemist, for example, who had regressed from elective affinity to simple affinity, must reasonably expect to be able to move back from this to yet another force of matter which had so far been unknown to him; which he had to assume all the more as the former expression tells us nothing about the cause of the combination of several different substances but only informs us that they are combined. The natural philosopher, who attributes seven or eight fundamental properties to matter, probably also had to assume that it must be possible to reduce these forces, to which he has reduced all the others, to even fewer. In spite of his need for the help of metaphysics, however, the empiricist has gone very far while the metaphysician has made no progress. He had to make a bold attempt to provide the empiricist with the advantages that he might claim from him, and we owe this attempt to Kant's perspicacity. However, his investigations are difficult, and he has not always succeeded in combining comprehensibility with his profundity although they are often found combined in him to a high degree. Those who have tried to elucidate and explain his *Metaphysische Anfangsgründe der Naturwissenschaften* are incomparably fewer than those who have commented on his other writings, and most of them have not produced any really new idea in consequence. I have therefore thought that an explanatory summary of this treatise, according to a somewhat different architectonic from that of Kant, accompanied by some new ideas which have occurred to me in this connection, would not be unwelcome to Danish readers, all the more so as little or nothing on this subject exists in the Danish language.

2. Generally, the word nature has two different meanings. When we talk about the nature of something, we associate with it a completely different meaning from when we talk about Nature. By Nature we mean the essence of all the things which are the objects of our external senses, while the expression the nature of something means the fundamental internal principle of all that is required for the existence of that thing. It is all of Nature, and not the nature of each individual thing, which will occupy us here.

3. Here we do not want to put forward a catalogue of such knowledge about Nature as experience has taught us and from that deduce laws which we presume to be valid for all of Nature without, however, being able to decide with certainty whether they are necessary and universal or not. On the contrary, we shall try to deduce general and strictly necessary laws of Nature from the nature of perception itself.

4. Nothing can be perceived by the external senses except in time and space. When we contemplate matter, which is what we call each object for the external

senses, in these two forms, the notion of motion and rest arises, the former by contemplating it at different times in changed relations to the rest of space, the latter by contemplating it as present in one space for some time. Here, then, we see the emergence of a pure *doctrine of motion*. However, in addition to this, the metaphysics of nature must also contain another part, which will teach us what properties we must necessarily presuppose in matter if it is to become an object of our external senses. This might be called a *pure doctrine of matter*.

Each of these main parts must be taken through the four classes (categories) of pure intellectual concepts so that each of them will consist of four parts. Thus, the whole metaphysics of nature must be divided into:

A. The *doctrine of pure motion* (Phoronomy)
   It contains
   a. The doctrine of the quantity of motion
   b. The doctrine of the quality of motion
   c. The doctrine of the relations of motion
   d. The doctrine of the modality of motion
B. The *doctrine of pure matter* (Hylology),
   which consists of
   a. The doctrine of the quantity of matter
   b. The doctrine of the quality of matter
   c. The doctrine of the relations of matter
   d. The doctrine of the modality of matter.

5. Anyone who is familiar with Kant's metaphysics of nature will see that this division is somewhat different from that of the famous thinker, but I hope that the following will demonstrate sufficient grounds for my deviation. Here I can state no other than the one I have just put forward, viz., the division itself.

DOCTRINE OF MOTION

Introduction

6. Here we consider matter only as that which is movable in space, and we do not permit ourselves to add anything to this concept from experience, but we now seek to deduce a priori what we can know with apodeictic certainty about motion. We initially accept the common definition, according to which motion means that matter is in different parts of space at different times.

7. Empty space cannot be perceived by our external senses, for it is in itself a kind of sense, the necessary condition for external sensation but not an object for external senses, not matter, not movable. Space can only be perceived by the senses in so far as it contains matter. Only in such space can we perceive changes because it alone can be perceived by our senses. Material space itself, like matter, must be movable, that in which such space moves must, in its turn, itself be movable, and so on ad infinitum. Thus people move on board a ship, the ship moves with them, the earth moves with that, etc.

8. Absolute space really means immaterial, immovable space, but as this cannot be an object of the senses, we usually understand by this the nearest material space whose motion we do not perceive, and which we then imagine infinitely extended beyond all its limits. From this we see that no motion can be perceived in absolute space, consequently no absolute motion, which should be that matter, at different times, was in different parts of absolute space, i.e., that it changes absolute position.

9. The common definition of motion, which we also used above, really only applies to the motion of a point, but not always to that of a whole mass. A ball can rotate about its axis and is then in motion, but it cannot be said to move from one part of space to another or to change its place. If, on the other hand, we contemplate what takes place in every motion, we see that sometimes it is the body itself that changes its distance from others, sometimes its parts. Therefore, it must always be regarded as motion when the body changes its external relation to space, which must occur when either the whole or the parts change their position. It might be objected that this definition did not include internal motion, but here it should be noticed that the object in motion must be regarded as an entity, not as a composite. Therefore, when I say that fermenting beer is in motion, I regard every individual particle of beer as an entity, not the entire mass of beer. When, on the other hand, the barrel which contains the beer is moved from its place or rotated, the beer might be said to be moved because here the whole mass was regarded as an entity.

10. Usually, rest is defined as the absence of motion, but this definition is not correct either, for if a point moved along a line AC, we would have to say that it was at rest at each point on this line as it did not change its location at the instant when it was at this point and, consequently, was at rest according to this definition, which, however, it would be unreasonable to assume since motion would then mean that the body was at rest every moment. Thus we would have to say that the point moves at every point on the line and does not remain at any point for even the shortest instant of time. However, we can present rigorous evidence against this assertion. If we assume that, instead of moving from A to C, the point moved from A to B, which is precisely at the mid-point of AC, and then back again from B to A, then B belongs to both movements, to AB as well as to BA, and the body is not at B for even the briefest period of time. Now it will make no difference in the quantity of the motion whether the point moves from A to B and back from B to A, or whether it moves directly to C, and if it now goes from A to C, it would be in motion at B, but if it goes from A to B and returns from B to A, it is obviously at rest at B if rest were the lack of motion, for two opposite motions at the same point and at the same time must cancel each other. In this way, the body would be in motion if it went from A to C but at rest if it went from A to B and then back to A, and yet the quantity of these movements would be the same, which would be unreasonable. Therefore, if the quantity of the movement is to be the same, the body must also be considered at rest at B in the motion AC, and as we could draw the same conclusion for every other point on the line, we would have to say that it was at rest at every

point on its path, which would be contradictory. If, on the other hand, we imagined that the point rose to[2] AC in uniform motion, but that this was delayed by gravity, we would undoubtedly say that the point was at rest at A[3] because its motion first had to cease before the opposite began. But why should we assume now that the point was at rest at B here but not in the previous case? Without any doubt, the reason for this is that when the line is vertical, the movement is imagined as decreasing and later as increasing, and consequently the movement does not stop completely at B but only to a certain degree, which is smaller than any statable size. Now if the movement was continued beyond B, without gravity having any further effect on it, its motion would continue uniformly, but it would still travel only an unstatably small distance in any statable time with an infinitely small degree of velocity and consequently not change position to any perceptible degree, or its presence in that place would become constant (*præsentia perdurabilis*), but because gravity acts on it incessantly, this constant presence is again neutralized. Therefore, the definition of rest must be this: Rest is the unaltered, constant presence of a body at a given place. I myself have added the predicate unaltered in order to anticipate the objection that according to this definition a body which turned around its axis would be at rest as it did not change place, but this difficulty is removed by means of the adjective unaltered because such a presence would not be unaltered.

The Quantity of Motion

11. After having established the concepts of motion and rest, it now becomes possible for us to proceed to the more specialized branches of the doctrine of motion. The only concept which arises here is velocity. As we cannot here pay any attention to the mass of the body which is set in motion, we can only measure the quantity of the motion by the size of the distance which it traverses in a given time. We say of a body that it moves faster the greater distance it travels in a certain time so that one which travels twice as great a distance as another, in the same time, is said to have twice as great a velocity; the one which travels three times as great a distance, three times as great, etc. The consequence of this is that *when the time in which two bodies traverse different distances is equally long, the velocities are proportional to the distances.* If we call their velocities V and v and the traversed distances D and d, then

$$V : v = D : d \ .$$

If, on the other hand, the distances are equally large, then the velocity is greater, the shorter the time in which they are traversed so that *the velocity is inversely proportional to time when the distances are equal.* If we call the times in which two bodies traverse a distance T and t, then

[2] [along.]   [3] [B.]

$$V : v = t : T \ .$$

Let us now suppose a third body which travels a distance $= d$ in a time $= T$ with a velocity which we shall call $\mathcal{V}$, then

because $T = T$ we find $\mathcal{V} : V = d : D$
and because $D = D$ we find $v : \mathcal{V} = T : t$

therefore, according to mathematical axioms,[4]

$$v : V = \left(\frac{d : D}{T : t} \right) = dT : Dt = \frac{dT}{Tt} : \frac{Dt}{Tt} = \frac{d}{t} : \frac{D}{T} \ .$$

Velocities are related to one another as products, directly proportional to the distances and inversely proportional to the times, or as distances divided by times. Therefore we can determine the velocity by dividing a traversed distance by the time in which it was traversed, and the velocity is always $= D / T$. The greater the distance, the greater the velocity, if the time remains unchanged. The velocity would therefore be infinitely great if the distance were infinitely great but the time finite, for the expression D / T (§11)[5] must become $= \infty / T$, which, according to mathematics, would be $= \infty$. When the distance remains unchanged, the velocity becomes greater the shorter the time is. If the time were to become infinitely short but the distance finite, the velocity would become infinitely great, for, according to the above, the expression for this case would be $D / (1 / \infty)$, which, according to mathematics, is ∞. Thus there are two circumstances in which the velocity is infinitely great, that is, when the distance is infinitely great but the time is finite, and when the distance is finite but the time is infinitely short.

13. There are also two cases in which the velocity is infinitely small, that is, when the distance is infinitely small and the time finite, and when the distance is finite but the time infinitely long. This can easily be seen from the expressions D / ∞ and $(1 / \infty) / t$;[6] for D / ∞ is $= 1 / \infty$ and $(1 / \infty) / t$ is similarly $= 1 / \infty$.

14. We have seen from the preceding how we must imagine infinitely great or infinitely small velocity. They are the limits on either side, within which a possible series of velocities must be encompassed, but these limits are really only figments of the imagination. They cannot possibly occur in experience, for an infinitely great velocity results when an infinitely great distance is traversed in a finite time, and an infinite distance is one the successive composition of whose parts cannot be

---

[4][The parentheses in the following expression indicate that the upper fraction must be multiplied by the lower one.]

[5][Ørsted has evidently forgotten to insert §12.]

[6][The expressions have been interchanged.]

completed in any finite time. The concept of infinite velocity, then, contains a contradiction as it requires that a distance which cannot be traversed in a finite time must be traversed in a finite time. The second concept which we advanced of infinite velocity is not at all different from this except in its expression, for it required that a finite distance had to be traversed in an infinitely short time, but at such a velocity an infinitely great distance had to be traversed in a finite time, which is composed of infinitely many infinitely small parts. The mathematical expression also gives us an easy proof of this theorem, for

$$\frac{D}{1/\infty} = \frac{D \times \infty}{(1/\infty) \times \infty} = \frac{\infty}{\text{finite time}} \quad .$$

An infinitely small velocity cannot be anything other than rest, for when no finite distance is covered in any finite time however long but only a distance which is smaller than any that can be indicated, then there is no change of position or of any other external condition, but the body remains unchanged in the same place, that is, at rest (§10). The second concept that we have given of an infinitely small velocity, viz., one at which a finite distance can be covered in an infinitely long time, is similar to the previous one, for when a finite distance is to be traversed in an infinite time, then no finite distance can be traversed in a finite time at the same velocity, and consequently only an infinitely small distance can be traversed in it. It is also easy to see from the mathematical expression that they are of equal size:

$$\frac{1/\infty}{T} = \frac{(1/\infty) \times \infty}{T \times \infty} = \frac{\text{finite distance}}{\infty} \quad .$$

15. In his *Versuch die Gesetze magnetischer Erscheinungen aus der Naturmetaphysik, mithin a priori, zu entwiklen*, Eschenmayer has made an amusing mistake by taking the velocity as the product of distance and time instead of the quotient. Thereby he deduces that since it does not matter whether one sets distance $= \infty$ and time finite or time $= 1/\infty$ and distance finite, then their product, which can be written as $S^{\infty}T$ and $ST^{-\infty}$, had to be of equal size. He has been able to prove that $S^{-\infty}T = ST^{\infty}$ with equal ease. Anyone who has become acquainted with even the basics of mathematics can easily see what consequences can be drawn from this. However, in order to facilitate the work for those who are not used to dealing with infinite quantities, I shall draw a couple of consequences.

$$S^{-\infty} = \frac{1}{S^{\infty}} = \frac{1}{\infty}$$

so that

$$S^{-\infty}T = \frac{1}{\infty}T = \frac{1}{\infty} \quad .$$

$T^\infty = \infty$ and $ST^\infty = S\infty = \infty$.

$$S^{-\infty}T = \frac{1}{\infty} ,$$
$$ST^\infty = \infty ,$$
$$\text{and } S^{-\infty}T = ST^\infty ,$$

from which it follows quite mathematically that $1/\infty = \infty$. The infinitely small = the infinitely great! And several more such things are to be found in the same book, from which certain logical minds would be able to draw the conclusion that mathematics is only a useless figment of the imagination, which leads to pure absurdities.

The Quality of Motion

16. With regard to quality, motion can be either positive, negative, or limited. The notion of limitation is composed of the notions of position[7] and negation. Thus it becomes necessary to demonstrate below how the former can be composed of the two latter.

17. We acquire direct knowledge of purely positive or negative rectilinear motion only through intuition. We get this knowledge not through *empirical* intuition (through the kind of sensible notion which is connected with *perception*) but through *pure* intuition (through one which is not connected with *perception*). What comes from empirical intuition must not be involved in a matter of *science*.

18. It is impossible to form a notion of negative motion without presupposing positive motion. We call a motion negative when it goes in the opposite direction of another which is regarded as positive.

19. It makes no difference whether a non-composite motion is imagined as if matter moves and the space in which it moves is at rest, or as if space moves in the opposite direction and matter is at rest. Kant adopts this principle as an axiom, i.e., a tenet which is evident from mere intuition, but it seems to me that it could very well be proved by means of the preceding. As it has been established that we cannot perceive any motion in absolute space, we cannot form an idea of any motion except when matter changes its relationship with the surrounding material space, and this happens whether matter moves and space is at rest, or whether space moves and matter is at rest, just as I get to the end of my room whether I walk there myself, or whether the room moves away beneath me, and I myself am at rest.

20. It is not so easy to combine two movements with each other at one point as it might appear at first glance. We are used to seeing that two opposite movements stop or cancel each other, that two movements in one and the same direction increase each other's velocity, and that two movements in directions that form an

---

[7] [In the sense of "affirmation" as used in formal logic.]

angle with each other produce a movement in the direction of the parallelogram of the forces. However, we cannot be content with such empirical principles here where it lies with us to deduce universal natural laws from the nature of intuition. Here we must express the combination of two movements with pure intuition as the mathematician does with his notions, i.e., we must construct the combination of movements.

21. There are three cases in which two movements can be imagined to be combined in a point, which are that either:

A. the directions of both movements are in the same straight line, and in that
   case they are
   a. not opposite each other
   b. opposite each other

Or

B. the directions of the movements form an angle with each other.

In none of the cases is it possible to imagine the movements combined in a point in a stationary space, but we must imagine the point itself with one motion and space with the other in the opposite direction. We shall present the proof for each case separately.

First case. When both movements lie along the same straight line and have the same direction. We shall designate the velocities by means of the straight lines *a*

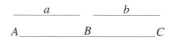

and *b*. Now if we wanted to combine the two motions *a* and *b* in one straight line *AC*, viz., *AB* = *a*, *BC* = *b*, so that *AC* was traversed in half the time required for the movements *a* and *b* combined, then the velocity *AC* would surely be the same as the sum of the velocities *a* and *b* (§11), but thereby we would not have presented the combination of the two movements with pure intuition, i.e., not constructed it, for if this were to happen, *AB* had to be traversed in the same time as *a*, and *BC* in the same time as *b*, but then *AC* would be traversed in the same, not half the time of *a* and *b* together. However, if one imagines, in accordance with the demands of the axiom, that the movable point moves in the direction and with the velocity *BC* from *B* and also imagines space in opposite motion *CB* with the velocity *AB*, then the point describes *BC* through its own motion and space *BA* through its motion. In this way, then, the point *B* completes two motions at the same time which correspond completely to *a* and *b*; therefore, these are constructed.

Second case. When both motions occur along the same straight line but in opposite directions. To imagine two opposite motions in the same space and at the same point is something that cancels itself. Therefore we only have to show how a point can have two opposite motions in two different spaces.

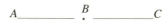

If the two motions *BA* and *BC* were to be combined at the point *B*, one of the motions, e.g., *BC*, had to be imagined as if it belonged to the point itself and took place in absolute space, but space, on the other hand, had to be imagined as moving with the velocity *BA* but in the opposite direction, that is, in the direction *AB*. Now if the velocity *BA* is = the velocity *BC*, then *B* has not changed its distance from any point in space, for every point in space has approached *B* through the motion of space by as much as *B* has moved away from it through its own motion. Thus *B* has not changed its external relation to space and therefore has not moved but kept its place in space, i.e., it has been at rest. If the velocity *BC* had been greater than the velocity *BA*, it would have moved farther away from the point at *A* than this would have approached. In this case, then, *B* must really change its external relation to space, i.e., move and do so with a velocity = *BC* − *BA*. If *BA* were greater than *BC*, the velocity would be = *BA* − *BC*. In the latter case, the direction would be *BA*, in the preceding one, *BC*.

Third case. When two motions which form an angle with each other should be imagined as being united at a point.

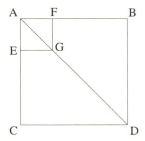

It is impossible to imagine that the point *A* moves through the same space with both the directions *AB* and *AC*, but it is easy to imagine that one motion produced a change in the other so that the point *A* now moved in an intermediate direction, but this would not be to show how a point could have motion in two directions at the same time but only to see the effect of two forces on one point. However, here we wish to construct the motion, so we must follow a different path. Let us imagine the point *A* moving in the direction *AC* but the space in the direction *BA*, the opposite of *AB*, so that the angle *BAC* remains unaltered. Thus *A* gets the motion *AB* and the motion *AC* at the same time. This motion, composed of two different directions, gives but a single direction for all possible experience, which is along the diagonal of the parallelogram described by the two directions. In order to illustrate this, one must imagine that *A* has moved from *A* to *E* in the direction *AC*, and then, during the same time, space must have traversed a proportional part of *BA*, such as *FA*. Using *AE* and *AF* a parallelogram can be drawn, the other two sides of which must meet at a point *G* on *AD*, which is the diagonal of *ABCD*, which can be proved mathematically. The moment when *A* reaches *E*, *F* must, through the motion of space, reach the point where *A* was initially, but *G*, on the other hand, will reach *E*, which can be seen from the nature of the parallelogram. Therefore the point *A* will be at the point *G*, consequently on the diagonal *AD*. From whatever point on the diagonal we might draw straight lines to *AC* and *AB*, parallel to the opposite side, we would always obtain the result that parallelograms were formed whose sides, cut from *AC* and *AB*, were proportional to the sides from which they were cut so that everything that was true of the parallelogram *AEFG* must also be valid here. During its motion, *A* reaches every point on the diagonal *AD* once, and when *A* reaches *C*, so does *D*, consequently *A* itself reaches *D* and has thus traversed the entire diagonal.

22. Consequently, three *laws of motion* result from the theorem advanced here:

1) *When two motions in one and the same direction are united at one point, the resulting velocity is = the sum of the velocities of the combined motions.*
2) *When two motions in opposite directions unite at one point, the result is a motion which is = the difference between the velocities of both motions in the direction of the stronger. If both velocities are equally great, the velocity of the resulting motion will be = 0, that is, the moved point is at rest.*
3) *When two motions, in directions which form an angle with each other, are united at one point, the result is a third motion, whose direction is the diagonal of the parallelogram of the two motions.*

These laws undergo some modification when they are applied to bodies, which we shall see below.

23. We have now learned how one motion can increase or decrease or change the direction of another. We immediately saw the possibility of positive and negative motion, and now we have learned how a limited motion can arise, that is, through a change in direction. *Circular* motion, whereby a point describes a curved line that falls back on itself, and *rotational*, whereby a body merely revolves around its own axis, are examples of limited motion. In general, it can be said that any motion becomes limited when it changes its direction, for even if it continues, it is really not the same motion but a new one which commences with the start of each new direction.

On the Relations of Motion

24. With regard to the relation of motion, we have the theorem that to every *motion there corresponds an equal and opposite one*. The preceding makes this theorem quite clear, for when it does not matter whether space moves and matter is at rest, or whether the latter moves and the former is at rest, it follows that when matter moves in one direction, space moves in the opposite direction, and when space moves in one direction, the point moves in the opposite with the same velocity as space. Thus the end of the room in which I walk up and down approaches me as much as I approach it, and it moves away from me as much as I from it. From this as well as from the definition of motion, it is evident that it is always relative.

On the Modality of Motion

25. In this part of the doctrine of pure motion we consider motion as an object of experience. Like all that is imagined by the senses, motion is only given as feeling. It is a still undetermined object which the intellect presents for intuition. This object must be specified by the intellect by means of a predicate which it ascribes to it. In this way the movable itself now becomes an object of experience, viz., when a certain object, in this case matter, is specified by the predicate motion. However, motion is a change of relations in space. Thus there are always two correlates here, such that in our sensation *we can equally well attribute the change to one or the other, and either one or the other can be thought of as moved because both are equally possible, or one of them must be thought of in experience as moved to the exclusion of the other, or both must necessarily be thought of by reason as moved.* Thus three notions arise: a) *Motion in relative* (movable) *space.* b) *Motion in absolute* (immovable) *space.* c) *Motion in any relative form,* which is distinct from motion in itself or without comparison to another.

26. *The non-composite motion of matter with regard to empirical space is, unlike the motion of space in the opposite direction, just a possible predicate. Rectilinear motion imagined without any relation to external matter is impossible;* for in experience (knowledge which determines the object validly to all sensation) there is no difference between non-composite motion in relative space and the body at rest with the motion of relative space in the opposite direction. Now the notion of an object with one of the two predicates, which are equally valid with regard to the object and only differ with regard to the subject and its mode of conception, is not a determination according to a *disjunctive* but only a choice according to an *alternative.* It is not according to a judgment which accepts one of two objectively opposite predicates to the exclusion of the other, but according to one which adopts one of two estimations, objectively of equal validity but subjectively opposite, for the determination of the object without the exclusion of the opposite, that is, it makes a specification of the object by mere choice. In other words, with the idea of motion, as an object of experience, it is in itself undetermined, i.e., equally valid, whether a body is imagined as moved in relative space, or whether the latter is imagined as moved relative to the former. Now that which in itself is undeterminable with regard to two opposite predicates is in a sense only *possible.* Consequently, the rectilinear motion of a body in empirical space, unlike the equal and opposite motion of space, is merely a possible predicate in experience. Furthermore, as a relation, and therefore also its change, cannot be an object of experience, except in so far as both the things which are imagined in relation to each other are material, no motion can be observed in pure space since this is not material.

27. *In composite motion, the motion of matter, unlike that of space in the opposite direction, is a real predicate, and the motion of material space in the opposite direction, regarded as the motion of the body, is mere appearance;* for in composite motion, space itself already has a motion in a definite direction, and as it cannot have two different directions at the same time, I cannot imagine the motion

of matter and the repose of space or the motion of space in the opposite direction as equally valid, from which it follows that the motion of matter must be considered real, and that the opposite mode of conception can only be appearance.

28. *In the motion of every body in which it is regarded as moved relative to another, an opposite, equally great motion is necessary.* We have seen above that every motion is *reciprocal.* As this must be imagined for every motion according to the evidence put forward, that is, as it cannot be absent, then it must be *necessary.*

29. Before I conclude this doctrine of motion, I want to show how motion can be divided into categories. Thereby I hope fully to prove the correctness of the way in which I have divided all the metaphysics of nature.

Motion is

A. With regard to quantity
   a. uniform
   b. difform
      α. regular
         aa. decreasing
         bb. increasing
      β. irregular
         aa. decreasing
         bb. increasing
         cc. now decreasing, now increasing.

In uniform motion equal distances are traversed in equal times. Here, then, we have the *unity* of the distances traversed. In regular difform motion, different distances are traversed in equal times, but their quantity is determined according to a certain rule. Here we have the *plurality* of the traversed distances in a certain time. In irregular motion, the distances which are traversed in a certain time can have all possible sizes. Here we see the *totality* of the distances in a given time.

B. With regard to Quality
   a. non-composite
      α. positive
      β. negative
   b. composite (of positive and negative)
      α. with a change in the position of the object in motion
         aa. rectilinear
         bb. curvilinear
            αα. periodic
               aaa. oscillating
               bbb. circular
            ββ. spatially expanding
         cc. with mixed lines
      β. with no change in the position of the object in motion, a rotating motion.

C. With regard to relation
  a. common
  b. individual
D. With regard to modality
  a. real
  b. apparent
  c. necessary.

Here we have a complete classification of motion. It will not be possible to mention any kind of motion which cannot be referred to one of the categories listed here although I fully realize that this division could have gone further, and that I could have made more subdivisions.

DOCTRINE OF MATTER

On the Quantity of Matter

30. As we have seen above, that which is movable in space is called matter, and we regarded it in this manner throughout the doctrine of motion, but the first notion we have of matter is that it is the object for our external senses. According to *general metaphysics*,[8] every object of the external senses must have extensive size, which must always be limited, for if it were infinitely large, the composition of its parts could never be completed, which, however, must happen during perception. If it were infinitely small, i.e., smaller than any sensible size, it would also be smaller than any material space, which can always be given, and consequently insensible.

31. As every part of matter itself is matter and, as such, movable, it follows that it can cease to be a part of it through motion; it can separate from it. Every such separated part again has parts and can therefore be separated or divided. This division cannot cease until one arrives at matter which has no parts, but then the space which was occupied by it would have no parts either, but a space without parts is impossible, and its divisibility never ceases, so neither does that of matter. Therefore, matter is infinitely divisible. By imagining the division of matter continued infinitely, one arrives at the conception of infinitely small parts, consequently of matter which occupies an infinitely small space, which seems to be at variance with the above, but this conflict is only apparent, for the conception of infinitely small parts is merely an idea which reason has thought up in order to arrive at a limit, but nothing like it can ever be found in any possible experience.

32. It is evident from this that the quantity of matter, i.e., the mass of the movable in space, cannot be determined by the number of its parts as this can never be determined.

---

[8] According to the axiom of intuition. *Kritik der reinen Vernunft*, fourth ed., p. 202.

33. Here we now have the complete doctrine of the quantity of matter, as the following table shows.

Matter is: with regard to

The totality; a finite whole		The plurality; composed of infinitely many parts
	Unity; an infinitely small unit	

I must anticipate the objection that might be raised here, which is that even though matter only has a finite size, it must still consist of infinitely many parts; but this objection falls when one considers that these parts become infinitely small. However, an infinite number of infinitely small parts is required in order to produce a finite size, according to the mathematical formula $1 / \infty \times \infty = 1$. The validity of the assertion that the parts are infinitely small hardly requires any proof. In order to convince ourselves, we only need the observation that the more parts into which a certain extent of matter is divided, the smaller each of them becomes; if, for instance, it is divided into 10 parts, then every part is $= {}^1/_{10}$ of it, if it is divided into 10000 parts, each of the parts is $= {}^1/_{10000}$, and if it is divided into ∞ parts, each of these, then, must become $= {}^1/_{\infty}$. We have now seen that matter must have extent, form, and divisibility. Matter, when confined within certain limits, is called a body, and thus matter must always manifest itself as a body.

On the Quality of Matter

34. In the doctrine of the quantity of matter, general metaphysics guided us to show the limitation of matter in that it had already shown that all material objects must have extensive size. In the present investigation of the quality of matter, another of its axioms is used as the basis of the following. In Kant's *Critique*, it is called the anticipation of observations, and it teaches us that in all phenomena the real, as an object of perception, always has intensive size, i.e., degree. In extensive size, one part is always disjoint from another, and we acquire a conception of the whole by first imagining all its parts. In intensive size, on the other hand, the parts are not disjoint, and we first acquire a conception of the whole and, through this, a conception of the parts which, for sizes of this nature, are nothing but degrees which can decrease until everything dwindles to zero. The real in phenomena, then, manifests itself to our senses through the intensive size of its quality, i.e., through a force. Now, matter is the real in material objects since space and time, which can make no impression on the senses, cannot be that, and consequently we must attribute force to matter. No change, and therefore no effect, can happen in sensible nature except through motion, so the force of matter can create nothing but motion, and consequently it must be a motive force.

35. Matter, then, has motive force. This can only be of two kinds; it can act in such a way on matter that it either approaches or moves away. In the former case, it

may be called an attractive, in the latter a repulsive force. It is not possible to imagine any other forces of matter. It now remains to investigate which of these forces matter has, or whether perhaps it possesses them both. Therefore we shall first examine how matter would behave if it possessed only one of these forces.

36. If we attribute repulsive force to matter, we must also attribute the same force to all its parts because every part of matter is itself matter. However, a repulsive force in every part of matter must cause it to spread through space, so it is an *expansive* force.

37. Matter is not possible through this force alone as it cannot be limited by it, for the force cannot limit itself as it consists only of an effort to occupy a larger volume. Space cannot limit it either, only cause the force to become smaller, the larger the space in which matter spreads, but the force must become = 0 if matter were to be limited by its diminution in this way, but a force = 0 is no force, and matter which has no other force is not matter. The expansive force alone, then, cannot give us limited but rather infinitely large matter.

38. Matter which only possessed an attractive force would not be limited but infinitely small, for because of this force which, for the same reason as the repulsive force, belongs to all the parts of matter if it belongs to matter itself, all the parts of matter must approach each other incessantly until space no longer permits it, and this cannot happen until it is so small that it cannot be diminished, which is = 0. It is evident that this force is no more capable of limiting itself than the expansive force.

39. Thus only the third possibility remains, which is to attribute both forces to matter. In this way the expansive force is limited by the attractive force, and the attractive force conversely by the expansive force. The expansive force prevents the attractive force from reducing the extent of matter to zero, and the attractive force prevents the expansive force from giving matter an infinitely large extent. They work in opposition to each other and produce motion in opposite directions so that one may be regarded as *negative* when the other is regarded as *positive*. The *positive* and the *negative* (reality and negation) combined give *limitation*.

40. The two fundamental forces which we have now discovered in matter distinguish themselves from each other in that the expansive force does not act except through contact, the attractive force, on the other hand, at a distance. Contact in a *physical* sense is the direct action and reaction of impenetrability. It is contrasted with mathematical contact, which consists in two bodies having a common boundary. The mathematical is part of the physical but does not constitute it alone; an active relation must be added, and as this cannot happen through the attractive force, it must happen through the repulsive one, i.e., through impenetrability. When, on the other hand, one body acts on another without contact, this is called *action at a distance*. This effect, which can also occur without the aid of any intervening matter, is called immediate action at a distance or the action of matter through empty space.

41. The *attractive force* has such an *action at a distance*, for this force is required if matter is to be limited and is therefore necessary for physical contact. As it thus contains the cause for this contact, it must also precede it, and its effect cannot be tied to the conditions of contact. However, if it is independent of contact, it is also independent of the filling of the space between the moving and the moved, i.e., it must take place even when the space between the two is not filled, thus as an action through empty space, for by insisting that the space between them was filled, one would also demand that there was contact, of which, however, this force was presumed to be independent. Consequently, the original attractive force, which is essential for all matter, is an immediate action at a distance and through empty space.

42. Much offence has been taken at the idea that matter should act at a distance as it was incomprehensible how an object could have an effect where it was not. However, this objection is without meaning. It is much more correct to say that matter cannot act at the place where it is, for if it acted at the place where it is, the object on which it acted could not be *outside* it, for *outside* means presence at a place where the other is not. Even if the earth and the moon touched each other, the point of contact would still be a place in which neither the moon nor the earth was, for the locations of these have a distance from each other which is equal to the sum of their radii. The point of contact would not even contain any part of the moon or the earth, for this point lies at the boundary of the space filled by them both, which does not constitute any part of one or the other.

43. The repulsive force does not act in this manner at a distance, but it is a *surface force*, which is such that matter acts only directly at the surface of contact. However, the force whereby one body acts only on the parts of another directly, also beyond the surface of contact, is a *penetrative force*. The repulsive force in one part of matter limits the sphere of action of the other, and the repulsive force cannot influence any distant part, except by means of the intervening ones, and a direct effect of the repulsive force which goes through these is impossible. On the other hand, an attractive force by means of which a body can occupy a space without filling it, whereby it also acts on other remote ones through empty space, and whose effect no intervening matter limits, must be a penetrative force. The original attractive force must be thought of in this way, so it is a penetrative force and is always proportional to the quantity of matter because it must become stronger as the number of points with which it can act grows, but since one point cannot prevent another from acting, it must work more strongly, the more points there are, i.e., the more matter there is.

44. This force extends to infinity, from one part of the universe to the other, for if we suppose that this force for immediate action at a distance, which every part of matter possesses, extends only to a certain distance, the cause of this limitation of its sphere of activity must lie either in the matter which is contained in this sphere or in the volume of space, but neither of these is the case. The former because this force is a penetrative force and acts immediately at a distance through every space as if it were empty. The latter because the attractive force is a motive force which

has some degree, of which it is possible to imagine smaller degrees ad infinitum, and the greater distance would thus only be a cause for the weakening of the force proportional to its extension over a larger space, but it could never make it disappear completely. Since there is now nothing which can limit the sphere of action of the attractive force, it extends beyond all limits to all other matter, and thus infinitely into the universe.

45. From the preceding, we have seen that matter is a result of the union of two opposite motive forces. According to phoronomy, two forces moving in opposite directions give us a motion which is equal to the difference between the forces. Consequently, the extent of matter must be equal to the difference between its two fundamental forces. It may be objected to this that both these forces are infinite and thus their difference $= \infty - \infty = 0$, but this difficulty is resolved by the observation that these forces may well be infinite with regard to their sphere of activity, but it certainly does not follow from this that they have to be infinite with regard to their intensity. In this regard they might very well be different and thus yield a finite difference. —Eschenmeier[9] has attempted another solution by declaring the expansive force $= \infty$ and the attractive force $= 1 / \infty$ so that these multiplied by each other, according to the axioms of mathematics, must be $=$ a finite quantity. Thus he sees matter as a product of these two fundamental forces. However, in part this way of looking at things is so obviously false that there is no need to refute it, and we must in fairness wonder that such an astute man, as he otherwise appears to be, could indulge in such a notion, and in part, if we accepted for a moment the notion that matter was to be regarded as the product of these forces, I do not know what justifies the assumption that the attractive force is $1 / \infty$ when the expansive force was assumed to be ∞; it should rather be assumed to be $-\infty$ since the other was assumed to be $+\infty$, but then their product would be $+\infty$ multiplied by $-\infty = -\infty \infty$, which is a negative quantity. According to this kind of reasoning, matter would acquire a negative extent, which would be contradictory. Therefore, all the conclusions which he builds on this collapse along with this foundation, and the chemistry and the theory of magnetism that he has given us are of no value even though they show him to be an astute man who would probably have found the right way if he had started out from the right point.

46. We cannot know very much about the mutual relationship between these forces but must content ourselves with the knowledge that these two forces are necessary conditions for the possibility of matter. Only this much is certain: The expansive force must always be greater than the attractive force, for if they were equally great, they would neutralize each other completely, and their difference, and therefore also the extent of matter, would be $= 0$, and if the former were smaller, it is just as clear that their result could not be extension.

47. All matter cannot be completely identical with regard to the fundamental forces, for then it would not be possible to observe any motion in material space,

[9] [Eschenmayer.]

where, however, all motion must occur, because space would then have to be homogeneous so that one point in it could not be distinguished from another, and consequently it would be impossible to observe that one point moved away from another or approached another, i.e., no motion could be cognized. It might be objected to this proof that the parts of matter could also be distinguished from each other through their shape, but where would one find a cause for the different shapes of matter when no difference could exist with regard to their forces? Besides, it would be impossible to observe form, and consequently also boundaries, in completely homogeneous matter unless one wanted to consider empty space, which philosophy rejects as something that can never be the object of any experience.

48. In this part of the doctrine of matter we have now proved a priori that *general attraction*, and consequently also *gravity*, and *impenetrability* are necessary conditions of matter. Usually cohesion is also included here, but that can hardly be regarded as anything other than general attraction modified by random circumstances. The cohesion of bodies appears to follow laws which are quite different from those of attraction, which has caused Kant to believe that this is a derivative force. However, in spite of all its deviations, it is still reasonable to assume that it is the same force, for good grounds can be found for all the deviations. The two things in which the cohesive force differs in its action from the attractive force are that it does not seem to obey the same laws as this, and also that, in contrast to attraction, it does not have any particular effect at a distance. In a treatise on this subject in Gren's new *Journal der Physik*, Vol. 4, p. 82, Murhard has made it comprehensible how the cohesive force can act according to other laws than the attractive force. I want to add to this that the pressure of the atmosphere modifies this force greatly, which is obvious from the fact that bodies in empty space, or rather in rarefied air, evince this force to a far smaller degree than when they are surrounded by denser air. In so far as air causes coherence, it must be determined by the surfaces on which the air can exert pressure. Likewise the force, in so far as it owes its existence to the air, becomes merely a surface force. In this investigation, we must also observe that the attractive force in and of itself does not have so considerable an effect as is generally imagined, for if we recall that the greatest effect known about the attraction of a mountain is that it can pull a perpendicular bob a little out of its direction, an effect which a force of less than one-half *quintin*[10] could produce, it is easy to understand that the effect of one of the bodies with which experiments on this topic are usually conducted must be inordinately small as these bodies do not constitute $^1/_{1000000}$ of such a mountain. Such bodies, then, can tolerate only an extremely small distance from each other if any interaction is to take place between them, for the strength of the attractive force decreases proportionally with the increase of the distance. As a result of this, two bodies which attract each other must have a stronger effect on each other, the more surface area they have in common, for in this way more of their parts are brought closer to each other. Thus the original attractive force itself must act partly in proportion to the mass (as a penetrative force) and partly in proportion to the surface. This makes it easy to understand that

[10] [One *quintin* corresponds to $^1/_{128}$ of a Danish pound of 500 g.]

the effect of the original attractive force on small masses must be extremely small and, when contact ceases, weak to the point of being imperceptible. This is also precisely what experience teaches us as the attraction of bodies, when contact is broken off, almost disappears though not completely, for when a metal plate is weighed with a very accurate balance just above mercury, it weighs a little more than it would otherwise weigh. Besides, chemical affinity often acts during contact so that this also causes them to cohere more strongly, in proportion to the surfaces on which they touch.

The distinctive shape of bodies, which they seek to assume at every opportunity, and which in dead nature gives us crystallization in living creation, also rests on these forces which we have here proved to be in the bodies. When bodies are in their liquid state, they cannot help but flow into a sphere unless other forces counteract that, for when the parts do not resist the force which wants to change their position, the attractive force must place them in various positions until an equilibrium is established between them, and this happens only when their position is such that the entire mass is spherical. In order not to enter into mathematical arguments, I merely want to state the following. If we imagine a point in a quantity of water which attracts all others equally independent of direction, everyone will realize that the circumference must be everywhere equidistant from this point, and this gives us a sphere, whose nature is that its surface is everywhere equidistant from a certain point. When, on the other hand, an external force like gravity acts on fluid matter, this will change its shape more or less. If the fluid becomes a solid while such a force acts on it, the solid thus generated will also deviate from a round shape, in different proportions according as its own cohesion and the externally acting forces influence it. Thus, experience also shows that when parts of a fluid become a solid (are crystallized), this does not happen in the middle of the fluid but at its edges or at the bottom because there it is in contact with another body which can exert attraction on it. Therefore threads or wooden sticks are often placed in the middle of a fluid in order that they may promote crystallization through their attraction. It is impossible to object that this may just as easily be explained by the surrounding solid parts attracting the caloric and thus promoting the transition to solidity, for in that case one would not use such poor heat conductors as one usually does. This tendency of bodies to assume certain shapes could be called a formative instinct, admittedly a figurative expression, like chemical affinity, but it is not always the concern of natural philosophy to abolish such expressions but to give them a proper meaning. So far the term formative instinct has been used only about the phenomenon in organic nature, but I think that the term must be extended to describe the phenomenon in the inorganic part of nature, too, for its cause is everywhere the same. What I have said here about this formative instinct is very incomplete, but on closer investigation, especially through the application of mathematics to this subject, it can no doubt be developed to a high degree of perfection.

Most bodies are composed of a great number of crystals which are, as it were, wedged in between each other, and from this it is also seen that bodies can rarely be made to cohere closely once they have been torn apart.

On the Relations of Matter

50.[11] In this part of the metaphysics of nature we must consider matter when brought into relation with other matter, both in their different states and with their different quantities and forces. Kant has called this part of science *mechanics*, but this expression does not characterize the entire science, which must also contain *chemical* action as this is nothing but a particular kind of the mutual action of matter through attractive force. Therefore, it would be better to call this part the *doctrine of action*, in contrast to the previous one, which might be called the *doctrine of forces*. Kant has not considered the action of one body on another when at rest, and this is undoubtedly the reason why he relegated chemical action to *dynamics* (the doctrine of forces), where, we must admit, it does not belong. In the same doctrine that famous philosopher and his successors have dealt with *elasticity, fluidity*, and *Mariotte's Law*, which, however, are all based on the interaction of matter and thus belong to the doctrine of the relations of matter. As the doctrine of motion is nothing but the doctrine of the external relations of matter, it is obvious that the doctrine of action will contain an applied doctrine of motion or rather become nothing but that. The difference between phoronomy and the doctrine of action, then, is that the former deals with the merely external relations of matter, without regard for its forces, i.e., its motion or rest regardless of its action. Here, on the other hand, we deal with active motion or rest. It is this part of the metaphysics of nature which is most closely related to empirical physics, and which must really give it its form, which I shall try to develop further on another occasion. First we shall seek out the general laws of all action.

51. *First law. The quantity of matter* (the quantity of the movable in space) *can be neither increased nor decreased.* In order to prove this, we presuppose the theorem from general metaphysics that the substantial in sensible nature can be neither increased nor decreased.[12] Here, then, we only have to prove that matter is substantial. By substance we mean a subject which is not a predicate to any other subject, that is, a final subject. Matter is such a final subject, for in sensible nature there is nothing but time and space and matter. It cannot be the first since that is only a form of sense according to which we must intuit, but it does not exist in itself. Therefore matter, as the movable in space, must be the substantial, which can neither come from nothing nor change into nothing but only alter its external qualities. We base so many of our judgments in empiricism on this law of nature that no natural science would be possible without it. Only because of that can we boldly claim that one body has given some of its mass to another when it has lost some of its quantity.

52. But how do we determine the quantity of matter? As we have seen above, this cannot be done through the number of its parts. Nor can it be determined according to the size of the space which the matter occupies because this presupposes that the bodies which are compared have the same density, so this method is not gen-

---

[11] [There is no §49.]   [12] *Kritik der reinen Vernunft.*

eral. As the quantity of matter, then, can be determined neither by the number of its parts nor by its volume, all that remains is to determine it by the quantity of motion.

53. In phoronomy we cannot imagine any other quantity in motion than velocity as we cannot there attribute any size to the movable. Here, on the other hand, where we can attribute a quantity to the movable itself, the situation is entirely different. According to phoronomy, it does not matter whether I attribute a certain degree of velocity to a single movable body or attribute smaller degrees of velocity to many equally movable ones so that their total equals the velocity of the former. From this there arises first an apparently phoronomic notion of a quantity of motion as composed of many motions distinct from each other but still united to form a whole. However, if these points are now imagined as having a motive force by means of their motion, a mechanical concept of the quantity of motion arises. The greater the quantity of matter, then, the greater also the quantity of motion when the velocities are equal, for the greater the quantity of matter, the more movable points whose motions can be added up. If we call the velocities V and v, the quantities of motion Q and q, the masses M and m, we then obtain:

$$\text{when } V \text{ is } = v, \quad Q : q = M : m$$
$$\text{and when } M \text{ is } = m, \quad Q : q = V : v.$$

If we now assume a third motion $\mathbf{Q}$, in which the velocity is $= V$ and the mass $= m$, then

$$Q : \mathbf{Q} = M : m$$
$$\mathbf{Q} : q = V : v$$

$$\text{therefore } Q : q = \frac{M : m}{V : v} = MV : mv.$$

The quantity of a motion, then, is $=$ the velocity multiplied by the mass.

54. The above makes it clear that motion with an infinite quantity is impossible, for that would require either infinite velocity or infinite mass, which we have seen above to be impossible.

55. According to phoronomy every motion presupposes an equal motion in the opposite direction. As we have now acquired a new concept of the quantity of a motion, this must also be applied to this theorem. The result of the application is that when A moves towards B, then B's mass : A's mass $= A$'s velocity : B's velocity, or the masses are inversely proportional to the velocities or the velocities to the masses because the quantities of the motions are equal. As times are always equal in such corresponding opposite motions, it follows that the velocities are directly proportional to the distance traversed, and consequently these are inversely proportional to the masses. If, for instance, in the adjoining figure, a mass A, which is

three times the size of *B*, approaches [*B*], then *B* must be imagined to traverse a relative distance, which is three times that of *A* because then *B*'s mass : *A*'s mass = *A*'s velocity : *B*'s velocity.

56. *Second law. No body changes its state* (passes from one motion to another or to rest or vice versa) *without an external cause.* Here we take it as proved by general metaphysics that every event has a cause, so it only remains to prove that every cause of the motion of matter must be outside it, i.e., an external cause. Matter, merely as an object of the external senses, has no other determinations than the external relations in space and therefore undergoes no changes except through motion. As a change from one motion to another or to rest and vice versa, this must have a cause. This cause must be external, for matter as such has no internal determinations or grounds for determination. Consequently, there must be an external cause for every change in matter.

57. It is this law which should be called *lex inertiæ* and be distinguished from the law which says that there is an equal reaction for every action, for this law has often been called the law of inertia and has thus attributed to bodies a force to resist another which would change its state. Such a positive force to maintain its state conflicts with precisely this law as it presupposed an internal force for its determination. The difference between life and lifelessness lies in the correctly understood notion of inertia. Life consists in an ability to act according to an *internal principle*, so it is exactly the opposite of inertia, which consists in a lack of this ability to determine itself. Without this law, no natural science would be possible. The opposite, which attributed life to matter (hylozoism), is the death of all natural science.

58. It is evident from the law of inertia that every motion which is not rectilinear must be composite, for every deviation from the straight line is a change in direction which cannot arise except through an external cause, that is, through a new motion. We have sufficient examples of such a composite motion, e.g., the stone in the sling.

59. *Third law. In all motion, action and reaction are equal.* If *A* is a body which in its motion approaches *B*, it must be assumed that *B* and the relative space with the same quantity of motion move towards *A*, and the distance traversed by *A* : the distance traversed by *B* = *A*'s velocity : *B*'s velocity, i.e., *AC* : *BC* = *B*'s mass : *A*'s mass.

A, then, must meet *B* at *C*, and as they have equal quantities of motion, one cancels the other so that *A* and *B* both come to rest, but relative space is not therefore at rest yet but continues to move with the velocity and in the direction BC. However, as the motion of relative space is the same as the equally fast motions in opposite directions of the bodies A and B, now at rest at C, they now continue to move in relative space, in the direction BD and with the velocity $= BC$. If A's mass is $= \frac{1}{3}$ that of B, then AC is $= 3\ BC$ (§55), and $AB = 4\ BC$. A, which previously travelled the path AB in relative space, now only manages to travel $BD = BC = \frac{1}{4}\ AB$ so that the quantity of its motion is now $= 1$, instead of being $= 4$ as before. B, on the other hand, which was at rest in relative space before will now move with the same velocity as A, and because its mass is 3 times as great, the quantity of its motion is $= 3$. Since the quantity of A's motion is now $= 1$, that of $B = 3$, then A, the quantity of whose motion was $= 4$ before, has lost as much of its motion as B has gained, so action and reaction have been equal. In this proof we have regarded B as being at rest, but anyone who understands the second argument can readily apply it to the case when B is in motion, for if B in its motion approaches A, a so much smaller motion must be attributed to relative space, and if it moves away, one so much greater. Here we have actually only proved that repulsion and counter-repulsion are equal, but with a little change it is possible to prove in the same way that attraction and counter-attraction are equal as attraction and repulsion differ from each other only in their direction. It is also obvious that pressure and counter-pressure are equal, for pressure does not take place without an effort from the pressing body to move towards the pressed one, and if an equal counter-pressure were not present, motion would really have to take place. Besides this mechanical action, in which one body transmits some of its motion to another, there is also a dynamic one, whereby a body can transmit motion to another, through its fundamental forces, without losing any of its own. This dynamic action must also be equal to the reaction, for when A forces B to approach it by means of its attractive force, this is the same as when it resists the force whereby B wants to move away. As it is the same whether A moves away from B or B from A, B resists A's recession from it to the same degree as A resists that of B, so attraction and counter-attraction are equal also according to this. The case of the action of the expansive force is the same, for when A repels B, it resists B's approach, but as A now approaches B as much as B does A, B consequently also resists A's approach as much as A resists that of B. As dynamic action can only occur as a consequence of the two fundamental forces of matter, it follows from this that dynamic action is also equal to the reaction.

60. From the laws of nature that we have developed here, there follows another law which is very important to all natural science, which is that a change never happens instantaneously but only gradually. In order to prove this law, we must first state a couple of other theorems.

61. The velocity which a body transmits to another in an instant is called the moment of acceleration. Once a body has acquired a moment of acceleration, it retains it until deprived of it by another. This follows from the law of inertia, according to which no body can change its state without an external cause. Now if it acquires an identical moment of acceleration in the next instant, it moves with a velocity which is composed of both of these as the previous motion continues, and an equal one is

now transmitted to it. If it thus acquires a new moment of acceleration in each of a series of instants, then it must acquire an infinite number of moments of acceleration in any finite time, which always consists of infinitely many instants. For this reason, each of these must be infinitely small, for if they were finite, an infinite quantity of finite velocities would thus be transmitted, which, combined in a body, would constitute an infinitely great velocity, which is impossible.

62. It is impossible to imagine a body which could transmit a finite moment of acceleration to another, for this would transmit to it an infinite velocity in a finite time, which is impossible.

63. From these theorems follows the law that no transmission of motion, and consequently no change in nature, happens instantaneously, but always gradually. This law is called *lex continui* and might be called physical, in contradistinction to the metaphysical one which teaches that all change, internal as well as external, must occur gradually.

64. Here we now have the general laws which must form the basis of every branch of natural science. It still remains to make a classification a priori of natural science, but the development and justification of this classification would make this presentation too extensive. I shall therefore confine myself to a few remarks on several subjects.

65. Kant relegates elasticity to dynamics, in my opinion incorrectly, as it can only be recognized when bodies are brought in relation to each other and thus act. Chemical action can be relegated to dynamics with even less justification as it cannot be cognized except for the relation of one body to another.

66. Kant mentions the expansive force as the force in bodies which gives them their elasticity, or rather he states that elasticity is only this force, but more careful consideration reveals that bodies can also, by means of the attractive force, regain the shape that has been taken away from them by external violence, or even that both the fundamental forces of the body can contribute to this. I shall give an example of each of these cases. When compressed air drives back the piston in a compressor, the elasticity is through the expansive force only, whereas a taut string rebounds by means of its attractive force, but both forces together compel a steel spring to return to its original straight line. It is evident that this is really so, for in the first case the air has been compressed into a smaller volume than before and therefore cannot be returned to its former space through any force other than the expansive force, in the second case the string is stretched from the straight line which is the shortest between its two fixed end points and therefore comes to occupy a greater space than before, so the attractive force must restore it to its former space, and in the third case the part of the surface of the steel spring that becomes convex is stretched and the part that becomes concave is compressed, so the attractive force on the convex side and the expansive force on the concave side force the parts to seek their former positions.

67. To indicate the causes of chemical action with strict philosophical precision is not so easy as it might seem at first glance now that we have the fundamental laws of the action of matter; on the contrary, the issue presents many difficulties in its execution. The two attempts to build a chemistry on the basis of the critical metaphysics of nature that I am familiar with are so unsuccessful that they have brought their authors into the most evident contradiction with its foundations. The first to make an attempt of this kind is, as far as I know, the above-mentioned Eschenmayer, who builds it on the doctrine of the relation between the fundamental forces of matter which we have seen above. In his *Ideen zur Philosophie der Natur*, Schelling has adopted the same doctrine and developed it more precisely. As the theory of the former philosopher is false from its first foundation, the chemistry which he has based upon it is also false and is in conflict with the basic ideas of dynamics. As Schelling tries to develop the same chemical theory on other grounds, I only want to demonstrate its incorrectness by means of a few observations. His definition of heterogeneous (p. 235) is at variance with dynamics and contains the basis of the falseness of all that follows. According to him, heterogeneous substances are those in which the ratio of the fundamental forces in one body is the inverse of that in the other. No such relation can exist, which is easily understood when the theorem is stated in mathematical form. Let us call the forces of attraction in two bodies A and a, the repulsive forces R and r, and if the bodies were heterogeneous, the following relation must exist: $A : R = r : a$, but this relation is impossible, for A can be neither greater nor smaller than R nor equal to it. If A were $< R$, then, as a consequence of mathematics, r would have to be $< a$, i.e., the attractive force greater than the expansive force, which is impossible (§46). Nor can A be $> R$, or $= R$, as this would be completely at variance with what has been proved before (§47). The entire chemistry of this author stands or falls with this definition, of which one can easily convince oneself just by reading p. 236 ff. In his *Vorlesungen über die metaphysischen Anfangsgründe der Naturwissenschaft* (p. 108 ff), Bendavid has advanced another theory of chemistry, which is no more correct than the one we just left. He distinguishes between chemical and mechanical separation in such a way that the former is supposed to be caused by the attractive force of the separating body, the latter by its repulsive force. This distinction is fundamentally erroneous, for, according to that, it would be a chemical separation when one put a finger into water and separated from it one or more drops which adhered to the finger, or when one had something torn apart by means of a rope, which, of course, is also due to the attractive force of the separating body. He gives the following explanation of how one body would be able to destroy the coherence of another by means of its attractive force. When the force by means of which the parts of a body A cohere is smaller than the one by means of which another body B attracts it, the parts of A will exist for a moment without coherence, that is, merely under the influence of its expansive force. This force would disperse them infinitely if B did not immediately evince its action on them, whereby it seeks to unite them in its centre, which would also happen if the expansive force with which A's parts are endowed did not cause them, in combination with those of B, to occupy a given space to a certain extent. How, it is fair to ask concerning this explanation, is it possible that A's parts can be deprived of

their attractive force through the attraction of another body? The attractive force of *B* would rather act on *A* as a compressing force which was strong enough to destroy its expansive force by compressing it infinitely, but as such a force in *B* is impossible, it follows that chemical action cannot be explained in this way. I dare not venture to contribute anything complete on this subject myself, but I feel it incumbent on me to make a few observations. There is no doubt that substances can unite for two reasons, either because they are so little different from each other that they can be made to mix by mere shaking, or because their fundamental forces are related to each other in such a way that they cancel each other's coherence. This does not happen in an instant but little by little so that the parts which are torn away are divided again into new ones, and so on indefinitely. This might be called chemical penetration as there is no part of one of the substances which is not also combined with some part of the other in this way, and both substances occupy one and the same space. Thus there is an infinite division, which seems to contradict what has been proved previously, but it must be remembered that here the division occurs in a series of small time segments, that an infinite number of parts work all at once, and that the points of contact are constantly multiplied during the operation, for which reason this chemical motion becomes accelerated. When a body neutralizes the constituents of another through the attractive force without the latter producing the same counteraction, the result is merely adhesion. In the development of a complete theory of our subject, it might be useful to observe that heterogeneous bodies can have three kinds of qualitative relation to each other, viz., either by having equal attractive but different expansive forces, or by having equal expansive but different attractive forces, or, finally, by having neither of them equal.

68. Until the time of Kant, one did not have the correct notion of *fluids* because one differentiated between them by means of the varying degrees of force with which their parts cohered. Thus a body was to be called *fluid* when its parts cohered with a small force and solid when there was a greater cohesive force. Not to mention that no sharply drawn border between solidity and fluidity was determined as it had not been decided how large the cohesive force must be in order for the body to be called *solid* or *fluid*. Besides, this distinguishing mark was not even correct, for several fluid bodies cohere with a not inconsiderable force, e.g., mercury. According to Morveau's experiment, a round gold plate with a diameter of one inch adheres so strongly to mercury that a force of 446 grains is required in order to tear them apart. If a mercury column with a diameter of one inch did not cohere with a force at least as great, it would necessarily have been torn apart during the experiment. However, though we do not approve of distinguishing between fluid and solid bodies in this way now, it still has some basis in truth, which is that fluids can very easily be drawn apart, but as we have seen, the weakness of their coherence cannot be the cause of this, so it must be sought elsewhere. We find this cause in the fact that fluids can be extended so easily that they can be drawn out into a thin thread by any force however small, and then this can be torn apart without difficulty. This extensibility is nothing but the property that the parts of the body can easily be moved backwards and forwards in between each other. Fluid bodies

possess this ease in having their parts moved around between each other to such a high degree that any force however small can change their position, and in this we have the true distinguishing mark of fluidity. Many phenomena are explained with the greatest ease according to this definition; for instance, the fact that a sewing needle which is carefully placed on top of some water does not sink but remains on the surface, where it forms a little valley. However, once it has got under the surface of the water, it sinks until it has reached the bottom. The explanation is easy. When the sewing needle is to pass from the surface of the water to beneath it, it must first overcome the coherence of its parts, for which a certain force is required, but once it has done this, it has only to thrust the water particles aside, for which any force is sufficient. It is clear that this explanation does not fit the old definition. The ease with which our definition can be used in hydrostatics and hydraulics is evident upon the slightest reflection.

69. In the same way as fluidity, hardness, softness, ductility, and brittleness can also be determined according to the relation of the bodies when the position of their parts is changed by an external force. A body is hard when a great force is required to produce this change and soft when only a small one is required. The bodies whose parts cannot change their position without also having their coherence abolished are brittle. Ductile bodies have the opposite quality.

70. Here I had wished to enlarge on a number of other physical subjects in order to demonstrate the usefulness of metaphysics to the empirical knowledge of nature, but I fear that I have already gone too far in presenting a part of philosophy which still cannot claim very many adherents. Elsewhere I have promised to say something here about the materiality of the source of heat. I dare not ignore this subject completely, but I shall be as brief as possible in what I have to say about it. The distinguishing quality of the cause of heat is that it expands all bodies, so one could easily be tempted to regard it merely as the expansive force, but when one observes another of its qualities, which is that it can pass from one body to another, it is obvious that it cannot be this fundamental force of matter, for to say that one body transmitted part of one of its fundamental forces to another would be the same as saying that some matter ceased being matter in order to enlarge another. The law of substantiality (§51) makes it clear that this is impossible. To assume that the cause of heat is a derivative force would probably be better, but eventually one would encounter the same difficulties when one had to show how this force was grounded in the original ones. So far no one has ventured any attempt to assign a place in metaphysics to this force which is supposed to cause heat.

Matter with Regard to Modality

71. Just as the doctrine of matter with regard to its relations was not much more than the application of the doctrine of motion to matter endowed with forces, the part of the doctrine of matter which we are going to deal with here also becomes nothing but the doctrine of the modality of motion applied to the doctrine of matter. This application is very easy. By applying the theorem that only a rectilinear

motion can be non-composite (§58) to the first and the second theorems of the doctrine of the modality of motion (§26, §27), it becomes clear that it can never be anything other than rectilinear motion which can be considered equivalent to the motion of space in the opposite direction. Therefore, unlike the motion of space in the opposite direction, the motion of a body, when it is rectilinear and also not produced by more than one force, is to be regarded as only a possible predicate, and, on the other hand, we always know that, unlike the motion of space in the opposite direction, the motion of a body, when it is not rectilinear, is a real predicate. This shows that it is not unimportant whether one assumes that the sun stands still, and the earth moves around it, or conversely, whether the earth is at rest, and the sun moves around it, for this motion is curvilinear, that is, composite (which we also know for other reasons) so that the motion of the body (in this case that of the earth) is a real predicate or different from the motion of space in the opposite direction. The third theorem in that part of our science (§28) can be expanded here with the theorem that to every motion belongs an opposite one in space of equal size.

On the Atomistic System

72. Here we have now seen the most important theorems in the metaphysics of nature, according to critical philosophy. We shall now consider the opposite system in order to see the arguments with which it defends itself. Our system is called *dynamic* because it assumes that bodies fill space with a force. The opposite is called *atomistic*, for reasons which will readily become apparent during its presentation. The atomistic system is distinguished from ours in that it does not assume the infinite divisibility of bodies but teaches that they are composed of many small particles, which are called atoms. These atoms are distinguished from each other only through their different shapes but not through any difference in forces or degree of forces. Their principal characteristic is that they cannot be compressed. These atoms also have an attractive force, whereby they cause the phenomenon that we call coherence. Between the atoms there are always empty spaces, which of course are different according to the difference of the atoms, and on which the differences of bodies are based. Besides its theory of divisibility, the atomistic system is also distinguished by the fact that it assumes the basic particles or atoms to be completely incompressible and bodies which are composed of these only compressible in so far as they contain empty spaces. The dynamic system attributes to all bodies a fundamental force with which they fill space, but this has a degree above which one can find a greater one which can overcome it in so far as it is confined to a smaller space. Thus, impenetrability is invincible according to the former system but not according to the latter.

73. The atomistic system has an advantage over the dynamic in that its explanations, wherever it can explain, are simpler as they are all based on difference in form and extent and therefore allow mathematical explanation, and as the dynamic does not easily allow such a construction because it still depends on the fundamental forces. On the other hand, our system has the advantage that it explains phenomena for which the former is not sufficient. Thus it is impossible to give a complete

explanation of elasticity by means of the atomistic system. It is true that we can give a fairly good explanation of the kind of elasticity which is based on the attractive force using this system as it assumes such a force, but it cannot explain the kind of elasticity which is based solely on compresssion and expansion, for the expansion following compression can no more be due to the empty spaces than to the small incompressible particles, which have no force at all except to attract and to resist. An attempt has been made to explain this by means of a fine, pervasive elastic substance, but then where does this acquire its elasticity? Indeed, the atomistic system does not even explain why bodies fill space, it merely assumes that this is so.

74. Many may believe that the atomistic system would have an advantage over the dynamic because it can be proved empirically, but this advantage is only imaginary. It is true that this system is always treated as a collection of empirical truths, but on reflection it is obvious that this procedure is incorrect. Neither the atoms nor the empty spaces, which are actually required to constitute bodies, are of such a nature that they can be presented to the senses. It may well be said that grey paper which absorbs water or leather through which mercury is filtered are proofs of porosity, but the pores which thus reveal themselves to the senses are of a completely different kind from those which belong to the constitution of every body. Consequently, it cannot be directly proved from experience that there are atoms and empty spaces, so they must be proved indirectly from it. Through these indirect proofs, the cause must be inferred from the effect, which is at variance with well-known logical rules unless it can be proved that there can be only one cause of this effect. Therefore, if an atomist wants to infer from a chemical solution the existence of pores in one or both of the active bodies, he is also obliged to prove that there can be no other cause of this phenomenon, but no one has even started on this yet. Besides, the generality and the necessity with which atoms and empty spaces are attributed to bodies are at variance with the principle that they are supposed to be drawn from experience. If the entire system of the atomists, then, is to have more than hypothetical value, they must seek proofs elsewhere.

75. The only attempt that has been made to prove the atomistic system a priori is the well-known proof of the impenetrability of bodies. It was believed that it was contrary to the principle of contradiction to say that a body could be in the same space as another at the same time. No doubt one would never have thought of this proof if this theorem had not been expressed in the following way: A thing cannot be and not be at the same time. If this is expressed in different words: A thing cannot have contradictory predicates, then it is not possible to see how to complete the proof as there is no contradiction in two bodies occupying one and the same space unless one presupposes the impenetrability that one wants to prove.

76. Consequently, the foundations of the atomistic system are shaky. Its principal axioms, that there are atoms and empty spaces, are unproved and therefore only hypotheses which we are entitled to reject if they do not fully explain the phenomena, or if we can explain them by means of other, proved theorems. With the foundation follows the collapse of the entire structure, which the perspicacity of many helped erect but could endow with no more than meagre consistency.

77. Our system also has the advantage over the opposite that it presents the laws of nature as founded on human cognition so that we know beforehand that there can be no exceptions to these because, in order to imagine that anything happened according to natural laws which were at variance with the ones we had proved in this way, we would have to change our cognition, that is, become other beings. Thus the question of whether there might be bodies with negative gravity can be answered with the greatest ease according to our system. We have proved that the fundamental attractive force of bodies acts in proportion to mass and is therefore not different from gravity, and as there can be no body without this force, the question is answered in the negative. The atomistic system does not have this advantage, for although it assumes theorems which cannot be proved by experience, nothing is more certain than that they are still aided by empirical proof. This has the harmful consequence that it is impossible to refute anyone who claims to have observed a phenomenon whose explanation required the assumption of a natural law which is inconsistent with those hitherto assumed. We saw an example of that when Gren believed that he should accept negative gravity in phlogiston; how many fruitless efforts were not made then to refute him.

78. I have now described both the systems in natural philosophy which still contend for precedence, one through its age, the other through the strength which the extension of critical philosophy gives it. Their strife is like the one Lavoisier's theory waged against that of Stahl, and it will probably also end in the same way, by the new theory suppressing the old. All that is needed to make the dynamic system that of all astute students of nature is a few more successful attempts at uniting the dynamic system with all of physics, like the one made by Gren, who died far too young. As it is now, it may not be sufficiently obvious to those who have not studied critical philosophy. I have taken pains to make this presentation so comprehensible that it could be understood by anyone with some knowledge of physics and mathematics, but I greatly fear that I have not achieved this goal throughout. Perhaps the doctrine of matter in particular lacks the comprehensibility that I wished to give it. If, like Kant, I had taken certain empirical ideas as my basis, it would probably have been easier for me to achieve this goal, but thereby I would have sacrificed something even more essential.

By taking empirical concepts as a basis and inferring the natural laws from them, one imparts to the natural laws thus proved only hypothetical validity instead of the rigorous generality which they should have. According to critical philosophy, all natural laws ought to be deduced from the nature of our cognition, which Kant has developed so excellently in his *Kritik der reinen Vernunft*, and I believe I have proved that this can be done by deducing them all a priori and taking only what that book has proved for my basis. Therefore I did not hesitate to deviate from Kant's letter in order to follow the spirit of critical philosophy. Many may find it superfluous to apologize for this deviation, but to those who know that one cannot, without sinning against sound logic, simultaneously assume a system and an opinion which is at variance with it, it will not be superfluous to point to the reasons why I have deviated somewhat from the creator of critical philosophy even though

I accept critical philosophy. To those who know how to separate the system from its originator, I have said enough; my defense must be found in the treatise itself. I realize full well that my presentation must have deficiencies, and when I have reconsidered this subject, I shall myself communicate the corrections which I then find necessary.

79. As the title shows, the inspiration for this account is our famous Lord Steward Hauch's textbook in physics, of which we have now got the beginning of a new edition. As this, the only book of its kind which we have in Danish, adopts the atomistic system, which I am convinced is incorrect, I believe that I have performed a useful task by comparing it to the opposite system. I dare say that I need not assure any man of science that, with this work, I have no intention of disparaging the man who, in spite of all the obstacles which his official duties and the diversions of the Court placed before him, has raised himself to a prominent rank among our best students of nature and won the esteem of foreigners by means of his talents and insights.

80. No doubt it would not be unwelcome to my readers to see the most important works on natural philosophy mentioned here:

Naturally, the book which may properly speaking be said to have laid the foundation of this science ought to come first: Immanuel Kant's *Metaphysische Anfangsgründe der Naturwissenschaft*, second edition. Riga 1787.

Lazarus Bendavid's *Vorlesungen über die metaphysische Anfangsgründe der Naturwissenschaft*. Wien 1798. This book is the most complete and the most informative commentary on that part of Kant's philosophy.

F. W. I. Schelling: *Ideen zu einer Philosophie der Natur*, Vols. 1 and 2. Leipzig 1797.

F. W. I. Schelling: *Von der Weltseele, eine Hypothese der höhern Physik zur Erklärung der allgemeinen Organismus*. Hamburg 1798. These two books certainly deserve attention because of the beautiful and grand ideas which are found in them, but the insufficiently rigorous method, whereby the author adds empirical theorems without distinguishing them adequately from a priori theorems, deprives the book of much of its value, in particular because the empirical theorems that he adduces are often completely false.

D. Johan Friedrich Christof Gräffe: *Kommentar über eine der schwersten Stellen in Kant's Metaphysische Anfangsgründe der Naturwissenschaften, das mechanische Gesetz der Stetigkeit betreffend*. Celle 1798.

C. A. Eschenmayer: *Sätze aus der Natur-Metaphysik, auf chemische und medicinische Gegenstände angewandt*. Tübingen 1797. By the same author we also have:

Versuch die Gesetze magnetischer Erscheinungen aus Sätze der Naturmetaphysik mithin a priori zu entwikeln. Tübingen 1798.

These two short works reveal an author with an original intelligence which is well worth following through many errors as he rewards the trouble with many excellent thoughts. Besides these, several people, who have written whole systems of critical philosophy, like Beck, Neeb, Buhle, as well as Mellin in his dictionary of critical philosophy, and Fischer in his dictionary of natural philosophy, have given summaries and explanations of the metaphysics of nature, though without thereby developing science any further. Kant's metaphysics of nature also serves as the basis of Gren's and Fischer's textbooks in physics, which is why there is a brief presentation of it there. In our Rafn's *Plantephysiologi*, we also find the main points of dynamics. Likewise in Gren's *Grundrisz der Chemie*, 1796. There are also things of some relevance in several of Kant's short treatises. In particular, it is interesting to see that this famous philosopher already had the ideas which form the basis of his dynamics in 1763, when he wrote his treatise *Versuch den Begriff der negativen Grösze in die Weltweisheit einzuführen.*

H.K.Ø.

7

Dissertation on the Structure of the Elementary Metaphysics of External Nature,[1]

WHICH HANS KRISTIAN ØRSTED,
GRADUATE IN PHARMACY, WILL DEFEND
ON SEPTEMBER 5 TO OBTAIN THE HIGHEST
HONOURS IN PHILOSOPHY

The faculty of philosophy in Copenhagen considers this work worthy of being submitted to public scrutiny with a view to earning the highest academic honours in philosophy.

Jacob Baden
Dean of the faculty of philosophy, pro tem.

TO THE BENEVOLENT READER

Before I submit these humble pages to the examination of learned men, it might not be inappropriate to point out that they were written within a short time although I had been thinking about the matter in question for quite a while. In fact, I had already obtained the permission to dispute for the highest honours in philosophy when I received the news that I would shortly be leaving this university, but I had not yet finished writing the dissertation, so I was forced to rush. Therefore, benevolent reader, I ask you to be so kind as to forgive me such errors as are due to the negligence of a man in haste, but as far as the principles are concerned, I must myself defend my theses.

This essay of mine was almost finished when Schelling's excellent Erster Entwurf einer Naturphilosophie *arrived here, so I could not use it in this place, which I certainly regret; in any case, his book contributes much more to the higher than to*

---

[1] [Originally published in Latin in Copenhagen, 1799 and printed by J. L. S. Winding. Together with the previous work (*Fundamentals of the Metaphysics of Nature*), this treatise was published with some additions under the title: *D. Joh. Christian Ørsted's Ideen zu einer neuen Architektonik der Naturmetaphysik nebst Bemerkungen über einzelne Theile derselben*, edited by M. H. Mendel. Berlin 1802. KM I, pp. 79–105.]

the elementary metaphysics of nature. What I have tried to establish in this disser-
tation about the force of cohesion is in accordance with the views of this philoso-
pher; I have not, however, derived these findings from his book, as may be seen
from the fact that I earlier put forward the same ideas in a review published in
Lærde Efterretninger 1799, p. 287. The same famous philosopher has promised to
publish a book on the same topic as mine; I am eager to read it.

THE metaphysics of external nature has been almost totally neglected while the remaining parts of metaphysics have flourished due to the efforts of eminent men. This is a fact that few would dispute. The merits of the immortal Kant, even in this part of philosophy, have attracted due attention; but his little book on the principles of this science has had very few readers, still fewer commentators, and next to none who really explored it. This neglect has been of no small damage to philosophy. For our science is not only needed to complete the system of philosophy, though this is a matter of great importance, but, by deducing the laws of nature—which we find in experience, too—from the very nature of our cognitive faculty, it confirms the principles of critical philosophy with an experiment, as it were.

Until now, no-one has attempted to deal with the nature of this science, its limits, and its relation to general metaphysics, but I think it is worthwhile to do so, in particular since no presentation of this doctrine has been published, at least not one which agrees with my views on its nature.

Therefore, please allow me to put forward my thoughts, for what they are worth, about these matters, together with certain objections to the views of others which diverge from mine. In so doing, I shall be as brief as possible and aim only at adumbrating our science, not at completing it. Consequently, I shall not repeat any demonstration already proposed by others unless this is necessitated by the train of the argument.

Since the laws of experience in general are deduced from the nature of human cognition, transcendental philosophy divides into two branches, as it were, one embracing the laws of external experience, the other those of internal experience. This gives rise to two doctrines, the metaphysics of external nature and the metaphysics of internal nature. The latter includes little or nothing which has not already been explained in general transcendental philosophy, the former contains the foundations of the whole of the science commonly called physics.

I assume that the transcendental philosophy about the highest principles of cognition, the one which Kant gave the name of ontology, is known to the readers, and no further presentation of its tenets is needed. To some extent, however, I want to deal with the universal laws of experience since, in my opinion, Kant listed them incompletely.

I. General Principles

A. With Regard to Quantity

All objects of intuition are extensive quanta.

B. With Regard to Quality

In all phenomena, that real thing which affects the senses has an intensive quantity, i.e., a degree.

C. With Regard to Relation

No experience can arise except through a *necessary link* between several observations.

D. With Regard to Modality

We cannot cognize anything except what is joined to our *cognitive faculty* by some link.

II. From the *General Principles* above, certain more specific ones can be deduced corresponding to each predicament (category).[2]

A. With Regard to Quantity

Any object of intuition is a finite whole; for if it were an *infinitely great* whole, its synthesis could never be carried through; if it were an *infinitely small* one, it would become smaller than any given thing; in either case an inconsistency arises.

Any object of intuition can be divided, and in such a way that its division never reaches a limit. It is an *extensive* quantum, so it has parts, and these cannot be inseparable since the force (intensive quantity) which is in it cannot be absolutely the largest but must be overcome by some other force. Consequently, there is nothing to prevent one part from being separated from another. However, any part can be considered a whole, and therefore it can be divided once more so that no part is absolutely the smallest. Thus, no part of an object is simple, or an *absolute unity*, but it is left to anyone's discretion which part should be counted as a unity. In every object of intuition, therefore, the *unity* is arbitrary,[3] as is the *number of parts (the plurality).*

B. With Regard to Quality

The force which is in a real thing is not simple but composed of two forces. For one force would not be capable of producing a limited whole unless it either limited itself or disappeared through extension, but neither is possible. It cannot limit

---

[2] [Ørsted here glosses the Latin *praedicamentum* with the Greek *categoria*. Later on, where he just says *praedicamentum*, this is translated as "category."]

[3] ["Arbitrary" renders *indefinitus* except where otherwise noted.]

itself since it is not capable of [4] acting against itself; nor can it disappear through extension since it cannot be converted into nothing through diminution because no degree of a force is so small that none smaller can be found. To be firmly convinced of the truth of this proposition, we need only consider that in human cognition every limitation is the result of combining two contraries (an affirmation and a negation), with the result that we cannot intuit anything which is not composed of two forces since a real thing can lack neither force nor limitation. We are therefore bound to admit two forces (a positive and a negative one), each of which keeps the other within certain boundaries, and also a third one, the *limited* force, which arises from the meeting[5] of the two.

C. With Regard to Relation

THE PRINCIPLE OF SUBSTANTIALITY

Substances can neither come to be nor pass away but are immutable. Therefore, their quantity can be neither decreased nor increased.

THE PRINCIPLE OF CAUSALITY

No change occurs without an antecedent cause.

THE PRINCIPLE OF INTERCHANGE

Phenomena, insofar as they are simultaneously perceptible in space, are in interchange, i.e., one acts on the other.

D. With Regard to Modality

Whatever is not inconsistent with the formal conditions of experience is *possible.*

What is consonant with the material condition of experience truly *is* (Danish: *er til*).[6]

What is connected with that which *is* through the universal laws of experience is *necessary.*

It is by now sufficiently clear that this presentation of the universal laws of experience is complete with regard to the categories, and this needs no further demonstration. The same laws must also apply to all objects of intuition, whether placed in time alone or in both time and space. Taken together, those objects of intuition which are placed in time alone constitute internal nature; those which occupy both time and space constitute external nature.

What is real and is located in space we call *matter*, and its changes, which likewise can only occur in space, we call *motion*. Thus, two doctrines arise, one of matter, the other of the motion of matter. To these a third might be added, viz., the *ap-*

---

[4] ["Since . . . capable of": *cum non valet.* The indicative *valet* must be a slip, cf. *neqveat* in the following *cum . . .* clause.]

[5] [Here and below "meeting" renders *conflictu[s]*.]   [6] [The gloss *er til* means "exists."]

plied doctrine of motion, which deals with motion arising from certain properties[7] of bodies and with the mutual action of moving bodies.[8] I beg, however, to be excused for touching only lightly on this doctrine as I fear that if I were to propose anything final in this matter, I might be caught trying to shoulder a burden too heavy for me.

Further, I feel that I must here point out that the parts which I have put forward above do not make up the whole metaphysics of nature since they only teach us what matter and motion are, and not how nature as a whole is to be formed out of those two. For this very reason these doctrines, taken together, are called the *elementary* metaphysics of external nature, and I believe that this was also the reason why the famous Kant gave his book about the metaphysics of nature the name *Metaphysische Anfangsgründe*.

The doctrines of matter and motion are each in turn divided into four parts corresponding to the four classes of categories.

The doctrine of matter must come before the doctrine of motion, for without knowledge of extension, shape, and several other properties of matter, no doctrine of motion can be established. It is true that a notion of motion must already be present in the doctrine of matter, but at least we do not posit anything to be known except this notion, which is also required for the division of the metaphysics of external nature.

A. The Doctrine of Matter

I. With Regard to Quantity[9]

All matter is extended through space and has the three dimensions of space. This is clear, if for no other reason, then because we cannot intuit matter as an object of an external sense except in the form of space so that any part of matter is thus placed in a particular part of space and has three dimensions like space itself.

The quantity of space which some portion of matter occupies is finite. For if it were either infinitely great or infinitely small, it could not become an object of intuition. If the former were the case, its synthesis could never come to an end; if the latter, it would become smaller than any given quantity. Any portion of matter, therefore, is bounded within certain limits.

Matter can be divided, and can be so ad infinitum, i.e., in the division of a portion of matter we shall never reach a simple part, for matter is something *real*, and this can be divided ad infinitum.

From the above, we can conclude that the quantity of a portion of matter cannot be defined through an enumeration of its parts since no part is absolutely the smallest but can be divided further.

[7] ["Property" renders *affectio* throughout.]

[8] [The often-used expression *corpora mota*, literally "bodies moved," is rendered as "moving bodies" throughout.]

[9] [In agreement with the notation adopted below, this should read "I. The Doctrine of Matter." and "A. With Regard to Quantity."]

No part of matter, then, is an absolute unity; which of them is to be taken as unity is arbitrary.[10]

Consequently: The unity in matter is indefinite and therefore arbitrary; the *number of parts (plurality)* is likewise indefinite and arbitrary. However, matter itself, considered as a whole, is finite.

We are now able to list the general properties of matter which arise from quantity, namely: *extension, shape, and divisibility*, and infinite divisibility at that.

B. With Regard to Quality

Matter is endowed with some kind of motive force. That matter possesses some sort of force may be gathered from the very fact that it is something real, but that this force is a motive one we conclude from the fact that every change in space happens through some motion. Therefore, the force of matter is not capable of producing anything but motion, so any force of matter fully deserves to be called *motive*.

Now, this force can equally well be attractive and repulsive. Indeed, a repulsive force alone would sever the parts of matter and thus expand matter, even infinitely, since no force can contain itself within boundaries, nor can the expansion of space diminish the force so much that it quite disappears as no force can be given so small that no smaller may be found. A similar consideration applies to the attractive force, which, if acting alone, would compress matter ad infinitum until it had reduced it to a point. Therefore, if matter were endowed only with a repulsive force, it would occupy an infinitely great space; if it were endowed only with an attractive force, an infinitely small space would contain it. However, neither of these can be the case. Therefore, neither of these forces is capable of producing matter on its own, but the real thing we call matter arises through their meeting. Due to this meeting, matter fills a definite space and resists any external force attempting to separate its parts. For the parts cannot be severed except through motion, and any such motion must thrust them in a direction opposed to that of one of these very forces. This property of matter is usually called the *force of cohesion*. The repulsive force of matter[11] acts only through contact, the attractive force also at a distance.

It follows from this that the force of cohesion can in no way act outside the surface of a body. For it owes its origin to the meeting of two forces one of which cannot act at a distance.

The most famous author of critical philosophy asserted that this force did not belong to the essence of matter but was rather to be known through experience alone. That things are different in this case will, I believe, be sufficiently clear from what has been demonstrated. For although I originally intended to follow in Kant's footsteps as far as this subject is concerned, when I thought it over more carefully I was forced to leave that trail.

---

[10] ["Arbitrary," here and just below, renders *arbitrarius*; elsewhere "arbitrary" is mostly used for rendering *indefinitus*. Ørsted seems to have used these terms interchangeably.]

[11] ["Of matter" translates *materiae*; the text has *materia*, which must be a misprint.]

The attractive force acts in proportion to the quantity of matter and is therefore in no way different from *gravity*. The repulsive force will always resist any external force which tries to penetrate the space it occupies, and due to its action no portion of matter can occupy the same space as another. The *impenetrability* of bodies follows from this.

If we posit that one of these powers is *positive*,[12] it follows without fail that the other one is *negative*. Each has an infinite sphere of activity, but due to their conjunction this turns out to be *limited*.

Thus we have now listed the forces of matter in accordance with the categories.

Matter, we saw, is generated by the meeting of two opposite forces, one of which restrains the other in such a way that the magnitude of the space which a body occupies becomes equal to the (mathematical) difference between the forces. At first glance, this may seem absurd since both forces are infinite, and therefore their difference ought to equal zero: $\infty - \infty = 0$. However, it should be considered that the forces are infinite only with respect to their *sphere*, not to their degree, which may vary. Eschenmeier,[13] a very astute man, has taken another road to remove this difficulty, claiming that the repulsive force $= \infty$ and the attractive $= 1/\infty$. To confirm this view he adduces the following reasoning: The attractive force is the opposite of the repulsive, which is infinite; but what is the opposite of something infinitely great must itself be infinitely small. This, however, strays far from truth, for it will easily be seen that only the negation of the infinitely great ($-\infty$) can be the opposite of the infinitely great ($+\infty$). He further claimed that matter is the product (in the mathematical sense) of the primitive forces so that matter ought to be $= \infty \times 1/\infty = finite$, but as this latter opinion is closely connected with the former, it falls with the refutation of the former. For if those forces were to be reckoned $= +\infty$ and $-\infty$ (not, as he says, $= \infty$ and $1/\infty$), their product would have to be $= +\infty \times -\infty = -\infty^2$, which is absurd. However, even if this be granted, it is still in no way possible to prove that matter is produced by the forces in the same way as is a mathematical product. This will be clear from the preceding remarks without any further demonstration.

In one and the same body, the repulsive force is always greater than the attractive; for if the attractive surpassed the repulsive, their difference, i.e., the extension of the body, would be a negative quantity. If, in turn, these forces were equal, their difference, and hence also the extension of the body, would be $= 0$.

Further, even as the forces of matter are[14] of unequal magnitude, so the bodies themselves must differ in the quantity of their forces one from another; for if they all had an equal quantity of forces, no body could be distinguished from another since they would all be of the same nature, with the result that no motion at all could be observed.

[12] [Ørsted uses the term "affirmative" belonging to logic. This has been translated into "positive" throughout.]

[13] [Eschenmayer.] *Versuch die Gesetze magnetischer Erscheinungen aus Sätzen der Naturmetaphysik, mithin a priori, zu entwickeln*, p. 39.

[14] ["Are" translates *sunt*; the text has *sint*.]

C. With Regard to Relation

The principle of the substantiality of matter, which says that no matter can either arise or disappear, has been so clearly demonstrated by Kant, I would say, that nothing is left for me but to add a few remarks in order to throw a clearer light on this proposition, which has often been the victim of misinterpretation and even more often of sheer neglect.

Matter is nothing but the product of the meeting of the primitive forces, so its quantity is the same as the quantity of the forces. Thus, if some force of matter disappeared, it would be just as if matter itself disappeared, which is inconsistent with the principle of substantiality. The same would be the case if a force were to pass from one body to another or from one part of a body to another, for in this way one part of matter would pass away in order that another should arise, which is likewise inconsistent with the principle of substantiality.

The principle of causality states that no change in external nature occurs except such as arises from an external cause (the law of inertia).

From this principle follows the principle of dynamic interchange, which asserts that the reaction of matter is equal to the action. As this principle is a different one from that which Kant demonstrated in his Mechanics, I shall have to support it with arguments.

a) If a body A were to act on some other body B solely with a repulsive force, its only effect would be to compress body B, i.e., the repulsive force of body A would restrain the same force in body B. However, in accordance with its nature, the latter must offer resistance and will restrain the compressing force no less than the compressing force restrains it. If this were not the case, equilibrium would never be restored, but the motion which had arisen from the compression would remain forever, and one of the bodies would be compressed ad infinitum.

b) If a body A were to act on a body B solely with an attractive force, the reaction would similarly be equal to the action. Here four cases may be distinguished:

   I) A and B are altogether equal, in volume as well as in density.
   II) They are equal in density but not in volume.
   III) They are of the same volume but of different density.
   IV) They are unequal in volume as well as in density.

In the *first* case [I] it is easy to see that action and reaction equal each other.

In the *second* case [II]: If we posit that body B only contains some part, e.g., $1/n$, of the space of body A, this will be the same as if that body contained $1/n$ part of the sphere[15] of activity of the attractive force of body A. Body A, then, will only be able to act on B with $1/n$ part of the force which it ought to have exerted on B if these two bodies had been equal in volume. Further, as, by our assumption, A and B are equal in density but unequal in volume in such a way that body B has only $1/n$ part of the volume of body A, it follows that B also has only $1/n$ part of the matter of body A. Thus B will act on A with only $1/n$ part of the force with which it ought to have acted,

---

[15] ["Sphere" translates *ambitus*, which is also Ørsted's word for "volume."]

everything being equal. From this it follows that A and B act on each other with only $1/n$ part of the force with which they acted in the case above; since the action and reaction of each body in that case were equal one to the other, they must be equal here, too.

III. Let the matter of body B be only $1/n$ part of the matter of body A. Here body B will act on A with only $1/n$ part of the force with which it acted in that first case, but as B is only endowed with $1/n$ part of the mass on which A is capable of acting, it follows that A acts on body B with only $1/n$ part of the force with which it acted in the first case. Consequently, action and reaction continue to be equal one to the other.

IV. Let body B be as above, but let its volume be changed in such a way that it makes up $1/m$ part of body A. This done, it will be obvious that with the decrease of space body A will act on B with $1/m$ part of the force with which it would act if the case were as just above; and further that, as the diminished space now contains m times as much matter as in the case above, the force with which A acts on B will be m times greater than in that case. One of these causes would have the effect that A acted on B with only $1/m$ part of the force which it exerted in the case above, whereas the other would have the effect that the same force would be m times greater than in the same case.

Thus, the mutual equality of action and reaction remains unimpaired.

That this demonstration is complete will become obvious to anyone without difficulty when it is noted that the expressions $1/n$ and $1/m$ may stand for any fraction, whether proper or improper.

Since the force of cohesion is the result of combining the attractive and repulsive forces, it is beyond doubt that the action and reaction belonging to this force are equal to each other.

This is the place for a division of bodies as regards their mutual relation. In this division, we only consider change of position of parts in a body, taking two things into account: The force which acts on the body and the condition of the body on which an external force acts.

I. With respect to the force capable of changing the position of their parts, bodies are:
  a. *fluid,* if the position of their parts can be changed by any force, however small.
  b. *solid,* if their condition is otherwise. *Solid* bodies may be further divided into:
    α. *hard,* if their parts do not change position unless a great force be applied.
    β. *soft,* if they only need a small force to change the relative position of their parts.
  However, this division only operates with an arbitrary degree of external force, so it pertains to empirical physics rather than to metaphysics.
II. With respect to the condition of bodies while the position of their parts is being changed, bodies are divided into:
  a. *fragile* ones, those which are easily destroyed by a change in the position of their parts.

   b. *ductile* ones, those whose parts can be moved relative to each other without
      the destruction of the body affected.
   A body which is very *soft* and *ductile* at the same time is called *viscous*.
III. With respect to the condition of bodies after their parts have changed posi-
      tion, bodies fall into two classes:
   a. *Elastic* ones, those whose parts (unless the body to be changed breaks) re-
      occupy the place from which they were moved by the external force when
      this force ceases to act.
   b. *Non-elastic* ones, which have the contrary property.

Even if this division is made according to rational principles, it cannot prove that
there really are solid, fluid, or fragile bodies, etc. Only experience can teach us that,
so this division of bodies seems to be located on the border, as it were, between rational and empirical cognitions. Thus, the whole applied doctrine of motion,
which cannot do without the notions of solidity, fluidity, elasticity, may justly be
said to need the assistance of experience.

B.[16] *With Regard to Modality*

We shall omit this matter as it contributes nothing at all to our cognition. For this
reason, investigation with regard to this category deserves a prominent place in
critical science but next to none in *doctrinal* science.

II. THE PURE DOCTRINE OF MOTION

A. *With Regard to Quantity*

The quantity of motion is composed of magnitudes of space and time, in such a
way that the longer the space traversed in a certain time, the greater the quantity,
and the shorter, the smaller. The quantity of motion is here called *velocity*.

Now, let the velocities of two bodies be indicated by the letters V and v, the
spaces which they traverse by S and s, the times in which their motions are completed by T and t. Then, if the times are equal, $V : v = S : s$. If, on the other hand, the
spaces are equal but not the times ($S = s$; $T \neq t$), then T will be $= nt$, and as the moving body traverses s in time t, it must traverse space ns in time nt. Thus, $V : v$
is $= S : ns$ because times T and nt are equal to one another. Further, by hypothesis,
S is $= s$, and so $V : v = 1 : n$, and also, since nt is $= T$, $t : T = 1 : n$. For this reason,
$V : v$ will be $= t : T$ if, that is, the spaces are equal. Employing standard mathematical method we can now demonstrate:

$$V : v = S \times t : s \times T$$

and

$$V : v = \frac{S}{s} : \frac{T}{t}.$$

Eschenmeier,[17] taking motion to be the mathematical product of time and space, thought that the following series, in which S^0T is $= 1$, should be placed before the whole of phoronomy:[18] $S^{+\infty}T \ldots S^{+n}T \ldots S^{+2}T, S^{+1}T, S^0T, S^{-1}T, S^{-2}T \ldots S^{-n}T \ldots S^{-\infty}T$. If, he says, the first and the last term of this series are multiplied by each other, the middle term will result, i.e.,

$$S^{+\infty}T \times S^{-\infty}T = S^0T.$$

This is as far from the truth as possible, for according to the most basic rules of mathematics

$$S^{+\infty}T \times S^{-\infty}T = S^0T^2.$$

From his principles it no doubt follows, as he himself admits, that

$$S^{-\infty}T = ST^{+\infty}.$$

This entails the following conclusion:

$$\frac{1}{S^\infty} = \frac{1}{\infty}T = \frac{T}{\infty} = \frac{1}{\infty} = ST^{+\infty} = S\infty = \infty \; ;$$

and so

$$\left(\frac{1}{\infty}\right) = (\infty) \, ,$$

the infinitely small equals the infinitely great, which is absurd.

Further, at this juncture we take leave to point out that what we just said about the quantity of motion is only valid for the motion of bodies if their masses are equal, or, equivalently, if they are considered as mere points, but not for the motion of bodies of unequal mass. The quantity of the latter motion is composed of velocity and mass, in such a way that the quantities of motion of two bodies whose masses are arbitrary have the same ratio as the products of the masses and velocities, or, in mathematical terms,

$$Q : q = M \times V : m \times v \, .$$

The proposition about the quantity of motion which asserted that it could become neither infinitely great nor infinitely small should be taken to pertain to this part of the doctrine.

[17] *Versuch die Gesetze magnetischer Erscheinungen aus Sätzen der Naturmetaphysik, mithin a priori, zu entwickeln*, pp. 239 ff.
[18] [Phoronomy: see p. 94 below.]

No part of motion can be absolutely the smallest if we posit that a greater veloc-ity is composed of smaller velocities. This is sufficiently clear both from the laws of universal experience and from the fact that no motion in itself is the smallest.

If the spaces traversed by a continuous motion in equal parts of time have the same quantities so that there is *one* velocity, this motion is usually called *uniform*; otherwise it is called *varied*. Varied motion, in turn, may be either *regular* or *irreg-ular*. The latter has a *universality* of velocities, the former a *plurality* of velocities.

B. With Regard to Quality

The direction of the moving body constitutes the quality of a motion. The direc-tion is a straight line along which the moving body travels; where it deviates from the straight line a new motion begins, so a non-rectilinear motion becomes limited.

A rectilinear motion which follows a direction exactly opposite to another recti-linear motion may be considered the negation of the latter; and therefore, if we posit that the former is positive, the latter must be negative.

Here it should be pointed out that several motions, whether they have the same or different directions, cannot be united at one point and in the same space. This proposition is a part of Kant's phoronomic theorem.

With regard to quality, motion is divided into positive, negative, and limited.

C. With Regard to Relation

Motion and rest are accidents.

This would seem the most suitable place to indicate that rectilinear motion of a body is the same as the motion of relative space in the opposite direction. For if mo-tion were not an accident, this condition would not obtain.

With regard to causality, it is our duty to demonstrate that several different mo-tions can become the cause of one motion. This is the second part of Kant's phoro-nomic theorem.

In the demonstration of this proposition, it is shown that no composite motion can come about unless one of the motions is attributed to space. Thus it is clear that the motion of a body cannot be the same as the motion of space in the opposite di-rection if, that is, the motion is composite.

One body cannot move towards another unless the other body together with rela-tive space assumes an equal but contrary motion. For motion is nothing but the change of a body's relation to what is outside it, and this relation does not only per-tain to the moving body and to space but also to the two of them mutually. E.g., the more the distance of body *A* from *B* increases, the greater the distance of body *B* from *A* must become. This, at least, will be the case if the bodies are equal; if they are not, then the ratio between their velocities will be the inverse of that of their masses so that the quantities of the motions become equal.

From this it is possible to deduce the Kantian theorem that "the action is equal to the reaction."

With regard to causality, motion can be divided into *simple* and *composite* motion. If, on the other hand, we consider its influx, we may distinguish between *proper* and *common* motion.

Motion is usually also divided into *absolute* and *relative* motion, but less correctly since all motion must be relative.

D. With Regard to Modality

Motion can be divided into *true* and so-called *apparent* motion. This division corresponds to the category of being and is commonly used in experimental physics because experience very often confuses the composite motion of a body with the opposite motion of space. Motion can also be divided, with regard to the categories of possibility and necessity, into *possible* and *impossible*, *necessary* and *accidental* motion. However, none of these divisions will be of any use to us, as is sufficiently clear.

III. The Applied Doctrine of Motion

A. With Regard to Quantity

To this part pertain all such effects of moving bodies as are produced either by their shape or by their definite extension, together with the quantity of motion. This part of our science deals with the effect of the motion of solids, of elastic fluids, and also of non-elastic fluids. Out of this, three doctrines arise: *mechanics, hydraulics,* and *pneumatics.*

The divisibility of bodies together with motion cannot produce any effect. Thus no part of the applied doctrine of motion can arise from this property of bodies. The same holds for extension, which can have no being unless it is circumscribed by limits and confined to a shape.

B. With Regard to Quality

The repulsive force of bodies is the cause of impulse and pressure, and also of the resistance of the *Medium.*[19] Consequently, this is the origin of three doctrines.

From the attractive force of bodies, which is usually also called gravity, a tripartite doctrine comes into being, the first part of which deals with motion due to the gravity of solids, the second with motion due to the gravity of non-elastic fluids, whereas the third deals with motion which comes from the gravity of elastic fluids. *Statics, Hydrostatics,* and *Pneumatostatics.*

The force of cohesion has the double effect of making one body resist another when this tries to separate its parts, and of reducing the velocity of another moving body through contact with it. The same force also seems to be the cause of chemical action. To demonstrate this, we shall have to present here the fundamental theory of chemistry, which has not yet been established a priori.

True, some, including some very prominent men, have attempted to construct a

---

[19] [*Medium*: cf. the last two paragraphs of this dissertation.]

rational chemistry. However, I shall try to prove that they have erred in the very principles of the metaphysics of nature.

The chemical theory of the most astute Eschenmayer, as he presented it himself, is built entirely on his false views regarding the nature of the primitive forces, so it would scarcely need to be refuted if it were not for the fact that the ingenious Schelling has given it his support, though founding himself on other principles. The entire rational doctrine of chemistry of this great man turns on the notion of heterogeneous bodies, so if we can reveal the falsity of that, the whole thing must collapse. He thinks that the bodies called heterogeneous are those whose primitive forces are inversely proportional. However, such a proportion cannot possibly come about. Let the attractive forces of two heterogeneous bodies be signified by the letters A and a, and the repulsive forces by the letters R and r. From his definition, it follows that $A : R = r : a$. If so, however, neither $A > R$ nor $A < R$ nor $A = R$ is possible. For if $A < R$ were the case, then it would also be that $r < a$, i.e., the attractive force would be greater than the repulsive, which cannot happen. That A can neither be $> R$ nor $= R$ is completely obvious from the preceding (see p. 85).

Another great philosopher, Lazarus Bendavid, thought that the union of heterogeneous bodies, which is what chemistry deals with, should be explained solely in terms of the action of the attractive forces. If we believe this, then a chemical action between two bodies, A and B, ought to arise when the attractive force by which the parts of A cohere is smaller than the force exerted by the parts of body B on the parts of body A. Therefore, when the parts of body A are close to B, then, for an infinitely brief moment of time, they will be endowed with only repulsive force, losing all cohesion, so they cannot but be dispersed in an infinite space unless the parts of body B were to try to unite them in its centre. However, the repulsive force in the parts of body A prevents this from happening, so together with the parts of body B they fill a definite space to a certain degree. Moreover, in the same way that B acts on A, A in turn acts on B, and thus they interpenetrate each other.[20]

This doctrine not only has the defect of positing, without any demonstration, that the attractive force disappears; in fact, it is not even intelligible how one body should be able to take this force away from another. Further, it is grossly inconsistent that the attractive force in two bodies should at the same time disappear and act, but according to this doctrine, this is a condition without which they cannot be joined into one.

Moreover, if the attractive force, which can even act at a distance, were the only cause of chemical effects, nothing could prevent bodies at a distance from one another from entering into chemical combination, which is contrary to experience.

Finally, even if this doctrine were not false, it is not capable of proving a priori that a chemical action must happen but only of explaining it. Of course, if it had done this correctly, it would have achieved something. I think the primary error of this doctrine is that he believed that the only cause of cohesion was the attractive force, but this view is quite refuted by what has been demonstrated above.

---

[20]Lazarus Bendavid's *Vorlesungen über die metaphysischen Anfangsgründe der Naturwissenschaft*, Wien 1798, §§237 & 239.

If, however, we pursue this argument, taking experience as our guide, we shall, perhaps, find Ariadne's thread. Anyone who has but the slightest acquaintance with chemistry will know that no body can act chemically on another except through contact, and a very close one at that. From this we gather that the force of cohesion is the cause of chemical action. Indeed, the attractive force cannot function as cause since it acts at a distance, too; neither can the repulsive force since it would rather cause bodies to be separated than to be joined.

Now it only remains to explain how the force of cohesion is capable of producing such an effect.

If two heterogeneous bodies, *A* and *B*, touch each other, they must cohere; for the attractive force of body *A* together with the repulsive force of body *B* will bring about cohesion, and so will, vice versa, the attractive force of body *B* with the repulsive force of body *A*. If such cohesion is strong enough to conquer the forces by which each body coheres, they[21] ought to enter into chemical combination and produce a new body.

Here we have only aimed at demonstrating the possibility of an a priori chemistry. However, through a more thorough investigation along these lines, I hope we shall one day be able to know the whole doctrine of chemical affinities.

C. With Regard to Relation

With regard to substantiality and causality, there exists no motive property, and therefore no applied doctrine of motion can be put forward corresponding to those categories. With regard to influx, communication of motion is dealt with between solids, between solids and fluids, between hard bodies, etc.

I have now presented a sketch of the metaphysics of nature, from which it may be seen what, in my judgment, its structure ought to be. However, in order that no-one may think that I have without good reason abandoned the structure given to our science by the famous Kant, I will present an outline of the metaphysics of nature as it ought to be if we were to follow him and subject it to critical scrutiny.

PRESENTATION OF THE METAPHYSICS OF EXTERNAL NATURE ACCORDING TO KANT

No changes occurring in the external world can happen without motion. For this reason, the fundamental determination of any object of external experience is motion of some kind, and human reason refers all other attributes of matter to this one determination. Thus the metaphysics of external nature is properly nothing but a doctrine of motion. This is divided into four doctrines, corresponding to the four classes of categories:

a) On motion considered as a *pure quantum*. This doctrine is called *Phoronomy*.

---

[21] [Reading *debent*. With the *debet* in the edition, one would have to translate "it"—i.e., the force of cohesion—"ought to . . . ," which does not make sense.]

b) On motion insofar as it is treated as a quality connected to matter. *Dynamics*.

c) On the relation of matter endowed with motive force to other such matter. Hence, *Mechanics*.

d) On motion and rest with respect to human cognition. Thence, *Phenomenology*.

The foundation of each of these doctrines consists in the attribution of a single empirical determination to matter, but with this exception nothing whatsoever belonging to experience is added.

Phoronomy

In phoronomy, motion is the only empirical determination attributed to matter. In this doctrine are put forward:

A distinction between relative and absolute space.
    A definition of motion.
    A division of motion into:
        a. rotative.
        b. progressive.
            α. expansive.
            β. recurrent.
        and the rest.
Some remarks about the notions of direction and velocity.
A definition of rest.
A definition of the construction of composite motion.
An axiom stating that the rectilinear motion of a body and the motion of space in the opposite direction are altogether equipollent.
A definition of the composition of motion.
A theorem by which it is demonstrated that the composition of two motions at one and the same point is unthinkable unless the mind conceives of one of them as occurring in absolute space and the other by means of the motion of relative space in the opposite direction.

Dynamics

That matter fills a space is the only empirical determination which we attribute to matter in this part of our science. This determination is followed by:

What it is to fill a space.
A definition of empty space.
A theorem by which it is proved that matter fills a space by means of a motive force.
A definition of the attractive and repulsive forces.
A demonstration that there are no motive forces except the attractive and the repulsive ones.

A theorem showing that matter fills a space due to the repulsive force of its parts.

That the repulsive force is also called elasticity.

That in no body can the repulsive force ever become absolutely maximal.

When matter is said to penetrate other matter.

A theorem teaching that matter can be compressed ad infinitum but cannot be penetrated.

The difference between absolute impenetrability and relative impenetrability.

A definition of material substance.

A definition of the separation of parts of matter from the matter itself, and of physical division as well.

A theorem about the divisibility of matter ad infinitum.

A theorem by which it is proved that an attractive force is necessary for matter.

A theorem declaring that an attractive force cannot by itself be the foundation of matter.

Definitions of physical contact and of action at a distance.

A theorem about the attractive force, showing that it is also an action on things at a distance.

Definitions of surface force and penetrating force.

That the repulsive force is a surface force, whereas the attractive is a penetrating force.

A theorem proving that the attractive force of matter acts throughout the universe.

That the attractive force is the same as gravity.

Some remarks on the construction of matter.

A corollary to dynamics, showing that the repulsive force corresponds to the category of reality, the attractive to the category of negation, and that together these two forces constitute limitation.

What body, volume, density are.

Some remarks on the force of cohesion.

What fluid, solid, friction, viscosity, fragility are.

What an elastic body is.

Some remarks on chemical action.

In addition, in the places most suitable for this, he has adduced not a few points to refute the atomistic system.

Mechanics

In this doctrine motive force is attributed to matter. Here are put forward:

The notion of quantity of matter.

What it is to act as a mass.

What a body is in the mechanical sense.

The difference between quantity of motion determined mechanically, and
   quantity of motion in a phoronomic sense.
A theorem showing that the quantity of matter can be determined only from
   the quantity of motion.
A corollary about quantity of motion.

Laws of Mechanics

a) The quantity of matter can be neither increased nor decreased.
b) Any change of external nature has an external cause.
c) In every communication of motion reaction is equal to action.
A corollary showing that the very smallest body can communicate with the
   very largest body.
A corollary teaching that the dynamic reaction is equal to the action.
A general note particularly designed to demonstrate the law of the *continuum*.

Phenomenology

Matter ought here to be regarded as that mobile which is an object of experience.
Here we have three theorems.

1) Rectilinear motion of a body, if considered as distinct from the motion of
   space in the opposite direction, is a merely possible predicate (attribute).
2) Composite motion, seen as a predicate distinct from the opposite motion of
   space, is a *true* predicate, whereas the opposite motion of relative space,
   when taken as a motion of the body, is by no means a true predicate, and if
   anyone thinks it is a true motion, he is deceived by *appearances*.
3) In any motion of a body, which in comparison with some other body is held
   to be *moving*, we must attribute an equal but opposite motion to the other
   body.

Having thus briefly presented the metaphysics of external nature according to
the structure given to it by its famous author, we hasten to its critical scrutiny.

The first point which I think deserves censure is that this most astute man did not
deduce the laws of external experience altogether a priori but, on the contrary,
judged that empirical properties ought to be attributed to matter, one in each part
of the science. For the laws of nature are universal and necessary and thus are not
due to experience, a fact which has been sufficiently demonstrated by Kant himself
and by many other great philosophers as well. Therefore, Kant offends against his
own principles when he states that the notion of matter must be derived from experience, for in this way its universality and necessity will be lost. Further, if the very
notion of matter is contingent, the deduced[22] laws of nature cannot become necessary. If, e.g., the determination of matter as space-filling were given only a posteriori, there would be no obstacle to the possibility of finding in external nature

---

[22] [Reading *ea* for *eo*. If the latter is kept, one should probably translate: "if the very notion of matter
is something contingent, the laws of nature deduced from that [contingent thing]," which seems
implausible.]

something without this determination, the result being that the primitive forces of bodies would have no universality except for a hypothetical one, and that one could easily admit of cases excepted from the laws built on those forces. If, on the other hand, it is demonstrated from the nature of human cognition that we can intuit no thing unless it be endowed with those forces, everything else will easily follow with the necessity and universality without which there can be no science in the strict sense of the word. Neither is the assertion altogether true that a new determination is added to matter in each of the parts of the metaphysics of external nature. It holds for phoronomy and dynamics, but not for mechanics and phenomenology; for the principle of mechanics, that matter has a motive force, has been proved in dynamics, and the so-called principle of phoronomy, which asserts that motion is here regarded as an object of experience, is not merely a new determination of matter but rather a definition of phenomenology.

With regard to the division of our science I have particularly many points to make. We have already said above that the metaphysics of external nature is not only the doctrine of motion but also contains the doctrine of matter. However, I would like to add some remarks here to prove this. That Kant's division is not carefully made may be understood from the fact that in the phoronomy of this great man the quantity of *motion* is dealt with; in his dynamics, the quality of *matter*; in his mechanics, *matter* with regard to substantiality and causality, and communication of *motion* with regard to influx; in his phenomenology, motion once more. Thus, he did not carry through his treatment of either motion or matter with regard to all the classes of categories, but he dealt with motion only with regard to the categories of quantity and modality and with regard to the category of influx, and he dealt with matter only with regard to all categories of quality and two of relation. For this reason we find many things presented in places which are not the most suitable. Thus the direction of motion, which, since it can be either positive or negative or limited, indubitably belongs to quality, is dealt with in phoronomy (the doctrine of the pure quantity of motion). And I cannot see for what reason the axiom stating that the rectilinear motion of a body and the motion of space in the opposite direction are altogether equipollent can be assigned to the quantity of motion. For my part, I have put this axiom next to the proposition teaching that motion is an accident, for if motion were not an accident and therefore something changeable, it could not take place. The composition of motion seems to me to belong to the category of causality because the joint action of several motions produces[23] one single motion which differs from those that are its cause, not only in quantity but often in quality as well.

What is presented in dynamics concerning divisibility does not belong to that doctrine but rather forms a part of its own. Eschenmeier, whom we cited above,[24] was the first to notice this, and he called this part *arithmology*; for this reason he

---

[23] [Reading, tentatively, *actione* for *actio qua*.]

[24] *Versuch die Gesetze magnetischer Erscheinungen aus Sätzen der Naturmetaphysik, mithin a priori, zu entwickeln*, pp. 231 ff. This very astute author was, I think, the first to show publicly that the Kantian division is not accurate.

held that the metaphysics of external nature ought to be divided into five parts, phoronomy, arithmology, dynamics, mechanics, and phenomenology. Several of my
objections to the Kantian division could obviously be adduced against this division, too, though it is to be preferred to Kant's. It hardly needs demonstration that
one should not define what body and volume are in the doctrine of quality but in the
doctrine of quantity. The notion of density arises from the relation of mass to volume. What a fluid, a solid, etc., are does not pertain to quality but to relation, as may
be seen from the definitions put forward by the author himself. I have judged that
chemistry should be assigned to the applied doctrine of motion as it is motion due
to the quality of matter.

The propositions regarding the quantity of matter and motion have no place in
the doctrine of relation; to demonstrate this will be quite superfluous. We have already shown above that with regard to influx Kant carried through his treatment of
the action of moving bodies (mechanical action) but not of the action of the forces
(dynamic action). True, he tried in a corollary to show that a dynamic reaction is
equal to the action, but he did not derive this equilibrium of action and reaction
from the nature[25] of the forces but from that of motion; consequently, this equilibrium is only relevant to motion, which arises through the effect of the primitive
forces, but not to the action of the forces of bodies at rest.

I have already earlier put forward my remarks about the remaining points.

APPENDIX

By presenting the doctrine of matter and motion, we have put forward the elements
of the metaphysics of external nature, but this by no means amounts to having explained how, from these elements, a science may be composed that will be capable
of teaching us the order which must constitute the whole of external nature. However, to my knowledge, no one has attempted to present such a science, at least not
along these lines. Neither my powers nor the brevity of time has allowed me to
achieve full insight into all of this science; nonetheless I cannot but offer some remarks which I think may throw light on how to proceed from those elements to this
more sublime doctrine.

From what we have so far set out, it is quite obvious that indubitably there is motion in nature; without motion nature would become a lifeless lump. If, then, motion is something necessary for nature, it must happen in such a way that it does not
cease again by itself, and it must take up a finite space. A motion of this sort would
occur if a body travelled along a curved line which always recurred in itself; by this
definition we have already established what the motion necessary for nature ought
to be like. Any curvilinear motion has to be composite (produced by several
forces), for some force is always needed if a body is to deviate from the direction in
which it has begun to go, or, in other words: if it is to be deflected from a straight
line. Forces capable of bringing about this sort of motion must be of two kinds, one
of which incessantly attempts to sustain the motion along a straight line while the

[25] [*indoles*, what a thing is in itself.]

other incessantly attempts to deflect it from it. Further, such a line must be regular, and therefore it must have some fixed point from which it can be thought to have been described according to the method of mathematics. However, since the present question concerns not only a mathematical description but also a physical one, we are compelled by necessity to treat this fixed point belonging to the line as a physical point (which might be called the central body) and to attribute to it some force; this should be such that it acts at a distance, for there must be some distance between a curved line and the fixed point belonging to it.

Now it would not be enough for this force to have exerted its action once and for all if it were to leave off after that; on the contrary, it must remain in action forever. Indeed, since a vacuum can by no means exist, then a body in motion must always be pushing some matter aside, thus communicating some of its motion to it, and this would eventually make the body stop moving altogether, unless its motion were incessantly being incremented by some external force. Thus, the force we are to attribute to the fixed point of the curved line is bound to be perpetual and active at a distance.

Most conveniently, we have already found such a force. We know it well—it has been demonstrated a priori—under the name of the *attractive force*.

The other of these necessary forces must also be perpetual, for a perpetual force can by no means produce a perpetual motion if it is combined with a non-perpetual force; the latter must dwindle little by little until it completely disappears. The proximate cause of this fact cannot be the repulsive force of the central body, for this force does not act at a distance, and besides it is altogether contrary to the attractive force, whereas the force we are dealing with here intersects that force at an angle, i.e., it is a tangential force. Nonetheless, we dare not take this to be a primitive force, but we are compelled to derive it from others. This could be done in several ways, I admit, but the following appears to me to be both the easiest and a short-cut, as it were. Some sort of matter emanating from the central body hits the moving body in such a way that the impulse is resolved into two forces; the direction of one goes through the centre of the body, and at the same time it becomes a *tangent* to the curved line along which the body to be moved will go. In fact, I think I have discovered such a matter in nature itself, but allow me to keep this opinion to myself until I have supported it with the lengthy mathematical reasoning required for this purpose.

The result of what we have put forward here is somewhat different from what has so far been believed about the motion of the heavenly bodies, namely that God the Creator from the beginning gave these bodies a thrust in the direction of the *tangent* of the curved line along which they were to travel in their motion. However, this opinion suffers from all the defects that ever beset an opinion since it does not explain by means of a natural cause but by means of a supernatural one; and even if this cause were to be admitted, it would not suffice to make us understand the problem, for it assumes one non-perpetual cause of an incessant motion, and we have already demonstrated above that this is inconsistent. There is hardly any need for more arguments to refute this opinion. I am aware that it was held by a man who easily deserves to be numbered among the very sharpest minds of all time, and that many men of great name have rallied to this same opinion; not, indeed, because it was founded on suitable reasoning, but rather because it was inserted into another opinion which is counted among the most eminent manifestations of the human mind because of the cleverness of what was found and no less because of the rigorous demonstration. A much more probable, though in my judgment false, view would hold that the tangential force is a primitive force of matter. If we posit that a mechanical cause brings about the tangential force, we gain the advantage that we can at the same time explain the daily motion of the celestial bodies, which would be a most difficult task with the contrary view.

To make the utility of our theses perfectly clear, I shall use them as a foundation for proving that the *Medium* in which a perpetual motion will take place must be filled with some fluid. That a space is filled with some sort of matter is already understood from universal metaphysics. What needs to be demonstrated here is that this medium is of such a nature that it offers infinitely little resistance to a force which tries to change the position of its parts. For if the resisting force were finite, it would in any moment take away a finite part of the velocity of the moving body, and from this it would follow that in any finite time, composed of infinitely many moments, it would infinitely diminish the velocity of the body, i.e., it would do away with it altogether. The way we recognize in nature a *medium* capable of resisting a moving body with only an infinitely small force is that it yields to any force, however small. But this is nothing other than the definition of fluidity which we put forward above (p. 87).

From all these facts it further follows that the vast space of the world is filled with a fluid. Since motion must take place in it perpetually, and it is anyhow filled with matter, the matter in question must be a *fluid*.

8

Experiments and Observations
Concerning Galvanic Electricity[1]

FROM the moment when English chemists had observed the gasification which the voltaic apparatus was capable of producing from water and the strange phenomenon that this generation happened in such a way that oxygen gas was generated at one conductor while hydrogen gas appeared at the other, this apparatus, and the galvanic electricity thus produced, became the object of every chemist's attention. However, when Ritter, by repeating these experiments with certain changes, made it likely[2] that water was not composite, but that each of the generated gases was a product of water combined with one of the electric substances or forces generated by the voltaic apparatus, it had to invite even more reflection on the part of every researcher as there were now the greatest prospects of a rich harvest in the field of this science.

Even before Ritter's experiments were reported, I had repeated some of the English chemists' experiments, but with a small apparatus and without new results. Since then, however, I have repeated Ritter's experiments with some changes, and, partly through this, partly through my own reflections, I have had occasion to make several more or less important discoveries.

Almost all the experiments I intend to mention here were performed in the presence of several scientists, who made the same observations as I.

I shall describe a series of my experiments starting with the one whose influence on science seems most significant to me, together with the reasons which led me to it.

As it is possible to produce a galvanic effect not only by means of two metals of different oxidizability, but even with oxidizable metals and graphite, which is not a metal, it seems very likely that the generation of a galvanic effect is merely a matter of bringing bodies of different oxidizability in contact with water in an appropriate manner. I thought that I found this idea confirmed by the well-known operation in

---

[1] [*Nyt Bibliothek for Physik, Medicin og Oeconomie*, Vol. 1, No. 1, pp. 91–97. Copenhagen 1801. Also to be found in *Nordisches Archiv für Naturkunde, Arzneywissenschaft und Chirurgie*, Vol. 2, No. 1, pp. 173–81. Copenhagen 1801. KM I, pp. 106–9. Originally published in Danish.]

[2] Not even Ritter himself denies that they do not have all the strict evidential weight that it was possible to give to them. Davy, however, seems to raise the matter above any further doubt.

which nitric acid, water, and alcohol, poured over each other in layers according to their specific weight without mixing, generate such an interaction that the alcohol is thereby changed into naphtha. I had already decided to see if it was possible to produce a galvanic effect in this way, but several other considerations brought me on to an easier track. The thesis that water is not composed in the way we have hitherto believed stands quite isolated and, as such, is of little or no use to science until it is possible to ascertain that the generation of hydrogen gas which appears when metals are dissolved in some diluted acids is the consequence of a galvanic operation. I found this so evident that I no longer doubted the possibility of generating galvanic electricity in several new ways through the dissolution of metals. For this purpose I chose 7 glass tubes, curved in the approximate shape of a V, in these I placed some lead amalgam, and over this I poured diluted sulphuric acid in one of the tubes. I now joined these tubes with conductors, each of which consisted of 2 parts, one of silver, the other of iron; the iron conductors were in contact with the amalgam, the silver conductors with the acid. Naturally, this device produced no perceptible effect as no perceptible oxidation takes place here, but when I placed a small piece of zinc in each of the tubes which contained sulphuric acid, I obtained the galvanic effect, which appeared both in the generation of gas from water in a glass tube, connected by means of conductors, and in the feeling it created on the tongue. Although I do not think that I used 20 grains of zinc in each portion of acid, and although there was scarcely one *lod*[3] of diluted acid in each tube, the effect lasted for more than 14 hours, at which point I emptied the tubes, and could have lasted longer, so I think this method could be used with advantage for the generation of galvanic electricity. As soon as my other commitments permit, I shall also make these experiments with other analogous substances. There is no denying that this experiment sheds light on a part of the theory of gas generation which was dark before. Thus it was not possible previously to explain why oxidizable metals did not produce hydrogen gas with water in liquid form without the addition of acid, nor was it possible to explain why sulphur and phosphorus showed no perceptible effect on water even though they, as it is said, decomposed it in combination with alkali. Now we are also closer to answering the question of why pyrophorus needs moisture in order to ignite, and perhaps we shall be able to answer many more questions by means of the clue which has been given us. Mme Fulhame's apparently very accurate experiments, in which water plays such an important role in oxidations and deoxidations, also belong here.

Galvanic experiments have already changed two of the substances which we considered the most perfectly elementary into water. Perhaps nitrogen (azote) will meet with the same fate. We now have two kinds of electricity, the one generated hitherto by friction and the galvanic. Would it not be possible for us to discover more with the help of chemistry? Would the heretofore known electricity not generate nitrogen gas from water? The very frequent phenomenon that nitric acid is generated when a mixture of hydrogen gas and oxygen gas is changed into water by means of electricity seems to me to support this view. Furthermore, on several

---

[3] [Danish weight = $^1/_{32}$ Danish pound.]

occasions water has been seen to produce nitric acid through electricity. Within a few days I hope to make experiments which will throw light on this, and I shall then present my results as speedily as possible.

My voltaic battery consists of graphite plates, instead of the more usual silver plates. These plates are not pure graphite but of a tile-like material, that is, mixed with clay. They are not as effective as the ones made of pure graphite, but it was almost impossible for me to get these even in small quantities. However, 60 layers of these generated both shock and spark though the latter was not always perceptible.

The thickness of the plates is not immaterial. I was convinced of this by the fact that I obtained more effect from thick than thin lead plates, and our now sadly deceased Abildgaard found no effect at all from some lead plates even thinner than the ones I had used.

That there are two different galvanic electricities was settled by the fact that each end of the battery produced a different gas from water. That these two were opposite and neutralized each other was certainly highly likely given the other analogies with frictional electricity, but it had not been determined by any experiment. This, however, is easy: one merely joins the two conductors through which the galvanic electricity is conducted to the water by letting them touch each other, and the generation of gas will then cease.

I once made an experiment to see whether several conductors from one side would increase the generation of gas from water, but contrary to my supposition I obtained no perceptible quantity of hydrogen gas with 9 conductors, while on the other hand I obtained not a little hydrogen gas with one conductor. Altogether I have observed that the smaller the mass of the metal wire used to affect the water, the more gas is obtained.

In rarefied air galvanic electricity acts on water. I have convinced myself of this by placing a battery of 8 layers under the bell of an air pump and duly conducting the effect generated by this to a glass tube filled with water. The generation of gas took place with more than usual speed, became weaker when air was let in, and increased again when the air was once more rarefied.

I shall reserve for another occasion news concerning the examination of the gases which are generated from water by means of the voltaic apparatus.

9

An Addendum by Dr. Ørsted to
His Remarks on Galvanism[1]

SINCE my first report I have often repeated the experiment to produce galvanism by dissolving zinc in dilute sulphuric acid. Only simple conductors, consisting of plain iron wire, were used, and the result was always successful. So far I have not tried more than 30 tubes. The battery thus composed seems to me to produce a stronger effect than 30 plates of graphite and zinc. I base this opinion primarily on the intensity of the associated generation of gas from water. The effect on the organs of taste was very pronounced, but there were no or only extremely minor vibrations when the force was conducted through the whole body. However, in order to convince myself that the length of the path weakens the effect greatly, and that it is not only due to the delicateness of the skin that the tongue and the face are so sensitive to the galvanic effect, I placed a conductor on each side of a finger and received the same stinging feeling which I had noticed on my face and in my mouth.

A battery of 30 tubes had not ceased functioning after 14 days. A tube battery therefore retains its effect longer than one made with plates. It is merely necessary to add a few drops of water every second or third day in order to replace what has been transformed into air. Everyone will easily discover for himself the many other practical advantages connected with this device. Lead conductors are without any doubt to be preferred to iron conductors because they are little or not at all corroded by dilute sulphuric acid. I shall try them some time soon. When I placed two conductors instead of one between each tube, the effect seemed to me to double. I convinced myself that it was much bigger by the fact that the battery generated gas when two gold wires were employed even though their thickness and the distance between them did not allow a battery to have any effect on water when single conductors were used. The effect on the senses was also much stronger.

I am now having an apparatus made for the measurement of galvanism. This apparatus is supposed to measure in two different ways at once. It is based on two different principles: The first, that the bigger the distance between two metal wires,

[1] [*Nordisches Archiv für Naturkunde, Arzneywissenschaft und Chirurgie*, Vol. 2, 2, pp. 60–63. Copenhagen 1801. An abbreviation of part of this paper can be found in *Nyt Bibliothek for Physik, Medicin og Oeconomie*, Vol. 1, No. 2, p. 288. Copenhagen 1801. KM I, pp. 109–11. Originally published in German.]

the stronger the galvanic effect which is required to generate air from water; the second, that the quantity of gas generated from water is proportional to the strength of the galvanic effect if conditions are otherwise the same. Following these principles, the apparatus consists of a glass tube which will be filled with water and connected to a brass case at each end. Two very thin gold wires, which are cemented into delicate glass tubes, pass through these cases. One of the wires can be pushed back and forth in a piece of leather so that the distance can be increased or decreased as desired. A delicate glass tube with a narrow opening rises from the lower end of the tube. It is easy to understand that the water in this tube must rise when air is generated in the bigger one. I shall present an accurate description, accompanied by figures, as soon as I receive the instrument from the hand of the artisan.[2] If the unrest of war, which at this moment disturbs me in my work, has then ceased, as I hope, I shall probably be able to perform a fairly long series of experiments.

[2][A short description of the apparatus, accompanied by a figure, is found in a letter from Ørsted to Manthey, April 28, 1801, published in *Breve fra og til H. C. Ørsted*, ed. by Mathilde Ørsted, Vol. 1, pp. 13–14. Copenhagen 1870.]

10

Continued Experiments on Galvanism[1]

I.

Diluted syrup of violet was brought into contact with a galvanic battery by means of two gold wires, whose points were approximately 2 inches apart. After 5 or 6 minutes the syrup around the negative galvanic needle showed a highly perceptible green colour, around the positive needle, however, a not quite so perceptible red. After some hours both colours were much more perceptible and widespread: yet the red was not nearly so strong as the green, a phenomenon which could be foreseen as the green colour which alkalis give to syrup of violet has far more intensity than the red which it acquires from acids. When the tube in which this so differently coloured syrup was contained was shaken slightly, both colours disappeared as the different parts, each of which possessed one of the colours, were thereby blended. This is completely analogous with what happens when a quantity of syrup of violet that has been coloured green by alkali is mixed with another that has been coloured red by a proportional quantity of acid, which is due to the well-known fact that the alkali and the acid saturate each other and neutralize each other's effect.

2.

A glass tube filled with sweet oil and equipped with a gold wire at either end was brought into a galvanic circuit without any change being observed. A small glass bell into whose top a fine gold wire had been soldered was now filled with the same oil, suspended over water, and thus exposed to the effect of the battery before so much oil had been removed from the bell that the tip of the gold wire came into contact with the water. Now a milky cloud appeared in the water, and it soon spread in such a manner that it resembled water in which soap had been beaten into lather. This effect lasted until the newly produced mixture had become so thick that it prevented the contact of the water with the needle. This substance was now mixed with distilled water, in which it proved to be completely soluble. When shaken, it foamed like soapy water and did not have any oily taste. Nitric acid was added to the soapy mass, which had been dissolved in a considerable quantity of distilled water, and it was set aside for evaporation.

[1] [*Nyt Bibliothek for Physik, Medicin og Oeconomie*, Vol. 1, No. 3, pp. 408–10. Copenhagen 1801. KM I, pp. 111–12. Originally published in Danish.]

11

A Review of the Latest Advances in Physics[1]

AT this moment, it is most delightful to take a look at the recent history of physics. The lively eagerness, the courageous contempt for scientific prejudice, and the profound sense of higher things which inspire not all and not even the majority of physicists, but which still emanate from some great investigators with warming rays over all, show us the beginning of a new creation. The tendency towards chaotic formlessness still struggles in vain with the light that begins to spread creatively over all.

We shall seek to communicate the most important products of this glorious *Zeitgeist*. Light is the soul of creation. Without doubt, expanded knowledge of its manifestations deserves the first place in the history of the most recent progress in physics. Therefore, we wish to report in the simplest possible manner what the last years have taught us about this.

Violet light is the most deoxidizing among the light rays, which Scheele's experiments have taught us. Herschel demonstrated to us that *red light is accompanied by the greatest warming*, and at the same time he proved that *next to the red light there are invisible rays which possess an even greater warming ability*. However, these discoveries were still quite isolated, without any connection to the remaining phenomena until Ritter discovered that *there are invisible rays on both sides of the spectrum; that those on the violet side cause deoxidation, those on the red oxidation, and that the rays promote oxidation more, the closer they are to the red; similarly, they promote deoxidation more, the closer they are to the violet.*

This great discovery was soon joined by a second, that of the effect of galvanism on the eye. If *the nerves of the eye* have been put into the positive[2] state, all objects are seen with a *red* colour (in darkness) and *larger than they are otherwise seen,*

---

[1] [*Europa. Eine Zeitschrift*, ed. by Friedrich Schlegel, Vol. 1, Part 2, pp. 20–48. Frankfurt A. M. 1803. In his bibliography of the literary works of H. C. Ørsted, J. G. Forchhammer writes, "This compilation is signed O., and in H. C. Ørsted's notes it is stated that he worked on such a review for Schlegel's *Europa*, so there can be little doubt that Ø. is the author of the treatise quoted here." KM I, pp. 112–31. Originally published in German.]

[2] Actually, if the negative pole has been kept in contact with the eyeball for some time. The liquid in the eye, like any other liquid, must polarize, and therefore, if it becomes negative on the outer surface, the inner becomes positive. This explanation stems from the astute Dr. Reinhold in Leipzig, who has also repeated Ritter's experiments and found them completely confirmed.

but if they have been put into the negative state, all objects appear *blue* and *smaller* than usual. If we recall that the positive pole of the battery is the oxidizing one, the negative the deoxidizing one, and that the blue colour lies closest to the violet in the spectrum, the connection between this and the previous discovery becomes very clear to us. Oxidation and the red pole of the spectrum, deoxidation and the violet pole are associated with each other.

This student of nature has also brought us enlightenment concerning the other senses, but so far it has yielded less conspicuously satisfying results. The fact that the *positive* pole leaves an *acid*, the *negative* an *alkaline* taste on the tongue had previously been noticed, but that the positively galvanized ear perceives sounds as *softer*, the *negatively* galvanized as louder, is a discovery of the same nature. The negative pole evokes an ammoniac smell, the positive seems to deaden the smell. As far as *touch* is concerned, galvanism has a *hot* and a *cold* pole, about which we may soon expect more detailed enlightenment.

If we now take a look at previously known facts from this vantage, we see once more a great many phenomena converging towards a focal point. The positive pole generates oxygen gas from water, and this transforms combustible substances into acids or acid-like substances; the negative generates hydrogen gas, and this is a primary component of the few alkalis which we have so far been able to decompose. This yields enlightenment about the effect of galvanism on both taste and smell, but in the latter regard, we need to note that oxidizing substances (like *gas muriatique oxygéné*) also suppress smell and cause catarrhs. As far as hearing is concerned, we recall that, according to Chladny's[3] discovery, the notes of a flute sound far higher in hydrogen gas than in oxygen gas.

However, it will forever remain a major discovery concerning the effect of galvanism that the *positive pole causes expansion, the negative contraction*. This law, in its nature so simple, in its application so fruitful, already explains why the eye sees everything larger in the positive state and everything smaller in the negative. At this moment, it would be too daring to establish all the important conclusions which can be drawn from this discovery. Instead, we want to recall only two secondary discoveries by the same student of nature which give cause for much thought. If the tongue is positively galvanized (of course continuously), a swelling appears at the affected spot, whereas a depression is produced by the negative pole. Positive galvanism makes the pulse big, negative makes it small. (Here, fast and slow should not be confused with big and small.)

In connection with these facts, it is beyond doubt that mercury is caused to move by galvanization, apparently like a frog's leg. In France, it has been found that the fibrous part of the blood contracts as a result of galvanization, and this has been presented as an *experimentum crucis* because it proves that contractions can be achieved without nerves.

The chemical effects of galvanism are already so well known that we need not begin by repeating the main experiment regarding its effect on water. The important conclusion which Ritter drew from it against the antiphlogistic theory of water

---

[3] [Chladni.]

has also become very well known, but the arguments have never been shown in their full extent.

It is true that we cannot promise to achieve this at the present time, but we may certainly hope to come closer to the completeness which is required for conviction than has so far occurred elsewhere.

The discovery that the water in the galvanic circuit generates *oxygen gas* on the *positive* side and *hydrogen gas* on the *negative* should show even one who is not used to finding as much in one primary experiment as in a hundred modifications of it how all basic experiments on water can be explained without the antiphlogistic theory. Nevertheless, most have occupied and still occupy themselves with the possibility that the antiphlogistic theory *could* be true after all though they do not do this in order to settle the argument with the help of new and specific experiments but in order to prove their talent for scepticism towards all innovation. They merely forget to treat the old just as sceptically, a practice which would transform their entire knowledge to nought and leave them with nothing but assertions. Some, however, have followed a better path, but their good intentions are more to be praised than their success. Gruner was among the first to step forward to save the antiphlogistic system by claiming that he had not observed any weight loss in a quantity of water from which he had produced several cubic inches of air. Gilbert showed immediately that this experiment was not reliable, and Simon has found him completely wrong. Fourcroy, Vauquelin, and Thenard[4] endeavoured to show that hydrogen passes from the positive to the negative wire of the battery by claiming that horn silver, placed between the two, was blackened (i.e., reduced) in the direction from the positive to the negative. Ritter has found the experiments on this to be incorrect and shown that exactly the opposite happens.[5] Therefore, it must be assumed that these chemists have been confused about the polarity. After these two failed experiments, the antiphlogistians found themselves forced to resort to mere arbitrary assumptions. They have done this in two ways.

Either they assume that the part of the water which liberates oxygen gas remains oversaturated with hydrogen gas while that which liberates hydrogen gas gains an excess of oxygen gas, or they believe that hydrogen or oxygen, or both at the same time, passes from one pole to the other. The first of these quibbles actually comes from the English chemists and has been put forward without pretensions in order to have an explanation. Others have later wanted to use it against Ritter but with little success. With observation of the most perfect cleanliness, gases can be liberated from water with the aid of platina wires without the rest being changed in the least. However, a hyperoxidation or a hyperhydrogenation of water could hardly take place without changes in all its chemical conditions. Whether we take water containing some vegetable or animal matter or connect completely pure water to the pile with silver wires (probably also with others which oxidize easily), acid is liberated on the side which produces oxygen gas and alkali on the opposite side. Thus water which, according to this hypothesis, should be *hydrogenated* is *acid*, and

---

[4] [Thénard.]

[5] The author of this essay has repeated this experiment and found Ritter's account to be entirely appropriate.

water which should be *oxidized* is *alkaline.* Some say that the impossibility of these two cases has not yet been proved, and we will not bother to prove it here; *habeant sibi.* The hypothesis of the migration of oxygen or hydrogen from one side to the other is not consistent with the facts either. If we put the hydrogen wire in sulphuric acid (of course in a closed circuit), the sulphuric acid decomposes, and sulphur is precipitated, so if hydrogen passed from one pole to the other, the sulphuric acid should also decompose if we placed it between them, but this does not occur. If the oxygen wire is placed in a solution of liver of sulphur, the latter also decomposes, but if the solution is placed in water between the two wires of the battery, this does not happen, so no oxygen passes from one side to the other.

Still other facts could be quoted concerning this if it were necessary, but we will content ourselves with pointing out only two more experiments.

The oldest is by van Marum, who obtained nothing but hydrogen gas when he electrified water only with the negative electric conductor and had placed the positive above it in the air. The second is by Hauch, who obtained gases from water in a ratio which was entirely different from what the antiphlogistic theory permits.

Only one who is able to refute all of these results completely will be capable of saving the antiphlogistic theory.[6]

Concerning the chemical effect of galvanism, Ritter has further discovered that metals, galvanized negatively in water, hydrogenate just as well as they oxidize on the positive side. These hydrates can be in different states just like the oxides.

R. had long presumed that the so-called vegetation of metals is of galvanic origin and expressed this supposition in his treatise on galvanism in inorganic nature. Through the later discovery of the voltaic pile, we are now able to prove this. Using negative galvanism, we have already obtained many metallic calces from their solutions in a metallic form and with the branched structure of a Tree of Diana. R. has extended this further by finding that the soot of a tallow candle also acquires a treelike form when in contact with the negative conductor, but either irregular or conglomerate forms in contact with the positive. He has also produced electric figures on the surface of mercury by means of galvanism.

Similarly, it has not escaped the attention of this researcher that the galvanic process might also be oxidizing and hydrogenating in the dry way. Thus, the galvanic spark ignites only on the positive side while it melts metals on the negative.

Concerning the relation of galvanism to electricity, we have also received the greatest elucidation from Ritter although in this area he does not stand alone as he does in all others. Volta, whose great merits are rightfully recognized everywhere, has already achieved much here; Ermann[7] also has true merits as far as this point is concerned, and Reinhold, too, has made experiments on this with insight and skill. Yet R. has found all relevant results by himself, and for the most part his discoveries have either preceded similar ones by other researchers or been made simultaneously, and everywhere he has exhausted the facts more than anyone else

---

[6] A very speculative fellow has told me that something unexpected could still happen here. He is certainly right, but so is the necromancer when he reads the cards and prophesies, "You will receive a letter in the near future."

[7] [Erman.]

although Reinhold must also be highly commended in this regard. After this historical remark about the originators, we will now list the discoveries themselves. The voltaic pile displays the phenomena of repulsion and attraction, just like bodies which have been electrified by friction, and in both cases they follow precisely the same laws. The voltaic pile is electrically positive at one end, negative at the other, but it is at equilibrium or zero in the middle. If the positive pole is connected to the earth with a conductor, the pile loses the attractive and repulsive phenomena of positive electricity and those of negative electricity remain. Just the opposite happens if the negative pole is treated in the same way, and if both poles are simultaneously connected to the earth with a conductor, all attractive and repulsive phenomena disappear, but the pile does not cease to act chemically or to cause shocks. Therefore, all the phenomena which with Volta we wish to call phenomena of tension are not inseparably connected to chemical ones. This is also confirmed by the fact that one pile made with salt water and another made with distilled water show equally strong tension phenomena but have very dissimilar chemical effects and shocks.

Several chemists also claimed to have found that plates of a large diameter which give very strong sparks have chemical effects and shocks that are equal to or only slightly stronger than those from smaller plates. However, the whole issue is reduced to the fact that conduction by large surfaces is much greater than by small ones, and that therefore only very good conductors can carry all the force while inferior ones only conduct what a moderately large pile of small plates could yield.

Just as the tension is strongest at the poles of the pile and from there decreases towards the middle (where it becomes = 0), the chemical effect also reaches its maximum at the poles and decreases according to the same law as the tension.

The rapidity with which the galvanic process takes place depends on the perfection of the conductor. An ordinary pile made with zinc, copper, and salt water charges an electric jar or even a large electric battery so quickly that we cannot measure it, but if the cardboard has not been moistened at all (but has only its own hygrometric moisture), the pile requires several hours to charge a jar. The same result is also obtained if, instead of a single piece of wet cardboard, a glass disc covered on both sides with wet cardboard is placed between each zinc-copper layer. No chemical effect was observed in such batteries.

We see from these facts that the chemical effect requires far more perfect conduction than the tension, and that the latter is not increased by the perfection of the conduction like the former. Moreover, we find both functions in electricity as well as in galvanism, although in proportions which can be changed arbitrarily.

Nor does the generation of sparks follow the other effects precisely since, on the one hand, it becomes proportionally stronger as the conduction becomes more perfect, but, on the other, it becomes astonishingly weak, indeed it ceases, if the isolation is imperfect, so it shows an ability to follow poor conductors as well. This would be explained most easily by the assumption that this function is the result of the other two as it presupposes the conditions of both. It is sufficiently well known that this function is also common to electricity and galvanism.

The fourth function of electricity and galvanism is the effect on the animal body, which, in spite of the manifold ways in which it expresses itself, can always be attributed to the shock. The electric shock is usually experienced with dry hands, the galvanic with wet ones. This difference between the conductors had the effect that a difference between electric and galvanic shocks was believed to exist even though it did not. If the shock of a very weakly charged Leyden jar is experienced with wet hands, no difference will be felt between this and a galvanic shock. Conversely, if the galvanic shock of a very large pile (or several piles) is experienced with dry hands, the feeling will be the same as that of electric shocks. It has not yet been determined whether this function is the same as the chemical.

After this enumeration of the phenomena of galvanism, it might seem a little late to return to the conditions for galvanism, and we would indeed have started with this if we could have presented something complete on the subject, or even if we wanted to include everything, even the better-known facts, but since we want to draw attention only to some unusual and less known facts, we may as well elect to append this discussion here.

The ordinary manner of producing galvanism through two solid conducting bodies of different combustibility and a liquid conductor is not the only one; circuits can also be constructed with two liquids and one solid. Davy has constructed entire batteries in this manner. However, it has not yet been determined whether a circuit can be constructed from three liquids or from three solids. The production of nitrate of naphtha from a layer of nitric acid, a layer of water, and a layer of alcohol seems to indicate something of that nature, as Ørsted has remarked. It would be very important for the theory of galvanism to determine this.

The use of pure water as the liquid conductor in the battery yields a very weak effect, but if it contains other dissolved substances, the effect is far more perfect. Ritter has made the general observation that liquid conductors are better, the more composite they are, but naturally this applies only to compositions in each chemical category, for acids enhance conduction more than salts, as Ørsted and Davy first discovered simultaneously. It should not be concluded from this that oxidation by acids causes this as alkalis increase the conductivity of water as well.

Ørsted first observed that the pile continues to have an effect in air rarefied by pumping. This has admittedly been contradicted by Pfaff, but Pfaff himself has been given the opportunity by van Marum to convince himself of his error; however, it must be noted that this student of nature also observed a cessation of the effect in rarefied air, for which he could not discover a cause. Anyone used to conducting galvanic experiments will not be surprised by this as the effect of the pile often begins to weaken in the ordinary atmosphere as well and even stops altogether, for which a laborious investigation often reveals some very insignificant mechanical cause. Undeniably, these experiments merit repetition. The experiments in which no effect of the pile was discovered in non-respirable gases have not been conducted with sufficient care either since the piles were first brought through water, and they also have to be repeated.

Ritter has made the remarkable discovery that heating causes metals to behave like more oxidizable ones so that, e.g., one piece of heated zinc and another at a lower temperature cause galvanic spasms in a frog, with the stronger spasms occurring on the side of the heated metal. The causes of error were avoided.

The same investigator has found the law of conduction for homogeneous substances: *The conduction of a cylindrical conductor is directly proportional to the cross-section of the base and inversely proportional to the length.* Of the many applications of this law, some important ones can be seen in his *Beiträgen zur nähern Kenntnisz des Galvanismus*, Vol. 1, Part 4. He has also found that the same body conducts far better if it is straight than if it is bent, and an iron wire, which conducts very well, can almost be turned into an insulator if it is bent in a zigzag.

The understanding of magnetism has also made great progress in Ritter's hands. The phenomena of attraction and repulsion which magnetism has in common with electricity and the effects of atmospheric electricity on steel, &c., &c., had already drawn the attention of physicists to the relationship between these two effects, but R. has demonstrated this far more accurately. *If an iron wire has been magnetized, the south pole shows greater affinity to oxygen than the north pole.* He has established this fact through many experiments of very great precision. In addition, the magnet causes *spasms in frogs*, where the south pole plays the role of the zinc, the north pole that of the silver. Therefore, if we exclude the generation of light, whose cause is yet to be discovered and might be found in the *aurora borealis*, we see the effects of electricity repeated in magnetism. However, in the magnet they exist in an entirely different manner since they cannot be educed as easily.

In what we have seen so far, almost everything had an equal degree of certainty as most facts had been observed and confirmed by several, and as those for which this was not the case acquired the most complete credibility both because of the thoroughness with which they had been presented and because of the proven observational skills of their discoverer. What we want to report now does not have this complete authority, but certainly not because we might have reason to doubt the credibility of the discoverer. On the contrary, we must grant him all the virtues of a love of truth, but the form which he chose for the publication of his discoveries has not allowed him to describe each experiment individually in such a manner that a mere reading of his writings could lead us to regard these as being as safe from error as the discoveries mentioned above. In spite of this observation, we shall have to admit, once we have taken a look at the whole, that he has had many excellent insights into nature. Many may already have guessed that we speak here of Winterle[8] and his *Prolusiones in chemiam seculi decimi noni*. Here we want to present the fundamental principles, so far known, of his system without restricting ourselves to his order and without being completely content with his proofs, especially when we can add others.

The fundamental principle is: *The cause of heat is the product of the two electric principles.* By rubbing two heterogeneous bodies, of which one is connected to the earth, these principles are distributed so that one receives +, the other −. Without using the important observation that friction between homogeneous bodies

---

[8] [Here and below, Winterl.]

(where, consequently, no basis for induction exists) produces heat and not electricity, we need only remember that a merging of the two electricities which is somewhat delayed (which must be distinguished from being hindered) produces very intense heat. Thus, when the poles of a strong electric pile are brought into contact with each other by means of a very thin iron wire, it heats up to such a degree that it can bring water to a boil, indeed, it is finally able to melt.

In any event, this is also proved by the fact that the combination of acids, especially the strong ones, with alkalis causes considerable heating. However, the acidity and the alkalinity, which merge on this occasion and are cancelled, seem to stem from the electric principles, which is suggested by the taste produced by electricity and further confirmed by the acidity and alkalinity of the electrified water.

However, if acidity and alkalinity stem from $+$ and $-$, it is easy to see that these two qualities must cancel each other, but this also happens if an alkali combines with an acid, and we call this *neutralization*. In a neutral salt, acid and alkali are therefore inert, and the reason why we do not produce them as such if we separate them again is that the means usually adopted for the separation reinstates the lost quality of the separated substance. For instance, if sulphurous acid is driven off from its compound with potash by means of sulphuric acid, the latter gives so much acid principle ($+E.$) to the former that it regains its lost acidity. The sulphuric acid itself, on the other hand, passes into the inert state. If, instead of this, a solution of sulphurous acid potash is subjected to distillation, the sulphurous acid will separate without regaining its acid principle. In this state, it has neither the sharp smell or taste of ordinary sulphurous acid, nor does it react so strongly to vegetable juices; it has also become more soluble in water and takes up a larger volume. All this also applies to the other volatile acids which do not decompose in fire. Nitric acid and several others cannot be driven from their salts by mere heat without decomposing. The gas-like mixture which is produced in this process is very much less acidic than the acid itself. However, the mixture of azote and oxygen gas which is obtained from the nitric acid needs only to be electrified in order to regain all that it lost and be transformed again into an acid.

The greater the quantity of alkali which is combined with an acid, the more acidity it loses, and if this is taken very far, the acids which do not decompose in fire are often found to decompose.

The inert acids regain their integrity through glowing heat (presumably red heat), from which it can be inferred that the acid principle should be sought here. This should be compared with Ritter's discoveries about light.

Alkalis are also made inert by acids, but it is difficult to produce them in that form because they seem to reinstate themselves with the slightest heat. However, potash can be made inert by nitre glowed with magnesium, whereby this metal is transformed into an acid with which the inert potash is then combined. It is separated from it when the solution is brought into contact with zinc, which decomposes the magnesium acid, and the insipid (inert) potash remains in the solution. Other alkaline substances can be made inert by the insipid potash. There are other methods for this as well, but we cannot embark on a detailed description of them here. Alkalis also become more volatile when inert so that, e.g., potash can be distilled with water and is less soluble in water.

The so-called acid matter[9] of the antiphlogistians is therefore not the acid principle, rather oxygen gas itself is an acid and can be hyperoxidized and deoxidized. The pure oxygen gas which is generated from metallic calces by glowing is hyperoxidized, but as it exists in air, it is not. In the so-called gaseous nitrogen oxide, the oxygen is half deoxidized, and this gas can be deoxidized even more completely if it is passed over glowing iron. By this it is deprived of so much of its ability to support combustion that it can be distinguished from nitrogen gas only in that it can deflagrate with hydrogen gas. It is evident that oxygen gas would be an acid of an entirely unique kind. It distinguishes itself from other acids by the fact that it retains its acid principle more firmly and, precisely because of that, is less inclined to combine with alkalis, whereas it easily forms synsomatic compounds (which is what Winterle calls the combination of homogeneous materials) with other acids.

The concept of *acid* is taken by Winterle in a much broader sense than is usually done. He believes himself justified in this because the previous classification of these was too uncertain to build anything firm on. Several characteristics of acidity are stated, but none is specific to acids, so nothing prevents the presence of all these insignificant qualities in a substance without its being acid at all. The reddening of certain plant juices by acid, as well as the green colour which they are given by alkalis, is not sufficiently reliable to serve as a significant characteristic. In any event, we cannot persuade ourselves that this property should constitute the nature of acids, but if this cannot be assumed, it would be possible to imagine that a substance could combine insolubility in water with the nature of an acid. In this case, however, it could not redden plant juices. This also applies to any judgement based on taste. Even the neutralization of acids by bases (which is what Winterle calls everything that is an alkali or behaves like an alkali with respect to acids) and of bases by acids cannot be used with certainty, for even if the tendency to combine is present, it must nevertheless compete with many other forces, and if these are sufficiently strong, the effect will not take place. Much enlightenment about the forces which strive to prevent combinations is to be found in Berthollet's *Recherches sur les loix des affinités chimiques.*

Phosphorus and sulphur are classified among the acids; metals generally among the bases, but gold and platina come much closer to the acids.

Sulphur is an inert acid. We have already seen that acids are often decomposed when rendered inert. This also seems to be the case with sulphur. It has probably been only one part of a complete acid, and it lacks the other which would restore its integrity. We find this lacking component in hydrogen. Combined with this, sulphur regains its full integrity as an acid, and the difficulty that the combination of sulphur with a substance striving against oxidation generates an acid is thus removed.

Carbonic acid is also decomposed when robbed of its acid principle. With

[9][*Sauerstoff*, oxygen.]

warming or even glowing of carbonate of lime, only one part of an incomplete carbonic acid, which clouds lime water but cannot dissolve the precipitate again, is
driven off, and the other part remains in the caustic lime. When the glowing becomes so intense that this part leaves as well, it has also lost its causticity although
not, thereby, its integrity as a base, which is quite independent of that. This part of
the carbonic acid is also the cause of causticity in other bases and, therefore, deserves the name of causticity principle. If the acid has been entirely separated from
nitrate of baryt by heating, this remains both insipid and deprived of its causticity.
However, baryt has more affinity to the causticity principle than potash but less to
the acidity principle, and thus a caustic potash can be made weak without being
either saturated with carbonic acid or made insipid. Here we cannot enter into all
of the details which chemists find so very interesting; we will only remark that the
part of the carbonic acid which is attached to the caustic lime can be driven off by
nitric acid in the form of a gas which supports the burning of a candle almost like
vital air, but it redintegrates into carbonic acid through contact with atmospheric
air, with nitric acid &c.

These are the fundamental principles of the first prolusion of Winterle's book.
We must pass over the infinitely rich detail here in the interest of brevity. A comparative look at these and the galvanic discoveries shows us a conspicuous similarity
in their results, which have been found in such different ways. In fact, the comparison must begin with a difference, for Ritter's discoveries, which play a principal
role in this, place light in the position which Winterle seems to accord to heat. We
need hardly assume on this account that one of these excellent researchers has been
mistaken. It is very likely (not to say more) that light and heat are results of the
same principles. Winterle considers the heat principle to be both material and mas-
sive and notes in this connection that an electric spark always travels from top to
bottom whether the positive or the negative conductor is placed on top. This materiality cannot be rejected by mere philosophy, for even if this had a priori proved
the existence of some immaterial principle, it would still not have determined that
the phenomena of heat, light &c. stem from it, but this must be left to experimental
investigation. This remark should not be taken as a defence of the materiality of the
heat principle but merely as a necessary remark regarding the conditions of this
point of contention.

The connection between magnetism and electricity has not been ignored by
Winterle. In various questions he seeks to draw attention to the fact that the different temperatures of the earth's zones and of the seasons, as well as the fiery meteors, can be explained by this. Finally, he also asks whether light can be considered
the general cause of magnetism. He derives gravity from magnetism as well.

The weakness of contemporary chemistry is undoubtedly carbon and nitrogen,
two substances which seemed almost as impossible to combine as they were to decompose, and which could only form compounds with oxygen and hydrogen while
they opposed all mutual union. Winterle seems to have succeeded in overcoming
this difficulty. Carbon and nitrogen are not opposite chemical elements, but combinations of one and the same element with oxygen, except that in carbon the acid

principle but in nitrogen that of alkalinity is essential. This new element is called *andronia* (in contrast to another, of the opposite nature, called *thelyke*). It is of acidic nature, of gelatinous consistency, and insoluble in water. It combines with bases (except for ammonia), forms synsomatic compounds with acids, and also combines with metallic calces, which it reduces with glowing, whereby it is itself transformed into azote or carbonic acid gas. Coal contains a great deal of andronia as do the ashes which remain after its burning, so the potash extracted from this is thoroughly impregnated with andronia, and it can be separated from it with ease.

Andronia dissolves in sulphuric acid and imparts to it the ability to dissolve all metals, gold and platina not excepted. This androniated sulphuric acid is so closely related to metals that earths and alkalis must yield to it. It is therefore called a metallophile acid. This acid also has more affinity to earths than to alkalis. The sulphuric acid which is obtained from iron is therefore free of andronia because the iron retains it, and the residue of this distillation is an *androniate d'oxide de fer*. However, the sulphuric acid which is prepared by burning sulphur is never free of andronia because common sulphur contains this element. Andronia also combines with muriatic acid and several other acids and forms its own acids thereby. It goes without saying that andronia is the basis of nitric acid.

When glowed with potash, andronia transforms itself partly into silica, and therefore much of the siliceous content which people believed to have found in minerals was perhaps merely a product of the glowing of the mineral with potash. It is clear from the total context that all other earths contain andronia as well although this has not yet been proved.

That all vegetable and animal matters contain andronia, that steel is an *androniate de fer*, and much else can be passed over as obvious consequences of the preceding observations. With lead, andronia forms metallic baryt, which becomes baryta when calcined. Winterle produces this substance synthetically, by precipitating a solution of lead in nitric acid with androniated potash. The precipitate dissolves in nitric acid and crystallizes into cubes which do not have a sweet taste. It captures the acid from sulphate of potash, when glowed with soot it is not reduced but only leaves a metallic carbon behind, and finally, when dissolved in water with zinc, it yields a metallic precipitate which does not behave like lead.

With copper, andronia forms molybdenum. Andronia is a component of several other metals, zinc and lead among them. It is therefore understandable that Hauch obtained a mixture of approximately 36 parts of oxygen gas and 64 parts of azotic gas when he passed steam over these metals (in a glowing state).

If one does not find the completeness of a system in this very brief description of Winterl's main ideas but still discovers a marvellous beginning, one will be happy to hear that Winterl will soon deliver the continuation of these discoveries, where the table of contents leads us to expect to find all that is missing here. Those who cannot easily extract the results of complicated chemical investigations from a treatise in Latin, like Winterl's (who, in any event, uses the old chemical terminology almost exclusively), have now been provided with a description which, according to the notices, will soon be published, and which will constitute the beginning of a book on the investigation of Winterl's system. The title is *Materia-lien zu einer Chemie des neunzehnten Jahrhunderts* by Dr. Johann Christian Oersted. Part 1.[10]

The theory of the chemical affinities, on which so many capable men have wasted their energy without achieving more than an accumulation of facts, has now taken a great step forward due to the efforts of the perspicacious Berthollet. Here, as so often happens, the old sandy soil had been so well protected that no solid construction was possible. Berthollet begins by destroying the old from the foundation and justifies this boldness completely by what he builds in its place.

So far it was regarded as certain that when a mixture is composed of several substances *A*, *B*, *C*, &c., and *C* has more affinity to *A* than to *B*, it would then have to combine with *A* and leave *B* completely empty-handed. This is a fundamental error; *a distribution takes place, whereby B acquires as large a part of C as the intensity of their kinship dictates.* If, for instance, *A* had an affinity to *C* ten times greater than that of *B*, *B* would obtain only one tenth of *C*. (We shall soon see how other considerations modify this.) It is easy to see that this must be the case if it is remembered that *B*, like *A*, has an attraction to *C*, only weaker, and that *C* is not a mechanical unity but can be divided. How can we assume other than that *C* must divide itself between *A* and *B* in proportion to the attractive forces? *The effects of the affinities are in proportion to the combination of intensity and mass.* What a quantity = 1 cannot achieve, a quantity = 2 is often able to accomplish, and so forth. This has not previously been taken into account in chemistry although examples of it existed which people, however, took great pains explaining in a different way. The earlier conception of saturation, as if the affinities contained a limit (could be satisfied to a certain degree), is founded on inadequate knowledge. There are many other forces which limit the affinities, e.g., cohesion, the tendency to crystallization, the tendency to assume the form of light, and so forth. For example, the more of a given salt which is dissolved in a quantity of water, the greater is the tendency to crystallization, and finally this tendency towards solidity becomes so great that the water can no longer overcome the cohesion of the salt. However, this solution is often capable of dissolving another salt because of the affinity of the dissolved salt to the one to be dissolved, whereby its cohesion is usually reduced.

Insolubility, which is to be viewed as a resistance to liquefaction, very often changes the affinities as well. Therefore, baryt attracts sulphuric acid very strongly and forms with it a salt which is very difficult to dissolve. However, that the affinity between these two substances is not a quantity which cannot be overcome by some simple kinship can be seen from the fact that when sulphate of baryt is boiled with a large amount of caustic potash, it attracts some sulphuric acid, and some caustic baryt is obtained.

The effects of this resistance to liquefaction have often been taken for the effects of chemical attractions. Therefore, e.g., it was believed that potash would have a greater inclination to combine with much rather than with little tartaric acid, but it

---

[10] [Cf. the following article.]

is nothing more than the greater insolubility of acidic tartaric potash (*cremor tartari*) which determines this.

Heat does not alter the affinities as was hitherto believed, rather it merely modifies the circumstances which disturb the affinities. If one of two bodies is liquid, the other gaseous, heat reduces the affinity of the liquid to the gas because it increases the expansion of the latter, that is, its resistance to liquidity; however, it frees the heated gas more from the liquid than if it is cold. Because its expansion is thereby increased, it is more assimilated with the air. When one body is solid but the other liquid, the solubility of the solid is increased because heat opposes cohesion.

This is merely a description of the fundamental principles of the affinities which Berthollet has discovered, the application of which offers itself in rich measure. Part of it is found in his *Recherches sur les loix des affinités*, and we can expect even more in his *Chemische Statik*, which will soon be published.

O.

12

Materials for a Chemistry of the Nineteenth Century[1]

FIRST PART

If, Benevolent Reader, you find some things unpalatable, I ask this one favour of you, that you do not cut the knot by refusing to lend credence to the experiments.—It is only fair to lend credence to a writer until you are convinced of the contrary by means of a counter-experiment.

Winterl, *Prolusiones*, p. vii

PREFACE

The *chemical system* which, a few years ago, after much dispute, hence after much testing as well, was acknowledged almost everywhere in Europe as a collection of completely confirmed facts and as a concatenation of principles based on the most accurate investigations, has already begun to waver again following the discoveries made by means of the *voltaic pile*. A courageous and meticulous researcher has already stated this, more cautious investigators are at least in doubt,[2] and only lethargic indolence still believes that its structure rests unshaken on its old pillars. One important step has already been taken, and we can dimly perceive those that will follow. We have already seen several brilliant results, and others shine through the darkness of their wraps, but much is still to be sought laboriously, and the completed whole will appear unveiled in all its splendour only after *prolonged* effort.

However, several paths lead to the shrine of nature, and another researcher offers

---

[1] [Originally published in German (Regensburg 1803, at the Montag- and Weiszischen Book Shop). KM I, pp. 133–210.]

[2] Several meticulous chemists, and almost all those who have seriously been engaged in aqueous experiments with the voltaic pile, doubt the correctness of the antiphlogistic theory and do not rely on it, primarily out of caution. We refer to what Davy, Erman, Simon, and *several others* have said about that in Gilbert's *Annalen der Physik*. Even the penetrating Berthollet finds an obscurity in this matter (as an enthusiastic advocate of the antiphlogistic theory). See his letter to Pfaff, in Friedländer and Pfaff's French *Annales*, Vol. 1, Part 1.

us a new light from a distant country. One seems to wish to meet the other on the way, and their friendly union will soon prepare a path for us which we could otherwise have trod only after great and prolonged effort. While the former observed nature in the phenomena where it reveals its laws most immediately, the latter discovered manifestations and circumstances in the complicated processes of the chemical workshop which only few might have suspected and created views which will cling to the results discovered elsewhere with other tools as one truth to another. He wishes to tell us what he has learned from Nature as her devout disciple for forty years. Without self-interest and without pride, he only wants us to help him examine whether he has really understood her teachings correctly. Every naturalist is his friend, from whom he accepts reprobation in the same spirit as he receives confirmation of his observations. It would reveal shameful ingratitude and blameworthy self-interest or indolence if we refused to pay attention to this noble researcher, and if we did not consider what carries the seal of the purest intention and the most diligent observation worthy of a serious investigation. Since this has been neglected too long already, it is all the more appropriate to begin this investigation without delay if, perhaps, a future era is not to say that most of us let the chaff pass through when a rich seed beckoned to harvest.

If I have not yet mentioned the work whose examination I invite, it is because I hoped that every reader would guess that I had in mind *Prolusiones in Chemiam seculi decimi noni* by the Hungarian researcher JAKOB JOSEPH WINTERL.[3] Yet I know that few will agree with my judgement. On the basis of inessential matters and external circumstances, most chemists believe that they must reach the verdict that Winterl has been mistaken in his observations and wrong in his conclusions.[4] In his work, they miss the experiments which have been performed to almost the highest perfection by Lavoisier, they find or believe to find such great disagreement between Winterl's observations and the ones which we have already

[3] The following notes serve as evidence for that in the above-mentioned remarks which will not otherwise be documented in the pages below. W. published his book before he could have known about Volta's pile and even less about the chemical discoveries made with it, so whatever he has in common with these has been found by a completely different method, which the entire context of his principles also proves.—He has distributed many copies of his work among chemists, or had them distributed, for nothing. In the preface, he says that in this way he deserves that readers should send him the confirmations or corrections which they might find in his work instead of publishing them in the few of the very best-known journals which are available to him there.

[4] I have been pleased to find that this prejudice is not completely general, and that at least a few very estimable men have not only desired the investigation of W's system but also undertaken it. Among these, Chief Commisioner of Mines Westrumb, of such outstanding merit in chemistry, has obtained *andronia* while examining potash and announced this in Crell's *Annalen*, Vol. 1, 1802. Among the unfavourable opinions of W's theory, I have not yet heard or read anything which was based on a serious investigation. However, in order to ensure that this does not appear to be an empty assertion, I shall examine the most detailed and apparently most thorough public judgement of this kind as representative of all the others. For this purpose I select the review of Winterl's *Prolusiones in chemiam seculi decimi noni* in Trommsdorff's *Allg. chemischer Bibliothek des neunzehenten Jahrhunderts*, Vol. 2, Part 1. Actually, all we needed to do was to list the confessions of the reviewer (loc. cit., p. 65) in order to show that he has not understood the author, for the reviewer *"does not understand what the author intends with his terreis et salinis basibus fatuis, (which, however, are distinguished from the mild alkalis),"* and he finds that *"the greatest difficulty about the author's mode of conception is to understand acidum fatuum and*

made, and, finally, many probably believe to find the key to all of nature in Lavoisier's system, and they regard it only as a sign of vanity, of ignorance, or of fraud if someone believes that he can show entirely unknown treasures to those who are so rich. Therefore, if I do not want to fail in my purpose, I must seek to correct these judgements, and the most important part of this intention will undoubtedly be achieved once I demonstrate that the comparison between Lavoisier and Winterl will not turn out to be so unfavourable for the latter as might have been expected at first glance.

There are two very different ways of penetrating into the depths of nature, of which the one is often more reliable, but the other also leads us further.

Whoever chooses the first takes an important and far-reaching fact which has not yet been understood, traces it through experiments in all their nuances, and thus establishes a chain of facts which is valid as a statement of nature. In this way, we obtain a natural law which always has a far more profound impact than was assumed at the beginning but in the end often deceives the astounded discoverer with the hope that he now possesses the standard with which he can measure all the forces in nature; so he pauses at the goal of his first course, from which he could still, with luck, begin a second. Lavoisier made his discoveries in this way, and what he did will never be erased from our grateful memories. However, we should not forget that only a secondary natural law can be found in this manner, and that only something very incomplete will be accomplished if we attempt to construct a natural system from such. The French chemists made this mistake, and the incompleteness of their system immediately attracted the attention of the first astute opponents, but the bright, new ideas that it contained gained it a victory which was more complete than it deserved. For what was missing was soon quite forgotten in favour of what had been achieved or was believed to have been achieved. The ease with which an open mind learned the theory of chemistry in a few hours gained many disciples for the new theory which was the only source of their chemical knowledge, and the world knows well enough what all of them wanted to make of the antiphlogistic theory.[5]

This blind enthusiasm has disappeared for the most part. In the orthodox church,

---

pottassa fatua." Anyone who must make such confessions cannot possibly understand the entire first prolusion. We hope that the following pages will prove that it is not difficult at all to understand what the author means by insipid acids and alkalis. The reviewer attributes (loc. cit., p. 62) to W. an opinion which he advances at the beginning of §16, in order to disprove it at the end *of the very same paragraph.*—The reviewer says that W. has given the names *symplectic* and *dialytic* to the bodies which are otherwise called *undecomposed* and *compound.* If we read through §10 *very superficially,* such an error is *barely* possible, but if we also read §11, it must become obvious that it is not meant like this at all.—The reviewer is surprised at Winterl's statement that metals do not calcine in *pure* water *isolated from air.* However, this is precisely the case (see, e.g., Ritter's *Beyträge etc.,* Vol. 1, Part 3, p. 184 footnote), and we can only wonder at someone who considers himself entitled to object to experiences with the results of an incomplete theory. It has been my intention to prove that the most detailed and apparently most thorough judgement which has been formulated against W's system so far has been written without the requisite insight. I believe that this example serves my purpose completely, and I shall save the space which I could fill with many more examples for better things.

[5] Who does not recall the attempt to derive principles of light, of electricity, of irritability, etc. from oxygen and hydrogen so that everything was turned upside down in favour of these substances?

there are already many disputes over the main points of the system, and hopefully it will not be long before it is generally understood that Lavoisier's *theory is nothing but an exposition of the chemical behaviour of vital air and its relation to other substances, performed with exceptional completeness and exactitude*, for what further occurs to hydrogen, azote, and carbon is obviously nothing but work of secondary importance which is not at all decisive. Therefore, this theory leaves everything else as it is or can be.

Let us therefore ask:

Why do acids and alkalis neutralize each other? —*Why* is electricity needed for the combination of several types of gases with each other?

Or indeed:

How are the electrical phenomena related to the chemical ones, in whose company they are so frequently discovered by accident but would be discovered far more often if we proceeded according to principles?

Why does water absolutely require an additional substance in order that the calcination of a metal, etc. may take place?

Does the antiphlogistic theory of combustion lead to a theory of self-ignition?[6]

Is there a thorough explanation of the manifestations of light and the changes of colour during chemical processes?

What is the common principle of metals?

What is that of alkalis and earths?[7]

I claim with certainty that no thoughtful supporter of Lavoisier's theory will find in it *satisfactory* answers to these questions; indeed, he will admit that several of them are not even addressed in it, and also that there is no hope *in this system* of ever seeing these questions answered. And yet, all these questions, *carefully weighed*, are of great importance, and how many others could not be raised if we wanted to *criticize* Lavoisier's system here, and not, in accord with our intentions, restrict ourselves to pointing out that it has very many and extremely significant gaps.

Winterl has not, like Lavoisier, summoned up all his abilities to solve a *single* problem; he is one of those *more exceptional* men who observe every noteworthy manifestation of nature which appears before them with a clear eye and pursue it until they understand it. In their hands, a system of facts is not derived from a *single* experience, but their genius creates a *true theory* from *all* the facts which they have found in nature by means of a profound understanding. Therefore, in their works, experiences are to be found compiled in *great numbers*, and, enriched by the view which they carry with them, they proceed from one to the next. The longer we follow them attentively, the more we begin to understand, and once we have accompanied them undaunted to the end, we see, as a reward, the clear light of day where once we found only twilight or darkness. At the beginning, therefore, we find entire

---

[6]Those who would at least like to have the antiphlogistic theory regarded as a complete theory of combustion must still fully answer this and part of the following questions as well as several of the other points.

[7]Whatever constitutes a class must have a common principle. The demand which can be made on a theory in this connection seemed at first to be satisfied by Lavoisier's system with regard to acids, but later investigations have also shown difficulties in this respect.

collections of paradoxical phenomena in Winterl's work, which are only gradually joined to each other to form a whole. We are often quite amazed to see the presentation of a great many facts which, if they occurred, should be found daily by chemists, but which they have simply failed to notice because they sought only that which was connected with a certain subject and did not consider anew the process itself as a unity which, as such, has a connection to a whole. This also explains why, in the writings of *good* chemists who describe their works *in great detail, phenomena* are *often* reported which in Winterl's work belong to a *system of facts* but which are presented there merely as *rare phenomena* originating from *accidental causes*.[8]

Precisely because Winterl, in his experiments, wanted to understand not only the connection of one particular to another but the *connection of the details to the whole* (he did not consider it atomistically but *dynamically*), it will also be found that he has performed many experiments with the raw materials needed for the preparation of the reagents in order to examine what in their precipitation has otherwise been discarded as impurities. This scrupulousness did not go unrewarded, for he found in the potash of commerce a substance (his *andronia*) which will become one of the most important substances in chemistry and give us unexpected information about the composition of metals, of earths and alkalis, of air, etc.[9] However, as much as Winterl and his like do for the whole, as little do they permit themselves to become involved in the pursuit of experiments to their extreme. They are *observers* rather than *experimenters*. If, in their reports of individual facts, they are far from the accuracy of a Lavoisier or a Volta, and if, therefore, we wished to reject the whole without further examination, we would reveal only that we did not possess the measure of men like Winterl. Besides, it would be desirable if Winterl, who like Priestly has examined only qualitative and very rarely quantitative relations, *might also find his Lavoisier*, who could add what was still missing and correct what was erroneous.

However, far be it from me to want to disparage the art of the experimenter. Whoever is capable of encompassing the universe with his spirit and still takes the laborious path of performing very detailed experiments, to him his creative genius shows, in every single phenomenon of his experiments, the laws according to which creative Nature has connected it to the whole, and what others have only half accomplished on this side or that, he presents in the highest perfection. He who sees in his task only petty effort forgets that the smallest part also contains the laws of the whole, and he who recognizes the laws of the organism in the frog must also understand them in man.

---

[8] I have mentioned some of these and could have done so more often if the conditions under which I lived during this work had not left me lacking in time and literary aids.

[9] I know that precisely because W. obtained his andronia from impure potash, many chemists were deterred from looking into this discovery as they were afraid of not getting a clear result, but if a substance has been obtained which dissolves all metals without exception when combined with vitriolic acid and binds them so strongly that they do not precipitate alkalis, if it makes muriatic acid capable of dissolving silver, if it evaporates in the atmosphere and still produces annealed silica with potash, if it changes lead into baryt, copper into molybdenum, etc., we have certainly obtained something from the impure potash which is worth all the attention, and we are certainly not misled by the silica content of its potash, as some believe.

So much for Winterl's work and merit. I have praised him as I believe he deserves, and if, after future investigations, it should seem as if I had spoken too warmly of him, it should not be forgotten that I have rather recommended for *examination* a matter which in itself was very important than intended to make a definite judgement of the value of his experiments, which I have had too little opportunity to repeat myself.

Now let me explain the *purpose of these pages*. It is none other than to *initiate a serious investigation of Winterl's theory*. In subsequent issues I would like to publish the records of this. The first step must be to make Winterl's *ideas* more generally known, for which purpose his *Prolusiones* themselves are not well-suited both because of the language in which they are written, with which not all chemists are sufficiently familiar,[10] and because of their arrangement, which does not provide a clear overview. As the *first part*, I hope to deliver a comprehensible *description of Winterl's system*. In the following, *experiments, criticism, comparisons, and applications* will be found. Some of the best German chemists have already promised to support this work with reports of their experiments. On my part, I shall not fail to continue the experiments which I have already begun though I do not want to publish them until they have produced a completely decisive result.[11] I would be most appreciative if many researchers would become involved in these investigations, and every student of nature is hereby invited to inform me of the elucidations which he can provide in this matter for the continuation of these pages. Every thorough examination, every accurate experiment, and every relevant observation will be accepted with due thanks and gratitude, and they will always reach me if sent to the publishing house of this first part.

In this presentation, a significant difference between the treatment of the first and the second of the prolusions will be noticed. This arose quite naturally from the fact that the *first* in a sense *constitutes a whole*, which made it easy to emphasize the system of natural laws in it without going into details. On the other hand, the second can only constitute a whole in combination with several prospective ones, and yet it touches on so many interesting topics, which actually make it the more interesting with regard to experimental issues, that its promulgation should certainly not be postponed.

If I succeed in describing Winterl's ideas with the proper clarity, I hope that I shall be excused for any rashness which may have crept in, and that, as a foreigner, I shall be forgiven for any imperfection of expression.

Jena, August 1802.

[10] A good translation of the *Prolusiones*, which will not be rendered unnecessary by my presentation, would, therefore, still be desirable, and if more attention were to be paid to W. than hitherto, we might also expect such a translation from a German chemist of exceptional merit.

[11] The reason why I have not reached the point where I can publish the results of my experiments is merely that I have been on a study tour for a year so that I have naturally not had much time and opportunity for such work. Therefore, I would also have preferred to postpone the publication of this work until my return if I had not been convinced that it was important that a matter which can provide so much enlightenment was raised as soon as possible.

ON THE ACID PRINCIPLE

The well-known fact that *the acid and alkaline substances* (in the broadest sense of the word) *completely lose their acidity and alkalinity*[1] *in their combinations* once the appropriate proportion is found would have attracted the attention of physicists a long time ago if they had not been content with the statement that a combination of two different substances must have other properties than its ingredients. As correct as this statement is, it does not explain anything, and it does not satisfy the inquisitive physicist, who wants to understand the reason for this fact. And yet, the simplest expression of this phenomenon easily leads us on the path to its explanation. I say: *Acidity and alkalinity neutralize each other*, and now I ask: What is simpler and more natural than to conclude from this that *alkalinity and acidity are opposite each other*? In fact, we draw conclusions like this everywhere in natural science, for how would we know that the different electricities, magnetisms, etc., are opposites if they did not neutralize each other? It might be objected that it would be far more obvious in the case of these forces because the substances with which they are associated did not have to be chemically combined in order for the forces to be able to neutralize each other. This objection, which in any case is of little consequence, is invalidated by the following, in which it will be demonstrated that *acids and bases can be separated from each other in such a way that they do not regain their stimulating forces,* from which it will follow at the same time *that the cause of acidity and alkalinity does not lie in something ponderable, but that they possess their character due to a stimulating principle of their own.*

With the usual method of decomposing neutral combinations, we restore to the separated part the force which was exhausted in the combination by using agents for the decompositions which contain the stimulating principle of one of the elements. Acids are such agents of decomposition, and their mode of action will be explained in the following. Here we are content to indicate a means whereby many combinations of acids and bases can be decomposed in such a way that the separated element is not restored but exists separately in an inert state. Such a means is found in heat which has not yet become glowing heat. If such neutral compounds as those which are composed of a volatile and a solid component are treated with this, one element will be separated in an *inert* state, and this often to such a degree that it is no longer capable of affecting even the most sensitive plant pigments, of neutralizing an opposite element, and generally of reacting as an acid or a base in any way.

If *sulphurous acid potash*,[2] dissolved in water, is subjected to a distillation, the *acid* for the most part leaves its base in an aëiform state, but *with properties entirely different* from those it had before the combination with the potash. Its distinctive

---

[1] With this, I refer to the property of rendering acids inert without decomposing them first. W. calls it basicity because he contrasts bases (bases of salts) with acids. That which has often been called the base of an acid, he calls the substratum. In this respect, I shall also follow his terminology, but I prefer the term alkalinity to the term basicity.

[2] W. had obtained his sulphurous acid by the distillation of 9 parts vitriolic acid with one part sulphur.

effects on eyes, nose, lungs, etc., have vanished, it no longer reacts on plant pigments, it dissolves in water in an even larger quantity than before, and could in this form be regarded as an entirely new kind of gas. If it is again combined with the remaining base, it yields the same neutral salt as before, whereas it would not neutralize a pure, non-inert potash.

With the aid of heat, *carbonate of lime* produces a gaseous carbonic acid which combines with water in an even larger amount than the ordinary one but without transmitting to it any taste or the property of making blue plant pigments red. This *inert carbonic acid* is also distinguished by the ability to precipitate lime from lime water in such a way that the precipitate cannot be dissolved again later.

If *boracic acid ammonium* is exposed to a temperature of 160° Fahrenheit, part of the ammonium disappears without the remaining part acquiring any effect on syrup of violet. At a temperature of 300° Fahrenh. (but not driven to glowing heat), the ammonium leaves the compound completely, but the remaining *acid* has *not* recovered *its previous properties.*

Indeed, many acids have the property that they are driven from their combinations by heat, though not in their uniform state but decomposed. These are called acids with a *dialytic* substratum (from δια and λυω), whereas those which retain their uniformity during expulsion are called acids with a *symplectic* substratum (from συν and πλεκω).

At the beginning of distillation, acids with a *dialytic* substratum, which are driven from a combination with an alkali or earth base without glowing heat, are expelled in part with their complete acidity, but in part decomposed into their constituents, *adiaphorically* or with a small part of their acidity.

Nitric acid belongs to this category of acids whose combinations with salts, when heated, yield only partly their acid, partly vital air and azotic gas.

Acetic acid belongs here, too, because the salts combined with it, in addition to complete acetic acid, also produce fixed and combustible gas as well as the impurities of the vinegar, viz., oil and ammonium (Winterl's claim).

Muriate of magnesia yields fixed and azotic gases at higher temperatures. A portion of the former remains combined with the magnesia, in addition to some undecomposed muriatic acid.

Sulphur is also an *acid*, although a *very weak* one. When diluted with three times as much water, its compounds with bases (the livers of sulphur) produce azotic gas during distillation, a little hepatic gas, and a *liquid sulphur*, which precipitates most metals from their acid solutions. Thus, *gold* is precipitated from *aqua regia*, *silver* from sulphuric and nitric acids, *mercury* from nitric and muriatic acids, *lead* from nitric and acetic acids, *copper* from vitriolic acid, all as black deposits. *Tin* is precipitated from vitriolic acid in a small quantity as a whitish, in a larger quantity as a black-grey deposit. *Mercury* is precipitated from sulphuric acid as an orange-coloured precipitate which, however, very soon changes to white. This occurs only with a small quantity of the precipitate, a larger amount also turns black. *Zinc* always yields a white precipitate. *Iron* is scarcely precipitated because a readily soluble salt is formed, which is also why an aqueous solution of iron vitriol merely turns milky with this and, only when completely saturated, produces a sediment which, however, is dissolved again by even more liquid sulphur. Nitrate of iron turns white with a little and milky with more sulphur, and with even more it becomes transparent, with a dark-blue reflection, because of the dilution.

Just as acids can be produced in an *inert state*, so can *bases*, but this is more difficult because only a single alkali can be separated from its combinations by volatilization, and because the fixed ones from which a volatile acid has been separated retain some of it and, therefore, cannot be regarded as pure. Temperature also seems to redintegrate alkalis far sooner than acids, which might be the best way to explain why *ammonium*, driven from a neutral salt by means of heat, loses only part of its alkalinity.

As the easiest method for bringing *potash* into the inert state, Winterl suggests combining it with metallic acids and adding to the solutions of these acid metallic salts metals which extract from the acids a part of their basis of vital air, whereby they cease to be acids and, therefore, must release the bases with which they were combined, and in the same state in which they existed in the compound, i.e., either *insipid* or *inert*. For this purpose, he particularly suggests *magnesium oxide*. Magnesium oxide is treated with saltpetre in fire. At the expense of the nitric acid, the magnesium oxide is transformed into a *metallic acid*,[3] and the *potash* which is released from the saltpetre combines with the new acid to form a neutral salt which, as mentioned above, is only decomposed by a metal.[4]

Inert potash is volatile and, distilled with water, yields a gas which can neither be ignited or feed a flame nor be dissolved in water, and it can no longer affect a solution of sal ammoniac, on which it had a strong effect before the distillation. The

[3] W. considers this acid to be chromic acid and does not regard chrome as a metal in its own right.

[4] The complete procedure is as follows: One part very finely pulverized black magnesium oxide which has been purified with odourless nitric acid and three parts pulverized pure saltpetre are placed in a Hessian crucible with a tightly fitting lid and kept at a medium temperature for three hours. Hereby, the nitric acid is resolved into its components. The azote leaves the mass, which has become liquid because of the water of crystallization, with effervescence and noise (therefore, the crucible should be only two-thirds filled), but the basis of vital air combines with the magnesium, which thus becomes acid and changes together with the potash into a neutral salt. Once the ear no longer detects the effervescence, an increased temperature can neither melt nor reduce the sponge-like mixture which has been produced. Therefore, glowing heat is applied in the fourth hour. If, during these four hours, live coals are brought in contact with the rim of the crucible, it still remains dark, which proves that no vital air escapes, but with a premature temperature increase the coal indicates escaping vital air, and the experiment does not succeed. If everything is now left to cool by itself, a black mass remains in the crucible, but the lid is found to be covered with insipid potash in the form of white crystals which have sublimed. Both masses are extracted with water, whereby the latter produces a reddish solution, beautiful beyond all description, while the former produces an equally beautiful dark-green one. However, every agent which communicates acid principle [*] changes these colours, as does the agent which extracts the basis of vital air. By evaporation and cooling, green, red, or blue crystals are obtained, according to circumstances. An added piece of metal (silver, mercury, lead, tin, zinc, but iron much less) changes the colour of the solution, and in the end it becomes quite colourless. During this process, the magnesium oxide (or, more correctly, the magnesium acid) is deprived of its basis of vital air, and it leaves its combination with the potash, which now remains insipid as it was in the compound. [Cf. "A Review of the Latest Advances in Physics" (Chapter 11, this volume).]

[*] Here and elsewhere, *acid principle* [*Säureprincip*] denotes the cause of acidity, but not what is usually called oxygen [*Sauerstoff*].

residue of the distillation is only a few grains of mineral chameleon. If the insipid potash has been exposed to open air for some time, it has gradually resumed all the properties of common potash.

Lime can also be produced in an insipid state. If pure, insipid potash is mixed with lime water, a quantity of insipid lime is separated which can be collected on the filter in the form of a gelatine. This insipid lime dissolves more weakly in water than caustic lime but can nevertheless be precipitated by the appropriate reagents. This lime is without taste, but the potash used has become caustic. Insipid lime is volatile and follows steam, which is why it cannot be separated from water by distillation. Lime which is too strongly burnt (dead-burnt) is also insipid; at least Bergman claims, with regard to this kind of lime, that it certainly dissolves in water but does not thereby heat and decompose.

If the acid has been driven off from nitrate of *baryt*, the remaining baryt still retains its neutral taste. This cannot be the result of a small amount of residual acid, for this could neutralize only a little, not all, of the baryt. Insipid baryt is distinguished from insipid potash and lime in that it combines with carbonic acid if it comes into contact with it while evaporating with water.

If an *insipid acid is combined with an insipid base*, the same neutral salt is produced as if a complete acid were combined with a complete base. For instance, if insipid sulphurous acid is combined with insipid potash, sulphite of potash is produced just as if the complete acid had been combined with complete potash.

A *complete acid* is made only slightly inert by an *insipid base*, and a *complete base* reacts in the same way towards an *insipid acid*. This explains the acidity of the combination of tartar with borax. The excess soda in borax[5] is insipid and therefore can take little or none of the acidity from the excess acid of the tartar. This combination certainly seems to be more acid than the tartar itself, but this stems from the fact that it is more soluble, due to which it has a stronger effect on the sense of taste. *The more base is combined with an acid, the more completely its acidic properties disappear.* An acid with a small amount of base is still acid, with as much as is necessary for neutralization, it is insipid, half deoxidized, and combined with even more of the base it will be even more deoxidized.[6]

By means of heat, an *acid with a dialytic substratum* is expelled in a completely decomposed state from a compound with a double portion of base, and without any trace of the complete acid. If expelled from a compound with a triple portion of base, however, even those parts of the substratum which would otherwise have retained a partial acidity have lost this.

Acids with a symplectic substratum which have been expelled by heat from their compound with a double portion of base have a far larger volume than if they had been separated from the amount of base which would just have been sufficient for neutralization and also attract water much more strongly in the same proportion. When produced from a triple portion of base, they exhibit all these deviations to a

[5] In natural borax and in the kind which is produced from it by purification without the addition of soda.

[6] This is why an acid with symplectic substratum which is extracted from its neutral compound by heat first becomes less, but later more, deoxidized.

much higher degree. If an acid in the highest degree of deoxidation is combined with alkalis, it can no longer be observed that they become inert, rather they colour blue plant juices as completely green after the combination as before, whereby it is only remarkable that these become orange shortly thereafter. Aëriform acids deoxidized in this way also enter into such a combination, even with hot water, that it is impossible to produce sulphurous acid, expelled from the triple portion of base, in any hydro-pneumatic apparatus. There are also acids among these which, in their combination with a base in the neutralized state, have a symplectic substratum which, combined with a larger amount of base, cannot be separated from it by mere heat without being decomposed.

Acids which are deadened by bases in the neutralized state will be called *deoxidized in the first degree*, or simply *insipid*. However, if they are deadened by a *double* portion of base, they are called *deoxidized in the second degree* and by a triple portion of base *deoxidized in the third degree*.

If *one acid expels another from its combination with a base*, this occurs because the weaker acid, which is combined with the base, deprives the stronger of its acid principle and again becomes acid itself, whereby it is less inclined to be combined with the insipid base. This can be concluded from the fact that the acid which expels the other now changes into the *insipid* state, whereas the expelled, previously insipid acid is now *redintegrated*. Therefore, the terms *stronger and weaker acid* must be understood in the sense that the *stronger* acid is the one which relinquishes its acid principle *easily*, and the *weaker* is the one which does *not relinquish it so easily* but retains it more firmly. The quantity of the various acids which are resolved in combination with a base is also directly proportional to the intensity of the taste and the strength of the mutual attraction of the acid and the base.

As a result of this, it is understandable *why a strong acid often expels a larger quantity of a weaker one from its combination than it should proportionally* if the separation were caused merely by the new salt combination. If, e.g., a small amount of a strong acid is added to a solution of liver of sulphur, most of the sulphur is precipitated, and a large portion of the alkali remains in the solution in a caustic state.

A small amount of *vitriolic acid* which is added to a solution of *borax* in hot water causes a precipitation of boracic acid, which is found after cooling, despite the fact that the vitriolic acid would not even be capable of saturating the excess soda. If the borax is produced artificially from its acid and soda, more acid is required first to saturate the excess portion; afterwards, the vitriolic acid also produces a disproportionately large precipitate of boracic acid in this case. In natural borax, the excess soda is insipid and, therefore, the vitriolic acid is made inert to a very small extent. The acid thus obtained has little acidity, but if it is precipitated with more vitriolic acid, it has far more acidity, which, however, cannot be the result of the added sulphuric acid as it does not lose this acidity even under a fusion to glass.

The *driving off*[7] *of bases by means of other bases* is now easily explained. It is only necessary to consider the principle of causticity since many bases cannot be produced without it.

---

[7] [Ørsted writes *Ausbreitung* but evidently intends *Austreibung*.]

If a volatile acid is to be separated from its combination with a base by means of a weaker but at the same time more fixed acid with the aid of an elevated temperature, various remarkable phenomena can be seen which are best explained with an example. When mixed with an excess quantity of boracic acid which has been dissolved in a little water and distilled by means of a pneumatic device, common salt certainly yields some undecomposed acid, but part of it is passed in a resolved state. The excess boracic acid is hereby sublimed and can be driven from one place to another with a coal. At the same time, *this unusually volatile boracic acid is insipid*, very difficult to dissolve in alcohol, and does not redden syrup of violet but rather colours it slightly green. The residue consists partly of undecomposed common salt and of borate of soda, which, however, is distinguished from that obtained in other ways by being more difficult to dissolve in water since the other deliquesces in air. It has different chemical affinities towards salts as well.

Insipid acids only separate others which are much weaker from such combinations where the base is stronger than the one by which it was made insipid, i.e., in so far as it is capable of rendering it even more inert.

Precisely the same law also applies to *bases*. Though it might seem at first glance as if the ability of acids and bases to render each other inert, even after they have been separated from each other, might result from the fact that the acid still retained or acquired a small quantity of base, and the base a small quantity of acid, this can easily be disproved, for:

a) *Bases can make inert only a proportional quantity of acids, acids only a proportional quantity of bases*, and, therefore, one cannot be rendered completely inert by a small associated quantity of the other.

b) Acids with a dialytic substratum pass partly intact, partly completely deoxidized. In a neutral combination, however, it is not possible that one part is completely deoxidized, the other not at all, rather all acids contained in them must be in an intermediate state between completeness and perfect deoxidation, i.e., must be partially deoxidized. The acid principle, however, is no longer capable of maintaining the integrity of its related substratum, so it binds as much as it can, and this part passes in a complete state while the other is completely deoxidized.

According to the usual view, this last argument lacks sufficient justification. Therefore, I believe that this subject must be examined more closely, and I hope to clarify the entire issue with an example which the opponents themselves might like to choose. If I decompose saltpetre by means of heat, I obtain oxygen gas and nitrogen gas, and these contain the components of nitric acid, which again produce acid if we combine them. Therefore, it might be said that there is neither less nor more cause for acidity in the separated acid than it had prior to its combination, for the cause of acidity resides neither in oxygen nor in nitrogen but in the combination of the two. However, I answer that your claim that nitric acid is nothing but the combination of the basis of vital air with azote is not only an unconfirmed assumption but also an incorrect one. So far, no-one has succeeded, and certainly no-one ever will, in transforming azote combined with the basis of vital air into nitric acid without interposing a new principle, that of electricity, and it will be proved below that this is where the acid principle is to be found.—

Let us summarize in a few words the result of our investigations so far. *Acids and bases owe their distinctive property to their stimulating principle.* The principle of one class is the *opposite* of that of the other. If acids and bases come into conflict with each other, their properties therefore *neutralize* each other. The greater the power of one, the more the other is exhausted, and conversely. If an acid is deprived of its stimulating principle beyond a certain degree, it can no longer exist by itself as a qualitative unity, but it dissolves into two opposites, one of which contains more, the other less of the remaining principle.[8] (Or perhaps more correctly, one part absorbs all the existing principle, the other is entirely depleted, indeed, it may change into the opposite state. This has not been investigated adequately.)

Now an attempt should be made to find out whether we cannot get closer to the nature of these *principles, of acidity and alkalinity* (*basicity*, as Winterl calls it), but before we proceed to do this, we wish to undertake several secondary investigations which might have an important influence on the entire issue.

WHAT IS VITAL AIR?

This is certainly one of the most important questions which can be asked about what has already been said. It is *no longer a principle of acids, for these can become weaker and stronger without losing or gaining the basis of vital air.* However, so many bodies become acidic following combination with this substance despite the fact that they contained no perceptible trace of acidity before. All this will be explained once it is proved that *vital air is an acid.* Perhaps, the most striking proof of this is that this air can be produced *in a more or less acid state.* It has often been produced in a *semi-deoxidized state,* but its nature has been misjudged by chemists in that it was considered to be a modification of saltpetre gas and listed under the name *dephlogisticated saltpetre gas* or *gaseous nitrous oxide.* Semi-oxidized vital air can be produced when pure vital air is brought into contact with liquid ammonium containing copper filings. Here, the vital air is deoxidized during the calcination of the copper. However, this deoxidation continues to a point at which it loses even more of its essential properties and could easily be taken for azote if it did not permit combustion with hydrogen gas, although with a weaker bang than complete vital air. If the semi-deoxidized vital air is driven through glowing iron pipes, a kind of *completely deoxidized* vital air is obtained which Milner regarded as *nitrogen gas.* Deoxidized vital air can again be transformed into oxidized (acid) air by electricity as well as by glowing heat. Vital air does not combine with all bases, but it is generally inclined to combine with metals (which are also bases). This acid

---

[8]The extent to which bases are also decomposed by being made inert has been less thoroughly investigated although there are indications of this, p. 128. However, it would seem that bases redintegrate more easily by means of heat, and this may be the reason why the ammonium which is precipitated by heat from compounds with fixed acids appears to be so little inert.

forms synsomatic compounds[9] with saltpetre gas, with muriatic acid, with phosphorus, sulphur, arsenic, etc., whereby several among them become more but others less acid. Semi-deoxidized vital air does not combine with metals, nor does it form synsomatic compounds with the above-mentioned acids. On the other hand, it now combines more easily with hydrogen.

Vital air can also become *hyperoxidized*, which happens if it is driven off from metals by means of glowing heat. A mixture of this air and nitrogen gas does not produce true atmospheric air but a mixture in which the nature of the hyperoxide is still recognizable.

WHAT ARE FUMING ACIDS?

Experience will readily instruct us on this subject. The same salts which are produced through the combination of an insipid base with a complete acid are also obtained if a complete base is combined with a fuming acid. However, complete bases extract more acid principle from acids than incomplete bases, so fuming acids, which retain as much acidity as ordinary acids after combining with the former, must contain even more acid principle in their combination with the latter. Consequently, they can be called *overacidified, hyperoxidized* acids. This is further confirmed by the following experiences. Sulphurous acid which is deoxidized in the second degree and in this state combined with a complete base is not redintegrated and separated by common dilute sulphuric acid, but this can be accomplished by means of fuming oil of vitriol as well as by other fuming acids.

There are *four* ways of *hyperoxidizing acids.* 1) By combining them with metallic calces and driving them off again with heat. 2) By combining the concentrated acids with an insipid base below the saturation point. As one part of the acid is saturated with the insipid base, it gives off part of its acid principle to the rest. 3) If an acid is driven from a dry neutral salt by means of a fuming acid, the excess acid principle of the latter also passes into it. If oil of vitriol is added to a concentrated solution of liver of sulphur, gas develops with an unbearable bituminous smell which affects the brain. If fuming oil of vitriol is poured over chalk, the only result is the production of a white vapour which no-one can bear. 4) Incandescent temperature (glowing heat) can also communicate acid principle to a very high degree, but more of this later.

If vital air forms a synsomatic compound with sulphur or phosphorus, the increased acidity can be attributed to the acid principle attracted from the adjacent bodies.[10] In other cases, this increase of acidity could result from the fact that the acid principle in the product was not so strongly bound as in its constituents.

It seems to be unfavourable to the theory presented so far that acids which are

---

[9] A combination of two bodies of a similar kind (acid with acid, base with base, etc.) is called a *synsomatic compound.*

[10] As an example, W. mentions that moistened sulphuret of potash extracts not only vital air from saltpetre gas but also acid principle and transforms it into semi-deoxidized vital air.

driven from combinations with metallic calces become hyperoxidized, for metallic calces are for the most part bases, so they should make acids inert and not make them stronger. In this respect, however, it must be taken into consideration that the metallic calces are already combined with vital air, i.e., they already have acid principle, so that the new combinations with acids can be regarded as supersaturations, which rarely cause any perceptible deadening. In any case, many metallic calces are already of an acid nature and form synsomatic compounds with other acids, whereby there is even less tendency to cause inertness. However, that does not explain this phenomenon but solves only some of the difficulties. It appears that, even though they are of a *basic nature, metals* have, as it were, a *sphere of acid principle*, for if *base* metals, especially iron, are added to non-fuming vitriolic acid, they transform their adjacent parts into a crystalline mass, it fumes, and it finally rises to the upper part of the glass as flakes and forms a salt which is completely like the crystalline, volatile part of oil of vitriol, yet the metals do not dissolve in this process. *Noble metals* do not have this property, rather they seem to be *endowed with the opposite force* since it took a portion of crystalline oil of vitriol 6 minutes to dissolve in a silver spoon, whereas a similar portion in an iron spoon remained solid for a long time despite the moisture which had been attracted from the atmosphere. Experience indicates that several base metals do not dissolve in fuming acids, indeed, they resist being attacked by other acids after contact with fuming acids. Lead which has been in contact with fuming nitric acid for a few minutes and then is quickly washed does not dissolve in common nitric acid. Keir has also observed this in the case of iron. W. noticed that iron which had thus been changed by fuming nitric acid did not lose its power in an isolated silver spoon. However, when several individuals whose hands were in contact with each other formed a chain, and the person at one end held a silver spoon while the person at the other end placed a piece of iron in it, this lost the power to resist acids.[11] However, even if acids[12] have an atmosphere of acid principle, their inner nature is still basic, which can be seen from the fact that they attract acids and, directly, vital air. It is remarkable that *water, as well as bases, extracts the supersaturating principle from hyperoxides* but cannot make them more inert. This can be compared with the fact that acids which are supersaturated with a base release it if water is added. For instance, this is the case with acetic or nitric acid which has been in contact with litharge for a long time. In other words, water renders inert that part of the acid which attracts bases with a smaller force, for the supersaturating part feels the weakest attraction.[13] The fact that *water is of a basic nature, albeit weakly so,* is therefore understandable because water is composed of the completely deoxidized basis of vital air and hydrogen, but the latter is of a basic nature. However, a base, even if it is very weak, can still have retained some of its nature if only it is combined with a

---

[11] W. does not regard these experiments as completely illuminating and satisfactory, but he assumes quite correctly that these phenomena are related to *galvanic* phenomena.

[12] [metals.]

[13] In my opinion, it seems more likely that the base would still be supersaturated with acid here, especially because the bismuth calx which is precipitated from nitric acid in a similar manner is a nitrate of bismuth calx, although with very little acidity.

highly deoxidized acid. It can also be added that though water is certainly *indifferent in itself, any change of temperature in it causes a change* so that water is more *basic at a higher temperature* but more *acid* at a *lower,* below zero. We shall return to this principle later.

An experience which agrees beautifully with this theory, although it may not be taken as proof, cannot be passed over here. If fuming nitric acid, which is flaming red and has the highest degree of oxidation, is mixed with water, it turns yellow, with more, green, with even more, blue, and with a large amount of water, it finally passes into a state of colourlessness or, more correctly, into such a dilute state that colour can no longer be perceived. However, from Ritter's experiments it is well known that the red in the spectrum is the highest expression of oxygenation, and that the other colours follow in the same degree as here in the dilution of nitric acid.[14]

What Is the Cause of Causticity?

This is a question which appears to have been answered a long time ago, and there may be few questions about which we are now less concerned. The salt bases are caustic in so far as they exhibit a strong power to dissolve other bodies, especially animal matter. Therefore, this property belonged to them originally, and thus we should not ask about the cause of causticity, but about the cause of the *non-causticity* of the original caustic salt bases as such. This cause of non-causticity is to be found in all conditions which can weaken the original strength of these bases, especially in the combination with acids, even weak ones, e.g., carbonic acid. This is the general view, and it seems that we have no reason to stray from the path which we have already chosen since we have just tried to shed more light on how bases are made inert by acids. However, if we do not require such a principle for our system, it may be an even greater proof of our attention if we discover one all the same. As is well known, caustic salt bases become weak in air, and this change is due to the carbonic acid which they have attracted. If we deprive them of the carbonic acid again without combining them with another acid, they recover their causticity, and nothing, therefore, might seem more unlikely at first glance than the assumption that carbonic acid could become the cause of causticity. However, we do not wish to be deterred by this but intend to investigate thoroughly if this appearance is based on a truth. Such a serious investigation is beneficial to science to the same extent that blind assumption destroys it.

The first objection which we encounter, and which we have already noted, is that carbonic acid makes caustic alkalis weak. This is perfectly true. However, nothing follows from this except for the fact that carbonic acid, if it should be the cause of

---

[14]Richter found a nitric acid in which, for every 1000 parts of liquid, there was the following pure acid content: 302.0 slightly bluish; 331.7 bluish; 372.8 greenish blue; 307.7 green; 368.7 dark green; [*] 524.6 yellowish green; 577.6 yellow green; 608.9 yellow; 678.4 high yellow; 723.2 flaming red; 812.7 dark flaming red; 853.0 the darkest colour of fire.

[*] [Instead of 307.7 and 368.7, the text should read 407.7 and 468.7, respectively.]

causticity, must exist in a state which is completely different from the one in which it renders bases inert. As long as only one state of carbonic acid was known, it was impossible to come to this conclusion, but now that we know how much it can be changed, the path towards this investigation has been opened to us.

We want to append an example to our investigation. This will make things easier, and the application to the general case will not be difficult. If carbonate of lime is to be made caustic, a quantity of carbonic acid is driven off from it by means of heat; but now two questions are to be answered, viz., Is *all* the carbonic acid expelled from the lime? And, if carbonic acid remains, what is its *nature*?

From caustic quicklime which has been slaked with water, nitric acid still expels quite an amount of carbonic acid gas, so there is still carbonic acid in caustic lime. However, it must be greatly changed, for we have demonstrated above that an acid is all the more inert, the more base is combined with it, so that an *acid which otherwise reveals itself to be symplectic can become dialytic in combination with a very large amount of base.* This is precisely the case with *carbonic acid.* Several experiments prove this. At the end of the process of the combustion of lime, Priestley obtained a gas which could support a flame. At the beginning, the carbonic acid became symplectic, but once only a little remained, in comparison to the large amount of lime, it was resolved.

When small quantities of odourless nitric acid were added to a pound of burnt lime slaked in water, which dissolved almost completely in the water although it had been preserved for several years, a perceptible effervescence developed, but the water which blocked the bottleneck, or a bladder attached to it, was not repelled but attracted. This attraction diminished gradually, and finally the bladder was inflated with fixed air, and some chalk was left behind in the flask, frequently a twelfth part of the lime used. If, however, a large quantity of nitric acid is added at once, a gas is produced in which a candle burns with an increased flame, but this gas is absorbed again in the following instant. Undoubtedly, the explanation is: If nitric acid is added to lime, incomplete carbonic acid is produced, which, however, redintegrates in the vessel upon contact with atmospheric air and absorbs part of it. If this has continued for some time, so much carbonic acid has been released that the vessel becomes filled with it. If a large quantity of acid is added at once, much incomplete carbonic acid develops, which only after some time can attract sufficient atmospheric air to make the absorption perceptible.[15]

Now we understand the reason why lime does not become caustic but inert by excessive burning because the resolved fixed air which is still combined with it is also expelled, so there is a very fundamental difference between this and the kind of lime which is burnt in the usual way.

The fact that caustic lime dissolves much more slowly in acids, especially weak ones, than weak lime is also unexpectedly elucidated by this.

The air which yields *minium* with the help of vitriolic acid is not vital air, as is commonly believed, but *resolved, fixed air.* If we add nitric acid instead of this

[15] Is this resolved carbonic acid not the same that Morazzo [Morozzo?] obtained when he calcined mercury or lead in carbonic acid and, by a reduction of these, transformed them into a gas in which a flame burnt vigorously?

acid, we obtain a complete fixed air because this acid redintegrates the incomplete fixed air. That the carbonic acid contained in the minium is not complete can be assumed from the fact that ammonium is expelled without a trace of carbonic acid by means of this metallic calx. Atmospheric air is also capable of redintegrating incomplete fixed air, although not as well in lime as in lime water, which also develops a crust if it is brought in contact with a large amount of atmospheric air which has previously been washed with lime water.

It will be objected that *potash becomes caustic* as it transfers its fixed air to the lime. However, this can easily be answered according to what has been said above. It is certainly not necessary that the *potash* give off all its fixed air to the lime; it can still retain a minimal amount,—or, more correctly: *It redintegrates the incomplete fixed air of the lime, whereby its own is transferred to the state which that of the lime was in before.*

It seems to be more difficult, however, to answer the objection that nitrate of lime leaves caustic lime when decomposed by fire, but it has not yet been proved that nitric acid expels all the carbonic acid from the lime. Therefore, it would certainly be possible that it retained exactly the portion which makes it caustic. This is also supported by the experience that nitrate of lime still has the taste of quicklime. Muriatic acid combines with lime with greater difficulty but is also more difficult to separate from it. Therefore, it has a weaker attraction than nitric acid before combination and a stronger one after. This could be explained by the fact that muriatic acid had to overcome a resistance before the combination, which would not be the case with nitric acid. The matter could also be considered from another point of view since both carbonic acid and nitric acid contain andronia, as will be proved later, and therefore a change from one to the other could occur.

Incidentally, for the confirmation of the first view it is not inappropriate to remark that *acids frequently do not expel all carbonic acid from a combination with a base*, no matter how common this belief may be. By means of nitric acid, Wiegleb expelled 0.35 carbonic acid from carbonate of potash, by means of muriatic acid it yielded only 0.30, and by means of vitriolic acid 0.25. Therefore, an amount of carbonic acid still remained in the last two compounds. From carbonate of strontium, Buchholz[16] obtained 0.30 parts carbonic acid by means of muriatic acid but only 0.25 by means of nitric acid. From witherite he obtained 0.20 carbonic acid by means of nitric acid and muriatic acid, but with sulphuric acid only 0.14. After the witherite had been properly treated with vitriolic acid, and free acid was still noticeable although carbonic acid was no longer expelled, B. added muriatic acid, whereby a new effervescence ensued. Tartarite of lime (made with tartar and chalk) still effervesces with vitriolic acid. Priestley obtained fixed air from neutral salts. According to Kartheuser,[17] acetate of potash becomes crystallizable by means of fixed air.

In order to demonstrate a specific principle of causticity, we are now left with *the production of a non-caustic and yet non-inert base*. This is accomplished in the following manner. Baryt which is separated from its combination with nitric

[16] [Bucholz.]   [17] [Cartheuser.]

acid in fire is *insipid and non-caustic at the same time*. If this is dissolved in water and gradually added to a solution of caustic potash, a sediment appears which derives partly from the sulphuric acid, which is almost always combined with potash, but partly from the principle of causticity. This latter part of the precipitate is readily soluble in acids which contain hydrogen, but it is sparingly soluble in others. Caustic potash treated in this way, which could not previously be put into the mouth without the most painful sensation, can now be *kept in the mouth without any difficulty although it has not lost its alkaline taste*. On the other hand, this has certainly made the precipitated baryt *caustic though not completely so*.

The behaviour of alcohol towards acids is sufficiently remarkable to occasion an investigation as well.

If an *acid* is boiled *with alcohol*, it loses some of its acidity. If the acid is fixed, and the alcohol is distilled, the acid is left *with less acidity*, and the *alcohol is partially converted into naphtha* and thus brought *closer to the acids*.

(In order to explain this, let us imagine that all combustible liquids could be arranged in a series according to their properties. We want to start with the fatty oils. These are distinguished from the rest by their absolute insolubility in water, by their low volatility, and by the ease with which they dissolve bases. Due to this last property, fatty oils actually come close to acids. Upon distillation they also yield an acid which contains vital air without the addition of vital air, from which it can be concluded that these liquids do not consist merely of carbon and hydrogen but also of the substance of vital air and may be composed of oxide of carbon and hydrogen. Empyreumatic and, even more, ethereal oils deviate from this acid nature in varying degrees, and the most volatile among them can be combined with bases only if their nature is completely changed. The naphthas stand between the ethereal oils and alcohol, so the last of these comes closest to the bases in this series of combustible liquids.)

Tartaric acid was boiled with ten times as much *alcohol*, and this was distilled afterwards and transformed into a *gummous mass*, which was certainly very acidic and had a metallic taste at the same time, but it did not affect syrup of violet.

Boracic acid, treated with an equal quantity of *alcohol*, turned into a *grey, absolutely non-acidic* powder.

In order to transform alcohol into genuine naphtha, the acid must have an excess of acid principle, must be a fuming acid, or must at least slightly attract the acid principle, which occurs if a large quantity of the basis of vital air is combined with the acid.

With the addition of more acid principle, *alcohol* which turns into *naphtha* can be transformed into *oil*, the oil into *carbon*, the carbon into *fixed air* by means of added vital air. If it contains much acid principle, even atmospheric air is changed into fixed air.

It is remarkable that *alcohol expels part of the acid principle*, in the truest sense of the word, *without absorbing it*; for if oil of vitriol is placed in a confined space in which naphtha is produced from alcohol by means of a fuming acid, it turns into a crystalline acid which is hyperoxidized.

WATER IS NECESSARY FOR THE PRODUCTION OF MANY GASES

So far, this has been almost entirely overlooked in antiphlogistic chemistry, and yet it cannot be denied that this is a truth of the *utmost importance*. *Nothing* can be *driven off from sulphurous acid by distillation in the absence of moisture*. If four parts *sal ammoniac* are distilled with 3 parts lime, *only one part ammonia* is obtained, but if *steam* is introduced after the production has ceased, *a new quantity* of it is obtained. This can be repeated many times. *Carbonic acid requires a great quantity of water for its gaseous state*. During burning, lime does not release all this acid because the water required for the formation of gas is absent. If water is added, the lime, which has already been burnt once, produces gas again. As is well known, *witherite* releases its carbonic acid only very imperfectly in glowing heat, whereas this proceeds without difficulty if steam passes over it. The accuracy of a Priestley vouches for the correctness of this fact, and the later investigations of the profound Berthollet have demonstrated even more clearly the presence of water in carbonic acid gas.

SULPHUR

Sulphur plays the role both of an acid and of a combustible substance, although not with the same energy. Who would deny sulphur its acid property since it combines with basic substances and deadens them like an acid. However, it is an *insipid acid*, which can certainly be even more deoxidized, but, conversely, does not seem able to absorb more acid principle. However, it must be mentioned here that just as an acid is often resolved and therefore loses a component when deoxidized, conversely, it must reclaim a new component when redintegrated. This is exactly what occurs in the case of sulphur. We need only combine it with hydrogen, and it easily adopts more acid principle and displays all the properties of an acid.

If *liver of sulphur* (with a base of fixed alkali or earth) is *distilled with three times as much water*, the result, apart from *azotic gas* and a small amount of *sulphuretted hydrogen gas*, is a liquid which contains *deoxidized sulphur* and precipitates most metals with a black colour. With another triple, an even more *altered sulphur* is obtained, which in a larger quantity dissolves the precipitates again, especially those of nitrate of *mercury* (prepared at cold temperatures), of acetate of *lead*, of vitriolic *tin and copper* as well as of *gold and silver*. At first, these solutions are grey, later they become colourless (those with copper become blue), and a black sediment appears. In this process, the *acids* have released *acid principle* to the sulphur which is deoxidized in the second degree in order to restore it to the first degree; this becomes clear from the fact that several among them, especially *muriatic acid, now freeze as easily as water*, a property which they were far from possessing in their complete state.

ON ANDRONIA

I

THE PROPERTIES OF ANDRONIA

In all realms of nature there is *a hitherto unknown substance*, which has the following properties.

A. It is *acid* and combines with all bases except ammonium.
B. Unless it is in contact with vital air or another acid, it is *fire-resistant* and can be heated to incandescence in a crucible without evaporating if only it is covered with fully burnt soot.
C. *It forms synsomatic compounds with acids*, and
    a. *Thereby it takes away part of their capacity for bases* so that acids which have been combined with it can dissolve less of the bases than before;
    b. It is *not separated from acids by any base*;
    c. *In sulphuric acid* and perhaps also *in other acids, it reverses the affinity towards bases* so that the metallic calces are at the top of the order, followed by the earths, and, finally, the alkalis. This affinity is also that of pure andronia.
    d. *Combined with vital air*, it produces either
        α. *Azotic gas*, or
        β. *Carbonic acid gas*.
The former is obtained if *less*, the latter, on the other hand, if *more vital air* enters into the combination. In the latter case, however, there is also more *acid principle*, and *water* may also be an essential component of it as several of the above-mentioned circumstances seem to indicate.
        γ. *With a large quantity of vital air and acid principle*, it seems to produce *nitric acid*.
D. *With hydrogen it produces the substances which constitute the largest part of organic bodies*. The richest nutrients, milk, egg white, etc., contain it.
E. If it is glowed with *metallic calces* which contain little oxygen, it produces either *azotic* or *carbonic acid gas*. If more of this substance was present than was necessary for the reduction, a metallic calx seems to remain which, however, is a compound of this substance and metal, and which again produces azotic gas after having attracted vital air.

This substance may be given the name of *andronia* as we might later be able to discover another substance of a basic nature which would correspond to andronia in the sense that it would also be a component of most bodies and thus—as *thelyca*—constitute the contrast to andronia.

II

THE PRODUCTION OF ANDRONIA

According to the preceding, *andronia* must be contained in *carbon*. In fact, it is obtained if carbon is deflagrated with saltpetre. The remaining potash contains so

much of it that a portion remains on the filter.[1] However, this andronia is not completely pure but contaminated with sulphuretted hydrogen. It can be obtained in
larger quantities and in a pure state from *potash*. When this is extracted from wood
ashes by means of water, it contains the andronia of carbon. Very pure andronia can
be precipitated from this combination, perhaps even purer than from most other
substances.[2] First, siliceous earth, which is contained in all potash, must be separated from it, and this is done by placing the clear solution of this, prepared with
distilled water, in a cellar in order to saturate it with carbonic acid; thereby the siliceous earth is precipitated and forms a deposit on the sides of the vessel. The fluid
is decanted, filtered, and placed in a cellar so that all the siliceous earth is precipitated. When this has been completely precipitated, there are *two possible ways to
obtain andronia*; viz., either *by cooling* or *by adding acids*. If the *first* method is
chosen, the solution[3] is diluted with 4 parts distilled water (because it will not
freeze otherwise) immediately after the first filtration and, once it no longer absorbs any carbonic acid, which can easily be tested by means of an inverted flask
containing carbonic acid, it is brought from the cellar into air at a very cold temperature, and, as soon as the solution has assumed the temperature of the air, it is
placed in another equally cold vessel and surrounded by alternating layers of snow
and ice. Andronia is precipitated even at the initial temperature decrease, and the
more it decreases, the stronger the precipitation. However, the most complete precipitation occurs only when most of the water is frozen. The ice is then removed
with a cold, perforated spoon, and the remainder is passed through a cold filter,
where a large quantity of andronia is obtained although the solution which has
passed through the filter will be purer. If the *other* method is chosen, the solution is
allowed to remain in a completely concentrated state, in which case the precipitation of the siliceous earth is accomplished much more rapidly. During filtration, it
should still be kept at cellar temperature so that the andronia, which is best precipitated by a decrease in temperature, does not hinder the filtration by blocking the
pores of the filter. If the solution is poured into an acid, no andronia is precipitated,
but if an acid is added to the solution, it is obtained immediately. In the first case,
the free acid immediately dissolves the precipitated andronia again, and therefore
andronia cannot appear as a sediment. In the second case, the precipitation actually occurs not because of the added acid but because of the carbonic acid which is
thus removed, for caustic potash, with which as much andronia is combined as can
be found in ordinary potash, does not restore andronia either by freezing or by
means of acids, but as soon as it has attracted some carbonic acid, it behaves like
ordinary potash. Therefore, care must be taken not to lose the carbonic acid. Consequently, we must 1) select a cold temperature for precipitating andronia; 2) dilute
the acid so much that a thin stream of acid falling continuously into the solution
does not cause an effervescence; 3) a glass funnel which ends in a capillary tube is
suspended above the solution in such a way that the tip almost touches the fluid,
and that it can easily be lifted little by little as the volume of the fluid increases;
4) the diluted acid is poured into this funnel. 5) Meanwhile, the fluid is slowly but
constantly kept in motion with a glass tube because otherwise the acid would

---

[1] Scheele regarded this as silica.   [2] Pelletier regarded this as silica.
[3] This must probably be completely saturated.

simply remain in the upper part of the vessel; 6) once the lixivium starts to become noticeably cloudy, and andronia is precipitated as a gelatine, we cease adding acid as further separation will then take place by itself within a few minutes; 7) the filtration is to be done neither sooner nor later than when the upper part of the solution has become clear again, for if it is left longer, the carbonic acid starts to disappear again, and part of the andronia returns to the potash leaving another product behind, but the filtration cannot be started earlier either because then andronia could still be deposited in the pores of the filter and thus impede the filtration; 8) acid should again be added to the filtered fluid immediately in the manner described above in order to see if more andronia can be precipitated. When left undisturbed, a sediment can certainly be seen, but this is by no means pure andronia but a salt of this and *glucinum*;[4] 9) andronia does not allow water to pass through the filter and, consequently, cannot be washed in the usual way but must be ground in a glass mortar with pure water and then filtered as long as the water flowing off still precipitates saturated metallic solutions.

III

ON THE SYNSOMATIC COMPOUNDS OF ANDRONIA

Although *andronia* does not combine with *simple sulphur* (*sulphur simplicissimum*) at normal temperature, this combination can be achieved by digestion and produces a *solid sulphur* which melts into a transparent mass in fire. However, this sulphur dissolves *lead calx* and, when sublimed with this, produces *ordinary sulphur*. *Sulphuric acid* which is still occasionally prepared from sulphur is not quite the same as *vitriolic acid* but consists partly of a *quite special acid*, which we shall now investigate.

If *pure*, still moist *andronia is dissolved in oil of vitriol*, the former heats up with the latter and forms a *synsomatic compound*. It has a *mild taste almost like a concentrated vegetable acid* and attracts a portion of atmospheric air. If the acid is now distilled, azotic gas is obtained, and the acid is partly transformed into pure vitriolic acid again.

This acid *dissolves all metals* even if it is diluted but does so more vigorously in its concentrated state. However, saturation (neutralization?) is not accomplished, not even with the addition of metal calces. During dissolution, no gas develops at atmospheric temperature, but much more air is absorbed than by the acid alone; however, at a higher temperature, it produces gas during dissolution, but that is azote. Therefore, this dissolution of metals occurs without their prior calcination by vital air. Nevertheless, *this acid retains metals so tenaciously that not even caustic potash is capable of precipitating anything from it.*

Most *metals* which produce *crystals* with *vitriolic acid do not* produce any *with this acid* (gold, zinc, and bismuth are exceptions), but the addition of alcohol causes crystallization to occur.

---

[4] This salt can be decomposed either by separating the glucinum from it with nitric acid, or, even better, with caustic potash lye which dissolves a quantity of andronia in a couple of weeks if it is applied in very small quantities and assisted by motion, and which leaves behind that earth which can be dissolved in nitric acid and precipitated with caustic ammonium.

Those *metals which*, like iron, *emit vitriolic acid in fire retain this stubbornly*; *colcothar vitrioli* contains this acid.

However, *metals which*, like lead, *fix vitriolic acid, follow this in distillation*, to a certain extent.

Prussiate of potash certainly precipitates all metals from this acid, but if it is diluted, they dissolve again in the added water.

Metals calcined by vital air certainly *dissolve in this acid*, but at a moderate temperature they produce a large amount of *azote*, and the acid is changed into *ordinary vitriolic acid. Metallic calces which are produced by andronia*, however, for the most part form *the same compounds with vitriolic acid as if the andronia had been combined with it previously*. But the andronia which is a component of a metal, with the exception of steel and tin, and belongs to it *as a reguline part*, does not absorb vitriolic acid as far as we know, and the precipitates of prussiate of potash from such metals are not dissolved again by water.

On the basis of its indicated properties, this acid deserves to be regarded as a very special acid, and its preference for metals entitles it to the name of *metallophile acid*.

Among the phenomena which *this acid produces with metals* in solution, we emphasize the following:

The *gold solution* has a double colour, by reflection it is an opaque yellowish grey and by refraction a transparent green. By means of evaporation it first produced crystals of the same colour (Could it be platina?), but afterwards the solution turned clear as water, and the crystals also became colourless.

The *silver solution* is not precipitated by liver of sulphur.

The *copper solution* leaves a metallic coal with a volume larger than that of the copper itself, and it has no colour.

The *solution of tin* is obtained more easily than in vitriolic acid, it is colourless and can hardly be distinguished from that in vitriolic acid; it is also precipitated by a solution of caustic potash (even a very dilute one), and it thus provides an exception to the general rule.

Next to metals, metallophile acid is most closely related to calcareous earth, and its solution cannot be decomposed by caustic potash and hardly by weak potash. This combination is soluble in muriatic acid and can again be precipitated from it by weak potash, whereby it absorbs carbonic acid but does not lose its metallophile acid. When edulcorated and dissolved in boiling water, this salt produces inert prismatic crystals. It reddens syrup of violet because lime, like all bases, is not absorbed by the androniated acid in the amount necessary for neutralization. Saccharic acid alone decomposes it.[5]

It is easy to determine the manner in which *vitriolic acid can be separated from metallophile acid*. A mere distillation is rarely sufficient, in spite of the fact that a large quantity of andronia is driven off in the form of nitrogen gas. A distillation with an appropriate quantity of iron filings is better. The *surest* method, however,

---

[5] *Stalactites* contain *andronia*, which changes the added vitriolic acid into metallophile acid. *Gypsum* which has been precipitated from tartarate of lime by means of vitriolic acid also contains andronia; the metallophile acid portion of it can be separated by washing and leaves only 0.4 undissolved substance behind.

is *to saturate the acid with iron* and allow it to *crystallize*, which does not occur if only a small amount of the metallophile acid is contained in it. If the *crystals are distilled*, they produce *pure vitriolic acid*, and the *metallophile acid is retained by the iron*, so acid which has been distilled from vitriol is always free of andronia, but acids which have been prepared from sulphur in metallurgical and in pharmaceutical laboratories are equally unacceptable.

Nitric acid can be distilled with *andronia* and then produces a *characteristic acid* which combines with potash to form a salt without becoming cloudy, but which again deposits andronia during digestion.

Muriatic acid willingly combines with *andronia* and produces with it a *new acid* with a sweet-sour taste which, when boiled, partly changes into a stable gaseous state and leaves the water.

It crystallizes with *calcareous earth*, and during a slow digestion a very large portion of this compound is lost due to evaporation.

The combination of this acid with *potash* can be dissolved in alcohol and changes into a gaseous state at a higher temperature provided only that water is present. This gas can be dissolved very easily in cold water.

It seems to attract *vital air* from *atmospheric air*, for in a confined space with atmospheric air it causes a strong absorption.

Zinc dissolved in this acid and thereby produced a combustible gas which burned with a green flame whose lower edges were blue, but the upper ones yellow. When left over lime water for several days, it yielded a precipitate which was insipid potash, so the andronia must have passed from the air into the lime water.

As *modifications of this acid* are to be regarded:

1) The *acid of the blood*. If blood is mixed with potash, boiled down, and charred without being made to glow, a neutral salt can be extracted from it with a little water; the salt contains this acid combined with a great deal of andronia, a part of which is precipitated by alcohol as it dissolves the salt. This blood acid contains so much andronia that it is precipitated from the neutral salt with the consistency of cheese.

2) *With prussic acid gas, blood acid produces genuine Prussian blue acid,* which Scheele confused with this gas itself. These *two acids can be separated from one another.* Prussiate of potash or Paris blue (prepared without alum), with a large amount of vitriolic acid, produces muriatic acid, which comes from the blood acid. Vitriolic acid now replaces the muriatic acid and, combining with andronia, it yields metallophile acid, and now the Paris blue becomes soluble in water. If the potash in the prussiate of potash is completely saturated with vitriolic acid, *prussic acid* is liberated without being resolved, but it remains like that for only a short time, for it gradually allows its prussic acid gas to escape and thus also loses its capacity to precipitate metals, and with the vitriolic acid potash only blood acid remains, which releases its excess andronia partly as a gelatine. The *synthetic* production of this acid (by Clouet) was made with charcoal, and this contains pyroligneous acid, which partly remains in the ashes despite the volatile nature of its neutral salt.

3) *Pyroligneous acid* is obtained especially from bark which has no aroma. When Göttling had saturated 4 ounces $1^1/2$ drachms of potash with this acid, he obtained only 3 ounces 1 drachm after evaporation and crystallization, from which it can be concluded with fair certainty that half has escaped as gas. (See Crell's *Journ.*, Vol. 2, p. 55.)

4) *Sebacic acid.*

5) *Gallic acid* is modified both by the associated benzoic acid and by the pigment which turns iron black. Extraction from gallnuts by a solution of weak potash produces a *neutral gas* during distillation and a *neutral salt*, which can assume the same form as soon as it is dissolved again in distilled water and distilled once more. The saccharic acid which Scheele obtained from this with nitric acid did not derive from the volatile acid but from the pigment. After the dissolution of gallic acid in open air, benzoic acid, together with the pigment, remains as small shiny crystals.

Vinegar does not dissolve *andronia* but merely extracts potash, which cannot be separated from it completely in any other way.

Vital air seems not to have so much effect on andronia as atmospheric air, for when andronia was enclosed in it and kept for 8 days at a temperature of 130° F, only $^1/10$ was absorbed. When this experiment was made over mercury, the reduction was even smaller, but as soon as lime water was added, it became cloudy, and in this case the reduction also grew to approximately $^1/10$. Nor did a candle burn quite as well in the remaining air as before the experiment, and it also extinguished sooner, so a little nitrogen gas must also have been produced. The andronia seemed to be only slightly modified on the outside, but in open air, where previously it would have evaporated except for the portion saturated with potash, it was now stable.

IV

ANDRONIA WITH HYDROGEN AND OTHER
EASILY COMBUSTIBLE BODIES

Hydrogen gas is modified by andronia. If it has been in contact with andronia, it burns with a perfectly *green flame.* This phenomenon has also been observed in the case of hydrogen gas which was mixed with nitrous acid. When the gas which had been expelled from a bladder was ignited, the flame forced its way into the bladder itself and caused an explosion. This can be explained by the fact that the water with which the bell jar was blocked while the andronia was exposed to the influence of the gas attracted air from the atmosphere at a lower temperature, which it had to give off again when the temperature increased. The *andronia had been changed significantly,* for it had *lost its semi-transparency* in addition to the fact that it had now become *stable in the atmosphere,* but it had *not become soluble in water.*

Alcohol does not dissolve *andronia* but makes it *opaque* and *stable in the atmosphere,* in which respect it is similar to hydrogen gas, but it communicates more hydrogen to it, whereby it obtains *an even weaker solution in water.*

Dry *sugar,* ground with *andronia* which has been dried to a caseous consistency, produces a transparent *honey* which, when dissolved in water, makes a *genuine milk* which also passes through a filter as such. This milk did not change when in contact with the atmosphere for 14 days, but in the third week it changed to a *caseous fermentation* and coagulated into a *genuine cheese,* as the smell indicated. Part of it separated immediately when freshly treated with vinegar, and the remaining part became gelatinous after some days. Now we understand why mineral acids precipitate less cheese from animal milk than vegetable acids since the former dissolve part of the andronia which the vinegar cannot dissolve. Freshly precipitated and purified *silica,* ground with *sugar,* yields a *similar honey* and a milk which the filter, however, does not pass; rather it extracts from it a fluid clear as water, from which alcohol dissolves some sugar and precipitates some genuine *India rubber.* The silica is retained on the filter and yields some insipid potash with added acids.

Oil ground with andronia yields a water-soluble mass, whose solution is white, passes through a filter, and is *not dissimilar to the milk described above. One ounce of andronia distilled with 100 drops of oil* until the complete glowing of the residue produced pure water and approximately 40 cubic inches of combustible gas, an amount which was far less than that which would have been obtained if this quantity had been passed through a red-hot iron pipe. The residue was the *purest carbon,* with respect to smell and form it corresponded completely to soot which has been calcined in closed vessels, it produced nitrogen gas when combined with vital air before glowing, but it changed into carbonic acid gas during glowing without any residue. Therefore, *carbon is not a simple substance* but a combination of *hydrogen* and *andronia, oxidized* by glowing,[6] and in this state it is more inclined than usual to produce carbonic acid gas. The composite nature of carbon is further confirmed by several facts. We have seen from what has already been said that carbonic acid cannot be expelled from its combinations with bases without the presence of water, so we were forced to assume that water is inescapably necessary for the gaseous state of carbonic acid, and that, in this state, half of it consists of water. However, combined with vital air pure carbon gives carbonic acid gas, and therefore it must contain so much hydrogen that the necessary amount of water can be produced. From what we have now discovered regarding the nature of carbon, it is easy to explain the differences between the various forms of carbon. There is *charcoal,* which is either smothered or burnt too much or extinguished in air; burnt *mineral coal (coke)*; *anthracite* (barren coal), which does not burn; *animal coal,* which can scarcely be reduced to ashes; *graphite*; Priestley's *metallic carbon*; and his *white carbon*; and, finally, the *diamond,* which is not true carbon only because of the lack of acid principle. These are not distinguished so much by admixed foreign parts but rather by the different quantities of saltpetre which they can decompose, i.e., *by the different quantity of the principle which attracts vital air,* which cannot be andronia because the acid in saltpetre is already saturated with it. If we begin the series of carbons with graphite, which requires 10 parts of saltpetre for its deflagration, then step down to the kind of coal which is quenched in air, and which can

---

[6] Perhaps this excess of acid principle is the reason why carbonic acid gas is so much more harmful than most of the other non-respirable gases.

change sulphur into an acid, and finally end with diamond, which does not deflagrate at all, then the different hydrogen content of carbons becomes sufficiently obvious.[7] The production of metallic carbon during the dissolution of several metals in dilute acids, whereby water is decomposed (this is why it does not happen in nitric acid or oil of vitriol), also strongly supports the theory presented here.

V

NEUTRALIZATION OF ANDRONIA

A solution of caustic *potash* absorbs a very large amount of *andronia*, but this process takes a long time. This salt has a very mild taste, freezes almost as easily as water, and dissolves in alcohol while depositing part of the andronia. It does not crystallize when boiled down but yields a crystal-clear gelatine; when dried and glowed, this becomes a white mass which is insoluble in water and in acids, a *genuine siliceous earth*.

If we shake *andronia* with *lime water* in a stoppered bottle for a long time, it loses its transparency, and its volume increases. If the liquid is poured off, and it is treated with lime water again, the result is the same until finally the andronia cannot change any further. The portions of lime water which have acted sufficiently on the andronia do not develop a skin when exposed to air and are not made milky by oxalic acid or by complete carbonic acid potash. When nitric acid is added to this liquid, beautiful *saltpetre crystals* are obtained in an amount too large to derive from the potash present in the andronia (prepared according to the method described above); therefore, *potash has been produced.* After this, irregular crystals of another kind appear which still taste sour even after edulcoration and contain *insipid potash* as a basis, which can be seen clearly from the fact that complete carbonic acid potash produces with them a precipitate which is soluble in water as well as in vitriolic acid with effervescence, but which can be extracted only incompletely from this compound by potash and not at all after being boiled down.[8] From the residue on the filter, nitric acid extracts *insipid potash* and a small quantity of unmodified lime and leaves something insoluble, which is *genuine silica.*

Muriatic calcareous earth dissolved in a small amount of water (*oleum calcis*) changes *andronia* as well but also absorbs some of it and thus becomes crystallizable. After proper examination, however, the altered andronia produces *insipid*

---

[7] The *ability of coal to smolder* is not taken into consideration here, for this phenomenon seems to derive neither from the purity nor from the hydrogen content of carbon but from a stimulating principle (of a basic nature?), for when Monokiensian anthracite was deflagrated with an equal portion of saltpetre, and the edulcorated residue had been placed to dry on an oven, it began to smolder at this moderate temperature despite the fact that the deflagration must have diminished the hydrogen content. The observation by Guyton that coals which do not smolder are distinguished from those which do in that they behave like the less oxidizable metals in the galvanic circuit may be of even greater importance.

From the above, it can be seen that W. cannot be satisfied with Guyton's theory of the varieties of carbon. He has demonstrated this further in a separate paragraph on diamonds, and there he raises several objections which show that he, as well as Berthollet, wishes this topic to be investigated far more precisely.

[8] Because insipid potash is redintegrated when boiled with oil of vitriol.

potash, lime, and *silica*. The situation is slightly different if the *muriatic calcareous earth* is precipitated by means of *androniate of potash*. The precipitate, edulcorated and glowed, dissolves in muriatic acid with effervescence (because of the carbonic acid produced) and leaves only a little silica. It also dissolves in potash saturated with carbonic acid, but the saline nature of this precipitate is clearly revealed by its perfect solubility in water. Therefore, *with andronia the lesser part of the calcareous earth has transformed into silica*, whereas *the larger* has become *insipid potash*, but it is remarkable that this insipid potash was so *decomposable* before glowing that some added oxalic acid began to make it cloudy after some minutes and later produced a precipitate in it.

It would now be particularly interesting to see whether *siliceous earth could be changed back into potash*, and this has also been accomplished to a certain degree. Although only *an extremely insipid potash* is obtained during the operation which is to be described, it is nevertheless sufficient to confirm the theory offered. For this purpose, it was necessary to employ a substance which kept atmospheric air out and, on the other hand, supplied hydrogen, and soot was chosen for this purpose. Pure siliceous earth precipitated from gelatinous silica was washed with vinegar, mixed with soot, dried well, and placed in a crucible where it was covered with a paste of clay and then heated gradually to incandescence, which was maintained for half an hour. A very black coal remained, which was now reduced to ashes. The soluble part was extracted from the ashes by water, but it did not react like potash with reagents, except in so far as it coloured syrup of violet green and expelled some ammonium from its combination with muriatic acid at a higher temperature (something which it could not do at a lower temperature). Muriatic acid was poured over the residue of the aqueous extraction, whereby a noise was produced which persisted for two hours. The muriatic acid was not changed perceptibly, but nevertheless complete carbonic acid potash precipitated from this as a white coagulum which clearly dissolved in water to some extent without any taste other than that of the added digestive salt, so it was *insipid potash*. The edulcoration had to be performed by adding very small portions of water so that not all of it was dissolved. When this was completed, dilute vitriolic acid was added, whereupon it effervesced but dissolved only partially. The undissolved substance was gypsum, but it had the special property that its solution coloured an ear of corn dipped into it a rosy red. Consequently, the *insipid potash* had already been *partly decomposed* in this case. The dissolved part was boiled down and produced an irregular salt. This was boiled with distilled water, whereby part of the salt was dissolved; this, together with complete carbonic acid potash, in turn produced a precipitate which was soluble in water and therefore had to be insipid potash. The undissolved part was found to be silica. Therefore, the potash was insipid to the extent that it would not only not allow itself to be redintegrated by boiling with vitriolic acid, but it was even decomposed so that first a part of the *lime* precipitated and gave off its andronia to the potash, which then, however, partly produced silica with it. This is an order of transformation which we would not have expected since andronia otherwise passes from potash to lime, which happens if the former is made caustic by the latter, for caustic potash saturated with carbonic acid does not yield andronia.

Could it be that the andronia of the potash is merely used to redintegrate the carbonic acid of the lime? This could almost be assumed from the large amount of potash which is necessary for the redintegration of the resolved carbonic acid in lime.

Although *baryt* certainly combines with *andronia*, no remarkable combinations are obtained thereby.

Magnesia is precipitated from its solution in vitriolic acid by *andronia* as the andronia combines with it to form *a salt which is almost insoluble in acids but readily soluble in caustic potash*. The magnesia gradually settles from this solution as the potash assumes control of the andronia, from which it can be concluded that potash has more affinity to andronia than magnesia.

In addition to several phenomena which *alumina* produces with *andronia*, it is particularly remarkable that it is required for the *crystallization of alum*, for a small amount of caustic lye mixed with alumina gave nothing more than a gummous mass with vitriolic acid, but when this was mixed with andronia which had been purified of its potash by vinegar, it yielded beautiful alumina crystals after proper digestion. Since alum always contains andronia, it is natural that it is made to crystallize by means of ordinary potash in factories.

In an *acid solution of gold, androniate of potash* produced a strong precipitate which was separated by means of a filter, whereupon, within a few days, the solution deposited a new precipitate, which was mixed with the first. The solution was concentrated and acquired a greyish-red colour like that of platina, so there was some hope that *gold might be changed into platina*. The portion which had been separated by means of the filter had the semi-transparent appearance of andronia and had a faint golden yellow colour. This was then dissolved in caustic potash and precipitated once more with gold solution, whereby a new precipitate similar to the previous one was produced; this was again treated in the same way, etc., until no further precipitation occurred. Everything had now turned into a liquid, which was first boiled down and then glowed; however, it did not produce any platina but really *reduced gold*. A very large quantity of neutral salt was to have been expected, but only a small amount of white and compact matter remained, from which water extracted a yellow tincture merging into green. The residue did not dissolve easily in water until it boiled, but the solution yielded hardly anything but saltpetre crystals and no digestive salt, which had probably escaped as a gas.

Andronia completely purified of potash was added to a *platina solution*, which caused the same precipitation as otherwise with potash or ammonium, of a beautiful orange which later turned minium red. This again gave rise to the speculation that platina might turn into gold, but it was not confirmed by the second investigation.

From a solution of *nitrate of silver* in water, *androniate of potash* separated a large quantity of *precipitate*, which was chalky *white* and was *not blackened by sunlight*, but which dissolved at cold temperatures in dilute muriatic acid and in caustic potash lye by means of digestion. The solution in muriatic acid was not precipitated by copper. The solution in potash is first red and later pale yellow, and it also yields crystals of this colour. Glowed in a glass retort, the same precipitate gives *azotic gas*, but if the solution of nitrate of silver has first been boiled down for crystallization, it produces *carbonic acid gas*. In this, it undergoes no further change in colour except that the part which is in contact with the retort turns red. When glowed with soot, it is not reduced but yields a *metallic carbon* which is somewhat shinier than the ordinary kind. When dissolved in caustic potash, placed in a retort, and glowed until the retort begins to melt, it does not reduce but turns into a *hard, white mass which adheres firmly to the glass* although it remains soluble in water. When linseed oil is poured over it, this mass is finally *reduced* by glowing. Now nitric acid easily extracts the metal, and muriatic acid easily precipitates it again from the solution, but this does not occur with dilute sulphuric acid, and copper precipitates it with a white, non-metallic colour.

If *mercury* is ground with slightly moistened *andronia* in a glass mortar, it first assumes the colour of ashes. During the grinding, it dries and becomes *almost white*. This powder dissolves in water and makes it slightly milky, but it passes through a filter unchanged. If this powder is *distilled* into a receiving vessel containing *lime water*, it is covered with a *reddish skin* with an almost metallic lustre. Later on, when the mercury rises into the neck of the retort, the *lime water becomes quite milky*. However, despite the fact that part of the mercury is forced upwards, most of it remains in a fixed compound with andronia as *a white mass with a reddish tinge*, but if the andronia has not been completely purified of potash, the colour is *a more saturated greenish-red*. If a mercury solution prepared at a cold temperature in nitric acid is added to the water with which the andronia has been edulcorated (see above p. 142, No. 9.), and which therefore contains andronia and weak potash, then far more of this is absorbed in successive precipitations than could have been expected from the small quantity of base contained in it. The first precipitate is black, the following ash-coloured, the third whitish, the fourth white though it becomes yellow by digestion, but the fifth becomes yellow immediately. All these precipitates are soluble in water. The black gives no precipitate when dissolved in water, neither with nitric acid, caustic potash, or liver of sulphur, nor with prussiate of potash, not even with acid added, nor with nitrate of silver or lime oil (dissolved muriatic calcareous earth). Left undisturbed, this black precipitate becomes lead-grey once the water has evaporated, and later it reduces even more until finally a very small amount of a non-metallic black residue remains. The yellow precipitate behaves in the same way except that it can much more frequently[9] be dissolved in warm water, but the solution precipitates part of it again unchanged with cooling. With caustic potash it is turned into a black precipitate. When boiled with water in a retort, it is partly reduced, but the *reduced mercury is far more volatile* than the ordinary kind, despite the fact that otherwise it does not differ noticeably from it; however, it can be *distilled with water*, indeed, it is even *more volatile than water*, for it is evaporated again by the steam passing over it and settles between the neck of the retort and that of the receiving vessel and for the most part probably passes into the air. The part which remained in the receiving vessel *floated in the water* and even settled after being shaken, but with an inclination of the vessel and when brought into contact with the atmosphere, it floated again as

[9][easily.]

often as the experiment was repeated. All these *mercurial precipitates* have a very insipid, slimy taste and may be *profitably applied in medicine* if very mild mercury preparations are needed.

A *solution of the sublimate* saturated with *androniate of potash* gives a *precipitate of a greyish-red colour*. After caustic potash lye was poured over this in a retort, it soon changed to a *yellow* colour and yielded some *reduced mercury* as the temperature was increased to red heat. The *residue* in the retort was very white, little soluble in water, and almost without taste. It differed from silver treated in a similar manner by its *brittleness* and by the fact that it did *not* yield *even the smallest amount of reduced mercury* when glowed with linseed oil but merely produced *a carbon* which was shinier than others so that the part of the retort which had been covered with it on the inside gleamed like a *mirror*.

Lead yields various *products* with *andronia, sometimes acid, sometimes basic,* depending on the different nature of the stimulating principle.

If *potash saturated with andronia* is added to a *solution of nitrate of lead,* a precipitate is obtained which 1) dissolves in nitric acid and crystallizes with it whilst becoming cloudy. This salt does not taste sweet and is not precipitated by muriatic acid; 2) it is only changed by caustic potash like pure baryt by its principle of causticity; 3) it extracts vitriolic acid from potash and leaves pure potash, which precipitates magnesia from its sulphuric acid solution in just as gelatinous a form as ordinary (non-caustic) potash; 4) glowed with soot, it produces no trace of reduced lead despite the fact that the carbon becomes so shiny that it makes the retort like a mirror wherever it is in contact with it; 5) dissolved in muriatic acid, it covers an added piece of zinc with a metallic, lead-like crust which dissolves only very slowly in nitric acid (in 8 days during summer), and concentrated muriatic acid added to this solution causes no precipitation, but it remains perfectly clear for several days; however, one quarter of an hour after the addition of some dilute sulphuric acid, it begins to yield a precipitate, which is somewhat increased after several hours. This precipitate, however, is somewhat soluble in water because of the incomplete saturation of the metal with andronia. Because it could not be precipitated by muriatic acid, it was not lead but *baryt*, and the metal which was deposited by the zinc was genuine *metallic baryt*.

As a *control experiment*, muriatic baryt was brought into contact with zinc, but no precipitation occurred until it had digested for 6 days, and some free muriatic acid had gradually been added, but now a crust formed which, when dissolved in nitric acid, was precipitated neither by muriatic acid nor by sulphuric acid.

Minium and *andronia*, which had a caseous consistency when ground together, dried, and distilled in a glass retort, first produced *azotic gas* as pure andronia usually does and later *carbonic acid gas* for an entire hour, but during glowing only pure *vital air* was produced, probably because no more water remained. The *residue* (A) had a white colour, but the surface was red; the mass did not melt. With caustic potash lye, part of it dissolved and crystallized, but another part remained and again assumed the red colour which it had lost during glowing although it was no longer as bright. This part (B) dissolved partly in nitric acid and was again precipitated from it by muriatic acid. Consequently, it was *lead calx, whose andronia had been extracted by the potash.* During the dissolution of this residue in nitric acid, something remains which is more insoluble and has a black colour, and there are also small grains of metallic lead in it which remain after the lighter lime has been washed off. From this part of the lead calx, therefore, the andronia had removed the part of vital air which is required for calcination, but with the lead, on the other hand, it had produced *a different variety of calx*, from which the potash, however, extracted andronia, and thus the lead regained the reguline state. The *residue* (A) does not dissolve in dilute *nitric acid*, but with *concentrated acid* it produces a *synsomatic acid of a particular kind*, which fumes vigorously, is soluble in water, and is not precipitated by vitriolic acid. With caustic potash or ammonium it produces salts which are soluble in alcohol. It expels carbonic acid from a solution of complete carbonic potash and precipitates andronia. Prussiate of potash gave the concentrated solution an olive colour, but the dilute one was blue. *Vitriolic acid* dissolves *residue A*, and the solution has less colour than the acid used. Water decomposed it into *two parts*, the first of which remained liquid and had all the properties of *metallophile-acid lead*, whereas the second was solid and had all the properties of *heavy spar.*

Black iron calx, produced by grinding iron filings with water, first gave *azotic gas* with *andronia* in fire, then a little *carbonic acid gas*, and finally *azotic gas* again. The *residue* was mostly soluble by *vitriolic acid* and formed *metallophile acid* with it. At first, prussiate of potash yielded no precipitate, but when more was added, a white precipitate was obtained which was immediately dissolved by additional water. Upon dilution with water, it assumed a green colour which became blue in contact with the atmosphere but without a sediment. Caustic potash caused no precipitation. Most of it also combined with *nitric acid*, and the *residue* seemed to be merely an excess of *andronia*. From this solution complete carbonic acid potash deposited a precipitate of a pale reddish-grey colour, from which, however, the excess of added potash again extracted the colouring substance, which could not have been there in a large quantity because the residue did not lose much of its volume during the process, but it became entirely white and was completely insoluble in potash. Glowed in an iron spoon, it assumed a rust colour. (Could it be that it was *magnesium*?)—

If *steel* is subjected to the same experiments as *androniated iron*, the *results* will be the *same*, but it must be pointed out that *even the best English steel contains a little pure iron*, which is revealed by the small precipitate which is produced by its solution in oil of vitriol with caustic potash and with prussiate of potash. The same can be seen when steel is dissolved in nitric acid, where the added carbonic acid potash again dissolves more than from androniated iron, but it hardly amounts to 0.05. The remaining 0.95 is genuine steel, which is absorbed in oil of vitriol without the generation of sulphurous acid (the very small amount which can be observed comes from the pure iron), does not impart any colour to it (except for a very faint tinge from the iron contained in it), and is precipitated with a white colour by prussiate of potash (the very pale blue is a result of the admixed iron), but this precipitate is soluble in water (apart from a small remainder which comes from the pure iron). This solution is not turned cloudy by caustic potash (except for the 0.05

iron). Dissolved in nitric acid, it leaves no andronia; this passes to the iron. Saturated with carbonic acid potash, it also gives a white precipitate, which is not soluble if more is added, but it is dissolved without any colour by all acids. (The greyish-green, which can be dissolved again by an excess of potash, and which changes into a bright red colour in air, comes from iron.)

In all chemical circumstances, therefore, steel corresponds to androniated iron, so only the actual *decomposition* of steel still remains. For this purpose, steel is dissolved in muriatic acid, and the solution is precipitated with zinc by digestion. *Andronia* is produced, which makes the solution cloudy, and it settles abundantly when left undisturbed.

If *androniate of potash* is added to an acid solution of *copper,* a white coagulum appears which has a tendency towards a bluish colour. The edulcorated *precipitate* is distilled with 3 times as much *nitric acid* until only very little fluid remains. The liquid part, which also yields blue crystals, is merely nitrate of copper. On the bottom, however, *large, irregular crystals* are found which have the colour and the texture of dental enamel, do not dissolve in nitric acid, but are absorbed in warm water, though only in the smallest amount in which any known soluble acid can be absorbed, so that it makes solutions of nitrates of baryt, silver, and mercury only very slightly milky and eventually produces a mere trace of precipitate. From muriatic calcareous earth this solution does not yield any precipitate at all. These crystals easily dissolve without any colour in caustic potash and ammonium, but they are precipitated again by all acids, even vinegar. The combination with ammonium, without an excess of the latter so that it has hardly any odour, is not precipitated by the sublimate of mercury or complete muriatic iron calx, so it is prepared for precipitations with a little additional ammonium. Both solutions precipitate gold white with a tendency towards the yellowish, and also magnesium though with a faint tinge of grey, but lead, zinc, tin, and baryt with the whitest colour. Similarly, lime, magnesia, and alumina, which were gelatinous and semitransparent. The solution from which magnesia was precipitated by the ammoniacal solution was blue but lost its colour in air. The sublimate of mercury is also turned white by the ammoniacal solution but yellow by the potash solution and greyish-red in a larger portion. Mercury, dissolved in cold nitric acid, is precipitated with an ash-grey colour, which turns black soon after, but the precipitate is disinclined to settle. Iron was precipitated ash-coloured from a muriatic solution by the potash solution, but the next day it became yellow; the ammoniacal solution precipitated it grey. Green vitriol, however, was precipitated blue by the potash solution, but when the liquid of the precipitate was poured off, and it came into contact with air, it turned white. Silver was precipitated yellow from nitric acid by the potash solution; it did not become cloudy with the ammoniacal solution but only pale blue, a colour which it soon lost again in air. Afterwards, however, it was turned cloudy neither by the potash solution nor by a saline solution. Baryt which was precipitated from nitric acid either by pure acid or by the ammoniacal solution dissolved in water, but with the potash solution a precipitate was obtained which was dissolved only by means of some added vinegar. Among the other precipitates, vinegar dissolved those of calcareous earth and alumina, but not that of magnesia;

that of completely oxidized iron, but not that of green vitriol, from which it absorbed only the more calcined part with an orange colour while leaving behind the less calcined part, which was entirely white. The same was the case with mercury, for it dissolved the precipitate from the sublimate with the potash solution without any colour although it was dark reddish-grey. On the other hand, it only partly dissolved the precipitate of a solution of nitrate of mercury, prepared at cold temperature, and left something which was blacker than the rest. However, it could not dissolve the precipitate from the sublimate obtained with the ammoniacal solution because of the *sal alembroth* which had developed. The partial dissolution of the precipitates of lead, copper, silver, and zinc seemed to have the same causes, for the residues were whiter. The precipitates of gold, tin, and magnesium were not noticeably diminished by vinegar, but that of tin became milky, that of gold yellow (with a very white residue), that of magnesium grey, and thus it revealed its activity as well as by a weak precipitate with potash. The part of green vitriol which did not dissolve in vinegar dissolved abundantly in water without any colour. From this, it becomes sufficiently clear that *androniated copper calx is genuine molybdic acid*, but *purer than Scheele's acid* which, prepared by distillation with nitric acid, still contains copper and, therefore, shows a slight difference in colours.

Molybdenum is also obtained *from copper* in the following experiment. *Blue vitriol* is dissolved in water, the solution is precipitated with potash, the sediment is edulcorated, a solution *of potash completely saturated with carbonic acid* is poured over it, and a vessel filled with the same carbonic acid is inverted over it; this now absorbs such an immense quantity of carbonic acid that, if the vessel is filled with *new carbonic acid gas almost daily for 3 months*, it is still found to be entirely absorbed by the following day. During this protracted work, however, the potash assumes a darker blue colour every day so that it finally appears *almost black*. If this tincture is boiled, it assumes an *inky black* colour for a moment, and immediately afterwards it deposits a *sediment* above which the liquid remains clear as water. This sediment is a genuine *molybdenum regulus*; it has a metallic lustre, is fusible in fire, dissolves in nitric acid with a green colour which changes into blue in air. *Distilled with a large quantity of nitric acid*, it becomes *molybdic acid* and is insoluble in nitric acid. The liquid, clear part is potash completely saturated with andronia. It was easy to understand that carbonic acid could produce andronia here, but whence came the basis of vital air, whence the acid principle? It is to be hoped that these questions will find their answers in the future.

Among all the acids, only *vitriolic acid* is capable of dissolving tin at normal temperature, but in order to obtain a *saturated solution*, it is necessary to distill off the excess acid. During this process, the vitriolic acid is for the most part transformed into a solid sulphur although it is completely free of andronia, which provides sufficient proof that *tin contains andronia*. Another product of this dissolution is *metallophile acid*, which remains after the vitriolic tin is dissolved in water and coagulated by boiling, another indication of andronia. So is the *ammonium* which is produced during this process. It is true that so far this has been found only in nitric acid derived from a two-fold decomposition of nitric acid and the water contained in it. However, acid and water are not decomposed simultaneously in any metallic solution. In strong nitric acid and above winter temperature, a strong nitric acid is decomposed only by tin. At a low temperature, however, and with a dilute acid, only the water is decomposed, and the hydrogen produces ammonia with the andronia of the tin.

In order to *decompose tin even further,* *vitriolic tin* was precipitated with complete *carbonic acid potash,* the edulcorated *precipitate* was mixed with water, and it was then exposed to a flow of *sulphuretted hydrogen gas* (prepared from pyrite and muriatic acid) for 24 hours. At the end of this time, a *heavy granular substance* appeared at the bottom and a *lighter* one floated in the water. The heavier one dissolved with relative ease in water, but the *lighter* was almost insoluble. In order to see whether the *first substance* was tin, it was glowed with soot for $^1/_2$ hour in a crucible which had been well-sealed with a paste of clay. After two days the crucible was smashed in such a fortunate manner that the heavy granular matter was knocked off completely by itself. However, it had been in contact with the atmosphere for hardly half a minute when it ignited spontaneously without any explosion and produced a flame in the shape of a fountain, and the part of the crucible which had contained this matter was left completely empty. The glowing pieces flying about were found to be pure soot. Therefore, the heavy part, which was soluble in water, *was not tin but only one part of it.* This part was produced once again and distilled with red mercury calx in order to precipitate all the andronia. The mercury calx was reduced, and carbonic acid gas and vital air developed. The *residue* was white and soluble in a large quantity of water, whereby, however, a red precipitate, previously unnoticed, was deposited. It is even more soluble in alkalis, which proves its highly *acid nature,* and it leaves the same red residue. The aqueous solution was sufficient to precipitate lime water and several metallic solutions, and with the former it yielded a white precipitate which dissolved only partly in boiling muriatic acid. Gold was precipitated white with a tendency towards yellow, silver white, nitrate of mercury prepared cold greenish changing to black during digestion; mercury sublimate, however, precipitated greyish-red in a small amount, white in a larger amount, copper from vitriolic acid white with a tendency towards greenish-blue, but from nitric acid even more so, iron from vitriolic acid white, but strongly calcined, from muriatic acid grayish, and zinc and baryt white from muriatic acid. The precipitates which had a colour other than green were turned white by vinegar, and their colour was transferred to it. However, the precipitates from gold solution and cold-prepared mercury solution became violet when digested with vinegar. The latter solution, however, became white again with extended digestion.

Diluted *vitriolic acid* combined by digestion with the *above-mentioned* residue, which was identified as an *acid* because of its behaviour towards bases, and when cooled it produced *large,* colourless *crystals.* Approximately the same happened with muriatic acid, but it did not crystallize and had a yellow colour which disappeared again when a grain of tin was added. The undissolved *residue* remained white.

From all this, it can be concluded that *this acidic substance or acid which has been precipitated from tin is tungstic acid,* but even *purer than Scheele's acid* as it precipitates bases far more vigorously than the acid obtained from tungsten or wolfram.[10]

Using *nitric acid* diluted with twenty times as much water, one part of granular *tin* was dissolved at very low temperature in 8 days, but another part remained undissolved. This undissolved part consisted of an ash-coloured powder and semi-metallic grains. The tin could be separated from the latter by nitric acid as a corroded powder; the unaltered part was soluble in fuming nitric acid. This was found to be *tungsten* which did not yet have sufficient vital air to be an acid. It was *not possible* to produce *tin again from andronia and tungstic acid*, so it is to be assumed that *even a third principle* is required for this.

We have seen now how the *principles of acidity and alkalinity*, under various conditions, *pass from one body to another and weaken or neutralize each other*, but we have only identified them according to their *relation* without examining their *nature* more closely. This will now be our object, whereby many other things will become clear at the same time.

So far we have seen that acids and bases mutually deaden their stimulating principles when they combine with one another in such a way that *both* become *inert*. This was the *internal* change, but *outwardly* a phenomenon can be noticed which no less merits our attention, viz., *a very noticeable temperature increase takes place which is directly proportional to the attraction between the acid and the base and inversely proportional to the period of time which the combinations require.*

If a *fuming acid* is combined with a *complete base*, a *large change in temperature* occurs, but if a *complete acid* is combined with an *insipid base*, the same aciduulous salt is certainly produced, but the *change in temperature is not nearly as significant*. If, finally, this *fuming or complete acid enters into a synsomatic compound with another, no such change* will be produced unless other factors cause it.

If a *complete acid* is combined *with a greater quantity of base than is required for neutralization*, the acid principle is even more completely exhausted, and *heat* therefore appears *even more strongly* during the combination.[11]

An *insipid acid* combined with an *insipid base* produces *no change in temperature* despite the fact that both form the same intermediate salts as if they had been complete.

The combinations of *acids with metallic calces*, during which no significant exhaustion of the acid principle occurs, also proceed *without any noticeable changes in temperature*.

In addition, *the temperature does not change* if two *neutral salts mutually decompose each other*, nor does any binding of the relevant principles occur here.

If a *strong acid* is combined with a *neutral salt* which contains a *weak acid*, a *temperature increase* certainly occurs, but it is *far weaker* than if the former had been combined with the base in its integral form.

---

[10] On closer inspection, it is found that this contains some molybdic acid.

[11] As an example W. mentions fulminating powder. With heating, the sulphur from the acid of the saltpetre is burnt, but not so much sulphuric acid is produced that all of the released alkali could be neutralized, and this explains the great heating and the consequent expansion of air.

Ammonium not infrequently produces *cold with acids* by communicating hydrogen to them and thus causes them *to retain their acid principle more strongly.*

Therefore: *Whenever the acid principle and that of alkalinity combine with one another and are thus liberated, heat is produced; if one of them is bound, cold results.* It seems that no more rigorous proof is required to justify the assumption that *the cause of heat consists of these two principles.*

In addition to the *changes in temperature* which derive from the *generation of heat from its principles*, changes can also occur as a result of the *increase or decrease of the already existing caloric.* Such changes arise if the *capacity of bodies is changed*, or if *bodies with different capacities are placed in a medium with a temperature higher or lower* than that which they had before. The *capacity is increased* with every *expansion*, and with every *contraction* or *compression* the *capacity is decreased.* However, the capacity is determined not only by the densities, so something still *unknown* must also be present which we cannot yet determine.

It is also quite generally assumed that *heat* can be *bound*, whereby *cold* develops, and that, on the other hand, it causes *warming* when it *is released.* However general this idea may be, *much* can be said *against it*, for it would be wondrous indeed if an absolutely elastic substance should lose this its only property without any reason in order to join the collection of bodies which act only because of the constitutive components of the caloric, whereby it would cease to be what it was. At any rate, the *caloric*, which *cannot be without weight* (as we shall prove later) *would necessarily increase the weight of bodies* if it were *chemically bound in them.* In general, the phenomena which are to be explained by the binding or the liberation of heat derive from changes in density. However, in order not to ignore the point which above all seems to support the fixation of heat, I note that the heat which is liberated if a uniform gas transforms into a more solid form or the heat which must be applied to change a more solid body into a gaseous state by no means proves what it was intended to prove, for wherever any substance is changed into a true gas, either the electrical spark or red heat or an electrical arrangement following the pattern of a galvanic circuit must concur, or, to sum it all up in one precise expression, *wherever a gas develops, conditions must exist which activate one of the principles of the caloric.* Vital air can be produced by glowing, by sunlight, and by electricity, but it can never be produced by heat alone. Hydrogen gas is also produced by electricity, either from an electrical apparatus or from bodies which form an electrical circuit with each other, or from water by glowing with iron, but if the fixation of heat had validity, I might well ask: Where does the large amount of heat come from which is necessary for the gaseous state of hydrogen if hydrogen gas is produced from zinc and muriatic acid or the like? If heat were bound by the developing gas, cold should be produced, but the *opposite* occurs. This can be seen even more in the generation of vital air and hydrogen gas in the voltaic pile; since both gases develop here between each layer, the *temperature of the pile* should *decrease* constantly, but this does not happen. What is demonstrated here concerning these two gases could also be claimed for all the others. For our purpose, however, this may suffice; a detailed study would lead us too far.

Just as *heat can be composed of its two principles*, it can also *be decomposed into them again*. One method to accomplish this is by *friction* which *gathers heat* in the rubbed bodies and *decomposes them* if they are of a heterogeneous nature, like *glass* and *metal*, so that the *former* attracts the *principle of acidity*, the *latter that of alkalinity*.[12] If a metal has a conducting connection to earth, it constantly receives alkaline principle from it as the glass attracts acid principle. The former becomes negative, the latter positive. The acid taste aroused by the positive conductor and the opposite by the negative conductor are quite well known. The combination of both electricities is the transition of the principles of heat from difference to indifference. If this transition occurs through air or another poor, non-solid conductor, the *spark always proceeds from the upper conductor to the lower,* no matter which is the positive and which the negative. This is a proof of the *weight of the caloric* which may not appear evident because of its novelty.[13]

If *fuming nitric acid* or *saltpetre gas* is brought into a conducting connection with the *positive conductor*, they *weaken its effect*. Freshly prepared *saltpetre gas,* blocked with mercury, *prevents electrical activity* for a long time. Once it has lost this power, it can be restored by the addition of some lime water, and thus it can repeatedly lose and regain its strength until finally nothing is left but azotic gas, which constitutes only 0.3 of the total. The lime, however, has been transformed into *nitrate of lime*. Here some *basis of vital air* in a combination (perhaps in the water or perhaps also in the azote) is certainly *transferred from the inert to the more active state* and now combines with the *saltpetre gas*. So long as the saltpetre gas has not lost the capacity to attract the acid principle, the spark that leaps across is *blue,* but once this has quite disappeared, it is *red*. The blue spark, however, is a sign of deoxidation, the red, on the other hand, of complete oxidation.

The most telling facts in support of the role which electricity plays in acids and bases are provided by the phenomena of the water in the voltaic pile. As is well known, water becomes acid wherever it is oxidized or gives off vital air, whereas it becomes alkaline wherever it gives off hydrogen gas, and the saline solution between the plates decomposes and liberates its soda on the hydrogen side, but on the opposite side its acid, which combines with the zinc calx that has developed. Although acid and alkali have not always been found in pure, galvanized water, and although the conditions under which these supplements appear have not been established precisely, this much is certain, that they do not derive from impurities in the *water* since (according to Simon's experiments) we obtain *acid and alkali* in a *very short time* with silver as the conductor, whereas *none* was generated from the same pure water after several days of gas production with *platina wires as the con-*

---

[12] This certainly contradicts the statement by W., who assumes E to be the principle of alkalinity, $-E$, on the other hand, the principle of acidity. His main argument is that fuming acids have the effect of a positive conductor and are changed into non-fuming acids, which, according to him, is a diminution of the acid principle. I think that I have sufficiently justified my opinion with the above explanation of the effect of E on saltpetre gas, without referring to the *galvanic* facts which support this view, but which could not have been known to W. at that time.

[13] Ritter has arrived at the same result in another way and will answer those who might find W's assertion far too paradoxical.

ductor. Perhaps all the more acid and alkali are produced, the less electricity is used for the production of gas, but acid at least is always (as far as I am aware) found in water where one of the metals oxidizes.

Therefore, it is *electricity which activates everything and constitutes the principles without which, as we have seen, the most active substances that our chemical art has ever discovered would be dead and inactive.* We now understand why friction is able to promote all chemical actions and even create combinations which would not occur spontaneously. There is hardly any need here to look for examples; to mention many in one, we need only recall the many deflagrations which are produced merely by a weak or a strong vibration.

The similarity of magnetic and electrical phenomena has long been recognized and has lately been carried even further by important experiments.[14] Here we may be allowed, with W., to suggest a few principles as contributions to a theory of magnetism.

1) A magnet behaves towards the earth as towards another magnet.
2) Steel behaves towards the earth as towards another magnet in that it takes magnetism from it, especially if it is held in the direction of the magnetic line of the earth while glowing.

Therefore, is the earth a magnet?

3) In that region of the earth which is affected by winter, we see the disappearance of a large quantity of caloric.
4) However, heat moves very slowly even through the best conductors.

Is it likely, therefore, that the caloric moves from one hemisphere of the earth to the other along with the change of the seasons? Or should it not rather be assumed that the caloric would be decomposed by the oblique rays of sunlight but would be composed by more perpendicular ones?

5) In winter heat leaves the surface of the earth sooner than the parts beneath and, conversely, reaches them again later than the surface in summer.
6) The layer of clouds in the sky tempers the cold of winter, the heat of summer.

Could it be that, in winter, the same cause which weakens the rays of sunlight prevents the resolution of caloric, whereas it prevents the generation of caloric in summer? The warming in winter, therefore, would derive from the caloric of the earth, and the cooling in summer would be caused by the dissipative force of the earth.

7) In the electrical machine, the principles of acidity and alkalinity seem to act strongly only on living bodies and but little on inorganic ones (except for those which already possess one principle or the other).
8) The seasons have the same effect. Could it be that the separated principles

[14]Ritter's *Contributions to Better Understanding of Galvanism*, Vol. II, Part 1.

of heat are bound in the earth without any perceptible indication of their existence?

9) As we know, the higher regions of the earth are colder, and the snow on the highest mountains is melted only from below.

Could it therefore be that the separated principles of heat were again combined at a very low altitude? Or is there a lack of the principles for the restitution of heat at higher elevations?

10) The magnetic needle constantly adjusts itself towards the poles and glowing steel becomes magnetic by its orientation alone.
11) The separated principles of heat combine with one another with the greatest rapidity even through the longest conductors (unless they are interrupted by poor conductors).

Could it be that each part of heat moves to one of the poles in winter, in order to proceed to where it is summer? Heat itself does not seem capable of such rapid motion.

12) Summer is also distinguished from winter by fiery meteors.

Are these perhaps phenomena of general magnetism? As they are not constant, however, what secondary causes are also present?

13) A magnetic needle consists of many other magnets with their own poles.
14) Refractory[15] bodies like iron are magnets by external influence, not by independent power.
15) The solar system extends as far as the heat-generating power of the sun.
16) Light alone has no weight because it is emitted by the sun in the direction opposite to gravity.

Are all heavy bodies magnets?
Are there different degrees of magnetism?
Is the *first* degree a phenomenon common to all heavy bodies?
Is the *second* the phenomenon of bodies which have been polarized by friction?
Is the *third* the phenomenon of polarity which is determined by that of the earth?
Which secondary cause establishes the relation of magnetized steel to iron?
Is light the cause of magnetism and thus of weight?

The *composition and decomposition of heat* plays a major role in nature and in the phenomena of magnetism, and they are also constantly at work in organic life.[16] It would certainly be of the utmost importance and of the greatest interest to trace

[15] [Ørsted writes *retractorischen* but more likely intends *refractorischen*.]

[16] A very well-known fact in chemistry follows from this, viz., that vinegar is better preserved by boiling. The *infusoria* in it are thus killed, and they could contribute to making it insipid only as living organisms, for the touch of a living hand also makes vinegar insipid even though it is itself capable of preserving dead meat.

this back to its origin, but here we must be satisfied with attributing to it some remarkable chemical phenomena which we have in part experienced before.

We have seen previously that *inert acids redintegrate by means of red heat.* The *explanation* of this now follows of itself. Here, the acids which have been partly deprived of their stimulating principle again find the substance which they desired, and if they do not redintegrate, another cause must be at work. In order to add several other examples to the ones we have mentioned above, we want to cite the following. When heated, *nitric acid combined with weak bases* produces first partly complete, partly insipid acid, then only insipid, and finally the last part, which can be expelled only by glowing, again complete. *Carbonic acid,* which exists only in the highest state of deoxidation in minium and in vegetable bodies, is driven off in a complete state by red heat.

Vinegar is only partially expelled in a complete state from acetate of copper by heat, but by means of glowing it is expelled totally in a complete state.

Ignited with alcohol, insipid *boracic acid* produces only a blue light initially but later a green one, which proves that it has been redintegrated by the flame.

That sulphur does not become more oxidized in red heat is due to the fact that sulphur can appear acidic only in combination with hydrogen. However, how much effect red heat also has here can be seen from the fact that sulphur vapours do not combine with hydrogen gas but dissolve at the focal point of a burning mirror and become sulphuretted hydrogen gas, which is an acid. It is almost the same with *boracic acid,* which does not become complete when it fuses to glass, but when extracted from its compounds by very strong acids, it appears complete because it acquires the basis of vital air from them. This is revealed by the fact that copper which was free of moisture calcines when melted with it, whereby a gas is released which has not yet been investigated.

The complete acids become hyperoxides with luminous heat. Thus *sulphuric acid* which has been deprived of its volatile part by distillation can become fuming again to some extent, the limit of which is set by the admixed water. The intense light with which *phosphorus* burns immediately transforms the produced acid into a hyperoxide so that it cannot leave this state until water is added. In the atmosphere, vital air certainly exists in a complete state, but it can still be oxidized even more if it is expelled glowing from metals, in which state it is capable of combusting iron and causing several phenomena of which the vital air of the atmosphere is incapable.

The *decompositions of several vegetable acids by fire* also seem *to be the result of such a hyperoxidation,* and as a consequence we find that certain acids tolerate only an intermediate state of oxidation, and that they must decompose with too great an increase or decrease. This idea is also confirmed by the behaviour of several metallic oxides in fire as they release all or at least part of their vital air during glowing, whereby it is proved that they can contain it only in a less oxidized state.

It is remarkable *that bases can be redintegrated with far less heat than acids.* We have had examples of this above and need only add here that if *insipid potash* is boiled with fuming sulphuric acid, we still obtain only the usual salt from it, sulphate of potash, a redintegration which can be produced only by means of heat.

Letter to a Friend Regarding Winterl's *Prolusiones*

You write to me, Dear Friend, that you cannot agree with my opinion of Winterl's *Prolusiones*. You see in this work merely a collection of not very accurately described experiments, of daring conclusions, indeed, probably even of gross offenses against previous experiences. You also confess to me that, therefore, you have not deemed the *Prolusiones* worthy of further study and wish even less to have a series of his experiments repeated. However much I usually value your opinion, I nevertheless believe that I have an important advantage over you here as I have not only read the *Prolusiones* but also studied them diligently. Without such an effort, it is not easy to grasp the idea of the whole, to compare it with the compass of previous experiences, or even less to assess it philosophically. Therefore, I dare hope that we shall still be able to reach agreement on this important matter if I explain to you quite briefly what I consider to be W's theory and illustrate the fundamental principles using only experiments which were not *originally* performed by W. In so doing, I shall not commit myself to the greatest scientific stringency but arrange the various parts as they offer themselves most easily at the present standpoint.

The principles of electricity are also the causes of acidity and alkalinity.

The acid taste which is stimulated by the positive and the alkaline taste which is stimulated by the negative conductor must already draw attention to this. Experiments with the electric pile have confirmed this even further as the water which has been properly connected to it becomes acid on the positive side and alkaline on the negative. It is true that some have not been able to obtain this result with platina wires and pure water, but from the same water which only produced gas when treated in this way, Simon obtained acid when he connected it to the pile by means of silver wires. When there is not the slightest calcination, it does not appear that such an effect of electricity occurs in the water which is necessary in order to cause a generation of acids and alkalis.

As you will see, the fundamental principle set forth here is very useful. From it follows:

Acidity and alkalinity are opposite one another.

Therefore, they must mutually neutralize one another by their combination.

Is this conclusion not confirmed most beautifully by experience, for what is neutralization but the indifference of acidity and alkalinity?

In a *combination of an acid and an alkali*, therefore, both exist in an *inert* state.

Acids and alkalis do not contain their principles in the free state as electrified bodies do, but their stimulating principles are bound, more in some acids and alkalis, less in others. The *acid* or the alkali in which the *stimulating principle is more free* appears to be *stronger* (more active) than those in which it is *more tightly bound*.

Hence, the mutual *neutralization of alkalinity and acidity* by one another is never absolute, only *relative*. Therefore, an acid or an alkali can be *deadened in various degrees*, by the *dominant quantity or strength* of either one or the other.

You will tell me that these conclusions are certainly not contradicted by experience, indeed, you will admit that they could easily convince you if just one more experiment were added without which, as a precaution, you do not wish to accept any theorem in physics. I could easily satisfy this demand by quoting new experiments from the *Prolusiones*, but, true to our intention, we shall stay with old ones. If it is true that the acid and the alkali are inert in their neutral combination, it must also be possible to separate them in the inert state if only a means is applied for this purpose which neither gives nor takes acid or alkaline principle. Heat is such a means. Therefore, if nitric acid is expelled from its combination with potash by heat alone, this acid must be obtained in a *less acidic state*. This also agrees perfectly with experience if only all circumstances are considered correctly, for we obtain nitrogen gas and vital air, and is this gaseous mixture not less acidic than nitric acid? —You will perhaps suppose for a moment that this might be due to the nature of the mixture, but you must abandon this opinion if you consider that we can transform this mixture of nitrogen gas and vital air into nitric acid merely by means of electricity, i.e., merely by imparting acid principle to it.

It is no objection to this that sunlight also decomposes nitric acid, for it also reduces horn silver and shows everywhere that the negative principle plays a major part in this, something which Ritter also found in his experiments and actually was the first to understand clearly.

From this well-known experiment, which can easily be compared to others that are equally well known, we learn not only that acids exist in an inert state in neutral salts, but also that several substrata of acids can only exist with a certain quantity of acid principle, and that the parts separate with less. The same might also be the case if more stimulating principle were added, and we actually find this in the metallic calces which release their vital air when heated red, for the acid principle dominates in red light as Ritter has shown.

It is natural that there are inert alkalis just as well as inert acids though we cannot prove it with such striking facts, but we have an example in the case of the soda in natural borax, which has less alkalinity than required for complete integrity. This can be seen from the fact that tartar does not lose any of its acidity if it is saturated with borax, which, however, should happen if the excess soda in the borax were not already inert. Tartarate of borax even seems more acid to the tongue than before, but this is merely the result of its greater solubility.

The complete series of facts by which these laws, or, rather, this *law* is confirmed was first discovered by W. and passed on to us.

Now we come to a fundamental principle in Winterl's system, indeed, to the one which in a strictly synthetic presentation should be placed ahead of all others. It is nothing other than:

The principle of heat is composed of the principles of acidity and alkalinity.

We learn this in part from the fact that the combination of a sufficiently strong acid with an alkaline substance generates heat. Here we are not aware of any other cause of heat except that acid and alkaline principles combine with one another. Elucidation of all the apparent exceptions to this law would require a very detailed discussion, for which some of W's peculiar experiments would also have to be used. Therefore, I prefer to strengthen the proof with several facts, and I shall be content if I have offered enough to engage your full attention even if I should not have the opportunity here to treat everything in such a way that the rigour of a proof becomes quite transparent. A major fact is that all frictions generate either heat or electricity, and if everything were carefully investigated, both would certainly appear simultaneously but under different conditions. However, this much is certain, that in general conductors of a homogeneous nature rubbed against each other primarily generate heat, heterogeneous conductors, on the other hand, electricity. Therefore, where there is no cause to separate the principles of heat from one another, a temperature increase appears, but where one body is more inclined to become +, the other − , a separation occurs, and electricity appears. Thus heat is frequently produced when it is believed to have been merely separated, but we do not deny that temperature changes would also be possible due to a changed capacity, which occurs if the volume of the bodies increases or decreases; on the contrary, we shall affirm this. We refuse to acknowledge only the *binding* of the heat principle, nor do we find evidence which could be explained *only* in this way. In the formation of gases, which should provide the best evidence for this opinion, there are circumstances which are not completely consistent with this theory but are in complete agreement with ours. For example, if carbonic acid gas is produced from chalk by means of another acid, a significant cooling of the surroundings should follow according to the theory of the binding of heat, but we discover none. We explain this very simply by the fact that the added acid, which combines with the calcareous earth, gives its stimulating principle to the carbonic acid and now appears as a gas. It is the same with the generation of hydrogen gas, where there is just as little of the cooling which should be expected when a portion of the water assumes a volume many thousand times greater. Doubly justified by the galvanic discoveries, we declare any such generation to be an electrical process, and we know that hydrogen gas contains alkaline principle while oxygen gas contains acid principle.

The constituent principles of heat, which are important in alkalis and acids, in electricity, and in light, are also the principles of magnetism, and thus we would have the unity of all the forces which act together to govern the entire universe, and previous physical knowledge would therefore unite into one coherent physics (even with this, our physics is not yet complete), for do friction and impact not produce both heat and electricity, and are dynamics and mechanics not thereby perfectly intertwined? (If it should be necessary, this will become even more obvious once we are able to survey at a glance all Ritter's beautiful and relevant discoveries, some of which were made a long time ago.) Our physics, therefore, will no longer be a collection of fragments on motion, on heat, on air, on light, on electricity, on magnetism, and who knows what else, but with one system we shall embrace the entire world. Everyone must do what is in his power to nurture the great task of completion.

I have here acquainted you with the essential features of what Winterl's system has given me, but I should certainly not entirely neglect to say a few words about *andronia* since some have believed that W's entire system is based upon it. As I cannot offer well-known experiments for this part either, I shall content myself with pointing out to you that this ubiquitous *acid* substance has its counterpart in *thelyke*, a similar *alkaline* substance, that both together seem to constitute *all earths* and *alkalis*, and that also many *metals*, perhaps *all*, contain andronia.

With vital air, water, and acid principle, *andronia* yields *carbonic acid gas*, with vital air and alkaline principle *nitrogen gas*. This is why von Hauch obtained nitrogen gas and vital air (in the proportions 64 to 36 and 65 to 37) by passing water over glowing tin and lead, whereas the metals evaporated. According to other experiments by W., tin contains andronia, and so must lead. The part of the metal which was not andronia had undoubtedly combined with the hydrogen in the water, and the so-called oxygen had partly formed nitrogen gas with the andronia.

These few words might suffice to show you how much we may expect from the discovery of andronia and thelyke in the more specific chemistry. You will find the experiments described, in part explicitly, in the *Prolusiones*.

Farewell, and let me truly recommend Winterl's chemistry to you.

Ørsted.

13

Correspondence[1]

Copenhagen, September 10, 1804.

MY work in physics runs in part parallel to my lectures, and therefore I have only recently returned to the investigations of Winterl's chemistry. When I was in Berlin, I had already begun a series of experiments on Winterl's inert or insipid sulphurous acid in the laboratory of Privy Councillor R. Hermbstädt, but my trip prevented me from continuing this series. However, during this work I found a new example to be added to the many older ones which prove how very often our chemical knowledge, considered established, is lacking in reliability. The sulphite of potash with which I worked had been prepared according to the instructions of Fourcroy and Vauquelin and agreed completely with the further description which these chemists have given. However, it was not a neutral salt but an acidulous one, whose solution reddened blue vegetable juices and became effervescent with carbonate of potash, etc.; the neutral compound could scarcely be crystallized. In addition, the sulphite of potash did not turn into sulphate of potash under the influence of air as Fourcroy and Vauquelin think, but a triple salt is formed from sulphuric acid, sulphurous acid, and potash. This salt is deposited rapidly in flat hexagonal prisms, is less soluble in water than the sulphite, but more so than the sulphate of potash, and, when treated with an acid stronger than sulphurous acid, gives off the odour of sulphurous acid. —I shall very soon be able to give you further information about Winterl's inert sulphurous acid.

Concerning the claim by the astute Rumford that a small portion of a liquid body cannot transmit heat to another, I have conducted experiments which, due to their simplicity, might be the most decisive ones that have so far been performed on this subject. I heated mercury to above the boiling point of water and then placed a drop of water on its surface, and the water came to the boil instantly; a drop of alcohol or ether evaporated immediately, etc. Heated oil or water also causes a drop of alcohol or ether to boil and evaporate instantly. Each time, I caused the drop of liquid to fall in the middle of the surface of the heated liquid, where it immediately boiled

[1] [*Neues Allgemeines Journal der Chemie*, ed. by A. F. Gehlen, Vol. 3, pp. 322–24. Berlin 1804. Also partially in *The Philosophical Magazine*, ed. by Alexander Tilloch, Vol. 23, p. 80. London 1805. It is stated there that the paper has been taken from *Van Mons's Journal*, Vol. 6. KM I, pp. 211–13. Originally published in German.]

and evaporated without approaching the walls of the container. It was therefore not possible for the added drop of liquid to receive heat from anywhere but the warming liquid.

These experiments provided an opportunity for several other observations. It has been known for a long time that water, dripped on to very hot metal, does not evaporate as easily as it does when the metal is less heated. Precisely that occurs with other liquids. The same heated mercury on which a drop of ether evaporated instantly did not cause this fast evaporation when heated to an even higher degree, rather the drop of ether ran around on the surface of the heated mercury for some seconds and disappeared only gradually. When the mercury was heated even more, the same phenomenon took place, but the ether also turned quite black and a pleasant vinegary aroma was emitted. Alcohol displayed the same phenomena, except for the vinegary odour, only at a somewhat higher temperature. Therefore, the slower evaporation of a liquid at an elevated temperature seems to be the result of an internal change which is certainly different at different temperatures. The experiments on the transformation of water in nitrogen might be mentioned once again.

I shall not yet venture a complete explanation of these experiments as I do not deem them sufficiently mature for that. I wish to add only that the slower evaporation of ether at an elevated temperature can also be demonstrated on the surface of heated oil or a concentrated salt solution. As soon as time permits, I shall seek to continue and expand these experiments.

J. C. Ørsted.

14

Galvano-Chemical
Observations[1]

RECENTLY, I have discovered a galvano-chemical phenomenon which has not previously been observed. The reason for this was the following: Several years ago Ritter told us about the discovery that the conductors of an electric pile are covered with sooty shapes when held in the flame of a candle. The sooty shapes on the hydrogen side are vegetal, but the ones on the oxygen side have a different shape. It was to be assumed that each oxidation would be associated with the same formation. I wanted to investigate this. To this end, I put a solution of acetate of lead in contact with the poles of the pile. The dissolved lead calx should be oxidized more strongly on the oxygen side and be precipitated as brown lead calx, but on the hydrogen side it should be reduced and thus be precipitated. This indeed happened. On the hydrogen side I obtained a beautiful metallic lead vegetation but on the oxygen side a brown lead calx which formed shapes comparable to the positive soot figures. I would prefer to compare these shapes with plant roots.[2] Could it be that oxidation and deoxidation are associated with definite forms which occur if no external causes oppose them? Could the organic forms be necessary products of the internal chemical process? I shall soon endeavour to provide more information about these and several related issues in a treatise dedicated to the subject.

In connection with his experiments on the vibration of mercury in the galvanic circuit, Ritter reports that the mercury seemed less fluid on the oxygen side but more fluid on the hydrogen side. I have repeated this experiment and found it completely confirmed. However, in order to convince myself, or rather others, even more completely, I chose a lead amalgam which remained fluid in warm water. I let the water with the amalgam cool down gradually in the galvanic circuit, whereby the amalgam solidified sooner on the side which produced oxygen than on the side which produced hydrogen. Therefore, this provides further confirmation of Ritter's discovery. In addition, this phenomenon agrees completely with two other

---

[1] A letter from the author. [*Neues allgemeines Journal der Chemie*, ed. by A. F. Gehlen, Vol. 3, pp. 578–80. Berlin 1804. Also in *The Philosophical Magazine*, ed. by Alexander Tilloch, Vol. 23, pp. 129–30. London 1805. KM I, pp. 213–14. Originally published in German.]

[2] Compare these remarks to the ones by Mr. Ritter mentioned above.

discoveries by the same physicist. That is, he found that the small metal leaves in spark experiments combusted on the oxygen side but melted on the hydrogen side. And, in his galvano-physiological experiments he found that the hydrogen pole arouses a sensation of warmth, which is not the case with the oxygen pole, which often produces an opposite sensation.

15

Criticism of the So-Called Eudiometry
with Regard to Medicine[1]

FROM the moment when it was discovered that atmospheric air was able to sustain life and combustion solely through its oxygen gas, or vital air, people also began to think of ways in which to discover how much of it air contained. Thereby they thought to be able to measure the fitness of air for health and thus gave the name of eudiometry to the art of determining the content of oxygen in air. It was soon discovered that the method which had first been used for this determination was full of errors. Consequently new methods were invented, but they were no less imperfect and therefore equally soon rejected only to be replaced by others which, in their turn, suffered the same fate. Thus, the matter was seen to occupy the physicist and the doctor for a long time as one of the greatest importance to mankind. I believe it is now time to draw a conclusion from these many experiments, even if it were only that eudiometry has never fulfilled and will never be able to fulfil what was expected of it. At a time when chemistry is so proud of its great progress, it is important to display its weak points and to awaken it from the slumber which trust in a supposedly infallible and unique doctrine has caused. According to human nature, hasty application of new discoveries is unavoidable, but now the experiences of several decades should finally lead us back to the right path. Guided by these, I shall endeavour to show that eudiometers, by their very nature, are unfit to fulfil their purpose, and also that it would not be as important as it is thought even if it were possible to determine most accurately how much oxygen gas the atmosphere contains, to which I must add some remarks on the various causes which change the effect of air on the animal body and on how to proceed in order to become familiar with them.

All eudiometers are based on the fact that a combustible body absorbs oxygen gas from the atmosphere. Thus, when such a body has been confined in a quantity of air, it is believed that it will absorb all its oxygen gas more or less quickly. This opinion presupposes the natural law which states that when three substances are in

---

[1] Read at the Royal Society of Medicine in January 1805. [*Nyt Bibliothek for Physik, Medicin og Oeconomie*, Vol. 8, pp. 52–79. Copenhagen 1805. Also to be found in German in *Neues allgemeines Journal der Chemie*, edited by A. F. Gehlen, Vol. 5, pp. 365–92. Berlin 1805. KM I, pp. 248–61. Originally published in Danish.]

contact with each other, the one among them which has the greatest affinity to one of the others will take complete possession of it. If this is not a law of nature, it follows that this way of determining the content of oxygen in air is incorrect, and this is indeed the case. Berthollet has shown that when two bodies, through their affinity, act on a third, each of them absorbs some of it, in proportion to both their affinity and their mass. Suppose, for instance, that nitrous gas had an affinity to oxygen gas 20 times greater than that of azotic gas, then the former would certainly absorb 20 times more oxygen gas than the latter, but the nitrous gas would still keep $^1/_{21}$ of it and therefore not be completely deprived of its oxygen. It would also show some affinity to the nitrous gas itself so that, after having mixed atmospheric air with nitrous gas, a mixture of azotic gas, oxygen gas, and nitrous gas would remain. Only two main objections can be made to this. Either azotic gas can be denied any affinity to oxygen gas and nitrous gas, or the validity of Berthollet's discoveries can be doubted. The first suggeston is obviously impossible, for if the two gases which constitute our atmosphere had no affinity to each other, they would soon separate because of their different specific weights, and all the oxygen gas would sink towards the earth in still air. Admittedly, it has been suggested that atmospheric air, left in a long tube for 18 months, had separated into its main constituents, but this experiment has not been confirmed, and if it were true, it would only prove that their affinities were somewhat weaker than the difference between their specific weights, but it does not follow from this that it is $= 0$. To assume this is contrary partly to the recently mentioned common experience regarding our atmosphere, partly to the truth that there are no two bodies which are completely without any affinity to each other. When we encounter two bodies which are not at all compatible with each other, it only proves that the force with which they strive to maintain their own quality is larger than the one with which they strive to combine, or to use an older expression, that the mutual attraction of their parts is smaller than the sum of their cohesive forces. If, on the other hand, the correctness of Berthollet's ideas about the affinities were to be doubted, it would be necessary to contradict reason and experience at the same time, for who can believe that A can stop acting on B because C also acts on it? Is it not more reasonable that A would just as well affect B with its force as would C? And must 2 C not have a stronger effect than 1 C, just as a mass of 2 pounds has twice as much effect as a mass of 1 pound, with the same motion? Besides, Berthollet has confirmed these laws of affinity through so many experiences that whoever dare not trust reason merely needs to look at the experiments. No matter how much doubt this discovery by Berthollet raises about all chemical analyses, none is more subject to these doubts than such an analytical method which, like the eudiometric, is supposed to give people who are not familiar with all the resources of analysis a means to determine the constituents of the atmosphere, even down to hundredth parts, indeed thousandths.

Therefore, Berthollet's discoveries force us to conclude that not all combustible bodies deprive the atmosphere of the same amount of oxygen, no matter how much time the one with the weaker effect was given for it, that no combustible body deprives atmospheric air of all its oxygen, and, finally, that some part of the combustible body itself must be able to combine with the residual air so that it is impossible to determine how much of the air remaining after the experiment belongs to the atmosphere, and similarly how much the combustible body has really increased in weight through its oxidation. The only thing that remained for the eudiometers, then, was that the same eudiometric substance was always used in the same form and in the same quantity, preferably an immensely great one relative to that of the air. It is true that the exact amount of oxygen in the air would remain unknown, but from the amount it gave off to one and the same substance, it would still be possible to infer an altered or an unaltered proportion of the constituents of the atmosphere, but only on the condition that a great many circumstances which might change the result were observed and taken into consideration, such as the change which heat produces in the affinities and in the volume of the products, the influence of humidity, changes in the atmospheric pressure, &c. We shall come to know all of these circumstances better through the best eudiometric methods suggested so far and see whether what we have observed about them in general is not also true of each of them in particular, and on the same occasion a great many other obstacles to the eudiometric investigations will confront us and strengthen the evidence of the inadequacy of eudiometry.

The eudiometric substances can properly be divided, according to the three main categories, into gaseous, liquid, and solid.

The gaseous depend on the fact that a gas combines with oxygen and contracts into a liquid fluid. As it is impossible to know beforehand the proportion of oxygen gas in the air under investigation, it is necessary to know the exact proportion of oxygen gas and eudiometric gas which comprise the new compound. Only two gases have been used for eudiometric purposes, nitrous gas and hydrogen gas.

Nitrous gas has been used most often, and great efforts have been made to improve the eudiometric investigations based on its use. Hereby all the difficulties which present themselves have been demonstrated so clearly that it is now easy for us to see that these are insuperable or, at least, far outweigh the advantages. When nitrous gas comes into contact with oxygen gas, they combine to form vaporous nitric acid, which is easily absorbed in water, so that nothing remains except nitrous gas and the irrespirable part of the air investigated. If it were known with certainty how much oxygen gas a given quantity of nitrous gas absorbed, it would be easy to calculate the volume of the oxygen gas consumed, but this is not the case. When combined with oxygen gas, nitrous gas not only changes into a red vapour, but part of it may possibly even change completely into nitric acid. In other words, the proportion of the absorbed gases is not constant. This explains the disparities in the results of physicists. That uniform results have been obtained under the same conditions is not at variance with this since inconsistent ones have been obtained far more often. Humboldt himself has observed that pure oxygen gas produced a relatively greater absorption of nitrous gas than did atmospheric air. The proportion of oxygen gas in the investigated air thus changes the results. Many other circumstances have the same influence. When the tube of the eudiometer is narrow, the absorption is smaller than when it has a bigger diameter, but in the latter case, on the other hand, measurements are less accurate unless the very greatest caution is exercised. If there is no air in the water used for sealing, it absorbs air, and if it contains air, it can very easily release some when moved. If it contains atmospheric air, this can also affect the nitrous gas, and if it contains nitrous gas, this affects the air under investigation. If the water contains carbonates, the generated nitric acid produces carbonic acid gas. If the temperature is raised, some of the sealing water is turned into vapour, and if it is lowered, the vapour is destroyed, and thus the volume of the air changes with every change in temperature, besides the specific expansion which even dry air would experience with such a change in temperature. Barometric changes require the same kind of double correction as temperature changes. Furthermore, Humboldt wants the purity of the nitrous gas to be taken into consideration. If the proportion of the absorbed gases were constant, this would be superfluous, but now it must be considered necessary, for it is true that the same results are always obtained with nitrous gas of different purity when all other circumstances are equal, but we have too little control over this. However, if Humboldt's correction is to be of any use, the purity of the nitrous gas must be measured before the experiment (with a solution of sulphate of iron, as is well known), and after the mixing of the nitrous gas with the atmospheric air, all the nitrous gas remaining after the absorption must be absorbed by the sulphate solution. It is easy to see that each of these measurements is subject to new errors or requires new corrections so that it is necessary to correct corrections ad infinitum. In addition to this, the sulphate solution which Humboldt used to test how much azotic gas the nitrous gas contains itself decomposes nitrous gas and produces azote in it, which Berthollet and Davy have shown. From what I have put forward here, I think it appropriate to conclude that the investigation of atmospheric air by means of nitrous gas leads into a labyrinth from which even the most practiced experimenter can hardly extricate himself.

The investigation of atmospheric air with hydrogen gas is undeniably much easier. An electrical spark completes it in a moment. We know exactly how much oxygen gas is required to change a given quantity of hydrogen gas into water and could therefore boldly infer the quantity of each of the vanished gases from their volumes, but this method is far from flawless. Experience teaches us that some azote can be oxidized in this operation and form nitric acid. Not all the oxygen gas is dissociated from the azote in this operation, indeed, it is even found that hydrogen gas cannot be ignited at all in air which is depleted of oxygen gas, but this error can be remedied if the bad air is mixed with good, whereby it is enabled to maintain the combustion of the hydrogen gas, which now also consumes most of the remaining oxygen. Hydrogen gas is rarely without carbon. This upsets the results greatly. The inconveniences which result directly from the sealing water also attend the investigation of air with hydrogen gas. It has not yet been determined whether mercury could be profitably used as a sealing agent. It is at least highly likely that the aqueous vapours which are generated by the combination of the two gases do not all condense when cooled, but that as many as can remain in the elastic state at the original temperature of the air really do remain in that state and thereby increase the volume of the residual air.

The liquid and solid eudiometric substances seem at least to have the advantage over the gaseous ones that the air remaining after the combustion consists of nothing but the irrespirable part of atmospheric air. However, it only seems that way, for it cannot be denied that azotic gas is able to dissolve part of any solid or liquid substance, and even more so since the eudiometric substance can never completely deprive the air of all its oxygen, and through its affinity with the eudiometric substance, this must naturally contribute to the absorption of even more of it. Still, it is true that solid or liquid substances cannot contribute so much to the expansion of the residual air as the gaseous ones.

Davy has suggested a eudiometric substance which was to combine the great absorptive capacity of nitrous gas with the perfect ability of liquid bodies not to change the volume of the air greatly. He uses a solution of muriatic iron or sulphate of iron saturated with nitrous gas. However, as he himself must admit that this gas separates from the metallic salt with diminished air pressure or at an elevated temperature, and that the gas is also readily decomposed by the imperfect iron calx in the salt, it is easy to see that this method will not give correct results except in the hands of the most skilled experimenter.

Sulphurated metals are hardly used by anyone to determine the oxygen content of air, whereas sulphurated alkalis have met with more approval. They are undoubtedly among the best eudiometric substances we know, but the objections to them are not unimportant. There is no doubt that some of the sulphurated hydrogen in these solutions is dissolved by the air, which thereby changes its volume. Even the smell of sulphurated alkalis, as soon as they are moistened with water, convinces us of this, not to mention that the general principles of the theory of affinity force us to assume it. Sulphurated alkalis work slowly, and there are no signs to indicate when all the oxygen gas has been consumed. In addition, Marti has found that freshly prepared sulphurated lime absorbs nitrogen gas as well as oxygen gas. Furthermore, liver of sulphur eudiometers have the imperfection that they can easily saturate the air with moisture.

Phosphorus is among the most perfect and most used eudiometrical substances. Its sudden combustion has often been used to measure the oxygen content of air, but this violent operation, which can also easily destroy the instrument, has now been largely abandoned, whereas the slow oxidation of phosphorus is used. For this Parrot has suggested a suitable instrument which can be sealed with mercury. There is no need to fear that the air is mixed with other vapours than it contained before, nor is there any need to calculate the absorbed carbonic acid gas, and it is obvious when the operation is finished as the phosphorus then ceases to be luminous in the dark. However, there is much to criticize about this eudiometer, too. Water is precipitated by the decomposition of air. The residual air is not pure azote but a combination of this with phosphorus and a little oxygen. Parrot himself has observed that when phosphorus, under certain circumstances, has been in contact with air for a somewhat longer time than is required for the absorption, the residual air begins to expand again. Corrections have to be made for all this. As in all other experiments with eudiometers, changes in the barometer and the thermometer necessitate a correction, which, however, is a little easier for the barometer, for if this had risen by an inch, it would only be necessary to place the tube in the mercury in such a way that this was an inch higher inside than outside, and more or less the other way round in the opposite case.

Lead amalgam, which absorbs so much oxygen when mixed with air, seems to be among the most commendable eudiometrical substances since it is likely that air can only dissolve very little of it and therefore will not change its volume very much. However, it must be noted that it has not hitherto been investigated so much that we can evaluate all the attendant circumstances, that such an easily oxidizable metallic mass is likely to decompose the moisture of the air to some extent, and that there is no indication when the absorption has ceased.

This must be enough about the individual eudiometers. It is true that there are many more, but as I have already shown the origin of their imperfection by means of general principles, I think it is sufficient here to apply these principles to the most important and best-known eudiometers. It is not my intention to infer from what has been put forward that it is completely impossible to obtain instructive results from eudiometrical experiments, but I believe that I have shown so many difficulties about them that none but the most skilled experimenters can be expected to find truth in them. It is sufficient to bear in mind that, in order to obtain an accurate eudiometric result, changes in barometric height, in temperature, and in the humidity of the confined air must be observed, that each of these circumstances requires its own observations and its own corrective calculations, that care must also be taken that the experiment is interrupted at the right time, and that it must be investigated whether the liquid used for sealing has given off air, in addition to other investigations which each individual eudiometric substance in particular renders necessary, and then the correctness of my assertion can hardly be questioned.

However, suppose that the eudiometer were used correctly in the hands of many doctors, would the advantages accruing from this be very great? I believe I dare deny that. The best investigations show us that air always and everywhere contains the same amount of oxygen. Marti found the air in Spain to be equally rich in oxygen, under similar circumstances, as the liver of sulphur always gave him considerable absorption. Even close to stagnant water he found no perceptible change. Berthollet has convinced himself of the same fact, both in Egypt and in Paris. Even at a time when Egypt was flooded by the Nile, at a temperature of 30°, Berthollet did not find the quality of the air diminished. In Paris and in Egypt he found the air equally rich in oxygen. Cavendish obtained the same result, already in 1783, when he made very accurate experiments with nitrous gas. Experiments with air which had been sent to Beddoes from the coast of Guinea confirm the same thing. In short, atmospheric air is the same everywhere, and how could it be otherwise as it is always in motion, and as the reservoir from which the earth and its inhabitants draw oxygen is so large that one or more oxidation processes cannot easily be imagined to be strong enough to change its components noticeably. This becomes particularly obvious when it is recalled that every oxidation or deoxidation process changes the volume of air and therefore must produce currents. Only in confined air can the amount of oxygen change perceptibly, but this can be ascertained without any eudiometer, and generally we are sufficiently well informed of the quality of such air when we know whether a candle can burn in it or not. It can be seen from this how limited the value of eudiometers would be even if it were possible to make them fit for general use.

Furthermore, even if the amount of oxygen in the air really were different in different places and at different times, it would still be very rash to infer from this a proportional acceleration or retardation of the processes of combustion which take place in it. It is well known that oxygen gas, mixed with a great quantity of carbonic acid gas, facilitates oxidation and respiration less than it would have done, had it been mixed with azotic gas, whereas hydrogen gas does not impede respiration nearly as much. We also know that oxygen gas and azote can be combined with very different degrees of affinity. We even know that electricity, in a very effective way, can change the form under which these two substances exist in combination. Would not the electricity of the air be able to produce such a change which, merely because of its smallness, has not hitherto been observed by us although we may often have felt it? A circumstance which produces very little change in the degree of affinity between the elements of the atmosphere, among which water and carbonic acid gas must also be reckoned, can cause the respiratory process to be slower or faster without the amount of oxygen gas in the air therefore having been decreased or increased. I believe the cause of the noxiousness of the air near bogs and any other place where much exhalation of vegetable and animal matter takes place to be found in this circumstance, for, being combustible, these substances have affinity to oxygen, and therefore they resist its separation from the air. In eudiometric operations, on the other hand, these vapours are suppressed, and the eudiometer will indicate approximately the same amount of oxygen as usual.

The humidity of air must contribute to the determination of the speed of the process of oxidation.[2] Who, for example, does not know that pyrophorus burns more poorly in dry air than in humid? And has Madame Fulhame not proved the influence of humidity in a myriad of oxidations and reductions? It is known that respiration is associated with sufficient moisture, but is it also known that the air has time to become saturated with it during the short period it stays in the lungs before it has to make room for new?

Air does not always have the same density. Hence it follows that, when breathing, we do not always get the same amount of air, that is, with an unchanged proportion of constituents, not the same amount of oxygen unless the speed of breathing was precisely inversely proportional to the density of the air, a tenet which cannot be accepted without further proof. Besides, the affinities between the various substances change according to the difference in pressure.

Within certain limits, heat promotes respiration. Spalanzani[3] found that several cold-blooded animals consumed more oxygen gas, the higher the temperature of the air, and that at 0° they completely ceased breathing and thereby became torpid. He found that they consumed even more oxygen while digesting than when they had taken no nourishment for a long time.

Yet another objection, which may eventually be considered the primary objection, against the views of most contemporary physicists on the influence of the amount of oxygen in the air, but which many now would believe it easy to dismiss because it is based on a principle which has not yet been accepted among current truths, is this: Oxygen gas undeniably owes an essential part of its properties to electricity, so we are forced to assume that it has an electric charge or, as many

[2] Parrot has proved this in very unequivocal experiments. See Gilbert's *Annalen der Physik*, Vol. 10, p. 168 footnote.
[3] [Spallanzani.]

would call it, bound electricity. This can be weakened somewhat without oxygen gas ceasing to be oxygen gas, so we have not yet measured its activity because we have measured its quantity. If Ritter's view on water is adopted, there can be no doubt that oxygen gas owes its entire nature to electricity, but if one still sides with the antiphlogistians, one certainly does not want to hear about electricity in chemistry, but after the experience of recent years it hardly seems possible to refuse to do so any longer. It would really be too rude a mockery of experience to wish to deny this now that we see water decompose into its constituents (in order to speak their own language) through an electrical process, and that there are compounds, like the one between oxygen and azote, which can hardly be produced without the help of our most powerful electrical machines. However, if electricity plays an important role for oxygen itself, it must be admitted that a very slight change in its charge (bound electricity) can sufficiently change the force with which it acts to render all our quantitative determinations of oxygen useless. Winterl has understood this situation correctly and has tried to confirm it in several experiments.

A long time ago the observation was made about eudiometers that the salubrity of the air had not been determined simply because the quantity of its oxygen gas was known. No matter how correct this observation is, it could still be useful to know how much oxygen gas the air contained and its relation to the intensity of the respiratory process. Therefore, I have not wished to mention this objection to the usefulness of oxygenometry. On the contrary, it is perfect proof that there are other investigations of changes in the air that are far more important than those of the amount of oxygen. One thousandth part of oxymuriatic acid in the air will certainly have a far greater influence on our health than a change of 2 to 3 percent of oxygen gas would have. Schmeisser has assured me that he has more often found both this acid, as well as sulphur and other such substances, in air when he has investigated large quantities of it. The most dangerous miasmas, rotting effluvia and the like, constitute perhaps not even one millionth part of the air, yet they make it very harmful to us. However, everyone must realize this, and therefore I wish merely to mention it. My main object here has only been to show that the speed of the respiratory process does not depend on the quantity of oxygen gas in the inhaled air. By this I certainly do not wish to deny that some day, when science has progressed, it could be useful to know the amount of oxygen in the air and if it really changed, which, however, it does not seem to do, but I think that I have given sufficient proof that so many other data are required for such a determination that science is still very far from being able even to attempt it.

Perhaps it would lead us to far more important results for animal economy if we investigated whether combustion always occurs at the same speed in atmospheric air. However, this would not be easy as it is very difficult to make all circumstances of oxidation completely identical every time. I, at least, have not yet been able to find a method for these investigations which so satisfied me that I dared recommend it at once. I hope that a series of experiments will soon give me the necessary information.

In all these investigations it is important that all the circumstances which have any kind of connection with the object under investigation are also investigated most scrupulously. It is advisable not merely to be satisfied with making investigations of the ability of air to sustain combustion, but to accompany experiments on this with other meteorological observations, for as these might have an effect on the condition of the body, there is no doubt that they might produce changes in the body which would cause decreased or increased respiration, whereby effects which have their origin in completely different causes might be attributed to the oxidation-sustaining quality of air. Consequently, the humidity of the air, its quantity of carbonic acid, its temperature, and its pressure must be very carefully examined and their effects observed as far as possible. Similarly, large quantities of air must be washed with distilled water to see what water-soluble substances it contains. This investigation would undoubtedly be very instructive. The intensity of light has to be determined every day as it probably has a very great influence on the well-being of animals. The free electricity of the air must not be forgotten. I want to suggest some further investigations which must be of the greatest importance, but which have hitherto been quite neglected: The conductivity of heat and electricity of the air should be measured every day. In order to investigate heat conduction, it would only be necessary to suspend a thermometer, heated to a certain temperature, freely in the air, until it cooled down. Or perhaps it would be even better to enclose the ball of the thermometer in a cylinder of warm water in order to have an even bigger mass to cool. That an accurate watch must be used to determine the cooling time, that the thermometer always has to be suspended in the same place as far as possible, and many other precautionary measures, I pass over here since anyone who knows how to make experiments will easily find them. The ability of air to conduct electricity could be determined by means of Coulomb's electrometer. When an insulated and electrified metal ball is suspended in this, it always loses power in each unit of time in proportion to the conductivity of air and its own electricity, which Coulomb has demonstrated admirably. As animals always generate heat and, as it seems, electricity, which is then conducted away by the air, it will be important for the medical investigator to know to what degree the air possesses this ability at any given time.

The magnetic needle never stands still but moves, often in a seemingly irregular way. Perhaps even this would not be completely unimportant to the physician. At least it seems to be connected with the greatest phenomena on earth.

Primarily I would recommend investigations with the voltaic pile, however paradoxical this may seem at first glance. Ritter has observed that its power changes every day and every hour of the day, in a way which has much in common with the variations of the magnetic needle. For reasons which it is not appropriate to mention here, he presumes a great sympathy between these changes and the effect of the earth's most secret forces; indeed, he has even made observations of this in sick people, in the announcement of which I dare not anticipate him here.

Perhaps water changes just as much and as often as air and deserves a parallel investigation. Daily experience hardly permits us to doubt this. First and last, I must warn that the experiments mentioned would only be instructive if all or at least most of them were made together, and that a single one teaches little or nothing. A new Hippocrates may be needed to make proper use of them in medicine.

I do not doubt that the difficulty of performing all the investigations suggested here will strike everyone, and this gives new evidence that they are not the task of the physician but of the proper experimental physicist. Indeed, I am even convinced that this kind of investigation will never achieve any great perfection until we obtain physical observations, just as we now have astronomical ones. However little comfort there may be in the fact that we are still so far from the goal, it is, however, better to know this, when it is really so, than foolishly to abandon ourselves to an imaginary wisdom which is as discreditable to the researcher as it is harmful to the practitioner.

16

A Letter from Dr. Ørsted of Copenhagen to Mr. J. W. Ritter of Jena, Concerning Chladni's Acoustic Figures in an Electrical Context.[1]

Copenhagen, October 5, 1804.

I HAVE conducted several experiments on Chladni's acoustic figures which may offer important insights into the theory of sound. I believed that I would also be able to discover electrical phenomena in the production of the acoustic figures and therefore chose *semen lycopodii* to strew on the glass plates instead of sand, in the hope that this dust would adhere to the positively charged places and would easily fall off the negatively charged ones. The first thing I observed during this experiment was that a number of small waves or nodal points developed with each stroke of the violin bow, and they all moved towards the larger quiescent lines and finally coalesced with them.

Therefore, each acoustic oscillation is composed of a number of smaller ones. The nature of each tone might therefore depend more on the relationship between the secondary oscillations and the primary oscillations rather than merely on the number of primary oscillations. Then each tone in itself would be an organisation of oscillations just as any music is an organisation of tones. I have performed many experiments on this and shall publish these completely in the near future.

When I wanted to shake the powder off the plates again, I found that it adhered more to the points which had been at either absolute or relative rest than to other places. However, if I now produced a different tone so that a new figure was forced to develop, it became loose and was easily shaken off.

I have often found signs of electricity in certain places on the plate with Coulomb's electrometer, too, but it is so easy to be mistaken in this regard that I do not yet wish to regard this as fact.

The phenomena which can be observed in these experiments are numerous and beautiful and certainly merit being seen.

Ørsted.

[1][*Magazin für den neuesten Zustand der Naturkunde*, ed. by J. H. Voigt, Vol. 9, pp. 31–32. Weimar 1805. KM I, pp. 261–62. Originally published in German.]

17

A Letter from Mr. Ørsted,
Professor of Philosophy in Copenhagen,
to Professor Pictet on Acoustic Vibrations[1]

Copenhagen, May 26, 1785.[2]

SIR,

The impartial interest which you take in all that can further the progress of science has, for a long time, made me desirous of establishing a correspondence with you. I hasten to take the opportunity given to me by a traveller who is going to Geneva and is willing to deliver to you some results of my researches in physics. As the subject of my present letter, I have chosen the experiments which I have made, and repeated many times, on the effects produced in the interior of solid bodies during the propagation of motion. I have been led to these researches by both ancient and modern observations on sound. Everyone is now acquainted with the interesting discovery of the celebrated Chladni, who knew how to produce certain figures by covering thin plates or strips of metal or glass with sand and then stroking them on the edge with a violin bow. The sand is set in motion by the effect of the oscillations of the sonorous body, and its grains leave certain parts of the surface in order to collect in others, forming lines and figures which are sometimes strange, sometimes regular. The lines on which the sand collects are without doubt the points of rest between the portions of the surface set in vibration by stroking the edge with the bow.

Modern physicists think that they see, in these curious results, the confirmation of the opinion which tends to exclude those internal vibrations which were formerly supposed to be the cause of the phenomena of sound, and to refer the latter exclusively to those oscillations which are visible and appreciable in their mechanical effects, of which the experiment with sand provides an example.

It seems to me, however, that the ancient hypothesis has been abandoned too easily. At least, my experiments have rather led me towards this hypothesis than

---

[1] [*Bibliothèque britannique*, Vol. 30, pp. 364–72. Genève 1805. This letter was published in English in *The Philosophical Magazine*, ed. by Alexander Tilloch, Vol. 24, pp. 251–56. London 1806. KM I, pp. 262–66. Originally published in French.]

[2] [The date 1785 is probably due to a misprint; it should no doubt read 1805.]

away from it. The following are the observations which my experiments have suggested to me.

It is agreed that in air, an eminently elastic fluid, a sudden compression is followed by a reaction which produces an expansion, which immediately compresses the neighbouring parts, and so on. The effects of an initial concussion propagate themselves in this manner to an indefinite extent. However, I do not see why things should not take place in the same way in an elastic solid when it is made to produce a sound. Chladni's experiments, although in other respects very interesting, are not appropriate to give evidence of the small vibrations which combine to form the undulations from which the figures in question result, and for the knowledge of which we are indebted to him. The grains of sand which he uses are too coarse to indicate the nature of the movement of the solid molecules which tremble under them, and, what is even worse, these grains are elastic so that they do not remain where they fall but bounce around. This prevents them from following, and consequently from showing, a regular progressive movement, like that of sonorous undulations. It was on account of this that, when trying to vary these experiments, I thought that I would substitute for the sand a much finer material which did not have the inconveniences that I have just mentioned. I have found this in *Semen Lycopodii* (seeds of club moss). After having covered a metal or glass plate with this dust, I try to produce a sound in the manner of Chladni, and in an instant I see the dust distribute itself in a great number of small, regular heaps, which start moving towards the points where the figures discovered by this naturalist are formed. They always arrange themselves in the form of a curve whose convexity faces the point touched by the bow or a point which has an analogous position. The closer each of these small heaps is to these points, the higher they are, which gives a rather remarkable regularity to the whole. Mr. Chladni's experiments are astonishing to anyone who sees them for the first time on account of the regularity of the figures which are produced by a single stroke of the bow, as if by magic. My experiments do not have the same charm, but perhaps they are more instructive because of the comparative slowness of my process, which allows their effects to be studied more advantageously. The interior of the small elevations which are formed in my experiments is seen to be in continuous motion while the sound lasts. The duration of these vibrations, though very short, is still appreciable on plates of three or four inches in diameter, and it is longer when larger plates are used. In these experiments I have often made use of a metal disc six inches in diameter. In this case, I always saw the small elevations change their appearance at different times during the duration of the sound. At one moment the height increases, at another it diminishes, and the dust has the appearance of arranging itself in small globules which roll over each other. It is easy to see that all these phenomena are still very complicated. The movement of the grains is in part vertical, in part horizontal. The horizontal movement is composed of two others; one force impels the grain forward, the other drives it to the two sides. I was eager to examine each of these movements by itself as minutely as possible, and the following is a brief description of what I observed. I hold a square glass plate in such a manner that two or three of its edges are in contact with my fingers. I strike the edge which I do not touch with a piece of smooth wood, in such a manner that every point receives the impact at the same instant. In this case, the powder arranges itself in lines parallel to the edge struck. These lines are rarely straight, no doubt because these edges always have irregularities. The straightest I ever saw had been produced on an iron ruler, struck with a smooth piece of the same metal. In order to obtain the best result in this experiment, it is necessary to rest the square or the ruler on a smooth table. If I do not strike the blow on the whole edge but only at some of its points, other lines are formed parallel to the direction of the blow and perpendicular to the edge struck. These lines seem to be composed of many small elevations, less regular than those produced when plates covered with dust are stroked with a bow. This experiment succeeds with difficulty when the entire plate is allowed to rest on a table because this arrangement prevents us from striking a clean blow. Finally, if I strike the plate in a direction perpendicular to its plane, the dust accumulates in small, fairly regular heaps. In the first case, the undulations only proceed forward; in the second, they occur both in a forward and a lateral direction; in the third, they cross in all directions because the compression produced by the blow perpendicular to the plane must propagate itself horizontally as well as vertically. The tone produced in the first case is very low, that of the second is higher, but that of the third is even higher, and it is the only one which deserves to be called a tone or a sound which can be placed in the musical scale.

One would suppose that the change which is produced in elastic bodies by the communication of motion could scarcely be limited to the simple mechanical displacement of their parts, but that there ought to be some other more intimate action in this modification. All friction produces not only heat but also electricity. De La Place and Biot have already drawn the attention of physicists to the first of these phenomena. It is my opinion that the latter deserves even more attention. In my experiments I have always found that sand or powder adhered much more to those parts to which the movement of the sonorous body fixed it than to others. I have very often strewn fresh sand on a plate on which I had produced a figure. I shook it gently after having inverted it, and I always noticed that the sand which formed the figure adhered while the rest fell off. The adherence of dust which is finer than sand is very remarkable. With the aid of Coulomb's electrometer, I have also discovered indications of electricity in the plates which had emitted a sound, but I have not yet repeated these experiments often enough to be able to give details about them. On these occasions, I have discovered that the edges and the corners of bodies almost always have an effect on Coulomb's electrometer, and I propose making a new series of experiments on this subject. The celebrated Ritter, to whom I had communicated my experiments on the role which electricity must play in the phenomena of sound, had long ago discovered that Volta's electric pile is capable of producing sound when the ear receives a shock from it. In a work which is about to appear under the title of *System der electrischen Körper*,[3] this great physicist makes it clear that a body acquires positive electricity by compression and negative by dilatation. Therefore, we may say that in each sound there are as many alternatives of positive

[3] [*Das elektrische System der Körper*, Leipzig 1805.]

and negative electricity as there are oscillations, but the union of the two electricities produces a shock. Thus there are in one sound as many extremely weak electrical shocks as there are oscillations. Each of these individual shocks would be totally imperceptible, but when received in very great numbers, in a period of time
too short to distinguish one from the other, they always produce a noticeable effect,
especially because positive electricity makes the organ more susceptible to the
negative than it was before, and vice versa. The perceptible effect of the union of all
these imperceptible shocks is sound. I confess that these ideas of Mr. Ritter's appear contradictory to all the received opinions about the auditory organ, but it must
also be confessed that our knowledge of the organs of sense is still quite imperfect.
We must take care not to refute new hypotheses with old ones. In the present case, I
am of the opinion that all the facts known to the ancients will agree perfectly well
with Mr. Ritter's theory, but in order to study and appreciate this theory, it is necessary to wait until he has published all his discoveries concerning the senses, discoveries which will throw light on each other. As for my own experiments, they are
easy to repeat, and those who take the trouble will see more in them than I have
been able to describe, and perhaps they will discover things that I have not been
able to see.

I am, &c.

J. C. Ørsted, Dr. of Philos.

18

On the Harmony Between Electrical Figures
and Organic Forms[1]

THE remarkable similarity which the figures that electricity produces on dusted surfaces have to the forms of organic beings has often attracted the attention of students of nature, but they have only been able to suggest the full connection between them imperfectly because some of the most recent electro-chemical discoveries are needed to present this connection in a clearer light.

The primary form of positive electricity is the radiating point, of the negative, on the other hand, the circle so that one seems to form the internal, the other the external, one the point which radiates from its centre in all directions, the other the enclosing periphery. The natural symbol of electricity, then, is a circle with its radii, the symbol of positive electricity the radiating point, and of negative electricity the point surrounded by concentric circles. These symbols undoubtedly deserve our fullest attention, for they reappear everywhere, and who knows whether all of Nature's mathematics does not lie hidden in them! At least it is obvious that they underlie all electrical forms, and I hope that the following observations will clearly demonstrate that these are the primary forms of nature.

If we draw a line on a poorly conducting surface using the positively charged conductor of a Leyden jar and then dust it, we obtain a line radiating in all directions, which has a most striking resemblance to vegetation, but if we draw a line with the negatively charged conductor in the same way, we obtain a collection of parallel lines;[2] what was the circumference of a circle for the individual point becomes parallels for the line. These parallels unmistakably depict the internal form of the plant, the parallel fibres.

If a dissolved metallic calx is duly brought into contact with the electric pile, the metallic calx is reduced (deoxidized) at the pole which generates hydrogen from the water, and it also assumes a vegetative form, whose beauty is often a most pleasant spectacle for the eye. If, for this experiment, we choose a metallic calx which can be combusted even more completely, for instance, lead as it is found in acetate of lead, we obtain, on the oxygen side, a highly combusted (even maximally oxi-

[1] [*Det Skandinaviske Litteraturselskabs Skrifter,* Vol. 1, pp. 1–22. Copenhagen 1805. KM III, pp. 96–105. Originally published in Danish.]

[2] It is true that we sometimes get a series of perfect circles, but this has not been the general rule in my experiments.

dized) lead calx, which precipitates from the solution though without assuming that vegetative form, its surface being smooth or bounded by parallel lines.[3] Sometimes it is also deposited in well-defined figures, but these are not ramified like the positive ones and have much greater similarity to the negative ones. Such vegetations are also obtained when the conductors of an electric pile are held in the flame of a candle, where the deposited soot appears ramified on the conductor of the hydrogen pole, while it forms more conglomerate figures on the conductor of the oxygen pole. They have a greater resemblance to the roots of the plant than to its branches. Thus, in the simplest, purest experiments, which actually serve as a basis for all the chemical discoveries of more recent physics, we find the process of reduction (deoxidation) united with the external form of vegetation, whereas the process of combustion is accompanied by a form whose boundary is the circumference of a circle when it radiates from a central point or parallel lines when it radiates from a central line, that is, we see in it the norm of the internal form of a plant. Therefore we should expect to find the same formations everywhere in nature, assuming that the same form must follow the same force unless the effect of foreign forces changes it.

We need only glance at nature to find our assertion confirmed. The plant lives solely by the influence of sunlight, and thereby it constantly generates oxygen gas and is deoxidized or reduced. The same form and the same chemical process which were united in the electrical effect are so here, too. Internally, however, the plant must oxidize. This follows from the fact that it deoxidizes externally, whereby it would otherwise reach the maximum of deoxidation and cease to be an organism, for which an interplay of forces is necessary. There is, however, another reason, which can be found in the nature of the plant juices themselves. These are acidic, and those that are not to any noticeable extent still have a tendency in this direction so that they are always acidified by fermentation. Thus we discover the same agreement between form and force in the interior of the plant as in its exterior, and in both the most perfect similarity to what we have seen in electricity. We could add that plant fibres appear as parallels only when viewed from one direction, that is, lengthwise, but when viewed crosswise, the circle is the dominant figure and forces us to acknowledge the negative in the interior of the plant, in every direction.

The same grand design by which the electrical forms appear in vegetative nature is also found in animal nature. However, the animal is the antithesis of the plant. As the latter turns its ramifications outwards and its parallel lines inwards, the exterior of the animal, conversely, consists of parallels or circles while its interior is a web of ramifications. Yet, is the chemical state of the animal not also the opposite of that of the plant? With its breathing, the animal is engaged in a continuous process of combustion, whereas the circulation of the blood reduces it again at every instant and in this way prepares it to be combusted again in the lungs. The life of the

---

[3] The negative pole of the electric pile is the one which generates hydrogen gas, the positive the one which generates oxygen gas. It might seem strange that the positive pole of the electric pile behaves exactly like negative electricity, and vice versa, but we need only remind ourselves of the fundamental law that one kind of electricity always induces the other, its opposite, and all doubt will be resolved. Hydrogen must then be regarded as the positive pole of water precisely because it appears at the negative conductor of the pile, and conversely oxygen is the negative pole of water because it appears at the positive conductor of the pile.

animal and that of the plant are thus both an unceasing struggle between combustion and reduction, with the only difference that in the animals the process of combustion moves from the outside in, and the process of reduction from the inside out, while exactly the opposite takes place in the plant.

If we proceed to consider the distribution of the processes of combustion and reduction in organisms, we readily discover that the process of combustion is dominant in the breast of the animal and the other parts which most directly get blood from it, whereas the process of reduction is dominant in the extremities. However, are the extremities not an incipient ramification, the vegetative part of the animal? Conversely, is the lower part of the plant, the stem and the root, not the one in which the process of combustion is dominant? Thus the plant is the animal reversed top to bottom. This contrast between the two organic realms of nature is not a completely new observation; a long time ago, perceptive students of nature declared the root to be the reversed stomach, the stomach turned upside down and inside out.

Animals and plants, then, are contrasts, and this contrast appears not only in their forms and in the chemical processes which take place in living organisms but even in their isolated constituents. Thus fermenting plant juices generate acid, fermenting animal fluids alkali.

Another contrast, which is most easily illustrated with examples from the vegetable world, is the sexual difference. If we dare agree with Linnee[4] and regard the pistil as the continuation of the pith of the plant, and the stamen as the continuation of the ligneous part, the sexual difference is obviously a contrast between the general + and − of the plant. Fertilization, then, becomes a discharge of the same nature as the electrical, a statement with which I do not pretend to have explained fertilization, in which all the forces of nature must necessarily play. The application of this to animal physiology is easy and can be made by anyone who has understood the preceding observations.

I hope that we have now discovered the main differences between animal and plant forms, but under each of these there are infinitely many subsidiaries, each of which again has its subsidiaries, and so on ad infinitum. It is quite impossible to make an exhaustive list of these; even a more detailed version than the one given here would constitute an essential part of physiology. As one of the very striking contrasts in the details, I merely want to mention hair as animal vegetation. Of course, such contrasts must confuse one who is not used to finding his bearings in that infinitude of contrasts, of + and −, which itself appears in every greater + and −.

Perhaps objections are more likely to come from another direction, which is with regard to those animals and plants where the indicated contrasts seem to disappear, that is, the least complete animals and plants, but if we just remember that animals and plants are in contrast, it is easy to understand that the most incomplete animals as well as the most incomplete plants must be found at the point of indifference.[5] Therefore, neither the positive nor the negative form can be predominant, but the lack of definite form will be characteristic of them both.

After what we have demonstrated so far, we are now entitled to regard the pro-

---

[4] [Linné.]   [5] Steffens's introduction to his lectures on philosophy.

cesses of combustion and reduction, which pervade every organism in an eternal interplay, as the formative process of nature.

However, a formative process presupposes a substance, for pure oxygen or hydrogen shows nothing but a tendency towards all forms, the gaseous form, their combination, water, nothing but indifference to all form, fluidity. This substance must continually be supplied to the organism as nourishment. In other words, the formative process presupposes a nourishing process. What this nourishing process consists of is easily discovered at a glance. In the animal it is obviously nitrogen, in the plant carbon, which constitute the primary substances, an assertion of which chemical analyses convince us completely. Even if chemistry succeeded in proving that nitrogen and carbon are not what they have so far been thought to be, nothing would result from this which could be at variance with the real difference which is found between the constituents of the animal and of the plant, or rather their chemical relation. Such a discovery could only give us more perfect insight and perhaps force us to adopt other scientific terms than the ones we use at present.

As this nourishing process is not the object of the present work but merely needs to be mentioned as necessary for an understanding of the formative process, I shall be content with adding a few observations here.

For those parts of the animal body in which the nourishing process takes place, every substance seems immediately serviceable, that is, nourishing, to the extent that it contains nitrogen, stimulating to the extent that it contains hydrogen. At least, the most animal foods are the most nourishing, while the substances in which hydrogen prevails, like alcohol, naphtha, ethereal oils, etc., are among the most stimulating that we know. Phosphorus also belongs among the most stimulating and most combustible. The fact that chemists have not yet found hydrogen in this substance cannot prevent us from ranking it among the other highly combustible substances. Oxygen, at least as represented by the acids, is depressive to the digestive organs. On the other hand, alkalis, as their opposites, are directly stimulating. As far as carbon is concerned, what we most lack are experiments. However, we may assume that carbon, as the opposite of nitrogen, has a retarding effect on the digestive process in animals. In so far as its quality could become dominant (in so far as it could become the characterizing constituent), it would also become a poison to the animal. Hence, the vegetable poisons. It is strange that animal poisons, in which azote is surely dominant, in Fontana's experience, only act externally, that is, in open wounds, but do not kill when administered internally. The external effect of vegetable poisons has not been investigated thoroughly, but it is hardly important. It goes without saying that plants can produce poisons in which carbon is not dominant and animals poisons in which nitrogen is not, and therefore what has been said here cannot be applied to all animal and vegetable poisons.[6]

---

[6] The fact that plants also provide nourishment is not at variance with the views on carbon stated here partly because plants contain more or less nitrogen and partly because the carbon in all the nourishing parts of the plant is neutral with respect to oxygen. It is only our chemical experiments which liberate the reactive carbon from it. We can also convince ourselves of the correctness of this statement by observing that the meat of graminivorous animals is very rich in nitrogen, which it can only have obtained from vegetable matter.

The opposite of what happens in the digestive organs must happen in the respiratory organs. In these, oxygen is stimulating, hydrogen depressive, carbon (in the vascular system), which belongs here, is precisely the nourishing element, and at the same time it continually provides the substance of respiration (to form carbonic acid gas). Azote, on the other hand, must be regarded as retarding, and animal poisons[7] now become evidence that nitrogen here is the same as carbon in the digestive organs. The fact that both carbon and nitrogen are fatal when inhaled as a gas or in gaseous combinations must partly be explained by their privative effect. However, carbon there seems to display the effect of a poison, which is easy to understand since the purpose of breathing is to liberate carbon, whereas the inhaled carbon is immediately ingested. Consequently, oxygen and hydrogen, carbon and nitrogen are also here revealed as the 4 chemical elements, the two former corresponding to the contrast in electricity, the two latter to that of magnetism, as our great natural philosopher Steffens first proved. Carbon and nitrogen appear in chemical action, like magnetism in nature, in internally determined forms; oxygen and hydrogen, like electricity, as eternally mutable, striving towards new forms. The life of nature could not exist through any of them alone. One of these contrasts would fossilize into a single, fixed, unalterable form, the other, finding no opposition, would rush from one form to another without permitting any of them to have even the shortest duration. Form, then, emerges only through the interaction between the chemical representatives of electricity and magnetism, yet in such a way that carbon and nitrogen must be regarded as that which is formed, oxygen and hydrogen as that which forms. These forms, which we have here shown as the companions of combustion and reduction, reappear in inorganic nature. Steffens has already pointed out how the granular fracture occurs primarily in the minerals which, in accordance with his discoveries, he places in the silicon series, whereas the radiant fracture is dominant in the ones which belong in the lime series. However, the lime series is alkaline while the silicon series is closer to the acids,[8] or the former approaches the hydrogen pole, the latter the oxygen pole. The same philosopher has also pointed out how the radiant fracture is prevalent in coherent, brittle metals, whereas the granular is found only in the more coherent ones, and that the radiant fracture corresponds to the positive electrical figures and the granular to the negative. The origin of this great formative process, which thus acts throughout the globe, must undoubtedly be sought in the nature of its construction. Alternating layers of substances containing carbon and nitrogen form its surface as far as we have been able to penetrate it. Anyone who knows the construction of a so-called galvanic battery must see that such an alternation serves to induce an electrical process. The substances containing carbon form the negative, the substances containing nitrogen the positive elements. The remains of prehistoric vegetation are found in the layers containing carbon, those of prehistoric animals in the ones

[7] The fact that Fontana, in his chemical analysis, found nothing but rubber in animal poison speaks against the analysis but not against the opinion stated here.

[8] This may be compared with what I have said about earths and alkalis in my review of Gadolin's *Indledning til Chemien*, Skandinavisk Museum. [For the year 1800, pp. 177–90. Copenhagen. KM III, p. 51.]

containing nitrogen.[9] This strange construction of the earth not only goes from its centre to its periphery, but in another form it goes from north to south. To the north, carbon is prevalent, which is indicated by the enormous number of forests, peat bogs, coal, etc., but to the south, nitrogen is found more often, which is demonstrated by the many coral mountains. To treat all this in greater detail here would be quite beyond my intention, but on that subject, I may refer to Werner's and Steffens's great discoveries, on which further information can be found in the latter's *Beyträge zur inneren Naturgeschichte der Erde*. Here we must be content with the result that carbon and nitrogen appear on the earth itself as the representatives of magnetism and form the basis of an electrical process, which, however, is far too bound by the rigidity of magnetism to appear as a formative process until some external force liberates it.

This external force is light. It is well known that sunlight has a deoxidizing effect on our globe, and if we had no other proof of this than that plants liberate oxygen in daylight, this would already be sufficient. With the departure of light, on the other hand, the process of combustion which the oxygen in the air is constantly seeking to initiate is given free rein. The day, then, is deoxidizing, the night oxidizing. The same relation reappears on a larger scale between summer and winter. Briefly, a constant process of combustion and reduction proceeds from east to west, the same electro-chemical process which we have demonstrated in the animal and vegetable kingdoms.

Steffens's glorious idea to regard oxygen and hydrogen as representatives of east and west, and carbon and nitrogen as representatives of north and south is then confirmed in the most perfect way, however paradoxical it might appear to all those who are not informed about recent physics.

There remains for us the firm result that the electro-chemical process is the formative process of the entire globe, and that the shape of the earth, the four corners of the world, exists through the struggle between this formative process and the pre-assumed shape of the earth, magnetism.

In order to shed even more light on all this, I want to add some information.

Shelling[10] has shown that three instances must be distinguished in the construction of matter by the attractive and repulsive forces. The first, in which the contrast between these two forces merely assumes the form of the line, the second, in which it is in the form of the surface, the third, in which both these interpenetrate and thus form the final dimension of space and matter, depth. Every time one body produces an internal change in another, whereby matter is really reconstructed, one or more of these actions must reappear. Thus, the longitudinal function manifests itself as magnetism, the latitudinal function as electricity, and the depth function as penetration or a chemical process. Each of these dynamic processes is the interaction of opposite fundamental forces in a different form. The transition to form occurs when a magnetic, electrical, or chemical plus and minus are roused in a homogeneous substance, or, in other words, when indifference becomes indifference.[11]

---

[9] It must not be forgotten that the negative pole of the electrical pile induces the positive process and vice versa.

[10] [Schelling.]   [11] [difference.]

One of numerous examples of this is the change of water into hydrogen and oxygen and the return of hydrogen and oxygen to water. The phenomenon of the process of indifferentiation is light, which is seen clearly when the chemical or electrical + and − neutralize each other, and the northern lights even seem to provide the empirical proof that the indifferentiation of the magnet also produces light. However, if the appearance of light is the phenomenon of indifferentiation, destruction of form, the disappearance of light must be accompanied by differentiation, formation. The correctness of this assertion is seen clearly in the best-known experiments. It would be superfluous to describe here in detail how colourless light, by its differentiation (refraction) through the prism, changes into colours, how it causes combustion and reduction, how the growth of vegetation is dependent on it, and how crystallizations are promoted by it. All this is dealt with in many places. Here I only want to point out how light, in its differentiation, assumes all the forms of the construction of matter in three different processes. The first fundamental law of light is that its effect spreads along straight lines, or that it is in the form of the first dimension. When light strikes an opaque body and thereby partly disappears, it necessarily excites an internal effect in the body, for, in order to neutralize the effect of light, another force must act against it, as its opposite. Thus, a real act of darkening inside corresponds to the act of illumination outside. This is the + and − of the first line. If the straight ray of light is refracted through the prism, a new effect arises, whose phenomenon is the spectrum. The direction of this effect is precisely perpendicular to that first straight line. The colour process is that of the second dimension or the surface. Finally, if this light, which has been changed by differentiation, falls on a chemically susceptible substance, a combustion is produced at the red pole of the spectrum, as Ritter has proved, a reduction at the violet, or, in other words, where the first and the second actions of light interpenetrate, the third arises, the chemical.

Thus, light appears as the formative principle in nature, and the one which reveals all forms to us is the one which gives all form and colour.

In this short outline I have merely wanted to suggest the connection which we find in all nature between force and form. Thereby I wished to prepare for more detailed investigations of this important subject. The ideas put forward here might serve as an introduction which provides a survey of the whole question although its further development might well go in a different direction from these preliminary observations.[12]

---

[12] A bibliographical note may serve to point out that Ludolf Christian Treviranus, in his *Untersuchung über wichtige Gegenstände der Naturwissenschaft und Medicin*, Vol. I, has made investigations of animal and vegetable forms which quite agree with the ideas put forward here. The most superficial comparison will show, however, that I have not borrowed from him but found my ideas in a completely different way, from which a far wider perspective also opened to me.

19

New Investigations into the Question:
What Is Chemistry?[1]

AFTER so many talented men, through several centuries, have endeavoured so eagerly to systematize chemistry, it might, at first glance, appear ridiculous to wish once more to pose the question: What is chemistry? It seems unreasonable to assume that so many clever men should not only have occupied themselves with chemical investigations but also sought to fit all chemical knowledge into a system, without first having asked and, through accurate measurement of the scope of the science, answered this question. I know all too well that many who hate all radical changes in science and, like inanimate objects, prefer to remain in the state in which they were once placed, would find sufficient cause to condemn the investigation which I intend to make here. Such people would follow me only half-heartedly when I show that it is in the nature of things that the fundamental concept of a science develops with the science itself, and that only complete truth can crown the complete science. They would fail to see when I show that chemists have so far disagreed on what to include in their science, and that their definitions of it contradict each other either openly or in more subtle ways. Least of all dare I hope to satisfy them when I try to demonstrate that the boundaries of chemistry are far wider than has hitherto been assumed, and that, in its new form, it will constitute one main part of physics while the doctrine of motion constitutes the other. However, I have nothing to say to those who *will* not investigate. It is not my wish to drown out their cries but, if possible, to use clear and strong arguments to convince those who do not believe that *opinions can gain prescriptive right*, or that *the sciences can be classified* by convention. Perhaps they are more likely to follow me if they know beforehand that my intention really is to unify the results of the various investigators, as at a focal point, in order to elevate science to a higher level.

Our sciences have not come into being through philosophical contemplation of the nature of things and an overview, based on this, of all the branches into which human knowledge could be divided. They have rather begun with a decision to undertake specific tasks. The solution of these led to others, and so forth. In this way,

[1] [*Det Skandinaviske Litteraturselskabs Skrifter,* Vol. 2, pp. 240–63. Copenhagen 1805. KM III, pp. 105–16. Originally published in Danish.]

systems of truths arose naturally. In these systems it was discovered, sooner or later, that each part pointed towards a common problem, the solution of which would provide the key to all the others. Thus the sciences came into being. However, it is easy to understand that no-one on such a course could be certain of discovering the full scope of the science, and that the conception of the science could not extend beyond the totality of knowledge acquired. This is not to deny that a secret urge has often led men of genius beyond these fixed limits, but they have not always arrived at the philosophical consciousness of their real objective which is so necessary in order to influence the masses. Chemistry is eloquent proof of this. It started with desultory experiences. Before seeking a science, people endeavoured to solve practical problems for which chemical forces were required. The fermentation of wine, the melting of metals, the dissolution and crystallization of salt, and the like, early occupied humans without thereby suggesting the creation of a science. The variety under which Nature hides her unity did not at first allow them to perceive the coherence of all phenomena. What human perspicacity could have discovered at first glance that it was the same forces which revealed themselves in different guises in combustion, in respiration, and in the calcination of metals? Chemistry, then, could not begin from its true point of unity, but the knowledge of chemical phenomena had to be gathered little by little, from a variety of scattered observations into larger and larger coherent units. The bodies which, being the least susceptible to interference, are most easily restored to their original form after each transformation were necessarily the ones about which some coherent knowledge was first obtained. Therefore, metallurgy was the first chemistry. Throughout the Middle Ages there was, properly speaking, no chemistry other than that of metals. The main problem, which is probably the most important question for the chemistry of metals, was the production and transformation of metals. All other problems which alchemy addressed revolved around this, as around a common centre, which is why it was generally assumed that the solution of this problem would lead to that of all the others. With increasing experience it became clearer and clearer that the same forces which are active in the transformation of metals also play a role in all other cases where two bodies seem to combine into one, or where one body seems to separate into several different ones. Thus, chemistry began to mean: the laws of the combination and decomposition of bodies. However, even this definition has not satisfied all chemists in more recent times. They seem to have felt the necessity of a more general expression, and several have tried to provide one, each in his own way. Fourcroy[2] calls chemistry a science of the internal and mutual effects of bodies.

In order to determine what chemistry is, Hildebrandt[3] begins by showing that all effects can occur either through a spatial change, in which case it is mechanical, or

---

[2] *Système des connaissances chimiques* [sic], Vol. 1, p. 4. Fourcroy uses expressions that show how little the definition of chemistry has hitherto been determined by philosophy. He says, "The true definition which can be given in the present state of the sciences must be more general (than the old one). This is the one that I adopted twenty years ago." Thus, he talks about this definition in the same way as about a knack that one has long been comfortable with.

[3] Hildebrandt's *Encyclopædie der gesammten Chemie*, Vol. 1, p. 40.

through a change in the properties of matter, in which case it is chemical. According to these various definitions, the boundaries of chemistry had to be set very differently, for it cannot be denied that many changes could take place in the nature of bodies, or in their interior, without the constituents being changed thereby. However, this difference in definition has not had much influence on the boundaries given to chemistry. After all, it has always been tacitly assumed that chemistry was the science of the combinations and decompositions of bodies,[4] and it can be justly claimed that no chemical textbook has yet gone beyond this concept. Therefore, since chemists in reality agree on the definition of their science, perhaps their conflicting statements are rather the result of a misunderstood attempt to abstract than the fruit of some notion about the extension of the science? This would seem likely if they agreed as much on the boundaries of their science as they tacitly seem to agree on its definition, and if this definition really permitted a consistent treatment of the science, but its acceptance will create a state of unavoidable uncertainty about the permissibility of including several important matters in chemistry. If the cause of heat is material, its theory must be included in chemistry according to the accepted definition, but if it is not, it must be excluded from chemistry since temperature changes, in that case, are not due to decompositions and combinations. The same is true of light, of electricity, and of magnetism. Therefore, Scherer[5] has declared that heat and light should not be dealt with in chemistry as he could not consider their causes to be material. So far no-one has ever thought of including magnetism in chemistry even though so many physicists considered its cause to be a substance. Nor have chemists so far wanted to take possession of electricity. Not until recently have two Spanish chemists included electricity in the chemical system, and in his textbook Trommsdorff has dealt with galvanism in a very limited way.[6] This difficulty regarding the present definition of chemistry can be illuminated even more fully. In order to agree with several well-informed physicists in denying that the cause of electricity is a substance and with Ritter in assuming that water is not composed of hydrogen and oxygen and claiming that these two substances are only water in opposite electrical states, it would be necessary to deny the theory of the transformation of water into oxygen and hydrogen a place in chemistry. Then the combustion of hydrogen would not be a chemical process. In short, there would in the end be nothing to which the name of chemistry could be given.

I believe that all of this urges us strongly to make a serious investigation of the boundaries of chemistry. It was easy to predict that these could not be found correctly without our going beyond the science itself, for we discover the boundaries of a thing by looking not at its parts and its contents but at its relationship with other things. Moreover, we must look not at what chemistry is but at what it ought to be. By ascending to the investigation of the necessary parts of all natural science, we would certainly also encounter that of which our present chemistry is only

---

[4]Hildebrandt, who has given the definition its most general expression, immediately restricts it by saying, "All chemical effects are either combinations or dissociations."

[5]See the first treatise in his *Nachträge zu seinen [den] Grundzügen der neuern chemischen Theorie.*

[6]*Chemie im Felde der Erfarung [Erfahrung]*, Part 6.

a fragment. Our natural science is obviously divided into two large primary parts. One teaches us to recognize objects, the other endeavours to make us familiar with the laws according to which they behave. Nobody will deny that chemistry belongs to the latter class, or to so-called natural science (physics) proper. Consequently, its parts must be examined more closely. Bodies are only able to act on each other in two ways, either in such a way that one forces the other, or its parts, to move, which is called *mechanical*, in the broadest sense of the word, or in such a way that one produces a change in the nature of the other, or that properties which do not depend on shape and motion are altered. Here, then, bodies act on each other through their properties, but an active property is a force. Consequently, the second part of physics must investigate the forces of bodies. These might be composed of others. Therefore, those last or fundamental forces, on which all others depend, must be discovered, if possible, and once they have been found must serve as the starting point. This part could justly be called dynamics, a word which, it is true, thereby acquires a completely different meaning from the one it has had so far.

It is easy to see that we have found in dynamics the part of natural science to which chemistry belongs, for no-one can deny that all the changes which have been called chemical are changes in properties. However, a considerable difference appears between what has so far been called chemistry and our dynamics in that the former examines only the effects, the latter, on the other hand, primarily the forces. And so it must be in the nature of every science. In experience, effects may put us on the track of causes, but it falls to the sciences to grasp the causes and to deduce from them all effects, which must then be rediscovered by experience, so that this becomes the touchstone of science.

Though I have now tried to establish the concept of chemistry as synonymous with dynamics, I would have accomplished but little towards a determination of the boundaries of this science if I did not apply it correctly, for our definition of chemistry is not very different from several of the usual ones. In order to determine more precisely what is to be treated in chemistry, we will first examine the chemical forces. However, in this investigation we cannot go into great detail but are forced to abide by results of experiments about whose validity confirmation can be obtained elsewhere or in the following treatises.

But what are the fundamental chemical forces? And what are the forms in which they work? Using philosophical arguments I could easily demonstrate that two opposite fundamental forces are at work throughout nature, in alternating expansions and contractions; I could show how their effects have as many fundamental forms as space has dimensions, and finally I could point out that all these forms, to varying degrees, must be discernible in every effect. However, those who do not fear the effort that an investigation of this kind entails would find themselves sufficiently satisfied by modern philosophers. It is my intention to persuade chemists, who so far have wanted to build their science only on experience, at least to look at this experience with less bias and not to believe that a limited concept of science can be maintained by any convention.

In seeking the chemical forces, then, we will look to experience, which teaches us that any friction rouses forces which before lay dormant in bodies. These forces show themselves not only in attractions and repulsions, but when their effect is concentrated, they also generate light and heat, change water into air, promote combustions, etc., that is, they interfere in the strongest possible manner with effects which we have hitherto considered to be chemical. Through these forces solid bodies can change their internal state so that they seem to be completely different from before, even so that metals, at will, can be made more or less combustible than they otherwise are according to their nature. However, when combustibility or, to adopt the usual chemical language, the affinity to oxygen can be arbitrarily increased or decreased by the application of such forces, these forces are seen to influence the property upon which virtually all of our chemistry depends, and their operation must be investigatively pursued until we see their transformation into what we usually call chemical effect.

It is strange that chemists have considered the affinities to be efforts towards combination but never thought that these efforts must also cause some change even when they are not able to produce a combination, for a force does not stop working because it encounters resistance which does not allow it to have its utmost effect. Recent experience has also shown us how to prove this experimentally as we can see that a more combustible metal, placed in contact with one which possesses this property to a lesser degree, becomes even more combustible. Very beautifully, proof of the same truth can also be seen in an experiment by Ritter, in which a wire of platina, gold, or copper, placed in dilute sulphuric acid or muriatic acid, liberates gas as soon as a piece of zinc is placed next to it in such a way that one end touches the acid, the other the original metal. Here the combustible metal evidently produces a changed chemical property in the more[7] combustible one without any combination, which is otherwise considered necessary for all chemical effects. This experiment can be varied in numerous ways, and I myself have very often repeated it in the most different forms.

The fact that we have many excellent investigations of the chemical forces can hardly escape notice, for what are the forces that we have been talking about but the opposite electricities? However, so far these have not generally been acknowledged as chemical, and therefore these investigations have not been useful to chemistry.

However, the same forces which manifest themselves in electricity also manifest themselves in magnetism, although in another form. Attractions and repulsions are the same in magnetism as in electricity, opposite forces attract, like ones repel each other. Through magnetism two pieces of iron can be made to produce the same effect on a prepared frog as two different metals. If an iron wire is magnetized, the end which becomes the south pole will become more combustible than it was before, but the one that becomes the north pole will lose some of its combustibility. Ritter has convinced us of this through many experiments whose validity can easily be ascertained through experience. Consequently, the same forces are at work in electricity and in magnetism.

[7] [less.]

Heat also seems to be produced by the same forces, for where the two opposite electricities combine, both heat and light are generated, depending on the specific circumstances under which the experiment is made. Similarly, friction generates both heat and electricity, and in particular the former when the conditions for an electrical indifference[8] (the separation of the two opposite electricities) are not present. However, if heat is nothing but the phenomenon of the struggle to combine[9] of the same forces which are found separated in electricity and magnetism, and this will be further proved in a treatise on heat, we are compelled to assume that these forces lie dormant in every body and in each of its parts so that they must be considered absolutely necessary for their constitution. We need only try to beat a metal wire or a metal bar, and it will soon reach an appreciable temperature. We remove this by cooling it in water, and renewed hammering will give it new heat, and so on, as long as any part of the metal bar or wire remains. Thus we can deprive a body of as much heat as we wish, there will still be dormant forces left which need only be awakened to produce new heat. It is as if the whole body could finally be dissolved in heat. And as heat is nothing but the conflict between the same forces which are at work in magnetism and electricity, we see from the experiment on heat what role these forces play in bodies. We could at least presume with justification that the forces considered here are the last into which any experiment has penetrated. Philosophy shows even more: They are the ultimate at which any construction of matter can arrive.

A brief outline of what we know about the effects of these forces is sufficient to show us the possibility that all the different forces of nature can be traced back to those two fundamental forces. How could there be three more different effects than heat, electricity, and magnetism! Yet, all of these are due to the effect of the same fundamental forces, only in different forms. Magnetism acts only in a *line* which is determined by the two opposite poles and the intermediate point of equilibrium.[10] Purely electrical effects only follow *surfaces*.[11] Heat works equally freely in *all directions* in a body. It cannot be denied that this difference actually exists. Only a detailed investigation can truly convince us that it is essential. However, it can hardly fail to attract the greatest attention that these three effects assume forms which correspond to the three dimensions of space and their realizations: line, surface, and body. It seems obvious to me at a glance that there can be no additional forms for the effects of the fundamental forces, but philosophy alone can provide complete certainty in this matter.

However, when heat reaches its greatest manifestation of strength, it changes into light, just as light is changed into heat when it loses its intensity. The best-known facts speak so strongly in favour of this that it must be attributed to theoretical confusion that it has not always been recognized. Through this change heat is made to radiate in all directions in straight lines, but in such a way that it fills space.

---

[8] [difference.]   [9] [*Forenings-Kamp.*]

[10] Of course a body can be magnetized along several lines, but then we rightly consider these to have been produced by several different magnetic processes.

[11] When it seems otherwise, it is already changing into magnetism.

Heat, which previously spread according to the form of bodies, is now seen at a *higher level* to move according to the form of the *line*. The spectrum is the dispersion of light onto a *surface*, but in a *higher form* than that of the electrical *surface*. The proof of this is not that it is primarily surfaces which show colours, but that a beam of white light is *dispersed* in the sense that it is transformed into coloured rays. Finally, during its elevation to a surface, light awakens a process of *combustion* and *reduction*, which Ritter's experiments with light demonstrate. It is easily seen that this is a process of the same form as heat but of a higher order.[12]

In our investigation we have now arrived at the first chemical process proper. It would be easy to proceed from this point, but this should be sufficient to demonstrate their relationship to the most basic forces. It should be obvious from the above that all of the science which has hitherto been called chemistry must have these investigations as a necessary prerequisite. It is also abundantly clear that what used to be called chemistry constitutes only a chapter of dynamics. Even if we wanted to limit it to this, it could still not be considered an independent science, and its limits could not even be determined until as much of the contents of physics as we have indicated here had been determined. Our investigations of the boundaries of chemistry have also shown us a new example of how uncertain all external determinations of the boundaries of a science are as long as its true principle has not yet been discovered. It seems quite obvious from this that the true definition and construction of science are inseparable, and that the speculative approach to science is the very opposite of that of experience. The latter merely comes up with scattered objects which invite reflection, and from this an arrangement into coherent elements emerges. The former seeks the first principle of everything, sees what constructions must result from it, and prefers to offer the most fundamental construction of science as its definition.

Now that we have demonstrated that what used to be called chemistry is only a fragment of a far higher science, we could in some sense be said to have destroyed it rather than determined its boundaries, for we have really shown that so far we have only collected fragments, under various names, for this science, and that chemistry constitutes only one of them. On the other hand, it cannot be denied that chemistry has influenced virtually every chapter of the dynamical part of physics, and that the old chemistry consequently became its main constituent. If we desired, we could therefore retain the venerable old name of Chemistry for this entire part of physics. For most people, this name expresses something hidden and mysterious,[13] and that is certainly appropriate for the internal forces of bodies.

It is not at all necessary for chemistry to be only experimental. On the contrary,

---

[12] On another occasion I shall demonstrate that electricity and magnetism cannot produce a chemical effect directly but only after having been intensified.

[13] The correct derivation of the word chemistry is completely uncertain, but most people agree that it means something mysterious. That people in old times understood chemistry in a broader sense can be seen from a passage in Zosinius Panapolita [Zosimus of Panopolis], in which he recounts the myth that the angels taught the women they loved *all the effects of nature* (ἐδίδαξαν αὐτὰς τὰ τῆς φύσεως πάντα τὰ ἔργα), and called the book about it *Chema,* the science *Chemia.* Boerhave [Boerhaave], *Elementa chemiae,* 1732, pp. 5–6.

it it evident that experiments can make us familiar with only a small part of nature. The existence of each body undeniably depends on the effects, both past and present, of others. The effect and existence of these are again tied to those of other bodies, and so on, until the complete circle of possible effects has been formed, that is, the universe. This makes it clear that a universal construction is necessary to complete science, and it stands to reason that such a construction cannot be given through experience but can only be expected from speculation.

Incidentally, it is far from being our intention to lower the value of experimental physics. Nature, or the universe, is the construction which must be reconstructed by thought. However, nature is infinite, so we cannot construct it differently from the way we construct an infinite series, that is, by presenting a certain part of it and from this deducing the laws for the whole. In this way the construction of a finite part of nature can be extremely instructive. In experiments we force nature to make a construction or rather a reconstruction in front of our eyes, and what reconstruction could be more instructive for us than the one Nature itself shows us if only we know how to see it. Experimentation, therefore, is the true art of the physicist, and if he has thus, with open eyes, really seen part of nature reconstruct itself, he can, from this point, survey or, at least, sense the coherence of all nature.

These considerations should suffice to ensure the value of experimental physics and of experimental chemistry as well. On the other hand, we are forced once again to limit them somewhat. We do not know nature because we know the forces of nature. We must also see how nature uses these forces in its entire economy, or, to perceive it from a higher and truer point of view, we must acknowledge all of nature as an expression of these laws. For instance, it is not enough that we know the laws of the processes of combustion and reduction. We must also know their movement across the entire globe, indeed, if possible, through the entire solar system, through the entire universe. However, as none of the physical processes is completely isolated but is connected with others, it follows that the science which we are discussing here cannot be divided into two parts, like physics itself, but that it must constitute a single, organic science, in relation to which experimental physics only serves as a means. We have fragments of such a science, for example, physical astronomy, geology, and meteorology, but the complete science does not exist yet and can never be reached by the path of experience. It is well-known that Schelling, through speculation, has produced an attempt which, as such, is of incalculable value, but the combined efforts of a great number of blessed geniuses are probably required for the accomplishment of this task.

20

An Attempt towards a New Theory of Spontaneous Combustion[1]

EXPERIENCE teaches us that there are bodies which have the property that, under certain circumstances, they are able to burst into flames without the application of any external fire or heat. These bodies have been called *spontaneously combustible*. The observations of chemists have familiarized us with several of these, and sad experience has made us aware of others. The bodies and their combinations which have proved to be spontaneously combustible have been diligently recorded. Both the intrinsic interest of the phenomenon and the importance of the matter in practical life urged this. However, in spite of all the diligence applied to collecting experiences of this subject, researchers have not been fortunate enough to provide a general explanation of it, a general theory of spontaneous combustion. They have tried to indicate the cause of some spontaneous combustions in particular, but they seem not to have thought very much about a general cause of spontaneous combustion,[2] and yet it would be very important to know that as it often produces such dangerous effects which could be avoided if foreseen.

I think that I have been fortunate enough to find a general theory of spontaneous combustion which makes it possible, if not to determine rigorously then at least to conjecture with good cause what bodies are spontaneously combustible.

However, before we progress any further, it will be necessary to specify the concept of spontaneously combustible bodies somewhat more precisely. The process which in daily life we call combustion, which has such an intense effect on our senses and compels their attention, can also occur so slowly and be so weak that our senses do not even perceive it, and that it can be detected only by experiments and particularly careful observation of all circumstances. Such a slow process of combustion takes place on the surface of all combustible bodies which are in contact with air. On the least combustible among these, it may happen so slowly that it will not be possible to notice any trace of it for several hundred years, but on others

<hr />

[1] [*Det Skandinaviske Litteraturselskabs Skrifter,* Vol. 2, pp. 487–517. Copenhagen 1805. KM III, pp. 116–30. Originally published in Danish.]

[2] The few attempts which have been made to provide such a theory are so incomplete and have won so little approval that it would be useless to try to refute them here.

it is so vigorous that we notice a light on their surface, especially in the dark. If we wanted to talk about this kind of slow combustion, we would have to assert that all bodies are spontaneously combustible in so far as they have not been burned, but it is evident that this cannot be the subject of the following investigation. However, we must go even further because in this general spontaneous combustion we see very different degrees, among which that of phosphorus is so high that it emits light which is perceptible to the eye and heat which is detectable by means of a thermometer. In the series of these general spontaneous combustions, there is nothing to prevent us from imagining some bodies that actually burst into flames. Not even spontaneous combustions of this kind could be the subject of the present investigation as they are to be regarded only as degrees of a general natural process, and we would immediately possess their theory if we had a general theory of combustion.

On the contrary, our investigation aims at discovering why many bodies which are very little combustible in themselves can become much more combustible than before when combined with others which do not possess this property to a high degree either, or, in other words, how compounds can manifest a greater tendency towards combustion than any of their constituents. Here we consider not only compounds which burst into flames by themselves but also those whose combustion is slower, provided only that it is more intense than the individual combustions of the constituents, for as combustion with flames is only a higher degree of combustion than that without, it is impossible for the theory to give a correct notion of one without considering the other, too.

In order to prevent all misunderstanding, we must make one more observation. There might be circumstances under which several bodies could be together without contact, and where one still had the effect of causing the other to burn. One such circumstance would be when burnt lime was moistened with water, whereby it generates strong heat, and then a combustible body was placed very close to it, whereby it must ignite. Now such a combustion would really be no different from any other caused by heat and consequently not a subject for separate investigation. The combustible body could have been mixed with the lime and then moistened, and now the combustion would seem to come from the chemical interaction of the mixed bodies. It is easy to see that such a combustion could be mistaken for the ones we are investigating if one did not observe all the circumstances carefully.

The discoveries of recent times have shown us a new natural law for the combustibility of bodies, which is that when two bodies are in contact with each other, the more combustible of them becomes even more combustible while the less combustible becomes even less combustible. This law applies especially to good electrical conductors but can also be demonstrated in the poorer ones, though not to such a high degree. According to this law, any combination of more and less combustible bodies must have a stronger effect on air and produce greater combustion than they could have done separately. It is true that the less combustible body loses some of its combustibility, but as it is most often very little combustible, it would have had little effect on air, whereas the more combustible body, which has now become more combustible than before, will react even more strongly. This, however, would not be sufficient. The effect mentioned here would not be able to cause perceptible combustion, but this mutual influence of two bodies is only the germ of a new effect. When two such bodies come in contact with water, this will be noticeably changed thereby, and at the points where it is in contact with the less combustible body, it will liberate hydrogen but with the more combustible oxygen, which then combines with this. If we imagine two bodies of unequal combustibility so finely combined that they almost constitute a homogeneous unity, indeed, even so finely that they cannot be perceived as separate by the senses, and then just as perfectly mixed with water, it is easy to comprehend that the effect must be even stronger. If we further imagine that the hydrogen liberated from the water decomposes a portion of one of the substances, that the resulting product can, to some extent, be dissolved in the water, that this substance again can be roused to even greater combustibility by the other substances from which it was produced, and, finally, that air can come in contact with it, then we have an idea of how the product of two bodies can become so much more combustible than they were individually.

In its compounds with alkalis, earths, and metals, sulphur presents a remarkable example of this law. Sulphur belongs to the bodies which prove to be extremely combustible at a high temperature, but which must be regarded as among the least combustible at a lower temperature, such as the one at which we usually live. We need only recall that acids have no effect on it in the cold, whereas they affect most metals in the same way as combustion otherwise does. This property of sulphur results in a very interesting phenomenon, which can best be illustrated by means of an example. Let us imagine sulphur and iron at a low temperature in contact with humid air or even moistened with water. In the first moments, the iron is considered to be the more combustible while the sulphur possesses this property in the smaller degree. As a consequence, the iron becomes even more combustible, changes the water into oxygen and then combines with it, while the sulphur becomes even less combustible and changes the water into hydrogen, which at the instant of generation immediately decomposes some sulphur and liberates sulpheretted hydrogen, most of which stays in the solution. Hereby two new products have been created: on one hand, a combination of iron and oxygen, or iron calx; on the other, sulpheretted hydrogen. Now, then, we have four links in the chain, among which sulphur remains the least combustible, then follows the iron calx, then the iron, and finally the sulphuretted hydrogen as the most combustible. However, the sulphuretted hydrogen will begin to combust through the action of the air, whereby sulphuric acid is created, which, as such, would represent the least combustible link in the entire chain if it did not combine with the iron calx to form sulphate of iron, but even in this state it should be considered the most negative (least combustible), especially the part of it which might have found perfectly calcined iron to combine with. Obviously, this process must not be imagined as thus divided into phases as we have been forced to describe it here, for as soon as the first atom (if I may so express myself) of iron calx and sulphuretted hydrogen is created, the combustion of the latter begins at once and, immediately after that, the action of the liberated sulphuric acid on the iron calx, and thus one effect constantly merges with another so that together they form a continuum. In the manner described here, the whole mass remains active until all the iron has been transformed into iron calx, and all the sulphur into sulphuric acid, that is, all of it into sulphate of iron.[3] So much activity was enclosed in this mass until it reached its goal, and how many forces did not work together in order to bring about the combustion of the sulphuretted hydrogen, but there remains an important one which has not yet been mentioned: heat. This has two sources here; partly it must appear with the combustion of the more combustible part of the mass, and partly it must arise from the internal activity in the mass. How the latter generation of heat is to be explained more precisely, we shall leave undecided here and simply observe that the internal action in the galvanic pile, which rests on the same foundation, is also accompanied by heat, or, if one prefers, one may agree with Berthollet that every intense chemical action is accompanied by heat. This generation of heat increases the combustibility of the sulphuretted hydrogen a great deal, and it is easy to understand that this can compel it to burst into flames.

If we now consult our experience, we find that sulphurated iron, when moistened, is able to burst into genuine combustion, in complete agreement with our theory.

The same theory also applies to the action of sulphur on alkalis and earths, but several points here require elucidation. At first glance, it might be assumed that the sulphur in these compounds must be considered the more, the alkalis the less combustible because alkalis are generally considered not to be combustible at all. It would be easy to answer that sulphur proves to be one of the least combustible bodies at the usual temperature of the atmosphere, whereas alkalis seem to experience a greater effect from the air. This answer could not be considered invalid. Ammonia shows considerable combustibility; it will be found that caustic potash, confined in oxygen gas, combined with some of it thus undergoing a weak combustion and then became carbonate of potash; alkalis and alkaline earths greatly promote the reduction of metals, etc. However, we could go further. We have stated the natural law on which we based our entire explanation far too one-sidedly and only in the language of chemistry in order not to have too many phenomena to explain at once. We could now present this natural law in more general terms. So far we have spoken only about a chemical relationship between the bodies in contact and have therefore had nothing but chemical evidence, that is, combustibility. However, we know that this increase or decrease of combustibility takes place parallel to an electrical change so that the more combustible body, whose combustibility increases, becomes positive while the less combustible body, whose combustibility decreases, becomes negative. If, conversely, we begin with the familiar electrical conditions, there is hardly any doubt that sulphur must become negative with alkalis, for though we have no definite experiments on this, we do know that sulphur is among the most negative bodies, whereas alkalis are far closer to the positive. Therefore, the result here is also that sulphur becomes negative and in combination with water produces sulphuretted hydrogen, that alkali, on the other hand, becomes positive with sulphur and negative with sulphuretted hydrogen, and

---

[3] Of course, this only applies to a certain ratio of these constituents.

consequently that the latter substance also becomes the one to combust here while the other substances only serve to intensify this combustion.

It is well known that sulphur and alkalis form compounds which combust very easily though without flame and strong heat. All that we have seen concerning the compounds of sulphur and alkalis also applies to alkaline earths, which really should not constitute a separate class, as I have previously proved in the publications of this Society.

One of the most extraordinary examples of spontaneous combustion is pyrophorus, which takes its name precisely from its ability to ignite on contact with air. Everybody knows that it has been made in many ways. The simple way in which an alchemist first found this compound has now been completely abandoned, and the detailed and apprehensive instructions then given for its preparation are no longer considered necessary. We now know that a salt which contains an incombustible alkali, especially potash combined with sulphuric acid, produces pyrophorus when it is heated strongly with a certain amount of carbon or a carbonaceous body. Through the heating, the sulphuric acid in the salt releases its oxygen to the carbon, whereby a portion of the sulphur remains in combination with the alkali. Furthermore, we know, from Thenard's and Lampadius's experiments, that sulphur, distilled with charcoal, forms a highly volatile and inflammable substance. The same must happen here, where glowing carbon is in contact with the sulphur liberated from the sulphuric acid. Also here, then, we have discovered sufficient grounds for an intensification of the chemical activity. Here carbon is the most negative, after that follows sulphur, then alkali, and finally the volatile product of sulphur with carbon, which undoubtedly consists of sulphur, hydrogen, and carbon. This compound does not ignite in dry air, but in humid air or when breathed on; if the air is dry, it burns spontaneously. Here, as in the previous examples, bodies of very different combustibilities and water provide the condition for spontaneous combustion.

Phosphorus reacts with alkalis like sulphur.[4] It releases hydrogen from water, and, through its decomposition, it liberates phosphoretted hydrogen gas, which, as is well known, is highly inflammable and does not even require contact with the mixture from which it is produced in order to ignite on contact with air. From the moment when phosphoretted hydrogen is produced, that is, from the first moment of contact between phosphorus and alkali, the latter is the more combustible, and part of it really combusts and becomes acid so that the final product becomes a phosphoric alkali. Of course, all this also applies to alkaline earths.

It is strange that phosphorus increases in combustibility when combined with sulphur. The same is true of resin and phosphorus. In a certain proportion, these mixtures are found to be fluid, and this has been thought to explain their greater

[4]There might be some question as to whether phosphorus should not be considered more combustible than alkali at a rather low temperature, but the answer is that phosphorus appears little combustible in solution, and it must be borne in mind that we have introduced another conception of combustibility, according to which phosphorus must become negative with alkali. Besides, alkalis are created where the positive pole is aroused in fluids, and therefore it is natural that they play the role of the positive.

tendency to enter into chemical processes, thus also to ignite, but many other related phenomena cannot be explained in the same manner. In order to be convinced of this, we need only remind ourselves that an amalgam of mercury and gold is calcined more quickly than pure mercury. The same applies to all other liquid amalgams which have been investigated so far, but as the same explanation could be given for most of them as for the combination of phosphorus and sulphur, the example of gold amalgam, that is, of a dissolved metal which is less combustible than mercury, is the most suitable to point towards the correct explanation of the phenomenon.

When sugar of lead (acetate of lead) has been decomposed by distillation, the retort contains a residue which often ignites on contact with atmospheric air. It can hardly be doubted that this residue consists of an oily substance, lead, a little acetic acid, and carbon. Therefore, it ignites according to the same procedure as pyrophorus. The same applies to several residues from the distillation of acetic metallic salts.

Burnt beans, burnt flour, and a host of other roasted vegetable substances, very tightly packed, often become spontaneously combustible, too. In order to understand this, it is necessary to know that all vegetable bodies are decomposed when intensely heated, whereby empyreumatic oil, water, acetic acid, and carbon are released. Such a substance cannot be completely without alkali either since potash is discovered in all vegetable substances that have been burnt to ashes, and some ammonia is obtained in most distillations performed with these. Some of the oil produced is certainly extremely volatile and inflammable and can almost be considered a liquefied carbonaceous hydrogen gas. Thus, in these burnt vegetable substances, as in the other spontaneously combustible bodies, we have a contrast between more and less combustible constituents accompanied by moisture. Carbon black, moistened with boiled linseed oil and wrapped in a mat, also ignites spontaneously. Oil and carbon are also present here, but there is good reason to assume that the litharge with which the oil is boiled plays a role in this. It could be further added that the oil is decomposed somewhat during boiling so that some volatile oil is released, as well as a little acid and water, which may still be present in the mixture. Wool, hemp, cowhair, and the like, impregnated with oil, also ignite spontaneously when they are tightly wadded. The compression has the effect that the heat which is generated by the incipient combustion is not carried away by the air but serves to promote further combustion.

During the preparation of boiled oils, which is done by boiling a fatty oil over vegetable matter, it is sometimes found that the mass, which has been almost completely freed of the oil by pressing, ignites spontaneously in the filtering cloth.

The kind of spontaneous combustion which the theory stated here might least be expected to explain is undoubtedly the one which takes place in quantities of moist hay, flour, malt, grain, and other seeds. However, it is not so difficult as it might seem at first glance to explain these spontaneous combustions, too. They only take place when these substances begin to ferment. Fermentation, however, always presupposes a decomposable (that is, a compound) substance and moisture. This is particularly easy to show when plants begin to ferment. We have careful investigations only of fermentations which produce wine, but these can be quite instructive to us here. Wine fermentation requires a mixture of sugar, ferment, and water. Ferment is, strictly speaking, the same as the so-called organic substances, which besides hydrogen, carbon, and oxygen also contain some azote. Sugar, on the other hand, contains only the first three constituents. The contrast between the sugar and the ferment seems to cause all the motion which occurs in this mixture, so much more so as it does not begin until the ferment has precipitated from the chemical solution and is only mechanically mixed with it.[5]

The sugar and the ferment here seem to arouse a mutual polarity in each other, the result of which is that the sugar is changed partly into a more combustible body, i.e., spirits, and partly into a less combustible one, i.e., carbonic acid. On the other hand, acetic acid and tartaric acid seem to be liberated from the ferment. We do not understand fermentation well enough to have full certainty about all this, but here it must suffice that we have reason to assume that the very lively motion which takes place during fermentation is caused by the contrasts between the closest elements of the fermenting substance and between the new products released from these during fermentation. What happens during wine fermentation also takes place in sprouting vegetation, when the seeds are said to *germinate*, for it is not difficult for chemistry to show the same constituents and the same internal motion as the ones in which wine fermentation has its origin. By analogy we may conclude that fermentation in hay happens in the same way, though we are less able to present immediate proof of this. If we are right about this, we are entitled to assume that fermentation is a process which is completely identical to the one that starts the other spontaneous combustions with which we have acquainted ourselves.

It is easy to see that the same explanation can be applied to the heat and the combustion which often take place in heaps of manure.

Among the most curious spontaneous combustions are the ones which take place when oils and concentrated fuming nitric acid are combined. However, it will hardly be possible to get the right idea about this without considering the relationship between acids and combustible liquids in general.

So far we have traced all spontaneous combustions back to one primary cause, which is that when heterogeneous bodies come in contact with each other, they are able to produce changes in each other's combustibility, and we found that this change was really an electrical process. That is also the case here. Perhaps we shall be able to shed more light on the previous explanations by making this connection clearer.

The law which we have previously referred to is actually a fundamental law of all electrical effects. From the simplest electrical experiments, we know that a positively electrified body induces negative electricity in the nearest part of an unelectrified body, and, conversely, that a negative body induces the positive. If we now assume that all bodies, in their internal nature, have a greater tendency to become

[5]That it is not the precipitation of the ferment from the liquid put aside for fermentation but merely the change in the manner of contact which causes fermentation can be seen from the fact that this does not take place at all when the admixed ferment is removed by means of a filter.

either negative or positive, it also follows that when two bodies of different natures come in contact with each other, the one whose tendency to positive electricity is the greater must induce its opposite, that is, the tendency to the negative, in the other body, which is also the more negative of the two. Conversely, the negative body induces a greater tendency in the more positive one to become even more positive. We also know that the part of an unelectrified body which is farthest away from the electrical one enters the state which is opposite that of the nearest part, and therefore like the electricity of the electrical body itself. We even know that in poor conductors the positive and the negative can alternate more often and remain in this state. This can be applied marvelously to the action of acids on combustible liquids, which are all poor conductors, provided we do not forget that the same conditions which appear externally in the usual electrical experiments are repeated internally in all contacts. We must also recognize that electrical transmission produces the opposite effect of polarization, and that chemical union in contact interactions serves as a substitute for transmission.

Some examples will serve to illustrate this phenomenon. When spirit of wine is brought into contact with acids, it changes into naphtha, which is more volatile and more inflammable. At first glance, it might suffice to explain that the spirit of wine, being a very inflammable substance, must increase in combustibility due to its contact with the acid, which is already a combusted body, but we must consider the whole process more closely in order to determine the true facts. With this object in mind, we want to compare two ways of producing naphtha. One, though an old pharmaceutical method, seems ideally suited to illustrate our theory. Concentrated fuming nitric acid is poured into a glass, water is poured over it, and spirits over this again in such a way that they do not mix. If these three layers of fluid are left undisturbed on top of each other, a great change will soon be seen taking place. While the colour of the nitric acid changes from red to blue at its surface of contact with the water and also farther down, the spirit of wine begins to become milky and cloudy and mixed with stripes, which are found to be naphtha.[6] It is clear that an induction, an alternation of opposite effects, is taking place in the spirit of wine, but it has not yet been established how this induction is otherwise constituted. However, we have other experiences of the production of naphtha which leave us in no doubt about this. If spirit of wine is combined with concentrated sulphuric acid, part of it will change into naphtha, but some of it will also be found to have decomposed so that, instead of spirit of wine, we obtain naphtha, a black resin, acetic acid, and some water. From this we may conclude that what produced the cloudiness in the experiment with nitric acid and spirit of wine was a liberated resinous substance, and we also have reason to assume that some of the spirit of wine has changed into acetic acid. We also have real experiments on what takes place in a simple mixture of spirit of wine and nitric acid which quite confirm our assumption. Now we can easily pass on to the well-known experiment on the ignition of oils by means of nitric acid, which we owe to one of our most famous compatriots.

[6] A very accurate description of this entire process is found in Crell's *Neuesten Entdeckungen etc.*, Vol. 5.

All concentrated mineral acids, but most of all nitric acid, have the property that they produce resin from ethereal oils. As this resin is released, some acetic acid is also created, which can be noticed from the smell, together with an oil which is much more volatile than the one on which the nitric acid was made to act. This latter product has not been studied adequately, but it reveals itself clearly enough by its odour. It deserves to be distilled off for closer investigation. It is to be assumed that it will have the same relationship to the oils as ether to spirit of wine, and that it might even become an excellent medical substance.[7]

The effects described here take place quite commonly in the action of relatively strong acids on oils, but more is required to provide a strong and intense spontaneous combustion. Only concentrated fuming nitric acid and only certain oils are suitable for this because their interaction takes place with such great speed. The mixture must involve relatively large masses because otherwise the air can cool the already warm mass too easily and carry away the oleaginous ether (if I may be permitted this term) already liberated. Of course, we shall not mention other precautionary measures which concern the experimenter more than the experiment. It is enough for us that in this spontaneous combustion, too, a less combustible body creates a greater combustibility in another even more combustible one, and that, during this action, new compounds of different combustibility arise which serve to intensify the total effect. Yet it remains to be mentioned that the heat which is generated by the interaction between acids and combustible fluids is so considerable that many have thought that all spontaneous combustion should be attributed to that alone, but we have clearly shown several causes. Since, strictly speaking, we cannot explain the origin of the heat which accompanies the interaction between acids and combustible fluids until we know their relation as positive and negative, we seem entitled to regard it merely as a means for the promotion of combustion, as we have explained in the mixing of sulphur and iron.

Of the remaining spontaneous combustions, we shall only recall one more, which is that moistened nitrate of copper wrapped in tin foil heats and begins combustion. Here, the tin is probably calcined at the expense of the nitric acid, whereby nitrous gas is liberated and probably ammonia as when tin is decomposed in nitric acid. The nitrous gas is not liberated at once but undoubtedly remains in contact with the copper calx while it draws oxygen from the atmosphere and combusts. When it comes in contact with the other substances, it becomes even more combustible, and the heat which is generated during the combination of the oxygen in the nitric acid and the tin promotes this even more.

From all that we have now investigated, it seems that we can rightly deduce the following results:

1) That spontaneous combustions are caused by the increased combustibility which a less combustible body produces in a more combustible one.
2) That the presence of water is one of the conditions for spontaneous combustions.
3) That vigorous and strong spontaneous combustions are only produced

---

[7] I have only now, after a long time, recalled to my memory the wealth of new investigations which this gives rise to. I dare say that it will not be long before I make the necessary experiments.

when the first fundamental constituents are or have been able to liberate new products with an even greater difference in combustibility.

4) That spontaneous combustions consequently result from an internal chemical motion necessitated by the composition of the bodies.[8]

5) That this chemical reaction is accelerated because each tendency in it serves to intensify the other.

6) That a many-sided contact between constituents promotes this increasing intensification of the tendency of the parts towards spontaneous combustion.

7) But that complete chemical homogeneity in a body cannot cause spontaneous combustions like the ones mentioned here.

Several precautionary measures can be deduced from the theory advanced here. If these are less definite than many could have wished, this is only due to the nature of things, for only the intense spontaneous combustions are dangerous while the less intense ones are not to be feared because they generate neither flames nor heat sufficient to ignite others. However, it can be said unequivocally that there is only danger to fear from accumulations of heterogeneous substances, but this class is larger than might be thought since the parts and residues of all organic bodies belong to it. Only bodies that are moistened or otherwise associated with water are susceptible to spontaneous combustion, but very little moisture is required in bodies which contain much free carbon. To this class belong all organic bodies which have been exposed to heat whereby they have been burnt. Such bodies must be aired and dried before they are stored away. Bodies in which burnt oils are in contact with carbon, flax, hemp, and most other substances of an organic nature, must not be piled up in great quantities. Therefore, painted and tarred canvas or rope should not be rolled and wrapped until it is completely dry. Completely dry or completely wet bodies present no danger of spontaneous combustion, so only the moist ones need be feared.

The reader might still wish to see the explanation of the phenomenon that people have sometimes ignited spontaneously and have been found almost completely consumed by the fire which had broken out in their own bodies. According to the facts mentioned above, no-one can doubt that all the conditions for such spontaneous combustion are found combined in the human body. However, they are found in the healthy as well as in the sick body, and they continually cause a process which is similar to combustion, so the question is only where that singular intensity comes from with which it sometimes, though rarely, breaks out. This can hardly be answered unless we know much more about all the circumstances which accompany this phenomenon than we do now. Besides, the explanation of this kind of spontaneous combustion is more a subject for the specific investigation of the laws of the animal body than for the study of natural laws in general. In the latter we have already taken many steps towards certainty while we are only beginning to see possibilities in the former.

---

[8] The analogy between all these reactions and fermentation is remarkable. The name of fermentation might very well be extended to all of them, and it would be possible to cite many of the old chemists' examples in favour of this expression.

21

On the Manner in Which Electricity Is Transmitted[1]

(A FRAGMENT)

As FAR As I know, no-one has yet attempted to penetrate into the internal mechanism of the transmission of electricity. It might also be very difficult to reveal the full secret of this process, but some interesting conclusions about this can certainly be drawn from the nature of things and from several well-known facts.

The first effect of an electrified object on one which is not electrified is, as is well known, to create an electric polarity in it. If we designate the charged object by A, and BC represents a conducting cylinder, then B receives the electricity opposite to A, whereas C receives the same as A. Everyone knows that this is called an *induction* of electricity. It is equally well known that, if the end C loses its electricity,

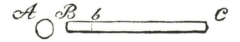

only that of B remains, but if B is brought in contact with A, the difference is cancelled, and the similar electricity of A and C remains, which is then usually called a *conduction* of electricity. Obviously, the induction is the *first act* of the spreading of electricity, the conduction *the second*. We can therefore call the former *electrification of the first degree*, the latter *electrification of the second degree*. The former is a *polarization*, the latter an *identification*. These designations will now serve to avoid the false notion of a conduction also in terms of language.

The electricity of the object A cannot possibly spread from B to C without taking a certain time to do so. In order to come to an understanding of this, we want to think of time and space, in the manner of mathematicians, as being divided into infinitesimally small parts. We want to think of such a small part of space as represented by Bb, where, in the first small part of time, an infinitesimally small electric

[1][*Neues Allgemeines Journal der Chemie*, ed. by A. F. Gehlen, Vol. 6, pp. 292–302. Berlin 1806. Also in *Journal de physique, de chimie et d'histoire naturelle*, Vol. 62, pp. 369–75. Paris 1806. A shortened Danish version is to be found in *Nyt Bibliothek for Physik, Medicin og Oeconomie*, Vol. 9, No. 3, pp. 268–75. Copenhagen 1806. KM I, pp. 267–73. Originally published in German.]

polarity is aroused; if, e.g., A becomes positive, Bb becomes negative at B and positive at b. However, in the next instant, A will seek to enlarge the negative zone, whereby the positive will be enlarged as well, while the positive in b will strive to produce a negative one even further towards C. And the entire process continues in this manner until the negative extends over the front half, the positive over the back half, and the middle remains indifferent. Everyone will now easily recognize this process, which we have represented as discrete for the purpose of description, as a continuum and visualize the internal life in each spreading of electricity in this way.

Once it is assumed that each electricity brings forth its opposite, it is quite natural that the inner mechanism of the transmission of electricity must be constituted as we have described it here. However, it serves to provide greater reassurance to the physicist when he sees his speculation confirmed at every point by nature itself. We shall search for this confirmation.

In good conductors electricity spreads over a mile in less than a second's time. At this fast pace, it is impossible to follow the changing waves of positive and negative by means of the electrometer. The situation is more favourable in poor conductors. Just hold a rod of glass, or resin, or sealing wax opposite a charged object and then examine it with the electrometer; alternating zones of opposite electricities are found. The experiment is known to every physicist. It is hardly necessary to mention that in this experiment we do not claim to have described those infinitesimally small alternations of positive and negative of which we have spoken above; on the contrary, we have spoken about them in such a way that we can have no hope of ever finding them in our experience; the changes demonstrated here are nevertheless repetitions of the same scheme on a larger scale. It would even be possible to calculate the size and composition of these zones mathematically if it were necessary.

It is sufficiently well known that we can follow the spreading of electricity in all poor conductors in precisely this way, and that we can show it in air, for example. Therefore, we are even more entitled to regard the transmission of electricity as *undulatory*. However, other experimental proofs are available to us. We cannot follow the rapid spread of electricity in good conductors with the electrometer, but it often leaves traces, and here we always discover confirmation of what we have previously found. It is sufficient to try to melt a long iron wire with a weak charge from the electric battery. It will soon be seen that only some parts melt while others remain intact, and that melted and unmelted parts alternate. If a stronger charge is applied, the wire does melt along its entire length, but in beads: with alternating *expanded and contracted zones*. It is also possible to find an intermediate charge for which the metal glows without melting. On such a wire very visible alternations of expanded and contracted zones are found. All these experiments are already familiar to most physicists and are the strongest evidence of the undulatory spreading of electricity.

If the charge of the battery is increased to the point where the metal wire evaporates, and the experiment is performed in such a way that the developing metallic vapour is able to condense partially on a piece of clean paper, an almost complete picture of the transmission of electricity is obtained, which —a cloud of vapour with regularly alternating expansions and contractions —speaks for itself. Even the density of the vapour, and often its colour as well, alternates so regularly that we actually have a coloured depiction of the oscillatory spreading of electricity. It can be seen from the constancy of the phenomenon that all this is not an accident. The experiment can be performed as often as one wishes and with whatever metals; no exception will ever be found. Or, if we wish to spare ourselves the trouble, we need only glance at the numerous and very accurate depictions in which van Marum has presented these experiments. The regularity of the pictures can be seen in yet another way. If the electricity merely worked expansively on the wire, all the vapour clouds would be parallel to each other and straight, but since each conductor has a repulsive effect on the end of the wire next to it, the vapour waves at the two ends, driven by two forces perpendicular to each other, must follow the corresponding diagonal or, more correctly, since the forces act continuously and unequally, form a curved line whose concavity is turned towards the wire. The farther a vapour wave is from one of the conductors, the less the repulsive force parallel to the wire will be able to act on it, and therefore its position will approach the direction perpendicular to the wire more and more. However, in the middle part of the wire there must be a perfect balance between the repulsive and attractive forces of the conductors, and consequently the position of the vapour cloud in that place will be completely perpendicular to the wire.

All this will not be observed as clearly if the force which is applied in the evaporation of the metals is disproportionately large; in this case the picture is bent into a zig-zag, where, nevertheless, clear traces of the order described above show themselves in each section.

Do we desire another proof? We need only consider the electric spark. If the conductors between which a spark jumps are brought sufficiently close to each other, nothing is observed other than that it has different colours at the two ends, red at one end, bluish at the other, but white towards the middle; if the conductors are farther apart, the same colours are seen to repeat themselves more often, with just as many alternations in which now the positive, now the negative has the upper hand.

All that has been said here about the spreading of electricity is also valid for magnetism. The action of magnetism also begins with a polarization and therefore must spread in an *undulatory* manner for the same reasons as electricity. In this case, a polarized zone must also attain its maximum expansion under certain conditions and then produce a second. Experience again confirms this, for if a long, thin steel wire is magnetized, it acquires alternating north and south poles along its entire length. In any event, we need only consider the operation of magnetization quite clearly in order to comprehend the manner in which magnetism spreads, for what is it when we draw the pole of a magnet across a piece of iron but driving the two poles forward so that the part which previously had $+M$ now gets $-M$, just as a wave on the ocean destroys a valley in front of it while it opens one behind it.

This mechanism for the transmission of actions through undulation is certainly common in all of nature, but it is quite difficult to demonstrate. However, it is possible to show it in the internal transmission of motion through bodies. It has been understood for a long time that an expansion must follow the compression of an air particle, whereby the neighbouring parts are compressed, which, upon subsequent expansion, must compress others, and so forth. It was assumed that sound is transmitted in air in this manner. However, the same mechanism has apparently not been assumed for the transmission of sound in solid bodies. At least, since the discoveries of the famous Chladni, almost all physicists rail against the assumption of a vibration of the smallest parts during the production of sound. Yet, nothing is easier than to understand the necessity of these vibrations from the nature of things and to prove their existence through experiments. Here, we can dwell only briefly on the theoretical aspects, since the same explanation which holds for the undulatory spreading of sound in air also applies to all bodies because in them, too, motion cannot be transmitted instantaneously, and therefore all parts cannot be affected in the same manner at the same time. If we wanted to convince ourselves of this by experience, a thick iron wire, resting on its two extremities, need only be powdered with lycopodium and struck quickly but not violently; it will be observed that the powder will distribute itself in small mounds which will form a line along the wire. Those which have formed closest to the points of impact are largest and the others smaller, in proportion to their distance. The essence of this experiment can be realized even more easily. A rectangular glass or metal plate with completely straight edges can serve this purpose. Its surface is powdered with lycopodium, after which it is held in such a way that two opposite edges are touched with the fingers while the other two are left free. Now the middle of one of the untouched edges is struck with the edge of a piece of wood, or something similar, and soon the powder will be distributed in lines which are parallel to the direction of the blow. Once again, various elevations and depressions can be distinguished. If, on the other hand, an entire edge is struck at once with a planed board or some other flat object, the powder will arrange itself in lines which are parallel to the edge struck. However, these lines are more or less wavelike, depending on whether the edge has been struck more or less equally at all points. If one of the surfaces itself is struck, a number of small mounds develop. These are without doubt the results of a composite oscillatory motion, i.e., one progressive and one moving up and down.

If the plate is held in such a way that the edges are not touched, but only a very small area of each surface is covered by the fingers, and it is then struck, not only do such mounds develop, but a tone is also produced.[2] The small mounds all move so that they finally meet and form a Chladni figure even if it is somewhat incomplete.

Now we proceed just as if we wanted to produce a Chladni figure, but in such a way that the plate is strewn with lycopodium instead of sand, and we can see the figure develop very slowly in front of our very eyes. At the first stroke of the bow, the first little mounds are immediately formed, and these, like so many smaller

---

[2] In the other experiments only a dull sound develops.

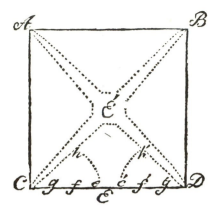

nodal points, move towards the place where the larger ones form. Let us imagine, for instance, that the rectangular plate *ABCD* is stroked at *E*, then little mounds would immediately rise at *e* and *e'*, as well as at *f, f ', g, g'*, etc. However, these little mounds will move faster at *e* and *e'* than at *f* and *f '*, and faster at *f* and *f '* than at *g* and *g'*. The dust at the point *E* and everything that lies on the line *EE'* will be driven straight out towards *E'*. But whatever lies at *e* and *e'* will be moved by two forces in the direction *EC* or *ED* and *EE'* and therefore describes the curved lines *eh* and *e'h'*, just as all the other points describe similar lines. The curved line *CE'D* is generated in this manner, and since each of the other quarters of the plate *AE'C*, *AE'B, AE'D* also vibrates, just as many such curved lines develop which together appear to form a cross or a star. However, it should be noticed that the nodal lines are not marked by the accumulated mass of dust but are enclosed by it. This cannot be shown in the usual manner by means of strewn sand because sand is elastic, and its particles are too large so that each small particle must jump around until it has found a suitable calm place. If a few sand grains are strewn on a glass plate, and this plate is stroked with a bow, this can be confirmed immediately. Therefore, the dust lines and the resting lines (nodal lines) should not be confused with each other.

After the experiment, the dust falls off the nodal lines very easily with the slightest disturbance; it is very difficult to remove it from the dust lines. Consequently, it seems that the bending back and forth has produced a kind of electricity which is undoubtedly negative at the nodal lines but positive at the dust lines because the negative lycopodium is attracted to the latter. What Ritter has said about this in Voigt's *Magazin* is probably known to all physicists.

22

Correspondence[1]

Copenhagen, March 4, 1806.
VOLTA's assertion that the storage battery merely resulted from the layers of alkali and acid produced during charging is not consistent with the experiments even if it were true that two charged metal wires lose their strength by being washed in water, which I dare to doubt. However, if it were true, I would say that this must originate in some other property of the electric charge, for if several tubes of water joined by metallic wires are connected with a galvanic pile but are separated from it again after a few minutes, the outermost wires of such a tube apparatus have a galvanic effect, even after the water in it has been shaken several times. The wires in this experiment can be made of platina, in which case no oxidation of the metal takes place. I have very often charged two platina wires between alkaline layers which were in contact with a pile and then dried them carefully, and I have still found them very effective. Often, many successive experiments with such charged wires were performed on frogs and on the tongue, and the wires were frequently wiped off without losing their strength. In any event, there must have been so much alkali on both wires in these experiments that the effect of the produced acid must have been completely insignificant. At the moment I have no frogs, so I cannot repeat Volta's experiment, but I shall do so as soon as it is possible.

Only now have I been able to communicate to you the description and drawing of the thermometer which H. Jürgensen has invented here. It is natural that the principle of it cannot be new, but the form and, especially, the internal mechanism seem excellent to me. From the outside it has precisely the form of a clock and is therefore entirely portable. Of course, it has this form in common with several thermometers manufactured in Switzerland. However, the manner of its regulation, the shape of the spring, and other similar features are unique. The particularly careful workmanship when it leaves his hands is yet another recommendation.

The most essential part of this thermometer is a curved spring (fig. 2, *accc*), which is soldered together from brass and steel in such a way that the former is the inner and the latter the outer curve. *AA* is the ring to which the whole is attached; *b*

---

[1] [*Neues Allgemeines Journal der Chemie*, ed. by A. F. Gehlen, Vol. 6, pp. 500–502. Berlin 1806. KM I, pp. 273–76. Originally published in German.]

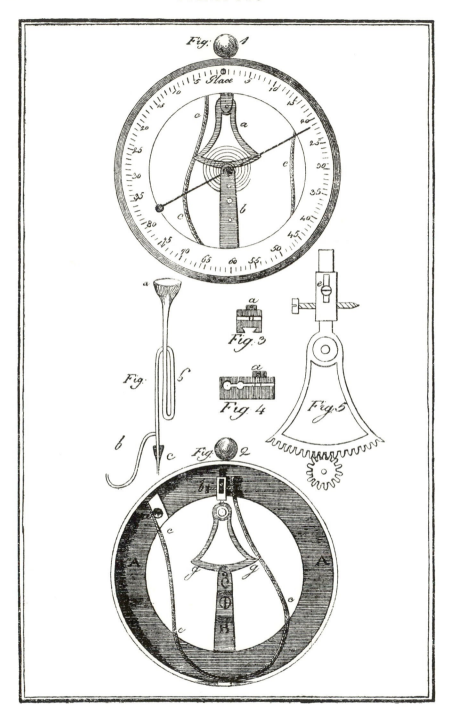

is a screw, on whose point one end of the spring can press, as can be seen in the figure. The inner curve will expand more than the outer with heat, whereby the curvature of the entire spring obviously becomes smaller so that it presses less on the point of the screw. Cold has the opposite effect. It is now easily understood from the figure how the dentated circular arc *gg*, which is shown enlarged in fig. 5, must change its position due to the changed curvature of the spring and thus turn the dentated wheel *C* (fig. 2), which can only be seen in fig. 5. The indicator is attached to this wheel, as shown in the first figure. The division of the degrees is easily understood. As the screw *b* is turned back and forth, the pressure of the spring increases and decreases, and this serves to regulate the instrument. However, the screw *e* (fig. 5) makes it possible to lengthen or shorten the balance beam on which the spring presses by moving the slide in which the screw is fastened back and forth. In this way the thermometer is made more or less sensitive at will. Fig. 3 shows the slide from the front, fig. 4 from the side.

The coil spring which can be seen around the indicator in fig. 1 is in contact with the gear which turns the indicator and is just as lively as similar coil springs in clocks. It serves to press the screw *b* (fig. 2) constantly against the large spring.

It is sufficiently evident that this thermometer is far preferable as a travel thermometer to common ones, which are so easily broken. It is also very pleasant and useful that the degrees are so sharply divided that there can be no doubt about one tenth more or less. Therefore it is certainly also to be recommended as a meteoroscopic instrument. It could also be easily arranged so that it would record its readings progressively; it is not very sensitive, but this is due to the polished surface. Therefore, the artisan will blacken it in future. This instrument would also be superbly useful for making observations in countries where mercury often freezes, and from which we do not have many thermometric experiments. Prolonged use will familiarize us further with its advantages and shortcomings.

Ørsted.

23

Experiments Prompted by Some Passages in Winterl's Writings[1]

1. So far nothing more than assertions, hypotheses, and among experiments only those whose superficiality immediately leaps to the eye have been set up against Winterl's writings. Nor have the works of this famous researcher given rise to many new experiments which would be able to confirm his theory, extend, or further determine its individual theorems.[2] I had early imposed on myself the obligation to contribute to this goal since I had recommended the investigation of Winterl's work with so much warmth, but at the beginning my scientific journeys hindered me, and after that I was occupied with other experimental work to which circumstances led me. At various times, however, I have also performed experiments on Winterl's ideas and constantly taken them into account during my investigations. Often this work occasioned me to correct opinions, often it offered me entirely new observations. I had hoped soon to acquire a coherent whole from these many individual parts, but the further I progress, and the more I seem to approach my goal, the more I become convinced that it can only be reached slowly. For this reason, I have decided to publish the most remarkable of my experiments although in fragments, nevertheless with the whole in mind. It goes without saying that each experiment which I report has been repeated frequently, and the principal experiments in particular have been carefully examined.

I. On the Behaviour of Acids Towards Alkaline Carbonates.

2. Among the experiments on obtaining andronia from potash, Winterl states that the alkali should be saturated with carbonic acid, and that a dilute acid should be added so slowly to the saturation that no gas generation occurs in the process. Since

---

[1] *Journal für die Chemie und Physik*, ed. by A. F. Gehlen, Vol. 1, pp. 276–89. Berlin 1806. This treatise can be found in Danish in *Nyt Bibliothek for Physik, Medicin og Oeconomie*, Vol. 9, No. 3, pp. 229–52. Copenhagen 1806. KM I, pp. 277–89. Originally published in German.]

[2] Mr. Jacobsen is the only one who (*N. allg. Journ. der Chemie*, Vol. 6, p. 605 ff.) has produced a coherent series of experiments in this respect.   Ø.

such a combination, of an acid with an alkaline carbonate without gas generation, has not so far been investigated more closely, I performed a series of experiments on this subject.

3. I poured a solution of potash into a not too narrow glass cylinder and then let dilute muriatic acid drip into it, with the aid of a glass funnel equipped with a filter, so that it fell in the middle of the surface of the potash solution from the least possible height. At the moment when the drop of acid touched the solution, a few air bubbles formed, and then the liquid curled downwards from the surface as usually happens when two liquids mix; a few air bubbles appeared again a little below the surface, and then the seething moved upwards. Soon, an entire layer of acid came to sit on top of the solution. If I now left everything undisturbed, hardly any gas production was observed; the few air bubbles which rose came from places where the contact surface of the acid and the alkali touched the glass. Further, this effect was only observed at the more uneven parts of the glass. The extent of the top layer increased gradually, and finally all was mixed. Now, some air bubbles did rise from the bottom, but this was so far from a lively effervescence that it rather resembled the air generation which weak galvanic batteries produce in water. The potash solution, as well as the muriatic acid, can be more or less dilute; it is important only that the former has a larger specific weight. Similarly, it should be noted that when completely concentrated muriatic acid falls on the surface of the potash solution, it produces an effervescence which is so lively that it roils the entire liquid mass, whereby the experiment is rendered impossible. This experiment has often been repeated with materials of varying purity. For different repetitions, a potash solution was used which had been saturated with carbonic acid at a low temperature, and which could be assumed to be completely free of silica. At times, I have also used ordinary potash, incompletely saturated with carbonic acid. This experiment can be particularly striking if the acid is coloured red and the alkali green with syrup of violet. However, the experiment, modified in this manner, did not teach me anything new.

4. The same experiment was also performed with other acids and alkalis; nitric acid, sulphuric acid, and acetic acid were tried with potash solution; carbonate of soda was combined with muriatic acid, sulphuric acid, and acetic acid; carbonate of ammonium was mixed with muriatic acid and nitric acid. All these experiments yielded the same result. Sometimes, a layer of water was poured over the alkaline solution and the funnel brought just in contact with its surface. In that case, it was not easy to see a bubble rise when the acid fell.

5. The experiment mentioned here was also reversed; a very dilute solution of carbonate of potash was poured drop by drop into the diluted acid. This experiment yielded the same result as above, viz., that gas was generated only at the moment when it fell and at places where both materials, mixed, were in contact with the glass. A similar experiment succeeds beautifully if water is poured over concentrated sulphuric acid in such a way that it does not mix with it, and then a solution

of carbonate of potash is allowed to fall drop by drop from some insignificant height in the middle of the surface of the acid; hardly any air bubbles are seen to develop.

6. On the other hand, as soon as I let the acid fall into the alkali, or the alkali into the acid, in such a way that they touched the glass while mixing, a strong effervescence developed.

7. Therefore, it seemed that an acid could not cause gas generation from an alkaline carbonate if the mixture of the two was not in contact with a solid body. Accordingly, I again mixed an acid with an alkali, in the manner indicated above (3), so that no air production took place. As soon as I dipped a solid body, e.g., a platina wire, a glass rod, a piece of sealing wax, a pen quill into it, gas was immediately produced which settled in many small bubbles on the solid body and was generated from there. As soon as the solid body was removed, the effervescence ceased, only to begin again the moment when it was put back in. This circumstance took place even several hours after the mixing of the two bodies. The experiment was successful with all the mixtures mentioned in 3–6.

8. The air bubbles appeared in extraordinary numbers at the sharp edges of a solid object; this is very easily observed if a knife with a broad blade is immersed. The very smallest, almost invisible solid bodies which may swim in the mixture result in a generation of gas which is assumed to stem from the liquid itself. I first detected this accidentally, and afterwards I repeated it intentionally, among other things with a small piece of leaf silver and with very small pieces of fine platina wire, which sank at first but were immediately covered with so many small air bubbles that they rose again.

9. Therefore, it was proved by many and varied experiments *that no generation of gas occurs in a mixture of acid and alkaline carbonate unless it is placed in contact with a solid body.*[3] One hypothesis offered itself very naturally which, however little I was inclined to accept it, still merited an investigation. As is well known, Rumford claims that liquid bodies are insulators against heat, and his experiments show them to be at least very poor conductors of heat. Therefore, the assumption could be entertained that liquid bodies, for lack of heat-conducting ability, could not supply the carbonic acid with the caloric which is considered to be necessary

[3] For the purpose of producing andronia, I made very similar observations in the spring of 1805 in experiments performed with Dr. Schuster in the manner indicated above in 2, for the completion of which and other experiments my situation has not given me the time and necessary aids until now. I only remark here what an attentive reading of Mr. Ørsted's experiments also makes sufficiently clear, that the necessary dilution of the acid and the carbonate of potash was not so great that, as might occur to some, the carbonic acid released could, under other circumstances, have been absorbed by the quantity of water present. In one of our experiments it was charming to see the role of the solid bodies used by Mr. Ørsted being played by several ice crystals which formed in the neutralized liquid placed in a cooling mixture while the rest, naturally, remained unfrozen.   G.

for their gaseous state. If this hypothesis could be given further confirmation by experiment, it would shed new light on the theory of the perspicacious Rumford, perhaps even provide a striking proof of its correctness, in its full extent.

10. If the stated hypothesis were correct, the solid body would probably provide the carbonic acid with the necessary caloric. Consequently, it should lose some of its caloric, or cool down, to the extent that air is produced at its surface. For this reason, I placed a thermometer in a mixture of carbonate of potash and muriatic acid; immediately, a number of air bubbles developed at its bulb and on a portion of the tube, and it rose noticeably. During the time when I was occupied with the repetitions of this experiment, the temperature of the air changed between $12°$ and $14°$ R, and usually the thermometer rose by a little more than $1^1/_2°$ if it was brought from the air into the mixture. Thus, I obtained a result opposite to the one expected. The experiment was repeated frequently in the presence of several people, with different acids and alkalis and with different thermometers, always with the same result so that no mistake was possible here.

11. However, the hypothesis based on Rumford's opinion was not completely disproved by this, for experience teaches us that alkalis and acids produce heat during their combination. The source of this is immaterial here. Now, following Rumford, it is further assumed that a liquid particle communicates heat to a solid one, and a solid to a liquid, but that, on the other hand, a liquid particle cannot perform the same service for another liquid. Hence, the solid body takes heat from several liquid particles and again communicates some of it to others, but in this case it receives more than it gives off, so the thermometer rises. In order to investigate this, I placed a glass cylinder which contained a mixture of alkaline carbonate and a stronger acid in a somewhat larger glass filled with water. No perceptible generation of gas took place; the temperature of the water contained in the outer glass increased until it reached a certain point, e.g., $14^1/_2°$ when that of the atmosphere was $13°$; then the thermometer did not rise higher. After this, it was placed in the mixture, where it was immediately covered with tiny air bubbles which soon broke free and rose in streams, but it did not fall in the slightest and several times even seemed to rise a bit, but so little that I can easily have been mistaken. I then wound silver wire around the bulb of the thermometer, whereby the generation of gas was promoted significantly, but it still remained at the same temperature. The glass cylinder which contained the solution had various bubbles from which the air frequently rose in streams, so I held the bulb of the thermometer in contact with these locations from the outside, but there was no change in temperature. The experiment was repeated often, also with the modification that the acid and the alkali were mixed more rapidly after they had been poured over each other. Notwithstanding, the generation of gas did not become significant as only so few liquid particles were in contact with the walls of the glass. The increase in temperature was not noticeably larger than before, and the thermometer had the same height in the mixture and in the water which surrounded it though, at the moment of the most violent generation of gas, it was at times $^1/_4$ degree higher in the mixture.

12. I communicate the following two experiments more as an addendum than as confirmation. I first dipped my finger in the water which surrounded the mixture on the outside, in order to feel its temperature, and then put it in the mixture itself. Immediately, gas developed on the finger, but I could not notice any change in temperature. After this, I put my tongue in the mixture and saw in a mirror that it was covered with air bubbles. During this, I experienced a very noticeable taste of champagne wine. After all, nothing is more natural since the fermentation of champagne wine has been interrupted before its completion, so it begins to generate gas as soon as the pressure which arrested the fermentation ceases. In fact, all the air bubbles which rise in a glass of foaming champagne are also seen to come from the bottom and the side walls of the glass. If it comes in contact with the tongue, gas must be produced.

13. The conflict between the outcome of these experiments and the usual theories of heat was not at all unexpected to me. For a long time, I have regarded the proposition that every gas generation was to be accompanied by a decrease in external heat as a hypothesis. I am convinced that the entire theory of heat as established by the antiphlogistians must undergo a great revolution, for which Ritter and Winterl, each in his own fashion, have cleared the way. One of the most striking experiences, which I wished the strict antiphlogistians would try to explain in their way, is the simple galvanic experiment on the influence of the voltaic pile on water. Everyone knows that gas is produced in the process, and yet heat is also produced at the same time because an immersed thermometer rises. Therefore, we have here a gas generation which is accompanied by an increase in heat. I know perfectly well that everything can be reconciled with any hypothesis if several other hypotheses are added, but in that way we finally build a castle of soap bubbles which gleams from afar with many borrowed colours, but which is destroyed by the slightest breath. Consequently, I do not believe that I venture too much if I state explicitly what my experiments have already said, *that carbonic acid requires no heat for its production.*

14. The most remarkable aspect of these experiments is undeniably the fact that contact between a solid body and the mixture containing carbonic acid was required in order for gas to be generated. I do not presume to give the kind of explanation for this that I believe an explanation must be, but it will already satisfy us somewhat to cast an eye on the connection between this and a number of other experiences. If a glass of water is placed under the bell jar of an air pump, and the air is rarefied, then, as is well known, gas is produced. These gas bubbles always rise from the points of contact between the water and the glass. To convince myself more completely of the analogy between what happens here and what occurs in the mixtures containing carbonic acid, I placed a platina wire in the water, from which I pumped out air. Now, this was covered with air bubbles on its entire surface. I poured diluted spirit of wine over a layer of fuming nitric acid. When I placed a glass rod in it, gas was produced as in one of the mixtures containing carbonic acid. However, I obtained a much larger air stream in this mixture if, instead of a glass rod, I immersed a platina wire, which incidentally was not at all attacked by the

acid. The platina wire seemed to produce more gas than the glass rod in the mixtures containing carbonic acid, but the difference was not sufficiently noticeable to be claimed with certainty. For this reason, the effect of solid bodies in mixtures containing carbonic acid cannot be explained directly through galvanism, according to which it is nevertheless certain that liquid bodies do not produce any effect on each other except when in contact with a solid; however, in that case the solid body must be a conductor, and here the bad conductors, even sealing wax, apparently produce as much gas as the best.

15. In order to convince myself further of this influence of solid bodies on gas generation, I poured a solution of sulphate of potash into a small cylindrical glass, water over it, and then let dilute muriatic acid drip into it. I obtained hardly any gas, and no significant odour of sulphuretted hydrogen appeared, but if I immersed a platina wire, it immediately produced a great air stream. Therefore, the outcome in the first case must have been a combination of hydrogen and sulphur, but it could not be examined closely enough at that time.

16. This influence of solid bodies expresses itself not only in gas generation but, as is well known, also in crystallizations. Everyone knows how much the crystallization of a salt is enhanced when a solid body is brought into the solution. Indeed, even chilled water can be made to change into a solid state through contact with a solid body, e.g., a piece of ice. Blagden found that it was very difficult to make completely clear water freeze, whereas water which contained particles which decreased its clarity changed easily into a solid state.

17. Various chemists have made the remark that a fluid which is to undergo wine fermentation does not begin to foam and emit air bubbles until it has become cloudy. If whatever makes it cloudy is strained away, the fermentation comes to a complete stand-still. It is possible, indeed even probable, that the substance which plays this role in a fermenting liquid acts predominantly through its chemical quality, but on the basis of our previous experiments, it cannot at all be doubted that the generation of gas is greatly promoted by this cloudiness.

18. These experiments also explain to us the great disparity which is found between the indications of various chemists regarding the content of carbonic acid in alkalis. Winterl has shown in his *Prolusiones*, and I in my *Materialien zu einer Chemie des neunzehnten Jahrhunderts*,[4] that many differences stem from the fact that different acids yield different results, but it is now clear that the manipulation has a great influence. Perhaps the best advice for avoiding this difference would be to use either the acid or the alkali in a solid form. Where this is not feasible, it would be appropriate to suspend a small net, made of gold or platina wire or even glass, in the fluid into which the other is to be poured in such a way that it would come into contact with the net immediately after it fell into it. Following the principles indicated here, others may find even better methods.

---

[4] [Cf. *Materials for a Chemistry of the Nineteenth Century* (Chapter 12, this volume).]

19. We now have the principle for the proper preparation of a drink which, in the old days, was called *potio Riverii*, where citric acid was poured into a solution of carbonate of potash, and the mixture was drunk while it was still foaming. The best way to prepare this medicinal drink would be to dilute the acid as well as the alkali significantly, e.g., with 6 parts water, to pour it suddenly into a glass which was just large enough, and to drink it immediately. Such a mixture hardly foams at all and still contains much carbonic acid, which shows itself in air bubbles when the tongue or another solid body is immersed in it. Such a mixture retains its carbonic acid for a very long time. Once, I let it stand in an uncovered vessel for 24 hours, and when I wanted to filter it after that, it produced a stream of air because it touched the linen through which it was to pass, and it still foamed for some moments after it had run through.

II. CONTRIBUTION TO A COMPARISON BETWEEN ANDRONIA AND SILICA.

20. In all the experiments in which an acid was combined with carbonate of potash in such a way that no noticeable gas production took place, I obtained a sediment similar to what Winterl describes as andronia. This sediment did not always develop equally rapidly. At times, the mixture turned milky after 15 minutes, but the sediment settled only after several hours, indeed, even after one day. I have found that the way to mix the acid with the alkali just described (19) yields an extraordinarily loose sediment which is extremely easily separated from the solution. This method for the preparation of andronia seems more convenient to me than any of all the others which Winterl indicates. If I used very pure materials, I obtained very little sediment; with less purified, on the other hand, I often obtained a large amount, but not always the same amount from the same materials. It was obvious that, if acid and alkali were mixed in such a way that a violent foaming occurred in the process, the sediment became weaker, indeed, was not even present with pure materials. In the case of the least pure materials, it was highly likely that the sediment was primarily that which chemists earlier called silica, but if I had used a potash solution saturated with carbonic acid or the solution of crystallized carbonate of potash, it seemed that the sediment, even if it resembled the other completely, could not be included in this class after all since chemists consider this material to be pure. Therefore, we would either have to contradict chemists in general or agree with Winterl. In the meantime, I had already observed that there had to be a relationship between silica and carbonic acid in the sediments, so I believed that it would be worth the effort to examine silica itself more closely as it seemed to me that not all its properties were sufficiently well known. Therefore, I decided to seek in silica the same properties which Winterl states to be those of andronia.

21. A concentrated solution of silicated potash was diluted with so much water that no acid produced any sediment in it, not even after several days had passed. Then, I let carbonic acid gas stream through this solution. One hour later, the solution was already strongly opalescent and gained gradually in opacity. Finally, the

precipitated particles separated in flakes and were then isolated from the liquid by filtration. This experiment proves *that silica is precipitated from the potash solution more easily by carbonic acid than by the best known liquid acids.*

22. Winterl says that andronia rubbed together with sugar yields a honey which forms a milky liquid when mixed with water. This was to be expected beforehand; however, it is still remarkable how the sugar deliquesces virtually upon contact with the moist silica. It is also noticed that the mixture of sugar and silica is far more fluid than the moist silica itself. I realize that this can be explained on the basis of the high water content of the silica, which is known to any practical chemist, but if the eye does not deceive, something more is involved here. In order to acquire a deeper understanding of this, I placed some pieces of gelatinous silica in a glass funnel with some lumps of white sugar. These were immediately permeated with water, for which reason I then sprinkled the silica with granulated sugar. This also vanished gradually, as did various portions added little by little. The gelatinous silica did not alter its form, but on the other hand, it assumed a light brown colour throughout, almost like amber, instead of the bluish-white it had before, and some drops of moisture flowed away. A hard body, harder than sugar itself, is obtained with a disproportionate quantity of sugar, e.g., 10 times the weight of the silica. If water is poured on one of these mixtures, a milky liquid is obtained which passes completely through a linen filter and remains for several days without changing. However, if vinegar is added, it curdles, just as Winterl reports about his milk of andronia. If I poured water on gelatinous silica which had been sprinkled with sugar, part of the jelly dissolved and produced a milky fluid without requiring any shaking or rubbing. The remaining jelly had changed the acquired brown colour back to the original bluish-white. All this seems to prove something more than a merely mechanical reaction between silica and sugar. And in any event, it is certain that the behaviour of silica in this respect does not deviate in the slightest from that which Winterl attributes to andronia.

23. It is also possible to rub oil with the moist silica, and this results in a mixture which resembles an ointment. This can further be mixed thoroughly with water and yields a milk-like liquid. Silica resembles Winterl's andronia also in this regard.

24. Winterl says that andronia is obtained when one part muriatic ammonium is mixed with three parts crystallized carbonate of potash. If I performed this experiment with potash, cleansed by flushing carbonic acid through it, I obtained only a very small amount of sediment. However, when the same experiment was performed with less clean potash, a significant sediment was obtained even if the proportions were very different. Consequently, it was easy to be led to the supposition that the sediment stemmed from the silica alone, for the muriatic ammonium must precipitate silica because muriatic acid combines with potash, and ammonium is not capable of dissolving silica. An experiment easily proves the validity of what has been said above as a sal ammoniac solution, added to the moistened silica, immediately produces a large quantity of sediment.

25. If a solution which consists of carbonate of potash and muriatic acid or nitric acid is boiled, a noticeable sediment is obtained during the boiling. However, no generation of carbonic acid gas is noticed. Here the question arises whether a decomposition of the carbonic acid takes place on this occasion, or whether a quantity of silica which remained dissolved in the more dilute salt solution is precipitated. It would be most convenient to assume the latter, which agrees quite well with what has been assumed about silica so far, but a closer investigation should first decide this.

26. I ask that no-one believe that I would presume to have determined anything about andronia with these few remarks. I have merely wished to make a small contribution which could induce others to work more with Winterl instead of being prejudiced against him. The experiments which I have reported do seem to give rise to some suspicion that andronia and silica are identical, but the most important points remain. At least it has been determined that Winterl has undoubtedly made his observations himself and is far from having fabricated his experiments, which some find it so convenient and soothing for their conscience to believe. It is also no less certain that silica has only been known very incompletely so far. It may even emerge that silica really undergoes a decomposition in the most common chemical operations, and that Winterl therefore, with complete justification, regards the silica precipitated from alkalis as a new product. However, all of this must be left to closer investigation. To wish to dismiss something in the work of a profound thinker merely on the basis of our fathers' beliefs, to wish to refute new hypotheses with old ones (following the legal principle of *beatus possessor*) reveals an intellectual and moral callousness which we can only hope would be banished to the barbarians.

24

The Series of Acids and Bases[1]

THE acids and their opposite, the bases, have not yet been organized as they deserve. The reason for this lies entirely in the previous treatment of chemistry. All attention was directed only towards the components of bodies. No independent comparison and organization of the substances were attempted, or only in an extremely superficial manner. In this way, however, only extremely limited knowledge of nature could be achieved, and furthermore, it must often lead to error, for our knowledge of the components of bodies is still very incomplete. First, there are a significant number of bodies whose components we do not know, and which yet form a barely assessable number of combinations with other bodies. Second, we do not sufficiently understand the manner in which several substances are combined in one body. It is hardly necessary to remind the chemists how many bodies are composed of carbon, hydrogen, and oxygen, or of these three and nitrogen, without our being able to say why the results are as they are and not otherwise, which is why chemistry is so frequently unable to assemble what it believes to have taken apart. If we further consider that our situation is similar to that of the results of the analysis of fossils, and if we compare all that we do not know here with what we do know, we cannot help searching for means and ways which lead us closer to nature. However, if someone had not made the requisite comparison, or done so only superficially, but wanted to claim all the more firmly that the previous theory of components would lead us slowly but surely to insight, we would ask him to pay some heed to the many conspicuous contradictions which spring from the contemplation of the mere components. We wish to introduce only a few examples which we can use in the following. With nitrogen, oxygen yields an acid, hydrogen an alkali, but both of these substances yield an acid with sulphur. Metals in combination with a small quantity of oxygen produce bodies which have the most perfect analogy with alkalis and earths, but if they can be united with a larger quantity of the same substance, they produce acids, which is to say that they become the complete opposite of what they were at the lower degrees of oxidation. Perhaps these examples will not appear particularly important to some because such phenomena, which are not

[1] [*Journal für die Chemie und Physik*, ed. by A. F. Gehlen, Vol. 2, pp. 509–47. Berlin 1806. KM I, pp. 289–315. Originally published in German.]

at all compatible with the theory, are very often regarded as inexplicable coincidences although this is not stated clearly. However, if we eventually succeed in explaining these circumstances, this prejudice may disappear altogether.

The direction which chemistry had once taken was also of great influence in the classification and arrangement of bodies, for they were united and separated, often forcibly, according to the knowledge of the components. Thus, in order to choose another example which will be examined more closely below, it was known that the metallic calces are composed of metal and oxygen, but this was not known about the earths, and even at this moment we cannot prove it; therefore, they were separated into two categories which are completely distinct from each other. Suppose, however, we discovered in the future that earths were actually metallic calces, only that they could not be reduced with our present means, or suppose we discovered that metals were composite and had components which, with an addition of oxygen, also constituted earths, only that, without being destroyed, they were as little reducible as the vegetable acids. If we assume this, our entire separation into earths and metallic calces would be more harmful to science than useful, for every such division is an obstacle to recognizing the general laws of nature, and yet this is the final goal of all our natural science.[2]

Thus, in order to know something general about the nature of bodies, we should consider their properties rather than their basic substances. A body will be sought which has a certain property to a very high degree, then others which have the same property will be compared with it and ordered, and thus we follow the same force through each manifestation of varying strength until we arrive at a point where it disappears entirely. If we then find the opposite of just this force and follow it in a similar manner, we again reach a vanishing point, and this will be exactly the same as the first, viz., the point of equilibrium of the opposing forces, and thus the two series fuse together into one. It is easy to see that, for the time being, we are not thinking of the series which occur so frequently with physicists, such as those of specific weights, heat capacities, and others of similar individual qualities. We are actually speaking here of certain manifestations which always occur together, which always decrease and increase proportionally, in short, manifestations which allow us to guess at an inner connection, without our being able to specify it more precisely. Examples of such properties are given to us by metalleity, acidity, alkalinity, etc.

The same bodies can occur in several such series at the same time, and one series will thereby shed light on the others. Therefore, it goes without saying that we will pay no heed to the arbitrary definitions and boundary determinations of the chemists. Almost all such divisions have been based on secondary properties, or determinations have even been made according to properties which do not even belong to the essence of the objects which are to be classified. It should be recalled how often the degree of solubility, of fusibility, of coherence has been used to determine

---

[2] It should not be forgotten that only a possibility is raised here; nothing is claimed. We will later encounter another possibility.   Ø.

boundaries, in order to remember at the same time how often that which God and nature have joined together has been separated most arbitrarily.

The inner *essence* of things is a unity which *expresses* itself through various phenomena. However, to distinguish these *externalities* from all other associated ones and to attribute them to the *inner unity* is the task of the true student of nature. No definition has ever succeeded in expressing this internality, but by staying with *individual manifestations*, it has always established boundaries across which the *inner unity* could still be seen in other manifestations. Winterl has demonstrated this most beautifully for the acids and bases, and we can assume, at least *de jure*, that every chemist has read and pondered upon what he has said regarding this at the beginning of his *Prolusiones* and his *Darstellung der vier Bestandtheile der anorganischen Natur*.[3] In general, in the course of this investigation we will frequently use what this profound researcher has done for this subject, but without meticulously indicating it each time, leaving it for the experts to find and to judge our agreements and deviations from him.

As stated, we do not wish to offer any definition of acidity and alkalinity, rather we select a body in this series, consider its most important properties, seek them little by little in other bodies where they are prominent, and make comparisons between them. We wish to begin with ammonium. Why? This will come to light in the context of the whole. Ammonium is hydrogenized azote, has a very strong characteristic odour and taste, is gaseous, has a very strong attraction to water, makes the syrup of violet turn green, and neutralizes the red colour of vegetable pigments caused by acids. Moreover, it merits mention that it attacks copper, forms combinations with oils and albumen. All of these properties are diminished by combination with acids, and at a certain point in the combination, it loses many of them completely. Everyone knows this though it does no harm to inspect it a little more closely. Our senses teach us that the smell and the taste of ammonium vanish altogether when it is saturated with acid. We may therefore suspect that it possessed these sensual properties by virtue of its alkaline quality. It does not follow from this that all bases should have the same properties. First, volatility is a condition for smell, but solubility for taste, so when these properties disappear, the effects on the senses must cease as well. Secondly, no body is *only* a base but, as a body, has many additional properties which modify the alkaline property. The gaseous form of ammonium also disappears in combination with acids. This manifests itself in a particularly striking manner in experiments when it is combined with an aëriform acid, whereby both the acid and the base lose their gaseous form. We shall be able to make even more remarks about this. Anyone who has performed the chemical experiments will have been struck by how much the chemical attraction of ammonium to water is reduced by its combination with acids. The most hygroscopic acids do not yield fluid salts with ammonium. The ability to change vegetable pigments, like taste, depends on solubility, and therefore what

---

[3] Compare also Schuster's introduction to his description of the *System der dualistischen Chemie*. Berlin, Frölich's Bookshop.   G.

has been said about the immediate effect on the senses applies here as well. It is fairly well known that ammonium loses this property under the influence of acids. That it is also deprived by acids of the ability to dissolve fatty oils and albumen requires no further discussion. The ability to dissolve copper seems to represent an exception to this. This is also attacked by salts composed of ammonium and acids, so this effect may not depend on the alkaline quality.

Among all the alkaline bodies none comes as close to ammonium as potash. We do not know its composition and thus cannot determine whether it is, like ammonium, hydrogenized azote, or whether it has a secondary component, e.g., carbon. Certainly, it will not be completely different from ammonium in its components. It is not aëriform, like the latter, indeed, it is even very little volatile, but it is far from being completely resistant to heat. In the chemical analysis of fossils, we often experience that potash evaporates during glowing, and potash must also transform into vapour during bleaching by vapours from a solution of caustic potash. Ammonium, placed at a lower temperature, loses its gaseous form, potash would assume this form at a higher temperature, so this difference is merely quantitative. The situation is similar for potash with respect to odour; potash is not completely deprived of it, at least if it is heated and moistened. If, however, ammonium smells so much stronger, this is not surprising since the temperatures at which they can become aëriform are so very far apart. The taste of potash is obviously closely related to that of ammonium. They have effects on vegetable dyes, on oils, on animal matter as perfectly in common as is possible for two different bodies. In addition, potash dissolves alumina and siliceous earth although the assistance of heat is required, particularly for the latter, if it is to act significantly. Therefore, it is understandable that ammonium cannot achieve very much in this regard even if it had this tendency.

It has probably not occurred to anyone to doubt the similarity of soda with potash. Therefore, we only wish to recall that this base has a lesser attraction to water, a taste not quite as pungent, a lower causticity, in short, that it presents itself as a weaker base. Potash and soda both lose their stated properties through combination with acids though, to some extent, more easily in the case of soda, for soda which is only slightly carbonated has already lost much of its taste, it does not extract any moisture from the air, and it does not form hygroscopic mixtures with silica as easily. And, what is even more distinctive, several salts which it forms with acids effloresce, i.e., emit their water of crystalization into the air. Potash and soda both attack copper, like ammonium, and lose this property as little as the latter upon neutralization.

Chemists of our time willingly grant baryta a place among the alkalis as well. In truth, it also possesses all the properties which we attribute to the above-mentioned alkalis as such. However, it possesses most of them to a lesser degree. Much could certainly be ascribed to its smaller solubility in water, but this is not the only thing; baryta is also less fusible. This diminished solubility, this greater resistance to melting indicate a greater tendency to be contained in itself, a greater concentration. It also has much less effect on silica than do potash and soda, and it has virtually no effect in solution. Nor does it act as completely on alumina as these alkalis.

We know that the other alkalis in combination with carbonic acid are very much less soluble in water. Rose has demonstrated that soda, completely saturated with carbonic acid, is very insoluble. This is true for baryta in an even higher degree so that carbonic baryta can be called insoluble. Here, too, there is only a quantitative difference. Several chemists have believed to find in this insolubility a boundary between alkalis and earths and have therefore ended the series. This would hardly occur to a thinking chemist any more. It would be all too easy to point out a larger number of analogies between alkalis and baryta to him. Thus, for example, it will be recalled that baryta dissolves sulphur just as well as the first three alkalis and, hence, is able to produce all the phenomena of liver of sulphur. Similarly, that baryta, boiled with phosphorus, yields phosphorated hydrogen gas.

It could be said of strontia that it is related to baryta approximately as soda is to potash. This remark, together with the above, must remind everyone of all the relevant facts, and therefore we need not add anything further.

In the case of lime, all properties of the two latter substances can again be demonstrated, although very much weaker. For example, the taste, the solubility in water, the ability to act on alumina and silica are even smaller here. In addition, lime is even less fusible, and this to the degree that it is not yet certain whether anyone has succeeded in melting it in its pure form.

If we were entitled to assume that the properties which we have listed here were all only manifestations of a *single* inner force, and if we were able to be encouraged by the fact that all these manifestations decrease *together* throughout the series, we might indeed believe that we were somewhat closer on the track of nature. But we have shed too little light on what is actually the primary phenomenon of this series, viz., the property of each of the listed substances to neutralize the acidity of acids as well as, conversely, being deprived by the acids of all their distinctive properties, in a word, their alkalinity. It is so striking that each property of the bases disappears to precisely the degree to which they are saturated by acids. Therefore, it is impossible to dismiss the thought that the degree of alkalinity of these bodies could be measured by the quantities of acids which are required for their neutralization. This also seems to be confirmed by the fact that the quantities of acids which are required for the saturation of a base increase in the same progression for each of the carefully studied bases. Conversely, the quantities of bases which are necessary for the saturation of acids follow the same progression for every acid.[4] However, this idea is disturbed when we observe that those bodies which are obviously the weakest bases require the largest quantities of acids for saturation so that we are thereby led to assume that the strength of bases is *inversely* proportional to the quantities which are necessary for the saturation of acids. Winterl remarks very beautifully that this is further confirmed by the fact that the metallic oxides lose their alkalinity through stronger oxidation but require more acid for their saturation.[5] However, this inverse relation should not occur in the series of acids, rather the direct one. Here, it should nevertheless be remarked that the most similar acids in the series

[4]Fischer in a comment to Berthollet, *Essai de Statique chimique*, I. 336.
[5]Winterl's *Darstellung der vier Bestandtheile der anorganischen Natur*, p. 8, footnote.

are far removed from each other, and that this is precisely what takes place in the series of bases, where the first three bases of our series, whose similarity is unmistakable, have been cast variously among the last three. If we thus wished to assume that the quantity of acidity and alkalinity could be measured in this way, we would at least have to acknowledge that we had progressed but little in our knowledge of acids and bases. In any event, the quantity of acidity is still very different from its intensity. In order to express ourselves more clearly, we would say to the adherents of the most common opinion that the principle of acidity or alkalinity was *bound* more in one substance than in the other, but to the friends of the dynamic view we would explain that the same force was either more concentrated towards the interior or striving to spread towards the exterior, depending on whether the intrinsic conducting ability was larger or smaller.[6]

The fact remains that we have neither a *definitive* external characteristic nor a measure of acidity and alkalinity, but rather that these properties lie hidden in the interior; still, we find ourselves justified in regarding each property of a base which is lost in conflict with the acids as dependent on alkalinity. Also, we may hope to have described the intensities of the bases correctly in the above series. Thus, we proceed. So far in our series, we have seen the gradual decrease of solubility in water, fusibility, and everything which depends on these. It can no longer surprise us if we must accept a body in our series which is completely insoluble, or more correctly, difficult to dissolve to a degree which cannot be determined by our experiments. Magnesia is such a body. Besides, it is impossible to deny its similarity to the other bases. It follows from this insolubility that it has no taste, cannot colour syrup of violet green, even as little as it can dissolve albumen and the like. It can no longer surprise us to see so many properties vanish along with solubility since we saw odour disappear with volatility so early on. In spite of the lack of solubility, magnesia is far from being without attraction to water, as calcined magnesia shows us. It also has sufficient chemical attraction for carbonic acid to deprive this of its gaseous form. Similarly, it neutralizes the acidity of other acids and can thereby restore the blue to vegetable colours dyed red by acids. It also fuses with silica and alumina to form glass in spite of the fact that it cannot melt on its own in our furnaces. Like the bases mentioned above, it also enters into a combination with sulphur. In short, all properties which a base does not lose along with solubility are still combined in magnesia.

Evidently, zirconia follows after magnesia. This does not have sufficient alkalinity to overcome the elasticity of carbonic acid. Nevertheless, it is capable of forming combinations with the other acids. It is very curious that it loses the ability to be dissolved in acids by drying and, even more, by glowing and can regain it only by glowing with potash. Glowing causes magnesia to become somewhat less soluble in acids, but not to this degree by far. On the other hand, we know that alumina, which is lower in the series, suffers even greater change by glowing.

In the case of alumina, it is very remarkable that it is dissolved by the stronger

---

[6]Concerning some of the points mentioned in this paragraph, cf. Schuster, *System der dualistischen Chemie*, Vol. 2, pp. 66–79.   G.

bases. Of the first four, we know for certain that they dissolve alumina.[7] The following two at least fuse with it, forming a kind of glass. Thus, alumina comes as close to the acids as to the bases and is, as it appears, at the point of indifference between these two categories. Glucina is at approximately the same level, as is yttria. From what we know of them now, it is difficult to decide whether they should stand before or after alumina.

Finally, silica appears to us more like an acid than an alkaline body, for it dissolves very completely in the stronger bases and with the weaker ones at least forms a glass, which is to be regarded as an insoluble neutral salt. It will not be found inexplicable that it is not dissolved by ammonium since the tendency of one towards the aëriform state and of the other towards the greatest solidity creates such a great resistance. On the other hand, it does not testify to its alkaline nature that silica is attacked by fluoric acid, for this can very well be synsomatic. Incidentally, in silica even more than in alumina there are phenomena which well merit new investigation and enquiry. Among these is the fact that silica, precipitated from silica jelly, appears to be somewhat soluble in several other acids as well. It might be desired to explain this by the neutralized coherence, but it speaks against this that such solubility is no longer found if siliceous earth has been dried, even though it remained extremely finely separated in the process, and also that siliceous earth, after it is truly dissolved, separates again under certain circumstances, e.g., during evaporation. Therefore, it should much rather be assumed that genuine chemical changes occur in the siliceous earth in these cases. Winterl's discoveries can probably shed some light here. Another phenomenon, viz., that a very dilute silica jelly does not yield any precipitate with the addition of an acid, is hardly explained adequately by the fine division of the siliceous earth, which even allows it to pass through a filter. The glowing dust particles which are produced by striking *silica* with *silica*, and which seem to be of a coal-like nature, might well also be explained, somewhat superficially, as dust particles from the air and leave an analysis of silica as something to be hoped for and desired. In this manner, we have arrived at the series of the acids by a gradual progression through the series of the bases. We know full well that this progression was not without gaps. We can certainly expect more discoveries of individual types of earths as well as of other bases. Perhaps the metallic calces should also be included in this series. We have already noted this possibility above. And we do not even need the reasons listed there for it. After all, it is self-evident that we have as much right to assign a base with a metallic substratum to the bases as to include acids with such a substratum in the series of other acids. It would be possible to make a fundamental objection to this only if it could be proved, or at least made probable, that the metallic oxides are bases due to a different principle than the earths.

From the point which we have reached, we could now ascend in the series of the acids just as we descended in that of the bases, but it is more advisable first to look at the bodies in which the character of acidity is very prominent and, therefore,

---

[7] We know this for baryta from Bucholz's *Beiträge zur Erweiterung der Chemie*, 3, Nos. 58 and 61.

Ø.

manifests itself more clearly. Accordingly, we will also begin with the strongest in the series of acids and finally arrive at the weakest bases in order to merge the two series into one.

After what has already been said about chemical quantity and chemical intensity, we can as little hope to discover a measure of acidity as of alkalinity. Again, we must make do with analogies and compilations. It has been assumed that one acid which drives another out of its compounds should be the stronger, but we now know how greatly the tendency of an acid towards the aëriform state, towards solidity, and the like influences this. For example, it does not at all follow that sulphuric acid should be stronger than sulphurous acid because the former separates the latter from its compounds with bases, for it is only necessary that the sum of the chemical attraction of sulphuric acid to the base and the tendency of sulphurous acid towards the aëriform state is larger than the attraction of sulphurous acid to the base. Such remarks can no longer be regarded as novel after the work of Berthollet and Winterl. We can as little assume that the same substratum with its maximum of oxygen must be more acid than when it is bound with less oxygen. Here, we can again invoke Winterl's work, but at least this much can be noted because of the context. In spite of the fact that we have no measure for acidity, there are still acids of such different strengths that everyone, through the most direct observations and the simplest experiments, will immediately declare one to be stronger, the other to be weaker. The weakest act very little on metals and only on those which are closest to the positive pole in the galvanic series. The stronger, on the other hand, also act on those that behave more negatively. Only very few act on gold. These are: oxymuriatic acid, nitrous acid, sulphurous acid.[8] It is therefore to be assumed that these are the strongest. In fact, they are all at the same level of oxidation. Everyone will immediately admit that this is the case for nitrous and sulphurous acids, but this must also be admitted for oxymuriatic acid if it is recalled that Berthollet and, after him, Chenevix have proved the existence of an even higher degree of oxidation for oxymuriatic acid, which must be called hyperoxidized according to current terminology. Nitrous acid, when not mixed with nitric acid, is vaporous or gaseous, if one wishes to call it that; sulphurous acid and oxymuriatic acid as well. However, hyperoxidized muriatic acid, sulphuric acid, and nitric acid are liquid. The three frequently mentioned acids destroy vegetable colours, but only nitrous acid adds a new colour, which the others do not do, and that is why they are suitable for bleaching. It is known that oxymuriatic and sulphurous acids destroy infectious poisons, but we have no experiments with nitrous acid proper. These would probably be fruitless because it attracts additional oxygen from the air so rapidly. The fact that the vapours of nitric acid destroy infectious poisons as well is not an objection, for we do not wish to deny the other stronger acids all similarity with those three. The experiments which we have before us do not suffice to determine the position of phosphoric acid because chemists have hardly even attempted the production of this acid in a pure form, which does present some difficulties.

[8] The *composite* acid which we call *aqua regia* need not be mentioned here.   Ø.

Nor do we have sufficient data to determine the position of hyperoxidized muriatic acid. If we might be allowed to conclude something from the fact that it does not destroy pigments, we would have to place it next after those first three. It is hardly to be doubted that it must be the first after the previous so-called incomplete acids. It is difficult to decide where common muriatic acid is to be placed for another reason. It must be presumed that it could act much more strongly if it could be obtained in a form as concentrated and as free of water as nitric or sulphuric acid. For now, it undoubtedly lags behind both. Accordingly, we wish to leave the position of muriatic acid undecided but otherwise let nitric acid follow after hyperoxidized muriatic acid and sulphuric acid after this. Every chemist will certainly agree with us regarding the effect of them both on metals. We place phosphoric acid after this. Its fairly strong acidity on the one hand, and its weaker effect on metals on the other, its lack of corrosive power, of which the previous possess so much, will justify this order. Arsenic acid deserves its place immediately after phosphoric acid if it does not deserve to be placed in front of it. Every chemist will acknowledge that it is weaker than nitric or sulphuric acid if he otherwise agrees with us regarding the method for estimating the strength of acids. With our present knowledge, it is not really possible to decide whether chromic acid should have a place here or deserves to be placed even higher. It would therefore be useless to pause for a long investigation whose result could only be shaky. Molybdic acid appears in all its effects as a weaker acid than the previous ones, and tungstic acid, in its turn, must be placed after this. The fact that we cannot assign positions to the remaining metallic acids is a natural consequence of our poor knowledge about them as we scarcely know more than their existence. It will emerge in the following why we have not also listed vegetable and animal acids here, but that we have passed over carbonic acid stems from the fact that, in spite of all investigations, we still do not know how to determine its intensity. Since it is bound to water even in the aëriform state, as Winterl and Berthollet have demonstrated, it cannot be concentrated to any significant degree, and in combinations with bases in the absence of water, it clings so tightly that it cannot be separated without being destroyed.

It might appear that, in the establishment of this series, we did not have as strong grounds for determination as in the series of bases. However, we have refrained from developing the reasons more elaborately only because we believed that everything would be more evident here. If it is only considered that, in addition to what we have stated above, the speed of all the effects decides so much about the intensity, if it is then recalled what significant effects the stronger acids have on oils, on alcohol, and on products of organic nature in general, and if it is taken into consideration how all these effects decrease along the series of acids established here, then approval of this ordering can hardly be denied. There is yet another important test which is valid for the strength of both acids and bases. Winterl has proved, and galvanic experiments confirm, that the opposite electricities are the principles of acidity and alkalinity, and that their combination produces heat. Therefore, if we could determine the heat which results from the combination of various acids and bases, this would probably place a measure in our hands. So far, however, we have only chance remarks on this subject, but what we know agrees very well with all that we have previously assumed. Every chemist will recall from his experience that precisely those acids which we have declared to be the strongest produce the most heat with bases and especially with the strongest bases. If we could list the general laws from all of physics here and thus prove that heat would be produced by indifferentiation,[9] the idea of using heat as a measure of the differences would perhaps not seem entirely unimportant.

An apparently compelling objection could be made to what we have established so far. It could be claimed that each body forms chemical compounds more easily, i.e., has even greater chemical effects, the less coherence it possesses. Therefore, it could be added that the established series of acids and bases is nothing but the series of their cohesions. This could certainly be answered briefly by noting that this postulate falls far short of explaining everything, but it could then be put to us that it explained enough to cut the main thread of our series if we wished to eliminate it. However, we can come closer to the heart of the matter. First, we may ask: Where does cohesion come from? Is this something independent of all other properties of substances, or is it not much rather the result of all the inner forces together? We can envision the situation in this way: The entire existence and being of a body is the product of inner forces. The more perfectly these are in equilibrium, the more completely they limit each other to a given volume, and the stronger must be the force which is to modify the existence of the body in this volume. Mechanical coherence and chemical resistance are equal to the extent that no other circumstances have any influence. The largest chemical difference will therefore be associated with the weakest cohesion. It is evident that this is the case in our series, and it is only a question of proving that coherence depends on the chemical quality and not the other way around. Experiments decide this in a most excellent fashion.[10] All aëriform acids solidify in combination with the only aëriform base. The aëriform acids condense with all the remaining bases, and even the non-aëriform acids are subject to a compression in their neutralization. Conversely, ammonium also condenses with all non-aëriform acids so that condensation really seems to be a necessary consequence of indifferentiation. It would not be easy to counter all possible objections beforehand; we merely wish to draw attention to the generality of this law. In fact, water is changed into hydrogen gas and oxygen gas by the opposite electricities, whereas the indifferentiation of them both contracts them to water again. Similarly, all combustions, which are also acts of indifferentiation, are associated with contractions. Even if the calcined body seems to be expanded in some cases, this is only an appearance because the body has been dissolved in the oxygen gas. This can also be recognized by the fact that a more intimate combustion frequently produces a true contraction. It is well known that some have concluded from this that oxygen is the contracting principle in nature.

At first glance, it might appear that we have pulled ourselves out of one difficulty

---

[9] That heat is not merely indifferentiation, but that there is a predominance of $+E$ in heat and of $-E$ in cold follows from Ritter's excellent investigations.   Ø.

[10] Here, we deliberately do not wish to call upon the results of natural philosophy, which agree with this so completely.   Ø.

only to become tangled in a new one, for we assumed initially that cohesion modifies the chemical effects, but we now assert that the chemical nature of the body determines cohesion. This is due to the fact that we first judged conditions by mere appearance and listed the phenomena individually without attempting any derivation of one from the other. We imagined that if *being* a body was one thing, *having* cohesion, *having* chemical forces would be another, and so it must be imagined if we wish to proceed from the outside to the inside. In fact, however, to be a body and to have these forces are one and the same. Or, to be even clearer, the forces do not have their existence because of the body, rather the body is nothing other than the product of these forces. Therefore, cohesion in itself is not a retarding force for the chemical action, but its degree is a mark of the intimacy of the act of indifferentiation, and consequently, it is simultaneously a measure of the resistance to every external force which seeks to disrupt the direction of the inner forces. However, no-one familiar with the spirit of natural research will be surprised that we, starting from a lower level, take for cause what appears as effect from a higher position.

So far, we have established the series of acids and bases without perceiving their principles. For a more complete understanding, it is necessary that we seek to penetrate their intrinsic nature to the greatest extent possible. It is an accepted fact, admitted by the more insightful antiphlogistians themselves, that in many cases the so-called oxygen[11] does not make things sour, and that, conversely, several bodies are sour in the absence of oxygen. Therefore, we are not averse to abandoning the theory of this substance as the souring principle. However, the facts which have given rise to the assumption of the souring power of the basis of vital air are too remarkable and too consistent for us to discard them now without obtaining some result from them. Carbon, nitrogen, phosphorus, sulphur, several metals, nine or ten, and any number of heterogeneous bodies become sour through their combination with oxygen. The common property which these bodies acquire through calcination is rather to be attributed to the common body added than to any other. On the other hand, we see that oxygen not only fails to produce any acidity in some compounds, but that it even causes alkalinity under certain circumstances. We see this in those metallic calces which, with a certain proportion of oxygen, act like bases and neutralize acids. With a larger proportion of oxygen, their ability to dull acids is reduced, and some actually become sour with their maximum oxygen content. Thus, oxygen causes totally opposite effects in one and the same category. It is not sufficient to conclude from this that oxygen does not deserve its name, for we clearly see here a regular progression of phenomena which cannot be without some law.

Nitrogen, which becomes acid with oxygen, yields a base with hydrogen, viz., ammonium. As isolated as this experiment is, it still causes hydrogen to be placed in the same relation to bases as we have long placed oxygen in relation to acids. We can do this all the more since we know that hydrogen is the only body by which oxygen is completely neutralized (this word is used in a broader sense than usual). Moreover, several have already understood this. However, precisely this hydrogen

---

[11] [*Sauerstoff.*]

yields an acid with sulphur, but here, too, we have only a single decisive example. In this case we can conclude, with only the aid of the analogous experiences with oxygen, that a similar contrast occurs in the series of hydrogenations as well. It is no wonder that we know so little about hydrogenations since they have been so little under our control.

The result of all this is that a substance in a certain degree of combination can cause precisely the *opposite* of what it does in some other state. However, we encounter such a situation again only in galvanism. It is therefore necessary that we establish a connection between this and our subject. This connection is already admitted by many physicists, but perhaps a summary of the relevant issues, although very condensed, would not be completely useless and might teach many something new.

It is known that, if the poles of a simple galvanic circuit or of a not particularly strong pile are placed in contact with the tongue, an acid taste is obtained on the part which is turned towards the positive pole but a strongly alkaline taste at the opposite pole. If the pile is very strong, e.g., 100 copper-zinc plates with a sal ammoniac solution, the opposite effect is obtained, i.e., the alkaline or basic taste at the positive pole but the acid taste at the negative. We find here a perfect analogy with electricity, in so far as this merely arouses its opposite at weak degrees, which we call induction, but at stronger degrees it not only eliminates this opposite but even produces the same electricity. Thus, it is not the positive pole (which arouses acidity at weak degrees) but the negative (which arouses acidity at stronger degrees) which is in possession of the acid principle just as, conversely, the positive pole contains the principle of alkalinity. That a pointed object which, as is said, emits positive electricity produces an acid taste on the tongue seems to me to prove nothing against this. The mechanism of the electrical actions is so constituted that the positive body first excites negative electricity in another body and then cancels it again. However, the indifferentiation of the small quantity which can be emitted from such a point probably takes place on the surface of the tongue and not in the taste nerves so that these remain negative during the entire process while what is located on the surface is positive.

The galvanic pile also transmits a polarity to water so that it transforms into oxygen gas at the positive pole of the pile, at the negative into hydrogen gas. This is perfectly analogous to another fact, also discovered by Ritter, that a solid body, like a metal, also becomes charged when in contact with a pile and then becomes more combustible at one pole but less combustible at the other. We may also assume here that the pile acts only by electrification of the first degree (induction), and that, therefore, oxygen must be the negative pole of water and hydrogen the positive. This is also confirmed by the fact that metallic reductions which arise at the hydrogen pole of the pile have vegetative shapes, which is entirely consistent with the dust figures of positive electricity.

Neutral salts are brought to acidic and basic polarity by the galvanic pile. All galvanic experiments with neutral salt solutions prove this. The acid appears opposite the positive pole, the base opposite the negative. Hisinger and Berzelius have reported a very beautiful series of experiments on this subject (*Neues allg. Journ.*

d. Chemie, Vol. 1, pp. 115–49.), from which this follows clearly. All their experiments give the same result, but the experiment with the sulphate of potash does not easily admit of more than one interpretation.

If we assume that positive electricity is the principle of hydrogen and of alkalinity, whereas the negative is to be regarded as the principle of acidity, it follows that the positive pole must attract oxygen and acids, but the negative must attract hydrogen and bases. —We have now arrived at a point where we recognize the principles of acidity and alkalinity as principles of electricity. These principles are to be found in all bodies and cannot be separated from their nature. We will certainly not claim on this account that all bodies are acids or bases, for it depends not only on whether these principles are present, but also on how they are present. Otherwise, we would be obliged to claim that even the coloured rays of light were acidic or basic. Now this, as a paradox, would not frighten us, but we would become entangled in a great many difficulties. Instead of calling a body with an excess of the positive principle a base, we could say with equal justice that it was violet internally, and that we should not be concerned merely with the outward appearance because there could be causes which impeded the manifestation of the colour. It is indisputable that we should not allow ourselves to be prevented by appearances from seeking the inner principle. Once we have found the principle and, at the same time, seen it revealed in the most varied forms, e.g., as light, as heat, as electricity, as magnetism, etc., it is then time to differentiate precisely between these forms and not to confuse them because of what they have in common. To be sure, the first abstraction must generalize everything, but he who possesses the general must separate everything again, in order to see a structure and not a void. The issue of what is an acid and what is a base is far from settled by the demonstration of their principles, which are general principles of nature. It does not depend on their presence in a body, for they are present in all bodies, but on the manner of their presence. Thus, we are still far from knowing the inner nature of acids and bases. So far, we have only seen that oxygen, in a certain proportion with metals, produces bases but, in a different proportion, produces acids, and then we have observed that not only oxygen owes its nature to one of the electrical principles, but that all acidity and alkalinity were created by them. However, the striking fact that the same principle can produce opposite chemical qualities has not yet been related adequately to other natural phenomena. We have made the remark that a weak electricity only excites its opposite while a stronger produces its equal, but this is not exhaustive. Ritter has not only shown that all sensory impressions are inverted by a stronger degree of galvanism, but he has demonstrated with numerous experiments that there are two kinds of irritability which are affected in two opposite ways. These two kinds of irritability are found not only in organic but also in inorganic nature, for the electricities which a liquid body produces with a solid one in the first moment of contact turn themselves into their opposites if the effect begins to become more intimate. In the former case, the activity is directed towards the excitability of the first kind, but in the latter towards the excitability of the second kind.

Thus, the effect of oxygen on metals in the first degrees of oxidation also proceeds, most probably, through the excitability of the first kind, in the later degrees through the excitability of the second kind. In fact, oxygen is in possession of the true acid principle. If it is caused to act on the excitability of the first kind, it arouses only its opposite, i.e., it yields alkalinity; on the second kind it produces its equal, and only then does it yield acidity. The fact that some metallic calces, at an intermediate stage of oxidation, are adiaphoral is easily understood from the above. It could be objected that Ritter regards every chemical action as an excitation of the second kind, and therefore the first degrees of oxidation, as well as the later ones, proceeded according to the same excitability. This is merely an objection in the letter, not in the spirit, for Ritter admits with adequate clarity that the excitability of the first kind is not overcome by that of the second kind in an instant, which would indeed be impossible. Therefore, it is quite possible that the excitability of the first kind can be dominant in the first degrees of oxidation. The fact that the excitability of the first kind is so quickly exhausted when metals are dissolved in acid could be due to the fact that the metal is not only oxidized but dissolved at the same time, in which double action it is not enough to see what oxygen does to metal, but also what metallic calx does to acid. It is curious that metals, which are such good conductors, lose much of their conductivity in the first degrees of oxidation but become good conductors again in the highest degrees, as Ritter first noticed in general.

If we now proceed to the combinations with hydrogen, the same situation appears. Since this substance owes its nature to the principle of alkalinity, it must produce alkalinity when it acts on the excitability of the second kind, and additional conditions are present. However, when it acts on the excitability of the first kind and thus produces only its opposite, it must produce acid. We have already mentioned that we know very little about the combinations of hydrogen. The only certain example which we have of acidification by hydrogen is that of sulphur, and we still lack other arguments for the claim that this combination is formed according to the excitability of the first kind, so we must be content with the analogy. It is very probable that prussic acid belongs to the same category; so far at least, it has not been possible to demonstrate its oxygen content. We could perhaps extend this supposition considerably, for it is not at all impossible that vegetable acids are acids because of hydrogen. It is at least certain that nitric acid, which changes the weaker vegetable acids into stronger ones, shows its effect more by de-carbonizing than by oxidizing. Hydrogen is also present in relatively larger amounts than oxygen; there is not even enough of the latter to oxidize all of the carbon contained, rather a very large part of the carbon disappears during the decomposition of these acids by hydrogen; another part remains in the form of coal, and yet another is transformed into an oil by hydrogen. Vinegar is also produced here, but its oxygen could not possibly cause a complete saturation of the carbon and the hydrogen of the other products. On the other hand, the components of this acid can accept even more oxygen, and thus it is no exception to vegetable acids in general, which are all more or less combustible. It can therefore be assumed that the oxygen in these acids would only be present together with a small portion of hydrogen in the form of bound water. All this can also be applied to the animal acids, as can be easily understood. If all this were so, we would be completely justified in assigning the vegetable and animal acids to an entirely different category from the others, which would then be acid through the opposite principle. Even if this were not the case, we could still not accept them in that series because they would not fit anywhere, neither at the beginning, nor in the middle, nor at the end. Moreover, they can form a series of their own. If we arrange them according to their volatility, the series would be roughly the following: *prussic acid, formic acid, acetic acid, benzoic acid, succinic acid, suberic acid, gallic acid, oxalic acid, tartarous acid, citric acid, malic acid, uric acid, lactic acid.* The first of these acids is aëriform, the following two are volatile with a strong attraction to water. After this follow several others which have less attraction to water but are still volatile at a temperature which does not destroy them. Even gallic acid does not have this property completely, oxalic acid is largely decomposed by heat before it evaporates, and all the remaining are almost completely destroyed before they can evaporate. The attraction of the last of these to water is also very small. We can also place hydrosulphuric acid at the head of them all as it is frequently produced in animal matter. However, it is very difficult to investigate whether the actual strengths of the acids in this series indeed progress in the same order as the property whereby we have determined the positions. Since our organs can endure tasting all of these acids, even in their highest concentration, this could provide us with a means for a determination, at least for lack of something better. We find, however, that the acids of the first category which possess the most acidity, like sulphurous acid, oxymuriatic acid, did not have the strongest *sour* taste, but they were rather distinguished by something very piquant. On the other hand, acids which were in the middle of this category had a proper sour taste which was very pronounced however much they were diluted. The last members generally have a weaker taste which almost vanishes in the case of tungstic acid. The situation is exactly the same throughout the second series of acids established here. The first distinguish themselves more by their irritating than by their sour taste, not excepting acetic acid if it is free of water. Oxalic, tartarous, citric, and malic acids have a purely sour taste in the order cited, and finally the last two have a very weak taste. If we desired to go even further (and why should we not pursue an analogy to the extreme?), we could even add tallow, fat, and fatty oils to the acids. In any event, they also dull bases, of which soap provides an adequate example. They dissolve metallic calces as well or otherwise enter into chemical combinations with them, whereby they represent a new analogy with acids. However, if hydrogen becomes too dominant in the oils, they leave the category of acids and *approach* the bases, *due to hydrogen*, as we see in the etheric oils. If we might assume what has become more and more probable during the investigation, that the second category of acids owes its nature to hydrogen, we would have to contrast this series with the series of metallic oxides, which are bases due to oxygen.

In the series of metallic oxides, the bases formed by combustible metals will be the strongest, just as the most weakly oxidized among several oxides of the same metal must have the greatest alkalinity, for the inner hydrogenous nature of the metal should only be excited by the oxidation, not exhausted. Everyone knows how much experience agrees with this both in that the oxidized metallic bases also give the more caustic, i.e., less neutral, salts with acids, and in that the oxides of very little combustible metals similarly produce more caustic salts. Since the facts which we must call upon here are not only very well known but are also described in the context we require so that only the interpretation might contain something new, we do not wish to make any attempt here actually to establish the series of metallic oxides and can feel all the more relieved on this account as, on the one hand, this discussion would be very lengthy, and on the other, we could and would at present only attempt to indicate rather than to elaborate.

In order to have a name for the investigation of the two categories of acids and bases which we have presented, it would doubtless be fitting to attribute a *direct acidity* to those acids which owe their nature to oxygen, but to bestow an *indirect acidity* on those which were made acid by hydrogen, and following the same principle, to designate bases by hydrogen as bodies with *direct alkalinity* but those by oxygen as bodies with *indirect alkalinity*.

In conclusion, the author may be allowed to make a historical remark regarding the development of this investigation. The series of bases with direct alkalinity presented here is approximately the same as that which Steffens presented in his *Beiträge zur innern Naturgeschichte der Erde*. Depending on their own convictions, people lacking insight into the subject, who nevertheless wish to judge scholarly matters, could easily either accuse this investigation of contradicting the other, or they could also think that the author has borrowed the idea for his series from that important work. The response to the first is already contained in the preceding, in that we assume that the same bodies can be members of different series. In fact, there is nothing which prevents the assumption that the nitrogen content of the bodies in our series decreases steadily from top to bottom, but that the carbon content increases. Thus, these two views of the same series cannot be in contradiction with each other. The author could easily counter the second accusation if he wished to confess directly that he had borrowed here, which, after all, is admissible. But, in truth, this is not so, and therefore it cannot be confessed. The author established this series before the publication of Steffens's *Beiträge*, in a review of Gadolin's *Indledning til Chemien*, which can be found in the Danish journal *Skandinavisk Museum* for the year 1800, p. 177. The issue in which it is printed is even the first of the year. Steffens's *Beiträge* appeared only in the year 1801. With this report, however, the author wishes to deprive his famous friend and countryman of something as little as he wishes to avoid giving something to him. The author knows with certainty that he was not familiar with the review mentioned when he wrote his *Beiträge*. Anyone who truly understands the *Beiträge* does not even need this statement since he must see that Steffens could have acquired his conception from the review as little as the author of that review could have acquired his from the *Beiträge*. This matter is not important in itself, for it is not the *idea* but the thorough, truly carefully reasoned *execution* of the idea which provides proof of its possession. However, one is frequently obliged to explain oneself about such matters in order to avoid being wrongfully suspected or, indeed, even insulted.

25

Reflections on the History of Chemistry,[1]
A Lecture[2]

It is an old, often repeated complaint that there is no agreement in the realm of science. Precisely where one expected an eternal peace, a unanimous effort towards one goal, there a continuous internal war reigns in ever changing upheavals. Who does not know how much occasion this discord has given the timid for distrust, the scoffer for laughter, and the enemy of science for declamations on the uncertainty of human knowledge.

What I said in the last lecture regarding the fall of antiphlogistic chemistry must occasion the same complaint. Once again we had to reject a theory which was advanced only a few years ago with the approval of almost the entire enlightened world, and not without bitter quarrel, in which the arguments on both sides were often weighed and tried again. Nor was the battle fought in the world of ideas, where reason might have strayed in its own infinite profundity, but in a sphere where everything could be brought to the old and tested touchstone of experience, and where everything, therefore, could reach an even greater certitude. This is the theory that must fall. As this philosophical system fell, so have many others, in every science, and have often had to make room for the exact opposite. Where do these vicissitudes cease? Is it likely that we are now in possession of the true theory, which shall stand firm against the attacks of future generations? We have no greater

[1][*Det skandinaviske Litteraturselskabs Skrifter*, Vol. 3, No. 2, pp. 1–54. Copenhagen 1807. Also published in *Journal für die Chemie und Physik*, ed. by Dr. A. F. Gehlen, Vol. 3, pp. 194–231. Berlin 1807. With some changes in *Samlede og efterladte Skrifter af H. C. Ørsted*, Vol. 5, pp. 1–32. Copenhagen 1851. In German in *Der Geist in der Natur*, pp. 371–428. Munich 1850. *Gesammelten Schriften*, Vol. 3, pp. 143–74. Leipzig 1851. *Der Geist in der Natur. Neue Ausgabe*, Vol. 1, pp. 290–314. Leipzig 1867. In English in *The Soul in Nature*, pp. 300–324. London 1852. KM I, pp. 315–43. Originally published in Danish.]

[2]When, two years ago, I was to give a lecture on chemistry to a gathering of men, most of whom were already familiar with older theories, I first presented three introductory lectures, in which I tried to remove several unfavourable ideas which might be conceived against a new revolution in chemistry. In the first of these lectures, I showed that arbitrary boundaries had already been set for chemistry in its definition, about which I have also contributed a paper to this publication. In the second, I gave a brief summary of the difference between antiphlogistic and dynamic chemistry, and then I concluded the introductory lectures with the present reflections.

likelihood of this than all our predecessors of the correctness of their ideas, which they held to be just as certain and true as we ours. Therefore, we must think it likely that we, too, shall fail.

For this reason, many men of experience have rejected all theory and found in it only a game unworthy of their attention. They ask us whether it is likely to be more than a foolish dream that reason should yet prevail if it has not arrived at agreement through the efforts of so many centuries, after the ruminations of so many wise heads. If we follow them, there is only one truth, the reality which firmly and stead-fastly has surrounded us during so many vicissitudes and, at every moment, forces new evidence upon our senses. Theories might serve to arrange our knowledge into certain categories where we could find it again more easily; they might be useful enough to sharpen the discernment of youth, which will later be used in a practical career, but it is quite impossible to believe that so many contradictions should con-tain any unity and truth and provide any real insight.

A nobler, though weaker nature rightly adds to this a very worrisome observa-tion. The entire worth of man lies in reason, and if his greatest endeavours have produced nothing but delusions, then man is the most imperfect and unhappy of all animals, for his race has misused the abilities that Nature gave it for its preserva-tion and through it tried to break away from the reality of which he is but a part. A revolt in all possible directions but punished by Nature with eternal unrest and countless weaknesses unknown to other animals. In an unhappy whirl of folly they are all carried away, but doubly unhappy he who realizes it, for he still cannot resist as he is only a limb of his entire race, which has been endeavouring to distance him from Nature for millenia. Even if he felt strong enough to approach it on one side, he would still have to tear himself away on the other. Thus man is hastening, ever more quickly, towards his own destruction, and even if that were to be the fate of the entire globe, of which he has erringly made himself the master, he is still the only one who is unfortunate enough to predict it.

The spirited youth goes another way. Courageously he cuts the knot. He says that cowardice justly leads to despair. Who has taught you that you shall win the greatest treasure, Truth, without toil? Do you usually count the votes to know what is right? Are you not, then, forced to denounce the multitude as unwise? Now apply this rule to those who pretend to be scholars. If there have been conflicting views, one of them must have been right? Can you wonder that they have changed as you must easily see that they have not all striven towards the great goal with the same zeal, with the same vigour, with the same love of truth? Besides, if Truth has not yet been found, it can still be found. A whole eternity remains. The force that tore you away from what you called reality must have been stronger than this reality itself. It can, it must some day lead us into the realm of Truth. And even to strive towards it is glorious. Our strength is tested, our spiritual life is maintained, our disposition is soothed by the happy prospect that our whole race is progressing towards better things. Is this not reward enough for our humble work!

Thus a new conflict arises as we seek unity. And we would gladly follow him with cheerful prospects to a future tranquility. However, we soon notice that he has just started to tread the path of science. He thinks of himself only as a rational being, and therefore he must be right, but if he glances back at his own individuality and its consequent limitations, he recalls the host of profound men who have lost their way, and for a moment, exhausted by the struggle of Life, he conceives the idea: Those men felt as convinced as you and failed. Are you stronger, then, or could you not also, unaware, be entangled in error? Then he will soon be consumed by doubt, not about reason but about himself. This doubt must be extended to all other human individuals, and consequently to the entire species, and we now see him at the same point where despair itself began.

However, I see the transfigured succession of heroes in the history of the human spirit looking calmly down at this confusion. They tell us: We have devoted more time and more effort than our brothers to find the profundities of Nature and Reason. Half-way we met darkness and doubt, but the deeper we penetrated, the more everything became light and unity for us. Each of us has learned from his predecessors and been a teacher to his successors, not just by accumulating a greater stock of knowledge but also by looking more deeply into the order of things. And have we not confirmed to you our truthfulness by giving you laws whereby you could understand and calculate many things that no eye had seen before, a sign for your disbelief. Be not deceived by an appearance of disagreement. The age, the country, the character of each led him in his own direction and gave his works their peculiar quality, but we are all united by a secret bond. If you seek more deeply in the history of science, you shall find tranquility where before you found only doubt and unrest.

We shall follow this voice, for it is the voice of truth. By looking closer at the conflict which caused our unrest, we shall discover the purest harmony, the most complete calm and certainty. —Indeed, it would be a subject worthy of the thinking man to seek that internal unity in all sciences, but such an extensive task would here take us too far from our goal. We must content ourselves with contemplating the evolution of the only science for the explanation of which these lectures are intended.

Just as the history of every science, to the less observant eye, seems to present nothing but a chaos of contradictions, a maelstrom of struggling forces, so, too, that of chemistry. From the time when we began to unite the scattered experiences which contained the first seeds of this science into a whole, we find a constant movement forwards and backwards, no calm progression. First it was shrouded in an impenetrable veil of mysticism during the Middle Ages, and it is easy to see that there were as many different opinions as there were dreamers in this period. To imitate the noblest metal that creative Nature offered, to find a universal solvent, to procure for mankind a remedy against all diseases, these once constituted the problems of science, which they attempted to solve in the most varied ways. However, almost all agreed that there was a profound similarity, hidden to the uninitiated eye, between even the most distant objects in nature, and that this could be found only through a special gift and used in the execution of the great schemes of mankind that they had proposed. Thus they sought the imagined characteristics of the planets in metals, and they organized their chemical tasks according to astrological combinations. They found themselves so much more easily convinced of the validity of this conception as they knew an equal number of planets and metals.

What would they now say in defense of their view when a more refined astronomy shows them that the sun is not a planet and the moon only a satellite like many others which we can see with the fortified eye? What would they think when they learned from a more ingenious chemistry to distinguish thirty metals and, in addition, had a probability, amounting almost to a certainty, that there are many more to discover? In short, who would any longer take pains to refute opinions which the whole rational world rejects though, in those dark times, they found friends among the most profound men, protectors among the mightiest princes? It must be enough for us to remember that to them chemistry was nothing but the science of the production of metals and their natural forces nothing but mystical similarities of characteristics. This, I say, is sufficient to show how different it was from our more comprehensive chemistry, how contrary it was to the efforts of our time towards a clear and transparent science. Nor are our minds set at rest by an attempt to comfort us with the thought that that great period was a dark night full of fantastic dreams now happily over. If they went so horribly astray, who can guarantee that we do not stray as far in the opposite direction and, from a misguided effort to understand, overlook what constitutes the essence of infinite Nature and incomprehensible existence? Or dare we, for historical reasons, regard a period as irrelevant to us, at a time when many of the most excellent intellects bring many of its doctrines back to light?

However, we shall avert our eyes from this period. It was the age of incomprehensibility, so why should we try to comprehend it? We shall not even pay any attention to the period of ferment between that time and the more recent; it is no wonder that this was an age of contradiction and struggle. We shall turn our full attention to the time when one sought the cause of natural phenomena in comprehensible natural forces and attempted to present each of these in the greatest possible purity through experiments. In this period, all acknowledged the same Reason as judge, so we ought to expect greater agreement among them. However, no expectation can be less fulfilled. To recite how opinions have appeared, disappeared, and regained control during this period would give material sufficient for an entire book and would, even if it were possible within the short scope of a lecture, tax your attention; hence only some main features here.

They began to classify chemical knowledge, and they invented a combustible substance, phlogiston, which was supposed to be contained in all bodies. Every combustible body, then, was a compound. Combustion was a decomposition, and metals were composed of a calx and phlogiston. The bodies which gave acid through combustion had to contain this along with phlogiston as their constituents. Every phenomenon was explained according to this theory, and it was believed to contain the key to the secret workshop of nature. However, they were far from agreeing about the nature of phlogiston, now it was assumed to be a sulphur, now a fine earth, now a part of the material substance of light. Indeed, finally it was even given a property which was opposite all other bodies, a force which destroyed gravity.

When this theory was believed to have been brought closest to perfection, it was upset by another which was so contrary to it that it even took its name from this contrast. By denying the existence of phlogiston, the antiphlogistic theory changed the old conception completely. Combustion was no longer a decomposition but, on the contrary, a composition, a combination with oxygen. This element had to be contained in acids, and these were by nature compound, whereas the substance which produced the acid through combustion could be simple. The grounds for assuming the compound nature of metals now disappeared, and they were regarded as elements. Water, on the other hand, which was an element in that theory, here became a compound. In short, everything was turned upside down, and not one stone was left on another in the old building.

However, no sooner was the controversy between these two systems over than a new one arose, which began by denying that water was a compound. It is true that this theory is not yet completed, but it is clear that the antiphlogistic theory, which seemed so well established, must be overthrown through its efforts. Indeed, it even goes much further as it seeks to establish an entirely new view for all of chemistry, according to which none of what we have so far called compositions and decompositions is to be regarded as such any longer. All that we have hitherto learned about chemical affinity, that which the phlogistians and the antiphlogistians still had in common, must now be regarded as useless!

We shall now attempt to shed light on this chaos which, at first glance, shows only a crude mixture of the remains of so many ages. However, in order to facilitate the outline for you, I shall begin by describing the course of the investigation that I intend to present to you. First, I shall show you that all those who have had knowledge of science, according to any theory, have also been in possession of some great and profound truth. In this investigation, it will also become evident that the course of science has been a true evolution. I further hope to prove to you that this process of evolution followed necessary laws. Finally, a closer consideration will convince us of the beneficial character of these laws.

I fully realize the great extent of this investigation to which I have introduced you, and I willingly admit that it cannot be accomplished with any degree of completeness within the brief scope of an oral lecture, but I believe that I shall be able to touch on the main points so that each of you personally will be able to apply them fully.

The mystical tendency of the Middle Ages is so opposite the striving of our time towards perfect clarity that it might easily seem impossible for them both to have a share in the truth. To deny the contrast between them would be against evident truth, but no contrast can exist where there is nothing in common. Perhaps the sharpest contrast has its origin precisely in the one-sidedness of both ages, and, on the other hand, there might be agreement where we had not suspected it. Every effort towards insight into nature aims at bringing separate phenomena under a common terminology, at discovering laws which everything obeys, in short, at bringing the unity of reason to nature. The mystical age had at least this endeavour in common with ours. It is natural that the first look at objects only meets the surface, which must be opaque to the untrained eye and hide the interior, but so much greater was the attention to the outside, which spoke to them in countless guises. They sought similarities and divergences in them and thought that they would learn the secrets of nature from them. Without doubt this course in an investigation can easily lead astray, but we cannot deny that an internal similarity lies hidden in the external because the same form must always be produced by the same force when no foreign forces disturb the effect. It is evident that excessive subtlety along this path can produce the most ridiculous monstrosities, but it is no less certain that an ingenious eye can also make the greatest discoveries there. It must never be forgotten that what is the surface of individual objects belongs to the heart of the universe. Some examples will throw more light on what I have advanced here. They compared the metals with the planets. At first glance, this seems mere fantasy, but if we consider the matter more closely, we find an underlying truth. Metals constitute the actual mass of the earth, and it is not unlikely that all earths are metallic calces. Just as all planets move around the sun according to the same laws, all satellites move around their planet. Therefore, it is also natural that all planets are developed in the same way and their satellites in relation to them. However, is it not likely that the same laws according to which the planets have been developed in the solar system have also ruled the development of the individual parts of the planets? If this could be assumed, each planet would find its corresponding element in the bowels of the earth, and the correspondence between the metals and the planets would no longer be fantasy. If we add to this that we have learned in recent times that metals are distributed on earth according to certain laws, the idea gains even more in probability. Finally, it is also worth noticing that gold, which was the sun of metals according to that time, is deposited primarily around the equator and also maintains its metallic nature most perfectly in all assays.

I hope that you will believe that I am far from defending the practical application of that astrological metallurgy. What is more: I admit that, in spite of all our greater knowledge, we are still not able to advance such a comparison between the metals and the planets, but the basic idea is hardly to be disdained. It is easy to answer the objection that the difference between the number of metals known to us now and the planets is so great, for why can certain metals which usually accompany others not be regarded as their satellites? Why could many of them not be compared with comets? I need hardly warn you not to go into too much detail in the application of these theories as the matter is probably not yet ripe for it. It must be sufficient for us to have seen that there was some truth behind the delusion of that time.

However, even regardless of this, that age has won some merit from science. Not to mention the many individual discoveries for which we are indebted to this age, we also owe it several coherent series of related substances, e.g., the mineral acids and the alkalis. The calcination of metals and various useful methods for this, as well as the increase in weight of metals with calcination, are also discoveries by the chemists of that time. They even contributed to the chemical knowledge of air by showing that there were several gases which were different from the one which surrounds the entire globe.

We see, then, that the mystical age acted neither without plan nor without fruit. We can arrive at a similar conviction about the phlogistic age with even greater ease. It is true that the founders and the supporters of this theory explained every-thing on the basis of a presumed element which they could not prove. Nevertheless, their concept of nature was not entirely wrong. The idea that combustion is, as it were, the centre of all chemical actions reveals an uncommonly profound insight into nature because, in order to conceive such an idea, it was not enough to find the outbreak of fire and the glorious brilliance of the flame worthy of their attention; they also had to see that nature often produced the same effects as combustion by means other than fire, and they had to discover that there was a common force in all these outwardly different effects. It requires a very astute and bold spirit to find traces of combustion where no flame, and often not even any heat, has announced its presence. However, it probably calls for centuries of preparation to see combus-tion even in the middle of a fluid and to find a certain similarity between respiration and the flame.

Only after such a great and yet deeply penetrating insight was it possible to ar-range bodies in series according to their combustibility, for now they knew what combustibility was. For this series, it was possible to determine the natural law that the more combustible body could restore the less combustible and already com-busted body to its initial state. They also saw that a body lost its combustibility to the very same extent that it was burnt, and thus the great and extensive as well as easily comprehensible law was established that combustibility and combustion are *opposite* each other, or, in other words, that combustion and reduction are two op-posite processes which pervade all nature. Such great ideas, the fruit of the endeav-ours of centuries, are contained in so few words. However, those who know nature know what these few words mean. I should like you to form a fairly clear concep-tion of this. Who does not know the role which metals play in the history of man-kind? They play a no less important one in that of the earth. They mix with stones and crystals, they cut through mountains, they form the foundation of enormous masses, and they appear everywhere in the most varied forms. However, that the-ory of combustion embraces them with almost unlimited generality. To whom is it not evident that the circulation of the blood is one of the mainsprings of animate nature? Who does not know that respiration is one of the elements in that great chain of actions? But has the phlogistic theory not incorporated it? How many ex-amples do we need? None of you is completely unfamiliar with the phlogistic the-ory, so you will find examples in abundance.

None of us will blame Stahl and his successors for assuming a common princi-ple in all combustible substances. The antiphlogistians themselves did that by at-tributing to all combustible bodies a chemical attraction to oxygen. The mistake of the phlogistians, then, can be only that they assumed a material cause of combus-tion, about which they were certainly wrong. However, we must remember the character of their age, this clinging to the material, from which only few of the chemists of our time have broken away. Phlogiston just became a cipher, an X with which they designated the unknown cause of combustion. Even if they added something to this symbol whereby it lost its purity, it must be admitted that it was possible, within certain limits, to make correct calculations with it.

In other respects, we are far from regarding the phlogistic theory as the perfection of science. We believe ourselves justified only in claiming that there was a true and great vision in it, a contemplation of a great natural law. However, the very clarity of this vision is darkened by every arbitrary assumption. There is always a point where this interferes substantially, and from there confusion spreads to all the rest. So, too, in phlogistic theory. The hypothesis made them blind to what nature showed them. Therefore they overlooked the real effect of air in combustion. It was granted to Lavoisier to discover this and to establish a new system, which, however, was claimed to be newer than it really was, for the basic idea of the old remained the foundation of the new and could not possibly be rejected. However, that combustion is a composition and not a decomposition, that every body absorbs oxygen in combustion, that this substance is a common constituent in many acids — these are among the characteristics of the antiphlogistic system. Only through these discoveries was the naturalist enabled to indicate and calculate in advance, not just generally but with the greatest precision, the products of many actions which we must now refer to the class of combustions; but the antiphlogistic theory did not become complete until the discovery of the constituents of water. That the most combustible of all bodies is found in water in combination with the one which is the condition of all combustion is a discovery whose importance can be doubted by no-one who will recall for just a moment the many bodies which receive the elements of water either separated in order to combine them or combined in order to decompose them. How many do not then present themselves to our eyes!

You can easily see that, in spite of its name, the antiphlogistic system is a continuation of the phlogistic. The fact that they contradict each other does not provide any proof against this, for you have seen yourselves that this was only in one respect, not in all. Therefore, we found the transition from one system to the other easy. The system which stems from the most recent experiences follows a very different course from those. It even has its origin in investigations of a completely different sphere. It was from investigations of electricity that new light would be shed on chemistry. The force which for a long time had been found only in rubbed amber was gradually found in many other bodies, and finally it was realized that all natural bodies must possess it. The fundamental laws of electricity were now found through Franklin's profound insight. Almost all old instruments for the investigation of electricity were improved, and many new ones were invented with the guidance of established principles. Electricity was now discovered where it had scarcely been suspected before, and finally it was found, in different ways, that even the mutual contact between bodies arouses this natural force. We gratefully recall Volta's merits in this respect. It had also been discovered that mutual contact changed the chemical forces of bodies, and Ritter's foresight already saw the connection between electricity and chemistry in these experiences. Volta went even further and found that a combination of several elements produces an increased effect, and now electricity is generally recognized as a chemical agent. Even if most continued one-sidedly to regard it only as a chemical agent and not as the revelation of a universal force of nature, they did not all allow their vision to be limited by an empty name. Now Ritter showed that the chemical changes in water were dependent on an electrical distribution of force, and thus he gave the whole theory of the composition of water another point of view. However, not only the theory of the composition of water but the whole chemical theory took a new turn because of this change. Through several discoveries, whose perfection is due to Ritter, it has been shown that all bodies constitute a series according to their ability to produce electricity. The first link in this series generates positive electricity in interaction with all other bodies, the second acquires negative electricity with the first but positive with all the others, and so on, until the last, which acquires negative electricity with all the others. Among bodies which, on the whole, are subject to the same conditions, it is found that this series is parallel to that of combustibility so that the more positive are also the more combustible, the more negative, on the other hand, the less combustible. This accord between combustibility and the tendency towards the positive state is confirmed even further by the discovery that good conductors can receive a charge through contact electricity and thereby become proportionally more combustible as they become more positive and proportionally less combustible as they become more negative. The so-called decomposition of water is such a charging, in which the only strange circumstance is that the differences generated are so great and appear so perceptibly to the eye. The positive pole of water is hydrogen, the negative pole of water is oxygen. Combined, the opposite forces again cancel each other and produce water. Of all bodies, water is the one in which there is the greatest equilibrium of all forces; hydrogen is the most combustible, and oxygen is the least combustible and itself a condition of all combustion.

Now, then, a new theory of combustion emerges. The combustion of hydrogen is only a uniting of its positive with the negative of oxygen. The flame is really a continuously renewed electrical spark. Or, more precisely, light and heat are generated because these effects have the same elementary actions as electricity. However, what applies to the combustion of hydrogen is true of all combustion as every combustible body is combustible through its positive and burns when it is combined with the negative. We thus assume an internal cause of combustion in agreement with the phlogistians and an external one in agreement with the antiphlogistians, but we differ from them in that we do not consider these to be material.

You cannot choose but feel that our whole conception here undergoes an essential change. Previously we assumed real combinations and decompositions in all cases where two different bodies became one. Now, on the other hand, we assume that a distribution of forces makes them identical. This applies not only to combustion but also to the interaction between acids and alkalis. In the course of these lectures, we shall see that acidity and alkalinity are also due to a certain mode of existence of the two fundamental forces which I have now mentioned to you so often, and that the many natural phenomena which the chemists relegate to neutralizations can also be regarded as indifferentiations.

During all these reflections, it cannot have escaped your attention that the view of all these natural phenomena may well have changed, but that the connection which had once been found between a large series of natural phenomena was not destroyed by it in order to create a new one. That combustibility does not merely consist in generating a flame under certain circumstances, that combustion has its opposite in another process, the one we call reduction, that an interaction between

oxygen and the combustible body is part of combustion, that water can be transformed into hydrogen and oxygen, as these again into water, remain discoveries of
the utmost importance, which we all use, but which we also subjugate to the necessity of a higher law.

I hope that the most difficult part of our undertaking, viz., to demonstrate an
eternal truth in the many contradictions which the history of science presents to the
untrained eye, has now been so far accomplished as the limited time permits. The
same investigation shows clearly enough that the course of science has led to development and real progess. However, I shall add some further reflections on this
subject.

Science has gradually gained not only with regard to the perfection of theory but
also with regard to its extent. In the Middle Ages, no chemistry was known other
than that of metals, and under such conditions all other tasks were dependent on the
great one, that of determining the composition of metals.[3]

The phlogistic theory already contained a far greater number of subjects within
its limits, such as fermentation, respiration, fire, etc. Although the chemistry of the
Middle Ages had touched on all these subjects, it had never assimilated them into
its proper substance.

The antiphlogistic theory may not have a perceptibly greater scope than the
phlogistic, but it cannot be denied that it was the first to adopt the theory of gases as
one of its principal components. The dynamic theory, however, extends the scope
of chemistry far beyond its old bounds. Electricity, magnetism, and galvanism now
become part of chemistry, and it is shown that the very same fundamental forces
which generate these effects also produce the chemical ones in another form. But
this is not all. We have found that electricity, especially in the form in which it appears in galvanism, is capable of producing the extremes of all sensations; in the
gustatory organ acidity and alkalinity, in the olfactory organ a similar contrast, in
the eye the two extreme prismatic colours, in the ear higher and deeper notes, in the
tactile sense change in temperature and expansion or contraction, in the nerves
changed incitability. The same effects are produced by different substances in relation to the fundamental force which is prevalent in them. In this way, the theory of
sensations can be made part of experimental physics. With the torch of chemistry,
we can even shed light on the organization of the earth and its connection with the
other globes. Only a couple of examples here; the full proof can only be given
below. Like each of the other planets, the earth is a large magnet. The two poles of
the magnet are of unequal combustibility, of unequal heat tendency. Must not,
therefore, the poles of these globes be that, too? The magnetic forces decrease according to the same law as gravity, so the globes could just as well act on each other
through the former as through the latter and thereby generate equally perceptible
changes in the weather, in plants, in animals, as the ones that one planet produces
in the motion of another by means of its attractive force. I hope that you already

---

[3] [In the German version this sentence reads, "In the Middle Ages, no chemistry was known other
than that of metals, which is quite natural as these bodies return most easily to their original state after
the most varied changes in their condition so that the first coherent experience was obtained from
them."]

find these examples worthy of some attention. In the more detailed lectures, it will become clear to you that some day chemistry will have just as much influence on astronomy as mechanics so far. Then it will be necessary to regard external motion as a product of internal forces, and all natural science will finally become a cosmogony.

You will become even more enraptured by the same hope when I present to you Ritter's great discovery that in every natural phenomenon there are certain period-ical alternations, on a small as well as on a large scale; a discovery which will teach us to cast many prophetic looks into the past and into the unseen future.

Much as chemistry has gained in breadth, it has gained no less in internal coher-ence and firmness. The so-called chemical affinities or attractions, these *qualitates occultæ* on which combustion depended, as did all chemical effects, are now re-solved into forces which we can liberate experimentally and thus come to know better. The contrast between the processes of combustion and of reduction become clearer to us now that we see that each of them is due to the predominance of one of two opposite forces. What we used to call neutralizations is no longer a mystery to us now that we know that they are due to the equilibrium of the same opposing forces, only in a different form. However, what in particular promises greater sta-bility and perfection to chemistry is that all questions about elements cease. This is of the greatest importance. As long as chemistry was only the theory of the constit-uents of bodies, it was only possible to ask about the ultimate among them, i.e., those which had no constituents themselves, the principal components, the ele-ments. But when can we know that we have reached the basic constituents, which the art of the future will never succeed in decomposing? Or how could we con-vince ourselves that we had really enumerated them completely? If, on the other hand, everything is due to certain fundamental forces and the forms in which they manifest themselves, it must be possible to find the principle for these forms and show which and how many are possible, more or less according to the pattern which Schelling has given us, by presenting them according to the three dimen-sions of space.

I feel that I should rather fear your mistrust because I promise you so much about more recent progress than your lack of attention because the subject might be too insignificant. However, I hope that you will not lose sight of the fact that every great epoch in history has been characterized by discoveries which have seemed to the eye, dazzled by the brilliance of novelty, to swallow up all older ones. I dare say I may also presume that it has not escaped your attention that the progress of science must always take place with increasing speed when there are no external interrup-tions, for who does not know that with every new discovery the means to make oth-ers are increased, and that, with the rise of science, the number of its students grows, and their eagerness intensifies? If this assertion about the growth of science required more elucidation, it could easily be found in a very rough view of the his-tory of science. Its first subject was the heaviest, hardest, most unalterable of all bodies, the most material matter, so to speak. The phlogistic theory still kept to solid matter, but it presupposed a fine substance which was considered by many to be imperceptible to the senses. The subject of the antiphlogistic theory, which is appropriately called pneumatic, was primarily the investigation of gases, in which it still assumed, with great firmness, imponderable substances and even determined the laws of motion of some. Finally, for the dynamic theory matter is only a product of the fundamental forces, whose laws it endeavours to discover. Consequently, from this side, science has attained the outermost limit which empirical investigations can reach.

Before I leave this subject, you will permit me one more comparison between the history of chemistry and that of astronomy. Everyone knows that before Copernicus the movements of the planets were imagined as they appear to an observer from earth. However, he succeeded in showing the observers of the sky that they had previously allowed themselves to be deceived by appearances, and that what they had taken for the centre only belonged to the circumference. Nevertheless, their work endured. Not only were their observations used by their successors, but also the form in which they had been handed down could rightly be kept, and from this, spherical astronomy was created. Through this introduction, by which the human spirit itself has been led to insight into the true structure of the world, every inquisitive soul must still make his way to the shrine of Urania. It seems to me that chemistry has taken a similar course. As long as chemical interaction was regarded as a combination or a decomposition of constituents, one kept only to appearances, like the observers of the sky before Copernicus. However, this observation of the surface was necessary before we could penetrate to the interior. Thus the past has given us an *elementary chemistry*, which we will use faithfully in order to penetrate to a *more advanced chemistry*.

I hope that these few reflections will be sufficient to convince you that the changes in chemistry have not consisted in a pointless drifting between two opposite extremes, but that its history is a true development from the first seed to a full organism which still puts forth new branches every day and will never cease to bear new fruit as long as nature, in which its root is planted, is inexhaustible in providing it with new nutrient fluids.

Actually, it is already implied by our investigation thus far that the course of this evolution was not accidental but followed necessary laws. Therefore, I can be so much briefer in the development of this matter. It is so completely in the nature of things that one field of knowledge always contains the seeds of another that this must already be sufficient for the philosopher to assume a process of development according to necessary laws, but as it is precisely doubt, caused by uncertain experience, that we wish to counter, we ought to unite its parts with a clearer eye so that more complete experience can destroy the nightmares of the inexperienced.

Chemistry has grown on the foundation of experience. Countless chemical phenomena surrounded man in so many forms that not even the sharpest eye could immediately discover their internal coherence. He began by finding the similarity between some of these phenomena which were very close to each other, between certain bodies which had much in common with each other, but to unite in a single vision all the natural chemical phenomena, as different as the dissolution of salt, fermentation, combustion, calcination of metals, and respiration are from each other, that was only possible after many experiences, which required several centuries to accumulate. It is in the nature of things that certain laws must first be discovered in those bodies which were least disturbed by chemical forces. Metals have this property. They may change their form through the agents which can cause combustion with more or less intensity, but they can also recover their old lustre and cohesion with great ease and again arise from their ashes. It is natural, then, that the first trace of a chemical theory developed around the metals. To produce metals, and above all to produce the most precious of all metals, had to be the great problem of chemistry at that time. To find a radical solvent which could decompose these otherwise immutable bodies into their basic constituents had to have the closest connection with this. If they had solved this problem, they would have released the most secret forces of nature and thus had the remedy against all illness in their hands. It must be admitted that they understood full well the goal to which their endeavours must lead if they reached completion, and we cannot but admire the canny eye which so accurately saw the absolute limits of their course and the strength of mind with which they fixed so distant a goal.

Even if they did not arrive at a philosphically clear awareness of all that was required for the great goals that they had set for science, viz., to know all the rest of nature, they still felt, and could not but feel, the necessity of drawing many nonmetallic substances into the investigation. Thus they gradually laid the groundwork for the expansion of chemistry and ended with the realization that all combinations and decompositions were objects of the same science as that of metals. In particular, they had to strive more and more to discover the laws according to which metals lose their lustre and their entire metallic nature under certain circumstances and regain it under others; i.e., in the experiments with metals they ultimately had to discover a portion of the theory of combustion. It was also along this path that the antiphlogistic theory was actually found, of whose merits I think that I have given you a sufficiently clear notion.

More is required in order to distinguish between different gases than between two metals or other solid bodies, for we cannot find any difference between them with our eyes, nor with our sense of touch, rarely even with our sense of smell. Thus, it is not possible to distinguish between them by direct sensation but only by observing their relationship.[4] Furthermore, many finer instruments were needed for their treatment, and, in particular, weighing and measuring them also required the finest instruments in addition to a great deal of knowledge. Therefore, complete knowledge of the gases could only be the fruit of a prolonged investigation and could not be obtained in earlier times. However, the entire antiphlogistic theory is based on the theory of gases and could therefore no more be discovered or systematically pursued and completed without this than it could fail to appear once gases became well understood.

The more bodies we know, and the better we know how to imitate natural phenomena artificially, the more we learn to reduce them to a unity, and the less willing

[4] [The German version here adds "to other bodies."]

we are to content ourselves with mere appearance. To penetrate to the heart of things, to discover the most elementary natural forces must then be the work of the most mature age of science. It is true that as soon as he opens his eyes, man already sees free manifestations of the inmost forces of nature here and there, but they are like miracles to him, distinct from the rest of nature. Thales might have seen electricity in rubbed amber, but he could not conclude from this that it was a universal force of nature. He had to regard it as a force that was peculiar to this substance because most bodies by far could not manifest it, except by being examined more carefully and by being placed in circumstances whose nature was not then known. As soon as they began practising the art of experimental investigation with greater ardour, the same force was found in many other bodies, but there was still an entire large class of bodies in which it was not found, and which were therefore considered completely unelectrical. Now it was discovered that some bodies allowed electricity to act more quickly through them than others, and that a body could therefore generate much electricity without showing any signs of it because it did not retain it. They now discovered the art of absorbing electricity by means of good conductors and of interrupting its progress by means of bad ones, and they soon learned to intensify the electrical effect to a previously unheard-of degree. Only after all this did it become possible for Franklin's genius to show completely that the relationship between the two different kinds of electricity was that of opposite forces. Now they could calculate electrical effects, and thus it became possible for Volta, through a series of conclusions, to invent the instrument which displays weak electricities intensified so many times that they could discover electricity, indeed, measure its quantity, where it had not even been suspected before. Only then did it become possible to demonstrate with complete certainty that bodies generate electricity by contact, and that this generation is governed by certain laws. The intensification of contact electricity was based on this, as was the transition from material to dynamic chemistry.

I am sure that you will not demand of me that I proceed from all the many starting points of our knowledge in this manner and traverse all the roads which finally meet in one point. Such an undertaking would lead us too far here. However, I anticipate a very natural objection, which I should counter. I claim that each step along the path of science has been made necessary by the previous one and, in its turn, makes a subsequent one necessary. Even if you would agree with me that this is the natural course of things, you might still object that very considerable deviations from this course might occur as speculation could hurry on ahead of experience, and genius, by means of weaker clues, could discover that for which the experience of centuries would otherwise have been necessary. It is also possible that chance might sooner place before our eyes natural phenomena which we would not ordinarily have discovered for centuries. You might mention the discovery of galvanism, which was actually due to a coincidence. To this I answer that no matter how interesting Galvani's discovery may be, it would have had little influence on chemistry during an earlier period. If we had not known electricity and not known how to arrange metals according to their combustibility, which presupposes many chemical insights, Galvani's discovery, even if it had been made, would have stood as a solitary phenomenon, without causing any upheaval in science. If Volta had not already discovered how to make weak degrees of electricity perceptible, it would hardly even have been possible to prove with certainty that the same fundamental force existed in electricity and galvanism. In short, galvanism might well have been discovered, but it would have remained an inexplicable phenomenon among other oddities, just as animal magnetism still is in part. Let us also suppose that a man of great genius seized such a discovery and pursued it through all the weak traces in which his sharper eye could see its connection with all else; he would still not have any effect on the ordinary body of thought. For example, it is seen very clearly in Ritter's earlier writings that he himself thought of creating a new chemistry from galvanic elements before the invention of the electric pile, but he would undoubtedly have found great obstacles if that invention had not facilitated his work, and even if he had completed his proposed work, as I think he would, it would not, though based on such fine exeriments, have persuaded any but a few great minds at most, without any great and radical influence on the course of science. We see an excellent proof of this in the older history of chemistry. Had Mayow not discovered the pneumatic theory a hundred years before Lavoisier and given a proof of it which now seems extremely clear to us? Had he not described his discoveries in a language that could be read by the entire learned world? And was his theory not nevertheless forgotten until it was again unearthed from the dust of the libraries after Lavoisier? It remains an eternal and glorious truth that, in the sacred hours of rapture, the genius can see far beyond the narrow horizon of his age, but it is just as certain that the higher he is above his contemporaries, the more difficult it is for him to raise them to his level.

You can easily see that I am far from wanting to deny the influence of genius on an age or on a country, but I only claim that it cannot affect the totality except by producing the link that fits into the last in the pre-existing chain. Therefore, it was just as impossible for Mayow's time to *accept* the pneumatic theory as it was for Lavoisier's *not to discover it*, either through him or through some other man with a bright mind. A genius who goes far beyond the intended goal may well be a marvel to his time and an object of admiration to posterity, but he does not play an important role in the history of the human spirit. Therefore, I am very much afraid that Winterl will not influence his time greatly. He has proved his genius through the comprehensive scope and the internal coherence of his system. He will also have an influence where he has already contributed so brilliantly, by making galvanic discoveries before the invention of the pile and because of the good fortune he has had in that Berthollet has made several similar discoveries after him. However, he would have had more influence if he had conducted a single series of experiments on andronia with the clarity of a Lavoisier than he has now with everyone scared away by the prodigious colossus which his genius has created. But here you must become aware of a new regularity. It is, of course, clear that Winterl could never have brought forth such a great system if he had wished to delve into the most minute detail. Nor does it depend on one man but on the age whether his great work shall receive the desired testing and refinement. If I had to give you another example of regularity in the progress of science, I would mention that Newton and Leibnitz both invented differential and integral calculus at the same time. Dynamics had achieved such perfection that the method of calculation which so excellently serves to express its internal mechanism had to be discovered. Philosophy required an infinite unfolding of time and space, mathematics had exhausted itself in finite formulas, and now these two excellent men were driven towards the same great goal at the same time. However, even as we found pleasure and consolation in having discovered an eternal truth in science and an inviolable law in its development, we still encountered a strange excess from one delusion to another opposite it, and here we found a source of conflict which would have been avoided if the true point of equilibrium had been struck immediately. At first glance, this may seem to be an exception to the progression that we thought we could advance as a law of the development of the human spirit, but on closer inspection, this would only be an objection against us if we assumed that this development had happened in a straight line. However, we have only assumed that a development takes place and leave it to closer investigation to determine the form in which it takes place. It is quite in the nature of the human spirit to work in alternating expansions and contractions. To substantiate this further here would be to go beyond our limits, so we shall content ourselves with illustrating this law by experience. The activity of our spirit divides itself into the two tasks of creating and organizing. These are not completely separate from each other, but rarely are they so fused that neither the creative force nor the organizing thought is predominant. Each need only recall what has happened within himself from time to time. It cannot have escaped anyone who is used to thinking that sometimes ideas so great have sprung from his internal creative powers that he has lost himself in their blissful contemplation, far from trying to give them definite form. It is true that the ideas had a form at their origin and often the best, but frequently something limiting from the individuality had crept in imperceptibly which disturbed the pure clarity of the images, and even more frequently the flood of enthusiasm had spread beyond all bounds. In more tranquil hours the organizing intellect now begins to function fully and cuts away, organizes, combines, and finally presents the product of creation in its pure heavenly shape. Therefore life, even in the most ingenious man, is divided between enthusiasm and reflection, without whose union perfection was never achieved. Now, I call the hours of creation expanding and those of reason contracting, and I think there are similar periods in the history of science. There are times which are rich in invention, when hosts of great minds appear as if with one accord and fill all sciences with great discoveries. They are absorbed in great quantity by the brighter minds of the day while the more slow-witted oppose them. Now a more tranquil period sets in again, during which the great ideas of the previous period are refined, organized, settled. At first this effort serves to organize the creation that has been started, but finally the settling goes so far that it kills all life and would change science into a fossil if new geniuses did not then appear to re-kindle the extinguished flame, and it seems as if it was precisely the fear of this universal death that most forcefully stirred the dormant creative power. Thus, throughout history, there is a creative and an organizing or an expanding and a contracting force, whose law undoubtedly is that one must decrease as the other increases. Consequently, they could not but be in constant conflict, and in their greatest clashes even war breaks out. At first glance, this might seem dangerous to the progress of our spirit, but does our own corporeal life not consist in a struggle between opposing forces? Can our spiritual life, in its final form, manifest itself otherwise? It is a law of material nature that one of two opposing forces always arouses the other, and it is no less so in the spiritual world. Every doubt, every contradiction of the truth rallies a defense and sets it in a clearer light. Even the forces employed by the dull multitudes to stop the progress of science only serve to raise to an even higher degree the forces which are to defend it. Thus, not only the struggle between the great minds of different ages but also the petty efforts of the wicked are links in the great chain. We must leave it to them to defend the choice by which they have taken on so vile a role.

So much is certain that nothing is more suited to form a spirit which is capable of great development than living during great scientific upheavals and taking part in them. I would therefore advise anyone to whom time did not offer this advantage to obtain it artificially, I mean by studying writings from periods when science underwent great changes. To read the works of the most contradictory systems and to deduce the hidden truth from them, to answer questions according to opposing systems, to translate the main theories of one system into another —this is an exercise which can hardly be recommended sufficiently to students. The greatest possible independence of the limitations of the age would certainly be the reward for this, in itself easy, labour.

By studying the history of one's own science in this way, one acquires insight into the development of the entire human spirit. Not just chemistry but all human knowledge has always, although with varying clarity, intervened in the essence of things. This has always developed through a continually renewed struggle, which, however, has resolved itself in perfect harmony. And it is not just science, not just human nature, it is all of nature that develops according to these laws. To show this to its full extent would be to present a complete natural science and a complete history. Therefore, here as above, I am forced to content myself with a simple account of a single observation. The evolution of the earth seems to me to be the most suitable for this. We are able to penetrate into the darkness which envelops the history of our globe as we penetrate into its bowels and compare the deeper layers with the older and the newer ones. By examining these layers and the fossilized or imprinted creatures which are found in them, we learn that the earth began with enormous creative forces but with little definite direction. Through alternating expansions and contractions, it has gradually killed and buried its earlier creatures in order to make room for the present chain of creations with man at the head. It is clear to every open-eyed observer of nature that the creative and organizing forces have alternated, with an ever increasing predominance of the organizing ones, however, and that it has arrived at the stage of development at which it now stands only after many struggles. In short, the development of the earth is like that of the human spirit. This harmony between nature and spirit is hardly accidental. The further we progress, the more perfect you will find it and all the more readily agree with me in assuming that both natures are shoots from a common root. Hereby, I hope at least to have made many of you aware of a rich material for further reflection. You will easily see that these hints are not unconnected with our subject. We have glimpsed a higher physics, where the development of science itself, along with all its apparent contradictions, belongs to the laws of nature. It shows us that all and everything has grown from one common root and will develop into a common life. However, where something must be and work and grow, there the forces must have abandoned pure equilibrium, and a struggle must have begun. One force must have won but only for a time. Another must since have achieved dominance, but that, too, must have yielded when it had created its product and threatened to continue and disturb the rest. While thus

> Everything 'twixt love and hate,
> To the last link must alternate,

while the scholar himself must take part in this alternation because his own human passions are swayed by the influence of external nature, he can still, if only he fixes his eye on the steady unity in this confusion, maintain a confidence and a tranquility, indeed, a blessedness that no might in the world can destroy.

26

On Acoustic Figures[1]

I HAVE almost completed an article for your journal, in which I have further elaborated on my discoveries on acoustic figures, which I have already mentioned in Voigt's *Magazin*, Vol. 9, p. 31[2] and in your *N. allg. Journ. d. Chem.*, Vol. 6. Through the most precise measurements I have now convinced myself that in all the cases where Chladni assumes triangular oscillating areas and states that the dust lines intersect, nothing but hyperbolas are to be found. Hence, the figure which Chladni depicts as two crossing diagonals of a square actually consists of two complete hyperbolas with crossing axes .

A four-inch square plate, which is subdivided into 1600 small squares, has served me in the measurements, though not without the aid of other artifices. As a consequence, I have made much more progress in the physical explanation of the phenomenon, but a description of this would make a letter too long. I am not surprised that Chladni, whose achievements in the physics of sound will earn him a lasting place in the history of science, has failed to see the actual circumstances. I, too, have only discovered it with lycopodium; it is not as clearly revealed with sand. To be sure, the correct figure can always be found in sand, too, but the stronger the stroke of the bow and the thinner the plate, the more pointed the hyperbolas become, and therefore they are all the more easily mistaken for lines which intersect at an angle, the greater the skill of the experimenter. The small heaps of dust, which I mentioned in the earlier reports, always occur with the greatest regularity. Their development and movement will convince any observer that each entire oscillation of a plate or a string is composed of innumerable overlapping partial oscillations, and at the same time he will find himself compelled to assume that these partial oscillations increase in rapidity as they approach the nodal points. With the help of these experiments, I have proved that the largest internal vibration

[1] From a letter to the editor. [*Journal für die Chemie und Physik*, ed. by A. F. Gehlen, Vol. 3, pp. 544–45. Berlin 1807. KM I, pp. 343–44. Originally published in German.]
[2] [Cf. "A Letter to Ritter Concerning Chladni's Acoustic Figures" (Chapter 16, this volume).]

coincides with the smallest external one and therefore takes place at the so-called quiescent points. In this manner I explain how the oscillations can propagate through the quiescent points. Finally, I also make a contribution to the theory of the attraction which I have observed between the dust and the plate in these experiments.

27

Ørsted on Simon's (Volta's) New Law for Electrical Atmospheric Effects.[1]

Copenhagen, Sept. 3, 1808.
THE printing of my textbook on physics, which was burned during the siege, has now been resumed and is progressing swiftly.

A year ago, I also made Simon's discovery that electricity spreads according to the inverse ratio of the distances, and not the squares of the same, but I made it in a different way. Before long, I shall publish a paper about this in which I shall not concern myself so much with this discovery but rather with some conclusions which can be drawn from it. The most important of these is that electricity which is spread from one to two, three, four, &c. equally large bodies is weakened in proportion to the squares of these numbers. I shall also reveal the reason for the incorrectness of Coulomb's electrometer.

[1] From a letter to J. W. Ritter. [*Journal für die Chemie, Physik und Mineralogie*, ed. by A. F. Gehlen. Vol. 7, pp. 374–75. Berlin 1808. KM I, p. 345. Originally published in German.]

28

Experiments on Acoustic Figures[1]

THE figures which appear on dusted surfaces of elastic bodies when tones are produced in them have already shed much light on the theory of sound, but in addition they present so many hitherto inexplicable phenomena, so many traces of undiscovered secrets that the student of nature cannot possibly contemplate them with equanimity. Through a series of many hundred experiments I have endeavoured to get somewhat closer to the inner mechanism of these remarkable phenomena, and I now believe that I have brought my investigations to the point where I dare submit them to the scrutiny of other students of nature. In fairness, I think I dare ask them not to content themselves with a mere perusal of the following investigations but to perform for themselves the most important experiments to which I refer, a request which I dare hope will be fulfilled the more as the experiments are neither costly nor difficult.

I connect this investigation directly to the discoveries which we owe to the astute Chladni, whose services to the physics of tones have procured him a lasting place in the history of natural science.

The first six figures on page 265 represent some of the simplest acoustic figures as Chladni himself has represented them in his *Akustik*. He produces these on glass plates which he strews with sand or ground marble, whereupon he strokes them with a violin bow so as to produce a tone, which is then always accompanied by a figure. The explanation of this is the following:

Certain parts of the plate are at rest while the others are moving, and the dust is thrown by the moving parts on to the quiescent. Thus the quiescent parts, or the so-called nodal points, are the ones which form the acoustic figure. If the plate is supported in the middle of each side line, which are designated in fig. 1 as a, b, c, and d, the straight lines between these points, viz., ad and bc, remain at rest whichever tone is produced on the plate. If it is stroked near the corner e^2, fig. 1 appears, but if it is stroked closer to b than to e, fig. 4 appears. One point, for example c, can be left unsupported, and the result will be the same because the quiescent line bc is already determined by b and the other two points a and d. If the plate (as in fig. 2) is

---

[1] [*Det Kongelige Danske Videnskabernes Selskabs Skrivter for Aar 1808*, Part 5, No. 2, pp. 31–64. Copenhagen 1810. Also to be found in *Journal für die Chemie, Physik und Mineralogie*, ed. by A. F. Gehlen, Vol. 8, pp. 223–54. Berlin 1809. KM II, pp. 11–34. Originally published in Danish.]

[2] [The point e denotes the lower right corner of the plate shown in fig. 1.]

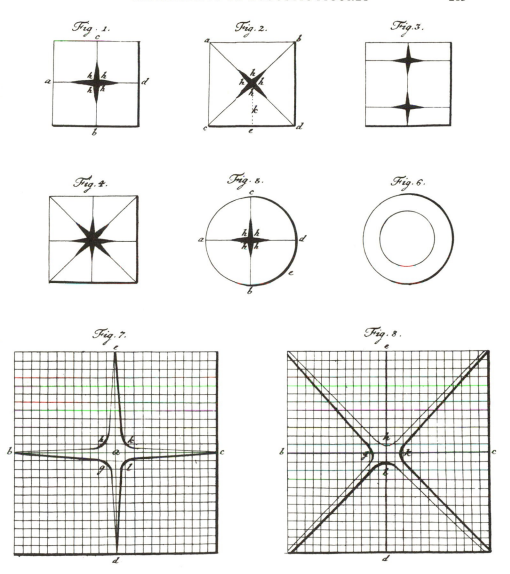

supported at the corners and stroked in the middle at *e*, the diagonals *ad* and *bc* are quiescent lines and form the acoustic figure. If, however, it is stroked closer to *c* or *d*, the result will again be the fourth figure. If the plate is held in the centre between two fingers so that the edge nowhere touches the hand, the result will be the first figure when the plate is stroked at a corner and the second when it is stroked in the middle. This clearly demonstrates the obvious fact that the entire plate cannot vibrate at once when it is held in the centre. A look at the second figure suffices. When this is stroked at *e*, the edge there flexes most strongly backwards and forwards but less and less, the closer we get to the corners *c* and *d*, where the vibration may be considered nil. Therefore, the vibration in this case is exactly the same as if the corners themselves were supported just as, conversely, the centre is at rest when the corners are supported. The same kind of argument can easily be applied to the remaining figures. If the plate in fig. 1 is stroked at *e* or close to it, the straight line between *b* and *d* must be compared with *cd* in fig. 2, and the straight line between *b* and *d* in fig. 5 must be regarded in the same way. Briefly, all these and the other figures are explained by the basic experiment that when an aliquot part of a taut string is supported at one of its extreme points and then struck to make a tone, each of the other aliquot parts will vibrate, too, as if they were also supported, but in such an alternating order that there will be a point of rest, a nodal point, between each vibrating part. This approximately has been the conception of acoustic figures so far. To this I propose to add the following investigations of my own.

It is inherent in the infinitude of Nature that no observer can discover all that is implied by an experiment. To understand an experiment completely would be equivalent to having found the key to all of Nature. Therefore, the perspicacious discoverer of acoustic figures cannot be blamed if he has not observed all that is really implied by his experiments. Did not Newton himself in his masterly enquiries into prismatic colours overlook several quite important phenomena, which also escaped the attention of his successors for a whole century until Hershel's[3] and Ritter's experiments enlightened us. I point this out with due gratitude to a man to whom science is indebted for such a considerable step forward, hoping that, by making such a statement, I dare contradict or rectify my predecessor more freely. I venture this all the more because, on the whole, this is an amplification rather than a mere correction.

The way in which Chladni draws his figures, of which the first 6 figures are a true copy, might give the impression that the figures 1 to 5 consist of straight lines which intersect. This is not the case in reality; *ahb, ahc, chd, bhd* (figs. 1, 2, 5) are not, as it seems, angles but hyperbolas which meet. Likewise, the angles seen in figs. 3 and 4 are really hyperbolas which have opposite vertices. These figures are seen drawn realistically in figures 9–16.

In order to produce these figures in their greatest purity, I do not support the plate at one of the points of intersection, as Chladni used to, but at the edges. In figs. 1 and 2, for instance, I put *a* on my thumb, *d* on my middle finger, and *b* on my ring finger, and then I stroke at *e*. If I put my finger in the centre, I only see that the acoustic lines are broken, but I do not discover the true state of affairs. Even in figures with several points of intersection, the size of the surface where the lines meet will be thought to derive from the fact that the corresponding surface which had been damped was of equal size.

I generally use metal plates instead of glass. These are not liable to break, they are more easily made uniform, and they hold the sound longer than glass plates. Thus, it is possible to obtain a very beautiful figure on a metal plate if sand is sprinkled on it after the tone has been produced. I do not find that sand is the best means of producing acoustic figures. It has considerable elasticity and therefore bounces

[3] [Herschel.]

rapidly from one part of the vibrating plate to another. As a result, it is possible to produce the figures with great speed as if they had been called forth with a magic wand, which may well please the eye but does not permit the investigator to observe the nature of the effect. For these experiments, therefore, I prefer to use fine iron filings, metallic calces, lycopodium, etc., according to the various purposes. Each of these powders has its distinctive advantages. As we shall see below, lycopodium shows each part of the phenomenon most perfectly, both by making perceptible vibrations which are not even suggested by other powders and by showing them so slowly that the eye can easily follow them. For quick experiments, on the other hand, and when I merely want to determine the nature of the total vibrations, other powders are to be preferred. The elasticity of sand makes it particularly suitable for the very quickest experiments, and above all when I want to please the eye. Fine iron filings give much more systematic results. Among all heavier powders, pulverized lead is the most excellent because it shows the lines with a sharpness like no other powder. Its weight and lack of elasticity cause it to stay exactly where it falls. Regardless of this advantage, however, I have not often used this powder because I clearly felt that I got some in my mouth and nose, which might have harmful consequences.

I shall start by showing what can be discovered by means of the coarser powders. Although the coarser dust is set in motion more quickly than the finer, the acoustic figure is not formed by one stroke unless it is exceedingly vigorous and the plate very elastic. Fig. 9 shows such an acoustic figure after the first or second stroke. In the next figure, on the other hand, it is shown completed. It can be seen, then, that the vertices of the hyperbolas approach each other more and more through repeated strokes though the curves can never be made to disappear completely. Fig. 11 shows another initial figure, fig. 12, on the other hand, a completed one. Fig. 15 is also an initial figure, fig. 16 a completed one. Figs. 13 and 14 are much too unclear initially and are therefore only shown completed. A glance will easily show to which of Chladni's drawings above each of them corresponds.

I have measured the acoustic lines which correspond to figs. 1, 2, and 5 particularly carefully. For this I use a square brass plate 4 inches long and wide and almost one line thick. It is divided into 1600 squares by lines parallel to the sides. When figures 9 and 10 are produced on this, the lines which divide two opposite sides into two equal parts are asymptotes, and their point of intersection is the beginning of these. This is shown in fig. 7, where ab, ac, ad, ae, are asymptotes to the equilateral hyperbolas bhe, bgd, cke, cld. The lines which are perpendicular to the asymptotes measure the distance between these and the hyperbolic lines. It is true that the divisions only indicate the distances in lines, but with some practice it is very easy to determine half and quarter lines, indeed even less, with the naked eye. Where the distances are less than one line, the measurement may not be quite accurate, but where this happens, the sides of the hyperbola are already so close to the straight line that they can be regarded as parallel to the asymptotes. Moreover, when the measurement achieves suitable accuracy, the lines which represent the ordinates are everywhere found to be inversely proportional to the abscissae as the nature of the hyperbola dictates. In fig. 8 bc and de are the axes of the hyperbolas, and the

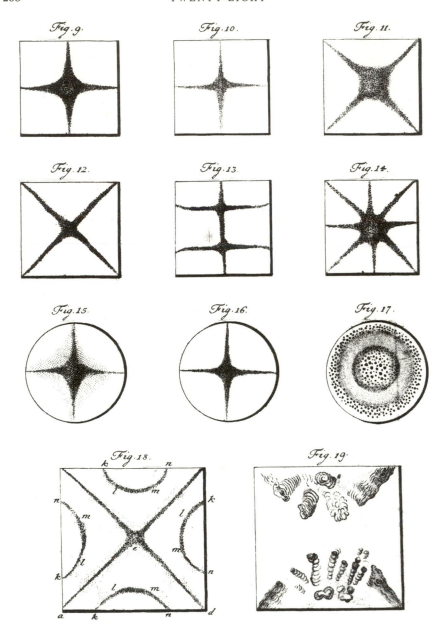

Fig. 9. *Fig. 10.* *Fig. 11.*
Fig. 12. *Fig. 13.* *Fig. 14.*
Fig. 15. *Fig. 16.* *Fig. 17.*
Fig. 18. *Fig. 19.*

lines which are perpendicular to each are ordinates for the corresponding hyperbolas. Therefore, it is easy to determine whether the squares of the ordinates are proportional to the products of the distances from the vertices of the hyperbola. I have produced figures 15 and 16 on glass and then measured them by putting the plate on top of the metal plate which had been divided into squares. I have examined the rest by superimposing on them hyperbolas which had been cut out in paper, and I have always found the hyperbolic form in the acoustic line. However, I have to admit that the latter method is less accurate. In the present treatise, I therefore intend to limit myself to those more accurately determined figures, which also represent the simplest cases, and I promise to provide future investigations of the more complex phenomena, which naturally must find their explanation in the simpler ones.

As I wish these experiments to be repeated by many, I shall add some further remarks to this already rather detailed description. In these experiments the dust lines are rarely formed so completely that each point in them can be determined. In particular, it very often happens that a little dust sticks in places on the plate where it is impossible to tell whether it is part of the figure or not. Likewise, it often happens that there is not sufficient dust on every part of the surface with the result that there are breaks in the figure, so it is important to clean the plate thoroughly before sprinkling it. It is most difficult to produce complete vertices of the hyperbolas when the plate is evenly covered, but when it is covered somewhat more thickly in the centre, a better result is achieved. As the metal plate blinds because of its lustre, especially as it is necessary to illuminate it well in order to observe the borders of the dust lines, the following artifice can be used: After the figure has been produced with a coarser powder, the plate is strewn with lycopodium, turned over, and tapped gently. The coarser powder will then fall off, and the finer lycopodium will remain. Thus the bright metal lines will show the acoustic figures clearly and relatively distinctly. Very often only one acoustic line will appear with complete sharpness, but this can be measured and the result applied to the part directly opposite, where the measurement will be found to match in the places where investigation is possible. Sometimes it is difficult to determine the vertex even though the figure is otherwise rather sharply defined. It is therefore best, before the investigation, to calculate the ordinates for various distances between the vertices which might appear on the plate. From the size of a few ordinates and their distance from the vertices, it will then be easy to determine the distance between the vertices. The pure mathematician may find what I have suggested here too elaborate, but I do assure that this method will feel comfortable in practice as it makes the observations of often ill-defined figures much easier.

After all these remarks, I think I dare proceed to the explanation of the basic phenomenon, for which I choose the figure presented in fig. 2. As the points a, b, c and d are supported, the lines ad and cb are at rest, and consequently only a triangular area, such as chd, is set in motion at once. How the motion of this triangle is repeated in the others will not yet be explained. When cd is now stroked at e, it will flex and vibrate like a touched string. As this side bends, so the whole surface chd bends. The closer the parts are to the nodes, the smaller their arcs of oscillation must be, and at the greatest proximity they will be so weak that they are not able to throw off the dust. If all the parts which are parallel to cd, in the surface chd, were bent equally much, the dust line thus produced would be equally wide everywhere, but as the arcs of oscillation must be proportional to the distances from he, the size of the parts which could not throw off the dust must also grow at the same rate as the distances from *h* diminish. In other words, around the quiescent lines a surface of dust must form, the distances of whose outer borders from these lines are inversely proportional to the distances from the intersection *h*. Thus, the borders of the dust surfaces are hyperbolas, and the actual nodal lines are their asymptotes. The same thing can be understood in another, if possible, simpler way. When bent, the surface *chd* becomes a conoidal surface. However, the part closest to the apex *h* cannot be bent so much that it can throw off the dust, so only the lower part of the conoidal surface is formed, of which *chd* is the cross section. According to the nature of the cone and the small size of the arcs of oscillation, this cross section must be a hyperbola.

In general, I have observed that the vertices of the two hyperbolas which are formed in a figure such as 8 are not equidistant, but that the line *gk*, for instance, is longer than *hl*. From this arises the apparent irregularity in the first figures on page 268, where, however, figs. 12 and 13 show it somewhat too large. This circumstance seems to be caused by a lack of uniformity of the surface, for in the same plate this difference always appears in the same position whether I stroke at *c* or *d*.

Hitherto we have paid no attention to the part of the vibrations which must necessarily produce some irregularity in the shape of the hyperbola even though the influence from this is not noticed in the smaller plates. Just as an elastic spring fastened at one end not only swings to and fro about its equilibrium when the other end is moved up or down but also bends, so *chd* in fig. 2 must also bend somewhat in the direction *eh*. The point *k* on this line, which at one instant is the place of the greatest convexity, will in the next instant be the place of the greatest concavity. Therefore, an interaction arises at this point which will accumulate larger quantities of dust. These are seen clearly at *abc* and the corresponding places in fig. 21, and also in every figure which is produced with a powder which has some finer particles, which will always collect there. This effect, however, is not transmitted to the quiescent lines, which can likewise be seen in fig. 21.

On square glass plates, with 8-inch sides, this interaction is revealed more strongly and forms a well-defined shape, whereas only a heap of dust collects on plates with 4-inch sides. This is seen in fig. 18, where the lines *klmn* represent this figure. They have a strong resemblance to ellipses but are generally more bent at *l* or *m* than this curved line. However, I have found that this irregularity is often very small. I think that it is really because such a large plate must be held by two persons while a third makes the stroke, which cannot but result in some non-uniformity. When one does not hold all four edges but has damped three of them, this non-uniformity is sufficiently large to produce very considerable distortions on the sides of the hyperbolas, such as are seen in the hyperbola *aed* and the others in fig. 18. When the plate is held most steadily, these distortions, as well as the irregularities in the elliptical figures, are noticed the least. Therefore, I have reason to believe that they would quite disappear if it were possible to get plates which were completely uniform throughout, and if the support were completely even everywhere. For further experiments, I intend to obtain a device made for this purpose, with whose help I can also determine the nature of the elliptical figures more accurately and decide whether they are perfect ellipses or not. Until then I shall also withhold the theory about this phenomenon, for it is indeed easy to see that the forms produced on this occasion must be conic sections; it is even very natural that these could pass through the axis of the cone and thus be ellipses, but there remain several observations which I want to investigate by experiment concerning the positions of these before I present them publicly. Furthermore, on the large plate I find the suggestion of a new figure which is likely to appear on an even larger plate.

It can be imagined that sound produces all the conic sections. If a plate could combine quite perfect elasticity with a flexibility so great that there was no resistance, the dust lines in such a plate would coincide with the completely quiescent lines and thus form triangles. Anyone can see that this case is only hypothetical, and that in reality only imperfect approximations to this can be achieved. The hyperbola is the section which is usually produced, but there is nothing to prevent the section from running parallel to the opposite side of the cone, thus forming a parabola, or intersecting the axis perpendicularly or obliquely, thereby producing a circle or an ellipse. Especially the latter cases could most easily occur in the figures which arise through the above-mentioned interaction. I assume that all the conic sections will be found at different distances from the centre on very large plates.

This is how far observations of acoustic figures produced by heavy and rather coarse powders will take us. If, however, such a fine powder as lycopodium is used, still more phenomena will be discovered which did not appear at all before. With the first stroke, many small elevations are formed, which all move towards the quiescent lines, and they move on with each stroke until they form a figure with the same outline as the one obtained with the coarser powders. Fig. 20 represents such a figure after the first stroke, and fig. 21 shows the same after completion. Anyone can easily see that they correspond to fig. 2 and figs. 11 and 12. It is clear that all the elevations of dust produced here form hyperbolas whose vertices, with each stroke, come closer and closer until their distance finally constitutes no more than $1/8$ or $1/10$ of the length of the whole plate.

Before I proceed to show all that lies in these experiments, it is first necessary to investigate the nature of the smaller dust figures.

It seems clear that they indicate smaller vibrations in solid bodies of which we have previously been unaware, but besides this we still know very little about them. The preceeding considerations have already shown that the action whereby the acoustic figures are produced is far from being as simple as it might appear at first glance. This action obviously produces motion in several different directions, or more correctly, it acts on all the dimensions of the body at the same time. For that reason, I have investigated the effects of simpler kinds of concussions. I take a square plate of glass or metal, sprinkle lycopodium on it, and hold it in the same way as in the production of acoustic figures. Then I take a ruler and strike the middle of one of the sides with its sharp edge as vigorously as I can and in such a way that the blow is completely perpendicular to the edge and parallel to the surface of the plate. Hereby the dust collects in lines which are parallel to the direction of the blow. In these lines several elevations and depressions are again observed. If the blow is gentle, the experiment does not have the same result but forms irregularly undulating lines, nearly parallel to the side which was struck. The experiment

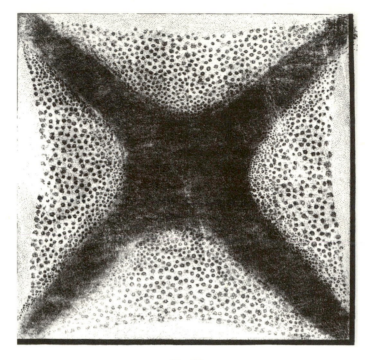

Fig. 20

gives the same result if an entire edge is struck at once with the blunt side of the ruler, except that everything now appears more regular.

The explanation of these experiments is without doubt the following: In the first, the blow was directed only against a point, but as it is not possible to impart this blow to the body at a truly mathematical point, a certain finite, though very small, part receives the impact, which, as everyone knows, cannot spread through the whole body instantaneously. According to the laws of elasticity, the compression of this part must be followed by an expansion, whereby the adjacent parts are compressed and then likewise expanded, whereby this motion is finally transmitted throughout the body, in a period of time which is generally so short that we are not able to distinguish its elements. However, when all the parts are set in motion in such a way that each must expand when the adjacent parts contract, the parts which lie between two such motions must naturally rest, and the dust will be thrown on these. Of course, the velocity is greatest in the direction in which the blow is imparted, and for this reason the alternations follow each other all the more quickly so that the intervals between them are observed to be smaller than between the ones that move sideways. Therefore the dust seems to be arranged in lines. The fact that, by striking the entire side of the square plate, we obtain undulating lines which are parallel to the side struck, hardly needs any further explanation. However, we might wonder that the lines are not straight if we did not know that the

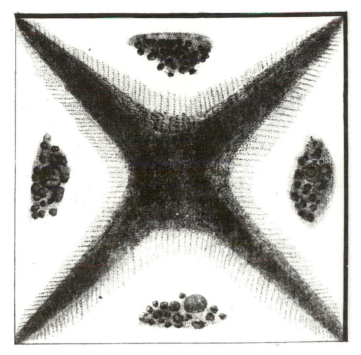

Fig. 21[4]

edges of the bodies are uneven. I must confess that I have not even tried to procure plates with perfectly even sides or perfectly even rulers with which to strike in these experiments. It is easy to understand that we got the same result by tapping rather gently at a point on the edge of the plate as when we struck the whole edge when we consider that the motion is then able to spread to the whole edge so that, without any significant error, it can be regarded as having received one single motion with which it then acts on all the other parts. If the direction of the blow is not in the dusted surface but is either perpendicular or oblique to it, we get nothing but small elevations which are no longer divided into lines. It is obvious that here a new, viz., a vibrating, up-and-down motion is added to the above-mentioned motions. It will generally be clear that these elevations must arise through a combination of the up-and-down vibration and the one which creates the dust lines, but I do not think that I am capable of giving a more precise explanation of this here.

When we want to produce the motion mentioned here, it makes no difference whether we strike the edge of the glass or somewhere else on the surface. We get the same heaps of dust, and in each of these the dust seems to be in undulating motion. This shows clearly that the point of greatest activity is the centre of the base of

[4][The points *abc* are not shown in fig. 21. The lowest isolated dust pile is so marked in the German edition.]

each of these elevations. It is also self-evident that when we strike a point on a surface, this point, which becomes the centre of the resulting pile of dust, will receive the greatest motion, and likewise for all the corresponding points.

It is interesting to observe the three different physical degrees which are found in the sound that is produced by an elastic plate. The sound which is produced when the blow is aimed at just a small part of the edge of the surface is nothing more than a dull thud, approximately as when a block is hit with a hammer. The sound obtained when the whole edge of the square plate is struck is a clatter. On the other hand, the sound produced when the surface itself is struck is a proper tone. Thus, when a tone is to be produced, the parts of the body must work in concert in all dimensions. This, however, is not yet a tone. Such will be produced only when the smaller vibrations arrange themselves into a symmetrical whole. This happens when the plate is held as described above and struck on the surface in the middle of one of the edges. In this way an acoustic figure is also produced, although not of such regular beauty as when the edge is stroked with a bow. Therefore we see that the most perfect and internally harmonious motion of bodies is also the one which, through the ear, produces the deepest impression on our internal sense of beauty. With this I believe that the first physical definition of the various kinds of sound has been given. By means of these remarks we can give more internal consistency and unity to the theory of the production of sound.

As we know, people have long agreed that sound in air is produced by a great many small compressions and expansions which follow each other with extraordinary speed. Something similar was found in fluids and therefore the same natural mechanism had to be acknowledged unless the effect was attributed to absorbed air because of prejudice. In solid bodies, however, it seems that no one has imagined the same mechanism with certitude. Now it lies in front of our eyes in an experiment, and I hope that what we discover there will also shed new light on sound waves in the air itself, but I permit myself to reserve these investigations for another treatise. Hereby, a dispute between some older and some more recent physicists can also be resolved. The former assumed that the sound which is produced in solid bodies was caused by the vibrations of the smallest parts, while the latter think that everything depends on the coherent vibration of certain primary parts. Through Chladni's experiments, it was demonstrated that the former, who attributed everything to the vibrations of the smallest parts, have been too one-sided. These new experiments show that those who attribute everything to the primary vibrations also go too far. Sound is the combination of both kinds of vibrations. In addition to its primary vibrations, even a taut string must have these subordinate ones, which I shall call subvibrations. This cannot be doubted if we just consider the possibility that the part which at a given time receives a blow can communicate it to others at the same time. Hence it follows that the part struck must be regarded as supported by the surrounding parts so that a number of small nodal points must appear which is equal to the number of such vibrating subparts contained in the entire string. Experience also confirms this assertion, for when a fairly thick metal wire is placed in such a way that its end points are supported, dusted with lycopodium, and then struck quickly but not violently, a great many small elevations of dust will form which are so constituted that variations of more- and less-moved parts can be observed in them clearly. A pipestem can also be used instead of a metal wire.

This has led me to a closer examination of the well-known experiment in which a pipestem suspended by two hairs is broken without tearing these. The speed at which the blow falls gives occasion to suppose that there must be several nodes and consequently also points at which the oscillation is at its maximum. At such points the pipestem, being of a fragile material, must break. Experience is in accordance with this, for a pipestem which is broken at the proper speed will always break into more than two pieces. The points which rest on the hair in this case become damped points, and for that reason the hair cannot break. It is false, however, to believe that the hair receives no motion at all. I have made the experiment in such a way that two persons held the hair while I hit the pipestem, and they always felt a jerk in their hands even though the hair remained intact.

After this somewhat detailed digression, I return to the experiments on the square plate. It is clear that each stroke with the violin bow must produce an undulating movement in all directions, and these sound waves are what we see in the experiments mentioned here. An internal motion from the centre out is seen in each of these small waves of dust. A kind of internal rotation is also clearly observed in it. All this is seen better on fairly large plates than on small ones. Coarser and heavier particles of dust are thrown out of the dust waves, whereby we are even further convinced of the presence of a centrifugal power inside. The waves grow smaller, the closer they get to the centre. Finally, they become so small that it is impossible to distinguish them from the specks of dust themselves, and here, then, the motion must cease because each dust particle comes to rest between points which oscillate in opposite directions. The size of the sound waves is also seen to diminish with the arcs of oscillation in the lines parallel to the edge. I cannot help thinking that the sound waves, as a consequence of this, must also follow each other with greater rapidity in these parts. This is also in complete accordance with our theory. The vibrations of the string must be compared to those of the pendulum, and therefore the parts in a taut string which are closer to the points of support have a tendency to vibrate more quickly than those farther away. This explains the decrease of the sound waves in the direction parallel to the edge. The size of the sound waves also decreases in the direction perpendicular to the edge. The explanation of this is not difficult. We can imagine that the plate is composed of nothing but strings which are parallel to one of the sides. In the triangular space *chd* (fig. 2), the length of these strings decreases in proportion to the distance from the sides. Therefore, the rapidity of their vibrations must increase proportionally, according to the law that the rapidities with which strings oscillate are inversely proportional to their lengths. Consequently, the internal motion increases in intensity as the external decreases in extent. In the quiescent parts, there is thus an extremely strong internal motion which continues beyond to the other side, where it decreases and changes into external vibrations, following the same laws according to which it was first created from these.

I hope that I have hereby made it comprehensible how the motion can be transmitted past the quiescent points, which could hardly be made clear without this.

It also seems possible to explain from this how a string or a plate, after having emitted the primary tone, emits a rising series of other less audible but higher tones in the reverberation. At the beginning, the higher tones are drowned by the lower ones, which have a larger amplitude of oscillation. The larger oscillation gradually becomes weaker, but the smaller ones still retain their vibrations because they were more violent. Thus, a higher tone is gradually perceived by the ear, though always with a smaller amplitude of oscillation and thus less audible.

It is understandable that the string is hereby gradually divided into new primary sections. I reserve this investigation for the future unless someone with more musical expertise than I will undertake it.

The mechanism of sound can be made even more understandable by the following experiment. We place a row of small heaps of lycopodium on one edge of the square plate and produce a tone on it. It will immediately be seen how these heaps of dust collect in small elevations which move in curved lines whose convexity turns away from the line to the point of the stroke and towards the places where the acoustic figure will be formed. If we put the dust in the place designated *abc* in fig. 21, it will be thrown both forwards and back towards the edge.

Both these experiments are shown in fig. 19. The lower part of the figure represents the former, the upper, on the other hand, the latter. It is easy to see from these directional lines that the plate must be well covered with dust, especially towards the centre, if the vertices of the hyperbolas are to be formed. This figure also shows, although not so clearly as in reality, that the sound waves come closer to each other, the closer they are to the nodal lines. It can also be seen that they decrease proportionally.

The same things which are observed in the various experiments described here in detail will appear, with easily predictable modifications, in all other experiments with acoustic figures. It would therefore be superfluous to give more depictions here, particularly as everything is seen more clearly in reality. I shall only draw attention to the experiment in which Chladni produces a circle on a round plate. Chladni's depiction is seen in fig. 6; the one produced with lycopodium is seen in fig. 17. In order to perform this experiment, the plate is held at a place close to the periphery, and then the edge opposite the finger is stroked. In my depiction, and even better in the experiment itself, it will be seen how the dust from both sides moves towards this point and actually forms a double circle, which finally coalesces to form one. If we stroke 45 degrees from this place, we get fig. 16. If we stroke somewhat closer to the point of support, we get 6, 8, 12, etc. hyperbolas, ever more the closer we get to it. The vertices of these hyperbolas are all more or less equidistant from the centre and are all the farther from each other, the more there are. Thus, when we stroke opposite the finger, they must form a circle. This transition is seen particularly clearly in the present figure because the many subordinate nodes still remain. However, it should be noticed that when the stroke starts to produce circles, the centre is also set in motion and rises and falls alternately. Without doubt, the inner part of the plate thereby forms a segment of a sphere, of which the generated circle is the circumference.

We have now seen the most essential part of the mechanical activity which the production of sound involves. There is yet another kind of effect in these experiments which I shall try to describe and explain here. The main point is that the dust adheres more firmly to the plate in the acoustic figures than elsewhere. I have repeated the investigation of this far more often than is usually deemed necessary in order to convince oneself of the correctness of an experiment, and I have always obtained the same result. Not only lycopodium but also sand shows this adherence. Metal filings, however, shows this property to a very small degree. In order to see this phenomenon, we only have to invert the plate on which we have produced the figure so that the dusted side faces downwards and then hit the back of it with a flat hand in such a way, however, that no tone is produced. When we now turn the plate over, we see that the dust has fallen off all the area enclosed by the dust lines so that we see a star-shaped space in the centre which is almost completely devoid of dust. The dust also falls off all the places on the plate where it has not been cast by the tone. Thus, a fine film of dust, in all its parts like the outline of the acoustic figure, covers the plate after we have removed what could easily be knocked off.

This property can be used to obtain prints of the acoustic figures. A sheet of black paper is coated with a solution of gum Arabic, and when this has dried to the extent that it is still sticky, we put the plate, from which we have knocked the superfluous dust, on top of it. When we have pressed it hard, we remove the plate and stick the paper, while it is still moist, on to glass; we can then be sure of preserving a print which, when successful, is more accurate than the best drawing.

It will readily occur to anyone to explain this adherence mechanically. At first glance, one might imagine that the dust falls most easily from the places where the most has collected, but this view would not agree with the experiments. When the experiments have been made with sand which has been sprinkled on very thinly, almost all of it adheres along the acoustic line, with a firmness which cannot but attract attention. When the experiment has been made with lycopodium, and the stroking is continued until the acoustic figure is very nicely delineated, the dust has also spread across the quiescent lines because it had been piled so heavily at their edge. When this is now knocked off, the quiescent lines will be completely devoid of dust, whereas the dust lines proper will retain most of theirs. Here it is quite true that the dust falls from the places which are most covered with it, but when an attempt is made to knock the dust off such a figure immediately after the first stroke when the area enclosed by the acoustic figure is not covered by thicker dust than before the stroke, the dust still falls more completely off the enclosed area than from the dust lines. Here, then, that mechanical explanation is not adequate either.

Therefore, this adherence of the dust seems to indicate an electrical effect as the one which could produce the phenomena mentioned here with attractions and repulsions. The most natural thought is that the dust, which we know from experience to be electrified when shaken, will adhere more firmly to the places which through vibrations receive the opposite kind of electricity and will, on the other hand, lie more loosely in the places which receive the same kind. It should not surprise us that the plate receives electricity through its internal oscillation since any friction, any shock, etc., creates electrical activity in bodies. However, if this conception were correct, such kinds of dust as become electrically positive when shaken would adhere to the places which least retained electrically negative kinds of dust, but this is not the case. The electricity of the shaken dust has no influence on the acoustic figure. Furthermore, it should be noticed that the same kind of adherence also takes place on metal plates, even on uninsulated ones. Consequently, the changes in adhesion found on the stroked plates could hardly be attributed to an electricity which we can detect by means of an electrometer.

In spite of all this, however, we need not quite abandon the thought of an electrical effect in these experiments. Through the investigations of the finer degrees of electricity which recent times have given us, we know that very weak electricities need no such insulation and yet can have considerable effect through their quantity. It is without doubt such an electricity which is aroused by the production of tones. It is a well-known experience that when two bodies of the same kind are rubbed against each other so that one experiences a stronger effect than the other, the most strongly rubbed will receive negative electricity, the most weakly rubbed, on the other hand, positive. Thus it also seems reasonable here that the parts where the weakest internal vibrations have taken place must have become electrically positive while the parts where the most powerful ones took place must have received negative electricity. In other words, those parts which have undergone the greatest external vibrations, and from which the dust has therefore been thrown, have become positive in relation to the nodal lines, which have become negative. I further imagine that, during the vibrations, the dust at each point will share the electricity of the place which it covers.

Thus, when the dust is thrown from one of the places which have a considerable arc of oscillation, it receives a weak positive electricity, whereby it adheres more to the negative parts near the quiescent lines. In the quiescent points proper, the dust receives the same electricity as these, but because it is not thrown off, it remains in contact with parts which have the same electricity as itself, from which a decreased adhesion must necessarily follow. The same contrast which occurs with regard to the whole acoustic figure and the whole plate also arises between the subordinate sound waves and the individual parts of the plate. The circumference of each of the small elevations of dust is the same for a small part of the plate as the large dust lines are for the whole plate. Neither does the adhesion need to be smaller there, for although the electrical opposition is smaller, the dust does not have such a long way to travel and consequently loses practically none of its electricity along the way.

The investigations which I have briefly described here allow a much more extensive application than just in the theory of sound. Every impact produces a vibrating motion in bodies, and this spreads in them according to the same laws as in the production of sound, or more precisely, every impact which is not too weak produces a sound, and even the weakest would produce one if the auditory organs had greater sensitivity. It is true that there are bodies of such poor elasticity that the vibrations in them must be extremely weak, but just as there are no absolutely inelastic bodies, there are none in which no vibrations take place. Therefore I think that, in the same way in which we have acquired knowledge of the acoustic figures, we must also be able to find information about the state of bodies during the internal transmission of motion and the relationship between this and elasticity. Chladni has successfully used longitudinal vibrations to determine the elasticity of bodies. I think that several of the experiments referred to here would already contribute something to the determination of the inner mechanism of an impact, but much more is probably still to come. However, we may hope that the development of this important chapter in physics will make greater progress once it has been clearly understood that the theory of sound and the theory of internal motion are one and the same.

It is also remarkable to see in the theory of sound how external motion is transformed into internal. A strong electrical effect seemed to be connected with the internal motion, which we could only discover by inferences but not perceive with our senses. More clearly than experiment could show us, we should be able to see from the nature of things that electricity must be generated under such conditions as surely as friction produces this effect. Therefore, would it not be possible that the motion of the whole, when transformed into a penetrating motion of the parts, also passed from purely mechanical motion to the production of forces? (I hope I may presuppose that I do not understand this question in the sense in which it would have been understood 20 years ago.) The idea is quite clear to me, and if I am not wrong, anyone who understands some of this must find it so.

Several important consequences for the theory of hearing can be inferred from the generation of electricity through sound. In this I avail myself of several excellent observations, which Ritter (he has arrived at a similar idea by a different path) published in Voigt's *Magasin für das neuste aus der Naturkunde*, Vol. 9, p. 33 ff, on the occasion of the first report of some of my discoveries mentioned here. He remarks that when a string is curved and thus extended more on one side than on the other, one side must acquire a greater tendency to become positive, the other to become electrically negative. At each oscillation whereby the part which before was concave now becomes convex and vice versa, the parts alternate their state. The same thing happens in air. The compression and the consequent expansion which take place in every sound wave produce a similar electrical alternation. Such an alternation between electrical states must also take place in the ear, but each transition from one electrical state to another produces a shock, however weak it may be. Each sound is accompanied by a great many such shocks following one after the other. Each of the preceding states prepares the organ for a greater sensitivity to the following, for we have learned from experience that a part of the body which has been in a state of positive electricity thereby has become more sensitive to the effect of negative electricity than before, and vice versa. The frequent alternation increases the sensitivity of the ear even more. This shows that the quantity of electricity generated by a shock, collected in an electrical jar, could not possibly produce the same effect as this series of small alternations, for, although they accomplish nothing separately, they form a whole and are perceived as such when they follow each other at intervals so small that each individual one cannot be felt. The more quickly they follow each other, the more perfect a continuum they form. Just as the fiery path described by a twirled brand constitutes an unbroken line more perfectly, the faster the rotation takes place, the tone also acquires more solidity, unity, and individuality, the closer its elements move towards each other. The less all this takes place, the more ordinary, indefinite, dissolved, even deeper, the tone seems to be. The latter pulls the soul down, the former up. Much could be said here about the use of higher and deeper tones in life; how sorrow and joy each has its own, the former in minor, the latter in major, while only the sublime, far above both, dares to proclaim the fullness of its expression in an all-embracing unity.

To these ideas, which may be the greatest ever said about tones, Ritter adds that the intervals between two sound vibrations in the end can become so short that they do not allow the ears the necessary rest, for which reason each impression itself becomes too short and too weak to have any effect at all. Here the tone gradually disappears in the ear, the whole effect turns from this to a higher organ, to the eye, and the tone becomes light. If we imagine a taut string making its slowest vibrations, we are able to distinguish each vibration with our eyes. Let the speed increase, and now we can no longer distinguish one vibration from the other; we see only the entire space through which the string vibrates filled by it. There is a gap between the point where the visibility of the individual vibrations ceases to the point where the deepest tone begins. Now imagine the vibrations proceeding with increasing speed and producing higher and higher tones; in the end the speed of the vibrations becomes too great to be perceived by the ear. The vibrations continue to increase, and after an interval like the one between the fastest individually visible vibration and the lowest tone, the vibrations here will rise to the production of the deepest colour. It appears in front of the eye as a faint blue twilight, and with increasing vibrations it clears to higher and higher colours and thus runs through all prismatic colours until they have reached the most vivid red. According to this conception, one sense would become an octave of the other on the grand scale of sensations, and all would be subject to the same laws. Thus all sensations spring from the same original force, which in light works *in puncto* but in galvanism spreads in space, where, however, it runs through all forms of vibration so that it becomes perceptible to every sense.

If we now combine into one all that the present investigation has shown, we cannot but be lost in the deepest admiration of the variety, the life, the harmony which are contained in a tone. Let us imagine an acoustic figure. Parts so small that they can no longer be discerned with the naked eye combine into a small orb which, in turn, proceeds to form part of a larger system, and thus new and larger combinations would continually arise if the size of our instruments did not finally set limits upon us. Let us further imagine how each of the smaller vibrations occupies its particular place and could not occupy another without disturbing the symmetry of the whole. Let us imagine how these harmonious vibrations travel throughout the air, in the same order in which they spread from the vibrating body. What a great and profound and in itself necessary harmony, what mark of an all-pervasive Reason. Here we clearly see that it is not the mechanical sensory stimulation which pleases us in the tone, but the mark of an invisible Reason which lies in it. And now a flow of notes which pervades our whole being with joy. What profundity unknown to the listener is not hidden in a single chord, what infinite arithmetic in a whole symphony! And now, joined with this, the invisible forms which appear before our soul in obscure intimations while the notes flow into the ear. In truth, we can repeat with joy and triumph at the nobility of our spiritual being that what fascinates and enraptures us in the art of music and makes us forget everything while our soul soars on the flow of notes is not the mechanical stimulation of tense nerves. It is the deep, infinite, incomprehensible Reason of Nature which speaks to us through the flow of notes.

29

First Introduction to General Physics[1]

A PROSPECTUS OF LECTURES
ON THIS SCIENCE

FOREWORD

The present treatise is a more extensive version of the introduction to the textbook which I published two years ago. Gradually I intend to treat more of its parts in the same way and thus provide my readers with a collection of pamphlets which in time might constitute a larger whole. Its more immediate purpose, however, is to serve as the basis for a small series of *prefatory lectures* with which I shall introduce my lectures on the various parts of general physics. These introductory lectures will start on Monday, November 18, and be given every Monday, Wednesday, Thursday, and Friday afternoons from 5 to 6 o'clock. They are also open to those who do not wish to attend the rest. When they finish, I shall lecture on *elementary chemistry* every Monday and Thursday afternoons from 6 to 8 o'clock, and on *advanced chemistry* Wednesday and Friday evenings from 5 to 6 o'clock. In the mornings, from 11 to 12 o'clock, I shall lecture on *general physics*, but particularly extensively on mechanics. Though each series is comprehensible in itself, the full series of lectures will provide as complete a survey of the entirety of experimental natural science (cf. §§11, 12, and 15 of the treatise) as is usually given at any university. Those who want to learn about the applied aspects of science, e.g., applied chemistry, require one or more of these general lectures as their necessary preparation. I shall gladly try to oblige those who, after having heard these, want some lectures on any of the applied subjects during the summer half. However, as the demands of such people, according to their different subjects, are so different that they cannot be satisfied in one or two lectures, and as I also want to be useful to the

---

[1] [Originally published in Danish (Copenhagen 1811, printed by Johan Frederik Schulz). KM III, pp. 151–90. The first 19 sections of this treatise are found in Schweigger's *Journal für Chemie und Physik*, Vol. 36, pp. 458–88. Nuremberg 1822. It contains the following note: "This is the introduction to the textbook on general physics [*Naturlære*] which Professor Ørsted is about to publish in Danish and German. From this fragment, here offered for public evaluation, the readers will realize what and how much they can expect from this new work by one of the most brilliant and, at the same time, most popular of contemporary physicists, who, through his works and ideas, has had such a momentous influence on the development of science. *The editor.*"]

regular students of science through something more than lectures, I offer to hold scientific conversations with my audience and other friends of science who want to attend on Friday evenings from 6 to 7 o'clock. During these hours I shall also, to the best of my ability, give each the directions which he might need in order to study those parts of the application of science that might be important to him. With this arrangement I hope to be more useful in many respects than through technical lectures. As usual, I willingly grant free admission to people of limited means, also those outside the class of regular students.

THE AUTHOR.

FIRST INTRODUCTION TO GENERAL PHYSICS

I. The Spirit, Meaning, and Goal of Natural Science

§1

With our reason we endeavour to encompass and penetrate all of nature and to present it in its complete context. We call the science whose aim this is *natural science* or *physics* in the broader meaning of this word.

§2

If we cast a critical glance at nature, we must be astonished by the magnitude of the task of including this infinite whole in one science. How immense is not the number of objects which are found spread across the globe we inhabit? What multitudes of animals are not known to us by name, and how many are not hidden from us by the ocean? How many do not live in territories on which no investigator has ever set foot, and how many do not escape our notice by hiding in the earth, in plants, or in the internal organs of other animals; not to mention those which can be discovered only by means of the artifice of optical instruments? No less marvellous is the profusion of species of plants, of which some 20,000 have been described. And now the minerals, dug out of the bowels of the earth, into which we have not yet penetrated as many fathoms as there are miles to the centre of the earth! How daring must not seem the intention to wish to acquire knowledge of all this! And yet it is only an infinitely small part of nature. If we rise in our imagination to our solar system, the earth already seems only a point in comparison, but nevertheless the solar system in its turn is as small compared to the part of the firmament which we can think to study, and what, finally, is all this compared to the infinity in which our increasingly speculative imagination finally loses itself? If we now turn in the opposite direction and endeavour to penetrate to the core of bodies by means of the cutting iron, we always discover parts there which, on closer examination, are found to be composed of others which, in their turn, have an elaborate structure, consisting of many parts, and so on indefinitely. In short, here too, we finally encounter something which can no longer be perceived by our senses. On the one side, we lose ourselves in the *infinitely large*, on the other, in the *infinitely small*.

However, this is still not sufficient. Throughout nature, we discover an activity which knows no rest. What seems rest to our eyes is only slow change. Through countless stages of development, every object hastens from birth to death. Not for one moment of its existence is it completely itself. Part of the full realization of this, then, is that all periods of time which it has traversed are united as at a focal point. In other words, this infinite chain of existence called the world, which already seemed so incomprehensible to us before, not only has to be seen as it *is*, but we must also find out what it *was* and determine what it *is to be*. Only when all this has been achieved, can natural science be said to have been exhausted.

§3

It is easily seen that here we have only presented some main features of an ideal. A science like this must always remain unattainable to a finite spirit. However, without setting ourselves a goal, we would have no direction for our efforts, and without an unattainable goal, the constant development for which mankind is intended cannot be advanced. The question is thus: *How is it possible, in our narrow sphere, to create a science which is even a faint image of that ideal?*

§4

A more penetrating look at nature shows us an admirable unity in its infinite variety. However different objects might seem from each other, closer scrutiny reveals a common essence in them all. Thus, throughout the animal kingdom, we find the same fundamental laws of organization in spite of the greatest and most varied diversity in external form and internal structure. By constantly directing our attention more and more to this basic unity, we have come so far that we need only know a few animals from each class in order to acquire valid insight into the nature of the entire animal kingdom. Thereby we can even form a fairly complete notion of animals which no longer exist, and of which we have only remains brought up from the depths of the earth. We find the same unity in the vegetable kingdom, where, similarly, a thorough examination of a few organisms is sufficient to provide a deep understanding of its entire nature. With continued investigation, we even find a point of unity for the animal and vegetable kingdoms, but even this unity, in its turn, is only part of a higher one, and so on, until the thought loses itself in a fundamental unity for all of nature. Wherever we turn our eyes, we find the same unity. The laws which rule the movements of our moon also apply to those which attend the other planets. The motion of these planets around the sun is again accomplished according to the same laws, and each newly discovered planet serves only to confirm the old law. However, we do not stop even here. We also have reason to assume that our sun, together with many others, repeats on a large scale what our planetary system shows us on a scale which is small by comparison. If, conversely, we turn from those planets to bodies here on earth, we find without exception that they all follow the same laws of motion and gravity as those great planetary bodies so that, on the basis of our experiments on motion here on earth, we are able to deduce consequences which apply to the whole. Now, if we understand these laws of motion correctly, we can also calculate what the positions of the planets have been, and what they will be at any given time. Astronomy offers us many examples of this. We have found a similar regularity, though still measured with far from the same accuracy as in astronomy, in the chronology of several other natural phenomena, such as regular, though not yet measured, periods in the development of the earth, in dynamic effects, and in the motion of the magnetic needle.

These examples show us what philosophy proves rigorously, *that every well-executed investigation of a limited object reveals a portion of the eternal laws of the Whole.*

§5

Now these laws and the force with which they are executed are the only unalterable things in nature. While every object moves incessantly, and the substances of which it is composed change incessantly, the laws according to which this happens, and only these, remain perpetually the same. It is also through them alone that one object is different from another, for we find that the most disparate objects are composed of the same substances, and the further our investigations progress, the more we are convinced that the substance of all objects and the forces whereby life and activity in nature are maintained are everywhere the same. However, that which gives objects their particular characteristics and produces the infinite variety in them is only the way in which the effects are realized in each object, the natural laws according to which everything in it is arranged and controlled. In other words, objects are in a perpetual transition from one state to another, in a constant becoming, everywhere of the same substance by means of the same forces. Matter itself is nothing but space filled by means of the fundamental forces of nature; what gives objects their unalterable characteristics, then, is the laws by which they are created. However, what constitutes the unalterable and also the characteristic aspect of objects can rightly be called their essence, and the part of them which they do not have in common with others, their distinctive essence. Thus, we may confidently establish *that the natural laws according to which an object is created together constitute its distinctiveness, and that knowledge of the actions of the natural laws is knowledge of the essence of objects.*

§6

The essence of an object does not consist in an individual law of nature which, as a concept, can be expressed in a theorem but only in a combination of many natural laws which all join to form a higher law which transcends what can be fully expressed in words. The essence of each object can therefore be regarded as composed of countless others. In its turn, it is itself only part of an even larger chain of essences and is thus combined to form a higher unity (as the earth, for example, cannot be fully understood except as part of the solar system), and this again constitutes only one part of an even higher one, and so on, until the thought loses itself in the infinite All. *Thus all natural laws combine to form a unity which, with regard to its actions, constitutes the essence of the entire world.*

§7

If we examine these laws even more closely, we find that they are in such perfect harmony with reason that we can truly say that the regularity of nature is that it follows the precepts of reason, or even more, *that the laws of nature and the laws of reason are one.* The chain of natural laws which through their actions constitute the essence of every object can thus be regarded as a *thought of nature* or rather an *idea of nature.* And as all natural laws together form a unity, *the entire world is the expression of an infinite, universal Idea which must be one with an infinite Reason, alive and active in everything.* In other words, *the world is merely the revelation of the combined creative power and reason of the Godhead.*

§8

Only now do we really understand how we may know nature with our reason because this is nothing more than that *our reason recognizes itself in objects.* On the other hand, we also understand why our knowledge is only a weak reflection of the great Totality, for our reason, though related to the infinite in its origin, is enveloped in finitudes and can only break away from them conditionally. Thus, it is not given to any mortal man to penetrate and embrace the Whole. With devout reverence, he must feel the limits of his powers in the happy realization that the few gleams which he is allowed to see are enough to raise him high above the dust. Moreover, it is not through clear, penetrating reason alone that we are connected to the innermost essence of nature. As we have been given a sense of the imprint of the spirit on forms through our taste for beauty and a sense of the imprint of reason on life through our conscience, thus we have also been given a sense of the imprint of reason on natural phenomena, through which we feel closer to it and, though lacking clarity in detail, acquire an impression of the majesty of All. This intuitive affinity with nature guides our reason in its investigation, and the latter rouses, intensifies, and purifies the former. Both stand in the most intimate union, but in such a way that the former must be dominant in life, the latter in science.

§9

We now feel quite vividly how unworthy it would be to make utility the purpose of the study of this or any other science, for when we ask about the usefulness of an object, we thereby reveal that we do not attribute any worth to it in itself but only with regard to something else, which must then be higher. Consequently, if science were to be studied merely for its usefulness, there would have to be something that was more worthy of a rational being than the use of reason or a better part of man than the spiritual, but if this is impossible, *then insight is good in itself*, and no external justification is needed for wanting to acquire it. Science, then, must be studied for its own sake, as the vital manifestation of our innermost being, as the acknowledgement of the Divine. The fact that this also produces the most glorious fruits in the lower sphere is a consequence of that rational harmony which inspires everything. These fruits are identical with what we call the utility of science, and to regard it in this way is the same as contemplating the glory of science from a lower vantage. This belongs to the completeness of the view and is thus of direct interest to the thinking being. Regarded in this manner, then, the utility of natural science is twofold, in that it both increases our powers and multiplies the means for their exercise. In addition to the ordinary development and growth that any science gives to our spiritual powers, natural science also contributes in a strange way to the enlightenment and fortification of our reason, which is wrapped in finiteness, by presenting to us, in an orderly sequence of contemplations, the eternal laws of reason as also dominating sensible nature. Filled with this insight, man proceeds with a sharper eye, a healthier confidence, and a purer joy to every task and completes it like one who acts on a conviction which springs from within and not merely according to some external precept. The soul also remains in internal tranquility and harmony with all of nature and is thereby purified of every superstitious fear, the cause of which is always the delusion that forces outside the order of reason should be able to influence the eternal course of nature.

This, in a few words, is the effect of the honourable pursuit of science, which emanates from the internal in countless ways. With regard to the external sphere of action of forces, the excellence of our science can be summed up in this one but great truth *that it teaches us to control nature*.

To primitive man nature offers the few necessities of life only sparingly and allows nourishment of only a few in a great expanse. Science extorts greater generosity. Through it, the soil is made more fertile, and its products are improved and prepared for uses whose possiblity the untrained eye scarcely suspected in the original substance. And thus it secures man an easy and pleasant existence where otherwise wretchedness would hardly have found soothing sustenance. Where many once had to work without thought like mere tools and spend their lives in bondage, there science liberates by replacing them with machines, which also fulfil the purpose more perfectly. —By nature alone, man would be limited to a narrow sphere; insight into nature expands it. Through science, he circumnavigates the globe, lowers himself to the bottom of the sea, flies through the air, and is thus no longer tied to the spot where he was born. A closer investigation has even been able to expand our sensory faculties so that, by means of artificial instruments, we find a world where the unfortified sense scarcely showed us an infinitesimal size, discover mountains and valleys on distant planets and solar systems, where even the boldest imagination hardly dared to place its creations before. In this way, then, man's entire existence is expanded and becomes more spiritual so that it is clearly seen that science and its consequences are in a mutually intensifying interaction with each other. —What science gave, it also protects. Without science, man would be mere sport for the wild struggle of the elements, intended for more ordinary natural purposes. With it, however, he learns to set one natural force against another and often to lead the most threatening one to a beneficial goal. In this way, science has taught us to conduct away the destructive lightning of the heavens, to tame the power of water so that it must serve our purposes, to control the consumption of fire and to force the most important services from it. Even when the ordinary forces of nature turn directly against the internal force through which our body exists, science teaches us to find the right counterpoise; against poison an antidote; against illness a remedy or even a protective power; against a widespread deadly pestilence, which could otherwise consume the people of a country and throw it back centuries in cultivation and development, a disruptive force which it cannot resist. The primitive human force, unguided by reason, can itself be regarded as a wild and hostile natural activity which has often destroyed the fruits of the enlightening industry of centuries. Natural science has made extraordinary contributions to the transformation of war into a scientific art which no people can any longer bring to perceptible perfection if it is not also at a rather advanced stage of development in other respects. And thus, at least in one sense, this ever dangerous manifestation of force has lost some of its atrocity. —In short, science facilitates, expands, and secures our existence in manifold ways and removes manifold obstacles to man's free activity and spiritual development.

II. The Division of Natural Science

§10

In our knowledge of nature we distinguish between something which comes more directly from the intellect and something else which rather has its origin in the senses. Both are in the most intimate connection with each other. It is the essence of man to present reason in an organic body, not merely in one particular form, but in its introspective totality. Literally, man's sensuous nature can only be regarded as the corporality of this reason. Therefore, the external sense organs already receive impressions in a manner which is in the most perfect harmony with it, and an unconscious reason in the internal sense impresses its own stamp even more markedly on these various abilities. Imperceptibly, they thus approach the conscious reason which organizes and combines everything into even higher units which, step by step, are finally transformed into the strange internal harmony of the independent reason. Thus, *the science of experience* (empirical science) comes into existence. Reason, on its side, is similar to the internal foundation and essence of nature. In a way, it contains the seeds of the entire world and must develop them through its necessary introspection. Consequently, it starts from the highest to which our spirit can ascend, from the essence of beings, the origin of everything. In itself, as a sign of this, it seeks out the various primary directions and through them the origin of the essential fundamental forms in the eternal unity. In its own laws it sees those of nature, in the variety of its own forms that of the world, and thus it develops and creates from itself the grand Totality. In this way arises *speculative natural science,* which is also called *natural philosophy.* The development of science in each of these directions has its characteristic difficulties and barriers. On the empirical path, we are stopped by the enormous profusion of objects which our senses offer, in which, however, there is no completeness. Even if they receive the stamp of reason because of the senses themselves, without our intervention, they do so only according to certain more obvious similarities, whereby large and more comprehensible units are formed, but in which this deeper connection, this internal unity, for which our reason strives, is not yet clear but more often misunderstood and placed in a disappointing light. If we did not understand this from the nature of things, history would show us sufficiently clearly that it is only through countless errors, through manifold vain attempts in contrary directions that our insight along this path has reached the point where it is at this moment, and from which it should proceed in the next. Consequently, speculative natural science seems to lead us more directly to the goal, but here we would do well to bear in mind that the reason which reveals itself in nature is infinite while ours, which must discover it there, is limited, trapped in finitudes. In innumerable sparks, reason spreads through mankind. Although a reflection of the whole in every individual, it has in each its distinctive direction which prevents it from spreading its light equally clearly and fully in all directions. Only recently shaped in its present form, speculative natural science will only approach significant perfection through the combined efforts of many thinkers. The closer it stays to the great principal forms, the more certain and pure it will be; the more it descends into the variety of nature, the more easily a link in its great chain is omitted, and the more easily it wanders off in some special direction. Only continually reminded and, as it were, inspired by experience can it progress without confusion. Each of these scientific directions, then, needs the other; the latter the variety and lively presence of the former, which our limited creativity cannot provide; the former the unity and comprehensive view of the latter, which can only be gained from a higher standpoint. To the empiricist, the idea of the Whole is to be regarded as a bright sun which shines into the pathless chaos of experiences, and to the speculative philosopher experiences are to be regarded as guiding stars, without which he could easily lose himself in the infinite depths of reason. The further they proceed in these two opposite directions, the more they meet each other, and, like different organs in the same being, they will finally be combined into a harmonious whole.

§11

Empirical natural science is divided into two main parts, the *descriptive* and the *investigative.* The first step is to notice, by means of the senses, the form, the structure, the composition, etc. of objects, in short, all that can be learnt by observing them as they present themselves in a given state and without being set in action. On the basis of certain principles, this enormous mass is arranged according to their similarities and differences. Thus arises the *description of nature,* which less correctly has been called natural history. To this belongs not only the description of

animals, plants, and minerals, but also the physical description of the earth and
the part of astronomy which merely describes the position, the shape, etc. of the
planets. Even anatomy and crystallography must be included in the description of
nature although they are already very close to the investigative part. —In this we
do not stop at that first acquaintance with objects but observe them in action and endeavour to discover the laws which they follow. This part of science is given the
special name of *physics*. In this again the *general* is distinguished from the *specific*. The former presents the laws according to which all bodies influence each
other regardless of any particular class to which they might belong. The latter endeavours to develop and present the internal nature and coherence in a separately
regarded class. This includes the theory of the laws of motion of the planets, *astronomy* in the stricter sense of the word; the theory of the laws of the development
of the earth, *geology*; the theory of the nature of organic beings, *physiology*. In reality, the same laws prevail in all of nature, only at a higher or a lower order, but this
difference in the power of the natural laws already makes a very considerable difference in the method and presentation of a science. This is all the more true as we
cannot connect one part directly with another and from the principles of one part
completely deduce those of another, but we must begin each single part from a new
point of origin, indeed, in yet others we must have several main points which cannot be deduced clearly enough either from each other or from a higher common
ground but must be connected through some feeling of unity. This lack of strict coherence is a natural consequence of the imperfection of science. The further we
have advanced in insight, the closer all its branches have come to forming a whole,
and through the collaboration of the speculative and the empirical natural sciences
this goal, for which we already strive considerably, will be attained as far as human
strength permits.

> The description of nature could also most suitably be called *natural history*[2] just as we
> call the person who possesses this knowledge a *natural historian*.[3] The person who
> works in the investigative part is called a *student of nature*.[4] Of course, no thinking
> man is *only* a natural historian even if natural history is his main purpose. Even in sci-
> entific presentation neither part is completely isolated, but one easily enters the prov-
> ince of the other.

III. GENERAL PHYSICS. ITS DIVISION, METHOD, AND RELATION TO THE OTHER PARTS OF NATURAL SCIENCE

§12

General natural science,[5] which is to be discussed here, is also called by another
name, *physics*, where the word is to be taken in its most narrow sense. Sometimes
it is also given a far too limiting name, *experimental physics*, because it uses experiments as much as possible to investigate the laws of nature and to confirm the theo-

---

[2] [*Naturkyndighed.*]   [3] [*Naturkyndig.*]   [4] [*Naturgrandsker.*]
[5] [*Naturlære*, which we usually translate as "physics."]

rems which have been found. We have seen that its object is the changes which may take place in bodies generally. It must then be divided accordingly. However, changes in bodies can be of two kinds which are substantially different: *external* and *internal*. A purely external change, which thus can have no effect on the composition of the body, can only affect its relationship to other bodies, i.e., its position, its location. A change in these is called *motion*, and the science about it the *theory of motion*. *Purely* internal changes can only affect the properties of bodies and can only be observed and investigated in so far as these can act on other bodies, so we always observe these properties in their actions. We also call an active property a *force*, and the theory of the internal changes of bodies must therefore be called the *theory of force*. This branch of science has developed naturally in two main directions. From one side, we observed the many compositions that take place between dissimilar bodies, whereby the most different ones together formed a physical unity perfect to our senses. The study of this has been called *chemistry* and treated as a science which is quite distinct from general physics since it was imagined that everything seen in effects of this kind depended on the substances which were to be combined. However, a closer consideration allows us to see that it is really the laws governing the forces which cause combinations and decompositions that really ought to be the objects of this science and not the substances which are combined and decomposed since knowledge of the substances, as such, either belongs to the description of nature or is a consequence of the investigation of the forces of bodies. By pursuing this further, it is finally found that all chemical effects can be traced back to the manifestation of two principal forces, widespread throughout nature, whose properties in their free state, however, cannot easily be found by chemical means. From another side, however, we have arrived at greater knowledge of these forces. In electric, galvanic, and magnetic effects two opposite forces have been found, widespread throughout nature, and it has been possible to investigate the laws which govern their freest manifestations and pursue them through the most diverse conditions to the point where they also produce chemical effects. This shows that we might very well expand the term *chemistry* to include the entire theory of force, in which case we should call all previous chemistry *elementary* as it has been restricted to the exterior of the effects (which, however, would not be as well-understood as it is now without the spirit, the insights, and the admirable efforts of many excellent scientists), and the other part, which by contrast shows us forces in their more immediate manifestations, should be called *advanced*. We could also distinguish these two parts as being the *advanced* and the *elementary theory of force*. The theory of force is also called the *dynamic* part of science, the theory of motion, on the other hand, the *mechanical*. In the same sense, we also speak about *dynamic* and *mechanical effects*.

In addition to the dynamic and mechanical portions of natural science, we could also imagine a third which dealt with the effects of the theory of force in combination with motion, but according to the present state of science this coincides with dynamics. Some day, when we have more perfect knowledge, particularly of organic nature, it may break away and become more independent.

Before all else in natural science there must appear, as a kind of introduction, a presentation of the general properties and conditions of bodies as the highest abstractions to which the description of nature has been able to ascend.

That the mechanical part of natural science deals with *forces* which generate motion and the dynamic part with *motions* generated by these forces can hardly lead to any doubt about the validity of our division as everything in one case is aimed at determining the laws of motion, in the other, the laws of forces.

We let the mechanical part precede the dynamic. At first glance this seems incorrect as the internal forces cause all external phenomena so that it would even be impossible for one body to set another in motion without these, but it is in the nature of the empirical method to start with the exterior and then penetrate into the interior, the conditional and the dependent, in order to get to the essence.[6]

§13

The foundation of general physics, according both to the concept of it which we have advanced here and to the manner in which it has developed over time, is experience. Nature shows us many of its changes so frequently, so intensely, and so sensuously that we cannot but notice them. These are *everyday experiences* to us. Others we do not discover without deliberately directing our attention towards them. Collecting information about these is *observation*. Finally, there are many which

---

[6][In the version in Schweigger's *Journal*, a new paragraph has been inserted here, which reads as follows:

4) PRECEPTS FOR THE TREATMENT OF NATURAL SCIENCE,
IN PARTICULAR GENERAL PHYSICS
§13

All precepts that can be provided for the investigation of nature must arise from the fundamental truth:

That all nature is the revelation of an infinitely *rational will, and that it is the task of science to apprehend as much of it as possible with* finite *forces.*

From this great fundamental truth follow a number of fundamental laws which the student of nature must keep before him as eternal guiding stars. *His primary task is to seek reason in nature.* Therefore he must assume *that the laws of nature agree everywhere with reason, and that any appearance of an exception to this rule must be due to the incompleteness of his insight.*

No contradiction is possible among the laws of nature; rather, they are in the most intimate agreement and together constitute a single indissoluble whole. If we direct our attention to the cause of deviations, we see that everything occurs in an entirely logical manner. If we then direct our attention to the result, we are compelled to admire a wisdom which exceeds all human imagination.

We can often see the strict consequence of events so clearly that it can become a basis for knowledge to us. The wisdom of the purposes is so superior to our ability that our thoughts about it may give us guidance in our investigation but no foundation to build on.

The laws of nature are immutable, like the will from which they arise.

The fundamental forces of nature are indestructible.

By fundamental forces we mean the simplest and most basic manifestations by which the creative force announces itself in material nature.

The same forces always act according to the same laws. Effects which are truly equal must stem from the same forces. In order to explore the laws of forces, we must strive *to recognize each particular force in its purity and the laws of its action in their simplicity,* but in doing this we must strive never to forget *that every force is part of the infinite whole and only exists to the extent that this exists.*

nature does not immediately show us in a very comprehensible way. In order to investigate the nature of these more closely, we must endeavour to bring objects together in such a manner that their effects become more comprehensible to us. In other words, in order to see the method of nature as completely as possible, we must know how to set it in motion at will and force it, as it were, to act before our eyes. We call this to *institute* or *conduct experiments*, to *experiment*. Nature forces everyday experiences upon us, and it invites observations, but we ourselves create the experiment; it is an act of our most complete freedom. It is easy to see that these are all degrees of the same kind of knowledge and merge into one another so that no completely sharp boundaries can be drawn between them. In a thinking being, every experience leads easily to a closer contemplation which, without any perceptible leap, leads him to actual observation. From the purely arbitrary direction of attention to the points which in particular constitute the present object of his curiosity, he soon proceeds to compare, distinguish, and arrange all the sensuous variety which might seem to have some connection with it. He seeks to sharpen his sensory organs through practice, he endeavours to measure their strength, to test, to determine, and, if possible, to correct their errors. Through habit, he acquires a proficiency in discovering the rare, the peculiar in natural phenomena, in finding their less conspicuous similarities and differences, and in distinguishing accurately what belongs to each part of it. When this is no longer sufficient, he tries to facilitate the observation, to expand its limits, and to make it more accurate by artificial means. He measures sizes by means of instruments devised for this purpose, he knows how to enlarge and to make clearer objects which would otherwise be too small or too far away in relation to the powers of his sensory organs. In short, he delves more and more deeply into nature by artificial and controllable means, and he gradually transforms himself into an experimentalist. The latter uses all the same tools but adds new ones and, above all, distinguishes himself by greater control. Where nature works through many combined forces, he tries to restrain some in order to allow others to work all the more freely, indeed, to restrain all others to allow just a single one to manifest itself freely. What nature achieves in great quantities, he must often try to produce on a smaller scale in order to bring it closer to the eye, and what nature offers so sparingly that it would escape even the most acute sense, he must learn how to make perceptible to the less acute with far more means than those of the mere observer. What nature shows only to one sense, he must endeavour to bring to the tribunal of the others so that the more acute can see what the more obscure could only suspect. Indeed, in order to discover the essence of things, he often places them in completely new conditions, never previously offered by nature, so that his preconceived assumptions are either confirmed or refuted. In short, he tries everywhere to make the most secret forces of nature reveal themselves, and he determines their course with scale and measure.

With limited abilities we strive to grasp the infinite reason of nature, *so we must always be suspicious of our own knowledge and only trust our notion about the thoughts of nature to the degree that it is clear, definite, and in agreement with all that is undisputed truth according to our absolute conviction.*

We should compare the pronouncement of reason with experience: We should endeavour to transform the pronouncement of experience into a pronouncement of reason.]

The treatment of the profusion of our sensuous knowledge, then, constitutes one great, coherent *empirical art*, whose development to heights never previously achieved constitutes the distinct characteristic of more recent natural science.

§14

This art presupposes many spiritual and physical natural talents and much skill acquired through long practice, but all these qualities would be in vain if they were not guided by a spirit attuned to nature. To have seen many natural phenomena is not yet to have insight into them. Our experiences become instructive only when they are placed in the proper context. To observe is to discover the actions of nature, but no-one will succeed in this without having some idea about its character. To make experiments is to ask questions of nature, but no-one can do this usefully unless he knows what to ask about. Throughout empirical science it is therefore necessary, on the one hand, that the investigator constantly maintains a view of the whole, without which it is impossible to have a clear notion of the parts, and on the other, that he considers no part unworthy of his attention because, after all, it belongs to the whole. He should never forget that the forces which maintain life and motion throughout nature are found in the smallest and most insignificant as well as in the greatest and most extraordinary objects. Then he will always approach his work with the strictest seriousness and attention, respectfully acknowledging that it is nature itself that speaks to him even in the most insignificant object. With this spirit, with this constant concern for the whole, often laborious and detailed tasks lose their pettinesss to him; he raises them up to him and does not let himself be pulled down by them. He is not content with a single one-sided experience. He always seeks to connect it to others, to deduce one from the other, and to organize everything in such a way that the whole series of observations or experiments presents a law of nature. The same object must therefore be subjected to all the most varied effects; the same effect must be tried if not on all bodies, which would be impossible, at least on many of each natural family, even on the most singular and the most dissimilar ones. Moreover, the same effect must be sought out through observation and produced by experiments in as many different guises as possible (the experiment is then said to *be varied*) in order to see, with greater clarity and certainty, all the conditions under which they take place. Only by giving the observations and experiments performed such a *coherence, extent,* and *variety* can his work give him insight and become more than an incomplete account of isolated phenomena. When, in one or several connected experiments, he has seen a certain series of phenomena which followed each other in a certain order, he begins the experiments from the opposite end in order to see if everything now follows in the opposite order; i.e., the experiment must be repeated in the two possible *opposite directions* (in chemistry we say that the proof is only complete when it is both *analytic* and *synthetic*). Where it is possible to follow the same development by observation, this should naturally be done. It is in this way that we are most clearly convinced that we have had the correct notion of the relation between natural phenomena. With all this, it would still have been easy to deceive ourselves if we had allowed circumstances foreign to the content of the investigation to interfere. If this happens, the observation or the experiment is different from what was imagined and is therefore said not to be *pure* with regard to its conception. In observation we must therefore be particularly alert to incidental circumstances. In experiments concerning the nature of substances, these must be chosen *pure*, free from foreign matter, and where form is concerned, we must obtain instruments whose construction for a given purpose is as mathematically correct as possible. If we add to all this a complete indication of all determinations relevant to the observation or the experiment so that every change is observed, its magnitude established, its connection with others and mutual relationships with external circumstances are not lost sight of, then the execution is *precise*. Finally, the student of nature must be sure that his senses have not deceived him and must *repeat* his observations and experiments frequently and in the most varied forms.

§15

In general physics, the experimental method is quite predominant, which is also why it has exclusively been called *experimental* physics, as mentioned before. It is true that everyday experience and observation constitute its first seeds and still provide its most important nourishment, but they have almost disappeared from its description and presentation. It is the spirit of general physics to transform every experience, every observation into an experiment, indeed, it even endeavours to express the insight acquired through reflection in this manner. The lover of nature delights in observing its effects, and therefore he will be able to re-create them often; he is desirous of understanding it as accurately as possible and from all sides, and therefore he wants to be the master of the investigation. Experiment must lead him to this. However, not everything depends on this alone. Empirical science would achieve its perfection only when it enabled us to make nature present to our eyes all its laws in a series of effects. However far our art may be from this ideal, it remains the ideal, that which it must imitate if it wants to be not only a collection of artifices but an art in its own right. Through this extensive experimental presentation, physics also gains a high degree of solidity, for mere inferences, however thorough they might be in themselves, all presume that the notion which we have formed of the object under investigation really agrees with it. However, it is easy to deceive ourselves in this respect. In nature so many circumstances work together at almost every juncture that we might easily overlook one or more of them and consequently form a conception which does not fully correspond to the object. If, on the other hand, experiment accompanies thought, delusion is not possible except in the case of several concurrent errors.

§16

However, we can consider the experimental art from an even higher level. It is not merely to contemplate the external world or to find its essence that it has thus limited itself as a peculiar art of imitation; it also wants to set our mind itself in creative activity in order to develop lively and vigorous knowledge which is more in harmony with the constant evolution of nature. The distinctive feature of this, then, is the *creative method* (the genetic method), and this does not take place only when we deal with physical objects but also belongs completely to all that is presented only to the mind. When we allow a point to move in our imagination in order to produce a line, or a line to rotate around one of its extremities to describe a circle with the other, what is this but a thought experiment? Differential and integral calculus consist of nothing but such thought experiments and considerations of them. When this procedure can take place, and that is far more often than might be believed, it is particularly well-suited to satisfy the striving for insight of a vital and forceful mind, for with other methods we generally learn why we must be *convinced* that something is a certain way rather than *why* it really is. Here, on the other hand, we are allowed to see every truth at its birth. The reason for its being and our certainty about it thus coincide so that, when it is presented in this fashion, it is also proved. If it is part of the essence of physics to let the development of our thoughts follow that of the objects in this manner, it is clear that we must often resort to these thought experiments, which we have so far overlooked too often. In his *Metaphysical Foundations of Natural Science*, Kant has given us the most beautiful examples of this kind of presentation without, however, drawing attention to it himself.

§17

Mathematics plays an important role in the presentation of physics, and this is quite in the nature of things. Every change has its magnitude, and so does each of its parts. These magnitudes, as well as the way in which they follow one another, can be determined only with the aid of mathematics. The theory of motion has been almost completely transformed into mathematics. The theory of force awaits the inventive mind which can lead it to the same point, for internal forces manifest themselves only in time and space, and their laws cannot be considered completely known to us until we can show all the concomitant circumstances with their correct magnitude. Many of the very best students of physics have tried far too hard to impress upon it the form of mathematics, or rather that of Euclidean geometry, whereby it has come to be regarded as applied mathematics. In this way, science is deprived of its natural form. The mathematician tries to deduce all his theorems from the least possible number of simple basic truths. All other considerations are sacrificed for the ingenious stringency of the proofs. Even in the applied branch of his science, where he must borrow certain fundamental experiences, he considers, after their clarity and certitude, only how to use as few as possible. The student of nature, on the other hand, endeavours primarily to find the most immediate connection between the effects of the various natural forces. He regards as his property the experiences which the mathematician may only borrow. Consequently, he is not afraid to use them abundantly in his proofs if only he can present them with the distinct stamp of their mutual connection. This is why he often deduces theorems directly from the nature of an effect while the mathematician only arrives at them circuitously from some basic truth on which he prefers to build. Applied mathematics and physics, then, deal with the same objects and also share the desire to display the rational connection between the same objects, but the former wants to show it by main force, so to speak, and is therefore content with an artificial connection when it cannot find a natural one, whereas the latter wants to see objects in their most natural or, if one prefers, most immediate rational connection and is not content with anything else. Therefore, it may be claimed that the two must coincide with a certain degree of perfection. Everywhere mathematics and physics must constantly approach a more intimate union. The former presents the natural laws of magnitudes, the latter the laws for objects which have magnitude, and which experience all effects with it. Consequently, one has always served the development of the other. If the former has given the latter some of its certainty, some of its inventiveness, the latter has given the former many essential new elements and will probably give it even more through the development of the theory of force. Physics has been brought sufficiently close to mathematics, probably even too close; perhaps it is time for mathematics to try to approach physics. Geometry in its present form will always be one of the most glorious monuments to the human spirit and through its inner perfection constantly serve to train and sharpen the intellect, but would it not be possible, in addition, to have another presentation, in which all geometric theorems were presented in a series of thought experiments? In this way, a much clearer and more immediate insight into the real source of every truth would be opened to mathematics itself, and physics would gain a much more intimate fusion with mathematics than ever before. The progress of physics, in its turn, will promote this fusion, for the further it progresses, the more will it succeed in tracing all effects back to a few forces whose strength and conditions in time and space will then constitute its primary object. Thus our science must be transformed into a mathematics of nature, which probably would so far exceed what we have hitherto imagined under the name of mathematics that this statement will easily seem to many to reveal a far too low and limited conception of the essence of physics.

> Some years ago the author attempted a formulation of geometry as described above. When he has had the opportunity to treat it more thoroughly, he will submit it publicly to the judgment of the experts.

§18

When we discover the general law of nature under which a phenomenon belongs, or when we trace a more limited natural law back to a more general one, we are said to *explain* it. This can also be regarded as the inclusion of a less widespread effect under a more widespread one, and in this way the explanation of an effect can be presented as the indication of its cause. When it is not clear under which law of nature an effect or a class of effects belongs, we try to fill this gap by means of a guess. Such guesses have been given the name of *conjectures* or *hypotheses*. They are really to be regarded as thought experiments through which we wish to discover whether something can be explained by a certain assumption in connection with other natural laws. If we find that everything in a rich and varied experience can be understood with the hypothesis, it is assumed to be true. If, on the other hand, some circumstance turns out to be at variance with the assumption, it is rejected, and we look for a new one, which may again be invalidated by a similar test, and so on, until we find one which is not destroyed by the test. When a conjecture cannot be refuted by experience but, on the other hand, does not explain all that ought to be explained, it is merely considered to be more or less likely, according to the degree of completeness of its explanations. In this case, we have not quite reached the object of the investigation, which is always to eliminate the hypothesis as hypothesis, either through a complete confirmation or through a complete refutation, but it remains either a properly posed question and, as such, a *tentative hypothesis* or a reasonable assumption which we endeavour to connect with the rest of the theory. In this state it is called a *hypothetical theorem*. Every hypothesis can really be regarded in either way, but generally there is a predominance to one of the sides so that either the invitation to further investigation is stronger, or the probability must almost exclusively be taken into consideration because the present state of affairs permits no further investigation. The former are vivid and active elements in the development of science, whereas the latter often prevent its progress because they take root in those who are so much under the sway of habit that they often defend them as though they were established truths.

§19

It is part of the complete confirmation or transition to certainty of a hypothesis that all the consequences which are deduced from the assumption actually occur in experience. If all possible consequences were deduced and found confirmed, the hypothesis would become certainty, for it is impossible for two different causes to produce effects which are similar to each other in every respect. However, as our insights and experiences are limited, we must content ourselves with going as far in this respect as our powers permit. First, the hypothesis must be presented in terms as simple and clear as possible. This circumstance is of the greatest importance, and its neglect has caused countless confusions. Next, as many immediate consequences as possible should be deduced from the assumption. If any of these contradicts experience in such a manner that we cannot expect to resolve the conflict, the fate of the hypothesis has already been sealed. If, on the other hand, the deduced consequences agree with the experiences with which they are compared, it must be examined further whether the consequences of the next conclusion also agree with reality, and whether this is found even under the most complex conditions. If this is the case so that the examined effects in all their parts not only take place but take place in exactly the order and display the magnitude which they should have according to the consequences deduced from the assumption, and if, finally, they do not take place where they should not according to the same consequences, then the probability has become certainty. If all this is to take place, countless circumstances really must coincide even though only a small number of them present themselves clearly to our attention, and thus the probability becomes infinite, i.e., it becomes certainty. Among these coincidences of ideas and experience, the agreement between calculated magnitudes and real ones is particularly important and in itself almost sufficient for confirmation because here, out of infinitely many equally possible cases, a given one coincides precisely with the calculation. In this way, it would even be possible to confirm an assumed cause or a conjectured natural law which had never been encountered in experience. However, in that case the most perfect and many-sided agreement between deductions and experience must also occur, and it may be that we could never quite satisfy the just requirements of science in this. As a tentative hypothesis, such a bold conjecture may be tolerated as it might lead to the discovery of what was previously unknown even though it should always be regarded as misleading, but as a hypothetical theorem, which is always more closely connected with the other elements of science, it is reprehensible. Therefore, a hypothetical theorem should only concern the connection between a cause or a more general natural law, whose existence is certain, and the effect or the more limited natural law which we wish to explain with it. Finally, we must, as much as possible, avoid mingling hypotheses with established truths of science. In this regard, we must distinguish very clearly the connection or relationship between several phenomena, which is almost always expressed as a hypothesis, from the view of the unknown cause of effects which it immediately seems to force upon us. If we can make this distinction clearly, we shall lose little by rejecting the latter and retaining only the former.

§20

The relationship between general physics and related sciences is partly implied by the preceding. Its interaction with philosophy and mathematics has already been indicated in such detail as is sufficient for our purpose (§§10, 16, and 17). We have also established the position it occupies among the various branches of natural science (§11), so only its interaction with these remains. It is obvious that physics presupposes the description of nature as it would be impossible to discover natural laws without knowing the objects in which they manifest themselves. However, it is also clear that insight into the laws of nature, which must apply to entire classes of objects, does not presuppose detailed knowledge of each but merely requires a general idea of what they have in common, which we can acquire by observing a rather small number of them (§§2–4). In our daily experience, therefore, we already obtain sufficient knowledge of nature to make great progress in physics. However, it cannot be denied that the scientific description of nature has offered many remarkable objects to physics and still discovers more every day which have prompted new experiments and new determinations concerning the laws of nature. On the other hand, physics is not without influence on the description of nature as it lends it many devices with which to discover objects which would otherwise be hidden, distinguish others (especially inorganic) which would otherwise be more easily confused, and preserve those which would otherwise be disturbed; not to mention that the means whereby physics has enabled man to extend his sway over nature have also given its describers the opportunity to find a profusion of objects which would otherwise have remained unknown to them. Our science is more intimately connected with astronomy than with any of the other branches of natural science. Through its discoveries, the latter has greatly expanded the ideas of gravity and motion of the former which in turn, through its investigations of motion and gravity, has made it possible for the latter to explain a variety of celestial phenomena which must otherwise have remained a riddle. The discoveries of the refraction of light, of the expansive power of heat, etc., have not been unimportant to astronomy either, and to this must be added the more perfect instruments that it owes to physics. The mechanical part of physics has had the most excellent influence on astronomy; chemical forces are particularly important in the effects which geology describes so that the latter will one day be regarded as a great chemical endeavour in the same sense in which the former can now be regarded as a mechanical one. Therefore, with the development of geology, that and the chemical branch of physics will enrich each other no less than astronomy and the mechanical branch do now. Meteorology, which in the nature of things must form part of geology, can already convince us of this through many examples. Our knowledge of organic nature still consists only of fragments whose connection may well be undeniable but cannot be presented with adequate certainty and clarity. General physics has given it many of these fragments while, on the other hand, the observation of living nature has offered general physics many extraordinary phenomena which have prompted new and extremely important discoveries. However, before any particularly close union between these two sciences can be formed, mechanical and dynamic physics must be united in a way which we can hardly imagine today. As the art of medicine, in its essence, is nothing but the application of science to living nature, we may omit its closer consideration here as above. If, after all this, we consider the whole, it seems that, in a certain respect, we could regard our science as the centre of all our other natural insights in that it contains the unities, or rather the unity, to which they can all be traced. These, on the other hand, contain a variety to which this ought to expand. Thus all the loosely connected branches of natural science must, as we have already mentioned (§11), merge completely into one great science.

IV. THE HISTORY OF GENERAL PHYSICS[7]

§21

General physics, in the sense in which it is used here, did not develop into a system until recent times. As far back as our history goes, ancient peoples have not lacked knowledge of certain laws of nature, but they did not form a unified whole. We know very little about the wisdom of the *Indians*, the *Egyptians*, the *Chaldeans*, the *Jews*, and the *Phoenicians*, except for *astronomy*, which of all the branches of natural science they brought to the greatest perfection. The taste of the *Greeks* led

---

[7] [The names in this section are given with Ørsted's spellings. Though occasionally somewhat idiosyncratic, the names are still recognizable. The dates are also Ørsted's and not always correct.]

them in particular to natural philosophy because they preferred embracing nature as a whole to busying themselves with individual details. Besides, Thales, Pythagoras, Democritos, Platon, Aristoteles, Epicuros, and others who have philosophized about nature also knew much of what we include in general physics. Consequently, we may look for many seeds of our present physics in what they left us although they themselves were far from regarding our science as a science in its own right. In that era, the most important preparation for its development happened through the refinement of mathematics, in which Thales, Pythagoras, and Platon greatly distinguished themselves. With the mathematicians of the *Alexandrian* school, among whom Euklides (300 B.C.) is the most famous, it then made extraordinary progress. This school flourished until the seventh century A.D. and also produced many of the meritorious teachers of applied mathematics, among whom the elder Heron and Ctesibos, as well as the astronomer Ptolomæos, who also developed optics, deserve to be mentioned. However, among all the Ancients, the greatest merits of our physics are due to Archimedes of *Syrakusa* (250 B.C.), who may be regarded as the creator of its mechanical branch. We owe him credit for discoveries concerning the equilibrium of both solid and liquid bodies which laid the groundwork for scientific *mechanics* and *hydrostatics*.

Natural science did not make any significant progress with the Romans, but from Lucretius, Seneca, and Plinius we learn something about how far it had progressed at their time. With the decline of the Roman Empire, all sciences also fell into decay, and among these natural science in particular. For a period of more than a thousand years, natural science declined rather than progressed. Especially from the third to the eighth century, the first part of this period, a profound barbarism prevailed, and even after this time Europe emerged from it only slowly so that increasing enlightenment was not able to benefit our science noticeably until the sixteenth century.

During this long period, science lost equally due to misunderstood philosophy and to primitive and superstitious mysticism, which were often mixed with each other. During this time, Aristoteles' writings rose to an unbounded prestige which suppressed all more independent investigation. Therefore, the excellent parts of these works did not prevent people from becoming lost in a confused technical language which offered only empty words instead of real natural laws. Their *qualitates occultæ* are sufficiently well known in this respect. From another direction, they sought the key to nature in the supernatural. Instead of clear empirical investigation or profound speculation, by means of which one should endeavour to understand nature, they now tried to activate nature by secret means. Thus, true natural science was replaced by *magic* and kindred *alchemy* and *astrology*. They wanted to discover and control the most secret forces of nature by means of mystical words and signs. The object of the endeavours of these times was to discover a universal solvent, to produce the noblest metal, and to find a remedy against all illnesses, but it was not possible to consider making a plan for an extensive investigation at a time when it was believed that the primal forces of nature could be seized with a stroke of good fortune. At such a time, then, it was no wonder that men with some clear knowledge of nature, like Albert Grot (known under the name of Albertus Magnus) and Rogert Baco (both in the thirteenth century), were regarded as sorcerers. However, in spite of all this, some knowledge of nature was preserved, and a little progress was made, partly through the writings of the Ancients and through the Arabs, who applied themselves particularly to mathematics and medical arts, partly through work done for alchemical and similar purposes, and, finally, partly through chance experiences. Thus we owe to this era the invention of the compass, spectacles, and gun powder. Nor can it be denied that, in the midst of the confusion of their sophistry or superstition, people of this era, precisely because they sought the highest to which man can rise, had many fortunate ideas about the relations between things, ideas which a more doubting age rejected completely, but which a better-informed one must again bring forth from the old darkness and into its proper light.

During the largely misunderstood endeavours of that dark age, however, they had gradually turned their attention to nature and in some respects learned to activate it at will. The empiricism of more recent times developed from these beginnings, but this development occurred slowly. Only in the sixteenth century, which already enjoyed the ripest fruits of the fine arts, did our science, like so many others, begin to flourish with notable vigour. The mechanical branch of physics, as well as the related theory of the motion of light, to which they applied the mathematics inherited from the Ancients in combination with experiments, was first imbued with the brighter spirit of the new age. The exposition of the chemical branch maintained much of the darkness of the Middle Ages for another century.

As mathematicians, Guido Ubaldi (whose mechanics was printed in 1577) and, with even greater success, Simon Stevin (whose principal work was published in 1596), concerned themselves with natural science. In his experiments, Johan Baptista Porta (1545–1615) made us better acquainted with several natural laws, especially for light. William Gilbert (whose principal work was printed in 1600) conducted many new experiments on electricity and magnetism. Willebrod Snellius (1591–1626) discovered the true law of the refraction of light. However, a new era had also begun in astronomy when Nicolaus Copernicus (1473–1543) transformed theoretical and Tyge Brahe (1546–1601) practical astronomy. At the same time, chemistry had also been given a great impetus by Theophrastus Paracelsus (1493–1541), who acted as its first public teacher and, though shrouded in the darkness of the Middle Ages, challenged the opinions of his predecessors with great force. And the metallurgist Georg Agricola (1494–1555) laid the foundations of a clearer and more certain treatment of one of the most important parts of chemistry. However, the experimental method became firmly established only with Franz Baco of Verulam (1560–1626) as its powerful advocate and Galilæo Galilæi (1564–1641), who, as an experimenter, also made important discoveries concerning the fundamental laws of motion and gravity. At the same time, Johan Kepler (1571–1630) also made his great discoveries of the laws of motion of the planets, and soon after that, science had excellent scholars in Peter Gassendi (1592–1655), who tried to renew Epicurean natural science, and René des Cartes (1596–1650), equally excellent as the subverter of scholastic philosophy and as an innovator in mathematics and its application to our science. The natural philosophy which he created ruled for almost an entire century. In addition to all this, that period was distinguished by the most important discoveries of the mechanical properties of air. Evangelista Torricelli (1618–1647) invented the barometer, and Otto von Guerike (1602–1686) the air pump. Christian Huygens (1624–1694), the inventor of the pendulum clock, also enriched the theory of motion with many important discoveries, especially concerning motion under a central force and impulse. Robert Boyle (1626–1691) worked successfully in practically all the branches of empirical physics. The chemical branch is also under great obligation to him. During the same century, the seeds for the perfection of chemistry were also planted by many other dedicated scholars, among whom Johan Baptista von Helmont (1577–1644) and Johan Kunkel von Løwenstern (b. 1630) may be particularly worthy of mention, but a relatively clear and pure view of chemistry was first offered by Johan Joachim Becher (1645–1682 or 1685), who made a consideration of fire the centre of this science and thus laid the foundation for a theory which, under the name of the *phlogistic theory*, endured for more than one hundred years, and of which essential elements have passed into more recent ones. Finally, one more scholar from this era deserves to be mentioned, Johan Mayow (1645–1679), whose theory was a hundred years ahead of his time, and who therefore did not have any noticeable influence on the course of science.

A new era began with Gottfried Wilhelm Leibnitz (1646–1716) and Isaak Newton (1642–1727). Their mathematical discoveries gave a new direction to mechanics, especially through the invention of differential and integral calculus. Thus arose a more advanced mechanics in which Newton even taught us to include the laws of the motion of the planets. His discoveries also started a new era in the theory of light. In philosophy, mathematics, and physics, these men laid the basis for the condition of science in the eighteenth century. Above all, there was now, in mathematical physics, a foundation on which only further construction was required. This was done by Johan, Jacob, and Daniel Bernoulli (a family that flourished in science at the end of the seventeenth and for the greater part of the eighteenth centuries), by Leonard Euler (1707–1783), Louis Jean le Rond d'Alembert (1717–1784), and our contemporaries Louis de la Grange and Pierre Simon de la Place, who have all contributed considerably to the perfection of advanced mechanics and thus also led us to solve many of the problems of physics. Of the discoveries in mechanical physics whose origins cannot be linked to those of Newton, one of the most extraordinary is due to Ernst Florenz Friedrich Chladni (our contemporary), that the motion of sound can be made visible to the eye. To this must be added that he separated the theory of sound, as an investigation of the internal vibrations of bodies, more clearly from other parts of physics than had been done before. The art of making mathematical and physical instruments has also been brought to previously unknown heights in our century. However, all the progress which the mechanical branch of physics has made since Newton hardly compares with what has been achieved in dynamical physics during the same period. During the transition between the seventeenth and eighteenth centuries, Georg Ernst Stahl (1660–1734) developed Becher's system in chemistry so that he deserved to be called its second founder. Stephan Franciscus Geoffroy (1672–1731) was the first to systematize the theory of chemical attractions and subsequently made tables of them. Thorbern Bergmann (1734–1784) then expanded this theory considerably, and even in our time Richard Kirwan, Guyton de Morveau, and C. L. Berthollet have brought this theory closer to other natural laws. Meanwhile, a number of excellent chemists investigated the constituents and mutual relationships of the most extraordinary bodies and thereby discovered new substances which had not been known to previous centuries. Thus the investigation of the constituents of bodies was transformed into a coherent art that would always give the same results in the hands of different masters. It would be too much to mention all those who have acquired great fame in this respect. Among the best, Bergmann, Karl Wilhelm Scheele (1742–1786), and Vauqvelin and Klaproth, who are still alive, deserve to be mentioned. In spite of all these efforts, this art still seems to have a long way to go before it attains the perfection of mechanics. A more profound knowledge of gases, in particular, is among the great conquests of empiricism in this century. Just as the seventeenth century made us familiar with the mechanical properties of air, so the eighteenth has informed us of its chemical properties. Stephan Hales (whose principal work was published in 1727) and Joseph Priestley (1733–1804) deserve great credit for this. Antoine Laurent Lavoisier (1743–1794) carried this investigation to even greater perfection by means of more perfect instruments and, primarily on the basis of the discoveries of various gases, constructed a new system which rightly bears the name of *pneumatics*.

Yet another important discovery approached its maturity during this period, viz., the theory of electricity. Already in the previous century, Gilbert had begun the investigation of this, and Otto Guerike had devised the first electrical machine, but far greater discoveries were reserved for this age. Stephan Grey (at the beginning of the 18th century) discovered the different conductivity of bodies. Du Fay (1698–1739) was the first to demonstrate that there were two different kinds of electricity. Even more important was the *discovery of the electrical charge* (1745), which was later refined so excellently by Benjamin Franklin (1706–1790), the inventor of the lightning rod, who proved that the two different electrical forces must be regarded as opposites. The theory of electricity received an extraordinary addition when Aloysius Galvani (1791) discovered the action of metals in contact with the animal body, and when Alexander Volta (b. 1737), who, among many other important discoveries, has taught us to use his condenser to measure electrical effects whose weakness made them otherwise imperceptible, demonstrated that bodies can become electrical by mere contact. His discovery of how to intensify this effect by means of the alternate connection of several bodies in contact has made the transition from the eighteenth to the nineteenth centuries a turning point in the development of science. Johan Wilhelm Ritter (1775–1810) may well have concluded on the basis of Galvani's discoveries that the same forces which generate electricity also produce chemical effects, but Volta's final discovery threw far more light on this truth. Ritter used this with rare spirit and power to show how the same natural forces manifest themselves in chemical, electrical, and magnetic effects, in light, in heat, indeed, even in the manifestations of life in organic bodies. He conducted his experiments, for the most part, with great diligence and often with quite scanty means. Those on the effect of electrical forces on the animal body he conducted in part on himself with a courage and a devotion which will not be easy to imitate. The experiments he conducted on light in connection with Herschel's discoveries are undoubtedly the most important since Newton's. Jacob Berzelius and Humphry Davy, our contemporaries, have with their work also contributed greatly to the confirmation of the connection between chemical and electrical activity, the former by showing how salts can be liberated by electricity, the latter by developing this even further and also by discovering that alkalis and earths are calcined metals, to which discovery the former has made excellent contributions in his turn.

The progress of philosophy in the eighteenth century has not been without influence on general physics. The perspicacity of Immanuel Kant liberated it from the atomistic system, which, though of speculative origin, was made the basis of experimental physics. F. W. J. Schelling created a new natural philosophy, the study of which must be important to the empirical student of nature and must both inspire many new ideas in him and also prompt him to re-examination of much that was previously considered unquestionable. Henrik Steffens has also contributed much to the perfection of natural philosophy, mainly through his investigations of the evolutionary laws of the earth, which are so closely related to chemical physics (§20).

V. On Studying Physics

§22

On the basis of the above, it is now easy to see what prior knowledge is necessary in order to study physics successfully. Besides the preparation which is required for all study, knowledge of mathematics is particularly necessary for him who wants to make progress in general physics, especially in all those parts which deal with motion. At the beginning, elementary mathematics, in which plane trigonometry must not be forgotten, may be sufficient, and beautiful progress can be made with that, but if the aim is to penetrate fairly deeply, it is also necessary to become familiar with more advanced mathematics. He who does not feel great ease in the manipulation of mathematical formulas will do well to begin with the study of the first principles of differential and integral calculus in a short textbook and then try the application of these to physics. Thus he will acquire a skill in their use and a familiarity with them which will cause him to proceed with enthusiasm. We have already seen (§20) that the scientific and detailed description of nature is not a necessary preparation for one who wants to study physics, but some knowledge of it is essential in order to acquire a more complete insight into our science. It cannot be sufficiently recommended to anyone who really wants to delve into general physics to turn his attention to all the other parts of natural science. This will greatly contribute to the expansion of his vision and prevent him from mistaking a small number of experiences for the whole of nature. In order to become more familiar with effects on a large scale and to observe the consequences which even small changes can have on the final result, it is important to become familiar with the application of science, which also has a useful reciprocal effect on its own practice in several respects. To acquire thorough knowledge of the history of science, to see the conceptions of different ages, and to attempt the explanation of some natural phenomenon in such a foreign spirit personally are excellent ways of fending off any tendency towards prejudice and of seeing the true relationship between science and nature. In order to form the broad and firm view which is so essential in every science, a free and independent study of philosophy must also be added, a study which should not be separated from that of any other science, the necessity of which cannot be mentioned too often.

§23

However, following all these precepts will still not provide an appropriate familiarity with nature if one does not add one's own observation to what has been learned from others. This must be as complete as possible. It is not enough to have seen the most remarkable effects once; one must have seen them often. And if one really wishes to make progress, it is even necessary to acquire a skill in producing natural effects at will by conducting experiments. Costly instruments and great preparations are not always needed for this, for the most insignificant things are often suited to present the most important laws of nature. It all depends on their correct application. It is best to begin by repeating well-known experiments whose results are considered unquestionable, and when some practice and proficiency have been obtained in this way, it is possible to proceed to the ones whose meaning is still in dispute, from which the final transition to free and independent experimentation is easy and natural. However, it stands to reason that it is also possible to go directly from basic exercises to completely self-devised experiments. The characteristics of the individual must decide the issue here. Strict rules cannot be given. As physics is both art and science, the best way to acquire the basic education is through the guidance of an experienced master, but in every case only personal diligence and practice will lead to greater perfection. Constant awareness of all the natural phenomena which take place around him, not only in the spontaneous acts of nature but also in art and in daily life cannot be recommended too strongly to the student of physics. He must try to explain to himself everything unusual that he encounters there, either by tracing it back to well-known natural laws or by discovering a new connection between that and the rest of nature, and those manifestations which he regards as being less in agreement with nature, he must try to bring into closer agreement with it. When he lives thus in harmony with nature and in a way makes it the centre of all his observations, he cannot fail to become its confidant and a naturalist in spirit and in truth.

§24

When the student has made himself familiar with the principles of our science from a textbook, he should turn his particular attention to such textbooks in which many natural phenomena are treated at some length and, either because of their remarkable connection with the rest of nature or because of their special relation to us, are brought closer to life. From there he proceeds to more detailed writings, in which some great part of science is presented fully and thoroughly. When relatively extensive knowledge has thus been acquired, he also begins to read papers on more limited subjects. The transactions of learned societies and scientific magazines, which are quite different from those, offer a sufficient supply of these. The study of many, especially less detailed, textbooks is not to be recommended until so much knowledge of nature has been acquired that an independent contribution can be made to the investigation of different opinions.

In order to familiarize oneself with the writings on physics, one can use: Jul. Bernh. v. Rohr's *Physikalische Bibliothek, mit Zusätzen und Verbesserungen*, ed. by Abr. Gotthilf Kästner, Leipzig 1754, 8°, which contains much useful information on the character and the mutual relation of the books, and also Hermanni Boerhave's *Methodus studii medici emaculata & accessionibus locupletata ab Alb. Haller*, Amsterdam 1751, 4°, Vols. 1–2. In the famous Chr. Wolf's *Elementa matheseos*, Halle 1742, 5 Vols., 4°, and in *Anfangsgründe der Mathem.* 4 Vols., 8°, by the same author, there is information about mathematical works, many of them physico-mathematical. A useful supplement to this is Joh. Georg Büsch's *Encyclopædie d. mathem. Wiss.*, second revised edition with the addition of a mathematical bibliography, Hamburg 1795, 8°. As far as the chemical part is concerned, *Repertorium d. chem. Litteratur, von 494 vor Christ. Geb. bis 1806, in chronologischer Ordnung, von den Verfassern der systemat. Beschreibung aller Gesundbrunnen und Bädern in Europa*, Vol. 1, Part 1, Jena and Leipzig 1806, Vol. 1, Part 2, 1808, is rather complete but covers only up to 1782. Information on recent literature can be found in *Allg. Repertorium der Litteratur* for 1785–1790, Weimar 1793–1794, for 1790–1795, Weimar 1799. Part 10 contains physical natural history; it is very complete for its period and has indicated the judgment pronounced on the works by means of symbols. A catalogue of all treatises which come under natural science and its application, found scattered in the transactions of learned societies, is given in J. D. Reuss's *Repertorium commentationum a societatibus literariis editarum*, Göttingen 1801–1808, 4°, Vols. 1–2, of which the third, fourth, and seventh parts, which contain physical, chemical, and mathematical writings, are most important to us. Each part can be purchased separately. Very complete references are found in I. C. P. Erxleben's *Anfangsgründe der Naturlehre*, 6th edition, with improvements and many additions by G. C. Lichtenberg, Gött. 1794, 8°, of which we also have a Danish translation by our Professor Olufsen.

We only have two very incomplete works on the history of general physics: De Loy's *Abrégé chronologique pour servir à l'hist. de phys.*, 4 Vols., Strassbourg 1786–1789, 8°, and Joh. Carl Fischer's *Geschichte d. Phys., von der Wiederherstellung d. Künsten u. Wiss., bis auf d. neusten Zeiten.* 8 parts, Gött. 1801–1808, 8°. Therefore, it is necessary to resort to Montucla's *Histoire des mathématiques*, Paris years 7–10, 4 Vols., 4°. I. C. Wiegleb's *Geschichte des Wachsthum u. der Erfind. in der Chem. in der neuern Zeit*, Berlin 1790–1791, 2 Vols., 8°, and *Histor. crit. Untersuchung d. Alchemie*, Weimar 1793, 8°, by the same author. Gmelin's *Gesch. d. Chemie*, 3 Vols., 8°, Gött. 1797–1799, is really nothing but a dry list of names, and the one that Trommsdorff has given in his *Taschenb. f. Aertzte, Chemiker, etc.* is the opposite, that is, empty declamation.

The following systematic textbooks are especially recommended for beginners: Michael Hube's *Faszlicher Unterricht in der Naturlehre, in einer Reihe von Briefen, etc.*, second edition, Leipzig 1801–1802, 4 Vols., 8°. J. B. Haüy's *Traité élémentaire de physique*, second edition, Paris 1806, 2 Vols., 8°, and Tiberio Cavallo's *Ausführliches Handbuch der Experimentalnaturlehre, in ihren reinen und angewandten Theilen.* Translated from English and annotated by D. Joh. Bart. Trommsdorff. Erfurt 1804–1809, 4 Vols., 8°. The first is very popular, the second more scientific but does not presuppose much knowledge, and the last, with all its comprehensibility, contains many quite profound sentences and also distinguishes itself by its practical spirit. The translation is not good, but the original is rare in this country.

In order to proceed to mathematical topics, the following deserve recommendation: 's Gravesande's *Physices elementa mathematica*, second edition, Leyden 1742, 2 Vols., 4°, which also contains accurate descriptions of experiments, and Muschenbroek's *Introductio ad philosophiam naturalem*, Leyden 1762, 2 Vols., 4°, which is a masterpiece for its time. For those who want to understand completely the mathematical theories which are connected with physics, we have an excellent work in our own language in Jens Kraft's *Forelæsninger over Mechanik*, Sorøe 1763 and 1764, 2 Vols., 4°.

In the chemical branch we recommend Lavoisier's *Traité élémentaire de chymie, présenté dans un ordre nouveau et d'après les découvertes modernes*, Paris 1789, 2 Vols., 8°, which deserves to be studied both as a major work of science, as a comprehensible work, and as instruction in how to conduct experiments. Concerning recent discoveries, one may consult Gren's *Grundrisz d. Chemie*, edited by von Bochholz, Erfurt 1810, 2 Vols., 8°.

As regards electricity, galvanism, and magnetism, the cited works by Haüy, Hube, Cavallo, Muschenbroek, Sigaud de la Fond, and Nollet may be consulted. In our own language, we have I. Saxtorph's *Kort Vejledning til Kundskab om Electriciteten*, Copenhagen 1807, in which the essential elements of the theory are presented, and F. Saxtorph's *Electricitetslære*, Copenhagen 1802 and 1803, 2 Vols., 8°, which gives many details concerning experiments.

If one wants to expand one's knowledge of individual topics which a textbook can deal with only briefly, one may use Joh. Samuel Traugott Gehler's *Physicalisches Wörterbuch*, Leipzig 1798–1801, 6 Vols., 8°, which is very well done, but which is becoming more inadequate every day because of the many new discoveries. More recent but less well done is Fischer's *Phys. Wörterb.*, Gött. 1798–1806, 6 Vols., 8°. For chemical topics one may use Klaproth's and Wolff's *Chemisches Wörterbuch*, Berlin 1807–1810, 5 Vols., 8°.

Concerning the description of the necessary instruments and experiments, we recommend, in addition to those which have already been mentioned in this respect, Nollet's *Leçons de physique*, Paris 1783–1786, 6 Vols., 8°, and *L'Art des expériences*, Paris 1770, 3 Vols., 8°. Also Sigaud de la Fond's *Éléments de physique, théorique et expérimentale*, second edition, revised and expanded by Rouland, Paris 1787, 4 Vols., 8°, and, connected with this, *Description et usage d'un cabinet de physique expérimentale*, Paris 1775, 2 Vols., 8°, by the same author. A similar work made complete with the discoveries of the last 36 years would be a true service to science. We may expect something like this from our renowned Lord Steward Hauch.

In order to keep pace with the times, one must make use of Gilbert's *Annalen der Physik*. That is a continuation of Gren's *Journal der Physik* and *Neues Journal der Physik*. It started in its present form in 1799 and now comprises 38 volumes, of which each group of 6 is provided with its own index. Besides this, one may use *Journal für Chemie und Physik*, edited by J. S. C. Schweigger, in collaboration with J. J. Bernhardi, C. F. Buchholz, C. v. Crell, A. F. Gehlen, S. F. Hermbstädt, F. Hildebrandt, M. H. Klaproth, H. C. Ørsted, C. H. Pfaff, T. J. Seebeck, and C. S. Weisz. This journal is a continuation of Scherer's *Allgemeines Journal der Chemie*, with which Crell's *Chemische Annalen* was later combined. Its main contents are still chemical.

He who wants to master the most difficult investigations of science and study the writings of Newton, Euler, the Bernoullis, d'Alembert, la Grange, and Laplace or wants to familiarize himself with the complete range of chemical, electrical, magnetic, and other experiences must seek advice in bibliographical writings.

30

View of the Chemical Laws of Nature Obtained Through Recent Discoveries[1]

INTRODUCTION

In the investigations of chemical effects so far, we have always stopped at the so-called affinities or attractions as the final limit which we could reach. This could not possibly mean that the ultimate reason for all chemical effects had been determined, but the unknown cause had to be given a name, and at the beginning the more figurative term, *affinity*, was chosen, and later the more characteristic term, *attraction*. The evolution of science, however, almost always caused attention to be directed towards the particular nature of each affinity. What has been accomplished in this respect in chemistry, thanks to the efforts of so many outstanding men, will remain an everlasting monument to the past century. The investigations, however, did not yet seem to be ready for research into the general nature of chemical affinities. Even though it produces the greatest and most wonderful phenomena before our eyes, it is also the nature of chemical action that sometimes invisible substances are transformed into solid bodies, and sometimes the hardest bodies again dissolve into liquids and even disperse into vapour and air, but the forces from which all these effects spring are concealed from our eyes. As a consequence, chemistry is still far from achieving the completeness of mechanical science and its ability to derive all its principles from a small number of interrelated, fundamental ones. Rather, it has been forced to discover almost every single theorem, every single law, by means of specific experiments. In many respects, all these special laws revealed so little internal coherence that anyone who was not already convinced of the necessity and the unity in all the works of nature could easily come to suspect that the regularity which was believed to be clearly visible at some points might be only accidental, and that, in general, the opposite of what was happening was equally possible. It would be appropriate to compare the previous state of

---

[1] [Originally published in German (Berlin 1812, at the High School Book Shop). An excerpt of this work (the section on the production of heat), with a short introduction, is to be found in *Journal für Chemie und Physik*, ed. by J. S. C. Schweigger, Vol. 5, pp. 398–440. Nuremberg 1812. KM II, pp. 35–169.]

chemistry with that in which mechanics stood before Galiläi,[2] Descartes, Huygens, and Newton taught us how to reduce the most complex motions to their most elementary components. Although many important facts, as well as series of facts, were known there, the grand unifying principle, from which everything was to acquire its intrinsic value, order, and strength, was still missing. For instance, circular motion was known empirically; the observation of the starry sky had also revealed a regularity in the large orbital motions of the planets. However, as long as they had not been reduced to linear motions, which may not be apparent but are in reality their essence, it was not possible to discover the unity of different motions. It seemed necessary not only to consider both linear and curved motion completely separately, but in the same way every kind of non-uniform motion had to be distinguished from uniform motion, and each kind of curved motion from every other, so that their connection could not be known but, at most, suspected. Until our century, the affinity of alkalis and acids, of earths, of metallic calces and the substances themselves, of combustible bodies and oxygen, etc., was likewise known in chemistry, without any attempt to deduce one from the other.

Through the reduction of all kinds of motion to their fundamental laws, mechanics rose to such perfection that it embraced all motion in the universe and regarded it as one great mechanical problem, and therefore it could calculate countless problems without waiting for empirical evidence. Through the reduction of all chemical effects to the primeval forces from which they arise, we must seek to prepare chemistry for a similar perfection. We shall then be able to deduce all chemical properties from the primeval forces and their laws, and as any substance, directly or indirectly, is recognized and distinguished from others by these properties alone, we shall be able to calculate all possible substances from them just as all possible motions can be calculated from the mechanical laws. Chemistry will then become a theory of force to which mathematics must be intimately conjoined in order to determine the proportions, the directions, and the modes of action of these forces, and a new branch of mathematics will most likely evolve just as the calculus of fluxions once arose because of the theory of motion. The most recent progress in our science has brought us so much closer to the realization of these great expectations that I think the time has come to collect all the scattered material in order to establish a new theoretical system. It is in the nature of such an initial attempt that it can only be incomplete, but the first step must be taken some time, and I think it not unimportant that it happens now. It can be of little benefit to science to construct a system from a few isolated observations, but it would be equally detrimental to let a great body of experience lie unused, for a student of nature will make great and far-reaching discoveries only if he has the examination of an idea before him. Therefore, the purpose of this work is merely to provide complete preparation and to raise important questions. Only through the combined efforts of many thinkers and after the course of several generations, will the science of chemistry significantly approach the perfection which many would find too bold to anticipate today.

[2][Galilei.]

As a first step, it will not be without benefit to summarize the entire course which is to be covered. We wish to begin our investigation with a classification and arrangement of all bodies according to their chemical nature. After that, we shall reflect on the ordinary and familiar chemical effects and thereby demonstrate that all chemical transformations investigated so far can be traced back to two ubiquitous natural forces. In addition, we shall show that these forces can act not merely in direct but also in indirect contact, and that, consequently, they can be controlled. This will lead us to the galvano-chemical effect, which has been known to us in galvanism for some time. And finally, this will bring us to the point where we can describe the chemical forces in their free state and, at the same time, make their identity with the electrical forces apparent. At this point, we shall turn back in our investigation and, by directing our attention to the electrical forces themselves, seek to discover how they, too, can be rendered chemically active. We shall first point out that there are only two electrical, just as there are two chemical, forces, which are similarly opposed, and that both are found everywhere and can be aroused by external forces from the relative calm in which they are found in bodies and be made active. To begin with, we shall observe these electrical forces in their freest state, but then we shall see, in the effects of the condenser, the charge, and in the electrophorus, how they become increasingly confined, trapped, and forced into greater coherence by their mutual attraction. Finally, we shall come to the point where we can show that electrical forces can be brought to such a degree of coherence that they produce permanent internal changes, and therefore chemical reactions. Once we have thus presented the broad connection between chemical and electrical effects and the identity of the forces which produce them in these two opposite directions, we shall, on the basis of an investigation of the nature of conductivity, attempt to show the conditions under which both forces produce heat, and those under which they produce light. Hereby, we shall see these great phenomena in a far closer relationship with the rest of nature than was possible according to the usual view. All this will pave the way for a more complete understanding of the nature of fire, whose manifold phenomena have so far been interpreted too one-sidedly. Finally, a look at magnetism and at sensory perceptions will confirm the universality of the chemical forces and their identity with the space-filling forces.

We are not without predecessors on the course which we are to follow. Ever since the theory of electricity was given a formal framework, we have found traces of similar ideas. Priestley already assumed that the substance which to him, as to Franklin, was the cause of all electrical effects was identical with the cause of combustibility, phlogiston. Wilke believed in the electrical forces as the cause of fire as well as of acidity. Kratzenstein assumed an acidic and a phlogistic force in electricity. Henry, Karsten, Forster, and Gren had ideas which approach these. Perhaps Lichtenberg, with his singular intellect, went further than all his predecessors, but he expressed himself only in sceptical questions. Hube states explicitly that, in his opinion, chemical affinities are based on electrical conditions. However, as these otherwise outstanding men, misled in part by the assumption of a characteristic electrical substance, regarded the particular mode of action which we call electricity as the basis for all other phenomena, instead of assuming it to be one of the various realizations of the universal natural forces, they limited their horizon and gave the entire grand theory, which should have originated from this, the appearance of a narrow hypothesis. On the other hand, it must be admitted that it was scarcely possible to develop this view more completely before our knowledge of electricity and several related effects had progressed to greater maturity so that this advance was naturally reserved for more recent times. Ritter can therefore be regarded as the creator of modern chemistry. His comprehensive ideas and his achievements, undertaken with such great vigour and exertion, spread a great light in all directions. To a certain extent, Winter[3] deserves to be placed next to him. His ideas on alkalinity and acidity, as well as on heat, are of great importance and have been confirmed many times by recent discoveries. It is not to be denied that the great minds of these men often led them too far into the realm of pure speculation. We shall try to avoid this by striving for clearer and more confirmed knowledge, but we must also content ourselves with the construction of a far more restricted theory than the one which they had in mind. Recent memory will make it easy for anyone to recall how much the profound investigations of Berthollet and the great discoveries of Davy, Berzelius, and some other experimenters have contributed to guide, to extend, and to confirm the views which we shall present here.

However, we would go too far if we wished to name everyone who helped prepare the way which lies before us. We gratefully acknowledge the merits of the natural philosophers of our time, also with regard to their views on chemistry. We also willingly acknowledge that we have received several felicitous hints from older philosophers, from mathematicians, and from individual experimenters. When using their work, we shall not neglect to acknowledge and honour their merits where we think they might be forgotten.

HOW INORGANIC BODIES ARE TO BE CLASSIFIED ACCORDING TO THEIR CHEMICAL NATURE

Because all chemical investigations proceed from the properties of substances, it will not be inappropriate to arrange them first in the most obvious and best order which we can discover. To be sure, attempts have always been made in chemistry to classify bodies according to certain characteristics and to give definitions of the classes thus formed, and due to this, an even more perfect arrangement has been prepared in a most fortunate manner, but this is not sufficient for the present state of the science. Those concepts whereby everything was to be ordered began to develop at a time when the bulk of chemical knowledge was still small, and when it had to be regarded as good fortune if it was possible to arrange subjects in individual groups. As long as these remained as isolated as they were initially, the concepts could easily be connected to the subjects in a precise manner, and therefore

---

[3] [Winterl.]

the limits of the concepts could appear natural. However, as the series of discoveries has gradually filled so many gaps between these groups, the limits formerly determined had to be, in part, extended and, in part, even eliminated. Thanks to their venerability and to the important service that they have rendered to science, however, some have retained a kind of permanence which must vanish with more careful examination. This will not prevent us from taking an unbiased look at the large series of bodies spread out before us and from seeking their similar nature without prejudice. We shall also subject these old concepts to examination wherever we think it necessary.

Less than one hundred years ago, in the days of Stahl and Boerhave, only 6 true or perfect metals were recognized. The other metallic bodies which were then known were excluded because of their brittleness until the discovery of several new ones gradually convinced them that the enormous gap between the elasticity of gold and the brittleness of arsenic is filled with a great many intermediate links and must be filled even more in the future. Likewise, they wanted to deny mercury its place among the perfect metals because of its usual liquidity until it was finally discovered that it could also become solid, and that its freezing point is far closer to that of tin or lead than the freezing point of these metals is to that of iron or even that of copper. A significant volatility also had to serve as a criterion for the exclusion of bodies from the ranks of the perfect metals. Now no limit can be set here either. To former generations the resistance of gold to heat may have appeared to be absolute; since the experiments with the burning-mirror and with electricity, we know that it resists heat as little as any other body. From this extremely low volatility to the rather significant one of arsenic, so many variations exist that no line can be drawn. To draw one here would represent the most impermissible arbitrariness even if we did not already have metallic ammonia as an example of a far greater volatility, perhaps even of an aëriform metal if it could be produced in isolation. However, even given that such production is impossible, it is certainly easy to comprehend that a gaseous metal is not self-contradictory, indeed, it is not even improbable. From the temperature at which gold evaporates to that at which this occurs for arsenic or mercury, there exists a distance of several hundred degrees of the thermometer or several units of our thermical measure.[4] However, in the event that the lowest temperature of the earth were 5 such thermical units above the present one, both of the metals last mentioned would be constantly aëriform for us while we would not yet notice any significant change in the volatility of gold due to the atmospheric heat.

[4]I find it far more convenient to indicate high and even undetermined temperatures in integer thermical units than in some arbitrary graduation. In addition, the difference between the freezing and boiling points of water is actually the unit of our thermical measure, which could simply be called one *thermical unit (mètre thermique)*. Consequently, all other degrees would have to be expressed either as fractions or as multiples of this; in this way all the different thermometeric scales would cease to exist. Everyone could measure with his thermometer, but 50° C, 40° R, 112° F would all be called $^1/_2$ or 0.5 thermical unit above the freezing point, or, even shorter, simply $+^1/_2$ or +0.5 Th. (since the freezing point could always be regarded as the starting point). Generally, these degrees are taken to mean longitudinal intervals of the measure, but the scale could also be divided according to intensity in the event that the appropriate determinations were available.

So far the large specific weight of metals has also been regarded as a very important feature, despite its variation over a very wide range between platina and tellurium. Discoveries in recent years have shown us metals which are lighter than water, and the gap between the lightest of those that have been produced in isolation, potassium, and the lightest known earlier has already been so well filled with other new metals, calcium and barium, that there is no reason to wish to separate these newly discovered metals from the others by a special name.

A more important characteristic for metals is opacity. However, it is not absolute as we see with thin sheets of gold, so it is also possible that it could be even lower, and who is to say what is the lowest opacity or the greatest transparency that a metal can have? The same applies to lustre. This is a consequence of the effect which bodies have on light through their combustibility and density. Because the density of metals decreases almost to the same degree as their combustibility increases, we have reason to expect that dissimilarities between these characteristics lie within rather narrow limits, but since this relation is not rigorous, we would not dare to exclude a body from the series of metals because of its smaller effect on light.

Even more than this, high electrical conductivity distinguishes metals from other bodies, but it is still undeniable that this ability is found in very different magnitudes in the most unquestioned metals, e.g., many times larger in copper than in platina. Therefore, the smallest degree of conductivity for a metal cannot be determined, and we should not regard it as impossible that poor conductors could be found among metals as well. If we further assume that aëriform metals can also exist, there must necessarily be an indicator which reveals metallic quality even in this state, but this can be neither opacity, nor lustre, nor conductivity. Hence, without wishing to deny the importance and significance of these properties for the recognition of metallic quality, we must nevertheless confess that they are not sufficient for the definition of a metal. The same can be said about the thermal conductivity of metals.

Let us, therefore, ask seriously: How do we recognize that a body belongs to the metals? The completely honest answer is: By its similarity to other metals. At least, this has been the practice so far because, if our starting point had been a fixed concept, it would have had to remain that way without all the extensions which have gradually been accepted. After all, it is the true method of natural science gradually to arrange similar phenomena next to each other without regard for the limits imposed by such concepts or, ultimately, to overcome them even if its progress is interrupted now and then. The concepts are only a means to understanding, not natural boundaries. If we assumed anything else, we would first have to maintain that nature represents certain concepts, and, secondly, that we have identified these with certainty. If we cannot do this, we must continue with our comparisons and compilations. If continued investigation requires otherwise, the limits of yesterday will not be those of today. Therefore, we do not proceed from concepts but from specifics. Next to the first we place a second, which has the greatest similarity with it, next to this a third, which is most similar to the second but somewhat more different from the first, etc., as long as we discover grounds to continue. By doing this, we must certainly end the series with substances which are most dissimilar from the first, even opposite to them, but we must not shrink from this as we wish to describe a relation and not a concept.

Among the metals already recognized, we must certainly place arsenic far from gold. The latter is malleable and can be stretched and hammered into finer pieces than any other inorganic body; the former, on the other hand, is so brittle that it can easily be pulverized. The latter is so heat-resistant that it will not evaporate in the fire of any furnace; the former is so volatile that it becomes perfectly vaporous even at 2.82 thermical units above the freezing point. The latter is so difficult to burn that we would readily consider it to be incombustible were it not for oxymuriatic acid and its compounds; the former is so flammable that it ignites with a moderate temperature increase; not to mention so many other dissimilarities. If we now make a comparison of arsenic with phosphorus or sulphur, we certainly find the dissimilarity here to be many times smaller. It is true that the last two are transparent and are poor conductors of electricity, which arsenic is not, but how many other similarities do they not have? All three are volatile, and they have a characteristic odour and a very penetrating effect on the animal body. All three can be combined with hydrogen and thus form gases. Their oxidation states are the same, at the lower level a volatile combination of an acidic nature and, at the highest, a very fire-resistant acid, among which that of sulphur is the least. Caustic alkalis cause all three to decompose water and to form rather similar products. In addition, all three combine with other metals, something which true non-metallic bodies that oxidize do not. We shall investigate the importance of this characteristic below; here we shall content ourselves with mentioning that so many highly malleable bodies become exceptionally brittle in combination with arsenic as well as with even the smallest quantities of sulphur and phosphorus. However, it is true that arsenic also forms some malleable compounds, in which it differs from the other two. Boracium and fluorine have so many analogies with sulphur and phosphorus, especially with the former, that there is no need to add them here.

Finally, if we compare the metals with regard to their combustibility, we find yet another example of the most varied gradations of one property in the same series. Let us first establish a really clear notion of the combustibility of the ammonium metal. In its oxide we find hydrogen, which was already known to us because of its high combustibility, but in the metal itself there must be something far more combustible than hydrogen; otherwise its amalgam could not possibly effect such a strong decomposition of water. We can only speculate as to what this combustible substance might be. If we agreed with several eminent chemists in assuming it to be deoxidized hydrogen, then what we call hydrogen would be an oxide which differed from other oxides in two very remarkable characteristics: the first, that it combines with metals, something an oxide usually does not do; the second, that it does not form a compound with any other oxide in its own right, a behaviour which is completely contrary to that of all other oxides. It could indeed be claimed that hydrogen has the opportunity to deoxidize in all the cases where it combines with metals so that the hydrides would be compounds of deoxidized hydrogen with metals, but, in addition to the fact that this solves only one difficulty, it is quite certain that mercury hydride is very different from the ammonium amalgam. We do not want to deny, however, that there may exist higher combinations of oxygen with combustible bodies which do not have the nature of oxides, but if hydrogen were such a compound, we would have no reason to separate it from the metals or to group it with the oxides. We have, however, several reasons to add hydrogen to the metals. We have already mentioned the first, the compatibility with metals and the non-compatibility with oxides. This is also the direct reason why we rank ammonium among the metals. In the solid state, the compounds of hydrogen with metals are also good electrical conductors, like the ammonia amalgam and all the compounds of metals which are themselves conductors. However, the similarity of its oxide with that of potassium, sodium, etc., may also have contributed significantly to the acceptance of ammonium as a metal. At first sight, it might appear as if hydrogen did not have such a great similarity with metals, but we want to take a closer look at this. Since it has been demonstrated by adequate experiments that lime, baryta, siliceous earth, etc., are true metallic oxides, no chemist can any longer refuse to consider the other earths to be so as well. We claim, however, that hydrogen oxide, water, has great chemical similarity to several of these, to silica and glucina, but especially to alumina. In order to see this quite clearly at once, we imagine our planet transported to a temperature 200^5 thermical units lower than the present. Then, water would not only be permanently solid but also appear to be almost infusible, in large fragments as solid crystals, powdered, like a white earth. We would find these to be equally soluble in acids and alkalis, approximately like alumina, and neither oxide would cause any great neutralization of either acids or alkalis. We could melt both by adding alkalis and acids (provided that nothing other than the fusibility would be altered at such a low temperature) and would then group them quite close to each other because of the similarity of their behaviour. Further, if it were possible to reduce both oxides through heating with iron or something similar, no-one would doubt that the combustible element in ice, like the one in alumina, is a metal. This would be all the less likely to occur here as both would certainly be found as solids. If we return to our temperature, however, ice is liquid, alumina is solid and, by itself, infusible; the ice metal will become a gas, but the alumina metal will retain its solidity, and now we would no longer acknowledge the affinity of these substances! —Whatever may be incomplete in this thought experiment, it will at least be obvious that the difference of hydrogen and its oxide from the other metals and oxides can only be a very relative one.

Since we have now proved the metallic nature of almost all undecomposed substances, it will not be inappropriate to investigate the few remaining ones in this respect as well. Among these, carbon is the most familiar. Its very close affinity with metals has already been recognized by many. Its ability to form a perfect metallic compound with iron, as we recognize it in steel, its highly metal-like state in graphite, and among other things its conductivity, as well as that of annealed charcoal in galvanical experiments, entitle it to this position. It can no longer surprise us that its oxide is sufficiently volatile to appear as a gas at normal temperature because we already know a gaseous oxide in the case of ammonia.

[5][2]

As nitrogen does not form direct combinations with any body other than hydrogen or oxygen, we could hardly assign it a place here even if its elementary state were not also so problematic.

Before we proceed to the consideration of oxygen, we wish first to cast a backward glance at the entire course which we have traversed. We find that all the bodies which we have listed here have in common that they are not decomposed by the strongest deoxidizing agent available to us, that each of them forms compounds with other bodies which are similarly undecomposed or uncombusted, and that only a few bodies at the extremes also combine with various combusted bodies. In addition, we find great chemical activity spread throughout the entire series of these bodies, though in the most different degrees; in all of them we find the ability to enter into combustion as an active agent. From ammonium, or whatever the most combustible metal may be, to gold, the metals form a quite numerous sequence whose members are already sufficiently well known, and which our objective here does not require us to arrange further. Here it will suffice to observe that combustibility, which has already fallen so low in the case of gold, is even lower for platina, but lower still for iridium and osmium in that these scarcely undergo oxidation by means of acid, rather by glowing with alkalis (perhaps only with hyperoxidized ones). However, once we have come so far in the decreasing order of the combustibility of metals that this property has almost vanished, nothing prevents us from proceeding to absolute incombustibility. This, however, can be found only in a body which in itself would be the necessary condition without which no body could be burnt; for any other body will either be burnt or have some relation to the substance which provides the external condition for combustion. In the series of bodies about which we have more detailed knowledge, oxygen is such an incombustible one. Even if we should find that oxygen itself could be burnt according to some even more remote principle, it would remain certain that only the body which represents the ultimate external condition for the burning of combustible bodies constitutes the principle of incombustibility. Thus, it can be seen clearly that oxygen is required to complete the series of metals, either as a weakly combustible substance or as incombustibility itself. That it is gaseous should no longer give offense. It serves as confirmation that it combines with metals but not with burnt bodies as such; if this seems to happen, it really only combines further with the combustible bodies in them in order to reach a higher degree of saturation. It is also quite remarkable that there are, among the metallic oxides which are combined with large quantities of oxygen, many good conductors as if they had a metallic character in this respect since the other oxides are poor conductors in comparison. However, it would appear that we could raise serious doubt from this about the metallic character of oxygen because it forms a new series of combinations with metals. Our answer to this is that this occurs according to a general rule which states that compounds of homogeneous substances remain in the series, but those of heterogenous substances form a new series. Thus hydrogen takes sulphur into a new series; similarly, an acid takes an alkali even though, as will be shown, they belong to the same series. However, two acids or two alkalis mixed with each other do not leave the series. The following will provide some further clarification of this subject.

The question now arises: Since there are so many combustible bodies, or such that attract oxygen strongly, should there not also be others of the opposite nature which would have a similarity to oxygen in their ability to attract combustible bodies and to combine with them directly? As we know, a combustible body is required for every combustion and another which is attracted by it and, in its turn, attracts it. We could call this quality, which is the opposite of *combustibility*, the *empyreal*[6] or, even better, the *igniparous*[7] quality. Both are equally necessary for combustion. It is somewhat arbitrary that the bodies which we call combustible have been given this name. This is due to the fact that everyday life has familiarized us with them, including bodies which also have some of the empyreal quality, but only in the context of combustibility because we usually see them only in interaction with air and heat but not with the most combustible bodies. In addition, the bodies which are empyreal in all ordinary cases act only as gases, and therefore their chemical combination with the combusting bodies remained unrecognized for a long time. If the atmosphere consisted of hydrogen gas instead of oxygen, bodies rich in oxygen would have been called combustible, and oxygen gas itself would have been regarded as combustible. If everything else remained unchanged, we would observe only that nearly all combustible bodies would undergo a decomposition during combustion. It would be interesting to consider how things would take place on a planet where the substances which we here call empyreal served as the basis of everything, just as those which we have called combustible do on our earth, and where, according to our use of language, the atmosphere would be of a combustible nature instead of the empyreal nature of our own. It would indeed be possible to imagine a complete organization there which would be completely contrary to ours. One has dared to speculate so much about the nature of comets; why, then, should we be afraid at least to ask whether they might not be such planets?[8] — However, be all this as it may, through these considerations we have, if we may express ourselves so, taken an impartial look at both classes of bodies which are involved in combustion, and we can now speak more easily without confusion and misunderstandings because we have clearly seen and acknowledged that the terms combustible and empyreal substances could, in principle, be interchanged at will but are best accepted for our immediate sphere of experience as they are now commonly used.

As, in the course of the preceding considerations, we have become familiar in particular with the decrease of the combustive force and have seen it in so many gradations that we eventually developed the idea of absolute incombustibility, it will be most convenient to continue in this vein for the time being. In so doing,

[6][Ørsted uses the term *feuernährend*, which literally means "fire-supporting." We have adopted Davy's term "empyreal."]

[7][Ørsted uses the term *zündungsfördernd*, which literally means "ignition-promoting."]

[8]If we might assume that the entire universe were filled with a gas containing oxygen, which, by the way, we could imagine to be arbitrarily dilute, we would also be able to explain the luminosity of comets because the universal gas would have to be compressed in the vicinity of comets due to their attraction and would therefore give rise to a combustion with the atmosphere of the comet. The light thus produced within the range of the atmosphere would not permit us to see the dark part of the gas but would only faintly reveal the center. Similar notions could be entertained about the sun.

however, we must take great care not to be misled by some very deceptive phenomena. In certain easily produced circumstances, some bodies are very combustible even though they exhibit very weak combustibility under the conditions which we most frequently employ for the observation of other bodies. Therefore, we must be particularly observant that the comparison is made only under similar conditions. For example, carbon, which restores most burnt bodies to the uncombusted state at higher temperatures, should not be regarded as a particularly combustible body; for at normal temperature, which is where we must compare all bodies because that is where we know them best, it is far less oxidizable than either gold or platina since it is attacked only weakly by acids which oxidize and dissolve those two. In galvanic actions, we also find carbon more on the negative side than any of the nobler metals, which indicates, according to general experience, that it must be less combustible. Sulphur is also fairly combustible at higher temperatures, but at lower temperatures it is less combustible than any of the noble metals. However, at higher temperatures it never attains a combustibility comparable to that of carbon. Therefore, sulphur at normal temperature must be numbered among the least combustible bodies. Regarding the increase of combustibility with temperature, this is certainly not a simple phenomenon but derives from several circumstances, as Berthollet's investigations also demonstrate. The weakening of cohesion by heat and the tendency towards the aëriform state of oxides produced by combination might be the most active factors. The deoxidizing force which carbon exerts on most oxides, and sulphur on a great number of them, would thus arise primarily from the tendency of carbonic acid and sulphurous acid to become aëriform, and in the case of sulphur also from its tendency to combine with metals. Another proof of the meagre combustibility of both carbon and sulphur is that they show so little contraction when combined with oxygen, for strong and intimate combinations always do so. According to recent investigations, the fact that sulphuric acid exists in a highly condensed state must be due to its combination with a small portion of water. In the case of sulphur, there is another very peculiar circumstance, which is that it not only combines with several metals, but that it seeks this combination with great intensity and thereby generates much light and heat without any addition of oxygen. This reveals a quite remarkable attraction to combustibility, or, in other words, a fairly considerable igniparous property. The fact that sulphur produces an acid with hydrogen provides a second analogy between sulphur and oxygen. However, the circumstance that sulphur and hydrogen form a gas might seem to contradict the assumption of a fairly strong attraction between the two, but we need not assume this attraction to be greater than that of oxygen and sulphur, which also produces a gaseous compound.

We also find a predominance of the igniparous property in tellurium, for when used as the positive conductor in the galvanic effect, it does not attract oxygen, but it hydrogenizes at the negative conductor. As Davy has found, it also forms an acid with hydrogen.

As we know, Davy has recently maintained that so-called oxymuriatic acid is a simple body, which he calls chlorine, and which produces muriatic acid with hydrogen. We cannot deny that he has made it very probable that his chlorine combines with many combustible bodies without being deoxidized, and he has shown that, without the addition of water, the action of these compounds on combustible bodies is not at all like that of bodies containing oxygen. In addition, this chlorine shows no tendency to combine with alkalis under normal circumstances. If its simple nature, deduced from these observations, should be confirmed by the inspection and further development of all the facts, we would have another highly empyreal body in addition to oxygen. However, the strong analogy which chlorine has with sulphurous and nitrous acids is in conflict with this. All three are aëriform but condense easily with water. Nitrous acid has such a weak attraction for alkalis that its compatibility with them has long gone unrecognized; but saturated with oxygen, as nitric acid, it reacts more strongly. The combination of oxymuriatic acid with alkalis may not be crystallizable and has, for that reason, been overlooked. Davy himself admits that dry ammonia gas combines directly with dry oxymuriatic acid. Saturated with oxygen, oxymuriatic acid is a more concentrated acid which can combine with alkalis. With heat, both saltpetre and hyperoxidized muriatic potash produce oxygen gas. Oxymuriatic acid destroys plant pigments and contagious vapours; sulphurous acid does the same, and so does nitrous acid, although that usually produces a yellow colour. All three of these acids also attack gold, though with very different intensity. So many similarities always make it very questionable to place these substances in different classes. However, if they were all regarded as simple substances, we would encounter other difficulties. We would then have to make sulphur from a combination of hydrogen and sulphurous acid and assume that, in the combustion of sulphur in the first degree, only its hydrogen would actually combust while the water produced would remain combined with the sulphurous acid, which already contradicts the analogy with the phenomena of muriatic acid. The issue becomes even more difficult in the case of nitrous acid. If this were non-composite, we would have to assume that, together with hydrogen, it would produce either nitrogen or a gaseous oxide of nitrogen. However, because of the manifestations of ammonia, we would further be compelled to regard hydrogen as an ammonium oxide. Then a serious question would arise concerning the following combinations. Either ammonium + 1 part nitrous acid would be = hydrogen, ammonium + 2 parts nitrous acid would be = ammonia, ammonium + 3 parts nitrous acid would be = nitrogen, etc., or we must assume that hydrogen is an ammonium oxide due to oxygen, but azote a similar one due to nitrous acid, so ammonia becomes a compound of two oxides, which again would not be in agreement with the analogy. Another question would be whether we should regard gaseous oxide of nitrogen as a combination of oxygen and azote or of azote and nitrous acid. We could also, on the one hand with a somewhat stronger, on the other with a far weaker analogy, regard gaseous oxide of nitrogen as the simple substance and derive the other compounds from it. Therefore, it is obvious that we cannot regard oxymuriatic acid, or Davy's chlorine, as a simple substance without becoming entangled in serious difficulties, but continued experimental investigations will show whether these difficulties can be overcome. The case cannot yet be considered closed. Even the important objections which Berzelius has raised on the basis of the quantitative relations of the components, and which are still to be added, cannot settle the matter completely, for through an extended analogy we might discover a regularity in the phenomena where now this new concept offers us nothing but contradictions.

Let us assume, however, that the substances mentioned here must be regarded, in the future as they have been in the past, as composite; the issue would still turn on the question of whether, in spite of this, they were not similar to the more combustible members of the proposed series. For it is all too likely that these are composite substances as well. Because of the predominance of their combustible properties, it will not be possible to extract their igniparous ingredient until we have found a substance which would be as well suited for attracting combustible substances as potassium or iron for empyreal substances. Therefore, it would depend only on whether the composition of the empyreal substances mentioned did not have something analogous to the composition of combustible bodies, or whether, under certain circumstances, they could not behave like undecomposable bodies even though, under different conditions, they would have effects which belong to a different series. In this case, they would also serve to complete the series. However, these are still only questions; also in this regard, much remains unresolved.

In any event, this much can be concluded from the foregoing:

1) No previous definition of metals provides a true natural distinction between metals and other bodies, but a transition through imperceptible degrees occurs in the established series from the most recognizable metals to those with scarcely more than a trace of metallic character. If it was desired to separate some bodies under the name of metals from the rest, e.g., to count only uncombusted bodies which conduct electricity and heat most perfectly, we would have no objection, provided that this restriction is regarded as an arbitrary limit and not as a natural one.

2) If we were to indicate something that is common to all the bodies listed here, it would be the property of combining with each other. And if we were to distinguish this category from others, we would have to start from one or several individual substances which were, so to speak, central to the category and then judge the others according to their ability to enter into combinations with them; e.g., the series of metallic bodies contains all those which either combine directly with gold itself or enter into direct combinations with bodies that are similar to gold.

3) So far, none of the bodies which we have included explicitly in this series have been decomposed. This should not be considered completely accidental, for it proves that they share a resistance to any means of decomposition yet tried. However, as this difference can just as well be merely quantitative as qualitative, it has significance only in so far as these bodies combine this similarity with others.

4) The activity which pervades this series is that which causes combustion. Each of the substances listed here contains one of the conditions for fire. However, the entire series is to be regarded as the series of uncombusted substances.

5) Most of the bodies in our series do not combine with burnt bodies. Sulphur and phosphorus, however, are exceptions. They could, without hesitation, be included among the bodies which are already burnt to some degree if they did not differ with respect to their ability to combine with uncombusted bodies. Now they represent a combining link at the transition between the two classes. Much remains to be explained regarding their compounds. We now know that alkalis, which were previously considered to be completely dry, still contain water. Perhaps this will always be decomposed by the melting of alkalis with those combustible substances so that the combination between them and the alkalis would not be a direct one. Or, vice versa, could the sulphur which evaporates if another portion combines with a metal perhaps be slightly less oxidized than ordinary sulphur? Could a yet unknown combustible sulphur exist?

6) Naturally, since only bodies capable of combining with each other can form a series of affinity, *the uncombusted bodies therefore form a series of affinity of their own.* This is certainly the most basic and important property, in a manner of speaking the *radical property* of our series.

The burnt bodies also constitute a coherent series of their own. This is also created from several previously distinct groups. Alkalis, earths, metallic oxides, etc., had been strictly separated according to specific definitions. For many years, the boundaries between them, which in their time served the useful function of embracing all that belonged together, have stood as divisive barriers, harmful to the further progress of science. There is hardly any need to tear them down now as we certainly no longer regard them as natural limits after the discovery of the components of alkalis and earths. However, we must make a few remarks regarding the concepts of alkalinity and acidity. It is not as if they were not assessed correctly by many chemists now, but we wish to avoid any offence which the following use of these very comprehensive terms might cause. According to the old practice, a sharp taste was considered to be the essence of alkalis, as was their ability to make certain blue plant pigments green, to promote the combination of oil with water, to draw moisture from the air, etc. Nearly all these properties presuppose that an alkali must be soluble in water. We have now found all these properties combined in ammonia, potash, soda, baryta, strontia, and lime in such a way that we cannot, without the greatest arbitrariness, distinguish between them and deny some of them the name of alkali. However, the solubility of these bodies in water decreases so considerably in this series, from ammonia to lime, that we could not, with any semblance of justification, claim that an alkali could not be even more insoluble. If such a substance exceeded lime in insolubility as much as lime exceeds soda, it would appear insoluble in all our experiments. Therefore, we must seek a property of alkalis which does not depend on insolubility, and this we easily find in the contrast between alkalis and acids. The well-known property of alkalis, to neutralize the striking properties of acids and, conversely, to lose their own through combination with them, is the property common to all alkalis which does not depend on solubility in water or on any similar circumstance, for even if certain particular conditions should prevent one alkali from combining with one or several acids, there will be others with which the combination takes place. If we further add the well-known experience that, with the disappearance of the neutralizing property of alkalis in their combination with acids, all of the remaining properties attributed to alkalinity disappear as well, we then have the complete theorem that the commonly assumed external characteristics of alkalis depend on the presence of a very widespread chemical property, the acid-neutralizing ability, but that this ability does not depend on any other external or secondary condition known to us. Among all the properties of alkalis known to us, then, neutralization is the only one that qualifies as a distinctive mark. However, if we want proof of the alkaline effect of insoluble substances on blue plant juices, we merely need first to colour them red with an acid and then add magnesia; as this substance dissolves in the acid, it will gradually restore the colour, in the same way as one of the soluble alkalis would have done.[9]

All that has been said here about alkalis can also be applied to acids. Their taste, their ability to colour blue plant juices red, etc., depend on their solubility in water and disappear if the acidity is neutralized by an alkali. Therefore, only the ability to neutralize alkalinity is permanent in acids and can be considered a distinctive mark. It must also be possible to apply to insoluble acids what we have said about the effect of insoluble alkalis on blue plant colouring. For this purpose, we want to mention an experiment which has probably never been performed in this way, but which is almost completely contained in well-known experiences. We dye syrup of violet green with caustic potash and then try to dissolve pyrolusite in it by means of potash. Once the alkali is as saturated as possible with the oxide, we add nitric acid. We then find that the green colour can be neutralized and changed into red with far less nitric acid than would otherwise be necessary for the saturation of the potash. This, however, could only take place if the pyrolusite had contributed something to the change in colour. Experiments of the same significance can be made with many other metallic oxides. Consequently, there is absolutely no reason to exclude a body from the acids because of its insolubility in water.

According to all this, we can perceive alkalis only by means of acids, acids only by means of alkalis, and, therefore, we find ourselves in a circular definition. This can be avoided only if we proceed in a manner similar to the case of metals described above. We must select several characteristic bodies with which we can compare all others if we want to determine their place. For instance, we could say: *Any body which, in its direct combination with others* (assuming no decomposition), *neutralizes the same distinguishing properties as nitric acid behaves like an acid; whereas one which, in its direct combination with other bodies, neutralizes the same properties as ammonia behaves like an alkali.* In some bodies we find an alkaline as well as an acidic effect, as in massicot and in so many other metallic oxides, which exhibit strong effects on both alkalis and acids. Not infrequently do we find both modes of action in fairly good balance, as in massicot, in alumina, etc.

[9]The author has further explained this, as well as the following, concerning the relation of alkalis and acids, in his treatise on *The Series of Acids and Bases* in the December 1806 issue of Gehlen's *Journal für Chemie und Physik*. [Reprinted in this volume, "The Series of Acids and Bases" (Chapter 24).]

From this, we see that only those bodies in which one of the forces is particularly dominant can be called alkalis or acids. Because acidity and alkalinity commonly occur together, it is also impossible to draw a sharp line between them. For instance, if we arrange alkalis according to the ease with which they act on other bodies by means of their alkalinity, then ammonia, potash, soda, baryta, strontia, lime, and magnesia follow each other in a rather coherent sequence. Many other metallic oxides, e.g., zinc oxide, could very easily be added. However, in order to remain with those which are quite irreducible, we mention zirconia, which still has sufficient analogies with magnesia. From this, we proceed to alumina, which already contains acidic and alkaline properties in relative balance. Next to this is glucina, which seems to possess even less alkalinity, and next to this, siliceous earth, which combines far more easily with alkalis than with acids and produces truly neutral compounds with the former. Next to this, we must place tantalum oxide, which does not combine at all with acids but well with alkalis; a fact which, according to Klaproth, is related to the irreducibility of earths. It is no surprise to find the saturated oxides of molybdenum and manganese next. It is not a big step from these to chromic and arsenic acids. Next to arsenic acid is phosphoric acid, to that sulphuric acid, etc. It would not be difficult to fill the gaps more completely by bringing in several more metallic oxides, but the ones mentioned here will certainly suffice for what we intend to prove. A complete list of all oxides, according to their alkalinity or acidity, will only be successful if we investigate all these substances with regard to these properties by using corrected principles.

However, even as we are now convinced that alkalis and acids form one continuous series, we must also admit that it is very useful, indeed, almost necessary for convenience of expression, to distinguish, by means of arbitrary limits, between groups of bodies which combine certain properties relevant for a number of applications. Thus, we can properly collect ammonia, potash, and soda under the name of the *readily soluble* alkalis. *Baryta, strontia,* and *lime* could be called sparingly soluble alkalis. Magnesia, ceria, iron oxide and other comparable substances could be called *insoluble* alkalis because of their very low, imperceptible solubility in water. We could also classify the acids according to their solubility, most profitably perhaps only as *soluble* and *insoluble* acids. For many other purposes it would be more convenient to divide them into *very volatile, nonvolatile,* and *heat-resistant* acids, where the boundaries would then be set rather arbitrarily. In the first category we would place those which readily appear in an aëriform state, in the second nitric and sulphuric acid, in the third phosphoric, arsenic, and boracic acids, etc. For other purposes, we could similarly divide acids into *readily decomposable* ones, such as hyperoxidized muriatic acid, nitric acid, oxymuriatic acid (if the old theory regarding this should be correct), *ordinarily decomposable* ones, like sulphuric, phosphoric, arsenic, and chromic acids, and finally, *sparingly decomposable* ones, like fluoric acid, muriatic acid (presupposing the usual theory), boracic acid, etc. In many cases, this last classification would be convenient if we wanted to apply it to burnt bodies in general. Insoluble, sparingly decomposable, burnt bodies would agree fairly precisely with the old notion of an earth.

The bodies which we now call salts derive from acids and alkalis. The meaning of this word has also been determined in different ways at different times in the most arbitrary manner. Initially, it always implied a taste; later, attention was directed primarily towards crystallizability; and even more recently, solubility in water was once assumed to be the characteristic of salts, and for this reason even spirit of wine was included among the salts. At present, there is more or less general agreement that we can define salts only according to their composition, and that all other determinations excessively separate what belongs together in many other respects. Therefore, we are no longer afraid of placing fluor spar, marble, and heavy spar among the salts even though they must also be regarded as varieties of true stone. However, if we now pursue this path consistently, we must give the name of salt to any compound that consists of two or more bodies in which the combination has taken place by means of alkalinity and acidity. In this manner, any compound variety of glass must be considered a salt as well, which also agrees very nicely with the brittleness, the transparency, and the insulating property of the other salts.

Salts as such hardly combine with the stronger alkalis or acids, but if this seems to occur, it must be due to the added alkali which, in addition to the alkali in the salt, binds to the acid in it, or due to the added acid which, in addition to the acid in the salt, binds with the alkali in it. However, combusted bodies in which alkalinity and acidity are in equilibrium appear to combine with salts as such. Among other examples, this is the case with water. In any event, salts distinguish themselves very little from combusted bodies in which alkalinity and acidity are in equilibrium. Nevertheless, everything taken into consideration, they will always constitute a class of their own, to which only oxides close to equilibrium represent transitional substances.

If we now compare the three classes we have established, we find:

1) In the first class, there are many ductile bodies, but in the second and third there are none.
2) In the first class, the great majority of bodies are opaque; in the second, they are, with very few exceptions, transparent only if they constitute a continuum; in the third, they are all transparent.
3) With very few exceptions, bodies of the second class are more difficult to melt than those in the first, and, at the same time, they are harder. Bodies in the third class, if they are composed of very strong alkalis and acids, are more refractory than they should be according to the average fusibility of their components, but otherwise they are often quite fusible. Generally, the situation here is very complicated.
4) Bodies in the first class are, for the most part, very good conductors of electricity, but those of the second class are, with very few exceptions, poor conductors as solids. When fused, they conduct quite well but are far from the metals. If the bodies of the third class are free of water and solid, they are poor conductors; combined with some water, they become better conductors.
5) We have mentioned repeatedly that the bodies of each class form an *affinity series of their own* in such a way, however, that the contrast between the

first and the other two is much sharper than the contrast between the two latter. It is easy to see that in this one difference lie countless others, e.g., that metals are insoluble in water, that they do not dissolve in acids without a preceding oxidation, that they have a curved (convex) surface when melted in earthen vessels, etc.

6) The three classes form not only different series of affinity but also different *series of compositeness*. Therefore, we could also refer to them as the series of the first, second, and third degrees of compositeness. If it turned out that sulphur, phosphorus, etc., which we have called transitional substances, had a composition similar to the oxides, they would be placed under them because, as transitional substances, they could be drawn to either side with almost equal validity. With regard to the remaining, undecomposed bodies, there is certainly no doubt that they are of a composite nature, but all analogies convince us that their composition must, at least, have a different form, a kind of coherence quite different from that of burnt bodies. However, the decomposition of the bodies of the first class would lead to the formation of an even higher order. It still cannot be determined whether some of the bodies known at present, e.g., oxygen, might not most appropriately belong there, but it is much to be doubted as we must rather expect that such a discovery would also reveal the principle from which oxygen acquires its essence. With regard to these degrees of compositeness, we must also mention that many compounds of bodies within a series, due to their homogeneity, can be formed by means of very weak chemical activity, so their inner chemical nature is not significantly altered either. Such compounds, which like Winterl we could call *synsomatic*, remain at the same level of compositeness, but those which are formed by strongly opposed forces, such as all compounds with oxygen and compounds of the most incombustible bodies with hydrogen, are transferred into the second series of affinity. Similarly, a compound of two alkalis or of two acids must still be considered an alkali or an acid and not a salt, but where the compound has been formed with a great heterogeneity as a result of the mutual attraction of alkalinity and acidity, only there do we find bodies of the third series, salt.

We began our present attempt at an arrangement of substances by rejecting the usual method of classification, according to which definitions were constructed following certain outward characteristics under which everything was to be organized. Even in the history of these concepts we recognized their arbitrariness as well as their inadequacy for the present state of science. We attempted to arrange bodies according to similarity by placing the most similar one next to any given substance, taking particular care not to make distinctions because of differences in degree. In this manner, we have now arrived at three series or classes which reveal significant internal and obviously important differences to us. If we wished to ask: *What is the main property of a substance whereby its chemical nature becomes most clearly visible?* the answer can hardly be other than: *The kind of chemical combinations which it forms.* However, we find a pronounced difference between these in the last summary of our series. The similarity of the chemical combinations which can be formed also causes the bodies to *form an affinity series of their own,* and in this expression we understand possibly even more completely that they belong to one chemical series. The third basic chemical difference imaginable could lie in composition. However, we have also noticed that the three series were based on three *degrees of compositeness.* We easily see that these three determinations are merely different views of the same subject, but therefore their importance becomes all the more obvious. Thus, we may believe that we have found a more natural and rigorous classification of the substances than the one which has so far been used in chemistry.

By looking at the entire process of nature, however, we can find an even more profound justification for this conviction. We see here not a mere accumulation of parts which can be arranged at will, but a constantly developing entity which expresses itself in each of its parts. Eternal in its laws, it permits objects to transform themselves incessantly. It begins each of its works from an outwardly simple basic structure and, from the forces slumbering within, gradually develops a creation which ultimately emerges with infinite variety yet tranquil unity. From this point on, however, it begins to destroy it again in order to present the eternal activity and the eternal law in new creations. In this manner, nature displays its forces in the most varied manifestations from change to change, and the space filled with such a particular activity, though it appears at rest to our senses, we call a body. Every organism reveals this process of nature to us. Thus every plant begins from a barely perceptible structure in the pollinated flower to develop from one stage to the next until it is finally transformed into a perfect plant, which again holds the seeds of new growth. In this manner, the substances now develop along with the forms. In the meantime, oxygen, hydrogen, and carbon must combine with each other in the most diverse ways to form the peculiar components which our decomposing art extracts from plants. None of these components, however, is permanent but must be regarded merely as a transition to a new one. Even in the plant which has progressed to the most complete development, each component is still in transition to a new one which will be formed during its decline.

Let us now apply this to our subject. Inorganic bodies belong to the earth. This, like the organic bodies on it, has evolved with time and continues to develop further. Rubber, sugar, resin, etc., are only different developmental stages of the same basic substances in plants, and this must also be the case for the inorganic substances in the earth. They, too, have certainly been formed by an evolution proceeding in all directions from a single basis and changed through continual transitions from one stage to another until finally the forming process, along with the great upheavals which in an earlier era transformed our planet in so many ways, fell to the present slow and feeble level where, to some degree, it even eludes the eye of the inattentive although it is far from finished. However, if all the substances of the earth are nothing but resting-points of the activity with which nature proceeds from work to work in the formation of the earth, then the substances are representations of the laws of this development. It is obvious, therefore, that the natural order of the substances forms a series which develops according to laws.

From this, we are now able to derive several laws for the formation of such series and to examine our own in their light.

1. The simplicity of a series requires that all its members are selected from the closest possible evolutionary unit. The earth is the closest evolutionary unit for all inorganic bodies. It would be more remote if we were to regard the earth and all its organisms as a small, self-contained world and then include all organic products in a single series along with the inorganic. We could establish such series, but they are more complex and therefore require the greatest perfection of science. We must first endeavour to establish separate series from each of these evolutionary units. Their parallels will be most instructive. Thus, resin, wax, and oils must form the series in organic nature which is parallel to that of metals. The relative quantities of their constituents, the transitions from one to the other, but especially the manner in which they decompose when oxidized may one day provide remarkable information about the nature of metals. The constituents which are common to organic and inorganic nature would also lead to numerous explanations, especially if we had even more accurate investigations of them as to whether they are produced by the organic process or merely taken from inorganic nature. Sulphur and phosphorus have a very great similarity to organic fat- and resin-like components and are perhaps among the most interesting connecting links between the two large branches of nature; likewise carbon. From the fact that carbonic acid can be found in large quantities in the limestone of primordial rock, which predates every trace of the formation of plants and animals, we can see that this also belongs to inorganic nature with as much right as any metal. Iron appears to belong to all periods and all realms of nature. So do quick lime and siliceous earth, but in such a manner that the former is found primarily in the animal realm, the latter in the vegetable. As is well known, nitrogen belongs primarily to animal nature, but it is also connected with inorganic nature because of its presence in the atmosphere. These and several other common substances certainly merit independent investigation solely as a consequence of this common occurrence.

2. The members of a simple series must be at the same level of development, e.g., we include in the series which we have established substances of the same degree of simplicity or compositeness.

3. These conditions can only serve to distinguish what does not belong together, and in this respect they can be regarded as guides to truth. However, they will be met immediately once we have discovered the correct radical property of a series whereby it is properly characterized and distinguished from others. In order to find this radical property, we must always exclude those which are merely quantitative because a property of variable quantity could also become = zero, indeed, even turn into its opposite. An exception to this would be some derived property, a function of others, which is of such a nature that its minimum has a finite magnitude. So

far there is no case in chemistry where we have seen this. Therefore, where we do not have such certain knowledge, it is far preferable to consider properties which cannot become larger or smaller, e.g., that chemical substances should form an affinity series with each other.

4. Once the radical property has been discovered, it is also easy to find a related primary activity which changes continuously throughout the series, e.g., combustibility in our first series. The body in which this force is most dominant is placed at one end of the series, beside it the next, etc. Such a series now behaves like a decreasing arithmetic series; the decreasing quantity not only vanishes gradually until it reaches zero, but it even changes into its opposite value. Thus, in our first series, the attraction to empyreal substances finally changes into its opposite, the attraction to combustible ones, and in the second, beginning with the alkalis, the attraction to acids passes through equilibrium and is finally transformed into the attraction to alkalis. Therefore, we should not be surprised to find opposing substances in the same natural series; on the contrary, it is part of their nature. In general, no contrast is possible without a certain similarity. For instance, there is a contrast between two lines drawn into opposite directions, but no such contrast is possible between a line and a plane. Because of this necessary contrast, the fundamental property in our series always depends on two opposite ones. In the first series, it is the *activity of causing combustion*, which is expressed through the combustible and igniparous property; in the second, it is the *activity of neutralization* through alkalinity and acidity. The third series has no really characteristic contrast except for the one between alkaline and acidic salts. Perhaps an investigation based on new views will find something hitherto unobserved.

5. Although the arrangement of substances according to the magnitude of their fundamental activity is the general principle for the inner structure of the series, we should note that the great common evolution has divided into different branches. It will be a very commendable task to investigate these divisions and their principles, although many data are still missing for this purpose. It would be highly desirable, however, that Steffens, who has accomplished so much in this respect with very limited facilities, would use the many added facts to make a new attempt.

Through the principles enunciated here for the arrangement of series, we may hope that we have not only explained and justified our procedure more than adequately, but that we have also provided some justification for the future arrangement of the remaining chemical objects. Here we cannot ignore an analogy which is found throughout nature. Just as it is a law of organic nature, from which there are but few and very limited exceptions, that only creatures of the same species join with each other sexually, there are also boundaries in inorganic nature which substances and their compounds do not generally cross. And just as it is only the opposite sexes which mate in organic nature, it is also only opposite chemical substances which form their unions with strength and vitality. Therefore, those who arranged chemical combinations on the basis of affinities had, in a certain sense, the right idea, and those who compared the chemical tendency towards combination with love in organic nature were no less right. This similarity should certainly not be regarded as accidental and superficial. It is the essence of things that opposite forces must seek each other everywhere in nature. Each force requires its opposite to justify its existence and, so long as equilibrium has not been reached, to approach this steadily and, as much as possible, to form a balanced whole. However, it is the great task of the student of nature to ascertain, wherever possible, the various forms of this separation, along with the requisite striving back to unity.

THE CHEMICAL FORCES

We now proceed to the investigation of the chemical forces and begin immediately with their greatest and most magnificent manifestation, fire, because in that we also find, as was to be expected, and as the following will clearly show, the freest and most vigorous action of the forces.

The only kind of combustion for which we have accurate investigations is the one which consists in the union of a combustible body with oxygen, so we must make this the basis of our investigation. The combustible body has a chemical attraction, a tendency to combine with oxygen, and vice versa. If the combustible body has been burnt to some degree, it has lost its ability to burn further under the present conditions, in which case we say that the body is saturated with oxygen. In other words this expression means: The chemical attraction of the combustible body to oxygen has now become so weak that, in the given circumstances, it can no longer neutralize the forces which resist combination. Under more favourable circumstances, the combustion of this body may continue, but here it will also find its limit, and so on, as long as a new combustion can be made possible. This generally known fact evidently shows us that the chemical attraction of combustible bodies to oxygen can be diminished and neutralized by a certain activity of oxygen. To be sure, there are several examples in which bodies appear to become more combustible through combination with some oxygen, but in each case this is nothing but the consequence of other circumstances, such as reduced cohesion in the combustible body as a result of the first degree of combustion, or an increased attraction to water; for it is a fact, confirmed through many experiences, that most oxidations are greatly facilitated in water. In aëriform bodies, the compression of the combustible body can also become a consideration. However, in general it is certain, and also assumed, that the attraction of combustible bodies to oxygen decreases through combination with it. On the other hand, this is also true for oxygen, which, as is well known, is also saturated with the combustible body; by combining with it, oxygen suffers a decrease in the chemical force of attraction. The attractions of both the combustible body and of oxygen, therefore, are such that *one neutralizes the apparent activity of the other*. Thus, we also find burnt bodies in which neither the combustibility of one nor the igniparous force of the other becomes apparent without the application of the strongest reagents, e.g., carbonic acid, in which it would be as futile to attempt to burn most bodies as to ignite the carbonic acid itself. As we usually call forces which neutralize each other *opposing*, we shall also use this expression here for the chemical attractions between combustible bodies and oxygen.

The force with which combustible bodies attract oxygen is not the only common feature whereby they are distinguished. There is already common experience of this, and as proof we need only mention the great effect on light which is always associated with combustibility. However, it will prove to be even more useful below. Therefore, the expression "attraction to oxygen" indicates only a very limited facet of the activity common to all combustible bodies. The negative aspect of this expression is that it has been quite harmful to a clear understanding of the situation. As we shall see, the antiphlogistians had by no means as vivid a conception of combustibility as the phlogistians. For this reason, we shall choose another expression for the action common to all combustible bodies and simply call it *combustive force*.[10] This expression, like any other with a certain generality, will be open to all sorts of misunderstandings in the absence of an explanation, but once this has been given as here, the expression will be found to be, if not exhaustive, at least sufficiently meaningful. Even if it were not necessary under the circumstances, its brevity would adequately justify the preference we have given to it. For the same reason that we have given the attraction to oxygen a name of its own, we must also do this for the attraction of oxygen to combustible bodies. For want of a better name, we shall call this force *empyreal* or *igniparous*, or, for the sake of brevity, *ignitive force*.[11] We have already acknowledged the relative nature of these terms and the possibility that they could also be used in quite the opposite way.

What has been presented so far can be expressed as follows: *Combustion is caused by the endeavour to combine of the combustive force and the ignitive force, which mutually neutralize each other and are therefore called opposing forces.*

It goes without saying that we do not presume to have explained anything with this definition. For the moment, we leave completely open what these forces are, and whether they are independent or merely modifications of other forces or of substances. We merely adopt the usage of referring to each active property as a force. In a sense, we do nothing but express well-known facts in new terms, but if these expressions are simpler and more accurate, such a change in terminology will have advantages similar to that common in mathematics.

As indicated previously, the neutralization of forces which we are discussing here is not an internal destruction, but rather a mere restriction whereby their manifestations become imperceptible. Opposing chemical forces are present in a burnt body in the same way that mechanical forces are in a body which is kept at rest by opposing forces. Therefore, we can say about chemical forces, in appropriate though figurative terms, that they *bind* each other. As we know, the force of one body can again be released from this bound state through the similar force of another which is more active under the given circumstances. For instance, the combustive force of the metal in copper oxide can be released from the bond of the opposing force in oxygen by the greater combustive force of iron. The outcome of all

10 [Ørsted uses the term *Brennkraft*.]
11 [This is a literal rendering of the term *Zündkraft*, which Ørsted uses here and below.]

elective affinities can be explained in a similar way. No decomposition would ever be possible without the neutralization of one of the forces, for the mere attraction of two bodies to a third would necessarily result in a combination of all three.

The combustion of bodies by oxygen is not only connected with a neutralization of forces, but *due to this combustion, bodies are also transferred into another class, into another series of affinity*, because, generally, no burnt body can combine with uncombusted ones, as we have seen earlier.

Through combustion, some bodies are transformed into alkalis, others into acids. We have already shown above what is meant by alkalis and acids and explained that their properties neutralize each other reciprocally. If we now adopt the newly stated terminology, we must say that *the activities in alkalis and in acids are opposing forces. Neutrality*, which could also be called chemical indifference, would therefore be the equilibrium of the two forces. However, the question now arises: How is it possible that one and the same operation can produce or develop two such opposite activities? We shall try to let the facts themselves answer this question, and therefore we shall compare the conditions under which alkalinity is produced with those which produce acidity.

1. Bodies which acquire significant alkalinity through combustion have the property of decomposing water easily and extracting oxygen from it. This is very striking in the seven metals recently discovered in the bodies which were formerly called alkalis and alkaline earths exclusively. Zinc, which oxidizes rather easily in water, also produces an oxide which has a not inconsiderable neutralizing power. Iron is similar, although to a lesser extent than zinc. Its oxide produced in water still acts on the relatively large amount of acid with which it combines with a rather large neutralizing power. However, the ability to decompose water consists in depriving the hydrogen in it of its oxygen, and this is clearly a proof of great combustibility.

Bodies which are transformed into acids through combustion display very little ability to decompose water in the absence of other agents. Sulphur can remain in water for a very long time without any indication of decomposition. Phosphorus also has little effect on water, as has carbon. The same is true of nitrogen if we dare mention such a problematic substance. Even the two recently discovered acid-forming substances, fluorine and boracium, which display such great combustibility in air, can be washed with water without being oxidized. Arsenic can be kept in water for a long time without losing its metallic lustre even though it is easily tarnished in air. According to Buchholz,[12] molybdenum can also be kept in water without oxidizing. Manganese does not oxidize, at least not perceptibly, when it is weighed in water. The remaining acid-forming substances of inorganic nature are not yet sufficiently known in this respect, but from what we have seen here of the ones we know, we would most likely find that chromium and tantalum cannot be particularly oxidizable in water either. It is remarkable that the oxidation of these bodies, which decompose water so little, occurs so easily in air, although the extraordinarily great influence of heat should be noted. Even phosphorus does not oxidize in air if the temperature has dropped to the freezing point.

[12] [Bucholz.]

2. If we compare the amount of oxygen absorbed by bodies which become alkalis or acids, we generally find that it is much larger for the latter than for the former. The most pronounced alkalis contain less oxygen than a quarter of their total weight, and not one of them contains half of it. It is the reverse with acids; in some of these we find almost three quarters of oxygen and in most of them not less than half.

3. The comparison between different oxides of the same body is even more instructive. There we find that the strongest alkalinity appears in oxides of the first degree. The most perfect oxides of several metals do not combine at all with acids if they cannot again give off part of their oxygen. We see this even in hyperoxidized potash. However, oxides of a higher degree which can still react with acids can be separated from them by far weaker forces than the less saturated oxides. Among other examples, we see this in incomplete and complete sulphate of iron, of which the latter can be far more easily decomposed by fire than the former. Acidity, on the other hand, occurs at the highest degrees of oxidation. Although acidity can also be found at lower degrees of oxidation, it is either very weak or disappears completely at the lowest degrees. Alkalinity and acidity are clearly found next to each other in the oxides of bodies with intermediate combustibility in which not much oxygen has been absorbed. We frequently see this in the oxides of the relatively reducible metals which easily dissolve in alkalis as well as in acids. It is unfortunate that we still have so few investigations on this subject. Very combustible bodies combined with a great deal of oxygen, as in water, can also show such an equilibrium.

Now it follows from everything mentioned so far that burnt bodies in which the combustive force still has some predominance are alkalis, but those in which the combustive force is so overwhelmed that the ignitive force must have significant predominance are acids. At a certain equilibrium of the two forces, we also find alkalinity and acidity in a state of equilibrium. However, we must consider not only this condition, but at the same time also the fact that, due to combustion, the forces have entered an entirely different mode of action. The combustive force, which constitutes alkalinity, does not act as a combustive force, and the one which constitutes acidity does not act as ignitive force; rather they act as alkaline or acidic forces in a completely different series of affinity, viz., that of the oxides. In a number of bodies the forces appear together in both forms in important phenomena, e.g., the combustive force along with alkalinity in ammonia, the ignitive force along with acidity in nitric acid. In some saturated oxides which are very deoxidizable, and in which the ignitive force is therefore only weakly restrained or limited by the opposing attraction, it appears entirely in its form as ignitive force without any sign of acidity, as we see in the saturated oxides of lead, mercury, and manganese. Therefore, one of the forces must be limited and reduced by the other to some degree in order to have an alkaline or acidic effect. For the moment, we shall not explain how this limitation, this reduction, or whatever it may be called, can transfer the forces into another series of affinity; we shall content ourselves with having demonstrated that such a change in the mode of action actually takes place. In order to have a name for these differences, we call the first mode of action that of the *un-*

combusted state, the second that of the *combusted state*. We can then establish that *alkalinity is a phenomenon of the combustive force, and acidity is a phenomenon of the ignitive force, both in the mode of the combusted state.*

If we want to assess the alkalinity and acidity in bodies correctly, we must clearly distinguish their intensity from their extensity. The former will be judged according to the ease with which the alkali or the acid overcomes obstacles or resists the decomposition of components in relation to the circumstances; the latter according to the quantity of the opposite substance which can be saturated by it. The intensity depends on the predominance of one of the two forces as does the chemical mobility, the weak coherence of the binding. The extensity depends on the quantity of oxygen required to balance the combustive force of alkalis or on the quantity of the combustible component required to balance the ignitive force of acids. We see this already with respect to the alkalis in Bergmann's[13] beautiful discovery, which was so magnificently extended by Berzelius, viz., that the quantity of acid which is to be saturated by an oxide is always proportional to the magnitude of the dephlogistication (as Bergmann called it) or to the quantity of the absorbed oxygen. As is well known, Berzelius has calculated the quantity of oxygen in ammonia according to this principle and given us a certainty about it which is really quite close to that gained through experiment. As far as acids are concerned, it also appears, in accordance with experiences so far and especially with Berzelius's experiments, that the neutral salts whose acids lose oxygen with heating and leave the alkali with an acid containing less oxygen do not lose their neutrality through such a loss of oxygen, and conversely, that the neutral salts which contain less-oxidized acids do not leave the neutral state if they attract more oxygen. However, if it were possible to combine the oxygen in a neutral salt with more of the basic combustible component, it would follow that the salt would cease to be neutral and show an excess of acid. In the same way, an excess of alkali would result if the combustible ingredient in the alkali were saturated with more oxygen. Consequently, the saturation capacity would always depend on the degree of balance of the combustive force or the ignitive force. Or the saturation capacity of an alkali is proportional to the amount of oxygen in it, that of an acid to the amount of the combustible component contained in it. The latter part of this law can be applied only for comparisons of the various quantitative relations of the same combustible substance. Now the question arises whether it could not be claimed that the saturation capacity of acids was proportional to the quantity of the combustive force in them. If this could be assumed, we would have a measure of the combustive force of various bodies which could become very important for us. The combustive force of the radical of an acid would always be the greater, the more alkali it needed for saturation and the smaller the quantity of combustible radicals contained in the acid. In other words, the combustive force of the radical would always be = the saturation capacity of the acid, divided by the quantity of the combustible radical. In order to make actual calculations according to this formula, we shall do well to await the completion of the beautiful series of quantitative determinations which Berzelius has begun

[13] [Bergman.]

because the current determinations have insufficient accuracy. For instance, if we wanted to apply the proposed law to alkalis and say that the ignitive force in the oxygen of an alkali is = the saturation capacity of the alkali divided by the quantity of oxygen, we would find very different magnitudes for the ignitive force of oxygen according to Richter's series of the various alkalis even though this formula, according to the general law, must always give the same magnitude because the saturation capacity of the alkali, which must be divided, is proportional to the quantity of oxygen, which is the divisor.

If we now compare the alkalis according to these principles, we find that ammonia combines a great excess of combustive force, a great chemical mobility, and a great saturation capacity, or, in other words, it has both significant intensity and extensity of alkalinity. The possibility of combining the two lies in the great combustive force of the basic component. We shall also find the effect of its alkalinity very strong if only we consider the great extent to which it encounters hindrances almost everywhere because of its volatility. Potash has a significant intensity more through the predominant activity of its combustive force than through its chemical mobility. In this respect, it is closest to ammonia among all the sparingly oxidizable alkalis, but it has only a very low alkaline extensity. Soda possesses the latter property to a higher, but the former to a lower degree than potash. Baryta has a very low extensity, but precisely therefore, and also because its radical is very combustible, it has a great intensity, which would be even greater if only its chemical mobility were not low. Strontia has more capacity and less intensity. Lime behaves much like strontia. And finally, among the sparingly oxidizable alkalis, magnesia combines the lowest intensity with the greatest capacity. In burnt bodies, where the amount of oxygen has become very large in relation to the combustive force, the alkaline capacity also increases quite considerably, but due to the diminished intensity they can no longer overcome the independence of acids (the essence of all forces by which acids strive to maintain their state) and can thus no longer act as alkalis. Therefore, there can still be an alkaline tendency in the strongest acids, and there must be, since a diminished combustive force is present, but its intensity has disappeared to our observation. For the same reason, there must also be an acidic tendency in every alkali because of the reduced amount of oxygen contained in it, but this tendency is often beyond all observation due to its weak intensity and the presence of strong alkalinity.

With regard to the acids considered here, we find in fluoric acid a very great capacity along with a most significant intensity. We derive the latter of these properties from the great force with which this acid is attracted by the weak alkalinity of water, from the great stability of its compounds, and from the power with which it dissolves siliceous earth, in which, after all, the coherence is so great and the alkalinity so low. We must conclude from this that its basic component is very combustible but also here combined with a large quantity of oxygen. Recent discoveries have not yet given us a satisfactory explanation of this. Similar conclusions must be drawn with regard to muriatic acid if it really is a combination of oxygen with a combustible substance. However, it is inferior to fluoric acid in intensity as well as in capacity. Carbonic acid has a large capacity, close to that of fluoric acid, but its intensity is low as we can see in its weak attraction to water, in the decomposability of its compounds by heat, and in its weak effect on solid substances. This cannot all be attributed to its aëriform state because the two previous acids are aëriform as well. Sulphuric acid has a low capacity, but a significant intensity. Phosphoric acid has almost the same capacity, but a far lower intensity. Nitric acid has an even lower capacity, but it is not so easy to determine whether its actual acid intensity is greater than that of sulphuric acid because a freer ignitive force than is typical of acids begins to manifest itself here. In hyperoxidized muriatic acid, the ignitive force is even freer and, therefore, has far less acidic nature. Sulphurous and nitrous acids, as well as oxymuriatic acid, have a lower coherence of binding than saturated acids of the same radicals. It is not our intention to try to explain this here but merely to observe that this behavior must be related to the fact that several of its effects occur with greater intensity.

Having now examined the compounds of uncombusted substances with one of the extreme members of the series, we may also turn our attention to their compounds with the extreme member in the opposite direction and to other combinations of uncombusted substances with each other. Before doing this, however, we want to make a general observation regarding combinations and their consequences. The most elementary kind of all combinations is the one in which two completely identical substances are mixed, e.g., water with water or oil with oil. Such a mixture cannot be called a mechanical one because a coherence which, at the same time, would be a continuity would never be achieved without the interpenetration of the characteristic forces of matter. Such a combination takes place without any resistance other than the extremely small one caused by the displacement of its parts, and without any change other than a spatial one. If we now mix other bodies of very low heterogeneity, e.g., warm and cold water, the mixing still proceeds without any great resistance (the dissimilar specific weight creates some) and without any great change (merely that of the temperature). If we take a greater difference, e.g., water and a dilute salt solution or one kind of fatty oil and another, the combination is still very easy, and it also produces only small changes. If we take two very similar metals, like tin and lead or gold and copper, the result is still very much the same. The similarity in the manner in which they fill space has the effect that they join easily with one another, or that no significant resistance prevents the two substances from spreading throughout the same space. Thus, these combinations occur without any strong tendency to unite because the forces are not much more satisfied by the combination than they were before, but they also occur without perceptible resistance and change. Such combinations could be called combinations of similarity. If bodies of greater heterogeneity are combined, far greater changes are obtained. For instance, the addition of approximately $1/900$ arsenic is sufficient to render gold brittle. Less distant metals make it brittle only in somewhat larger quantities, as we see with tin and lead. Gold is also made brittle by the other very volatile metals, though least with lead, which is closely related to it. The same can be applied to silver, but lead, which is far more similar to silver than to gold, produces even smaller changes in it. It is everywhere evident that arsenic is one of the substances which change other metals most, and at the same time we know from other comparisons that it is among the most extreme in the series. Antimony comes very close to arsenic with regard to its embrittling effect. Tin is again similar. Lead already produces a greater number of mixtures of higher ductility with other metals. In this respect, it is interesting that copper which contains so little lead that this shows no effect on it makes gold brittle when combined with it. We see clearly how the more distant metal becomes more brittle with lead than the less distant. Zinc is a similar case, only its well-known combination with copper produces a particularly ductile mixture. This might seem striking at first glance because zinc is far more different from copper than lead with regard to its combustibility and brittleness, but if we take into account that zinc has a much greater similarity to acids of higher coherence than to acids of lower coherence, it is obvious that this is not an exception. Sulphur changes metals even more and also forms combinations with greater force. As is well known, light and heat are produced the moment sulphur combines with several different metals so that the similarity of this phenomenon with combustion cannot be denied. The metals which are known to have such an intense reaction with sulphur are potassium, sodium, iron, and copper. There will certainly be several more of these, but how distant metals must be from sulphur in order to result in the production of light and heat has not yet been determined. It is also undeniable that metals are transferred into the second affinity series by sulphur, but at the same time they remain in the first series, and it cannot really be claimed that these compounds acquire a new primary chemical property, but only that the nature of sulphur is predominant in the mixture. However, it merits closer investigation to determine whether it might not be possible to combine metals with very small amounts of sulphur, precisely as we do with arsenic, and then it might turn out that sulphur shifts at least the most combustible into the second series.

Let us now consider hydrogen. This substance is currently the most extreme, reasonably well-known member of our series, at the end opposite to oxygen. It is too similar to the more combustible metals to cause a change in their affinity series. However, we know that two of its compounds with less combustible substances, sulphur and tellurium, are acids and combine with alkalis completely like other acids. It might even seem as if an alkali rather than an acid were to be expected from the combination of such a combustible body as hydrogen with sulphur, which has little ignitive force in comparison with oxygen. It appears, however, that precisely because of the great combustive force in one and the not insignificant force in the other, the combustive force cannot be restrained sufficiently to form an alkali. Thus, only the restricted ignitive force can reveal its second-order effect in it or appear as an acid. We do not wish to conceal that we offer only speculations here, but, in any case, such classifications are not without value to the scientist. The conviction that the cause of acidity does not lie exclusively in oxygen but in widespread forces through which bodies of very dissimilar combustibility can also form acidic compounds allowed us to foresee discoveries some years ago, such as the acidity of tellurium hydride.[14] Moreover, we do not at all deny the possibility of

---

[14] See the article about the series of acids and bases referred to above. The author will not repeat the hypothetical theory presented there. It is based on analogies which are far too weak, but he reserves the right to explain and to examine it further on another occasion.

finding oxygen in sulphur and related bodies, but if this should happen one day, we shall surely be on the way to finding it in tellurium and arsenic as well, and then it will certainly also be the case that it is contained there in a manner different from alkalis or acids. After such a discovery, however, we would also expect to find hydrogen or, better, an even higher, universal combustible agent in all metals, but it would also be all the easier to distinguish the difference between this mode of combination and that of the combusted state. —If we now look back at what has been said, we find that bodies of the first series, in which we acknowledge either the combustive force or the ignitive force as dominant, are also generally the more expanded ones with a lower specific weight. Thus, on one side, we find ammonium, potassium, sodium, etc., on the other, sulphur, phosphorus, carbon, and especially oxygen. However, the metals in which the forces approach an equilibrium, such as gold, silver, etc., are distinguished by their density and their heat-resistance. But as other circumstances also have an effect on these properties, it will be impossible to attribute every one to this law without further explanation. The same applies to the series of oxides, where, however, a larger number of bodies and several sharp distinctions make our task easier. Here we find that the acids which distinguish themselves through the strongest attractions to alkalis all number among the more volatile ones; indeed, they might all be in a gaseous state if the attraction to water did not condense them. It is at least certain that, among the acids with several degrees of oxidation, the most easily decomposed, i.e., those in which the components have attracted each other most weakly, are also the gaseous ones. Similarly, we know that the alkali in which the combustive force is particularly dominant is a gas in its water-free state. Even more important are the condensations which are the results of the chemical combination of opposite substances. Every combination of a body with oxygen produces a condensation. The smallest occurs when oxygen gas combines with the combustible agent without changing its volume, in which case the contraction is not larger than the volume of the burnt body. We see an example of this in carbonic and sulphurous acids. Otherwise, as is well known, it is most common that oxygen leaves its aëriform state and forms a solid with the combusting substance. We also find that bodies composed of oxygen and a combustible substance are almost always harder and more refractory than the solid bodies from which they were formed, i.e., even more than they should be according to the calculation based on the fusibility of oxygen (which is exceptionally large) and that of the combustible body. Similarly, alkalis and acids condense each other during their union. As is well known, ammonia and aëriform acids form solid salts in which the components are as much as a thousand times more dense than in their free state. In general, neutral salts are found to be far less volatile than their components. Usually, they are also harder and more refractory, but here there are several other circumstances which cannot be so easily determined with our present knowledge. With greater confidence, however, we venture to note that the point of neutralization, with its fairly precise equilibrium of opposite forces, coincides with the turning point of the formative forces. It even occurs at the appropriate mixing ratio, where virtually all salts crystallize. There are still several exceptions, which will certainly one day provide scientists with excellent information about the formative urge in inorganic nature, but they cannot be very important to us as objections to the generality of this rule if we consider that, among the infinitely many mixing ratios at which the formative force could have lingered, it missed the precise point of chemical equilibrium in only a few cases. We also find that the chemical forces at this point are usually incapable of overcoming any considerable resistance. For instance, we rarely find that an alkali is capable of condensing more of an aëriform acid than is required for neutralization. The exceptions here are the same as in crystallization. This gives us all the more hope that we shall soon discover the law which they all obey.

From all this, we now have reason to conclude that *each of the two chemical forces presently known is expansive in itself, but combined, they produce a contractive effect due to their mutual attraction.*

We find these forces in their freest state in the series of uncombusted bodies, where the greatest intensity of simple chemical action also occurs. We see them moderated, although still with a significant dominance of one force over the other, in the series of combusted bodies, but these forces are found extremely weakened in neutral salts. In none of the chemical compounds currently known to us do we find anything which would compel us to assume other forces. However, no chemical combination is produced merely by two opposing forces but rather by a twofold interaction between them. Even in the series of uncombusted substances, we cannot conceive of one in which one of the forces existed alone, for such a substance, with its unlimited strength, would immediately combine with any other in which there was the merest trace of the opposite so that it could never exist in isolation. Thus, we must assume both forces even in oxygen and hydrogen. If, consequently, a body, A, of the first series combines with another, B, of the same series, the combustive force of A will combine with the ignitive force of B, and the ignitive force of A with the combustive force of B, and through this double activity the combination is realized. This is even more clearly the case in the combination of alkalis and acids. We have already shown that a latent acidity exists in every alkali and a latent alkalinity in every acid. The combination, therefore, occurs by means of both. The predominant activities will always initiate the effects, but the latent activities will always have a significant influence on the intensity of the effect. The investigation of this subject will be of great importance for the theory of salts; in particular, it will certainly be helpful in the determination of taste, solubility, saturation beyond the point of neutralization, and crystallization. Here, we only wish to observe that every alkaline and acidic activity is already a consequence of the interaction between opposite forces and, thus, has a form of action which is different from that which pure combustive force or pure ignitive force would have. It is therefore understandable that the entire process of chemical activity in the second series of bodies must be very different from that of the first series. This will become even clearer when we consider that all the chemical properties of the constituents actually appear, though in a different form, in the chemical union of the substances in conflict with each other. Due to the great restriction of the forces, the tendency of neutral salts to combine must be far weaker than that in the previous series. The effect must also be much more complicated because of the combination of at least two alkalis and acids. For our present purposes, however, we require only the general observation that, in order to understand these combinations which are so similar to mere cohesion, no forces are necessary other than the attractions of acids and alkalis which are already limited by other, similar attractions.

In our entire investigation of chemical effects, we encountered no forces other than these two, the combustive force and the ignitive force. Both are expansive in themselves but contractive when combined. This contraction causes the most different degrees of cohesion so that, even if we cannot completely explain every single circumstance which arises, we have no reason to doubt that they are determined by these forces. Now we have come to a most important point where the mechanical and chemical forces appear in their unity. The mechanical existence of matter requires an expanding activity in order to fill space and a contracting one in order to limit the expansion. We find the former in both of the chemical forces, the latter in their mutual attraction. Cohesion is merely a manifestation of this holding together which is common to all matter. The more specific forms of cohesion, such as rigidity, liquidity, and the aëriform state, are only modifications which depend partly on the simplicity of the form of action of bodies in space, and partly on chemical mobility, whose manifestation, as we shall prove, is heat. Even the attraction and repulsion which we observe in electrical and related phenomena can easily be attributed to the forces stated here, for if we observe the various parts of a space in which an expanding activity takes place, they must necessarily appear to repel each other. Indeed, the expanding activity itself, propagating in space, must appear to have a repulsive effect on itself when observed with respect to different points in space, no matter how the very nature of this process is to be explained. Thus, each of the forces must appear to have a repulsive effect on itself but, for similar reasons, an attractive effect on the opposite force.

Therefore, the results of our entire investigation are:

1) *that combustive force and ignitive force are the ultimate chemical forces, to which all our experiments refer;*
2) *that they are also the ultimate mechanical forces, to which we are led by all our investigations of motive forces;*
3) *that, therefore, we can regard them as the universal and fundamental forces of the material world.*

The following portions of our investigation will establish this result even more firmly, and they will also provide more precise views of the nature and distribution of these forces. Moreover, when we speak of fundamental forces, we intend nothing more than the designation of the most elementary activity and effect which shines through our experiences. It is the task of speculative philosophy proper to discover the original nature of these forces. The highest which we can achieve along our way is to show that these forces, as opposites, nevertheless must possess a higher unity. This, however, does not lie on our course but determines it like an invisible centre. Whoever prefers to view the circle from some other vantage will still discover its regularity; or, speaking without images, whatever view someone may hold of the nature of these forces, even if he associates each of them with some particular, fine matter, he will still be able to join us in the investigation of the series of natural laws which we shall try to establish here.

GALVANO-CHEMICAL EFFECTS

Now it will also be easy for us to relate the so-called galvanic effects to our total view of chemical forces. We shall consider their most important facts and see how we can now understand them from the nature of chemical forces.

The best known of all relevant experiments is the following: If we connect two different metals of very unequal combustibility with each other on one side and with water on the other, the more combustible of the metals will oxidize more rapidly than it would without such a connection. If an acid is added, whereby the oxidation of the more combustible of the two metals is furthered, the reaction happens much faster, and now bubbles develop at the less combustible metal as well. These bubbles can be collected and investigated and are found to be hydrogen gas. Without such a connection where each body has a common boundary with the other, such an increase of power is not possible. This is the *galvanic circuit*, composed of bodies of an inorganic nature, but we could also call it the *chemical circuit* with equal justification and attempt to relate it directly to the theory of chemical forces.

This can be explained as follows: ACD (fig. 1) represents such a circuit, in which AC is the more combustible, CD the less combustible metal, and AD is the water. The attraction of the combustive and ignitive powers on the water is different.

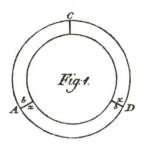

Therefore, the water will decompose. The ignitive force of water is drawn towards the metal in which the combustive force is dominant while its combustive force, repelled by this metal, will accumulate at the other. However, the metal itself will not remain undisturbed during this process, but the separation of forces, which already existed because of their initial dissimilarity, will increase further as a result of the attraction from both sides. The igniparous component of water will accumulate at A, together with the ignitive force of A, and there it encounters the more combustible metal with increased combustibility; consequently, it combines with it and thus, for a moment, cancels the conflict which, however, is immediately re-established through a new accumulation of forces from both sides. Similarly, the more combustible component of water will accumulate at the opposite metal, in which the ignitive force is relatively more dominant and will increase further through the attraction of the combustive force in water, and as it is released in the form of gas, it will always carry away part of the ignitive force of the metal, whereby the conflict, here as on the other side, will alternately be cancelled and restored.

If we now observe the true nature of this process more closely, we see that the combustive force starting at A is repelled throughout the liquid mass but is attracted by the z of the metal at D.[15] Here, the b of the water itself also repels the b of

---

[15] [In the following, z and b refer to *Zündkraft* (ignitive force) and *Brennkraft* (combustive force), respectively.]

the metal, whereby it must approach the point A in the direction DCA, where it is attracted by the z of the water and vice versa. In this manner, there is a constant flow of the combustive force in the direction ADC and a similar one of the ignitive force in the direction ACD. If this circuit of forces were interrupted at any point, every perceptible increase of forces would cease. Let us merely imagine the circuit interrupted at C, then the z, repelled by the b of the more combustible metal, would accumulate at C, and it would disturb the continuation of the effect due to its attraction to the b and its repulsion by the z in the water. A similar process, only with the opposite location of the forces, would occur in CD. The interruption of the circuit at any point will result in a similar interruption of the effect.

The interruption of the circuit, however, occurs not only because of an empty space but usually because of a space which is filled with other bodies or because of another body, which is the same thing. However, not all bodies interrupt the circuit.[16] If we separate the two metals by another metal or by carbon, graphite, or saturated oxide of manganese, the process continues, but if it is interrupted by glass, resin, or by most salts and metallic oxides, the effect ceases completely. The same is true if the water is interrupted. This can be done by using another mass of water, in which acids, alkalis, or salts can be dissolved; it can also be done by using the most concentrated acids we know as long as they are liquid. The separated masses of water can easily be connected again with siphon tubes filled with these liquids, and thus the effect can be re-established. The masses of water can be connected with metals as well, and the circuit can even be closed by using cylinders of metal far thinner than of any liquid. It will always be found that not all these bodies can be used to close the circuit with the same success, or, in other words, that not all bodies transmit the galvanic effect with equal ease. The metals can be drawn out and used in thin wires, and the effect is not significantly weakened by traversing a long distance in a wire. However, the siphon tubes enclosing the connecting liquids must be much wider, and their length has a large attenuating effect. We refer to bodies which transmit the effect of the chemical forces easily as *good conductors* but those which transmit it with difficulty as *poor conductors*. Several bodies lie between the two classes with regard to conductivity and can therefore be called *moderate conductors*. With all this, we have said nothing which would not already be known from galvanism, but it has become apparent here that nothing prevents us from regarding this as the effect of a purely chemical circuit. The concept of conductivity itself is not unknown in chemistry, so we should not hesitate to apply it directly to chemical forces. Moreover, given the circular course of the forces, it might be even more appropriate to refer to this circuit as the chemical *cyclic effect*.

For the formation of such a chemical circuit, it is absolutely necessary to have a conductor which decomposes and conducts at the same time so that the conditions are not the same on all sides, and an equilibrium is established. If everything except the combustive forces were equal, we could, e.g., in fig. 1, set the excess combustive force of $AC = u$, that of $CD = u + x$, that of $AD = u + x + y$, noting only that

---

[16]The experiments which are mentioned here can be performed particularly well with zinc, silver, and diluted muriatic or nitric acid. As soon as the circuit begins to act, gas bubbles appear at the silver. With pure water instead of acids, the process is too slow and often imperceptible.

u would be a negative quantity if the ignitive force of *AC* were greater than the combustive force. Then the effect of *AC* and *CD* on *AD*, due to their different combustive forces, will be $= u + x - u = x$, and as a result the attraction will be in the direction *AD*. Similarly, *CD* and *AD* will act with the force *y* in the direction *CA* or *AD*. The effect in this direction would therefore be $= x + y$. However, the effect of *AC* and *AD* is $= x + y$ in the opposite direction. According to this, everything would be completely balanced if some other circumstance did not occur. This is the decomposition of the conductor, which will be described below as the effect of a far poorer conductivity than that of metals. In this context, it might seem quite remarkable that water appears to be the only body which possesses this property of a conductor of the second kind (as Volta has called it) since all such bodies contain water. It must be noticed, however, that all burnt bodies which are not excellent conductors conduct very poorly in the solid state, but in the liquid state they either contain water or have a very high temperature. At higher temperatures, we might obtain a galvanic effect even without the presence of water, but these experiments are very difficult to perform.

If we place a solid conductor between two aqueous liquid conductors in which it can decompose the water, though in such a way that it oxidizes more easily in one than in the other, the part which is in contact with the more oxidizing liquid will act as the more combustible. However, if one of the liquids contains no water or nothing that can be decomposed under the given circumstances, the metal will act as the more combustible and the acids as the less combustible body, and the water between the two will decompose.

If two metals of very different combustive power are brought into direct contact in air, an altered relation of forces will be produced here, too. The two together can be regarded as a single body, in which the combustive force is not equally distributed. Due to its attraction, the combustive force itself will be pulled to the end of one side, and the opposite force will become dominant in the opposite part. As is well known, this effect can be produced by means of an electrical condenser and only through attractive and repulsive forces. For purposes of decomposition, air is a poor conductor without the assistance of heat.

We could derive many other remarkable circumstances from the nature of the simple chemical circuit if we did not see them even more perfectly in the composite circuit. As an introduction to the theory of this, we wish to consider one more fact concerning the simple circuit which, in a way, represents the transition to the composite. In fig. 2, *FG* represents a more combustible, *GH* a less combustible metal. *FE* and *HI* are two identical liquid conductors, *EKI* is a metal. Since the liquid conductors are connected to solid ones, the entire arrangement represents a functioning circuit. From the most combustible point *b* of the composite solid conductor, the ignitive force and the igniting component will be drawn towards *z'*, whereas the combustible component will be repelled by *b* and drawn towards *b'*. On the way, however, the more combustible component encounters the solid conductor *EKI*,

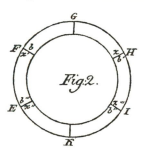

Fig.2.

where it is stopped, but where, at the same time, it encounters at z'' the ignitive force attracted from b. At b''', it will therefore experience the same conditions as at b', so it will seem like a gas. For similar reasons, combustive force will also accumulate at b'' and at z''' the igniparous component of the liquid, which in most cases combines with the metal at b''. This phenomenon will be difficult to observe unless zinc and a very weakly combustible metal are used as solid conductors and diluted muriatic or nitric acid as liquid conductors, but then the phenomenon will be sufficiently clear for an investigation.

If EKI were a solid conductor composed of a more combustible body KI and a less combustible one KE, but FGH a single piece of metal, the combustive force would also become predominant at b and the ignitive force at z, and the same generation of gas would occur as if FG were a more combustible, GH a less combustible metal. From this we see that if FG and IK were both more combustible, GH and EK less combustible metals, each of these elements would produce in the other the distribution which already existed in it by virtue of its composition, i.e., one element will reinforce the effect of the other. This, then, already constitutes a composite chemical circuit. The same conclusion can be applied to as many elements as desired. The effect of such a composite circuit now increases in proportion to the number of elements, subject only to the restriction of the decrease in conductivity due to the increase in the number of elements.

In our further investigation, we shall not deal with the more detailed determinations of all these effects. Their nature and their relation to the basic phenomena are known only too well. Here it was merely our intention to convince ourselves that the phenomena which we call galvanic can, in fact, be derived from chemical forces; and for this purpose, a short reference to the facts might have been sufficient once we had described the chemical forces as attractive and repulsive, even if we had not proceeded in the interest of providing a more complete context and in order to explain the closer relationship as fully as possible. Now, however, we shall content ourselves with a glance at the well-known action of the chemical circuit, merely to hold it up for closer inspection in order to see how well it confirms our view of the chemical forces.

As is well known, the true form of the chemical circuit is really a circle. To expose a body to the effect of such a circle actually means to incorporate it as an element in the circuit. We merely remove a liquid link similar to what is done in a conventional construction, where the circle is made directly rather than closed by means of a metal, and let the more combustible metal on one side and the less combustible on the other terminate in wires which we connect with a liquid of any kind. If the wires are made of another metal, this represents either a small disturbance or enhancement of the total effect which can be disregarded simply because of its insignificance. The nature of the liquid conductor also produces a change in the effect of the circuit which, however, does not affect the quality but only the magnitude of the total effect. If we now observe the total chemical effect of the circuit, we find, proceeding from the liquid conductor, that the substances in which the ignitive force is predominant are drawn from the less combustible to the more combustible metal, but in the opposite direction in metals in which the combustive force is dominant. If we use pure water in the link under observation, the igniparous component (the oxygen), as is well known, will separate towards the combustible side, but the combustible component (the hydrogen) towards the less combustible side. If we take a dissolved metallic oxide which does not belong to the most irreducible ones, the metal in its combustible state will be drawn towards the conductor with predominant ignitive force, but the oxygen, or a small quantity of metal with a great deal of oxygen, will be drawn towards the opposite conductor. Even stronger reductions are produced by very powerful circuits and only slightly moistened oxides. If, finally, we take a salt, the law will be confirmed here, too. The alkali is attracted by the conductor with predominant ignitive force, but the acid by the opposite conductor. This happens even across broad expanses of liquids, and even across different ones, so that the salt need not be in contact with both conductors, indeed, not even with one of them if only the circle be closed by other liquids. By means of this large circuit, it can even be closed by bodies which show a rather strong attraction to a substance. The connection will be largely or totally interrupted only if, in its passage, the circle produces an almost insoluble compound which is precipitated. Thus we see, on the one side, alkalis and combustible bodies, on the other, acids and igniparous substances attracted by the same forces as already seen in our general view of ordinary chemical processes. We see that these decompositions occur through attractions and repulsions which pass through even solid bodies, and that neither of the two separated bodies needs to enter a new combination with any other perceptible substance. Finally, we see here the chemical forces intensified to a degree which can hardly be produced in any other way, at least not in such a pure state; therefore, we are absolutely convinced that the forces which we have regarded as primary chemical forces are just that.

The circuit also gives us an opportunity to examine the nature of these forces far more precisely, for even though they appear as attractions and repulsions in a simple circuit, they display this property in an even higher degree in composite circuits, for as soon as we interrupt a circuit with a fairly strong effect so that only a highly imperfect connection remains through air, these attractions and repulsions appear with sufficient clarity. Their perfect similarity to the electrical attractions and repulsions produced by friction has already been adequately demonstrated. Therefore, we shall use them to assist us in our investigation.

ELECTRICAL FORCES VIEWED AS CHEMICAL FORCES

At present, we know that the same forces which produce chemical effects are also found in electrical effects, and there even in their freest external manifestation. We shall now begin our observation of this most independent mode of action in order to see how it can, by various stages, again assume the introversion of the chemical form. However, in order for the investigation to gain greater independence, we shall conduct it throughout as if we had so far determined nothing about the origin of the electrical mode of action from chemical considerations, and we shall there-

fore begin with the results of purely electrical investigations and gradually proceed to the chemical.

It is well known that two different properties are active in electrical effects
which are such that they cancel each other in their manifestations. As they are active properties, we also refer to them as forces, without any intention of further determining their nature by doing so, and because of their mutual cancellation, we
call them *opposite forces*. In this mathematical meaning, we shall adopt common
language practice and designate the force which is produced in glass by rubbing
with silk as *positive* and the force produced in the silk as *negative*. As demonstrated by familiar experiences, we further assume that each of the electrical forces
has a repulsive effect on itself but an attractive one on the opposite kind. Therefore,
the two forces may combine in such a manner that neither can be identified through
any external feature, so a body might long contain an immeasurable quantity of
them undetected. If such a body were brought close to an electrical one, that equilibrium would soon be destroyed by the attraction of the dominant force in this to
its opposite and its repulsion from its like. The positive force would become dominant in one part of the previously non-electric body, the negative force in another,
and between the two there must be a zone of equilibrium, but as soon as the electrical body ceases to exert an effect, the forces will, sooner or later, return to equilibrium because of their own mutual attraction. If, on the other hand, a body shows
such a distribution of forces by the mere approach to an electrical body but regains
its equilibrium, either immediately or gradually, when the electrical body ceases to
have an effect, we must assume that it already contained both forces, but in a state
of equilibrium. This phenomenon is common to all bodies, so we must assume that
all bodies contain electrical forces which, however, are imperceptible due to their
equilibrium.

If a body which has suffered the distribution of forces just described, through the
approach to an electrical body, is brought even closer to it, the contrast is intensified. The part closest to the electrical body acquires more and more of the opposite
electricity while the more remote part acquires the same kind as the electrical body
itself. Finally, at a certain distance, which is different for every conductor, part of
the electricity in the first body will cancel the opposite electricity which it induced
in the second conductor and leave behind only the one similar to its own in the
more distant part. This, along with the electricity remaining in the first body, will
be sufficient to maintain equilibrium, but before this can occur, the body initially
electrified will always attract even more of the opposite electricity and thus destroy
the equilibrium again. Therefore, it appears as if the electricity were flowing directly from one body to the other. Due to this appearance, as is well known, this
mode of transition has been called *conduction*, in contrast to the first which, not
without reason, has been called *induction*. In any case, it is legitimate to name
things according to mere appearance if only we acknowledge that the expression
shall have no deeper meaning. Perhaps, it would be more accurate to use the term
electrification through *equalization* for the so-called conduction. However, the
name is of no great importance if only we do not overlook the fact that a body
which seems to receive electricity from another is actually losing some of its own

and appears to be electric through that which it retains. If we wanted to illustrate this truth further, we need only consider a body with a point at one end, placed opposite the electrified body with the point turned towards it; then we know, and we can also see it in the dark, that the point discharges electricity which is opposite that of the electrified body, but the body with the point retains the same electricity as that of the body initially electrified. The Franklinist would express himself in an entirely different way; in fact, we do not discuss this topic with him. His theory has the merit of having provided the first formulation of the fundamental laws of electricity, and this will always remain, but the arbitrariness with which one of the activities is assigned a significance which is completely different from that of the other must appear even more daring than our formulation, which is not based on any hypothetical causes. Therefore, they must justify themselves to us, not we to them. Besides, on most of the following topics he will agree with us, especially as far as mere electrical laws are concerned, for according to either of the assumed theories, we must admit that an induction always precedes a conduction.

Consequently, if one of the electrical forces spreads through space, it happens in such a way that it immediately attracts the opposite force in the neighbouring zone, binds it, and is itself diminished as a result. Thus, the next zone has actually received a predominance of the same force which is spreading, but this itself excites a new zone of the opposite force, only to neutralize it again, etc. We could express this by saying that *electricity propagates in an undulatory manner.* In good conductors, these contrasts are cancelled so rapidly that no perception of them is possible, but in poor conductors they can be discovered with an electrometer, and many mutually induced alternations of $+E$ and $-E$ can be observed. If the electrical effect in a good conductor leaves traces behind, in which case, however, it has not behaved like a good conductor, we also see evidence of the alternation between force zones. For example, a metal wire which has been melted by electricity always appears in small pearls or beads in which expanded and contracted zones alternate. If a metal wire is vaporized by a strong electrical charge, and the vapour is collected on a piece of paper placed underneath, it condenses in such regular parts that the propagation of electricity in it is clearly depicted.

Therefore, it follows from the nature of things, and appearance supports it, that *the propagation of electrical forces consists only of an alternating disturbance and restoration of the internal equilibrium of forces.* Hence, it follows immediately that *the electrical forces are conducted only by themselves (viz., by other electrical forces).* Whoever shares our conviction that it is precisely these forces by means of which space becomes material must find this expression equivalent to the usual one, that *bodies conduct forces.*

According to this view, then, conduction is an internal change in the forces of the bodies themselves, but an internal change of forces deserves to be called a chemical change. The changes which have been described here differ from chemical changes only in that they are merely temporary because the contrasts are cancelled again through the attraction of the forces themselves. The more slowly they acquire electricity, the more slowly this cancellation occurs. Therefore, the poorest conductor remains longest in the state in which it has been placed. If there were a perfect non-conductor, and if there were a means to electrify it, a permanent internal change would be created. Admittedly, this seems absolutely impossible, in view of both conditions. However, it would depend on whether we might not find that, under certain circumstances, bodies could be non-conductors for the same quantity of force for which they otherwise possess some conductivity. In order to investigate this, we must examine more closely the nature of the propagation of electrical forces.

Since each of the electrical forces has a repulsive effect on itself, a single force cannot come to rest in a conductor until it reaches a boundary. We also see this law, which follows from the nature of things, confirmed by our experience, where it is expressed as follows: Electricity spreads only on the surface of bodies. However, as soon as both forces attract each other, they easily penetrate the interior of bodies as well.

In general, we must distinguish clearly between *free* propagation of electricity and *enforced*. The former happens through the independent activity of one of the forces, both through its expansive force and through the attraction of the opposite force, provided that it is in equilibrium and uniformly distributed over the entire surrounding space, or provided that the forces which have lost their equilibrium due to external circumstances can be restored to equilibrium by their own attraction. The latter, however, occurs where a free opposite force attracts the spreading force, or where the force itself excites such an attracting opposite force by means of resistances and non-uniformities in space. We see examples of enforced conduction in the electrical charge and discharge as well as in the condenser effect.

The conduction of electricity is determined by its degree and its quantity. The degree is measured by the strength of the attractions and repulsions, for which the electrometer is required. This property could be called the *strength* (the intensity) of electricity; it has also been called the tension. The *quantity* of electricity is measured by the surface which it can electrify to a certain level on the electrometer. If all else is equal, the quantity must be proportional to the surface, but the strength (provided all else, including the quantity, is equal) must be in some inverse proportion to the volume. Naturally, the *magnitude of the effect* of electricity has a composite relation to the strength and the quantity.

It is obvious that electricity propagates more in space, the stronger it is. We also see this by experience in an isolated body which in free air always loses an amount of electricity which is proportional to its electrometer reading. This loss, however, is the result of the conducting power of air, so we can also say that the conductivity of air is proportional to the strength of the electricity to be conducted. Such an experiment cannot be performed in non-aëriform media, but we see from countless experiences that a weaker electricity remains isolated or behaves only inductively where a stronger electricity would propagate freely, and equalization would occur.

Therefore, an infinitely weak electricity would be isolated in every body because no body is an absolutely perfect conductor. Consequently, for very poor conductors there will be intensities of electricity which they isolate completely.

The larger the quantity of electricity, the more difficult its complete conduction becomes; for to the extent that the quantity of electricity is increased in a given conductor, the repulsion in it must also increase and offer a resistance which must always be at its greatest for the given circumstances. The poorer the conductor, or the longer the propagating electricity must remain at each point, the less it will be able to conduct in a given time. If we now compare two different conductors, it must always be possible to find two quantities of electricity which they must conduct in the same time. Relative to these quantities, they are now equally good conductors, no matter how differently they might behave with respect to equal quantities. However, if the intensity of electricity is different in the two, less in the better conductor, it will be even easier to find a point where they are similar to each other. Therefore, for any conductor it is possible to find a sufficiently low intensity and a sufficiently large quantity of electricity that it becomes isolated. Moreover, it is quite clear that the internal activity in two such conductors, which are made equal by different circumstances, is very different and largest in the better conductor. However, an equality can also be effected in this respect, for it is obvious that the same intensity will create greater internal activity in the better conductor than in the poorer. It must therefore be possible to find for the latter an intensity of electricity which makes it equal to the former with regard to internal activity. These principles will be completely confirmed below.

If we now want to use enforced conduction to place bodies in a state of internal activity which they cannot leave again by their own conduction, or if we want to act on them chemically, then the quantity must be made in equal, but the degree in unequal proportions to the conduction. We find the smallest quantity with the highest degree in the spark of the usual conductor in an electrical machine. Consequently, this is applicable to the poorest conductors. In the electrical battery, the quantity of electricity is found to be larger in relation to its intensity, so it can be applied to larger conductions. Actually, there is no great leap between these two differences because we can approach the effect of the charge with a large conductor of a very powerful machine. However, there is a very great leap from the effect of frictional electricity to that of contact electricity, in which the quantity of electricity, with a barely perceptible intensity, is immense. As is well known, van Marum found that a voltaic pile charged Teiler's[17] large battery to full equilibrium in a very short time, and that Teiler's large electrical machine, the largest known, needed several minutes' rotation to achieve as much. The effect of an ordinary voltaic pile will therefore be zero for poor conductors, but all the more considerable for better conductors.

Let us now compare experiences with this. On marble, calcite, gypsum, witherite, and strontianite, the electrical spark, as Simon has shown, produces not only a mark but also makes the alkali dominant in them. Previously, this phenomenon was considered a mere rupture of the surface, but now we cannot choose but accept it as a chemical effect. We also find that electricity has a deoxidizing effect on metallic oxides; on the other hand, we find that oxidation is produced in metals by strong sparks or, even more, through discharges. The precise circumstances have not yet been thoroughly investigated.

[17] [Teyler.]

If electrical sparks are struck in turpentine oil, a gas develops immediately, and the oil gradually assumes a brown colour. However, the conductors must be very close in this case. Other oils produce similar results.

As is well known, ammonia gas, as well as aqueous vapours, is decomposed by electrical sparks.

The voltaic pile has little or no effect on all these bodies.

If frictional electricity is caused to act on water, the most instructive phenomena present themselves. If the conductor of the machine is perfectly connected to water by means of another conductor, no decomposition can be observed. On the other hand, if spark discharges from a charged jar are made to act on it, or very big sparks from the conductor of a machine, gases are produced. However, a decomposition is achieved with far less force if two isolated conductors, of which nothing is exposed but their tips, are placed opposite each other in a tube filled with water, for now the quantity of electricity applied is forced to act on a far more limited volume where conduction is no longer possible. Thus, oxygen and hydrogen are expelled from the water in gaseous form.

Contact electricity has this effect in common with frictional electricity, but it produces it without isolation and with such a low intensity that it can often be made perceptible only by means of a good condenser. It is understandable that, with such low intensity, it cannot have any effect on poor conductors. Even water is not a sufficiently good conductor for this kind of electricity to produce its greatest effect. Through added alkalis, acids, or salts, whereby conduction is increased significantly, its effect is also improved considerably. Only the highest degree of conduction of burnt liquid bodies, such as concentrated sulphuric or nitric acid, seems to be less suitable for it.

As is well known, the voltaic pile produces all chemical effects, oxidations and deoxidations, as well as attractions of alkalis and acids, which we have mentioned earlier in another context. However, we certainly cannot hesitate for a moment in calling the forces which produce all chemical effects chemical forces. From two different sides, then, from the observation of both chemical and electrical effects, we have finally arrived at the same conclusion, that chemical and electrical effects are both produced by the same forces. Since we also find that the mechanical filling of space may be due to these forces, and since the same forces will subsequently be seen in all other effects of bodies as well, we must take care not to confuse everything. Therefore, we shall not lump together the electrical effect, the chemical effect, the thermal effect, and light, but we shall regard them all as different modes of action of the two general forces. However, these modes are transformed into one another through a number of steps. Consequently, the basic electrical form is the free propagation of both forces through their expansive and attractive forces. In the electrical charge, this is already combined with the essence of the chemical mode of action. This happens even more in electrophorous effects, and, finally, we come to the complete binding of the forces in electrical explosions, decompositions of air, oil, water, etc. However, in order to understand more precisely the differences between the modes of action, it will be necessary to have observed the others as well. We turn our attention to heat next.

THE PRODUCTION OF HEAT AND THE
LAWS DERIVED FROM IT

Ever since we learned to deal with large quantities of electrical forces, we have also known that they often produce effects which are identical to those of heat. This phenomenon has led physicists to conceive many contradictory theories. Some have not wanted to accept these heat-like effects as truly identical with those of heat but have believed that electricity itself produces meltings as well as dissociations which could be deceptive because of their similarity to vapours. Others have expressed the assumption that electrical vibrations generated heat mechanically in bodies. Still others have held the opinion that electrical matter contained heat as a component. Here, the easiest explanation would be if we assumed two kinds of electrical matter which were to be regarded as two indescribably fine gases which condense through their mutual attraction and thus release the caloric contained in them. Winterl first voiced the felicitous idea that the caloric was produced through the combination of the principles of alkalinity and acidity, which, however, is not quite accurate but leads us most directly to the correct answer. We do not want to criticize all the untenable theories here; the description of the true circumstances, to which we shall add a few disproving remarks, must suffice to refute them.

We wish to begin with a series of facts.

If electricity is transmitted through a body which conducts perfectly, we do not observe any trace of heat production.

Thus, if metallic wires are used to transmit a certain quantity of electricity, it will always be possible to choose one which is so thick that it will not heat perceptibly during the process. If we use a thinner one, some heat will be produced by the same electricity, and it is easy to find a size such that the wire glows for a moment due to the electricity and is afterwards deprived of its elasticity as if it had been heated red-hot with coals. If we take an even thinner wire, it melts into small beads. Finally, an even thinner wire is transformed into vapour. This can be demonstrated most conveniently with an electric battery although sparks from very large machines can also melt wires. Not all of these gradations can be demonstrated for every metal with equal ease, but it is quite easily done with iron wires. For some metals, however, the point of evaporation seems to be so close to the melting point that the latter can be found only with difficulty, and the transition is almost always to the former.

The same gradations can also be covered with one and the same metal wire by using different quantities of electricity. If a given wire can be melted with a battery of a certain size, it will evaporate with the use of a larger one.

With a given battery charged to various levels, the amount of heat generated can be modified, but here we must emphasize that the increased intensity results in an increased quantity if the effect is then restricted to a certain volume.

From all this, it appears to follow that the more difficult it is to conduct electricity through a body, the more heat is produced, provided that electricity can penetrate into it.

This is further confirmed by the fact that a better conductor is not melted nearly as easily by electricity as we might expect from experiments with poorer conductors. Copper, for instance, which is a better conductor, is much more difficult to melt than iron. Lead and tin are also much more difficult to melt than we might expect from their melting points in an oven. However, as was first noted by Ritter, we must also take into account in our observations of these experiments that we are dealing with specific quantities of forces, so fusibility cannot be determined from the melting point alone but rather from the product of the melting point and the heat capacity. With these remarks, we are able to disprove van Marum's assertion that meltings by electricity and by heat were quite different because the order of fusibility is not the same for both. Similarly, it is obvious that the opinion that heat in electrical phenomena is generated by the combination of the two kinds of electrical matter is not compatible with these experiences, for otherwise considerable heat would also be perceived in conductors with a larger circumference during the flow of electricity. In order to take a specific example, we choose one of van Marum's experiments. A battery with an area of 45 square feet melted 84 inches of iron wire with a $^1/_{240}$ inch diameter, but only one-half inch with a $^1/_{75}$ inch diameter. Thus, the molten mass was approximately 16 times larger with the thin iron wire than with the thicker one. This experiment, which is only one among several similar ones, demonstrates sufficiently clearly that it depends not only on the combination of a certain amount of force but also on the nature of its combination.

Other experiments by van Marum further confirm our view. He first tried to conduct the electricity from Teiler's large machine through a copper cylinder in which he had placed a thermometer. It did not rise during the entire conduction. After that, he made the same experiment with a wooden cylinder and found that the Fahrenheit thermometer in it had risen from 61° to 88° after three minutes of continuous transmission. After 5 minutes, it increased to 112°.

By using Teiler's machine, considerable heat was also produced through the emanation of electricity from a point or a small sphere into the air. Thereby, the Fahrenheit thermometer was increased from 63° to 102°. In air rarefied to $^1/_{60}$, which is commonly considered a better conductor at every individual point than atmospheric air but, in any case, cannot conduct a large quantity because of its small mass, the thermometer even rose to $151^1/_2°$. Perhaps this had much to do with the fact that the air was enclosed. In his experiments on electrical emanations, Charles only succeeded in making the Reaumur thermometer rise one degree ($2^1/_4°$ Fahr.), and he wants to attribute even this small increase to an oxidation of the iron in the black substance with which he has coated the thermometer bulb. It should be mentioned, however, that the otherwise so excellent apparatus used by Charles is not very decisive here, for it is composed of two fairly powerful devices which are operated simultaneously in order to charge a battery, whereby it is not possible to obtain the extraordinary emanations of Teiler's large machine, which is so far unique of its kind.

Without such a large apparatus, we can convince ourselves of the generation of heat by electricity by means of Kinnersley's so-called electrical air thermometer. The inventor noticed that no heat was generated if he let one of the electrical forces act on the isolated instrument, not even if he brought both its conductors into contact with each other. If, however, electricity was forced to arc through air from one conductor to the other, the air was noticeably expanded. The momentary expansion caused by sparks must naturally be distinguished from the lasting expansion produced by heat. Confusion of the two effects has tempted honourable physicists into maintaining that any expansion caused by electrical sparks in air was merely temporary. Kinnersley also found that the air in his instrument expanded when the electricity had to pass from one good conductor to another through a linen thread, through a very thin metal wire, or through another quite imperfect conductor.

According to our law, significantly more heat is generated in better conductors through contact electricity than through frictional electricity because, with its large quantity and low intensity, it is not conducted nearly as easily. With very poor conductors, or those which are usually called partial conductors, we cannot expect any generation of heat from this kind of electricity until we learn to produce a far higher degree of it.

If water is connected with only a moderate voltaic pile, e.g., consisting of one hundred pairs of plates, an increase of the temperature can be noticed with a thermometer, and if the generated air is not allowed to disperse completely freely, the heat will rise by several hundredths of the thermical measure. When a zinc-copper pile, whose liquid conductor was a salt solution, was once used for this experiment, an increase of more than 0.10 Th. was measured. The water was in an open channel cut into white wax, approximately 3 inches long and 3 lines wide, but somewhat enlarged where the thermometer bulbs were immersed. The conducting wires were of platina. The air temperature was 0.10. As soon as the circuit was closed, the thermometers began to rise, and after several minutes the one on the oxygen side was at 0.205, the one on the hydrogen side at 0.18, but a thermometer in the middle showed 0.23. In better-conducting liquids, the heat reduction[18] was less, and the thermometer did not rise above 0.03 in a solution of sal ammoniac. In water which had lost some of its conductivity through the addition of alcohol, the thermometer on the oxygen side rose to 0.1875, on the hydrogen side to 0.1625, and in the middle to 0.205. In several of these experiments we neglected to use salt solution, which could have provided some additional insight. More experiments of this kind on a large scale were performed earlier by Buntzen. His circuit consisted of 1500 pairs of plates, and the liquid conductor was a solution of sal ammoniac. This powerful pile increased the temperature of water from 14° R (0.175) to 23° R (0.2875), i.e., by 0.1125, which is less than the result for a pile of 440 pairs. In a solution of sal ammoniac [sic], on the other hand, he produced an increase to 38° R (0.475), and when he allowed the air to escape only through a narrow pipe, the heat increase was more than $1/2$ thermical unit. Perhaps the differences between the two series of experiments derive partly from the better liquid conductor in the pile but partly also from the fact that Buntzen always had the liquids used in the experiments enclosed in a tube so that the heat from the escaping gases is better retained, and

18 [heat increase.]

the heat gradually reaches an equilibrium throughout, as he also observed in his experiments.

This generation of heat by conduction in liquids is not at all favourable to the concept of a mechanical production of heat through vibrations, for the vibration here is certainly not large, and, in any case, liquids do not emit any heat when subjected to mechanical vibrations. Besides, the fact that air is produced and heat generated here at the same time agrees poorly with the usual caloric theory in chemistry. It is hardly necessary to mention that it is possible to escape such difficulties if one is not afraid of becoming tangled in new inconsistencies.

In these experiments, we found the temperature in the middle highest, then that on the oxygen side, and finally that on the hydrogen side. The reason for this seems to be that no gas is produced in the middle, some on the oxygen side, but not nearly in such large volumes as on the hydrogen side. Because of the importance of the subject, this merits closer investigation. Metallic wires are also brought to high temperatures by contact electricity; but for this we require devices with very wide plates, whereby a very large quantity is obtained in proportion to the intensity. Gold leaf can also be melted by conventional batteries with small plates, but using batteries with large plates, Davy has not only made a thin metallic wire so hot that water which touched it began to boil, but with his large trough apparatus, he has also heated an 18-inch-long, but thin platina wire until it glowed. Conductors with a larger circumference do not become hot here either.

After all these experiences, we may boldly repeat what we have already established earlier: *That a body heats when it is forced to conduct a larger quantity of electricity than it would have conducted freely.* A glance at the nature of conduction will explain the meaning of this. We have seen that conduction consists of an interruption and a restoration of equilibrium at all points in a body. As long as the conduction is perfect, restoration will always be brought about by the mutual attraction of the forces which have been driven from equilibrium. However, as soon as a larger quantity of force than it can carry by itself passes through the body due to enforced conduction, the internally disturbed equilibrium will not be restored by its own forces. Because of poor conduction, this would happen very slowly, but because of the equal attraction which every point is subject to from all sides, and whereby every direction of force is blocked, equilibration is delayed indefinitely. *This state, in which the equilibrium is disturbed at every point in a body in such a way that no separation of the forces which can be perceived by the senses has occurred, now results in the phenomenon of heat.* This disturbance of equilibrium is naturally connected to a tendency towards its restoration which will not be satisfied but maintains a lively internal activity none the less. This also causes us *to regard heat as an internal conflict of opposite forces.*

Let us now consider the most important commonly known thermal phenomena from this point of view.

One of the most important of these facts is the generation of heat by impact and friction. We already know from electrical experiments that the equilibrium of forces can be disturbed by mechanical influence, and this does not really surprise us since we have also acknowledged both forces as space-filling. Through differentiation, we separated them in so far as each of the forces was attracted more by the surface of one body than by that of the other. In so far as such a separation does not occur, and it always happens only partially, an internal disturbance of the forces is produced, whose manifestation is heat. Whenever we rub two solid bodies against each other, they generate heat and lose it again through cooling. This does not agree with the usual theory of a caloric fluid, for there friction should always liberate caloric, and cooling should take it away. Therefore, two bodies should continue to give off caloric until they were completely worn away each by the other. However, the dust into which the bodies have been transformed must still contain caloric, and no experiment has given any cause to assume that these dust particles would not again possess the same ability to produce heat. Consequently, we must assume an infinite amount of caloric in every body. Rumford, who combined his wonderful talent for conducting experiments with truly scientific acuity, has turned this experience into an experiment to which an unbiased observer will not easily find objections. He investigated the heat produced by drilling a metal cylinder and accurately determined all the quantities which could have an effect on the results of the experiment. When 837 grains of metal had been drilled out, a heat had been produced which could have melted $6^1/_2$ pounds of ice or brought the borings to a temperature of 66360° F or $368^1/_2$ Th. Nevertheless, he did not find that these metal particles had a lower heat capacity than other small pieces of the same metal, which they should have had according to the caloric theory. Therefore, we have good reason to assume that the particles of a body, ground to dust during heat production, still have all the heat-generating power of the original body. In Rumford's experiment, we must assume that part of the heat derives from the iron drill and from the part of the cylinder which has not been bored out. However, we must not forget that the heat emanates from highly stressed points, and that it would be quite absurd to assume a flow of caloric towards the points of stress. Even if we only wanted to assume, which, however, would not be acceptable, that only one-tenth of the heat produced came from the borings, this would still be six times more than required to make these borings glow in a perfectly visible manner, and this without any sign of a change in the heat capacity.

Quite recently, there was an attempt to compare this with an experiment which shows that a coin which had attained its greatest possible compression under the stamp and consequently could not be compressed by additional strokes, did not produce any heat. This experiment, however, only proves that no heat is generated by impact where a disturbance of spatial relations is no longer possible, but this limit is only reached when the external force no longer causes compression. Through rubbing, filing, drilling, and the like, the vibrations do not impose any such limit. In his experiments, Rumford also found that continued boring produced the same amount of heat as it did initially, which would not have happened if the compression of the drill, as well as that of the adjacent parts of the metal, had had a significant influence.

Therefore, if we wish to postulate a caloric fluid as the cause of heat, we must actually assume an infinite amount of it in every body. Assuming that we should discover something which would cast doubt on the validity of this conclusion, we must in any case admit that every body contains an extraordinary amount of bound caloric, which is what we do when we say that the zero point for heat is far below the freezing point. It seems, however, that the consequences of this assumption have not been considered adequately, for the question then arises: What are the forces which retain such an immensely expansible matter within bodies? Due to its pressure in all other parts of space, the caloric could not remain in a given space, for if we seek any explanation of the propagation of the caloric not according to chemical attractions but according to the laws of mechanical equilibrium, then a vacuum would offer this matter the easiest passage and also show the greatest relative heat. If, however, we now want to retain the caloric in a body by means of an attractive force, this cannot, on the one hand, be the ordinary attraction of all bodies to each other, for in this case, the specific heat must be in proportion to the attractive forces, i.e., to the masses, which is not the case. On the other hand, it is also at variance with the analogy to let the caloric be retained by chemical attractions, for then we must assume that all bodies had such a strong attraction to it that they were able to compress this very expansible matter millions of times since, otherwise, it is unprecedented that *all* bodies would have a *very large* chemical attraction to any given body. Finally, if we want to assume, quite without analogy, a special attraction for heat inside matter, then, in the end, we support the fabricated caloric by means of a new fabrication. We must add, however, that even if someone would like to build hypotheses on hypothetical grounds in this manner, he would still face the difficulty that he must assume that the immense force required to produce such a compression as that of heat can be overcome by a mechanical force such as pressure and impact, and that so far there is no evidence that a body loses specific heat through such a release of caloric although Rumford had found one case in which this had *not* happened.

The production of heat by impact and friction not only serves to confirm our point of view, but it also shows us the immeasurable quantity of opposing forces which lie dormant in every body. We may divide a body as many times as we will, we may deprive it of as much heat as ever we can, *the capacity to obtain and to conduct electricity through induction, and to produce heat through friction, impact, and compression, is as inexhaustible as corporality itself.* If we now combine this inexhaustibility of the forces with their universality, which has been proved above both chemically and electrically, we feel even more strongly the conviction *that these forces truly constitute the elementary forces of material nature.*

We also see here that the forces which constitute the chemical distinctions of bodies and were previously acknowledged by us to be the dominant fundamental forces can only constitute a miniscule amount in comparison with the forces bound deep inside the body. Among other things, we can recognize this in the fact that even those bodies in which one of the fundamental forces is clearly dominant still generate enormous quantities of heat through pressure and impact, and in addition they can show very great conductivity if they are not aëriform. Therefore, the expansion of a body cannot generally be attributed to one of the dominant forces but rather to the expansive power of the bound forces, which have been more or less restricted in proportion to the strength of the coherence of their union. However, the more contracted a body is, the more coherently the forces in it are combined, or the more expanded, the less coherent this union is, thus the warmer body, with its disturbed equilibrium, is also more expanded than the colder body with its less disturbed equilibrium. *Therefore, heating must expand, and cooling must contract.*

Now that we have established the fundamental concept of heat, we shall seek to derive from it the most important of the other principles of heat production and propagation.

All bodies contain heat, for each body will suffer a continuously renewed disturbance through its interaction with all the rest of nature. In each body, the tendency of its own forces to combine fights against this and, therefore, creates a tension which determines the thermal condition of the body. If a body could balance its struggling forces in perfect equilibrium, it would also cease to manifest activity towards other bodies, and thus also to claim its space in competition with other bodies and to act upon our senses.

Heat can pass from one body to another. This does not happen through a real transition of the forces themselves but through equilibration. In a body which is uniformly warm, the separated forces do not achieve equilibration, not only because of incomplete conduction but also because of the equal attraction from all sides, whereby no direction for union can become dominant. However, when a colder body comes in contact with it, equilibration will occur, although only slowly, so that the inner equilibrium of the colder body will itself be disturbed because it promotes an equilibration in the warmer body, and this will continue until it can no longer produce an equilibration in the other body, i.e., until they have the same temperature. We could almost imagine this as if each of the forces were a continuum and would equilibrate itself with the forces in the colder bodies according to the law of electrical equilibration. In this regard, it is quite remarkable that smooth surfaces do not radiate or absorb as much heat as rough ones, for this could be attributed to the electrical law which says that every projection produces a stronger electrical effect than do the points on a smooth surface. In so far as the propagation of heat is sufficiently vigorous in relation to the mass of the body through which it passes, the separation of the forces may become so strong that they break through space due to their attraction and combine, exactly like very strong opposing forces in ordinary electrical experiments. In this case, heat will spread without leaving any trace in the body until it encounters a stronger resistance. In so far as heat spreads in this manner (this kind of propagation never exists in an absolutely pure state), it is called *radiating* and is close to the transition to light. The poorer the conductor and the smaller its mass, the more it favours radiation. Only gases favour it in such a way that we can clearly observe the phenomena. Both by radiation and by equilibration through contact, *heat always seeks equilibrium.* Investigation of the mechanism for the propagation of heat deserves to be pursued much further, but what we have found here might contain elements for a closer determination, and it might attract all the more attention as it is generally agreed that the caloric theory is not entirely satisfactory.

Heat decreases the cohesion of bodies, makes solid bodies soft to the point of liquidity, and makes liquids more fluid and, finally, vaporous. Cohesion is often confused with hardness. The former resists breaking, the latter a displacement of its parts. Where there exists perfect internal uniformity in all directions of activity, there can be no obstacle to displacement, for the location is not important, one point being no different from another. The actual resistance to breaking, however, can be fairly significant. Of course, resistance to the displacement of parts has the effect that a body breaks less easily than where it does not exist; this should not be considered a primary effect of the cohesive force but only a secondary one. As a result of the particular directions of activity in bodies, hardness must become increasingly weak with the disruption of the existing internal equilibrium; for, first of all, with an increase of the general internal activity of the forces, the activities in particular directions, even if they remain unaltered, become steadily less important to the whole. Then, however, they also become seriously disrupted by that internal conflict, and in most bodies there is finally a point where the remainder of its integrity suddenly disappears, and the body becomes fluid. In some cases, the transition happens more gradually. If hardness were merely a degree of cohesion, it must be zero in liquids, or at least, since no liquid conforms precisely to the concept of fluidity, almost zero, but this contradicts well-known experiments on cohesion. The disappearance of a definite shape at the moment of transition from solidity to liquidity provides sufficient proof that hardness, and rigidity in general, derives from that fixed shape and not directly from cohesion. What the usual experiments on the cohesion of solid bodies show could appropriately be called their *strength*. As already noticed, this is determined not only by the intrinsic cohesion and by hardness but also by ductility; for if the body is very brittle, the slightest beginning of a displacement may cause a complete fracture. Brittleness is also frequently reduced by heat, indeed, this is always the immediate result of heat, but the different volatilities of the components often cause exactly the opposite phenomenon since the resulting tendency towards separation further increases all non-uniformities. Several salts, which are said to effloresce, provide an example of a disintegration which is due to a genuine separation. In a liquid, brittleness as well as hardness disappear; it could be considered a minimum of both.

Good conductors of heat are also the best conductors of electricity. Conversely, we cannot state that all good conductors of electricity are also good conductors of heat, for carbon, graphite, and probably also the complete oxides of manganese and lead, which are good conductors, are exceptions to this even though they are among the poorest of the very good electrical conductors. It is quite natural, however, that the forces in a state as bound as heat would require even better conductors than usual.

The conductivity of bodies for electricity, heat, and chemical activity increases with heat, for it is obvious that the greater the disruption of the equilibrium with which the propagation of forces begins at every point in a body, the more easily it must proceed. It is unquestionably due to its smaller mass that a vapour cannot conduct as large a quantity of force at every point as its related liquid. However, the entire mass which has been turned into vapour certainly has greater conductivity

than the fluid from which it developed. Gases are all poor conductors, perhaps because of their rarefaction. This much, however, is certain: They all become better
conductors through heating. The forces are most strongly bound in their chemical
form. If a very weak but free electricity is already isolated by a very thin layer of
a poor conductor, we may imagine the forces in a chemical bond as in themselves
isolated at every point. Therefore, the excitation of the forces in heat is often necessary to bring about the complete union of adjacent substances. As the preference
for specific directions in rigid bodies must yield to the freely propagating internal
forces of heat, so must the individual modes of chemical action behave for substances in contact. A reduction of cohesion, which it was desired to attribute to the
promotion of chemical action by heat, cannot be assumed everywhere, e.g., where
two previously mixed gases still require heat in order to form a stronger union of
their components, such as a mixture of oxygen and hydrogen gas. Without this, in
any event, the union in combinations must be less strong, the greater the heat in it;
hence heat renders bodies more susceptible to direct decomposition. *That heat so
very strongly favours chemical combinations and decompositions*, even when a
mere change of cohesion would not be sufficient, agrees completely with our view.

Conversely, *heat is produced in every vigorous chemical reaction*. There have
been earlier attempts at deriving this fact from the associated contractions.
Berthollet, who certainly understood the difficulty of this explanation, formulated
the expression of this common fact almost as a law. We want to consider these circumstances more closely. We understand almost immediately that the strongest
chemical effects are only possible because of strong opposite forces. However,
these must produce heat when they combine because they are too strongly con-
strained in their chemically bound state to enter into a perfect union. In addition,
the otherwise static forces suffer a disturbance during such an internal motion of
forces. The result of a combination through opposite forces is generally a contraction. These two, heat and contraction, are effects of a single cause, the mutual activity of forces, but in such a way that contraction only reveals itself completely once
the generated heat is in equilibrium with its surroundings. *Combustion* gives us the
union of the most powerful opposite forces and at the same time *the greatest generation of heat*. If the union of oxygen and a combustible ingredient occurs in a liquid, a gas is usually produced, e.g., nitrous gas from nitric acid, hydrogen gas from
hydrated sulphuric and muriatic acids. Despite these gas emanations, heat is produced. According to the caloric theory, it was necessary to assume that the dissolved metal and the oxygen, which was already contained in the acid in compressed form, would release caloric through their compression sufficient to heat
the entire mixture and the escaping gas and, furthermore, to produce a quantity of
vapour which escapes with the air. However, we must consider that in most cases it
can be clearly demonstrated that the general result of such a decomposition is
expansion and less cohesion. If, for example, iron is dissolved in muriatic acid, hydrogen gas develops, and at the same time the iron dissolves to a readily soluble salt
where the great coherence of iron has disappeared in the liquid, and where it is volatile in the solid state. If we wanted to resort to more special compressions here,
like that of iron and oxygen, or of iron oxide and muriatic acid, and claim that these

compressions produced more heat than those expansions would absorb, then we would have to claim that the heat capacity of the metallic solution combined with that of the generated hydrogen gas is smaller than the capacity of the dissolved metal combined with that of the dissolving acid, an assumption which is at variance with all probability although a stringent proof would not be easy to construct because the capacities of gases are so difficult to determine.

When acids and alkalis combine with one another, heat must develop according to our principles, and, indeed, this occurs everywhere. The stronger alkalis and acids produce high temperatures, the weaker ones, naturally, lower. Even solutions which contain barely one-twentieth of the weight of water in acid and alkali, when mixed, still produce enough heat to make the thermometer rise noticeably. A particularly remarkable contradiction to the caloric theory, however, lies in the fact that heat develops even when carbonic acid escapes as a gas through the combination of an alkali and an acid. If, e.g., sulphuric acid combines with the lime in chalk, carbonic acid gas develops, and gypsum is formed. The gypsum and the gas combined should therefore have less capacity than the carbonic acid lime and the dilute sulphuric acid combined. We now know, however, that heat is produced when carbonic acid and lime are combined, so carbonic acid lime must have an even lower capacity than carbonic acid and lime together. Therefore, we are almost forced to derive all of the considerable generation of heat, together with the caloric required for the formation of gas, from the already fairly compressed sulphuric acid in the water. This, however, becomes even more striking in the generation of gas from carbonic lime by means of nitric acid, for Lavoisier and de la Place have demonstrated that the heat capacity of nitrate of lime is greater than that of the lime and the acid combined. Here the caloric theory is deprived of even its last excuse.

We mentioned earlier that alkalis combine with water as with an acid, and that an acid combines with water as with an alkali. The water which is retained by an acid with very great force also contains as much oxygen as the alkali which is required for the neutralization of the acid. However, acids still attract water vigorously beyond this degree of saturation, but at a higher saturation point they lose all their strong attraction to water, which would be capable of overcoming considerable barriers. This is also the case with very soluble alkalis. Even after having absorbed as much water as is necessary for their crystallization, they attract more water from air. For solid and liquid bodies we may assume that they have almost reached the goal of their stronger attraction when they are no longer capable of condensing more moisture from air, and for gaseous bodies the same may be assumed when they have condensed with water. Several salts also attract moisture from air. The first transition point of their combination occurs at crystallization; the second when they are no longer capable of condensing steam. There is no definite explanation of the origin of this strong attraction of various salts to water. We might assume that it is due to a less coherent combination between acids and alkalis in such salts, but this remains speculation. Therefore, we must be content with the realization that these combinations are not at variance with our views. In all these violent combinations, however, the existing equilibrium of the forces is disrupted, and heat is generated in the process.

Through combinations in which the disturbance of the equilibrium is greater than the effect of the contracting forces, a state is created which is equivalent to that of a higher temperature. We have really clear examples of this only in the combinations of water, but these are sufficiently numerous and noteworthy for the presentation of this law. If an acid or an alkali or a hygroscopic salt combines with water, its equilibrium is disturbed, as we have seen. The heat thus generated will soon equilibrate with the surrounding space. The attraction of the alkali to the bound oxygen and of the acid to the bound hydrogen still remains active and weakens the intensity with which the elements of water were otherwise combined. Through its combination with those substances, water will enter a chemical state which is similar to that which would otherwise have been created in it by a higher temperature. Thus, it will conduct better, be more decomposable, and require a lower temperature to freeze than would otherwise have been the case. All this is so thoroughly confirmed by experience that we need hardly provide any details. The greater oxidation of metals in a mixture of water with acids, alkalis, or salts, than in pure water is a phenomenon which is difficult to explain in any other way, and hardly any attempts have been made to explain the increased conductivity and the decreased ability to freeze in such cases. With regard to conduction, however, the matter does not arise independently, for water-free salts, acids, and alkalis are all, as far as they have been investigated, poor conductors, but all their solutions in water are better conductors than water itself. We shall immediately indicate the less obvious examples.—

When very powerful attractive forces act on one another to produce a combination, the ultimate result of this action is a more coherent union of forces, whereby the body produced behaves as if it had been placed at a much lower temperature once the initial heat has established an appropriate equilibrium with the surrounding space. Therefore, the metallic oxides are much more refractory than they should be according to the average fusibility of oxygen and the metal, indeed, they are usually even more refractory than the metal itself. In addition, they are poor conductors. Decomposability cannot be considered here since we know nothing about that in metals. We should also be careful not to draw conclusions in such an unknown sphere, where circumstances which will make all these bold conclusions useless can so easily prevail. The same law is confirmed in the combinations of the stronger alkalis and acids. We see it already in the fact that two aëriform substances form a solid if they combine coherently. As is well known, neutral salts are otherwise more coherent as well, and they have a higher melting point than the mean of their components. However, in salts we must not disregard the effects of the hidden alkalinities and oxidizing agents, which we have already mentioned, for they can also produce a disturbance of the internal equilibrium in salt and acid, which results in a weaker cohesion than that calculated. This must occur particularly in weak acids or alkalis, just as we find that lime and siliceous earth produce combinations more fusible than they are themselves. Likewise, it is certain that the components in those strong combinations are also less decomposable. This can be demonstrated most clearly in the most easily decomposable acids, which, when combined with alkalis, will not be decomposed nearly as easily as before by combustible bodies. The fact that they do not evaporate at higher temperatures due to neutralization further confirms our assertion; that they are often easily decomposable at such temperatures does not contradict it. It is not yet clear whether, on the other hand, the components in easily fusible combinations, other than those with water, are also more decomposable, but it should be investigated whether glass made of lime and siliceous earth, magnesia, and alumina, or the like, would not be more easily deoxidized by iron than its individual components.

When a body becomes a better conductor, it simultaneously becomes colder and acquires a greater heat capacity, but when it becomes a poorer conductor, the opposite occurs. The better conductor a body becomes, the less resistance the union of forces will encounter, so there must be a greater quantity of them in a given volume in order to generate the same tension, the same opposition. Therefore, if the body becomes a better conductor without any other change, it will exhibit a lower temperature with the same quantity of force and, consequently, receive heat from adjacent bodies for equilibration. For the same reason, it will also, in future, deprive other bodies of more of their heat than before for each increase in temperature it receives, but it will itself give to others with a smaller loss. If, on the other hand, a body becomes a poorer conductor, the contrast between the competing forces will increase relative to that in the adjacent bodies, so it will appear to be relatively warmer, but henceforth it will also manifest and receive activity as a less powerful body. It is possible to find an important objection to this view in the low capacity of metals since they are certainly good conductors of forces. We shall not conceal this difficulty, nor shall we maintain that we can solve it, but we wish to convince ourselves that what we are dealing with here is merely an unresolved difficulty and not a total contradiction. First, we must point out that we have referred only to bodies of a similar kind in such a way that we compared the change of heat capacity relative to the conductivity of forces in the *same* body. After all, it is possible that another condition might arise in dissimilar bodies whereby the law would become invalid. Then we must also point out that heat capacities must be judged not according to equal weights but according to equal cubic content of the bodies, in which case the capacities of metals are not all that small. Furthermore, it is well known that experiments on this subject are very far from being reliable. Thus, with regard to volume, the heat capacity of mercury is 0.666 that of water, according to Black's experiments, but according to Lavoisier and Kirvan[19], more than four times that of water, which is the largest yet found in any body. If we should obtain similar corrections for the other metals, the experience would so completely support our view that we could now be satisfied. Finally, however, we must point out one more circumstance which we should perhaps consider to be the most important. We have said that if a body becomes a better conductor of the *elementary forces of heat*, it will simultaneously lose external thermal activity as well, but if a body becomes a better conductor *of heat*, the exact opposite may be true. Conduction of heat and conduction of its elementary forces are not exactly proportional. A circumstance which hardly interferes with the conduction of the latter can interrupt the conduction of the former very strongly, and conversely, a circumstance which furthers the conduction of the latter significantly can be of little importance

[19][Kirwan.]

to the former. Therefore, we see that carbon and graphite, which still belong to the good conductors of elementary forces, are very poor conductors of heat, and water, which acquires a very significant conductivity of the forces at its transition from the solid to the liquid state, seems to have gained little in the process with respect to the conduction of heat. If we now wanted to assume, as seems quite natural, that a body would acquire greater thermal activity in so far as it becomes a better conductor of heat, we could further assume that the heat capacity in metals must certainly be very large because of the large conductivity of the forces, but, conversely, that it must be very small because of the large conductivity of heat, whereby one heated point could support another more in its effect and thus show a higher activity against foreign influence; thus, one of these circumstances could approximately balance the other. However, there remain difficulties here which can only be appreciated properly through a penetrating investigation of the entire theory of heat capacity. Actually, the heat capacity of a complete series of bodies should be determined experimentally according to several different reference substances. We have only one series of capacities determined from heat distribution with water. We now need a similar series based on the melting of ice, for which we possess few contributions as yet. We also need a similar capacity series on the basis of heat distribution with sand, another with linseed oil or some similar body, and one with metal filings or granules. Only then, if all these agreed, would we know for certain that what we call heat capacity is really something constant, and that no peculiar interaction is involved in the phenomena.

After all this, it would be wise to follow the broad context of facts, in which we see that bodies lose heat through the increased conduction of force and gain in heat capacity but gain heat and lose capacity through the reduced conduction of force. From this, we shall explain most of the laws for the binding and the release of heat which are usually proposed in chemistry.

If a body changes from the solid into the liquid state, it loses external heat (the heat is bound) and gains heat capacity. Since we now know that a body will also become a far better conductor of forces when it transforms into a liquid, this law is an immediate consequence of the more general one which has just been established.

If melting occurs as a consequence of chemical forces, heat should be generated due to the disturbance of the internal calm but cold due to the increased conduction. As we know, however, the alkalis, acids, or salts which are supposed to produce coldness with ice must first combine with some water, whereby they produce a significant quantity of heat. If they are saturated with water to a certain degree, a further addition will not cause renewed heat production because the increase in the conductivity of the water produces as much coldness as the disturbance of the internal calm produces heat. In this state, they can be regarded as cold-producing only in connection with snow and ice. This explains the paradox that the same bodies produce heat with water and cold with ice. If a crystallized salt dissolves in a dilute acid and produces cold, this can naturally be explained by the increased conduction of force of the dissolved salt and not by that of the liquid. Perhaps it is always the water of crystallization of the salt which causes the cooling here, for the acid is already saturated with it. In most cases, the effect of the acid on a completely dry salt would probably be sufficiently strong to generate more heat than the increased conduction would cold. Perhaps the acid will also lose more conductivity by absorbing some salt than the salt will gain by dissolving.

If a body changes into the vaporous state, the heat is also decreased and the capacity consequently increased. This, too, is merely an example of our general law, for the same mass can undeniably conduct far more as vapour than in the volume, often several thousand times smaller, which it occupied before. Each individual point may conduct far less in the much thinner vapour. However, we have already provided an adequate explanation of why heat is almost always generated during the release of gas, a fact which absolutely contradicts the usual theory.

It will be sufficiently clear that the release of heat and the reduction of capacity during the transition of bodies from vapour to fluid and from this to the solid state can just as easily be derived from our general law. However, our previous remarks have already shown that we explain some of the chemical phenomena which seem to belong here in a manner which is different from the usual theory.

According to our view, we might feel obliged to explain the cold which is produced by the rarefaction of air and the heat which is produced by its compression from the fact that air in the rarefied state is a far better conductor and in the compressed state a far poorer one, but on closer consideration this seems questionable. Actually, it is a great question whether rarefied air is really a better conductor than air with a higher density. We could explain the propagation of electricity by arcs, by sparks, rather as a rupture than as a conduction. If electricity could really be conducted through air as rapidly as it leaps in the form of sparks or propagates across large expanses of space in the form of emanations, it would certainly belong to the best conductors. It seems quite necessary first to distinguish between a *radiating and a conducting effect of electricity*, as well as a radiating and a conducting effect of heat, and then further to assume a *rupture* produced by means of opposite attraction, as again distinct from though related to radiation. The radiating electrical effect is what we wish to call the effect which instantly propagates by attraction or repulsion across a large expanse of air and thus merely produces a distribution in bodies which readily disappears again. The rupture can be understood only in terms of a previous radiation. Conduction has already been discussed sufficiently above. The expression *radiating effect* may not be the best, perhaps the expression *free propagation* would be more appropriate than *radiation*; the latter was chosen for the time being only because of the comparison with heat. According to all this, rarefied air would indeed be more suitable for the free propagation and the rupture of electricity than air with a higher density but not truly a better conductor. If this were the case, air would have to become cooler with rarefaction because the same activity would spread over a much larger volume and thus become much weaker; and, conversely, air would become warmer by compression merely because of the higher concentration of the activities.

It would be very interesting if, one day, it would become possible to determine the magnitude of the influence of the conduction of force on temperature changes. In itself, and also because it would enable us to measure the magnitude of chemical attractions by the heat which is generated in combinations, it would provide us with a foundation for mathematical-chemical investigations of the greatest importance.

Before we leave these investigations altogether, we must examine another relevant subject more closely. In some experiences, it seems that heat is associated with a certain predominance of combustive force, and cold with a similar predominance of ignitive force. For instance, the positive electrical emanations from a point are warmer, the negative colder. However, this phenomenon might merely be caused by an unequal evaporation on the skin. Schübler found that only bodies which evaporate will be cooled by electrical emanations. If we close a fairly powerful galvanic pile with a well-moistened finger, we shall feel cold if this happens on the positive side, heat on the negative. If the pile is very powerful, the reverse situation will occur, but for every such pile another can be found where heat is felt on both sides. Ritter, who noticed this first, also found that the taste which is produced by opposite conductors changes similarly with strength so that it can be determined that heat is always felt on the side which produces an alkaline taste on the tongue, but cold on the one which gives a sour taste. From this, we might almost imagine that this manifestation of heat is not a primary but only a secondary one so that predominant alkalinity always increases the fluidity of the blood among other animal fluids, and therefore also its conductivity, from which cold results, whereas predominant acidity, on the other hand, would produce the exact opposite. The non-uniform heat which we discovered in the water of a circuit also permits another explanation, as we have seen. In the galvanic circuit, a warm body reacts with a colder one of the same kind as the more combustible. However, this can be explained by the fact that the warmer one is a better conductor and, therefore, can have a stronger effect with its dominant force. More peculiar is an experiment by Ritter, in which a gold leaf dissolved in water mixed with a few drops of muriatic acid at the moment when the water froze; from this it should be concluded that the acid had become oxidized, and that the water, while freezing, gave off oxygen. This and several similar experiments certainly merit attention and must lead to further investigations. However, from all that has been said so far, we can safely maintain that the essential factor in heat is that conflict of the forces, indeed, we could even add that a genuine excess of one of the forces is not conceivable, but if a difference between positive and negative should occur in heat and cold, this must consist in a difference in the directions of the activities.

If we now compare the view of heat which we have proposed here with the two which existed previously, we find that we have succeeded in combining their opposite directions at one focal point, as it were. The natural philosophers of earlier times, who paid special attention to the force which is necessary for the production of heat and to the lively activity which becomes apparent in the manifestations of heat, believed that heat must be a vibration of the smallest parts of the body. More recent physicists, focusing their attention more on chemical reactions and, most of all, on that quietly transient activity which is so powerful in the production of cold, assumed the existence of a caloric fluid which could combine chemically with bodies, and which is sometimes bound and sometimes appears free. The former theory could be called *mechanical*, the latter *chemical*. Since we start from the forces, our theory could be called *dynamical*. If we now consider these theories with regard to their first principles, we must give some preference to the mechanical theory over the chemical, for the former begins with something quite empirical, viz., that all production of heat is accompanied by internal motion, whereas the latter begins directly with the assumption of a caloric fluid, for whose existence we do not have the slightest evidence. In the larger context, however, the mechanical theory has the great disadvantage that, as a consequence, all other chemical effects must also be regarded as the mere result of internal mechanical motion. This is the reason why the caloric theory was victorious at a time when the clearer understanding of all the chemical facts no longer permitted their mechanical explanation. In the fundamental principles of the mechanical theory, the assumption that the internal activity in heat is merely mechanical vibration is completely arbitrary. The dynamic theory does not start with some such arbitrary assumption, rather it first displays two forces, common to all of nature, which constitute the essence of all chemical as well as all mechanical effects, and after that it shows the law according to which the interaction of these forces produces heat. At the same time, this fundamental law of heat production offers the explanation of every kind of heat generation, both mechanical and chemical, as well as an interpretation of that lively activity which is visible in all manifestations of heat. At a glance, we also discover the reason for the expansion which accompanies heat. Indeed, we can sense, as it were, even the feeling created by heat and believe that we might even be able to derive it from this law if only we were able to resolve feelings into words. Moreover, both older theories clearly reveal their origin in what they explain and do not explain. Each explains what it has been invented to explain, or, in order to express the matter in a more respectable fashion, which is also closer to the truth: Each of these theories is a description of a regularity in the manifestations of heat but regarded only from one side. The mechanical theory really presents the inner vigour of heat, but with an attached one-sidedness which, in this context, cannot be removed, but which, in this very same context, also loses its otherwise repugnant aspect, as anyone will feel who will give more than half-hearted consideration to the mechanical view. The chemical theory really presents the regularity in the transitions of heat in all chemical changes, but also with a one-sidedness which results in a number of misleading interpretations. We cannot deny, however, that the caloric theory has served to present a more diverse and well-developed regularity than the mechanical theory, although it can be doubted that the concept of a caloric fluid actually brought about these discoveries. Precisely because of the beautiful laws regarding the binding and the release of heat, entirely new properties must be attributed to the caloric which are not at all contained in its original conception. It would not be difficult to apply the mechanical theory to the binding and the release of heat as well, and admittedly with apparent logic; it could be said that the oscillations would become stronger at the transition of vapours to liquidity or of the latter to solidity, and that they would become weaker in the opposite transition. Even some analogies to sound could be adduced in support. On the other hand, it would not be completely impossible, though more difficult, also to extend the caloric theory to explain the generation of heat; but we leave all this to those who remain content with such principles after the discovery of more solid ones. We only wish to offer a historical consideration of what they have accomplished. We hope that the dynamic theory will continue to prove its worth because of its specific correlation with all chemical and mechanical facts and because of the connection which it establishes between the two (e.g., the parallel which it provides between chemical and mechanical internal mobility). It is still only in its infancy; when it has been further developed by as many astute men as the older theories, it will undoubtedly be able to oppose them with greater advantage.

THE PRODUCTION OF LIGHT

When a body has reached a certain temperature, approximately 3.35 Th., it begins to be faintly visible in the dark; at a higher temperature, approximately 4 Th., it becomes very clearly visible; at an even higher one, somewhat above 4.6 Th., it will also be luminous in the twilight, and, finally, at approximately 5.4 Th., it will be seen to glow even in bright daylight. At lower temperatures, the light is red, but with the increase of heat, it will change through various degrees to white. This occurs not only in bodies whose surface suffers a weak combustion, but it also happens to those that do not undergo any such change, e.g., gold, which can be kept glowing for many months and even years without showing any trace of oxidation. Therefore, if we say: *Heat changes into light*, this is nothing more than expressing in words what actually occurs.

If, conversely, light falls on surfaces which reflect only a little of it, heat is generated. This can be seen with sufficient clarity in the well-known experiences according to which black and dark-coloured surfaces are warmed less[20] by light than lighter or white ones, and these again less[20] than polished surfaces. Whatever theory of light we wish to adopt, we are always forced to admit that light, as light, disappears to the degree in which it acts on the surfaces of bodies without being reflected. Similarly, we see that light, where it must pass through bodies, heats them up so much more, the less they transmit. The general expression for these experiences is: *Where light, as light, disappears, heat is produced.*

It should be mentioned that the experiences to which we are referring here belong to those universal ones which can be demonstrated in all bodies where only general conditions apply, and which always yield the same results. Therefore, the transformation of heat into light, or light into heat, cannot be denied; on the contrary, it must rather be regarded a great fact. However, as is obvious from the preceding remarks, it is certainly not our intention to claim that light and heat are the same thing, or that they differ only in degree; this would hardly agree with experi-

[20][more.]

ence. We conclude only the similarity of their generative forces and regard them as different modes of action of these forces.

The same law reappears in the electrical generation of heat, as could not be otherwise. The same effect which generates heat also enhances the activity to the point of glowing. The electrical spark is nothing but glowing air. In rarefied air this same glowing spreads over a larger volume, but nevertheless it is always the same. Even the so-called electrical emanations would not produce light if each of them did not attract its opposite from the air. Not only solid and gaseous bodies show this phenomenon but also liquids, e.g., water if we allow sparks to strike in order to decompose it, alcohol even more, but turpentine oil most splendidly. Through a strong galvano-chemical effect, even moist alkalis and salts can be made to glow.

Even though we have seen that heat can become light and light heat, and that heat can also be generated through the union of the same forces which produce light, we still need to be able to produce the two chemical forces from light by some form of analysis in order to be fully convinced of the equivalence of their generative forces. This, however, we have learned from Ritter's magnificent experiments, which were prompted by Scheele and Herschel and confirmed by Englefield, Seebek[21], and several others. The result is sufficiently well known, viz., that by passing through a transparent prism, light not only disperses into colours but also experiences a chemical distribution so that, beginning from beyond the violet end of the visible rays, a deoxidizing activity occurs which is weaker in the violet light itself, even weaker in the blue, and which finally, as it approaches the red, disappears completely only to change there into the opposite, igniparous, activity which continues to increase beyond the red into a region filled with an invisible activity. The experiments on which this is based cannot be questioned. The combustive force shows itself clearly at the violet end and in the violet light itself primarily by blackening muriatic silver very quickly, and at the opposite end, the ignitive force shows itself by preventing the blackening of muriatic silver, which otherwise occurs so readily in daylight. Especially the deoxidizing force of the violet end has been confirmed many times. The opposite force of the red would follow immediately from this fact even if we did not have any experiments, for if violet light has a greater reducing property than white, then there must be a force in the latter which neutralizes the one in the former, i.e., the opposite, in other words, the ignitive force. Wollaston found that guaiacum behaves in exactly the opposite way in prismatic light, becoming green in violet light, as it would otherwise in oxidations, and again brown in red light, just as with ordinary heat. This inversion of the facts is remarkable but does not in any way contradict the force distribution in light; for such inversions are very frequent in the free manifestations of forces due to contrasts between the distribution and the equilibration or the so-called conduction. The very low conductivity of guaiacum may be the reason for this inversion. Further investigations, however, may reveal this. We can always postulate that we have found combustive force and ignitive force in light, and that, because of such a con-

[21] [Seebeck.]

trast and distribution, we would have good reason to regard everything which lies within the limits of the greatest isolation of each as mere manifestations of various mixtures and activities. However, we readily admit that the analytic proof here is far from having the strength of a synthetic one, but being so well prepared and supported by this, it may well suffice for our complete conviction. However, for the artful perfection of the entire presentation, as well as for the more precise understanding of all the details of the process, a continued investigation into the production and cause of the internal activities of light is certainly one of the most desirable tasks of physics.

If we have now fully convinced ourselves that light and heat are produced by the same forces, the following question arises: To what extent are these forces, so dissimilar externally, also different internally? We noticed above that the contrasts must have attained a far greater intensity in the production of light than in that of heat. However, in our ordinary electrical experiments we notice, and it also follows from the nature of things, that when the contrasts have reached their highest degree, they overcome the resistance which the imperfection of the conductors offers them, discharge, and equilibrate. The same thing must occur with the smallest possible contrasts, as determined by the nature of the conductor, and cause a phenomenon other than the mere disturbance of equilibrium. The electrical spark, as well as any other glowing, is an extended discharge throughout space. In more solid bodies this discharge does not occur instantaneously at all points; on the contrary, they retain their luminous power for some time. This may be due to an oscillation which, during such a struggle to discharge, would be as natural as the well-known oscillations in the discharge of an electrical pile. However, propagation from the luminous body and outward through any given medium proceeds, as we know, without trace if the medium is not transformed into heat. Radiating heat has a similar mode of action, only the velocity of the action in it must be far lower because it is much more easily restrained by the resistance of the medium and transformed into heat which is conducted away. We must leave it undecided as a matter for further investigation whether something must be added to the mode of action of the forces in order to create light, or whether the unity of the mode of action which has already been indicated is sufficient for this purpose. The only question concerning the internal nature of light which we can raise here is the following: Apart from other differences from heat, should white light not also contain more combustive force than heat? As we know, in his thermometric experiments on light Herschel found that the oxidizing rays were accompanied by very high temperatures, whereas the purely deoxidizing ones did not generate any heat. We also find that white light promotes many deoxidations which heat cannot further. It also seems confirmed by the fact that the electrical light of positive emanations is much stronger than that of negative ones. The stronger refraction and reflection of light by combustible bodies also seem to support this. These indications are interesting and could be extended; they also agree very well with the opinion that the combustive force always increases with heat so that the transformation to light would depend only on a certain proportion of the predominance of the combustive force. All these indications, however, encounter a contradiction in the ordinary electrical generation of heat, whether it reaches glowing or not; we always notice the same expenditure of each force and not of one in particular. Whatever the length of an iron wire which has been made red-hot by a discharge, no difference can be seen between the positive and negative sides. Among other experiments, Davy caused an 18-inch long platina wire to glow with the discharge of his large galvanic battery without indicating any difference between the two ends, which he would certainly have done if he had noticed any. With all this, we do not wish to dispute that further experiments could reveal a cause of the dissimilarity in these electrical generations of heat which we do not yet know, at least a difference between the external and the internal. However, until this occurs, we must be content with knowing that *light is produced when the tension in the contrasts of the internal forces has reached its maximum and changes to an equilibration.*

According to what we have already seen, the propagation of light occurs through *dynamic undulations*, which is what we call the unbroken alternation of the opposite forces. This view stands between the theory of vibration, which was taught by Huygens and Euler, and the theory of emanation of the Newtonian school in approximately the same way as the dynamical theory of heat stands between the mechanical and the chemical theories. The possibility of such a view was previously acknowledged by Schelling in his *Weltseele*.

For the same reason that the poorest conductors are most easily made to glow, light is also most easily propagated in them, for this is nothing but a momentary glowing of a body which, and this is important to notice, does not begin with heat but with light and, therefore, leaves no heat after a long transmission of light other than that which is produced by the resistance to light. This law is undeniably very general; we need only consider that all fused oxides and all salts are poor conductors and transparent at the same time or are at least inclined to transparency. Sulphur and phosphorus are also poor conductors and transparent. Several good conductors are transparent as well, but they are still not among the best. As is well known, bodies may become opaque by being pulverized, fragmented, or mixed with other foreign substances which do not form chemical combinations. Moreover, it can hardly be doubted that there are other additional causes for transparency and opacity which have not yet been discovered. We know one of them but not its connection with the nature of light, viz., that combustible bodies refract and reflect light more strongly than others, but it is still an open question whether all the very combustible bodies are not also particularly poor conductors and in this state offer a somewhat greater resistance to light than is compatible with its most complete transmission. Hereby, however, the difficulty would only be shifted elsewhere. Even if we cannot resolve this at the moment, it should be noticed that this has not been accomplished anywhere else in experimental physics either, so, in this respect, we are not behind traditional theories, but we make all the greater demands of the new theory, and with good reason, for it places itself in closer connection with the whole. If we could assume that light contained a predominance of combustive force, either throughout or by virtue of a distribution from the inside to the outside, some of the difficulties would actually have been resolved.

An objection to our view will probably be made with regard to the velocity of light, and it will be claimed that such a propagation as we have described is not at all compatible with it. Our best response to this is experience, viz., that one side of a charged glass disc immediately changes its electrical activity as soon as the other side experiences a change, and this happens even though a poor conductor lies between them. Moreover, the induction through several alternations of good and bad conductors is established very quickly even though actual conduction through them is almost impossible. Therefore, should the very great surge in luminous bodies not transmit the same disturbance initially through a very rapid induction which is followed by an equally rapid discharge? However, if we want to pursue the nature of the propagation of light here, we shall become involved in the most difficult investigations and, at the same time, deviate from our goal. Let us return to the truly chemical generation of light.

In that chemical effect in which the greatest contrasts are neutralized, light also appears most frequently, viz., in combustion. Combustions which take place in air are particularly rich in the production of light. As oxygen gas combines with the combustible body, a very rapid union of opposite forces occurs, even in a medium which conducts only very imperfectly; so the union of the forces generates not only dull but also incandescent heat. The flame is only a glowing body composed primarily of vapour and air. It consists partly of the burning substance, partly of oxygen gas, and partly of the product of the combustion. In so far as vapour develops during combustion, the heat will be significantly diminished, partly because of heat absorption (as has been explained above), partly because of diffusion. If we put all this together, we understand why the slow combustion of phosphorus produces so much light and so little heat. This body begins to burn at a temperature which is only slightly above the freezing point. In this state, air is a very poor conductor, as is the insignificant vapour which is produced by the phosphorus, and finally, the product of the combustion, phosphorous acid, is also an aëriform body. Thus, the entire atmosphere which surrounds the phosphorus is a poor conductor with very little mass at each point, so it is easily made to glow by a combination of the forces, or, in other words, a glowing takes place with a very small quantity of heat. This quantity of heat is easily diffused by the gaseous phosphorous acid which develops and through the immediate transition into the air or into the solid mass of the phosphorus, where a portion of it is already expended at the surface to form new phosphorus vapour. That a diffusion of heat is a very important reason why phosphorus does not produce any heat during its slow combustion can be seen from the fact that stacked pieces of phosphorus gradually heat up internally during the slow combustion and are finally brought to a state of vigorous combustion. In general, we could say that luminescence appears without any significant temperature increase during the glowing of a very small mass if it is combined with a diffusion of heat.

The different colours of the flame at various points are also worthy of examination. They are most beautifully produced with burning alcohol. In a place where strong sunlight does not directly blind the eyes, alcohol can be burnt without a wick in a vessel made of completely white porcelain or faience, and to prevent any illusion, white paper or other uncoloured objects can be placed around it, and it will always be seen that the flame is a very beautiful violet blue at the bottom, a pure blue a little higher up, while the highest part of the flame is white mixed with some yellow. Once the alcohol has become quite hot so that the vapours develop more rapidly, a reddish-yellow colour and, at the uppermost edge, red itself can be noticed. At the upper border of the blue, some green can also be noticed now and then, but it never appears as clearly or as purely as the other colours and cannot in the least be compared with the violet blue and the pure blue of the lower part of the flame. In his work *Zur Farbenlehre*, Goethe declares this colour distribution to be a mere illusion because, according to him, it is supposed to derive from the circumstance that the wick in ordinary candle light, or else some dark object, can be seen through it. Therefore, he claims that no colour can be seen in an alcohol flame if a piece of white paper is placed behind it. However, such a result is not obtained in this experiment provided that the flame is not too small, and strong sunlight, which renders every flame unclear, is not allowed to shine on it. With a sufficiently large flame, this does not prevent us from seeing the colours, at least not their main features. The wick, which promotes the combustion of fatty bodies, and charcoal, which is produced by the combustion of other organic bodies, have no influence on the flame either; for if such bodies are heated in closed vessels, and if the emerging vaporous and gaseous bodies are conducted away through tubes, they can be ignited, and the same colours are obtained in them, against even the whitest background, as in the combustion of the body itself. In the part of the flame which is closest to the burning body, the colour is always found to be violet blue or pure blue but reddish-yellow or red in the most distant part. The strongest light of the flame is usually yellowish-white. In its centre, a dirty-green colour frequently appears. Consequently, it becomes obvious that the colours in which combustive force is dominant really stand out in the part of the flame in which combustive force is dominant, but where igniparous gas is in excess, there colours will appear which are determined by ignitive force; in the centre, an equilibrium of colours and forces will be seen. The old observation of chemists that metallic oxides are easily reduced in the blue region with the aid of bellows, and that, conversely, metals or little-oxidized substances easily attract oxygen in the red region, is in complete accordance with our view. We find similar confirmation in the combustion of hydrogen gas. The entire product of its combustion is a fairly good conductor: steam. Therefore, the activity will not increase sufficiently during combustion in atmospheric air to produce white light, but in the centre there will be a beautiful green, at the bottom violet blue, at the top yellowish-red, occasionally with a red rim if the flow of air is rapid. The green light is, however, dominant. If hydrogen gas is mixed with atmospheric air, in the ratio required to make water, and is then ignited by means of an electrical spark, the combustion occurs with a green light, but if mixed with oxygen, which, however, is dangerous in glass vessels, it occurs with a white light. If we take six parts of oxygen gas for each part of hydrogen gas, the combustion occurs with a strong red light. If highly combustible bodies combine with oxygen in a solution, e.g., a metal in nitric acid, significant degrees of heat are often generated, but they do not begin to glow. In order to explain this, we need not assume that some material substance of light has previously been extracted from the oxygen in the acid; everything can be explained quite easily according to the laws already established. The combination of the combustive force and the ignitive force in a medium which conducts many times better than air can rarely or never occasion the highest tension of the opposite forces according to the laws which we have already found. Only in a few cases do very dry alkalis and highly concentrated acids produce a glowing through their combination. In this case, the velocity of the union and the intensity of the attractive forces are obviously very high.

Now the fact that light produces effects which are entirely different from those of heat does not present any difficulty to us either. The difference lies primarily in the fact that internal[22] substances whose union is furthered by heat are separated again by light, as we see, among many examples, in the relation between oxygen and mercury. However, we now understand that once the tension of the contrasts in a body is at its highest, a separation must finally occur wherever the disposition towards it exists. On the other hand, where this separation of the contrasts is not so considerable, the increase in conduction, due to the internally disturbed calm, will promote the combination far more than prevent it. This may one day be realized by experiment. It can hardly be doubted that we shall succeed in constructing a galvanic apparatus whose power is sufficient to make an enclosed column of water glow immediately upon closing the circuit. Such a device must combine a large quantity of action, as in Davy's apparatus, with great strength. If the water in such a circuit could suddenly be made to glow, it would most likely decompose at every point, i.e., it would immediately turn into oxygen and hydrogen gas everywhere. Perhaps this could even be accomplished by means of an electrical battery if we conducted an electrical discharge through a thin, wet thread. If this had been soaked in a salt solution instead of water, we might have obtained images on coloured paper placed underneath, admittedly not of the ultimate and finest distributions but certainly of the larger ones, as in the experiments on the vaporization of metals by van Marum.[23]

REMARKS CONCERNING THE
MANIFESTATION OF THE GENERAL FORCES
IN ORGANIC NATURE

As we have now found, in all the effects of organic[24] nature, the two forces whose laws we have endeavoured to establish, we now come to the question of whether these forces are also found in organic nature, and whether they might be just as general there as in inorganic nature? In a sense, we can only answer this question by means of general chemical analysis, for this resolves every substance of organic nature to such elements as are also found in inorganic nature, and thus the ultimate

---

[22][Some? Ørsted writes *innere*, but *einige* seems more likely.]

[23]While writing this, the author is travelling and is therefore unable to conduct this experiment.

[24][inorganic.]

elementary forces of chemistry must be the same in all realms of nature. In fact, there can be no objection to this if the subject is viewed with certain limitations, but some might still counter with the objection that all this applies only to destroyed organic bodies, and that in living creatures other forces supervene which completely elude our chemical investigation. We would readily agree with such a person if it was his intention to claim that a higher law, a higher principle of unity prevails in organic nature than in inorganic nature, for this will result in an impartial look at what happens, but to claim that totally new forces are present in organic nature, and not merely new modes of action of well-known forces, is to advance something quite unproven. It is erroneously believed that this will maintain the dignity of organic nature, but the fundamental forces are only the material agency from which all manner of structures can arise just as, for instance, the most perfect example of human beauty and the most insignificant vessel can emerge from the same mechanical substance. Here, everything depends only on the idea of the creative artist, there on the idea of the thing in its endless reason, which we acknowledge in all our investigations as the essence of all the laws of nature which is contained in the nature of things, and which acts as a unity. In order to make this quite clear, we shall examine all the principal phenomena of organic nature and see how the common laws of nature already accepted are reflected in them. In this respect, we are forced to deal primarily with animal bodies because we have so few investigations of plants, but if only we can accomplish here what we have in mind, no-one will doubt that a better understanding will accomplish the same in the vegetable realm.

As is well known, one of the most common phenomena in animal nature is the contraction of muscles. This is certainly stimulated by electricity according to specific laws, so we need not assume the existence of a new natural force for this. Long ago, Ritter demonstrated that similar contractions and, consequently, also the relaxation of contractions, in other words twitches, can also be induced in inorganic bodies by means of the galvanic circuit. Erman, being a true master of the art of experimentation, made this even clearer. The living nerve is a most perfect conductor of electricity, and it can stimulate this force merely by contact with the muscles. Thus, the nerves can make the muscles move without any other force. All that a mechanical stimulus could accomplish here can be explained by the galvanic circuit changed as a result of the changing positions of its parts; on the other hand, the effect of the forces in their galvanic form cannot be derived from a mechanical stimulus. In this respect, any external contact must cause a change in sensations, and therefore, what we usually call the sense of feeling or touch does not require a particular force which is different from the general forces of nature but certainly a mode of action which has by no means been discovered yet, indeed, which may never be discovered completely. With regard to the sense of heat, this is obviously a product of the same forces that act in inorganic nature, for the thermal condition is not only produced in the same way everywhere, it is not only measured by the same means both for living animal bodies and for inanimate nature, but even the sense of heat itself, as we have mentioned earlier, agrees entirely with what we can imagine concerning a stimulation of forces at rest. Besides, it can hardly be doubted that animate activity, according to the laws established by us, does not have to produce a heat of its own. Thus, even animals, which enjoy an autonomous life, also have a more independent temperature, in the production of which we certainly do not wish to deny the role played by respiration; but it must be granted that this is not the only cause of heat. In order to mention but one fact among others, we need only consider how exceptionally quickly a thought can increase or decrease the feeling of heat through the stimulation of passion, which would be impossible if all heat proceeded from the respiratory organs. We noticed earlier that every body, including the inorganic ones, has its own heat. However, the latter, as well as many organic ones, have a far lower independent temperature than the one which they must have as parts of the entire planet, and for this reason, their characteristic temperature can never be revealed.

As long as chemistry has been studied, people have had the conviction that chemical transformations occur in our bodies, and after more definite knowledge of chemical laws has been acquired, it has also been discovered that some processes in organic nature can be derived from the most common chemical processes, especially if we consider only the conditions and the ultimate result, which should content us all the more as, strictly speaking, we would not see more in chemical experiments which we ourselves performed. Only with the discovery of the galvano-chemical effect have we begun to make progress in both directions. The wonderful secretions and absorptions in organic bodies, for which no open passage for the substances was to be seen, are no longer entirely incomprehensible since we know about the conduction of chemical forces and especially about Davy's excellent experiments on the transmission of substances. In future, anatomists may succeed in discovering, in the structure of the components, the conditions for various galvanic effects in animal and vegetable bodies. Already the single circumstance that the fluids which are removed directly from the animal body are acidic, whereas those which are consumed are predominantly alkaline, must encourage anatomists to perform new investigations and physiologists to suggest highly interesting combinations. However, we need not go further. The galvano-chemical circuit reveals a structure of inorganic bodies in which every element contributes to the general effect; and in this we have an instrument which has a certain analogy to the arrangement of organic bodies. However inaccessible the infinite art of organic bodies may remain to us, we shall always be able to reproduce it in the individual parts, and this is already of great importance for our understanding.

The peculiar effect of the galvano-chemical circuit on our sense organs also deserves to be mentioned here. As is well known, the very same conductor which calls forth the acid from a salt also creates a sour taste on the tongue, and the opposite conductor an alkaline taste. This was to be expected from the nature of things, as we know from experiments with inorganic bodies; but in any case we see that we are able to stimulate a force in inorganic nature which passes rapidly through all good conductors, therefore, also the nerves, and which is not accompanied by any substance, at least none apparent to the senses, and this force can activate the conditions of taste or at least two of the most remarkable extremes of taste. A similar though less conspicuous connection can also be found in the case of smell. With regard to the organ of vision, we find there the appearance of a bluish light produced by the conductor of the ignitive force and of a reddish one by the conductor of the combustive force, so here, too, we produce the extremes of visual sensations by means of the galvano-chemical circuit. The fact that the red colour is not aroused by the conductor of the ignitive force nor the blue by the conductor of the combustive force admits the natural explanation, as mentioned by Reinhold, that the retina does not come into direct contact with the conductor but only through the mediation of the fluid in the eye. Therefore, the optic nerve is actually to be regarded as the continuation of the opposite conductor; consequently, in the optic nerve, dominant combustive force is really associated with blue, dominant ignitive force with red light. The inversion of the effect, which we notice when we leave the circuit, that, for example, an alkaline taste follows the previous acid taste, can probably be explained by the fact that the nerves oxidize or hydrogenate while in the circuit and, upon leaving, therefore form new circuits in which the previous conductor of the combustive force, which is now oxidized, plays the part of a less combustible body, and the previous conductor of the ignitive force, which is hydrogenated, assumes that of the more combustible body. It is not easy to explain precisely why, according to Ritter, almost all effects on living bodies are reversed if the strength of the pile is increased considerably, but we can say in general that such an inversion must be based on the fact that, as a result of the stronger conduction, rapid and vigorous combinations occur where contrasts associated with very slow and weak combinations remained in the less powerful circuits.

Before we leave this subject, we must mention that the well-known experience of the exhaustion of sensitivity through increased stimulation can very easily be related to our view of things. This follows from a far more general law: The more dominant a force becomes in a body, the larger is the resistance in it to a force of the same kind, and the more the opposite force, through whose attraction the effect of the force is actually received or absorbed, is overwhelmed. For that reason, we must also assume the existence of stimuli of the opposite kind, without which the restoration of sensitivity would not actually be possible. In this way, for instance, it is easy to explain why, for one or two given colours, we see only the complementary ones in all white objects when our eyes grow fatigued; i.e., red if the eye has lingered on green (yellow and blue) for a long time, yellowish-red if it has become tired by looking at blue, etc.

These remarks may now suffice to demonstrate to chemists the general occurrence of the combustive force and the ignitive force even in organic nature. The physiologist will find proof of this throughout his science.

OBSERVATIONS ON MAGNETISM

In order to prove from every side the universality of the two forces which we have examined, we should not ignore magnetism. There has always been a tendency to compare its forces with those of electricity. The clear manifestation of the attractive and repulsive forces in electricity and magnetism and the perfect similarity of their laws must necessarily occasion this comparison. It is true that nothing was found in it which would correspond to the so-called conduction of electricity, but what was found was so perfectly comparable to induction that it was almost impossible to discover any difference; for example, if we want to compare these effects with those in a newly discharged Leyden jar, or even better, with a discharged disc condenser stripped of its coating. With such a device, too, conduction is never possible, only induction. We also observe an electrical induction of forces in each sheet if the disc is composed of several thin lamina, similar to a magnetic induction of forces in each piece of a magnet. A great difficulty appears in the circumstance that electrified and magnetized bodies have no attractive or repulsive effect on one another as a consequence of their condition. It would be of the greatest interest to eliminate this difficulty completely; but as we see no prospect of that given the present state of affairs, we shall content ourselves with drawing attention to the fact that even frictional electricity and contact electricity show similar and quite remarkable phenomena in their attractions and repulsions for each other. For instance, it is easy to cancel the repulsion of an electrical pile completely by approaching a rubbed glass rod and to induce an attraction or an opposite repulsion in its place, yet not only does the chemical effect remain undisturbed, but a long channel of water, such as a circuit containing a moistened thread, undergoes the same internal chemical change as usual, despite the fact that it has acquired a completely opposite electroscopic state due to an induction from outside. Thus, it seems that the forces in various modes of action can cross or meet without disturbing one another. The mode of action in the circuit, or the galvanic effect, stands between the pure electrical and the magnetic modes of action in that its forces are bound far more than in the former and far less than in the latter. Consequently, it is not unlikely that the electrical forces can also cross or meet the magnetic forces much more easily without disturbance. In the galvanic pile, it is actually the electrical state of the system as a whole and not that of an active electrical circuit which is changed by the approach of the rubbed glass rod; it is the same with magnetism. It is not the internal distribution, which constitutes magnetism, that is changed by it but the electrical state, which comes to the magnetic body merely as a body. We openly admit that this falls far short of explaining the matter, but we shall be satisfied if we can present this difficulty for further investigation as a subject which is not understood, but which does not contradict our view.

The remaining similarities between magnetism and electricity are so great that we need only remove the apparent contradictions in order to accept the identity of the forces in them. Even the fact that the magnetism of steel disappears with annealing is one of these similarities, for is this other than that, by being annealed, it becomes a far better conductor in which the opposite forces combine so much more easily? The change of the magnetic needle during thunderstorms is another example of the relation between the two. Ritter has also found that magnetized iron wire is less oxidizable at its northern end and more oxidizable at its southern end than iron, but iron or soft steel must be used here because harder steel produces less activity and, in fact, in the reversed order due to its poorer conduction and its correspondingly smaller quantity of force. Under similar conditions, muscular contractions are also induced in a prepared frog if two opposite poles of a magnetized iron wire are connected to it in such a way that a closed circuit can be formed. The wires must be magnetized by means of relatively strong magnets. These experiments are still somewhat disputed by physicists, but so many have been successful that it is not easy to assume a false conclusion. Ten years ago, Ritter speculated that the *aurora borealis* was of magnetic origin. Now, a Danish mathematician, Hansteen, has demonstrated with very important arguments that the earth, as claimed before by Halley, has four magnetic poles. He has determined their locations at various times and found that polar lights, whose centre he places, as do others, in the dark spot (of those lights which can be observed in their totality), are completely correlated with these magnetic poles. Therefore, all the functions which can be demonstrated in electricity can also be observed in magnetism: attractions and repulsions, chemical difference, effects on the living animal body, the production of light. Their similarity is unmistakable even though not all have been demonstrated with the same clarity. It would be worthwhile pursuing the recent discovery that all bodies are actually capable of assuming some magnetism, and that, therefore, the magnetic mode of action can occur in all bodies, although in highly varying degrees. At the same time, an attempt should be made to discover if it is possible to produce some effect on the magnet as a magnet in one of the states in which electricity is found to be extremely bound. This would not be without difficulty because the electricity would react on both magnetic and non-magnetic bodies, but perhaps it would be possible to obtain some information on this by comparing magnetic and non-magnetic needles.

GENERAL OBSERVATIONS REGARDING THE TWO FUNDAMENTAL FORCES

In our previous investigations, it became obvious that the two forces which we first introduced under the names of the combustive force and the ignitive force are spread throughout nature, and that all effects which we have so far been able to investigate and to establish as laws are to be regarded as different modes of action of these forces. This is true not only in a chemical, but also in a mechanical respect, and we also understand that nothing more than these forces is necessary even for the filling of space.

The thought that space is filled by repulsive and attractive forces has been expressed many times with various modifications. Some associated them with atoms which attract and repel each other, others imagined them as free forces distributed throughout space. The former are divided into those who still attribute an extension to these atoms although they assume it to be extremely small,[25] and those who regard atoms merely as points endowed with attractive and repulsive forces. They are also divided because some, like Knight, assume that similar atoms repel and dissimilar ones attract each other. Others regard some atoms as repulsive, some as attractive, and, finally, still others attribute to atoms a force which is attractive at larger distances but repulsive at very small ones. Among these views, Knight's comes closest to our own, which assumes that there are two kinds of atoms, with no extension and with the property that those of the same kind repel each other, and those of a different kind attract each other. We shall not attempt to criticize these different systems here. The atomistic system in general has long been subjected to fairly sharp criticism in Germany, and this criticism can also be applied to its variants. However, no objection can be made to Knight's concept because of the limited divisibility of matter, but only because of what can be said against an atomistic view in general, viz., that it makes a futile attempt to explain the whole on the basis of its parts, something of which Knight was actually less guilty than any other atomist.[26] In his results, however, as mentioned earlier, he comes exceptionally close to what we have found by experience.

In the dynamic view, in so far as it has been applied to experimental science, a particular repulsive or expansive force and another attractive or contractive force are assumed. It is true that we have also found two forces in our survey of experience, but at the same time, we have found that each is independently expansive and repulsive, whereas both are attractive with respect to each other. If we observe the matter more closely, it would also seem that it cannot be otherwise, for when we imagine a force, we must also imagine an outward striving which, in space, can only be described as expansion. A purely contractive or purely attractive force,

[25] Priestley, for instance, does not regard it as impossible that all matter, i.e., all the atoms in the universe, could be contained in a nutshell. The gap between this small extension and its complete elimination is not large. However, all students of nature who have observed chemical phenomena with a minumum of attention realize that, in any case, the size of atoms in bodies must be very small. For instance, let us imagine ammonia gas and carbonic acid gas, each under a pressure of only one line of mercury, i.e., very diluted. Let us further imagine that they are reduced to normal atmospheric pressure, then the space in which the atoms are confined has become over 330 times smaller. Finally, let us imagine these substances chemically combined, in which case they form a solid body of approximately the specific weight of water. Here, then, we have a concentration which is more than 25,000 times greater, and yet a space was filled in both cases. Nothing, however, prevents us from imagining an even greater dilution. We are also convinced that carbonic ammonium is still far from being a filled space. This goes even further if we imagine oxygen and hydrogen from their highest possible dilution as a gas to their greatest contraction in combination with sulphuric acid. It is easy to see that a space can be completely filled by atoms which contain no more than one hundred-thousandth of its volume, indeed not even one millionth; and there is still no limit. The caloric which, as might be assumed, escapes during the condensation need not be considered, for it weighs virtually nothing, so it can have only a very small volume of atoms.

[26] Knight's work is difficult to find. Its title is: *Attempt to demonstrate, that all the phænomena in nature may be explained by two simple principles, attraction and repulsion.* Lond. 1748. Through its elevation to the ultimate origin of things, even in its generality, it comes very close to modern natural philosophy.

then, would not be possible but must always be regarded as the product of two other forces. A possible answer to this might be that no such distinct independence should be attributed to the forces, rather a simultaneous presence in space should be assumed, and in general the attractive force should be regarded as originally a contractive force and the attraction of celestial bodies, for example, merely as a manifestation of the contracting force in the universe. However, no matter how inseparable the forces may be imagined to be, if they are merely regarded as two opposite aspects of a single, space-filling activity, the mind must still be able to distinguish between them as two activities and comprehend each in itself, but then the same difficulty returns in the imagination which we previously believed to find in existence. If this difficulty is to be avoided, the expansive force must be regarded as primary and the attractive force as secondary. However, the cause of attraction cannot lie in the expansive force itself, so it must be something else which determines contraction, and thus we would be brought back to our first view of two forces, each separately conceivable.

However, given our point of view, the question now arises: In what respect do these two forces differ since we cannot establish their dissimilarity through the direction of their activity alone? Although this is beyond the scope of our investigation, we shall nevertheless attempt an answer. Being well aware of how difficult it is not to err in this matter, we beg the reader to consider this answer itself merely as a more specific question which might elicit a real answer. —Let us imagine all of space pervaded by an elemental force, of which our two forces would merely be different modes of action. Within each force, each activity, there is a striving outwards in all directions which must manifest itself at every point so that space becomes filled. We could call this striving from a point in all directions the radiant activity, whereby, of course, we must not think of isolated rays. We also wish to call it the *positive* and the opposite the *negative*, invoking the accepted mathematical arbitrariness even though immediate experience also supports this choice. However, the radiating point also experiences an influence from all the other surrounding points; it reacts against this, and with the same strength against all equally distant points. In this respect, then, the force would act as a concentric force, forming infinite, merging boundaries. This would be the mode of the negative force. In so far as the activity of one point is merely due to itself, we call it *positive*, but in so far as it is determined by others, it is *negative*. If, however, these modes of action are the most primary in space, they will also form the basis of all the others, but they themselves will never cease to exist. If we wish to assume some separation of forces with regard to the whole of space, an assumption which any natural philosophy must make and must justify, then the distribution of self-determination and of being-determined becomes different, and thus the heterogeneity of matter becomes understandable. If we wanted to follow a very remarkable but perhaps misleading indication, we might imagine that the radiant activity was dominant in the combustive force and in positive electricity, but the opposite would manifest itself in the ignitive force and in negative electricity. This is our view of a topic in which we can hardly hope to have done more than increased the number of directions for investigation in a manner which is not without value.

Several chemists seem to be of the opinion that all diversity of matter may be explained from differences in the degree of the filling of space and from the dissimilar proportions of the forces. This is associated with the theory that chemical combination consists only in the combined substances together occupying space with continuity, and this is really implicit in the previous assumption. However, if there were no other differences, then all bodies must be able to combine with one another in all possible proportions. Hardly anything can be said against this conclusion, for he who wishes to seek the reason why this is not so in the difference between solids and liquids or something similar would first have to explain how only solidity is possible as long as one remains with this difference in density and proportion; not to mention that the example of several combinations of solid, liquid, and gaseous bodies provides ample proof that it is not this difference of form which prevents combination. According to this assumption, it would also be easy to induce the disposition for any polarity in every body; for in no combination would it be possible to say that two or more different substances were combined, but only that a certain proportion of forces existed in which any distribution was equally possible. Experience, however, shows quite the opposite. Completely different compositions can even be formed from the same elements and only separated again in a specific manner. For example, it is easy to prepare a combination of ammonium and water which has the same basic components and the same proportions of the basic components as a given dilution of nitric acid, but the same procedure will produce entirely different educts from one of these combinations than from the other. Consequently, we must admit that bodies retain their particular mode of action in their combinations with others, and that the uniform occupation of space by forces is therefore not sufficient.

Once the existence of two such fundamental forces has been assumed, as we have done, it follows immediately that any predominance manifests its effect by attracting the opposite force and repelling the similar one. In so far as the propagation of this effect encounters resistance, a more or less lasting internal difference will arise; indeed, we have seen that this can even become permanent. Because of the general interaction between all bodies, none will escape such a distribution of forces, but in each one, even if imperceptible to the senses, points of opposite activity will be separated from their point of equilibrium; for allowing the antitheses to merge ad infinitum means nothing but neutralizing them again. On the other hand, in any antithesis new ones can be imagined, in these again new, and so on, and here it does not seem necessary to set limits. These various depths of antitheses might offer us the best explanation of several dissimilar degrees of combination. If, for example, hydrogen gas and oxygen gas combine as gases, this should happen with less deeply interfering antitheses; if, on the other hand, both become better conductors through glowing, or if a sudden pressure produces an internal motion, they combine through the engagement of deeper antitheses, and thus water is produced.

However, this whole discussion of the distribution of forces still gives an inadequate picture of the internal variety of substances if we do not look at things in their entirety. We must always bear in mind that all these antitheses exist only because of the development of the whole and are supported by continuous interaction with it. This internal differentiation of substances must be imagined as an organizing process. And just as we would have described the internal structure of a plant only poorly if we said that it consisted of vessels which again contained other, finer ones, and so forth, as far as this artificial decomposition can be taken, so we have also spoken of the dynamic organization of bodies in the most inadequate generality by pointing to the internal antitheses. However, this may serve as the start of a more extensive investigation. In this spirit, we add here some observations on the differences between bodies. It becomes clear from what has been said already that bodies can differ in the degree with which the forces fill space, in the predominance of one of the forces, and in its mode of action. The last always constitutes the fundamental difference between substances. The simplest mode of action, perceived in our experience, manifests itself in the first of our series, that of uncombusted bodies. The flexibility and the conductivity which are so frequently found in them also seem to come from this. There is significantly greater complexity in the forms of activity in the second series. There, all the antitheses found in the bodies of the first series are combined with new ones. In the third series, the salts, the antitheses of the two previous series, are again associated with new ones. That the antitheses of the first series really still exist in the third is seen from the fact that they can be restored again by powerful agents, e.g., a less combustible metal can be obtained from a salt by a more combustible one. There are undeniably one or several levels of simplicity in the modes of action which are even higher than those in the series of metals. We shall not have a completely scientific chemistry until their nature has been investigated through some examination of the elementary forces. The experimental investigation may first have to perform much preliminary work by the discovery of the components of metals and perhaps by even more discoveries, for nature's clues can only cause the mind to produce what lies within it.

The more combustible bodies differ from the less combustible or the more igniparous ones, the stronger alkalis from the weaker and from acids, etc. because of differences of the forces. The character of every alkali or every acid is determined by the antitheses in the substances of the first series and by the coherence of the combination which, like the relative proportion of the components, can be different for the same substances. The character of the substances in the first series certainly has a similar relation to an even higher one. And so it must continue until we encounter a series in which we find the elementary law for the first separation of forces.

Wherever we discover different degrees in the filling of space as the only difference, as for instance in air of different compression or in bodies of a dissimilar temperature, we do not acknowledge a difference of substance, but combined with other differences these may, of course, help explain a greater dissimilarity.

It will be found that our view in a way approaches the atomistic theory, but we shall not take this as a reproach since this theory has also served to describe nature to profound men and therefore must, in some sense, contain elements of truth. The inner difference between the atomistic theory and the dynamic, however, remains;

the former wants to compose the whole from already-finished parts, whereas the latter allows the whole to develop together with its parts as different forms of an elemental force. This difference becomes clear only at the boundary of science, for every investigation of nature which is not purely speculative begins with pre-existing objects, studies their effects, and investigates the laws by which the effect is determined. From specific laws we try again to construct more general ones, and so on. The investigation of nature has been pursued in this spirit throughout the last centuries, and this is the way in which we have conducted the entire preceding investigation. Therefore, there can be no question about ultimate principles until the investigation has reached its end. However, another question is the extent to which a deeper insight can serve as the guiding idea, or a wrong metaphysics can lead astray. For instance, it is undeniable that pure atomism, where everything is to be explained solely on the basis of the size, the form, and the motion of atoms, has had a limiting and misleading influence on natural science. However, since this purely mechanical concept has had to yield to the contemplation of chemical activity, the assumption of pure dynamics or of a dynamic atomism will not create any great differences in the assessment of facts or the establishment of natural laws. What all true students of nature share is the conviction of the regularity[27] of nature, and that this cannot but agree with reason. This realization has great consequences. A clearer perspective soon adds to this that there is nothing dead and rigid in nature, but that every thing exists only as a result of an evolution, that this evolution proceeds according to laws, and that, therefore, the essence of every thing is based on the totality of the laws or on the unity of the laws, i.e., the higher law by which it has been created. Every thing, however, must again be regarded as an active agent of a more comprehensive whole, which again belongs to a higher whole so that only the great All sets the limit of this progression. And thus the universe itself would be regarded as the totality of the evolutions, and its law would be the unity of all other laws. However, what finally gives the study of nature its ultimate meaning is the clear understanding that natural laws are identical with laws of reason, so they are in their application like thoughts; the totality of the laws of an object, regarded as its essence, is therefore an idea of Nature, and the law or the essence of the universe is the quintessence of all ideas, identical with absolute Reason. And so we see all of nature as the *manifestation* of one infinite force and one infinite reason united, as the *revelation of God*.

To contemplate this is the highest in natural science, and we believe that an atomistic theory merging with a dynamic neither hinders the achievement of this view substantially nor alters the meaning and spirit of the study of nature, although it might be confusing at the most extreme point of the union of this system with speculation, and therefore only for those who attempt this unification. We are inclined to believe, however, that the system itself would change for one who did this. Even more confidently, we can assert that a dynamic system which approaches atomism in its more detailed description, even if it were in error, —that is, if it did not correctly describe the connection between the highest laws of experience and the

---

[27] [Here and below, "regularity" renders *Gesetzmäßigkeit*.]

prototypes of existence, would, in fact, like any other error to which we are subject, blur the great image of truth but in no way destroy the true spirit, we might say the religion, of natural science.

GENERAL OBSERVATIONS REGARDING THE STATE OF CHEMISTRY

Through the expansion which chemistry has experienced as the result of recent discoveries, the determination of its contents, its boundaries, and its relation to other sciences also becomes much easier in many respects. As long as chemistry was regarded merely as a theory of the constituents of bodies, it was always necessary to distinguish it very clearly from general physics because of the special nature of its contents. It also remained uncertain whether the theories of heat, light, electricity, etc., should be included or not, depending on whether it was desired to attribute an independent material cause to them or to deny it. Some even felt themselves forced into the inconsistency of always treating heat within chemistry while excluding electricity and magnetism from it even if they assumed a material cause for these, although it cannot be denied that the combinations and separations of an electrical or magnetic substance, by definition, had to be part of chemistry just like those of all other substances. Now that we not only see the intimate relation between the electrical and the chemical effect in so many facts but are even convinced of their identity, we have no doubt about the contents of our science from this perspective. However, in order not to appear to proceed arbitrarily, we shall begin at a higher level. Obviously, all science which can be gleaned from nature along the path of experience is first divided into two large sections, one which *describes what objects are like*, and another which *explains how bodies affect one another or change*. The former is the *description of nature*, the latter *physics*. To understand how bodies act or change is, in other words, to understand the laws according to which this happens. *Therefore, physics is the science of natural laws.* The laws which we can demonstrate in all bodies are called *general*, those which occur only in a certain closed entity, such as the large system of motion of the planets, the evolution of the earth, or the living body, are called *special*. The general laws are dealt with in the science which is usually called *physics*,[28] and which could now also be called *general physics*.[29] As we recognize all activity through its manifestations, we can also classify general physics according to possible changes. These are either merely external or internal. The purely external change is nothing more than a change of position with respect to other bodies, i.e., *motion*. A purely internal change must be a change in the properties of things. As these properties can never be recognized except through their effect, they can also, as acting properties, be called *forces*. The *theory of motion* and the *theory of force*

[28] [*Physik.*]
[29] [*allgemeine Naturlehre.*]

are, therefore, the two branches of physics, to which another could be added in which they both combine even though it has not yet been regarded as distinct. The theory of force is identical with what we must call *chemistry* according to our more comprehensive view. So far it has been believed that chemistry must be included in special physics, but in such a way that it was also accepted in the theoretical system which we now call general physics. This was considered necessary because chemistry dealt with substances, i.e., with something specific. Now we know, however, that chemistry deals with forces which are as general as motion, and therefore with very good reason, it is incorporated in general physics.

It is gratifying to see how chemistry has gradually ascended to such a generality in its laws, and how the devoted efforts of so many scientists, although they often moved in opposite directions, in the end still meet at one focal point, in one unity, as if guided by a higher principle (the general harmony of reason). Unfortunately, whenever the development of a new theory was discussed, it was always the custom to point out only the mistakes of our teachers, of our forefathers, without considering their merits and their true scientific worth. The inevitable consequence of this was that, even if we were delighted by the new truth, we could not, on the other hand, ward off a certain despondency thinking about so many successive generations who had apparently done nothing but err, and that we, therefore, could expect no other fate. It is also sufficiently well known that quite a few high-minded if not also influential friends of science, overwhelmed by this despondency, have doubted all knowledge. This thoroughly self-punishing ingratitude towards our forefathers seems to derive from a false notion of the goal of science, in that this was believed to be primarily the discovery of the causes of phenomena and not their laws. It is true that these are ultimately the same but not during the process of development. Thus, for instance, enthusiastic investigations have been made to determine whether or not the existence of a caloric fluid must be assumed, at a time, or rather in a context, where a decision was not yet possible. However, the complete review of the laws according to which heat is produced under the most different circumstances now shows us what we must think about this. The supporters of the caloric had partial knowledge of the laws according to which heat behaves and, to this extent, enjoyed the possession of some beautiful and comprehensive truth. The same is true about the theory of electricity. Only recently, an otherwise not unknown student of nature stated that we would deny Franklin his most substantial contribution to the theory of electricity if we refused to agree with him in accepting only *one* electrical substance. This is a totally incorrect view of the matter. It is not the feat of inventing such a substance which required the genius of a Franklin, or which constitutes the essence of his theory, but the great law of antithesis with all that flows from it; and it would be absurd to compare him with some inventor of a second substance or some denier of both if he had not discovered other natural laws of equal importance. The same is true for the theory of chemistry. It is both a ridiculous and an annoying spectacle to witness the contempt with which several writers on chemistry refer to the phlogistic era and scarcely want to grant it any credit other than for having accumulated facts for the illuminating perspicacity of the antiphlogistians. If phlogiston is rejected, it seems as if the entire phlogistic theory has been renounced, but basically its essence remains. Great merit is due to the creators of this system for having seen combustion as the centre of chemical effects. It is true that a certain feeling for nature already assigns it to this position since we see fire, on the one hand, act on the bodies exposed to it with the greatest decomposing and destroying force but, on the other, spread vivifying light and heat so that it produces the elemental forces of life and destruction at the same time. This, however, was not sufficient; it was also necessary to see how an effect, in its consequences completely like combustion but without an outward flame, spreads throughout nature. A mind of great penetration and daring is certainly needed to recognize combustion where the usual characteristics, light and heat, are absent. However, not to overlook combustion if it occurs, against every conception of everyday life, slowly and without perceptible heat in a liquid and to discover the *definite* similarity of respiration to combustion: This truly requires the preparatory investigations of several generations. This beautiful discovery, that *all bodies are related to combustion in a manner which determines all other effects*, constitutes the basis of the phlogistic theory. Now the principal law was established, *that bodies, when they burn, lose their combustibility, but that a burnt body can regain its combustibility through an even more combustible one.* Consequently, the phlogistians very clearly recognized combustion and restoration as two opposite general processes of nature and acknowledged their importance. Nor was it unknown to them that many bodies turn into acids upon losing their combustibility, and that this greatly depends on the magnitude of the lost combustibility. Thus they recognized the same half of the law of the transformation of bodies due to combustion that the antiphlogistians know, and in a certain respect even more vividly. Therefore, the essence of the phlogistic theory is not the invention of phlogiston as a cause of combustion but the discovery of some of the laws of combustion.

This justice towards the phlogistians still leaves the antiphlogistians with their true merit. The laws of combustion were not known well enough as long as a *material* principle of combustibility was assumed because it followed from this that combustion was regarded as a decomposition even though its clearly material part is a combination. It was undeniably of the utmost importance to realize this. Lavoisier taught us this and thereby shed light on many very complicated phenomena. The discovery of the composition of water, nitric acid, and ammonia is the result of this fundamental discovery. However, especially through our knowledge of the composition of water, we learned to assess more accurately a far-reaching regularity in countless phenomena, partly of inorganic but primarily of organic nature. On the other hand, a certain vividness in the conception of combustion was lost, which we cannot deny the phlogistians. It is clear that the antiphlogistic system is a continuation of the phlogistic, but with regard to its more specific form, a continuation in the opposite direction. In one, special attention has been directed towards the conditions of combustion from the perspective of the combustible bodies, in the other, from the perspective of the igniparous bodies, so each system has its one-sidedness as well as its merits.

Recent discoveries have taught us to combine the good parts of both. We do not regard combustibility as an empty capacity for an igniparous substance or the igniparous property as a mere capacity for a combustible substance, but in each we see an ever-active force, both of equal dignity. However, we have been able to understand the crucial point of our consideration more clearly than others because we have recognized the chemical forces in their free state. *Pure combustion* is really a combustion in which the combustive and ignitive forces combine directly, and in which, to remain within our scheme, the combustive force burns by means of the ignitive force. It will not be uninteresting for the philosopher to notice that this union of forces, if it happens throughout without disturbance, must be without any perceptible consequence, so this process appears to the senses only in so far as it is disturbed or interrupted. Therefore, light, heat, and combustion exist in that pure union of forces, but with a transparency and indivisibility inaccessible to our senses. Being-in-itself is here the actual form of light. Mutual engagement is the form of heat, the identification of activities that of combustion. However, we leave this investigation to the philosophers. Even in our time, they have eagerly tried to compare the archetypes of being and existence with their manifestations. It is to be hoped that a better arrangement of our experiences will facilitate their great endeavour. But we shall return to our original subject. From what we have called pure combustion or the union of forces, accompanied by the generation of light and heat, we possess our knowledge of the primary functions of combustion in general; viz., of light as a union of forces, of heat as their mutual interaction, and of the consumption of bodies in combustion as the neutralization of the forces by combination. We have already seen all this in material combustion although we have not discussed combustion specifically; but because the entire theory of chemistry can be regarded as a theory of combustion in all its consequences and potencies, only the completely developed theory can explain this phenomenon. In order not to overlook anything which might be useful for an explanation of our theory, we now wish to go through another series of phenomena which belongs to a theory of combustion and at the same time refer to the explanations already given.

Prior heating is necessary for a combustion in air. We have seen that the *general* reason for this is the increased conduction brought about by heat. Consequently, this preparation is not always necessary in good conductors, such as liquids containing oxygen.

Previously, it was said that combustion cannot occur in a gas in which there is no oxygen. We express the conditions for combustion far more generally when we say that we need a body with a predominant, chemically free combustive force and another with a corresponding ignitive force.

We have described the production of heat and light during combustion in detail and demonstrated its necessary connection with the nature of combustible and igniparous bodies without any new, unconfirmed assumption, such as that of an inherent substantiality for light or heat in bodies. It has not been difficult to derive from our basic concept the reasons why no light appears during some combustions, and why no perceptible heat develops during others. Finally, the different colours and peculiarities of the flame readily followed from it as well. With regard to these last points, both of the older theories, as is well known, offered little or nothing.

A body loses its combustibility through combustion, and the air in which the combustion occurs loses its igniparous property if it is not absorbed as well. In this behaviour, we have seen the effect of a binding of opposite forces.

Combustion is accompanied by contraction. We have shown the connection between this phenomenon and the nature of forces.

Both the older systems state that many bodies become acidic through combustion. Recent experiences add to this that many also become alkaline thereby. We have seen that alkalinity and acidity are nothing but muted combustive force and ignitive force.

From this we explain the neutralizing property exhibited by these opposite products of combustion towards one another.

It also follows naturally from our view that the union of acids and alkalis must produce heat, for we can, in fact, regard this as a combustion on a lower level, of a lower order.

It is the same with regard to the contraction which alkalis and acids experience when they combine. The greatest possible binding of the forces leaves nothing but cohesion. Salts do not have an absolute maximum of cohesion, nor can they reach an absolute chemical indifference, only because the binding is never complete, neither qualitatively nor quantitatively. Nevertheless, we see that chemists have rightly identified this final manifestation of all chemical forces with coherence.

Anyone who has followed the entire foregoing investigation attentively, and who is not unfamiliar with the older views, will easily understand how superior the theory resulting from the most recent experiences is to the earlier ones with respect to extent and internal consistency; this was to be expected given the increased number of facts which we have acquired since the establishment of the antiphlogistic theory. It will be seen that it was our intention throughout to discover laws, not to identify causes which, incidentally, are easily obtained from the laws discovered. As for the causal, it would not be difficult to translate our view into another, e.g., to assume two electrical substances instead of forces, indeed, to imagine that all bodies are composed of two such elements. Apart from some difficulties, such a view would make it possible to use all the regularity established here. Only at a point which lies beyond the presentation of experiences would the true difficulty arise. However, it is quite gratifying that the discovery and description of regularity, i.e., the manifestation of reason in nature, is possible for man to a certain extent even if he has not grasped the highest principles of knowledge. This reassures the friend of truth without dulling him, for he knows full well that the brightest light will not be found until he discovers the right point of view.

Such a conception of natural laws lay not only in the phlogistic or the antiphlogistic system, as we have mentioned often enough, but we find several profound insights into the laws of nature even in the mystic age of alchemy. If we ask ourselves what the most important problem of chemistry is today, the answer must undoubtedly be the decomposition and composition of metals, for these constitute the basic series of all our chemical substances. Without knowing what lies within them, we shall never be able to give a thorough account of the differences of other bodies. For a time, the hope for such knowledge of metals was considered absurd and ridiculous, but at present, its admissibility can hardly be doubted by experts if, on the one hand, we consider that so many similar substances which merge into each other through so many nuances must, according to all analogies with other chemical conditions, have an identity of composition; and then further consider that, with the discovery of ammonia metal, we must already either have a composite metal or expect the decomposition of bodies which are otherwise believed to be entirely simple. The alchemists certainly did not acquire their opinion about the importance of metals in chemical knowledge by accident, but because they saw metals constantly rising from their ashes and, therefore, regarded them as the most indestructible substances on earth, which is why they placed gold above all, as the most indestructible among them. This was obviously the scientific view of the matter; concerning the selfish view, we may only mention here that, in the end, this can also be traced back to the indestructibility of the noble metals. The hope for a universal medicine, a universal solvent, has given cause for the most foolish delusions, but it is based on the underlying truth that the decomposition of metals would give us the key to pharmacology and chemistry. However much this is buried in misconceptions, it is nevertheless a true presentiment which cannot be denied the alchemists, emerging from the contemplation of the antithesis between destruction and restoration. Even their comparison of the metals with the great celestial bodies might be a presentiment which is not yet sufficiently comprehensible to us. For if metals developed with the earth, and this with the other celestial bodies of our solar system (comets as well as planets and their satellites), how very likely, indeed, it could be said, how certain is it not that both developments proceeded according to the same laws, only on different scales! But it is not our intention to deny that we must regard their applications of this presentiment as often quite brazen, often even wrong. It is sufficient for us merely to see how, from different perspectives and in different eras, the striving for truth has never been completely fruitless. On the contrary, we see that these diverse attempts all lead to one goal. They must do so in so far as the entire civilized human race at once constitutes a rational whole, and the various directions, as parts of this whole, must necessarily dissolve in its harmony. We have seen many examples of this above, and an attentive eye will easily find confirmation everywhere. Even one-sidedness is part of this development, for without the one-sidedness of the individual, which is not that for the whole, there is no direction to the development, without struggle no awakening, no testing, and no strengthening of the forces. The development of the human spirit can well be compared with the development of an organic being. In both, there is an original unity in space and time in their individual functions, in both there is this separation of directions, this struggle of the forces for the development and generation of life, but in both we also see that this leads only to unity. If we may use a daring expression, strife provides the vital warmth of science, the union of antitheses is its light and love. One, as well as the other, is indispensable to it.

This parallel between knowledge and life courses through the entire history of mankind even if it is not recognizable everywhere with the same ease. Let us just look at metals. Science has now taught us that everything in the earth which is not uncombusted metal is combusted. Consequently, metals constitute the dynamic core of the earth, and this already indicates the magnitude of their significance in nature. Who does not know their significance in the history of mankind? Who does not know how the ability to work metals has kept pace with all other development and interacted with it? It was no wonder that the most indestructible on earth were chosen as the standard for all worldly, material values, and that the most indestructible of these indestructibles, the noble metals, were again given precedence among them. Gold was the metal which man had to discover first because, among all metals, it is the one which appears most frequently in its pure metallic form, not only in the earth but also in the sand of rivers. More like happy children than acquisitive men, to them this most noble of metals became merely a shiny toy. This was the childhood of mankind and its golden age together. Silver, which does not present itself so openly and often hides its metallic nature in foreign forms, was only discovered with greater attentiveness. Men had already begun to pay more attention to the value of things, they compared objects more and made greater efforts to acquire them; and thus the discovery of silver coincided with the transition of the human race from early childhood to a somewhat greater awareness. The age of gold and silver was followed by that of copper. This is truly the metal which had to be discovered after them.[30] It certainly had a more decisive influence on the condition of mankind. It was not used for jewelry and drinking vessels but for weapons which brought artifice to battles between men and gave some predominance to the more civilized over the barbarians. From then on, however, men also gathered in larger masses for combat, passions became more complicated, and the life of men less carefree. Iron was discovered far later. At least in certain areas, there is historical evidence of this. However, science also tells us that iron, which is found almost exclusively in a combusted state, can only be restored to its metallic nature by means of a laborious and artificial process. Consequently, a people which knows how to work iron must stand at a higher level of development than its predecessors, but the much-glorified carefreeness and the ignorance of many evils of the old days were gone, and therefore the poets have profoundedly disparaged this era, named after iron. We probably need not mention that all this applies only to the natural development of men and not to the knowledge which has been passed on from one earlier civilization to another. Probably, no one will object that we refer only to myth or poetry, for we can undoubtedly assume this much about the natural history of mankind in the old myths. Now the poetry stops, but the harmony between the history of mankind and the knowledge and use of the basic elements of our earth does not cease with it. Let us look only at all the degrees in the preparation of iron, from the rawest founding to the most perfectly hardened and polished steel; all that constitutes the convenience and ornament of life is connected with this. Even our own science, which requires all manner of assistance from the mechanical arts,

---

[30]Naturally, local circumstances can make a difference. Here, we are speaking only about the case where all the more important metals are found in the domain of a developing human community.

would still be in its infancy without the perfect working of this metal. At first glance, it is not so easy to see what has occurred because of the other, less important metals, and this cannot be described here in view of our primary goal. It was reserved for our era to describe the metals which were the most difficult of all to bring back from their ashes. They will certainly never be employed for mechanical purposes, but all the more for dynamic ones. They have already initiated a new era in our investigations of the composition of bodies. But what this means can be judged only by one who has seen in history what dominance over nature science has gradually won for us. The coming age will decompose metals themselves and thus loose the many bonds with which nature has tied us to our external circumstances and, generally, win for us a hitherto unknown dominance over nature.

> The secret's ancient, mighty spell
> By insight's hand alone is freed;
> When revelation does succeed,
> The dawn of freedom comes as well.

And thus the author, who otherwise cannot compare what he has done with what there was to strive for without an inner sense of melancholy, can comfort himself that he has participated in a work which is ever progressing and ever leading closer to a glorious goal.

___

31

On the Law of Electrical Attraction[1]

___

Copenhagen, Nov. 22, 1814.

I HAVE finished a very significant series of experiments on the decrease of electric forces with increasing distances, from which it turned out that the forces are in an inverse relation neither to the distances nor to the squares of the distances, but that the decrease can only be expressed by a series. Immediately after the New Year, you will receive a paper about this in which some not insignificant conclusions will also be drawn from this law.

___

[1] [*Journal für Chemie und Physik*, ed. by Dr. J.S.C. Schweigger, Vol. 12, p. 106. Nuremberg 1814. The same subject is considered in: *Videnskabernes Selskabs Oversigter*, 1814–15, p. 9. Copenhagen. KM II, p. 178. Cf. "The Law for the Weakening of Electric Effects with Distance" (Chapter 32, this volume). Originally published in German.]

32

Proposal for New Danish Terms in Chemistry

The Law for the Weakening of
Electric Effects with Distance[1]

PROFESSOR Ørsted, Knight of the Order of Dannebrog, has submitted to the Society a proposal for new Danish terms in chemistry. He remarks that, since the introduction of antiphlogistic chemistry, the chemical terminology in Danish, German, Swedish, and Dutch is merely a translation from the French. In these translations the mistake has been committed of designating the first, simplest elements we know by means of derived or compound words, from which it was then, naturally, impossible to form easy and convenient words for compound objects. As a consequence of this, native terminology had to be mixed with foreign words, which rarely complied with the grammar of the native language. The course which the author follows to find a name for an element is that he chooses a word which designates a particularly important property of the object and in that seeks out the first primary sound. This is always given a more precise linguistic determination, either through an addition or through a slight change of letters, whereby appropriate variations are formed to designate various related concepts. Thus *br* is a primary sound in *Brand*, *Brænde*, *Brynde*, *Brunst*,[2] and a minor change of letters determines their different meanings. The ancients also distinguished between *brinna* and *brenna*, as we still do between *sidde* and *sætte*, *ligge* and *lægge*, *synke* and *sænke*,[3] and this difference is still maintained in Swedish. Just as an ignited body is called a *brand*,

---

[1] [*Videnskabernes Selskabs Oversigter*, 1814–15, pp. 7–10. Reports of meetings held on April 21, 1814 and April 28, 1815, respectively. KM II, pp. 431–33. Originally published in Danish.]

[2] [These Danish words mean: fire or firebrand, firewood, concupiscence, and heat (or rut), respectively. They are all related to the English word "burn" with a metathesis of the "r".]

[3] [Similar distinctions are made in English between intransitive and transitive verbs, e.g., "sit" and "set," "lie" and "lay."]

the author thinks that a substance which is characterized by an extraordinary combustive power could be called "Brind" or "Brint."[4] Therefore he suggests this name for *Vandstof*,[5] whose current name he rejects, partly because it uses the name of a compound as part of the name of one of its components, and partly, and primarily, because no name can be deduced for the derived concepts, for example, French *hydrure*, "et Brinte"; *hydrogéner*, "at brinte"; *hydrogénation*, "Brintning"; *deshydrogénation*, "Afbrintning"; *protohydrure*, "Forbrinte"; *perhydrure*, "Fuldbrinte," etc.[6] Similarly, he creates for what has so far been called *Suurstof*[7] the name "Ilt," and from this: "at ilte," *oxygéner*; "et Ilte," *oxide* [sic]; "Iltning," *oxydation*; "at afilte," *desoxygéner*; "Afiltning," *desoxydation*; "iltelig," *oxydable*; "Forilte," *protoxyde*; "Tveilte," *deutoxyde*; "Fuldilte," *peroxyde*.[8]

On closer examination the author noticed that the roots he had chosen were common to Danish, German, Swedish, and Dutch, in short to all Scandinavian and Germanic languages. Because of this he took the opportunity to revise this paper and expand its subject. He did this in a Latin publication on the occasion of the celebration of the Reformation at the University in the year 1814. In this he proposes a common foundation for a chemical terminology in all these languages, in the same way as this is found in all languages which derive from Latin.

The same author has submitted to the Society a paper on the law for the weakening of electric effects with distance. With his electrometer Coulomb had tried to prove that the electric effect is inversely proportional to the square of the distance between the interacting points. This law seemed so natural that its validity could hardly be doubted. However, Volta had already made experiments which did not agree with this, and Simon's beautiful experiment, with a very cleverly devised instrument, seemed to refute Coulomb entirely. On the other hand, the experiments which Coulomb adduces are nonetheless important. Therefore, the author has repeated Coulomb's experiments, with the instrument which the latter has indicated, and found that the law advanced by Coulomb really is obeyed for distances which are neither very great nor very small. At small distances, Coulomb's electrometer indicates a decrease which is approximately inversely proportional to the distance, as Simon indicates. At very great distances, on the other hand, this reduction can even be proportional to the third power of the distance. The general law of the decrease of the effect with the increase of the distance is: that the weakening has an ever increasing ratio to the distance as the distance grows, and that it is only at certain points that the ratio of the decrease can be expressed as an integer power of the ratio of the distance. Moreover, the results do not progress only in accordance with

[4] [Danish for "hydrogen."]

[5] [The old Danish term for "hydrogen" means, literally, "water substance."]

[6] [The English terms are: hydride, hydrogenate, hydrogenation, dehydrogenation, protohydride, and perhydride, respectively.]

[7] [The old Danish term for "oxygen" means, literally, "acid substance."]

[8] [In English: oxygen, to oxidize, oxide, oxidation, deoxidize, deoxidation, oxidizable, protoxide, dioxide, and peroxide, respectively.]

such a law, but the way in which the electric forces are distributed in the air also seems to have a most decided influence on it. It still remains to be investigated whether some of these results could just be due to the torsion of the wire on which Coulomb's electrometer depends, or whether electricity really obeys such a law as has been shown here. This coming winter, the author intends to examine this more closely as it is impossible to make progress in the mathematical investigation of electric forces until this question has been decided.

33

Theory of Light[1]

PROFESSOR Ørsted, Knight of the Order of Dannebrog, has submitted his theory of light to the Society. As is well known, only two theories of the nature of light have received any considerable approbation. One of these, which bears the name of Newton, assumes that light consists of a fine substance which streams from the luminous body in all directions with extraordinary speed; the other, which was elaborated by Euler with much art, assumes that light is a motion in an all-pervasive ether. Although physicists now more or less agree on their preference for Newton's theory, they readily admit that this, as well as Euler's, is beset by significant difficulties. Therefore, the present author has tried a new approach. He has already, in the main, developed the theory he embraces in earlier publications, but he has now tried to develop it further. As a result of the discoveries with which the efforts of the last twenty years have enriched science, no-one will any longer deny that the forces which appear in electrical effects are ordinary forces of nature and no different from chemical forces. The author now assumes, as does Winterl, that the union of these forces gives heat as well as light, but Winterl had confined himself to offering proof of the validity of his assertion, without indicating the conditions under which the union of these two opposite forces produces light, and without applying this principle to the explanation of phenomena.

The author now finds that the union of the two opposite forces does not produce light unless it occurs with considerable resistance. If the two electric forces are combined against very little resistance, no change is observed other than that both forces neutralize each other. When there is a perceptible resistance, however, the body in which the union takes place is heated, and when the resistance becomes very great, the body glows, that is, it can be seen by its own light. The effect of the resistance is greater, the weaker the strength of the electricity as measured by the electrical repulsions. The resistance also grows with the quantity of the forces which, at each instant, act on the conductor, while the strength measured by the electrometer remains unaltered. Under the same conditions, therefore, the galvanic apparatus, especially with large plates, generates far more heat and light than

---

[1] [*Videnskabernes Selskabs Oversigter*, 1815–16, pp. 12–15. Report of the meeting held on March 21, 1816. KM II, pp. 433–35. Originally published in Danish.]

the electrical machine and its charged battery. The same force as in positive electricity is contained in all combustible bodies, and the same force as in negative electricity in all empyreal substances, but both are bound in such a way that they cannot show any repulsion. Therefore, they certainly cannot be conducted by means of spontaneous attraction and repulsion, but experience shows that one can set the other in motion by means of its attraction, especially when the conduction is perfect. The light which appears during ordinary combustion, then, is produced by the combination of the positive force, which is predominant in every combustible body, and the negative force, which is predominant in the empyreal component of air. At the combination of an acid and an alkali, the effect is rarely strong enough to produce more than heat.

The author compares the mode of action of the forces in light with that which occurs in the electrical spark. In the generation of the latter, each of the opposite forces accumulates in its own part of space, one near the other, and then they break through the intervening space and combine. The moment of union brings light. All these circumstances also occur during any obstructed conduction. The electricity which is to be conducted always begins by attracting the opposite and repelling the like electricity which is found in the conductor. Now, if we imagine conduction which has been freed of all resistance, the attraction which the electrical body exerts on the opposite electricity of the conductor and the repulsion which it exerts on the like will produce a disturbance and a restoration of the equilibrium, which permeate the entire body without interruption. However, to the extent that resistance is present, both the attracted and the repelled electricity will, within a few moments, accumulate each in its place, though very close to one another. When the accumulation has attained a certain strength, the opposite forces will combine in a rapid discharge, like a spark. If we now imagine that this effect exists everywhere in the conductor, and that the distance between the opposite points is extremely small, we have the idea of the generation and the propagation of light. The highest rapidity in the combination of the opposite forces gives the invisible rays which appear next to the violet light in the prismatic spectrum. The violet rays have the next-highest rate of combination after these, and so forth, according to the sequence of the colours, until the red, which have the lowest rate. An even lower rate of combination produces thermal radiation. The mutual transition between heat and light, together with all the attendant effects, is easily explained by this conception.

According to the theory advanced here, it is reasonable to view a ray of light as a series of immeasurably small electrical sparks, which could be called the principal elements of light. The line between the two most opposite points of such a principal element could be called its axis. The position of the axis relative to a reflecting or refracting surface will naturally influence the further progress of the ray of light. Therefore, this theory seems more consistent than any other with the polarity of light rays which has been discovered in our time. The profusion of objects to which a theory of light must be applied is too great for us to be able to mention them all. We must then confine ourselves to the remark that, on the basis of his theory, the author has tried to explain the production of light which is not accompanied by any perceptible heat, the colours of the flame, the chemical effect of the various rays of light, etc. The author believes that what speaks most in favour of his theory is the fact that it does not presuppose any force or matter whose existence has not been experimentally proved; that it follows the generation of light from darkness through all the instances in which it occurs and explains them easily; that it describes the relationship between heat and light without leading to contradictions with what we know about nature; and that it finally places the generation of light in the most intimate connection with chemical activity.

34

On Galvanic Trough Apparatuses and Spark Discharge in Mercury Vapour[1]

EVEN though galvanic trough apparatuses have many important advantages, especially when the object is to obtain very great effects, most of these devices have not quite satisfied the wishes of physicists. If the troughs are made of wood, they will soon be saturated by the acids in spite of the application of all kinds of varnish, and the connection thus created weakens the effect considerably. If instead one uses partitioned china troughs, in which the zinc and copper plates are suspended, a much greater and more certain effect is obtained, but this kind of trough will be very expensive when the apparatus is to have a considerable size. In order to procure, if possible, a more perfect device of this kind, Counsellor Esmarch and Professor Ørsted have collaborated on a project, whose results they have presented to the Society. Instead of using other materials for troughs or vessels, they turn the copper plates themselves into vessels which contain the requisite liquid. Each copper vessel is connected by a metallic strap to a zinc plate which is to be immersed in the liquid contained in the next copper vessel. It has since become known that Count Fredr. Stadion in Vienna has used a similar apparatus, which, however, is not quite the same as the former nor could have been known to us when the apparatus we made was shown for the first time. Moreover, the two Danish physicists have further pursued the idea they originally conceived. Even in its first version, the new galvanic apparatus had an excellent effect, but nevertheless they succeeded in improving it by using a hot liquid. Although the ability of heat to increase the galvanic effect had been known for some time, it was not easy to apply it because of the arrangement of the device. In the new apparatus, this is easy. However, in order to maintain the device at a fairly high temperature, they have later equipped the copper vessels with an arrangement which is rather similar to that of a tea urn, by making them cylindrical and by putting a chimney in the middle. This apparatus must be made rather large but will then produce the most glorious fusions. The one which was shown to the Academy of Sciences consisted of only 6 cylinders, but each of these could hold 18 Danish *pots*,[2] and the whole apparatus thus 108 *pots*.

---

[1] [*Videnskabernes Selskabs Skrifter*, 1816–17, pp. 7–9. Report of the meeting held on November 15, 1816. KM II, pp. 436–37. Originally published in Danish.]

[2] [One Danish *pot* corresponds to approximately one quart.]

When it is filled with water which contains $1/60$ of sulphuric acid and $1/60$ of nitric acid, and the chimneys contain enough embers to keep the water rather hot, it is possible to make a No. 1 iron wire, with a diameter of $1/24$ of an inch, glow, even melt. Also when the cylinders contain only a saline solution instead of acidulated water, they produce a very beautiful glowing of the metal wires. By the same means, No. 2 iron wires can be made to glow.

So far the aim of these galvanic experiments has merely been to find convenient arrangements of the necessary apparatus, but they suggest one experiment which may not teach us anything really new, but which presents a well-known truth in a new form. They filled a U-shaped glass tube, one leg of which had a very narrow capillary constriction, with mercury and boiled it in the tube. Then they introduced it into the galvanic circuit and now observed the formation of sparks in the narrow part of the tube, the mercury there being made to glow intermittently, its parts separating due to the mercury vapour generated but again combining after the liquefaction of this vapour, which follows quickly. If an experiment were needed to make it perceptible that the electrical spark is just a violent glowing of the substance which fills the space in which the spark appears, this experiment seems well suited.

35

Observations Regarding
Contact Electricity[1]

My TREATISE on the law of electrical repulsions was completed 2 years ago in the form in which it was read to the Society of Sciences. However, it was my wish to repeat these experiments with another apparatus, but this has not been constructed by the artisan quite in accordance with my wishes. This circumstance has caused a delay. This summer, I certainly hope to make this treatise more worthy of the attention of physicists with this addition.

During the last year I have occupied myself a great deal with galvanic apparatuses and their arrangement and have, in the company of Counsellor Esmarch here, performed many experiments on this. Even our first experiments turned out to be quite satisfactory. We sought a galvanic apparatus which would have the advantages of a trough apparatus, but in which we would need neither wooden troughs, which are so easily penetrated by the acid and thereby obtain a detrimental conducting ability, nor troughs made of faience or porcelain, which become expensive and, moreover, were not easy for us to obtain.[2] Therefore, we conceived the idea of using the copper element of the pile itself as the vessel for receiving the moisture. Fig. 1 depicts the cross-section of this very simple apparatus. AB is the vertical section of the rectangular zinc plates, $DFGH$ the cross-section of the almost cubical[3] copper box, ACD a soldered copper bow which is split at D in order to connect firmly to both sides of the copper, outside as well as inside, so that no oxidation of the solder might interrupt the conduction. At the bottom and on both sides of the zinc plates, holes are drilled in which there are small wooden stoppers which prevent contact between the zinc and the copper at these places. Only one of these stoppers can be seen at B. This apparatus can appropriately be called the galvanic

[1]From a letter to the editor dated April 16, 1817. [*Journal für Chemie und Physik*, ed. by Dr. J. S. Schweigger, Vol. 20, pp. 205–12. Nuremberg 1817. The same topic is treated in *Videnskabernes Selskabs Oversigter 1816–17*, pp. 7–9. Copenhagen. KM II, pp. 206–11. Cf. "On Galvanic Trough Apparatuses" (Chapter 34, this volume). Originally published in German.]

[2]Actually, porcelain troughs do not achieve what is hoped from them, especially if the acid solution is to be left in them.   *The Ed.*

[3]The cubic shape is due to a misunderstanding by the artisan, for it is obviously more advantageous to make the dimension FG smaller than the one perpendicular to it.

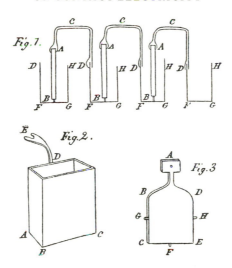

trough apparatus.[4] The vessels were normally filled with a mixture of 30 parts water, $^1/_{60}$ nitric acid, and $^1/_{60}$ sulphuric acid. The zinc plates had a size of approximately 9 □ inches. 6 elements of such an apparatus sufficed to cause the glowing of a thin iron wire, which is commercially denoted No. 13. When we heated the water, we obtained an even greater effect, whereby also thicker iron wires were made to glow while No. 12 and No. 13 wires melted. As is well known, No. 12 has a $^1/_{175}$ inch diameter.

I see from Gilbert's *Annalen*, 1817 (Part 4) that Count Friedrich Stadion has successfully tried a similar galvanic arrangement. We, Esmarch and I, knew nothing of this when we performed our experiments. I presented the results of these experiments in a lecture on October 1, 1816. On the first Tuesday of each month I deliver a public lecture on recent discoveries in physics and chemistry, whereby my audience is enabled to follow the progress of science. This galvanic apparatus was also shown in passing in such a lecture. I am not stating this in such detail because I consider the subject which is dealt with here to be important (Count Stadion does not seem to do so either since he has not even made his apparatus public), but rather because I cannot credit anyone with a share in our work which he does not have, and because I do not like to make a claim without proof.

Even though the trough apparatus described here performed very good service, we found cleaning it to be quite laborious because the soldering of the zinc and the copper very easily causes damage to the box during cleaning. A pronounced effect with respect to the glowing of a metal wire cannot be expected without a complete cleaning. However, the usual experiments on decomposition also proceed nicely after less thorough cleaning. As we intend to perform many more experiments on

[4]The reader can also find this device shown in Gehlen's *Journal*, Vol. 7, Tab. 5, Fig. 18, in which the main features of this device (viz., that the copper surface encloses the zinc surface), having also recently become very common in England, emerged as a corollary of my galvanic combinations.

The Ed.

the heat produced by galvanism in the future, we have separated the zinc and copper elements of the apparatus from each other. Fig. 2 shows the outline of the box, the base of which is a rectangle in which one of the sides BC is $2\,^1/_2$ times the length of the other AB. At E the bow DE has a notch for the zinc plate, which is depicted in fig. 3. At the top, this plate has a head A, which is equipped with a hole in order to attach a series of zinc plates to it. The broad part of the zinc plate $BCDE$, of 16 □ inches, is suspended by means of a narrow neck on this head. Small, wooden pegs are placed at F, G, and H in order to prevent contact between the zinc and the copper. It is now easy to see that the zinc plate will be mounted in the notch at E, fig. 2, in such a way that the head of the zinc plate comes to rest on the end of the copper bow. Therefore, when filled with water, the copper case must bear the weight of the zinc plates without tipping. I have already tested an apparatus made from 12 such elements and obtained the desired effect. At the moment, I am in possession of 48, which I have not yet tested. They will be arranged in two rows so that one box has the bow on the shorter side of the rectangle, bent so that the zinc plate can have the proper position in the next box. The two outermost boxes, which will constitute the two poles, have a device through which a conductor can be attached.

In order to exploit the beneficial influence of an elevated temperature on the galvanic effect, we have installed another device which is very advantageous on a large scale, but which would become too expensive and burdensome on a small scale. Fig. 4 presents an element of this pile. $ABCD$ is a copper cylinder which holds 18 Danish *pots*, each of $48._7$ cubic inches (old French measure), i.e., more than half a cubic foot. $EFGH$ is a chimney which is equipped with a grid at the bottom. The entire cylinder rests on three glass feet, of which two can be seen at I and K. The short glass columns which make up the feet adhere by mere friction to 3 tubes soldered to the bottom BC. DL is a copper bow which connects a zinc cylinder $LMNO$ with the copper cylinder. This cylinder is composed of several pieces because it was difficult to cast such a large cylinder from zinc in one piece, and because we could not obtain rolled zinc. At P and Q we see two of the three small wooden feet on which the zinc cylinder must rest. We have not made this apparatus with more than 6 elements, but these are sufficient to produce enormous effects. If this apparatus is to be used, it is first filled with hot water and the chimneys with glowing coals; the water can also be brought in cold, but then it takes a long time before the water becomes really hot. If nothing more than boiling water is used, beautiful sparks are obtained, but a thin iron wire barely glows. If, on the other hand, some common salt is added to the water, the effect is increased to such a degree that a No. 2 iron wire can be melted by it. If acid is added to the water in the proportion stated above, the effect becomes even more significant, and a No. 1 iron wire, which has a diameter of $^1/_{42}$ inches,[5] can easily be melted. It is not a good idea to add all of the acid to the boiling water at once. It is true that this produces a very large effect but also one which passes very quickly; however, if the acid is added little by little, a nice effect is obtained for quite a long time (approximately 2 hours).

[5] [In the Danish treatise in *Videnskabernes Selskabs Oversigter* (p. 8), the diameter is indicated to be $^1/_{24}$ inch. Cf. Footnote [1] above.]

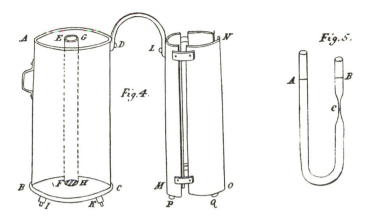

As we have so far concentrated on the improvement of the devices, we have not yet performed all the experiments on the glowing of metals and, in general, on the production of heat by galvanism which these devices invite. However, I must tell you about one experiment which, while it does not teach us anything really new, still presents a well-known law in an unusual manner. We have, in fact, produced electrical sparks in mercury. In order to manage this, we need a glass tube, which fig. 5 depicts. The space from *A* to *B* is filled with mercury, at *C* the tube is drawn into a capillary. Now, if the large galvanic circuit was closed by means of the mercury, numerous sparks developed in the narrow part of the tube. When the spark was formed, we noticed a separation of the mercury column in the capillary which disappeared again immediately. This happened in boiled mercury as well. Probably some of the mercury had transformed into vapour and condensed again as soon as the heat ceased. The mercury had become extremely hot in the narrowed part. The spark must naturally be regarded as a glowing of the mercury at the place of least resistance; just as, in my opinion, an electric spark must always be regarded as the glowing of some material (of air, of water, of oil). So far we have tried in vain to produce the same effect in water, salt solution, etc. The intensity of the electricity of our apparatus has probably been too low in relation to the resistance of these liquids to conduction. I consider the form of this experiment suitable for producing chemical decompositions in a new manner. *Perhaps metals will one day be successfully decomposed in this manner.* However, enormous devices will certainly be required for this. I see that Childern[6] adds some theoretical remarks to the description of his beautiful galvanic experiments which agree entirely with the theory of heat which I have stated previously. He draws the conclusion from his experiments that conductors are heated by the penetrating electrical forces to the degree that they offer resistance. I have known this natural law for a long time and stated it in my *Ansichten [sic] der chemischen Naturgesetze*,[7] as well as in your journal, and not without convincing reasons. On this basis, I have formulated a general theory

[6][Children.]
[7][*View of the Chemical Laws of Nature* (Chapter 30, this volume).]

of heat, in which I derive the facts from the principle in a natural way. How can it be that this is not considered in recent articles at all? I see to my astonishment that several purely theoretical speculations by English and French physicists are widely explained in German reports while complete silence is observed regarding the analogous investigations which I have mentioned in my *Ansichten der chemischen Naturgesetze*, even though corrections could still be obtained there for the theorems *later* postulated by these foreigners. I have a fair amount of material with which I can prove the basis of this remark.

36

On the Compression of Water[1]

Copenhagen, April 28, 1818

LATELY I have been preoccupied with the compression of water. The book by Zimmermann on this topic is filled with the strangest calculational errors. Once these errors are corrected, there is far more harmony among the results of the experiments than was previously believed. I have also conducted entirely new experiments on this topic, which show that the compression of water is proportional to the compressing forces, as Canton claimed based on very limited experiments, but which Zimmermann's experiments contradicted, in the form in which these results have been presented so far. I have found that the compression which Canton quoted is too small by almost a factor of three. This is important because, according to de la Place, the velocity of sound in water can be calculated from this compression. If Canton's results are used as the basis of the calculation, sound turns out to have a velocity in water which comes close to that in metals; according to my results it is almost three times smaller than according to Canton's. I am still working on a major experiment in order to determine most accurately the compression of water caused by a pressure equal to that of the atmosphere. Even though I am already rather certain that this quantity has to fall between 0.00012 and 0.00014 at 14° R., I still wish to obtain as accurate a determination as I possibly can.

---

[1] From a letter to the editor. [*Journal für Chemie und Physik*, ed. by Dr. J. S. C. Schweigger, Vol. 21, pp. 348–49. Nuremberg 1817. KM II, pp. 211–12. Originally published in German.]

37

On the Way in Which a Textbook in Physics Ought to Be Written

Investigations on the Compressibility of Water[1]

PROFESSOR Ørsted, Knight of the Order of Dannebrog, has submitted to the Society two papers, one of which was the first of a series of papers on the way in which textbooks in physics ought to be written; the other contained an investigation of the compressibility of water.

It is well known that many textbooks in physics have been published in various European countries in recent years. Although there are not a few differences of opinion about the causes of things, even in books from the same period, the differences in treatment seem far greater. It has not even been possible to agree on what should be included in this science. While some wanted to exclude both that which can be treated in applied mathematics and that which belongs to chemistry, there were others who not only included all this in physics but even added the science of the nature of the earth and a survey of the laws of motion of heavenly bodies. There was no less disagreement regarding the division and the order of subjects, but in particular there seemed to be great doubt about the way in which those natural laws which could be expressed mathematically should be put forward and proved in physics. All these matters give rise to a great many questions, which the author will attempt to answer in a series of papers. The aim of this, the first, is to show what ought to be understood by physics. Here, the author is trying to justify in detail the definition which he has already proposed in other writings, according to which physics is the science of the *general* laws of nature and is therefore called general physics. However, by using this name he does not limit this science to the narrow scope of several German writers, who have used the same name, but also includes the theories of electricity, magnetism, light, heat, and chemical combinations, as

---

[1] [*Videnskabernes Selskabs Oversigter*, 1817–18, pp. 8–12. Reports of the meetings held on December 12, 1817 and on April 4, 1818, respectively. KM II, pp. 438–41. Originally published in Danish.]

all these follow immediately from general natural forces. From this broader point of view, he wants to regard even the properties of various substances as distinctive manifestations of general natural forces, which in each of them are seen at a specific stage of development and strength. As the author has attempted to show in several earlier writings that the electric forces are the same as the chemical, only in a freer state, and as he has also advanced the theory that magnetism, light, and heat are effects of the same forces, then it follows that all things in physics which are not part of the *theory of motion* together form one coherent *theory of force*, or *chemistry* in the broadest sense of the word. The first of these parts of general physics, then, comprises all *external* changes, the second the *internal* ones. It is obvious that no third part can be added to these two main parts except for the science of the combination of forces and motion, e.g., in light and radiant heat. However, it may not be possible to determine whether this science should be separated from the rest as an independent part or included in the theory of force until physics has reached a far higher stage of perfection.

As soon as possible the author proposes to deliver the continuation of these investigations. Thereby it is his intention to initiate an investigation of this matter among physicists, and he thinks that it would be of considerable benefit to science if agreement could be reached as to the form and composition which a textbook in this science ought to have. Then all scholars in the field would work together towards its perfection, and thus, with time, a work would develop in which a complete picture of this science at a given time could be seen. Naturally, the author does not believe that all textbooks on this subject should have the same organization; this can be changed according to the different aims of the presentation. However, in textbooks which have no aim other than a thorough presentation of the science, he believes that the same organization and procedure should always be maintained, once scholars have agreed to recognize its correctness. But he considers such agreement on a theory of textbooks to be just as likely as the agreement which prevails among physicists concerning so many other theories.

Although we are already in possession of such proofs of the compressibility of water that this can no longer be doubted, the magnitude of this compression for a given force and the law which this compression follows have continued to be subject to doubts, even very considerable doubts. It is true that certain mathematicians and physicists assumed that the compression of water was proportional to the compressing forces, but the Abich–Zimmerman experiments, the only existing series of experiments on the compression of water with different forces, seemed to contradict this opinion completely, which is why Gehler, in the account he delivers in his dictionary of the experiments concerning the compressibility of water, expressly states that it has not been possible to discover any law for the resulting magnitudes. All other physicists have pronounced the same judgment on the results of the Abich–Zimmerman experiments.

Professor Ørsted recently reviewed these and found, to his surprise, that Zimmerman's calculations of the results of these experiments, although of the most elementary nature, were confused by mistakes which were fraught with consequences. When these were corrected, it appeared that the experiments described did not prove that unboiled water is less compressible than boiled, but that, on the contrary, it was, as should be expected, more compressible. It also appeared that the compressions were proportional to the compressing forces provided the experiments were not made with the greatest forces, which might compress the piston itself. This caused the author to have an apparatus constructed for the compression of water. This consisted of a wide but very thick brass cylinder and a narrow tube with a piston for the compression of the water. By means of this device it was now possible to compress the water with the application of a very small force and to measure very small changes in the volume of the water. In order to measure the compressing force, which, because of the frictional resistance of the piston, cannot be done directly with sufficient accuracy, an opening was made in the water cylinder, into which was screwed a strong, narrow glass tube. This was filled with air which showed the magnitude of the compressing force by the magnitude of its compression according to Mariotte's law. With this device he found that the compression of water is in reality similar to that of air, that is, proportional to the compressing forces. The magnitude of this compression at 12° of a hundred-division thermometer was almost 0.00012 for a pressure equal to that of the atmosphere, but the author proposes to continue experiments on this matter. The author has also extended his criticism to Canton's experiments and shown that a correct evaluation of the influence of heat in those experiments would give a greater compressibility for water than that which the English physicist has concluded from them. As De la Place has made Canton's experiments the basis for his calculations of the speed of sound in water, his determination of this must be corrected in the light of these more recent experiments.

38

On Piperine, a New Plant Alkaloid[1]

Copenhagen, February 15, 1820.

THE discoveries of new alkalis in plants have led me back to an old piece of work on pepper which I began several years ago.[2] By resuming this investigation, I easily discovered a new alkaline substance which we shall merely call *piperine* without searching for a more suitable name derived from its nature because our knowledge of the entire class of substances to which this belongs is still so new and incomplete. Piperine is obtained by extracting the resinous and oily substances from pepper by means of alcohol; piperine is also contained in the solution formed in this way. Muriatic acid is added, whereby a salt of piperine is produced which is soluble in water. The resin is now precipitated in water, the spirit of wine is distilled from the watery solution, and finally the piperine is extracted with potash. Magnesia can also be used for this but, it would seem, with less advantage.

Piperine is almost insoluble in water, soluble in cold alcohol, but even more so in hot. The solution has an exceptionally sharp taste, gives paper prepared with curcuma a brown colour, re-establishes the colour of litmus, forms salts with acids, and thus generally has the same properties which have been found in the other newly discovered plant alkalis. The saturated alcohol solution of piperine is faintly

[1] [*Journal für Chemie und Physik*, ed. by Dr. J. S. C. Schweigger, Vol. 29, pp. 80–82. Nuremberg 1820. The same topic is treated in *Videnskabernes Selskabs Oversigter*, 1819–20, p. 13. Copenhagen. A similar announcement, dated March 4, 1820, can be found in *Journal de physique*, Vol. 90, pp. 173–74. Paris 1820. KM II, pp. 212–14. Originally published in German.]

[2] [An announcement of this work can be found in the following communication in *Neues Archiv für medizinische Erfahrung*, ed. by Dr. Ernst Horn, Vol. 10, p. 333. Berlin 1809.

ON A NEW SURROGATE FOR CINCHONA

Mr. Oerstedt, Professor of physics in Copenhagen, has invited the local physicians to perform experiments with a surrogate for cinchona invented by him. In fact, he believes to have found that the sediment produced by a decoction of pepper, after its pungent substance has been removed by frequent digesting with strong spirits of wine and mixed with tincture of gall, is precisely the component of cinchona which possesses the antipyretic quality. He bases this assumption on the fact that a similar sediment is produced when a decoction of cinchona is mixed with tincture of gall, combined with the circumstance that the common man normally uses pepper against fever.

The same communication can also be found in *Allgemeine medizinische Annalen*, p. 1063. Altenburg 1809; and in *Medizinisch-chirurgische Zeitung*, Vol. 2, No. 43. Salzburg 1809.]

green but acquires a more pronounced green colour when nitric acid is added. Dry piperine is altered in the same way by nitric acid. However, if the effect of the acid is strong, or if it is continued for a long time, the colour of piperine becomes yellow and finally reddish. I have not yet been able to decide with certainty whether these changes in colour stem from a small amount of resin still remaining.

I am continuing my experiments on this subject with enthusiasm and hope soon to be able to communicate more complete news about this, from which it might be decided whether this substance is different from the other newly discovered alkaloids, or whether additional knowledge of the subject leads the multitude of these substances now presenting themselves to us back to something simpler. I harbour the suspicion that resins and volatile oils in general contain an alkaloid. First, I have directed my particular attention to common resin and camphor. The method which I have applied to pepper appears to me to be particularly convenient. A fair amount of oil was removed by the addition of acid to the alcohol solution of the pepper resin. Had the pepper alkali absorbed this previously?—Are the resins perhaps soaps of these new alkaloids and an oil? —

———

39

Experiments on the Effect of the Electric
Conflict on the Magnetic Needle[1]

———

THE first experiments on the subject which I shall address here were made during the lectures which I gave last[2] winter on electricity, galvanism, and magnetism. It seemed demonstrated by these experiments that the magnetic needle can be moved from its position by means of a galvanic apparatus, but by a closed galvanic circuit, not an open one, as several very celebrated physicists tried in vain some years ago. However, as these experiments were made with a less effective apparatus, and as the resulting phenomena therefore did not seem sufficiently clear, considering the importance of the subject, I joined with my friend Counsellor Esmarch in order that the experiments might be repeated and expanded by means of a large galvanic apparatus constructed jointly by us. The distinguished Mr. Wleugel, Knight of the Order of Dannebrog and Superintendent of Nautical Schools, was present at the experiments as a participant and a witness. Also present as witnesses to these experiments were the excellent Mr. Hauch, who has been decorated by the King with the highest honours, and whose knowledge of the natural sciences has long been famous, the perspicacious Mr. Reinhardt, Professor of Natural History, Mr. Jacobsen[3], Professor of Medicine and an astute experimenter, and the experienced chemist, Dr. Zeise. I frequently made experiments on the subject in question on my own, but I have always repeated the phenomena which I succeeded in discovering in this manner in the presence of these learned men.

[1] [This brief description of Ørsted's most important discovery was written in Latin under the title *Experimenta circa effectum conflictus electrici in acum magneticam* and sent to learned societies and scholars in all European countries on July 21, 1820. Translations of this article quickly appeared in many scientific journals, among which are the following: Thomson's *Annals of Philosophy*, Vol. 16, pp. 273–76. London 1820. *Journal für Chemie und Physik*, ed. by J. S. C. Schweigger, Vol. 29, pp. 275–81. Nuremberg 1820. Oken's *Isis*, Col. 57–60. Jena 1821. *Annales de chimie*, Vol. 14, pp. 417–25. Paris 1820. *Journal de physique*, Vol. 91, pp. 72–76, Paris 1820. *Bibliothèque universelle des sciences*, Vol. 14, pp. 274–84. Geneva 1820. *Annales générales des sciences physiques*, Vol. 5, pp. 259–64. Brussels 1820. *Annalen der Physik*, ed. by L. W. Gilbert, Vol. 66, pp. 295–304. Leipzig 1820. *Giornale Arcadico di scienze*, Vol. 8, pp. 174–78, Rome 1820. *Giornale de fisica, chimica e storia naturale*, ed. by L. Brugnatelli, Vol. 3, pp. 335–42. Pavia 1820. *Hesperus*, ed. by Rahbek, Vol. 3, pp. 312–21. Copenhagen 1820. The discovery was reported in *Videnskabernes Selskabs Overskrifter*, 1820–21, pp. 12–21. Copenhagen. KM II, pp. 214–18. Originally published in Latin.]

[2] [*proxime-superiori.*]   [3] [Jacobson.]

In reporting these experiments, I shall omit all those which have undoubtedly led to the discovery of the nature of the matter, but which, once this had been discovered, could not further illuminate the subject; we shall therefore content ourselves with those things which clearly demonstrate the nature of the matter.

The galvanic apparatus which we employed consists of twenty rectangular copper troughs, each of which is twelve inches in both length and height, but whose width scarcely exceeds two and one-half inches. Every trough is equipped with two copper strips, bent in such a manner that they can hold a copper rod which supports a zinc plate in the water of the next trough. The water in the trough contains $^1/_{60}$ of its weight of sulphuric acid and also $^1/_{60}$ of nitric acid. The portion of each zinc plate which is immersed in the water forms a square, whose side is approximately 10 inches in length. Smaller apparatuses can also be employed, provided that they can make a metallic wire glow.

The opposite ends of the galvanic apparatus are connected by a metallic wire, which, for the sake of brevity, we shall henceforth call the connecting conductor or the connecting wire. However, the effect which takes place in this conductor and in the surrounding space we shall call the electric conflict.

A straight portion of this wire is placed in a horizontal position above and parallel to a magnetic needle which has been properly suspended. If necessary, the connecting wire can be bent so that a suitable part of it assumes the position required for the experiment. When this has been achieved, the magnetic needle will move, and it will deviate to the west under that part of the connecting wire which receives electricity most closely from the negative end of the galvanic apparatus.

If the distance of the connecting wire from the magnetic needle does not exceed $^3/_4$ inch, the deviation will amount to an angle of approximately 45°. If the distance is augmented, the angles will decrease as the distances increase. Moreover, the deviation varies with the power of the apparatus.

The connecting wire can be moved, either towards the east or towards the west, provided that it maintains its orientation parallel to the needle, without any change in the effect other than with respect to its magnitude. Hence, the effect cannot possibly be ascribed to an attraction; for the same pole of the magnetic needle which approaches the connecting wire when the latter is placed on its east side should move away from it when the wire occupies a position on the west side if these deviations were dependent on attractions or repulsions. The connecting conductor can consist of several joined metallic wires or ribbons. The nature of the metal does not alter the effect, except perhaps for its magnitude. Wires of platina, gold, silver, brass, iron, ribbons of tin and lead, and a mass of mercury have been employed with equal success. A conductor which is interrupted by water is not completely without effect provided that the interruption does not extend over a distance of several inches.

The effects of the connecting wire on the magnetic needle pass through glass, metals, wood, water, resin, earthenware troughs, and stones; for it is by no means stopped by an interposed plate of glass, metal, or wood; nor does it disappear when plates of glass, metal, and wood are interposed simultaneously; indeed, it hardly seems to decrease. The same result is obtained if we interpose the disc of an electrophorus, a plate of porphyry, or an earthenware trough, even if it is filled with water. Our experiments have also shown that this effect is not changed if the magnetic needle is enclosed in a brass box filled with water. Needless to say, an effect which penetrates all these materials has never before been observed in electricity and galvanism. Therefore, the effects which take place in the electric conflict are very different from the effects of one or the other electric force.

If the connecting wire is placed in the horizontal plane below the magnetic needle, all the effects are the same as in the plane above the needle, but in the opposite direction; for the pole of the magnetic needle below which is the part of the connecting wire which receives the electricity from the negative pole of the galvanic apparatus will deviate towards the east.

In order that this may be more easily remembered, we may use this formula: The pole above which the negative electricity enters turns to the west, below which to the east.

If the connecting wire is turned in a horizontal plane so that it forms a gradually increasing angle with the magnetic meridian, the deviation of the magnetic needle increases if the movement of the wire is towards the place of the disturbed needle, but it decreases if the wire moves away from that place.

When the connecting wire is placed parallel to the magnetic needle in the same horizontal plane in which the latter moves, balanced by means of a counterpoise, it does not impel the needle either to the east or to the west but only causes it to tilt in the plane of inclination so that the pole close to which the negative electric force enters the wire is depressed when the wire is situated on the west side and elevated when on the east side.

If the connecting wire is placed perpendicular to the plane of the magnetic meridian either above or below the needle, the latter remains at rest unless the wire is very close to the pole; in which case the pole is elevated when the entrance occurs from the west part of the wire and depressed when it is from the east.

When the connecting wire is placed perpendicularly opposite the pole of the magnetic needle, and the upper extremity of the wire receives electricity from the negative end of the galvanic apparatus, the pole is moved towards the east, but when the wire is placed opposite a point between the pole and the middle of the needle, the pole is moved towards the west. When the upper extremity of the wire receives electricity from the positive end, the opposite phenomena occur.

If the connecting wire is bent in such a manner that it is parallel to itself on either side of the bend or forms two parallel legs, it repels or attracts the magnetic poles according to the various circumstances. If the wire is placed opposite either pole of the needle in such a manner that the plane of the parallel legs is perpendicular to the magnetic meridian, and if the eastern leg is connected to the negative, the western to the positive end of the galvanic apparatus, then the nearest pole will be repelled either to the east or to the west, according to the position of the plane of the legs. If the eastern leg is connected to the positive end and the western to the negative end, the nearest pole will be attracted. When the plane of the legs is placed perpendicularly at a point between the pole and the middle of the needle, the same effects will occur but reversed.

A brass needle, suspended like the magnetic needle, is not moved by the effect of the connecting wire. Likewise, needles of glass or so-called gum lac remain at rest when subjected to a similar experiment.

From all this, we may make a few observations towards an explanation of these phenomena.

The electric conflict can act only on the magnetic particles of matter. All non-magnetic bodies seem to be penetrable by the electric conflict, whereas magnetic bodies, or rather their magnetic particles, seem to resist the passage of this conflict; hence, they can be moved by the impetus of the contending powers.

It is sufficiently evident from the preceding observations that the electric conflict is not confined to the conductor but, as mentioned above, is dispersed quite widely in the circumjacent space.

From what has been observed we may likewise conclude that this conflict moves in circles; for without this condition it seems impossible that the same part of the connecting wire moves the magnetic needle towards the east when placed below the magnetic pole, but towards the west when placed above it; for it is the nature of circles that motion in opposite parts must have opposite directions. Moreover, it seems that circular motion combined with progressive motion in the longitudinal direction of the conductor ought to form a conchoidal or spiral line, but this, unless I am mistaken, contributes nothing to the explanation of the phenomena hitherto observed.

All the effects on the north pole presented here are easily understood if we assume that the negative electric force or matter moves in a spiral line bent towards the right and propels the north pole without acting in any way on the south pole. The effects on the south pole are explained in a similar manner if we ascribe to the positive electric force or matter the opposite motion and the power to act on the south pole but not the north. The agreement of this law with nature will be better understood by a repetition of the experiments than by a long explanation. However, the assessment of the experiments will be greatly facilitated if the course of the electric forces in the connecting wire is indicated by either painted or engraved marks.

I shall merely add this to what I have said: I have demonstrated, in a book published seven years ago, that heat and light are an electric conflict. From the observations just reported, we may now conclude that circular motion also occurs in these effects. I believe that this will contribute very much to the elucidation of the phenomena which are called the polarization of light.

Copenhagen, July 21, 1820.

Hans Christian Ørsted.
Knight of the Order of Dannebrog,
Prof. Ord. of Physics at the University of Copenhagen,
Secretary of the Royal Academy of
Sciences in Copenhagen.

39a

Experiments on the Effect of a Current of Electricity on the Magnetic Needle[1]

THE first experiments respecting the subject which I mean at present to explain, were made by me last winter, while lecturing on electricity, galvanism, and magnetism, in the University. It seemed demonstrated by these experiments that the magnetic needle was moved from its position by the galvanic apparatus, but that the galvanic circle must be complete, and not open, which last method was tried in vain some years ago by very celebrated philosophers. But as these experiments were made with a feeble apparatus, and were not, therefore, sufficiently conclusive, considering the importance of the subject, I associated myself with my friend Esmarck to repeat and extend them by means of a very powerful galvanic battery, provided by us in common. Mr. Wleugel, a Knight of the Order of Danneborg,[2] and at the head of the Pilots, was present at, and assisted in, the experiments. There were present likewise Mr. Hauch, a man very well skilled in the Natural Sciences, Mr. Reinhardt, Professor of Natural History, Mr. Jacobsen,[3] Professor of Medicine, and that very skilful chemist, Mr. Zeise, Doctor of Philosophy. I had often made experiments by myself; but every fact which I had observed was repeated in the presence of these gentlemen.

The galvanic apparatus which we employed consists of 20 copper troughs, the length and height of each of which was 12 inches; but the breadth scarcely exceeded $2^{1}/_{2}$ inches. Every trough is supplied with two plates of copper, so bent that they could carry a copper rod, which supports the zinc plate in the water of the next trough. The water of the troughs contained $^{1}/_{60}$th of its weight of sulphuric acid, and an equal quantity of nitric acid. The portion of each zinc plate sunk in the water is a square whose side is about 10 inches in length. A smaller apparatus will answer provided it be strong enough to heat a metallic wire red hot.

The opposite ends of the galvanic battery were joined by a metallic wire, which, for shortness sake, we shall call the *uniting conductor*, or the *uniting wire*. To the

[1] Translated from a printed account drawn up in Latin by the author, and transmitted by him to the Editor of the *Annals of Philosophy*. [Thomson's *Annals of Philosophy*, Vol. 16, pp. 273–76. London 1820. The text here is precisely as it appeared in *Annals of Philosophy*. KM I, pp. LXXXIX–XCIII. Originally published in English.]

[2] [Dannebrog.]   [3] [Jacobson.]

effect which takes place in this conductor and in the surrounding space, we shall give the name of the *conflict of electricity*.

Let the straight part of this wire be placed horizontally above the magnetic needle, properly suspended, and parallel to it. If necessary, the uniting wire is bent so as to assume a proper position for the experiment. Things being in this state, the needle will be moved, and the end of it next to the negative side of the battery will go westward.

If the distance of the uniting wire does not exceed threequarters of an inch from the needle, the declination of the needle makes an angle of about 45°. If the distance is increased, the angle diminishes proportionally. The declination likewise varies with the power of the battery.

The uniting wire may change its place, either towards the east or west, provided it continue parallel to the needle, without any other change of the effect than in respect to its quantity. Hence the effect cannot be ascribed to attraction; for the same pole of the magnetic needle, which approaches the uniting wire, while placed on its east side, ought to recede from it when on the west side, if these declinations depended on attractions and repulsions. The uniting conductor may consist of several wires, or metallic ribbons, connected together. The nature of the metal does not alter the effect, but merely the quantity. Wires of platinum, gold, silver, brass, iron, ribbons of lead and tin, a mass of mercury, were employed with equal success. The conductor does not lose its effect, though interrupted by water, unless the interruption amounts to several inches in length.

The effect of the uniting wire passes to the needle through glass, metals, wood, water, resin, stoneware, stones; for it is not taken away by interposing plates of glass, metal or wood. Even glass, metal, and wood, interposed at once, do not destroy, and indeed scarcely diminish the effect. The disc of the electrophorus, plates of porphyry, a stone-ware vessel, even filled with water, were interposed with the same result. We found the effects unchanged when the needle was included in a brass box filled with water. It is needless to observe that the transmission of effects through all these matters has never before been observed in electricity and galvanism. The effects, therefore, which take place in the conflict of electricity are very different from the effects of either of the electricities.

If the uniting wire be placed in a horizontal plane under the magnetic needle, all the effects are the same as when it is above the needle, only they are in the opposite direction; for the pole of the magnetic needle next the negative end of the battery declines to the east.

That these facts may be the more easily retained, we may use this formula—the pole *above* which the *negative* electricity enters is turned to the *west*; *under* which, to the *east*.

If the uniting wire is so turned in a horizontal plane as to form a gradually increasing angle with the magnetic meridian, the declination of the needle *increases*, if the motion of the wire is towards the place of the disturbed needle; but it *diminishes* if the wire moves further from that place.

When the uniting wire is situated in the same horizontal plane in which the needle moves by means of the counterpoise, and parallel to it, no declination is produced either to the east or west; but an *inclination* takes place, so that the pole, next which the negative electricity enters the wire, is *depressed* when the wire is situated on the *west* side, and *elevated* when situated on the *east* side.

If the uniting wire be placed perpendicularly to the plane of the magnetic meridian, whether above or below it, the needle remains at rest, unless it be very near the pole; in that case the pole is *elevated* when the entrance is from the *west* side of the wire, and *depressed*, when from the *east* side.

When the uniting wire is placed perpendicularly opposite to the pole of the magnetic needle, and the upper extremity of the wire receives the negative electricity, the pole is moved towards the east; but when the wire is opposite to a point between the pole and the middle of the needle, the pole is most[4] towards the west. When the upper end of the wire receives positive electricity, the phenomena are reversed.

If the uniting wire is bent so as to form two legs parallel to each other, it repels or attracts the magnetic poles according to the different conditions of the case. Suppose the wire placed opposite to either pole of the needle, so that the plane of the parallel legs is perpendicular to the magnetic meridian, and let the eastern leg be united with the negative end, the western leg with the positive end of the battery: in that case the nearest pole will be repelled either to the east or west, according to the position of the plane of the legs. The eastmost leg being united with the positive, and the westmost with the negative side of the battery, the nearest pole will be attracted. When the plane of the legs is placed perpendicular to the place between the pole and the middle of the needle, the same effects recur, but reversed.

A brass needle, suspended like a magnetic needle, is not moved by the effect of the uniting wire. Likewise needles of glass and of gum lac remain unacted on.

We may now make a few observations towards explaining these phenomena.

The electric conflict acts only on the magnetic particles of matter. All non-magnetic bodies appear penetrable by the electric conflict, while magnetic bodies, or rather their magnetic particles, resist the passage of this conflict. Hence they can be moved by the impetus of the contending powers.

It is sufficiently evident from the preceding facts that the electric conflict is not confined to the conductor, but dispersed pretty widely in the circumjacent space.

From the preceding facts we may likewise collect that this conflict performs circles; for without this condition, it seems impossible that the one part of the uniting wire, when placed below the magnetic pole, should drive it towards the east, and when placed above it towards the west; for it is the nature of a circle that the motions in opposite parts should have an opposite direction. Besides, a motion in circles, joined with a progressive motion, according to the length of the conductor, ought to form a conchoidal or spiral line; but this, unless I am mistaken, contributes nothing to explain the phenomena hitherto observed.

All the effects on the north pole above-mentioned are easily understood by supposing that negative electricity moves in a spiral line bent towards the right, and

[4][moved.]

propels the north pole, but does not act on the south pole. The effects on the south pole are explained in a similar manner, if we ascribe to positive electricity a contrary motion and power of acting on the south pole, but not upon the north. The agreement of this law with nature will be better seen by a repetition of the experiments than by a long explanation. The mode of judging of the experiments will be much facilitated if the course of the electricities in the uniting wire be pointed out by marks or figures.

I shall merely add to the above that I have demonstrated in a book published five years ago that heat and light consist of the conflict of the electricities. From the observations now stated, we may conclude that a circular motion likewise occurs in these effects. This I think will contribute very much to illustrate the phenomena to which the appellation of polarization of light has been given.

Copenhagen, July 21, 1820.

JOHN CHRISTIAN ØRSTED.

40

New Electro-Magnetic Experiments[1]

SINCE the publication of my first experiments on the magnetic action of the galvanic apparatus,[2] I have multiplied my researches on that subject as much as my other duties permitted.

The electro-magnetic effects do not seem to depend on the intensity of the electricity, but solely on its quantity. If a strong electric battery is discharged through a metallic wire above a magnetic needle, the latter acquires no motion. An uninterrupted series of electric sparks acts on the needle through ordinary electric attraction and repulsion but produces no genuine magneto-electric effect. A galvanic pile, composed of one hundred two □–inch plates of each of the two metals and with paper moistened with salt water to serve as a fluid conductor, likewise has no perceptible effect on the magnetic needle. On the other hand, we obtain an effect with a single galvanic element of zinc and copper which is provided with a strongly conducting liquid, such as a mixture of equal parts of sulphuric acid and nitric acid with sixty parts of water. The quantity of water can be doubled without diminishing the effect significantly. If the surface of the two metals is small, the effect is also small, but it grows in proportion to the increase of the surfaces. A zinc plate of six square inches, immersed in a copper trough which contains the aforementioned liquid conductor, is sufficient to produce a considerable effect. However, a similar apparatus with a zinc plate which has a surface of one hundred square inches acts

---

[1] [*Journal für Chemie und Physik*, ed. by Dr. J. S. C. Schweigger, Vol. 29, pp. 364–69. Nuremberg 1820. This work is also to be found in Rahbek's *Hesperus*, Vol. 3, pp. 321–27. Copenhagen 1820. — Thomson's *Annals of Philosophy*, Vol. 16, pp. 375–77. London 1820. —*Bibliothèque universelle des sciences*, Vol. 15, pp. 137–41. Geneva 1820. —*Journal de Physique*, Vol. 91, pp. 78–80. Paris 1820. — *Giornale Arcadico di scienze*, Vol. 8, pp. 343–53. Rome 1820. —Oken's *Isis*, Col. 62–65. Jena 1821. KM II, pp. 219–23. Originally published in German.]

[2] See the Latin article on p. 275 of this volume. [Cf. "Experiments on the Effect of the Electric Conflict on the Magnetic Needle" (Chapter 39, this volume).] It represents one of the most significant physical discoveries of recent times, which will have great consequences for science. The present more detailed commentary will enable the readers to repeat the experiments in a simple manner and to convince themselves of the importance of Ørsted's discovery. Since Galvani's first experiment, perhaps no experiment has been performed which is of greater importance for the theory of electricity and chemistry than Ørsted's. *The Ed.*

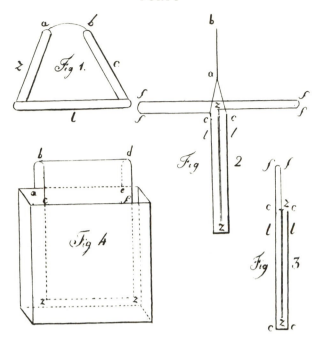

[The figures reproduced here are from the version of this
article that appeared in *Journal de Physique*. There are
minor changes from the figures that appeared in
Schweigger's *Journal für Chemie und Physik*.]

on the magnetic needle with such force that the attraction is very noticeable at a distance of three feet, even if the needle is not particularly sensitive. I have not produced greater effects with an apparatus of forty similar elements, rather it appeared to me that the effect was smaller. If this observation, which I have not
expressly repeated, is correct, I would assume that even a small diminution of the
conducting power, which can be attributed to the increase in the number of elements of the apparatus, weakens its electro-magnetic effect.

   To compare the effect of a single galvanic element with that of an apparatus
composed of several elements, we begin with the following observation. Let fig. 1
represent a galvanic element composed of a piece of zinc z, a piece of copper c, a
metallic wire ab, and a liquid conductor l. The zinc always communicates a portion
of its positive electricity as does the copper of its negative electricity. This would
create an accumulation of negative electricity in the upper part of the zinc and of
positive electricity in the upper part of the copper if the communication through ab
did not re-establish equilibrium by affording a free passage for the negative electricity from z to c and for the positive electricity from c to z. We thus see that the
wire ab receives negative electricity from the zinc and positive electricity from the
copper. This is in contrast to a wire which unites the two poles of a pile, or of an-

other compound galvanic apparatus, which receives the positive electricity of the zinc pole and the negative electricity of the copper pole.

By attending to this distinction, it is possible to repeat all the experiments which I first made with a compound galvanic apparatus using a single galvanic element which is arranged as I have indicated above. The use of such a single galvanic element gives a great advantage in that it enables us to repeat the experiments with little preparation and expense, but it presents an even greater advantage in that we can arrange a galvanic element which is sufficiently powerful for electro-magnetic experiments and still sufficiently light to be suspended from a thin metallic wire so that it can rotate about the extended axis of the wire. In this manner, conversely, it is also possible to observe the action of a magnet on the galvanic element. As a body cannot put another in motion without being moved itself, provided it is moveable, it is easy to foresee that the galvanic element must acquire some motion from the magnet.

I made use of different variations of the simple galvanic apparatus in order to examine the motion impressed on it by the magnet. One of these devices is shown in fig. 2, which represents a perpendicular section of it in the direction of its width. *cccc* is a trough of copper, three inches high, four inches long, and $^1/_2$ inch wide. These dimensions may doubtless be varied in numerous ways, but it is necessary to observe that the width is not too great, and that the walls of the trough are made as thin as possible. *zz* is a plate of zinc. *ll* are two pieces of cork which keep this plate in its position. *cffffz* is a brass wire with a diameter of at least $^1/_4$ line. *ab* is also a brass wire so fine that it is just strong enough to support the weight of the apparatus. *cac* is a linen thread which is tied to the metallic wire. The trough contains the liquid conductor. The conducting wire of this apparatus will attract the north pole of the magnetic needle when this is to the left of the plane *cffffz*, viewed in the direction *fz*. On the same side, the south pole will be repelled. On the other side of this plane, the north pole will be repelled and the south pole attracted. The needle must not be placed above *ff* and not below *fc* or *fz* if this effect is to be produced. If, instead of a moveable needle, one of the poles of a powerful magnet is presented to one of the extremities *ff*, the galvanic apparatus will be put in motion and will rotate around the extended axis of the wire *ab* according to the nature of the pole.

If a wide strip of copper having the same width as the zinc plate is used instead of the conducting wire, the effect differs from the one just described only in that it is much weaker. On the other hand, the effect is increased slightly if the conductor is greatly shortened. Fig. 3 represents the perpendicular section of this device in the direction of the width of the trough. Fig. 4 shows the same device in perspective. It is obvious that *aebdcf* represents the conducting plate and *czzf* the zinc plate. In this device, the north pole of the needle will be attracted towards the plane of *abc*, and the south pole will be repelled. *edf* causes the opposite effects. Thus, we have here an apparatus whose extremities act like the poles of a magnet. It should not be forgotten that only the two ends, and not the intermediate parts, offer this analogy.

It is also possible to construct a moveable galvanic apparatus from two plates, one of copper and one of zinc, twisted into a spiral and suspended in the liquid conductor. This apparatus is more mobile, but greater care is required to avoid errors in experiments made with it.

I have not yet succeeded in making a galvanic apparatus capable of pointing towards the poles of the earth. For this, the device would unquestionably require much greater mobility.

41

Note on the Discovery of Electromagnetism[1]

OVER the years, Professor Ørsted, Knight of the Order of Dannebrog, has submitted to the Society a series of investigations of magnetism, whereby it is shown that the magnetic effects are produced by the same forces as the electric. For a long time many students of nature have suspected this connection, but neither did they have conclusive evidence of it, nor did they guess the relationship between electricity and magnetism which experiments have now shown us. A long time ago the author himself adopted a system according to which all internal effects in bodies, such as electricity, heat, light, as well as chemical combinations and dissociations are due to the same fundamental forces. This system, which he has advanced in a few earlier treatises, has been developed more completely in his *Ansichten [sic] der chemischen Naturgesetze*,[2] which was published in 1812, and already then he arrived at the result that magnetism must be produced by electrical forces in their most bound form. For a long time he imagined that it would be far more difficult to confirm this idea experimentally than the outcome later showed it to be. He therefore devoted himself to other investigations until his lectures on electricity, galvanism, and magnetism in the spring of 1820 caused him to pursue this idea again. He now discovered that the conductor which connects the two opposite poles of a galvanic circuit, and in which all effects that can be detected by any kind of electrometer have disappeared, exerts a powerful side effect, whereby it sets a magnetic needle in motion. This effect takes place with equal ease whether the space between the conductors and the magnet is filled only with air or with other bodies. Water, glass, metals, resin, and different varieties of stone have been placed between the conductor and the needle without any difference being noticed. The effect penetrated all the bodies tested in the experiment with equal ease, except iron, which itself easily becomes magnetic to such an extent that this may contribute to the action, but iron is only an exception in so far as this happens. In this ability to penetrate all bodies with equal ease, this effect agrees completely with magnetism and distinguishes itself from electricity, galvanism, light, and heat, in short from all other effects which we are able to produce in our experiments, and approaches gravity, which

[1][*Videnskabernes Selskabs Oversigter*, 1820–21, pp. 12–21. Based primarily on the reports of the meetings held on January 5 and April 6, 1821. KM II, pp. 447–53. Originally published in Danish.]

[2][Cf. *View of the Chemical Laws of Nature* (Chapter 30, this volume).]

unceasingly penetrates all, but which can no more be produced by us than it can be strengthened or weakened.

This effect of the uniting wire does not consist in such an induction of forces as the one we know in the magnet, but it consists in a double circuit of magnetic effects around the uniting wire so that one of the two magnetic forces works from right to left while the other works from left to right. Although it is unnecessary to assume anything other than this double circuit in order to explain the experiments, the author has nevertheless believed that the combination of a progressive motion and a circular motion had to form a conchoid (a spiral). Given this, he assumes that the course of the electrical forces in the conductor is a conchoidal line which, with respect to an observer who stands outside the circuit, goes from left to right. In this line, each of the two electrical forces themselves moves from left to right when seen by an observer who faces the side from which the force comes. Given all this, the north pole of a magnetic needle is repelled by the negative electricity and attracted by the positive. Naturally, the south pole of the magnetic needle has the same relation with the positive electricity.

What we here for the moment call electricity is not really that, in the strictest sense of the word, for the force which in the open galvanic or electrical circuit acted in a particular way, in a particular form, which we call the electrical or galvanic, acts here in a completely different form, which we can most suitably call the magnetic. However, as magnetism works in the form of a straight line, that is, the opposite forces move in opposite directions, while the forces here, on the other hand, flow incessantly into each other and form a circuit, the author has called this effect *electromagnetism*.

After having made this discovery, the author performed experiments with easily movable galvanic apparatuses and found that they had precisely that relation to the magnet which the newly discovered law made it possible to predict. However, the instruments he procured proved neither movable nor sensitive enough to be put into a definite position by the earth's magnetism or to follow the effects which other uniting wires might have on it. While he was still occupied with this, Ampère, a member of the French Academy, invented a convenient device for this purpose by making a movable conductor which could receive the discharge of a powerful galvanic apparatus. Thereby he found that conductors through which the electric forces move in the same direction attract and, in the opposite case, repel each other. Ampère considered this to be the fundamental law for all electromagnetic effects; the author, on the other hand, regards it as an obvious consequence of the law he has discovered. If we imagine two parallel conductors which carry current in the same direction, the similar forces would not meet except at a couple of turning points precisely because they move in the same direction, but the opposite forces must meet and thus produce attraction. If, on the other hand, the two conductors are traversed in opposite directions, the similar forces meet and produce repulsion.

The author does not content himself with regarding this as a consequence of the law he has found for electromagnetism; he even tries to show that a circuit of forces is the only condition under which current-carrying conductors can exert the attractions and repulsions described here. If the forces flowed outward from the conductor in a straight line from its axis, no effect would be produced when both forces balanced each other and repulsion in all cases where one of the forces was domi-nant. This empirical relationship is possible only when the radiating forces act along the tangents to the circumference of the conductor, or in directions which can be decomposed in such a way that only effective tangential forces remain, and when each of the forces acts from any given point along opposite tangents.

It was natural to expect that the same force which had such a strong effect on the magnetic needle must also produce magnetism. Nor was it long before several ways were discovered in which a steel wire could be magnetized with the aid of the galvanic circuit. Arago found that a steel wire could be magnetized by winding the conductor around it in spirals.

Several other investigators found that the same effect can be produced by wind-ing the steel wire around the conductor in spirals. These procedures, however, leave considerable doubt with regard to the theory. One could either agree with Berzelius, the famous student of nature, in assuming that the conductor had a cer-tain number of fixed magnetic poles in its circumference and then, most reason-ably, assume two pairs, or one could agree with the author that there was a tendency towards magnetic opposition along the direction of the two opposite tangents at every point on the circumference of the conductor. The author has decided this question by placing a thin steel wire, which covered half the circumference of the conductor, close to the latter and thus magnetizing the wire. If there had been two pairs of magnetic poles, of which the similar ones were at the ends of the same di-ameter, the magnetized wire must have had the same poles at both ends, but the ex-periment showed that it had opposite poles. The magnetization took place in the same way when the wire covered larger or smaller portions of the conductor. It is therefore correct to assume that every point on the circumference has a tendency towards magnetic opposition in the directions of its two opposite tangents. With some modifications, Wollaston, Seebeck, Prechtel[3], and others share this opinion which they have tried to prove each in his own way.

It naturally followed that it had to be possible to produce magnetic effects through frictional electricity as well as through contact electricity, or the so-called galvanism, but it was not possible to distinguish the effect of frictional electricity on the magnetic needle from the strong attractive and repulsive effects it exerts as electricity. On the other hand, Davy and Arago have found that a steel wire can be magnetized by frictional electricity. As could be expected, the author finds this completely confirmed, and thereby the name of electromagnetism becomes all the more perfectly justified.

Ampère had found that a movable uniting wire is brought to orient itself along the magnetic east and west through the effect of the earth's magnetism. This sug-gested to him that magnetism consisted only of electrical circuits around an imagi-nary axis, situated in planes perpendicular to it. The author has tried to show that this theory, which has been devised with much perspicacity in order to explain the effect of the current-carrying conductor on the magnetic needle, is far from de-scribing all effects between magnets. He also agrees with Erman that this theory does not apply to the well-known magnetization by means of stroking. The author

[3] [Prechtl.]

believes that the difference between electromagnetism and magnetism is only that the former consists of a circuit in which the opposite forces incessantly move in opposite directions at every point, while the latter is to be regarded as a similar effect except that it does not form a circuit but is constituted in the same way as the electromagnetic circuit would be if it could be opened and straightened out into a line without the cessation of the ongoing effects.

The earth's magnetism was considered by Ampère merely to be a consequence of a galvanic circuit parallel to the equator, which he attributed especially to the earth's composition of layers which could be regarded as elements in a galvanic circuit, but on whose condition the sun must also exert influence. The author believes that the earth's assemblage of layers does not have such an orientation that it can produce a circuit from east to west, but he does believe that the sun alone can do this. Through the heat, evaporation, and chemical decomposition it causes, an electric circuit from east to west must necessarily be produced. This electric circuit from east to west, then, produces an electromagnetic effect in the direction between south and north. Thus it may be assumed that the earth is surrounded by an electromagnetic belt. On the upper surface of this belt, negative electromagnetism (corresponding to negative electricity) moves towards the north, positive towards the south. On the inferior surface, according to the nature of the circuit, the effect is the opposite. Since no body is completely without the ability to receive magnetism, though most only to a small degree, it follows that the earth's mass must be magnetized because of the electrical belt. This magnetization occurs by means of the lower or inner surface so that the earth's mass receives its magnetism in the direction opposite to the one shown by the upper surface of the belt and by the magnetic needle oriented by its electromagnetic forces. However, since a magnetic needle must orient itself in relation to another magnet in such a way that their opposite poles seek each other, it follows that the earth's magnetic mass and the surface of the electromagnetic belt strive to give the needle the same direction. The magnetic belt, which must extend approximately as far as the alternation between day and night, undergoes daily and yearly changes, which must also influence the magnetic needle: The earth as a magnet will not easily suffer anything but very slow changes. As all electrical discharges are accompanied by electromagnetic effects, it is now easy to understand the influence which lightning often has on the magnetic needle. The author suggests using the magnetic needle as a meteorological instrument, but to use only either very weak magnetic needles, on which the force of the earth is not considerable, or to suspend them in directions in which the force of the earth has no influence on their direction. He thinks that the electromagnetic theory could be useful even in the art of surveying, as it would be possible to observe what influence galvanic effects in mountains could have by means of magnetic needles suspended in various directions.

Finally, the author points out that the theory of heat and light which he advanced publicly a long time ago assumes that these two great effects should be produced by the interaction of the same two forces which produce electric phenomena. It has seemed unthinkable to many physicists that these forces could be united without completely destroying each other's effect. Electromagnetism, on the other hand, shows us an example of an effect produced by the same two forces which produce electricity, but under circumstances in which they are united in such a way that no electrometer detects their presence. Heat and light are generated by the same forces and by the same procedure as electromagnetism, but a larger quantity of active forces and greater rapidity in the interaction between them are required for this. Heat is produced only when the conductor has received so much electricity that it can only conduct it away very inadequately. Is it not the case that the force, pressing outwards in all directions, must make a path which itself must follow the same law as the motion on the surface of the conductor and consequently either form a spiral itself or spread in waves, in a constant circuit? However, the issue of constructing new laws for this effect, which is either more complex than the electromagnetic or perhaps does not propagate in curved lines of the same shape but even in a zigzag, must be left to further investigations to decide. The author has previously claimed that light could be compared to a series of innumerable electrical sparks of imperceptible size, and that the propagation of light must be regarded as an interruption and restoration of the equilibrium between the two opposite forces in space. This theory is similar to that of Huygens or Euler in that it assumes the propagation of light to be a kind of wave motion (undulation) but is different from them in that it does not consist of an alternation between compressions and compactions[4] but in a series of decompositions and compositions. Thus, the undulation became chemical, not mechanical. This was the author's earlier theory. The electromagnetic discoveries now seem to show us the way to deeper insight into the nature of these undulations. Perhaps the mutual distance between the windings or the circuits could determine the colour of the rays of light, and the shape of these windings or circuit elements may some day serve to explain the so-called polarity of light.

Even in the theory of chemistry, the motions which electromagnetism reveals to us could hardly be without value since we know that compound substances are decomposed by galvanism into their opposite constituents, and that these move from one electric or galvanic pole to the other, just like the electric forces in a metallic conductor. In opposite substances which neutralize each other's effects, then, there should be a tendency towards opposite motions. How wide are not the perspectives which open here, as a consequence of the discovery of one single primary quality.

The author himself feels that he will progress but slowly in this wide range of investigations. However, he may venture to hope that other students of nature will give the conclusions he draws from his discoveries some of the attention that they have given his experiments. In that case the many great subjects which seem to him to hover in a distant twilight would be brought forth into the clear daylight of truth.

[4][expansions.]

42

Observations on Electro-magnetism[1]

(A.) THE HISTORY OF MY PREVIOUS RESEARCHES ON
THIS SUBJECT

When I began to examine into the nature of electricity, I conceived the idea that the propagation of electricity consisted in a continual destruction and renewal of equilibrium, and thus possessed great activity which could only be explained by considering it as a uniform current.[2] I then regarded the transmission of electricity as an electrical conflict, and my researches into the nature of the heat produced by the electrical discharge, particularly led to the conclusion, that the two opposite electrical forces, which pervade a body heated by their effect, are so blended as to escape all observation, without however, having acquired perfect equilibrium,[3] so that they might still exhibit great activity, although under a form of action differing entirely from that which may be properly termed electrical. Notwithstanding my efforts to justify my idea, this complete annihilation of power indicated by the electrometer, accompanied with very considerable action of another kind, appeared to the greater number of philosophers to possess but little probability. This feeling may, perhaps, be partly attributed to the obscurity of the subject, and partly also to the imperfect manner in which I explained my theory; for it must be confessed that new views are rarely developed even to their authors with perfect clearness. A thorough conviction of the agreement of my theory with facts, inspired me, nevertheless, with so strong a persuasion of its truth, that upon this basis I ventured to form my theory of heat and light, and to attribute to these forces, apparently destroyed, a radiating action capable of penetrating to the greatest distances. Having for a long time considered[4] the powers which are developed by electricity as the

---

[1] Communicated by the author. [This article is reproduced as it appeared in Thomson's *Annals of Philosophy*, Vol. 2, pp. 321–37. London 1821. It also appeared in *Journal für Chemie und Physik*, ed. by Dr. Schweigger and Dr. Meinecke, Vol. 32, pp. 199–231. Nuremberg 1821. (KM II, pp. 223–45.) — *Bibliothèque universelle des sciences*, Vol. 18, pp. 3–29. Geneva 1821. —*Journal de physique*, Vol. 93, pp. 161–80. Paris 1821.]

[2] My treatise on this subject will be found in Gehlen's *Journal*, 1806, and the *Journal de Physique* of the same year. [Cf. "On the Manner in Which Electricity Is Transmitted" (Chapter 21, this volume).]

[3] See my *Considerations on Natural Chemical Laws*, pp. 132–234. Berlin 1812. [Cf. *View of the Chemical Laws of Nature* (Chapter 30, this volume).]

[4] See the letter added to my German publication *Materielen muner [sic] Chemie, & c.*, 1803; and also *Researches into the Identity of Electrical and Chemical Forces*, p. 127, &c. [The first of these references

general powers of nature, it necessarily followed that I should derive magnetic effects from them.[5]

In order to prove that I admitted this consequence to the utmost extent, I cite the following passage from my Researches into the Identity of Electrical and Chemical Powers, printed at Paris in 1813, "it must be determined whether electricity in its most latent state has any action upon the magnet as such."[6] I wrote this during a journey, so that I could not easily perform the experiments, besides which, the manner of making them was not at that time at all clear to me, all my attention being directed to the development of a system of chemistry. I still remember that I expected, though somewhat vaguely, the effect in question, and particularly by the discharge of a strong electrical battery, and also that I did not hope to obtain more than a weak magnetic effect. Thus I did not follow the idea which I had conceived with the requisite zeal, but the lectures which I delivered upon electricity, galvanism, and magnetism, during the year 1820, recalled it. My auditory consisted mostly of persons previously well acquainted with the science. On this account, these lectures and preparatory reflections, led me on to deeper researches than those which are admissible in common lectures.

My original persuasion of the identity of electric and magnetic powers were developed with greater clearness, and I resolved to submit my opinion to the test of experiment, and the preparations for it were made on a day in the evening of which I had to deliver a lecture. I there showed Canton's experiment on the influence of chemical agency on the magnetic state of iron. I requested attention to the variations of the magnet during a storm, and I mentioned at the same time the conjecture, that an electrical discharge might produce some effect upon the magnetic needle placed out of the galvanic circuit. I immediately resolved to make the experiment. As I expected the greatest effect from a discharge producing ignition, I inserted at the place under which the needle was situated, a very fine platina wire between the connecting wires. Although the effect was unquestionable, it appeared to me nevertheless so confused that I deferred a minute examination of it to a period at which I hoped for more leisure.[7]

At the beginning of the month of July, my experiments were repeated and continued without interruption until I obtained the results which have been published.

(B.) EXPLANATION OF THE FIRST LAW
OF ELECTRO-MAGNETIC EFFECTS

The electro-magnetic effect which I discovered by the aid of the galvanic apparatus, has since been produced by common electricity, so that the expression of

---

is to *Materialien zu einer Chemie des Neunzehnten Jahrhunderts*, cf. *Materials for a Chemistry of the Nineteenth Century* (Chapter 12, this volume). The second is to a section of the work mentioned in the previous footnote. In this edition on pp. 346–51.]

[5] Letter, pp. 234–38. [This edition, pp. 378–79.]

[6] Letter, p. 238. [This edition, p. 379.]

[7] All my auditors are witnesses that I mentioned the result of the experiment beforehand. The discovery, therefore, was not made by accident, as Prof. Gilbert has concluded from the expressions which I made use of in my first announcement.

electro-magnetic effect is perfectly justified by experiment. It is well-known that for the first experiments on this subject we are indebted to M. Arago, who has been equally successful in enriching both physics and astronomy with his discoveries; the illustrious President of the Royal Society of London has also made an important series of experiments on this subject.

I shall here state, rather more in detail than I have done in my first publication, the rule by which I think all electro-magnetic effects are governed. It is this: *When opposite electrical powers* [8] *meet under circumstances which offer resistance, they are subjected to a new form of action, and in this state they act upon the magnetic needle in such a manner that positive electricity repels the south, and attracts the north pole of the compass; and negative electricity repels the north, and attracts the south pole;* [9] *but the direction followed by the electrical powers in this state is not a right line, but a spiral one, turning from the left hand to the right.* Many philosophers, and some of them of great merit, have thought the spiral motion of electrical powers to be improbable. I shall endeavour in the sequel to show, that this supposition is less arbitrary than it may appear to be at first; but to prepare for this, it is necessary first to explain the meaning of this supposition, and then to prove that all electro-magnetic phenomena so completely harmonize with the rule given, that it will suffice even to anticipate those among them which were not known before experiment. I have not discovered so perfect an agreement with facts in any other theory which has been hitherto advanced. When I have shown that the rule is quite sufficient to comprehend all the facts under one point of view; that is to say, that it is a correct rule, I shall invite the reader to examine with me, whether it may not also be *a law*, according to which the phenomena are arranged in nature. [10] It is extremely difficult, and especially for those who are not much accustomed to the representation of complicated figures, to understand the spiral quite clearly. What I am going to state will be most readily understood in the following manner:

Upon a slip of paper, fig. 1, draw the line *AB* which is to be longitudinally divided into two equal parts; draw some small triangles so that the summits and the middle of the bases may be cut by the line. Put the sign + at that end towards which the summits point, and the sign − at the end towards which the bases are placed. This piece of paper is to be twisted round a quill, a piece of glass tube, or any other cylindrical body, in such a manner that the triangles, reckoning from the summit to the base, shall be placed from the left to the right hand of the observer. The cylinder enclosed in this mannner I call the electro-magnetic indicator. With this indicator,

[8] I here repeat what I have already stated in other works, that by *electrical forces*, I mean only the unknown cause of electrical phenomena, whether it belong to imperceptible matter or independent motion.

[9] In my first memoir, I grounded all explanations upon the repulsions only which are exerted by electrical and magnetic forces; but I soon discovered, that from the fear of assuming more than the phenomena required, I drew an unjust inference; for if magnetic forces are the same as electrical under another form of action, it follows, that opposite forces ought to attract each other reciprocally, and forces of the same kind to repel each other.

[10] I had intended to develope [sic] this matter in the present memoir, but the desire of stating all I am able to say upon so difficult a subject with all possible clearness, has induced me to defer it to a future opportunity.

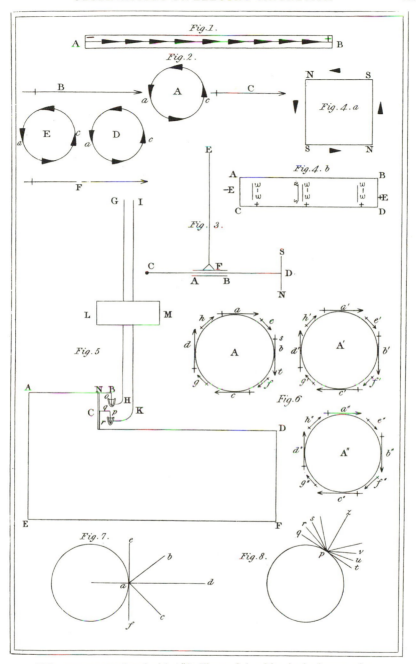

[The arrows associated with *A″* in Figure 6 should point in the opposite direction. In addition, the conductor *A″* should have been placed directly under the conductor *A*.]

that part of the connecting wire is compared whose effect is to be judged of, by imagining it to be put in the place of the latter in such a position that the end marked + may receive the electricity of the positive extremity of the galvanic apparatus, and the end marked − of the negative. This being done, it will always be found that the south pole of the suspended needle is repelled by negative electricity. For brevity's sake, we will designate positive electricity by +E, and the negative by −E. But when these forces have assumed their new condition, in which they possess no action upon the electrometer, and affect the magnet, we shall call them electromagnetic forces, and denote them by the Greek letter +ε and −ε.

It would be useless to repeat upon this occasion the description of all the experiments mentioned in my Latin memoir; it will be sufficient to say, that with the assistance of the electro-magnetic indicator, all the effects of the connecting wire in the most dissimilar positions which I have described, may be anticipated; I shall mention one example, as it may make the subject clearer. Place a part of the connecting wire perpendicularly opposite the magnetic needle, and let the upper part of the conductor receive the electricity of the negative end of the galvanic apparatus; that part of the needle, which receives the effect, will turn towards the east. A, fig. 2, represents the horizontal section of the conductor with the signs abovementioned to describe the direction of the electro-magnetic forces. B represents a magnetic needle, whose north pole turns towards the connecting wire. C is another which presents its south pole to the wire. Both will be directed towards the east; the north pole by −ε, which comes from the west; the south pole by +ε, which comes also from the west. It will be readily seen that the attraction caused by +ε upon the north pole, and by −ε upon the south pole, must increase the motion towards the east.

If the conductor be placed opposite a point in the northern half of the needle, −ε in a directed towards the east will not act upon the north pole of the needle; while the contrary will be the case with −ε in c directed towards the west. It is true that the attraction of +ε in a will draw the needle towards the east; but it will act upon a very weak magnetic point, and, consequently, with but little power; the north pole will then go to the west. I do not reckon upon the advantage which the lever −ε in c possesses over +ε in a; for it may be readily shown that this advantage is not necessary to produce the phenomenon in question; it requires only that a magnetic needle should be fixed perpendicularly to one of the ends of an arm of a torsion balance, and that it should be arranged with the galvanic connecting wire. When the conductor is so placed that E represents a horizontal section of it, −ε in c will exert some attraction upon the south pole of the needle, but +ε in a repulses it with greater force; this end will, therefore, turn towards the east. I have already shown by the effect of A upon C what will happen when the conductor is opposite the south pole. It may also be readily foreseen, what will happen when the conductor is placed on the western side of the needle pE with respect to the needle F.[11] When the conductor is at D, −ε of a will repel the north pole, but +ε of c will attract it with

[11] [The German version of this sentence reads: It may also be readily foreseen what will happen when the conductor comes to rest on the western side of the needle, e.g., here towards the needle F.]

more force, the north pole will then turn towards the west, whether the perpendicular conductor be placed on the eastern or the western side. In the same manner, the south pole will in both cases turn towards the west, which will be readily explained by referring to the figure.

If any one be desirous of seeing the contrary effect of the two sides of the connecting wire in a more direct manner, it is only requisite to give a moveable needle the same magnetism at both ends. It is not that I consider this experiment as necessary after so many similar ones, but it may, perhaps, be very useful to represent the thing in the most simple manner in an elementary lecture. With the same needle the experiment may be made with the connecting wire placed horizontally opposite the ends of the needle, and observe the movements from the top to the bottom, or the bottom to the top. In this way it is possible to show in a manner very easy to be understood, the direction of the electro-magnetic forces in the connecting wire; it is also very easy to perform electro-magnetic experiments by the following arrangement: Let *AB*, fig. 3, represent a small hollow cylinder, which may be made of paper, and in this a very small cylinder of wood must be made to turn with but little friction. The end *D* carries the needle *SN*, and the whole is fastened with a hair or a very fine metallic wire. When the effect of the connecting wire has been tried upon *N*, it is only necessary to turn *CD* in the cylinder, so that *N* may be uppermost; and then the effect of the other side of the needle may be immediately tried. It is also easy to place *SN* horizontally, and to try the attraction or repulsion exerted upon it by any given part of the connecting wire.

Many ingenious attempts have been made to explain electro-magnetic phenomena. The first which I am acquainted with is that of my illustrious friend Berzelius. This philosopher supposes, that the galvanic conductor possesses double transverse magnetism, so that a conductor in the form of a parallelopiped has a north magnetic pole at one of its angles, and a south at the other. Let *NS*, *NS*, fig. 4*a*, represent the transverse section of such a conductor placed in the magnetic meridian, and receiving the current of electricity from the positive end of the pile in the direction of south to north. The letters *NN* denote the two north poles; the letters *SS* the two south poles of the conductor. This theory explains many of the phenomena satisfactorily, and with surprising facility, as might naturally be expected in the hypothesis of so distinguished a philosopher, but it agrees nevertheless with only a part of the phenomena. The observation which I have so frequently had occasion to make in my experiments, that round conductors act in so equable a manner in every part of the periphery, that no distribution of poles is discoverable in them, excited some suspicions against this new hypothesis, and a direct experiment decided me absolutely against it. Twist a steel wire round one half of a square conductor, in such a manner that it may coincide with the semiperiphery *NSN*, or *SNS*, on which side soever it may be; according to the hypothesis, this wire ought to have either no magnetism at all, or equal poles at the two ends; but it will be found that the wire has always a north pole at that point towards which $-\varepsilon$ is directed, and a south pole at the point to which $+\varepsilon$ is directed. These directions will be explained in fig. 4.

As in these experiments very fine wire only should be used, a weak needle should also be employed: a small piece of the same iron wire fastened to a bit of raw silk is extremely convenient. In general, a steel wire may be magnetized by placing it across the conductor, although the latter be a parallelopiped, round or flat, and the wire may occupy a great or small part of the periphery; the point towards which $-\varepsilon$ turns always gains the property of turning towards the north. What is also remarkable is, that the magnetic pole produced in the steel wire applied to the conductor, is of the same kind as the pole of a neighbouring magnet repelled in the same direction. This proves also that the conductor cannot be considered as a body which has distinguishable poles on the surface; for in this case the poles produced and repelled would be of the same denomination.

In order to answer the question, whether the wire attached to the surface of the conductor might be considered as a part of that surface, differing only from others in its power of retaining the magnetism communicated, I put a piece of fine paper between the conductor and the steel wire: in other respects I performed the experiment as before. I had the same results, with this difference only, that the effect was rather weaker.

When a light magnetic needle is placed upon a large conductor, through which a strong discharge is passed, its direction is almost entirely determined by the electro-magnetic effect, and the magnetism of the earth causes but a very slight direction. Let $ABCD$, fig. 4b, represent the large conductor, and suppose $-\varepsilon$ to enter at AC, and $+\varepsilon$ at BD; the direction of the electro-magnetism may then be marked by the signs $+\varepsilon$ $-\varepsilon$. Place a magnetic needle SN properly mounted above the conductor, and let us call the end of it which turns towards the south pole s, and that which turns towards the north n; this being done, it will be found that the direction sn will coincide with the direction $+\varepsilon$ $-\varepsilon$. If the needle held always in the same horizontal plane be put towards one of the sides of the conductor, the north pole will be repelled from the side AB, but attracted by the side CD, only much more feebly than before. The cause of this phenomenon is undoubtedly that every point in that half of the needle which turns to the north is repelled by $-\varepsilon$ coming from the south, and attracted by $+\varepsilon$ coming from the north. In every point of the conductor, there is, therefore, an effort to act magnetically in two different directions.

M. Prechtel,[12] of Vienna, a distinguished chemist, has succeeded in representing the phenomena of the galvanic conductor by means of iron wire turned into a spiral form, which he touches with the magnet in the same manner as if he were magnetising a cylinder. This spiral thus gains transverse poles, but no sensible polarity from one end to the other. By employing the requisite means, each coil of the spiral has more than two poles given to it, and it will then produce the same effects as the connecting wire upon the magnetic needle. This experiment has led him to consider the connecting wire as a transverse magnet, having a great number of successive poles, which are alternately north and south. It will be observed that we have arrived by different routes at opinions which are almost entirely similar. I prefer, however, to keep the name of electro-magnetism for the state of the connecting wire; for in the first place, there is no distinct pole in such a conductor; and besides this, the continual production of fresh electricity in the galvanic apparatus requires

---

[12] [Prechtl.]

that we should suppose the electro-magnetism to be continually renewed, and an uninterrupted circulation of electrical forces in the conductor. In order that magnetism, properly so called, may be exhibited, it is requisite that the circulation should be interrupted, without the contrary effects of the activity which existed in the conductor, being suspended.

(C.) Explanation of the Attractions and Repulsions which Galvanic Conductors excite among each other

As soon as I had suspended a small galvanic apparatus in the manner of the torsion balance, I tried whether the connecting galvanic wire would act upon that of the suspended apparatus; but on account of the too weak action and great weight of this apparatus, I had no sensible effect. The same thing has happened to several other philosophers who have tried the same process, as I have seen in several treatises published upon electro-magnetism. M. Ampère selected a better process. He made a moveable conductor which he communicated with an apparatus of considerable strength, and thus he succeeded in discovering the attractions and repulsions of the galvanic conductor. His memoirs upon electro-magnetism are already too well known to render it necessary for me to say, that this distinguished philosopher has evinced the same extraordinary sagacity in the application of the discovery, as in his preceding labours, all of which evince great penetration; and if I adopt a theory of magnetism differing from his, I shall never cease to acknowledge the great merit of his labours.

My present apparatus for experiments upon the reciprocal effects of the galvanic conductor, appears to me to be sufficiently simple; and I shall now describe it. *ABCDEF*, fig. 5, is the moveable conductor, made of brass wire, of one-fourth of a line in diameter; *NC* is a small wooden cylinder, to prevent as much as possible any alteration in the form given to the brass wire. The two points, *o*, *p*, move in two conical iron cups, *q* and *r*, filled with mercury. In *q*, the point rests upon the bottom, and upon this, the whole of the conductor; in *r*, on the contrary, the point moves freely in the mercury. *GH* and *IK* are brass wires which support *q* and *r*. *LM* is a little bit of wood, in which these wires are inserted, and which, by means of a screw, are fastened to any support. When *G* and *I* are put into the requisite communication with the conductors of the galvanic apparatus, the wire *ABCDEF* forms a part of the communicating wire, and arranges itself in the direction of the magnetic east and west, as was discovered by M. Ampère, and may be subjected to the principal experiments upon the action which the connecting wires exhibit. But in order to render the effect imperceptible, which the conductors, designed only to convey electricity, produce upon the moveable conductor, it is requisite to make *GH* and *IK* a foot or more in length, and especially to prevent the conductors of the apparatus approaching the moveable conductor. It will be understood, that in more delicate experiments this apparatus may be inclosed in a glass case, provided only that the wires *GH* and *IK* are passed through it by means of a cork; but for the greater number of experiments this precaution is not necessary.

It is to the well-conducted researches of M. Ampère that we owe the law, *that the conductors or parts of parallel conductors attract each other when they both receive the electric current in the same direction, and repel each other when they receive it in contrary directions.* He does not endeavour to derive this law from the nature of electric forces, but considers it as a law which is independent of any laws already known.

I shall show that this law is necessarily derived from that which I have discovered.

Let us regard the thing at first as if the effect of electricity upon the magnetic needle had not been discovered. The attractions or repulsions of conductors of which we see no trace, unless they are pervaded by electrical powers, can be attributed only to those powers, and they must have such a direction in the conductors as to enable them to produce the effects discovered.

Let us consider the various modes of action which may be conceived, in order to discover that which agrees best with the circumstances demonstrated by experience. Fig. 6 represents the transverse sections of two conductors, which receive the current in the same direction. Neither of the electrical forces can be in any sensible excess, for such an excess would cause the conductors to repel each other mutually. The effects of the two forces cannot either, possess the same direction; for in this case, they would destroy each other. Still less can any inequality be suspected in the states of the two conductors, because they are supposed to be equally, and in the same manner, pervaded by the two forces. Thus the forces must leave each point of the surface in opposite directions; consequently their direction cannot be in the lengthened radius, but each of the forces must follow the direction of one of the tangents opposite to the point from which they set out, pe, to the point C on the conductor A, $-E$ will go towards t, while $+E$ will go towards s. Let us call also in this place $+E$ and $-E$, which act transversely, $+\varepsilon$ and $-\varepsilon$, in order to distinguish them from the forces in the longitudinal direction, as besides they agree absolutely with what we have before marked with these letters.

If any one should adopt the improbable idea, that the forces leave each point in two opposite directions, which will be found on contrary sides between the tangent and the lengthened radius, as ab and ac, fig. 7, each would nevertheless resolve itself into two directions, one of which would be ad, and would produce no effect in consequence of the union of the two forces, and the other would be for one force ae, and the other force af, consequently the effect would depend upon direction in a tangent.

I have observed that this supposition is improbable, but it is very likely that the forces may act at the same time according to the tangent, and in every direction between the tangent and the lengthened radius, so that they may form pencils from the point p, fig. 8, $+\varepsilon$ in the directions pq, pr, ps, &c.; $-\varepsilon$ in the direction pt, pu, pv, &c. but always so that $+\varepsilon$ remains on one, $-\varepsilon$ on the other side of the lengthened radius.

No one will readily believe that the effect of a force can proceed from the surface under a different angle from the other, for suppose one to act in the direction of the long radius, and the other in another direction, that which would have the advantage of acting directly, would produce greater effect than the other, and the conductors would repel each other. It would be almost the same as supposing that one force approached more in the direction of the lengthened radius than the other; for this force would resolve itself into two others, one longitudinal, and the other radial; and the latter would be stronger than the radial effect of the other force.

The only supposition then by which the electrical forces can produce the described effects is, that they proceed from every point in such a manner that the directions of the opposite forces are separated by the lengthened radius. But in order to show it clearly by a figure, let us represent only the directions according to the tangent, and those of some particular points, which may be exhibited as examples of what passes in others. When we consider the effects which take place in the longitudinal directions, fig. 6, where the analogous points are marked by the same letters, it will be seen that the direction of $-\varepsilon$ of a meets that of $+\varepsilon$ of a'. In the same manner $+\varepsilon$ of c, and $-\varepsilon$ of c' meet each other. This meeting of opposite forces which ought to produce attraction, occurs also in most of the points of the two peripheries, that of e in relation to h, f in relation to g. It is true that the points e and g, f and h, as well as the neighbouring points, repel each other; but on account of the small number of active points and the oblique direction, this effect must be much exceeded by the attracting effect.

A'', fig. 6, represents the transverse section of a conductor, in which the directive effect of the electricity is opposed to that of A and A'' [sic]. The points d and d'' repel each other on account of $+\varepsilon$, and the points C and C'' on account of $-\varepsilon$. Besides this there exists also here repulsion between all the points which are respectively in the same situation as the points which in the first case attract each other. The attracting effect which g and e'', f and h'' produce, as well as of the neighbouring points, is here overcome by the repulsive forces, in the same way as in the preceding case the repulsive effect, was overcome by the attracting forces.[13]

Although these conclusions are expressed only as it is requisite to do when treating of cylindrical conductors, they may nevertheless be readily applied to conductors of other forms; it appeared to me that the simplest representation should be preferred.

(D.) THE MAGNETIC NEEDLE

As fresh electricity is continually evolving in the galvanic column, the discharge of it must be regarded as a continual addition and subtraction. The peculiar state of forces which exists in the connecting wire, in which they act as electro-magnetic forces, appears to me to be a state of continual agitation. But in the magnet, the

---

[13] I have already given this explanation in a public lecture on Jan. 2, in this year. The first Tuesday in every month I give a public lecture, in which I announce new discoveries as they come to my knowledge. At the same time I demonstrate their relation to the system of our physical knowledge. It was in one of these lectures that I explained my opinion of the cause of the attractions and repulsions of the connecting conductor.

mode of action of the same forces differs from that of electro-magnetism, in their being almost entirely in a state of repose, and forming no close circle. Here we must alter the denomination $+\varepsilon$ into that of $+m$, and the denomination $-\varepsilon$ into that of $-m$. The pole of a magnet towards which $+m$ is directed ought then to produce the most marked effect of $+m$, and the pole directed opposite to $-m$, ought, in the same way, to exhibit the strongest effect of $-m$, supposing always that the extension and the state of the conducting property of the conductor occasion no exception. We are now speaking of the effect of each point, and not of the greatest effect of the whole half of a magnet, which evidently can take place only opposite the end. In a certain sense it may be said that the magnet is a body charged with electro-magnetism. This manner of considering the magnet agrees with that generally admitted from the point at which we changed the expressions from $+\varepsilon$ and $-\varepsilon$ to those of $+m$ and $-m$; it may now be left without further explanation.

But it will be necessary to explain in this place the reasons which prevent my adopting of M. Ampère's ingenious theory of magnetism. It is well known that this philosopher supposes that the line which unites the opposite poles of the magnet, is surrounded with electrical currents placed in planes perpendicular to the axis, so that these currents, and not the longitudinal magnetic distribution, are the cause of magnetism. According to this idea, two neighbouring moveable magnets, would have a tendency to turn in such a manner, that their circular currents would attract each other. If then we place two magnets, one of which at least should be moveable, one above the other with their axes parallel, it must happen that the moveable needle will turn, until the opposite poles are placed upon each other. A and A', fig. 9, represent transverse sections of two magnets placed upon each other in the same direction; so that thus we see only one of the currents of each magnet. The darts do not represent the circular movements of the forces in the conductors ($+\varepsilon$ and $-\varepsilon$), but the direction of the current, such as it is usually imagined to be ($+E$ and $-E$). The similar letters in the two circles represent those places, in which the direction of the current is the same. Now the parts which have opposite directions of the current approach each other the most; and thus they will repel each other, until the needle has made half a turn, by which the currents are placed in a situation in which they may have parts in equal directions opposite each other, which cannot take place, excepting when the opposite poles shall be placed one above the other. In every other parallel position of the axes, the same relation occurs, as may be satisfactorily conceived by imagining the circles A and A' to change their places in several ways.

SN, fig. 10, represents a magnet constructed according to this idea, and the darts have the same meaning as in the preceding figure, and will retain their meaning in the sequel. The direction of $+E$, or, according to M. Ampère, the direction of the electrical current, goes always to that side which turns towards us from bottom to top, as is indicated by the dart with a cross at the end, placed in the middle of the magnet SN. The right face of such a galvanic or electrical circle, always turns towards the north, the left side towards the south. SN thus representing a magnet, the end of the right side will endeavour to turn towards the north, or will be the north pole of the needle, and the end on the left hand will be the south pole. $S'N'$, fig. 10,

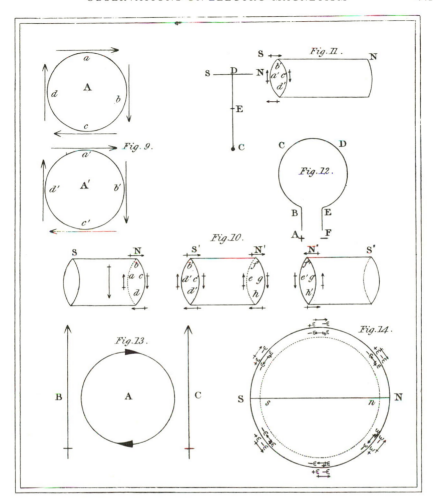

represents another needle like the former. When S', the south pole of this needle, is placed opposite N, the north pole of the needle, so that a, b, c, d, of the latter are opposite a', b', c', d', of the former, it will be readily observed that parts, the currents of which have the same direction, are opposite each other, and must reciprocally attract each other, as occurs with opposite poles. If, on the contrary, $S'N'$, fig. 10, be made to act upon $S''N''$, so that N' and N'' be opposite each other, and e, f, g, h, opposite the points e', f', g', h', opposite currents are placed together, and repel each other, as would occur with poles of similar descriptions.

To this point the explanation is perfectly correct; but when the magnets SN and $S'N'$ are placed beside each other so that the end N of the first shall be beside the end S of the other, it will be conceived that the point a will be put in relation with c or a' with c, or d with b', or at last d' with b; in all these cases, these currents meet, and must repel each other. Thus the theory is not in this case in unison with the phenomena. As that[14] has been said of the relation of N to S' may be applied to N' and N'', the requisite changes being made.

A true magnet acts in the same manner with respect to a magnet constructed according to this idea. Let SN, fig. 11, represent the latter, and $S'N'$ the former; but let SN be fastened to one end of a thin metal plate, DC, provided in E with a centre, and suspended like a magnetic needle. It will in fact be found that N is attracted by all the points placed in the interior of the circle a', b', c', d', but that it is repelled by all the points placed on the exterior of the circle, as ought to happen according to the theory which I have proposed, but it does not agree with the opinion, that $S'N'$ should be considered as a magnet. As there is not in this case any question as to the length of the axis of the magnet, in making the experiment it is only requisite to employ a wire curved like $ABCDEF$, fig. 12, which receives $+\varepsilon$ at A and $-\varepsilon$ at F. The moveable magnet SN may also be mounted in the manner shown in fig. 3, only it is necessary to observe that CD should be so turned in the cylinder AB, that SN should be in a horizontal position.

The experiment cited by M. Ampère as particularly favourable to his theory, that a magnetic needle properly suspended is either totally attracted or repelled by the connecting wire in complete conformity with his theory, cannot be considered as contrary to mine, and has not moreover been so considered by M. Ampère, especially as he did not intend to criticise the theory which I had in few words proposed, and but very slightly explained; but the explanation of this account will increase the facility of comprehending that which I am here going to describe.

A, fig. 13, is a transverse section of an electro-magnetic conductor; B and C are two magnetic needles. It will be readily conceived that C will be repelled, and B attracted; for $-\varepsilon$ in A will meet the north pole, and $+\varepsilon$ the south pole of C; on the contrary $+\varepsilon$ of A will meet the north pole, and $-\varepsilon$ the south pole of B. This phenomenon is, therefore, a consequence which results with equal facility from both theories.

One of the difficulties of M. Ampère's theory is the explanation of the common manner of magnetizing steel, as has been already remarked by the celebrated Erman. It is unintelligible how by touching or rubbing the side of a steel wire, a galvanic circle can be placed around the whole of the wire. As the theory here sanctioned differs from the common one only in the derivation of the magnetism, and not in the idea which ought to be formed of the distribution of the magnetic forces in the magnet, the common theory may also be adopted. That which neither one theory nor the other can yet explain is the peculiar power which some bodies, especially iron, nickel, and cobalt, possess of receiving a high degree of magnetism, while almost all others are susceptible of it only in a very slight degree.

I shall not explain more at length my objections against the theory of the celebrated Ampère. If I have perfectly understood it, what I have already said will be sufficient. If I am deceived upon any point of his theory, I flatter myself that I have given him an opportunity of shedding new light upon this important matter.

[14] [According to the German version, this should read "All that."]

(E.) Magnetism of the Earth

The daily course of the light of the sun round the earth produces warmth, evaporation, and chemical agency, from the east to the west. From this also proceeds an alternation of the destruction and renovation of electrical equilibrium, and the effect of it must be similar to that of a galvanic circle applied round the earth. It is true that this electrical effect will be weak at each point, but the great extent of the electrical surface will abundantly supply it. The length of the circle or electrical belt is that of the periphery of the earth. The width of this belt extends nearly as far as the vicissitudes of night and day during a revolution of the earth. The width of this belt varies every day, since the diameter of the circle around the poles of the earth, during night or day, changes continually during several revolutions. For at 66° 32' from the equator, there is once a year 24 hours of day, and 24 hours of night; at 67° 18' there is a whole month of day light, and a month in which the night continues without interruption, &c. The mean width of this circle will extend but little further than the polar circle; for the violent changes which impede the regular progress of the weather, and consequently the regular effect of solar influence, exert great controul [sic] in the countries near the poles.

M. Ampère supposes that there is also an electro-magnetic effect round the earth from the east to the west, but he is of opinion that it belongs properly to the construction of the earth, although its revolution is not without some effect; he supposes also that there is no other magnetism of the earth, besides the immediate effect of electro-magnetism. As to this last point, my opinion also differs from that of the French philosopher. A body capable of becoming magnetic cannot be surrounded with an electrical current without receiving a magnetic charge. All bodies are susceptible of magnetism to a certain extent, although generally it is very slight when compared with iron. It follows then necessarily from the electrical circulation round the earth, that the earth itself becomes magnetic.

Let fig. 14 represent a section of the terrestrial globe supported by its two poles. Suppose $+E$ in the electrical belt goes from the east to the west, and of course $-E$ from west to east, $-\varepsilon$ upon the surface of the earth goes then towards the north, and $+\varepsilon$ towards the south; but the contrary direction must take place at the inferior side of this belt. The globe $SzNz$[15] here represented as a nucleus surrounded by the crust which contains the electro-magnetic belt, will become magnetic, and at n it will possess the magnetic power which we find in that half of the magnetic needle which turns towards the south. It is thus that the magnetic needle receives its direction by the magnetism of the earth, and by the electro-magnetism of the surface. If the magnetic nucleus derived its magnetism from the electro-magnetism of the surface, its north pole would repel the same end of the needle that is attracted by the north side of the electro-magnetic belt; but as the lower plane produces opposite magnetism, the surface and the magnetic nucleus have the same effect with regard to the needle.

---

[15] [According to the German version, fig. 14 should include points marked z on the vertical and just inside the dotted circle.]

The intensity of action cannot be equal in the whole of the electro-magnetic belt of the earth, just as the effect of the sun is not the same upon the earth and the sea, and even differs according to the elevation of the country above the level of the sea. It is even possible that the different conducting power of different parts of the globe, particularly of land and sea, may possess great influence. It appears then that the new discoveries do not as yet furnish us with new facts sufficiently developed, to be useful in mathematical researches upon the situation of the magnetic poles of the earth. It is nevertheless to be wished that M. Hansteen, who has displayed so much solid learning in his researches into the magnetism of our globe (undertaken when he could not have recourse to the electro-magnetic discoveries), would resume his labour with the means which natural philosophy now offers. I shall confine myself to proposing some ideas to those who are willing to undertake a deeper examination of this subject.

According to the manner in which magnetism is produced on the globe *SzNz*, the strongest magnetic tension ought to occur in a circle round the ends *s* and *n* of the axis *sn*. We have hitherto supposed that the limits of the electro-magnetic belt were throughout equally distant from the poles of the earth. But there is reason to suppose that the electro-magnetic effect of the sun is but weak in those places which are covered with ice and snow during a great part of the year. It is then very likely that the electro-magnetic belt is very nearly parallel with the isothermal line of 0°. The form of the electro-magnetic belt determines also the form of the curve in which the greatest intensity of magnetism occurs round the poles of the globe *sznz*. But the points of this line which are nearest to us would act most strongly upon our magnetic needles, and would appear as magnetic poles.

It must be confessed that in these conclusions we cannot always support ourselves upon evident principles; but I will, however, cite, as a remarkable confirmation, that the two magnetic poles, indicated by M. Hansteen, in the northern hemisphere are under the same meridian as the celebrated Humboldt, (who has rendered such great services to natural sciences), places the greatest concavity, that is to say, the greatest polar distance, from his isothermal line of 0°. I likewise remember to have heard M. Hansteen remark, several years since, that the magnetic poles are distinguished by extreme cold. We are now speaking of the northern magnetic poles, as to the southern hemisphere, we are in possession of too small a collection of facts to fix the isothermal line.

The annual and daily variations of the magnetic needle are intimately connected with the relation of the earth to the sun, but they do not appear to depend upon any variation in the intensity of the magnetism of the interior of the earth, by the electro-magnetism which the sun produces; for these variations do not occur upon different parts of our globe at the same time, in such a manner as they must do, if the variation depended upon the increase or decrease of the magnetic powers of the poles.

It is more probable that the electro-magnetic state of the surface of the earth determines these changes. Not having a sufficient number of observations upon this subject to found principles upon, nor having sufficiently developed the principles of the electro-magnetism of the earth to be able to arrange the observations which we possess, we ought to content ourselves at first, with indicating the acknowledged analogy which[16] the diurnal and annual variations of the needle, with the periods of the day and seasons.

I have framed and examined many hypotheses as to the cause of the variations of the needle without satisfying myself. The different direction which the electro-magnetic belt receives by the united action of the annual and diurnal motions of the earth, the yearly and daily variations which occur in the figure of the electro-magnetic belt, the discharges which may occur when the electro-magnetic effect is at its maximum, the inequalities which are produced by the different effects of the sun upon the land and the sea, are considerations that have not yet given me sufficient agreement with the phenomena which have been observed in different parts of the earth. The frequent and unforeseen variations of the magnetic needle seem to depend upon electro-magnetic discharges, of which we have not at present any experimental knowledge. Among such discharges I particularly reckon polar light, known by the name of aurora borealis. I do not, on this account, oppose the opinion of the celebrated Biot; for I think it very probable that these discharges occur in certain clouds. Tempests have also a well-known influence upon the magnetic needle, which no longer surprises us after having found magnetism in every electrical discharge. It appears to me also very probable that several discharges occur in the air, and, perhaps, even in the earth, without our perceiving them. Among other irregularities of the needle are those which embarrass persons who have geometrical operations to execute in the hot days of summer in the open air; these seem to be owing to such imperceptible discharges. I hope that in future the magnetic needle will be used as a meteorological instrument. Weak needles seem to be preferable for this purpose, because the directing power of the earth produces only a weak action upon them, while a neighbouring electrical discharge has a marked effect. This agrees perfectly with the observations of Cassini, according to which a weak needle was subject to many irregularities, which he did not observe with a stronger one. But I would, above all, propose strong needles in meteorological researches, suspended, however, in such a manner that the magnetism of the earth may possess but little or no influence upon it. In order, however, to determine the direction of the discharges, it is requisite to have needles differently suspended. The mode of suspension represented in fig. 3 is one of the most important in this respect, as the magnetism of the earth has no influence upon its position; it is the bending of the wire alone from which the needle is suspended, which determines its direction. Another needle suspended in the same manner, but in a horizontal position, would be of great utility. By similar means, it may, perhaps, be discovered by subterranean geometry, whether the galvanic effects in mines do not disturb the magnetic needles employed.

In another memoir, I shall endeavour to show that the circular movement of electrical forces in the conductor, which I have admitted as an hypothesis, results from the nature of electrical forces; and I shall also endeavour to give a new explanation to the opinion which I expressed several years since, upon the production of light and heat by the conflict of electrical forces.

[16] [of.]

43

Correspondence[1]

In a treatise on electro-magnetism, Vol. 2, No. 2,[2] a printing error has slipped in which is insignificant in itself but might still bother the reader, viz., everywhere it reads $+F$ and $-F$ instead of $+E$ and $-E$. I wish to point out that printing error since, in my future communications, I always intend to designate the two electrical activities by the usual $+E$ and $-E$ while using the corresponding Greek letter ε for the designation of electro-magnetism, the latter having been correctly printed. An error has also occurred in the copperplate in Fig. 5, where *RST* has been displaced. The leg *SR* of the angle should lie in the extension of the horizontal wire with the end point *D*.

In a letter from Mr. Poggendorff, printed in Gilbert's *Annalen*, this young physicist, so favourably announced by the excellent Erman, made the remark that he finds the law for electro-magnetic effects established by me to be in general accordance with the facts, but the total deviation of 180° of the magnet needle seems to him incompatible with them. I have to remark here that, from the beginning, I have assumed the windings of the spirals in question to be extremely close to each other. In the Latin announcement of my discovery, I state expressly that it can be concluded from the facts that electro-magnetism performs a motion in circles (*hunc conflictum gyros peragere*). Then I say that it seems necessary to me that a motion in circles, combined with the progressive motion along the conductor, must form spirals, but I add that, in my opinion, this does not contribute to the explanation of the observed phenomena (*quod tamen, nisi fallor, ad phaenomena hucusque observata explicanda nihil confert*). Therefore, it is my opinion that the windings of the spiral deviate so little from circular motion that observation cannot distinguish the difference. But far be it from me to blame Mr. Poggendorff for this misunderstanding, which I rather ascribe to the brevity of my announcement.[3] Many have

---

[1] Letter to the Editors from Professor Ørsted, Dated Sept. 9, 1821. [*Journal für Chemie und Physik*, ed. by Dr. Schweigger and Dr. Meinecke, Vol. 33, pp. 123–31. Nuremberg 1821. KM II, pp. 246–51. Originally published in German.]

[2] ["Observations on Electro-magnetism" (Chapter 42, this volume). The printing error occurred in the German version in *Journal für Chemie und Physik*, not in the English version reprinted in this edition from *Annals of Philosophy*.]

[3] Since Mr. Poggendorff has performed a beautiful series of experiments with Schweigger's electro-magnetic multiplier (or condenser as he calls it), it would have been desirable if it had pleased Professor Ørsted to consider what is said about these amplifying devices in Vol. 1, No. 1 worthy of his examina-

shown a great aversion to the assumption of a spiral motion of the electro-magnetic effect because they viewed this assumption as an arbitrary fabrication. I hope that, after the statements made here and in my last treatise, my assumption will be judged more favorably. Hardly anyone will want to deny the electro-magnetic circular motion around the conductor, or rather around the axis of the conductor, especially after what the astute and profound experimenter Seebeck has said on this subject, with which my investigations from a different perspective agree perfectly. I am not completely sure whether Seebeck regards the electro-magnetic circulation about the conductor as circular motion, but I suspect so and believe that it is not at all possible that forces which always strive to cancel each other can be uniformly distributed at all points on a circle without cancelling each other; co-existing forces which always strive to cancel each other can produce a lasting activity only if an unceasing separation occurs. Consequently, I consider the circular motion of the magnetic forces in the conductor as established. Now, whether this circular motion turns into a spiral motion is a question whose answer requires a decision about another. The propagation of electricity can either, as has been done so far, be regarded as a progression, or contrary to our present concept of the subject, it can be regarded as a pulsation such as takes place in the motion of sound, and as Euler, whom some newcomers seem to follow, assumed in the propagation of light. If the propagation of electricity is a progression, then the circular motion cannot be considered proved by the facts unless it is allowed to turn into a spiral motion. Let AB be a piece of the connecting wire. Let the electricity at point a be about to move to point b, but it is led towards c by the circular motion in the time during which it should have reached b. The same happens at points d, e and generally at all points on the circumference of the conductor. However, electricity must distinguish itself in its magnetic mode of action through an important characteristic. Here, this follows automatically. The electro-magnetic effect takes place only under circumstances in which an enormous accumulation of electricity occurs; under these circumstances, it cannot proceed fast enough to give off a quantity of electricity at the same rate as it is

tion. One of these amplifying devices, Vol. 1, p. 13, is certainly compatible with the theory of vortices (whether circular or spiral), but the necessary conclusion that, with an appropriate change (Vol. 1, p. 15 and 16 and Vol. 2, p. 46 and 50), continuous motion of the needle must develop is not correct, as is already mentioned in the quoted passage (loc. cit.). This vortex theory could much less lead to the construction of multipliers with wires wound *tightly around each other* (in the same sense) because the vortices surrounding such tightly wound wires must disturb and hinder one another; an amplification of the force and a larger extension and vitality of the vortices could never be derived from this in a natural way. At least no-one who adopts this vortex theory could logically arrive at the construction of *such* multipliers while their construction, and the explanation of the same, as given in Vol. 1, table 1, figs. 3, 10, and 11 (shown in the cross-sectional drawing), offers itself if we merely regard the wire as being surrounded by a magnetic sphere. Any theory is admittedly only an image, and for the initial main facts, discovered by Ørsted, the image of vortices was certainly excellently chosen, but only a different picture could lead to the construction of Schweigger's multipliers. Still quite different and new images and similes (views and theories) will offer themselves when this important discovery by Ørsted is pursued further.

The editors.

received: a part of the electricity will therefore seek a way out, by moving side-ways. It is evident that this must occur with a speed which far exceeds that of the propagation, in which the accumulated electricity always offers resistance to that following, because this transverse motion is merely the consequence of a much heightened impulse, and the effect will not be able to penetrate all insulators with-out a quite extraordinary speed. Now, if the speed of this transverse activity is per-haps a million times larger than that of the propagation, then the spirals will also nicely approximate a series of circles around the same axis. However, if one wants to adopt the second new or renewed opinion that the propagation of electricity should merely be regarded as a pulsation, then the opposite tendencies of the two forces in definite directions must be assumed, so the hated spirals are avoided, but the most miraculous and the most inexplicable, the definite directions right and left still remain and will perhaps always remain the stumbling block for those who wish to explain electro-magnetic effects from already-known laws of nature. Moreover, it is obvious that the second conception deviates more from the present than the first. There is also a third one, viz., that the magnetic and electric forces not only differ in their mode of action but must be regarded as entirely different forces. Electricity should therefore liberate magnetism, in a manner of speaking. How this is supposed to happen is no less incomprehensible than the fact itself, and I doubt that anything could be stated in favour of this opinion other than analogies with other, quite unwarranted, hypotheses.

The propagation of substances in the galvanic circuit could be mentioned in fa-vour of progression in the effect of electricity. It can hardly be doubted that the acid, which moves towards the positive pole, and the alkali, which moves towards the negative pole, are driven by electrical forces. However, such a liquid in the process of galvanic decomposition has the effect on the magnetic needle which I have already described in my Latin announcement. Therefore, we are here dealing with a progressive motion of electrical conditions in combination with a circular electro-magnetic motion. However, I do not yet present this as decisive proof but merely as a strong indication. In Gilbert's *Annalen*, Pfaff has recently affirmed that an interruption of the galvanic circuit by even the thinnest layer of sulphuric acid cancels the electro-magnetic effect. The observation is certainly correct within certain limits but does not contradict my above-mentioned experiment, which

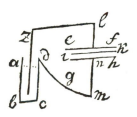

was performed with a strong pile. Meanwhile, I have succeeded in obtaining electro-magnetic activity from a simple circuit with an interrupting liquid. Let *abcd* be a copper box, *z* a zinc sheet, *ef* and *gh* brass plates ap-proximately 6 inches in diameter, *ik* a piece of filter pa-per soaked in sulphuric acid diluted with a solution of caustic potash or in a similar liquid. The conductors *zl* and *dmn* have an effect on the magnetic needle. Should *zlef* be the main piece here, or should the principal source of activity be assumed to be in the zinc and the copper separated by a liquid? Here, again, the question arises regarding the basis of the theory of galvanism.

The circular electro-magnetic motion in liquids which decompose galvanically also seems very curious to me in the sense that it leads to the idea that combustible and ignitive substances, alkali and acid, manifest tendencies towards opposite motions. What great prospects do not lie in this thought. I am almost afraid of expressing this thought because some will judge me summarily. They will condemn my view as unphilosophical because they do not see how it fits within our system of knowledge. However, I have thought more about this connection than those who will most quickly express their condemnation of me, and I hope to prove this at some point. For the present, this is still merely under investigation.

Mr. v. Yelin reports a curious experiment (in Gilbert's *Annalen*), according to which a conductor, bent back on itself, will behave like indifferent iron towards a magnet so that its ends are attracted equally by the two poles of the magnet. I have attempted to repeat this experiment. *abcd* is an elliptical ring of brass wire, 15 inches on the longest axis, *e* and *f* are two steel points which rest in iron hats filled with mercury. The whole rotates around the line *ef*. If this device constitutes part of the galvanic circuit, it is extremely responsive to the magnet but completely according to known laws. The

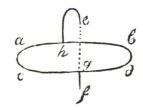

segment *hacg* behaves like a bent conductor *h'a'c'g'* and *hbdg* like a conductor of the shape of *h'b'd'g'*. Therefore, there is nothing new in all this. However, we may hope that Mr. von Yelin, if he has not made a mistake,will provide us with more details about the conditions of the experiment. Mr. Hill from Lund, from whom science has much to expect, took part in my experiments on this.

Finally, another remark regarding my spiral. Pick up my electro-magnetic indicator and observe the course of $+\varepsilon$ and $-\varepsilon$. It will be found that at each point $+\varepsilon$, thought of as moving towards the negative conductor, always moves to the left side of the point. In the same manner, $-\varepsilon$, thought of as moving towards the negative[4] conductor, goes to the left. Therefore, the whole activity in the

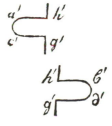

spiral can be imagined as a continuous endeavour of the effect to move around the axis of the conductor to the left and forwards. Should not a characteristic of all the small parts be imagined in the construction of each body, according to which a right and a left would be distinguished in inorganic as well as in organic nature?[5]

---

[4] [positive.]

[5] In his treatise on the twin crystallization of feldspar (cf. Vol. 10, pp. 229–34), Weisz has pointed out the right and left in the constructions of inorganic nature in an excellent and particularly interesting way, and yet, when similar views present themselves in the phenomena of electro-magnetism, they are precisely the most remarkable ones, whose further elucidation seems quite essential. However, if Ørsted seeks to deduce spirals or vortices in electricity from the construction of the conductors through which it moves and generally (Vol. 2, p. 203) ["Observations on Electro-magnetism" (Chapter 42, this volume).] believes to find the cause for the electro-magnetic phenomena observed by him in the meeting, *connected with resistance*, of the opposite electrical forces, then obviously no-one will misjudge the astuteness of this view; but it no longer seems admissible since it has been shown by Schweigger's experiments that the free electrical spark is surrounded by a magnetic atmosphere even if it leaps into a Torricelli void.   *The editors.*

44

An Experiment on Zamboni's Double Galvanic Circuit[1]

I HAVE recently conducted a series of experiments on Zamboni's double galvanic circuit. Even though my experiments are not as complete in all regards as I had intended, the results already obtained may nevertheless be of some interest and also serve as models for the application of electro-magnetism in other investigations. The principal experiment which I have conducted is the following:

Two zinc sheets of different size, that is, one narrow, the other wide, are immersed in a diluted acid and each connected to one end of the wire of Schweigger's galvano-magnetic condenser. The magnetic needle of the condenser will immediately indicate an effect. If the acid consists of the usual mixture of $1/60$ sulphuric acid, $1/60$ nitric acid, and 1 part water, the effect will be approximately as big as that of zinc, water, copper (2 □ inch area), where the wide part acts like the copper, the narrow like the zinc. In other words, the direction of the $+E$ in the conducting wire goes from the wide part to the narrow.

When two zinc sheets are used, but one of them is immersed in the liquid sooner than the other, the one immersed later acts like copper while the other is regarded as zinc.

If a strip of zinc has been wrinkled from being cut with a pair of scissors, this causes deviations. However, I have often re-established the original situation by merely bending the metal strip back into shape with my fingers. It seems that with similar metal sheets the situation can be affected by merely turning them to the right or to the left, but I have not yet performed enough experiments on this.

If the liquid contains significantly more acid, and especially if it is heated, the opposite results are obtained; the effect of zinc from the wide plate and that of copper from the narrow strip.

A circuit composed of 24 plates of 10 □ inches with the associated narrow extensions bent in such a way that the plates could sit in one vessel, the narrow extensions in another, was constructed in the manner of a trough apparatus. The acid was

[1]From a letter dated October 12, 1821. [*Journal für Chemie und Physik*, ed. by Dr. Schweigger and Dr. Meinecke, Vol. 33, pp. 163–65. Nuremberg 1821. The same subject is treated in *Videnskabernes Selskabs Oversigter*, 1821–22, pp. 5–6. Copenhagen. KM II, pp. 251–53. Originally published in German.]

the same as in the other experiments. The effect was not larger in this case, in fact, hardly as large as in the experiments with two zinc sheets.

A polished zinc surface placed opposite a dull one did not result in decisive success.

In these experiments some anomalies still occur which I shall investigate further. The sensitivity of the galvano-magnetic condenser is exceptionally great. I wish many physicists would make use of this in the investigation of galvanic conditions. Following Erman, Poggendorff has done this already in many cases. I wish he would publish his experiments.[2]

[2] This happened in Oken's *Isis*, No. 9.—*The editor.*

45

A Method to Facilitate the Generation of Steam[1]

In Gehlen's *Journal für Chemie und Physik*, Vol. 1 (Berlin, 1806), pp. 277–89,[2] I announced some experiments which showed that gas generation, which was supposed to take place in a liquid according to its chemical constituents, does not take place unless it is facilitated by contact with a solid body. Naturally, the same procedure can be applied to the generation of steam. If a metal wire is suspended in the middle of a boiling liquid, it will be seen that bubbles of vapour adhere to this as well as to the bottom of the vessel in which the boiling takes place. Consequently, a great number of appropriately coiled thin metal wires which are placed in a body of water will increase the rate of the generation of steam. I have tested this idea by placing about 10 pounds of brass wire of $^1/_5$ line's thickness in a still which holds approximately 20 quarts of water. The result was that with equal heat 7 quarts of distilled water were obtained in the same time as it would have taken to distill only 4 quarts without this wire. The same method was then used with a steam boiler such as was used in Siemens's experiment on the production of spirits from potatoes. Here there was no opportunity for such an accurate comparison, but the effect was obvious. In England, a procedure has recently been devised which depends on the same principle. When a steam boiler is quite filled with stones, the liquid in it will no longer boil with the appropriate speed. On the other hand, a rapid generation of steam will result if some of the dust which remains after the riddling of the malt, and which consists primarily of discarded sprouts, is thrown into the boiler. Here, then, a great number of small solid particles facilitate the generation of steam.

Ørsted.

---

[1] [*Tidsskrift for Naturvidenskaberne*, ed. by H. C. Ørsted, J. W. Hornemann, and J. Reinhardt, Vol. 1, pp. 299–300. Copenhagen 1822. Also to be found in Schweigger's *Journal für Chemie und Physik*, Vol. 38, pp. 511–12. Nuremberg 1823; and in Trommsdorff's *Neues Journal der Pharmacie*, Vol. 7, pp. 161–62. Leipzig 1823. KM II, p. 253. Originally published in Danish.]

[2] ["Experiments Prompted by Some Passages in Winterl's Writings" (Chapter 23, this volume).]

46

The Ørsted Experiment on the Compression of Water[1,2]

SEVERAL years ago, Prof. Ørsted submitted some experiments on the compression of water to the Royal Society of Copenhagen, and on this occasion he showed that this might be produced by a much smaller force than is generally assumed, provided the instrument was constructed according to the well-known principle that a pressure acting on a small part of the surface of an enclosed liquid has the same effect as an equally great force acting on each similar part of the whole surface. For the compression of water he made use of a wide brass cylinder, on which was screwed a narrower one equipped with a moveable piston. Therefore, he was able to demonstrate the compression of water with a small force quite as clearly as Abich and Zimmermann had done with many hundred pounds of weight. To measure the magnitude of the applied force, he used a tube full of air, blocked by mercury, whereby the air experienced the same pressure as the water from which it was separated by the mercury. As we now know that the compression of air is proportional to the compressing force, it was easy to calculate this force. However, in spite of the great strength of the brass cylinder in which the water was compressed, it was possible that its walls might have given way so that not only the compression of the water would have been measured, but the combined effect of that and the expansion of the vessel. Furthermore, in this as in all other experiments, except those by Canton, no change of temperature which might have occurred during the experiment had been taken into consideration, which, however, was doubly necessary as it is conceivable that heat was produced by the compression itself. Canton's

---

[1] [*Journal für Chemie und Physik,* ed. by Dr. Schweigger and Dr. Meinicke, Vol. 36, pp. 332–39. Nuremberg 1822. This version, supplemented by the figure, was taken from *Videnskabernes Selskabs Skrifter,* 1821–22, pp. 6–11. Copenhagen 1822. A version including the figure also appeared in Thomson's *Annals of Philosophy,* Vol. 5, pp. 53–56. London 1823. KM II, pp. 254–58. Originally published in German.]

[2] When Prof. Ørsted demonstrated this experiment to a party of natural philosophers during his recent stay in Halle, we were very pleasantly surprised by the certainty and the simplicity with which such a difficult task was solved in a manner which no longer permitted any doubt. This instructive as well as elegant experiment by Ørsted must not be neglected in any future chemical and physical lectures.

The editors.

excellent but too often ignored experiments were made with the pressure of rarefied or condensed air. However, every compression or expansion of air is accompanied by a considerable increase or decrease in temperature, so it was to be feared that this otherwise so ingenious experimenter had been deceived by this influence. He found the compression of water, at a pressure equal to that of one atmosphere, to be between 44 and 49 millionths of the volume of the water, which is only $1/3$ of the contraction produced by a decrease of one degree centigrade. Canton's experiments possessed a great advantage over all later ones which have quite displaced them, and that is that the vessel which contains the liquid undergoes the same pressure both internally and externally so that neither its form nor its size can be altered. Recently, Mr. Parkins,[3] the ingenious inventor of siderography, has conducted some experiments which have the same advantage as those by Canton in that he placed the metal tube, in which the water was to be compressed, in water

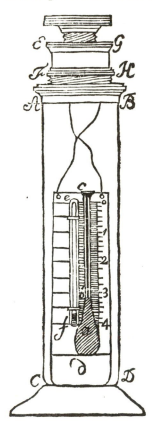

which was exposed to the same pressure. His ingeniously contrived experiments will always be of considerable importance because they were made with a force which an experimenter seldom has at his disposal, that is, a pressure several hundred times larger than that of the atmosphere, but the question of the generation of heat and its influence on the volume of the water remained unanswered. Therefore, the author endeavoured to devise an instrument which allowed an accurate measurement of the compressing forces, as well as of the compression of the water itself, and which at the same time made it possible to determine accurately the influence which heat might have on the effect. The water which is to be compressed is placed in a glass tube (a.), which holds about 4 lod[4] of water. It is closed at the bottom, and its upper part terminates in a very narrow calibrated tube (b.c.), 52 lines long, so that the vessel is like a bottle with a neck like a long capillary tube. On the upper end of this tube is a small funnel (c.), two lines wide. The bottle holds 709.48 grammes of mercury, but the mercury which fills 24.6 lines of the capillary tube weighs only 96 milligrammes, which gives 55 ten-millionths for the length of a line, or, to be more exact, 0.000005501 of the contents of the bottle. When the experiment is to be made, the bottle is warmed a little by being held for a moment in the hand, if possible only by $1/4°$ centigrade, and then a drop of mercury is introduced into the funnel. During the subsequent cooling, it will sink part way into the tube and block the water. The bot-

---

[3] [Perkins.]   [4] [1 lod equals $1/32$ Danish pound.]

tle is now placed in the strong glass cylinder (*A. B. C. D.*) with another, smaller cylinder (*E. F. G. H.*) on top, equipped with a piston. A pressure now exerted on the water in the glass cylinder by means of this piston will act on the mercury in the funnel and thus on the water in the tube. As soon as the water is compressed, the mercury in the capillary tube will sink, which happened in every experiment. To measure the compression, the author places the bottle on a stand (*d.*), equipped with a rule divided in $1/4$ lines. A small, calibrated glass tube (*e.f.*), closed at the top and filled with air, serves to measure the magnitude of the compressing force. All changes of temperature are easily perceived on the narrow neck of the bottle much more accurately than on any thermometer, for an increase of temperature amounting to one degree centigrade makes the water rise 27 lines at a temperature of about 15°. At a considerably higher or lower temperature, it will of course increase more or less for each additional degree. The scale being divided into $1/4$ lines, and $1/8$ being easily perceived by the eye, it is evident that a change of $1/100$ degree cannot escape observation, and that $1/200°$ is by no means difficult to observe. It is scarcely necessary to add that the temperature of the liquid is to be determined by a thermometer at the beginning of the experiment. As soon as the desired pressure has been reached, and a note has been made of how much the mercury has fallen in the narrow tube, and how much the water has risen in the air-filled tube, the pressure is removed. It will then be seen that the water almost always forces the mercury somewhat higher than it was immediately before the start of the experiment. If the experiment is made quickly, and no persons are present beside the observer, the difference in the height of the mercury before and after the experiment will generally be $1/8$ line, though frequently also $1/4$. In the former case, the change of temperature has been not quite $1/200$, in the latter hardly $1/100°$. If the experiment is performed slowly, the difference may be $1/2$ and even one line. In every case the mean height between these two observations is to be taken. In a long series of experiments, of which the most accurate were made at a temperature of 15°–16°, a pressure equal to one atmosphere has produced a compression = 47 millionths of the original volume of the compressed water. Several alterations of the pressure, from $1/3$ to 5 atmospheres, were tried and agreed in proving that *the compression is in direct proportion to the compressing forces.* The same result had been deduced by the author from his former experiments, which, however, were influenced by the expansion of the metallic vessel; this must likewise be proportional to the compressing forces.

It seems fairly evident that no heat is produced by this compression of the water, the boundary between the water and the mercury after the experiment being in almost the same place as before. The insignificant change in temperature must be considered a necessary result of the unavoidable contact of the observer during the experiment. Even after a pressure of 5 atmospheres, the difference of temperature was not quite $1/100°$, and in general neither greater nor smaller than when only a pressure of one atmosphere had been used. It was, however, possible that the expansion of the water when the pressure had ceased would absorb the heat produced before by the compression, so a Breguet thermometer, with which a difference of $1/10°$ may easily be perceived, was placed in the water in the cylinder and exposed to the greatest compression which could be produced, but no trace of a change in the temperature could be observed. The agreement between these experiments and those of Canton is truly remarkable. At 64° Fahrenheit = $15^{1}/_{2}$° centigrade,[5] the English physicist had obtained a compression of 44 millionths for a pressure equal to one atmosphere, and at 34° F = $1^{1}/_{9}$° centigrade, it was 49 millionths. This unexpected result can easily be explained by the difference in the effect of heat, but it does not deviate much to either side from the new determination, which is 47 millionths.

[5] [64° Fahrenheit = $17^{7}/_{9}$° C.]

47

On the Compressibility of Water[1]

ALTHOUGH Canton's experiments established the compressibility of water more than fifty years ago, people have not generally had the confidence in them which they deserve because the methods which this English philosopher used permitted changes in temperature to exert a considerable influence on the results. The rare ingenuity of this scientist was needed to avoid this influence, especially at a time when instruments did not yet have the perfection which they have acquired in our time. Therefore, a great many experiments on the same subject were published after Canton, but they were quite inferior to his, both because of the principles on which they were founded and because of the accuracy of their execution. Struck by the discrepancies between these many results, I engaged to submit this subject to new researches four years ago. I made my first experiments in brass cylinders with very thick walls, hoping thus to avoid dilatation of the vessel in which the compression took place. In that respect my apparatus did not present any advantage over those which had been used since Canton, but at least it did not require great mechanical forces to make the compression perceptible, and its measurement was accurate, being made in a narrow tube which communicated with another larger one. At the same time I used a tube filled with air which was connected with the water inside, in order to measure the compressing force, a method which protects the results against the influence of friction.

This apparatus was therefore very convenient to use, but it had the great imperfection that it dilated when yielding to the interior pressure, for it was not balanced by any pressure in the opposite direction. In order to remedy this deficiency, I employed the principle also adopted by the famous Mr. Parkins,[2] whose memoir was not known to me at the time, but as he makes his experiments with immense forces which are not at the disposal of many philosophers, I think that a description of my apparatus will be useful, especially because it can be used in very extensive researches on the compression of various kinds of liquids.

[1] [*Annales de chimie et de physique,* ed. by Messrs Gay-Lussac and Arago, Vol. 22, pp. 192–98. Paris 1823. KM II, pp. 258–63. Originally published in French.] We gave a very brief summary of this memoir in the issue of September last, taken from an English journal; but since Mr. Ørsted, during a short visit to Paris, kindly agreed to let us have the original paper, accompanied by a detailed description of the apparatus that he used, we hasten to enrich the *Annales.—The editors.*

[2] [Here and below, Perkins.]

The central part of my apparatus now consists of a small bottle, the neck of which is a capillary tube ending in a small funnel. After having filled it with water from which air has been expelled, a drop of mercury is introduced into the top of the capillary tube, where it remains due to adhesion. This drop serves as an indicator and as a piston during the experiments. This bottle is placed in a very thick glass cylinder, filled with water and equipped with a small pressure pump by means of which sufficient pressure can be applied to the water in the cylinder. This pressure is transmitted to the mercury in the capillary tube, which in its turn transmits it to the water in the bottle. It is evident that this water, receiving the same pressure from without as from within, will not experience any change in volume, but that, at the same time, the compression of the water will show itself in a very perceptible way, the capillary tube in which the liquid must descend being very narrow in comparison with the bottle.

In order to measure the magnitude of the compression correctly, I determined the capacity of the bottle and of the capillary tube by means of very accurate measurements of the weight of mercury which each could contain. I made experiments with apparatuses of slightly different sizes. Here are the measurements for one of them: the bottle filled with mercury weighed 709.48 grammes; a column of mercury, measuring 24.6 lines in a carefully calibrated tube, weighed 96 milligrammes, which makes 0.000005501 of the total volume for each line of mercury, or 0.000001375 for each quarter of a line, the value of one division of the scale.

In this apparatus, the measurement of the compressing forces is obtained by means of the volume of air contained in a well-calibrated tube, open at the bottom, which is placed next to the capillary tube in the bottle, on a graduated scale which serves to measure both the volume of the air and the position of the column of mercury. After careful observation of the diminution in the volume of air produced by the pressure on the water, it is easy to calculate the compressing force according to Mariotte's law. However, it must be observed that it is necessary to add to the pressure indicated by the barometer that produced by the column of water which acts on the air in the tube, and also the pressure of the column of mercury in the capillary tube. It is the sum of these three pressures which indicates that experienced by the water in the bottle before the piston was applied.

Heat exerts a great influence on the apparent volume of the water in the bottle. At a temperature of 15° centigrade, a change of a single degree makes the column of mercury rise or fall 27 lines in the capillary tube; each of these being divided on the scale into 4 parts, of which the eye can still clearly distinguish one quarter, it is possible to observe changes in temperature of $1/400$ of one degree centigrade. When making an experiment on compression, the body of the observer always transmits a little caloric to the apparatus. However, in re-establishing equilibrium with the air immediately after the observation has been made, one rarely sees a difference in the position of the mercury before and after the experiment which rises to $1/100$ of a degree, and this difference is not greater after a pressure of five atmospheres than after that of a single one, provided that the experiments have otherwise been performed in similar times.

According to the average of a great number of results, a pressure equal to that of the atmosphere produces a diminution of 0.000045 in the volume of the water. In all the experiments which I made with my apparatus, from pressures of $^1/_3$ to 6 atmospheres, I found that the compression of water was proportional to the compressing forces. In most of his experiments, Canton obtained 0.000044 for a pressure equal to that of the atmosphere, which only differs from my result by one millionth. It is true that he once found 0.000049 at a temperature of $+1°$, but since water condenses at that temperature through an augmentation of heat, this measurement could be excluded from the determination of the average value; at any rate, the difference between these results is very slight. Mr. Parkins's ingenious experiments, made with several hundred atmospheres, give 0.000048 for each atmosphere. I would be tempted to attribute this difference, otherwise very slight, to the compression which the material of the walls in Mr. Parkins's experiments must have experienced. At least there is reason to suppose that the walls of his apparatus, being of metal, were much thicker than the walls of my little bottle. I must also point out another circumstance which perhaps ought to be taken into consideration here, which is that water seems to lose its compressibility after some compressions. However, I dare not assert this as fact, not having submitted it to rigorous tests.

According to all my experiments, it is quite evident that there is no trace of an augmentation of temperature when the water has returned to its original volume after a compression, but one might think that the caloric liberated by the compression was re-absorbed during the subsequent dilatation.

It would not be possible to determine this question by means of an ordinary thermometer as the ball might undergo a compression and give a very inaccurate result, which is why I used a metal thermometer of Mr. Breguet's invention. This very sensitive thermometer did not indicate any change in temperature after a compression of water produced by five atmospheres.

In all these experiments on the effects of compression, it is necessary to pay particular attention to the variations in the temperature of the liquid, for a single one hundredth of a degree is sufficient to change the volume of water by as much as a pressure equal to that of three atmospheres.

DESCRIPTION OF THE APPARATUS FOR THE COMPRESSION OF WATER

ABCD (fig. 1) represents the vertical section of a glass cylinder, closed at *AB* by a copper stopper, into which is screwed the body of the pump *EFGH*.

JK is a screw which serves to lower and raise the piston *lmno*.

rs is a tube through which water is introduced into the body of the pump after the cylinder has been filled. *t* is a screw which closes this tube.

The lateral opening *u* in the body of the pump allows air to escape while water enters through the tube *rs*, but as soon as the piston descends, it closes this opening.

The other lateral hole *r* does not go all the way through and only serves to use the wrench (fig. 2) when the screw which joins the body of the pump to the top of the cylinder has to be tightened or loosened.

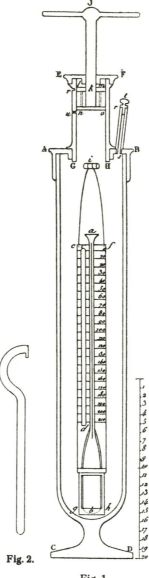

Fig. 2.

Fig. 1.

ab is a glass bottle, whose neck is a capillary tube; the whole is filled with water.

cd is a calibrated tube, open at the bottom and containing air which, through its compression, shows the compressing force which the piston exerts when it is made to descend.

efgh is a brass plate with a scale divided into quarters of lines or half millimetres. It serves simultaneously for both tubes, but it is a good idea to put the numbers on both sides of the scale, the lines of division being too long to follow with the eye with the desired facility.

i is a piece of cork connected to the brass plate by two silk threads. It is obvious that it only serves to remove the small apparatus from the cylinder when it is necessary.

Before placing the small bottle in the cylinder, a drop of mercury is put into the funnel, and the vessel is heated slightly with the hand. In the water of the cylinder, the bottle will soon lose the caloric transmitted to it by the hand, and the drop of mercury will descend a little into the capillary tube, where it will form a small column.

48

New Experiments by Dr. Seebeck on Electromagnetic Effects[1]

DR. SEEBECK, a member of the Academy of Berlin, has discovered that an electric circuit can be produced in metals without the interposition of any liquid. The current is established in this circuit by disturbing the equilibrium of the temperature. The apparatus for demonstrating this effect is very simple. It consists of two arcs of different metals, for instance, copper and bismuth, soldered together at both ends so that they form a circle. It is not even necessary that the metal pieces have the shape of an arc, or that their combination has that of a circle; it suffices that the two metals together form a circuit, that is, a closed ring of any shape.

In order to produce the current, the ring is heated at one of the two places where the two metals touch. If the circuit is composed of copper and bismuth, the positive electricity moves from the copper to the bismuth in the part that has not been heated, but if the circuit is composed of copper and antimony, the direction of the current is from the antimony to the copper in the non-heated part. These electric currents can only be discovered by means of a magnetic needle, on which they have a very perceptible influence. From now on, it will no doubt be necessary to distinguish this new class of electric circuits with its own characteristic name, and for this I propose the expression *thermoelectric* or *thermelectric circuit*; at the same time, the galvanic circuit could be designated the *hydroelectric circuit*.

The order in which conductors are arranged with regard to their hydroelectric effects is well known. Similarly, bodies can be ordered according to their thermelectric action. This series is far from coinciding with the hydroelectric series. Bismuth and antimony constitute the two extremes of the former while they are quite far from the extremes of the latter. Silver, on the other hand, which is at the negative extreme of the latter, is quite far from the limit of the former. In short, each of these two series seems to have its own principle of arrangement.

Dr. Seebeck has also succeeded in producing a thermelectric current in a single metal, but this succeeded only with metals that have a quite perceptible crystalline

[1]Note communicated by M. Ørsted. [*Annales de chimie et de physique*, ed. by Messrs Gay-Lussac and Arago, Vol. 22, pp. 199–201. Paris 1823. Translated into Danish in *Tidsskrift for Naturvidenskaberne*, ed. by H. C. Ørsted, J. W. Hornemann, and J. Reinhardt, Vol. 3, pp. 142–44. Copenhagen 1824. KM II, pp. 263–65. Originally published in French.]

texture so that the various parts of a crystal then seem to play the role of different metals. Two pieces of steel, one soft, the other hardened, also constitute a therm-electric circuit when joined together, and there are other similar cases where a difference in cohesion produces a current. However, a comparison between the different metals in the series clearly reveals that it is not cohesion that determines the thermelectric current, for the most different metals, as regards their cohesion, are close to each other in this series, and those whose cohesion is less different are often quite far from each other.

The next volume of the transactions of the Academy of Berlin to be published will report in detail on the many varied experiments of which we have given only a very brief outline here. There we shall also find investigations on the effect of alkalis and acids in the circuit, which will establish an even more marked difference between thermelectric and hydroelectric effects. Dr. Seebeck continues to pursue his important work, which will undoubtedly conclude with the establishment of an intimate relationship between thermelectric and hydroelectric phenomena although it has begun by showing us only the differences.

49

An Electromagnetic Experiment[1]

IMMEDIATELY after the discovery of electromagnetic phenomena, several distinguished physicists thought that they could explain them by assuming two magnetic axes in each transverse section of the conductor. In reality, it seems that the authors of this hypothesis abandoned it as soon as multiple experiments showed its inadequacy. Later, however, there have been other physicists who have tried to restore it to favour, which is why I thought that an experiment designed to show directly that every point on the circumference of the electric current has an equal action on the magnetic needle would not be entirely superfluous even now.

All that is needed to decide the question is to bring, in succession, to one of the poles of a magnetic needle each point on the circumference which would be obtained by cutting the electric wire by a plane perpendicular to the direction in which the electricity is transmitted.

In order to make this experiment, I arranged a circuit *ABCD* (fig. 1) in the shape of a square with sides of 10 feet; one of the vertical sides *AB* was fixed to a column *GH* with a support *J*, on which the magnetic needle was placed; the other vertical side contained the galvanic apparatus *kz*, composed of a copper vessel filled with acidulated water and a zinc plate *z*. It is clear that the two horizontal parts and the second vertical part *CD* could not have any very perceptible influence on the needle, being quite far from it. In order to achieve greater stability, the vertical part *CD* was fixed to a support. *L* and *M* represent two small vessels filled with mercury which serve to put the two horizontal parts in contact with the first vertical part. Through the action of this circuit, the needle was given a certain deviation which did not change when the mobile part of the circuit was moved so that it traversed three-quarters of the entire circumference. However, it is clear that the point

Fig. 1

[1][*Annales de chimie et de physique*, ed. by Messrs Gay-Lussac and Arago, Vol. 22, pp. 201–3. Paris 1823. The same experiment is described in Thomson's *Annals of Philosophy*, Vol. 5, pp. 155–56. London 1823. A summary is found in Gilbert's *Annalen der Physik*, Vol. 3, p. 278. Leipzig 1823. There is a short description in *Tidsskrift for Naturvidenskaberne*, Vol. 1, p. 301. Copenhagen 1822. KM II, pp. 265–66. Originally published in French.]

of the conductor *AB* which was closest to one of the extremities of the needle was not always the same physical point of this conductor during the experiment, and that it therefore changed constantly with respect to the circuit. Let us, in fact, assume that at the beginning this point was situated on the outer side of the circuit; when the apparatus had described a semicircle, it had to belong to the inner side while, after one fourth of a revolution, the point closest to the needle was a lateral point.

The experiment which I have reported here, then, directly proves the truth that every point on a transverse and circular section of the electric current has the same effect on the magnetic needle, which is contrary to the hypothesis that there are two magnetic axes in each of these sections.

50

On M. Schweigger's Electromagnetic Multiplier, with an Account of Some Experiments made with it[1]

IMMEDIATELY after the discovery of electromagnetism, Prof. Schweigger, of Halle, invented an extremely useful instrument for the purpose of discovering very weak electrical currents by means of the magnetic needle. The effect of this multiplier is founded upon the equal action which every part of a conducting wire when it transmits a current exerts upon the magnetic needle. When a part of this wire is curved as in *abc*, fig. 1, so that the two branches *ab* and *bc* are in a vertical plane, and a magnetic needle *de* is properly suspended in the same plane, it will be readily conceived that the needle receives an impulse double that which it would receive from one only of the branches. The impulse given by each branch has also the same direction, since it is in fact the same side of the wire which in both branches is opposite the needle. The effect is still further increased when the conducting wire makes several circumvolutions round the needle, as in fig. 2, and thus an electromagnetic multiplier is formed; fig. 3 represents an apparatus according to my construction, which differs, however, from M. Schweigger's in no essential respects. *AA*, fig. 3, is the foot of the apparatus. *CC*, *CC* are two stands which support a frame *BB*, which has a groove on the edge to receive the multiplying wire. *DD* is a stand to support the wire from which the magnetic needle is suspended. *EE* is a metallic wire passed tightly through a hole made in the upper part of the stand *DD*. To

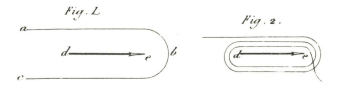

Fig. 1.

Fig. 2.

[1] Communicated by the author. [Thomson's *Annals of Philosophy*, Vol. 5, pp. 436–39. London 1823. This version is reproduced here. The same article is found in *Annales de chimie et de physique*, ed. by Messrs Gay-Lussac and Arago, Vol. 22, pp. 358–65. Paris 1823. —*Quarterly Journal of Science*, Vol. 16, pp. 123–26. London 1823. KM II, pp. 266–70. Originally published in English.]

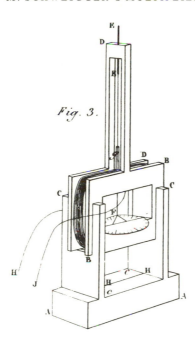

Fig. 3.

this metallic wire there is attached by a little wax a thread of raw silk *EF*, sus-
pending a double triangular loop of paper, in which the magnetic needle is placed.
EG is a tube which allows the suspension wire a free passage, and prevents the multiplying wire from touching it. Below the magnetic needle a divided circle is placed
to measure the deviations. The multiplying wire is of plated copper, and a quarter
of a millimetre, or about $^1/_{100}$ of an inch in diameter. It is covered with silk thread,
which prevents any communication between the different parts of the multiplying
wire; *H* and *J* are the two ends of this wire. The use of this apparatus will be understood almost without any explanation. In order to multiply the effect produced by a
galvanic arrangement upon the needle, it is requisite only to effect a communication so as to make the multiplying wire a part of the circuit. The effect of a disk of
copper and of zinc with pure water as a liquid conductor, was rendered perfectly
sensible by this apparatus, and it is even possible by its means to render those galvanic actions sensible, which are too weak to produce a marked effect upon the prepared muscles of a frog. When it is required to discover an action which is so extremely weak as to occasion a scarcely visible deviation, the circuit is interrupted
immediately after it has been completed, but it is again effected at each time that
the needle is at the point of terminating the preceding oscillation; the apparatus
may be rendered still more sensible by putting a small magnetic needle in *HH* in
the situation required to diminish the force with which the suspended needle tends
to preserve its direction.

When the multiplier is employed for moderately strong electromagnetic action,
thicker conducting wires must be used. If this precaution be neglected, the effect

may be diminished instead of increased, owing to the imperfection of the conductor. M. Seebeck has made some very satisfactory researches on this subject, in his memoir on electromagnetism, published two years since in the memoirs of the Berlin Academy.

M. Poggendorf,[2] of Berlin, a distinguished young philosopher, constructed an electromagnetic multiplier very soon after M. Schweigger, and made some striking experiments with it. The experiments of M. Poggendorf having been cited in a work upon electromagnetic phenomena by the celebrated M. Erman, published soon after the discovery of these phenomena, were known to several philosophers before those of M. Schweigger, which circumstance has given rise to different names for the same apparatus. M. Poggendorf has made a very useful application of this apparatus by employing it for the purpose of examining the order of the conductors in the galvanic series. An account of his labours, is contained in the German Journal, the *Isis*, for the year 1821. M. Avogadro, in Italy, has used the same plan, but without experimenting on so great a number of different bodies, his memoir contains some other observations which are worthy of being known. By the indications of the electromagnetic multiplier, he discovered that some metals at the first moment of their immersion in concentrated nitric acid, produce an effect contrary to that which is observed in a few seconds afterwards; but this alteration does not occur in dilute nitric acid. The metals which have exhibited this property are lead and bismuth, lead and tin, iron and bismuth, cobalt and antimony. M. Avogadro states, that the first effect which occurs in a concentrated acid is similar to that which happens in a diluted acid, and that it is afterwards that the contrary effect is perceived. I have repeated these experiments with lead and bismuth, and I have confirmed them by other means, excepting only that I have always had at the end of the experiment with concentrated acid, the same effect as that constantly produced by the dilute acid. I have also found that the bars of lead and of bismuth which have been acted upon by concentrated acid, gave in repeated experiments constantly the same effects as by dilute acid, unless fresh surfaces were given to them before they were again immersed in the acid; this renewal of the surfaces may be effected not only by mechanical means, but also by diluted nitric acid. It also frequently happened that the bars which had been in diluted acid, and which had been only slightly wiped, gave at first in the concentrated acid a momentary deviation in the same direction as in the diluted acid, very probably on account of the fluid which remained on their surface; they then gave for some seconds the contrary deviation; that is to say, the same as that observed when the experiment is made with bars well cleaned. At length the deviation became such as it would have been, if diluted acid had been employed as a fluid conductor. It is to be remarked that concentrated nitric acid acts much more strongly upon bismuth than upon lead; and, on the contrary, that the diluted acid acts strongly upon the lead, and scarcely at all upon the bismuth. Thus it follows, that the lead acts as the more positive metal in the dilute acid, but as the negative in the concentrated acid.

It remains only to explain why the deviation produced by the concentrated acid

[2] [Poggendorff.]

does not continue the same during the whole of the experiment. As I am travelling, I have not time to treat of this, or the analogous experiments related by M. Avogadro thoroughly, but I shall content myself with having contributed to call the attention of philosophers to this class of experiments which are equally interesting as regards the theory of solution and that of the excitation of the electric current. M. Avogadro mentions also that arsenic acts with respect to antimony as a positive metal in concentrated nitric acid and as a negative in dilute acid. This phenomenon appears interesting in relation to the chemical effect of this acid upon the two metals in its different degrees of concentration.

Among the experiments to which the electromagnetic multiplier gives rise, it may be stated that by its use, we may show, that when two pieces of the same metal are immersed in an acid capable of acting upon them, that which is first immersed acts towards the other as the most positive metal; this experiment is extremely well performed with two bars of zinc and diluted sulphuric or muriatic acid. It would be extremely interesting to examine the electromagnetic changes which take place during every period of the action of acids and alkalies upon the metals, and nothing affords greater facility for this purpose than the electromagnetic multiplier.

51

On Some New Thermoelectric Experiments Performed by Baron Fourier and M. Ørsted[1]

I HAVE had the honour of demonstrating to this illustrious assembly the remarkable experiments by which M. Seebeck has proved that an electric current may be produced in a circuit formed of solid conductors only by disturbing the equilibrium of the temperature. (See *Annales*, Vol. 22, p. 119.[2]) We are therefore in possession of a new class of electric circuits, which may be called *thermoelectric circuits*, thus distinguishing them from galvanic circuits, which from now on may conveniently be called *hydroelectric*. On this subject, a question arises which is of interest to electromagnetism and also to the theory of the motion of heat in solid bodies. The question is to examine whether thermoelectric effects may be increased by the alternate repetition of bars of different substances, and how it will be necessary to proceed in order to obtain such effects. It appears that the author of the discovery of the thermoelectric circuit has not as yet directed his researches to this point. Baron Fourier and I joined forces in order to examine this question experimentally.

The apparatus which we first used is composed of three bars of bismuth and three of antimony, alternately soldered together so that they form a hexagon and constitute a complex thermoelectric circuit, consisting of three elements. The bars are about 12 centimetres long, 15 millimetres wide, and 4 millimetres thick. We place this circuit in a horizontal position on two supports, taking care to give one of the sides of the hexagon the direction of the magnetic needle, and then we place a compass needle as close as possible under this side.

By heating one of the soldered joints with a taper, we produce a very perceptible effect on the needle. By heating two soldered joints which are not contiguous, the deviation is considerably increased. Finally, when the temperature is raised at all three alternate soldered joints, an even greater effect is produced.

---

[1] Notice Read at the Academy of Sciences by M. Ørsted. [*Annales de chimie et de physique*, ed. by Messrs Gay-Lussac and Arago, Vol. 22, pp. 375–89. Paris 1823. The same contents are found in Schweigger's *Journal für Chemie und Physik*, Vol. 41, pp. 48–63. Halle 1824. —*Bibliothèque universelle*, Vol. 23, pp. 50–62. Geneva 1823. —Thomson's *Annals of Philosophy*, Vol. 5, pp. 439–46. London 1823. —*Tidsskrift for Naturvidenskaberne*, Vol. 3, pp. 145–60. Copenhagen 1824. —In part in *Quarterly Journal of Science*, Vol. 16, pp. 126–30. London 1823. —*Videnskabernes Selskabs Oversigter*, 1822–23, pp. 9–10. Copenhagen. KM II, pp. 272–82. Originally published in French.]

[2] [Cf. "New Experiments by Dr. Seebeck on Electromagnetic Effects" (Chapter 48, this volume).]

We also used an inverse process, reducing the temperature of one or more of the soldered joints of the circuit to the freezing point by means of melting ice. It will be readily understood that in this case the joints which are not cooled are to be considered as heated with respect to the others. This mode of operation allows comparisons between different experiments, without which it would be impossible to discover the laws for this kind of phenomena.

By combining the action of ice with that of the flame, that is, by heating the three solderings which were not cooled, we obtained a very considerable effect: the deviation of the needle grew to 60 degrees.

Later we continued these experiments with an apparatus composed of 22 bars of bismuth and 22 of antimony, much thicker than those of the hexagon. Thus we satisfied ourselves that each element contributes to the total effect.

After having opened the circuit at one point, we soldered small brass bowls to the separated bars, and we filled the bowls with mercury in order to be able to produce at will a perfect connection between their extremities by means of metallic wires. A copper wire, one decimetre long and one millimetre thick, was nearly sufficient to re-establish the entire connection, and with two such wires next to each other, the connection was perfect. A wire with the same diameter, but more than a metre long, still transmitted the current quite well, but a platinum wire, half a millimetre in diameter and 4 decimetres long, produced such an imperfect connection that the deviation of the magnetic needle was not even one degree. When a slip of paper, moistened with a saturated solution of soda, was interposed, no effect could be observed.

It is worth remarking that an apparatus which was capable of giving such great electromagnetic effects did not produce any chemical action or any perceptible ignition.

We may also add that the effect of the complex electromagnetic circuit is much smaller than the sum of the isolated effects which the same elements could produce when used to form simple circuits.

Details of the Above-Mentioned Experiments and Further Observations

The bars which were used in the following experiments are parallelopipeds, the transverse section of which is a square, each side of which is 15 millimetres long.

First Experiment.

We formed a rectangular circuit *abcd* (fig. 1). One half *acd* was antimony, the other *abd* bismuth. These two halves were soldered together so that we had two contiguous sides of antimony and two contiguous sides of bismuth. The longest side measured 12 centimetres, the other 8.

Fig. 1

The circuit was placed horizontally on supports, with two of its sides in the direction of the magnetic needle, and the compass was placed on one of them. The circuit was then left for some time so that it could regain thermal equilibrium, which might have been disturbed by the hand of the experimenter. After some time ice was put on one of the two solderings which join the two heterogeneous metals. The compass then showed a deviation of 22 or 23 degrees at an atmospheric temperature of 14° centigrade. At a temperature of 20 degrees, we observed a deviation of 30 degrees, but as we had neglected to note the temperature of the atmosphere at the beginning of the experiment, we only compare results of experiments made at the same time, so to speak.

Second Experiment.

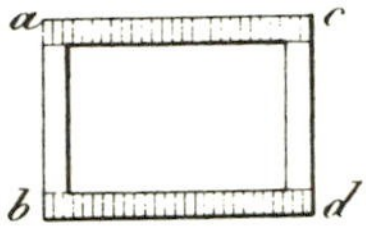

Fig. 2

A second circuit (fig. 2) was formed of approximately the same size, but in which the opposite sides were of the same metal, for example, *ab* and *cd* of bismuth, *ac* and *bd* of antimony. The apparatus was rendered active by placing ice on two opposite corners. This circuit produced a deviation of 30 to 31 degrees, under the same conditions in which the simple circuit produced only 22 to 23 degrees. Thermal equilibrium in this circuit is restored very quickly so that the thermoelectric effect appears weaker than it would without this circumstance.

Third Experiment.

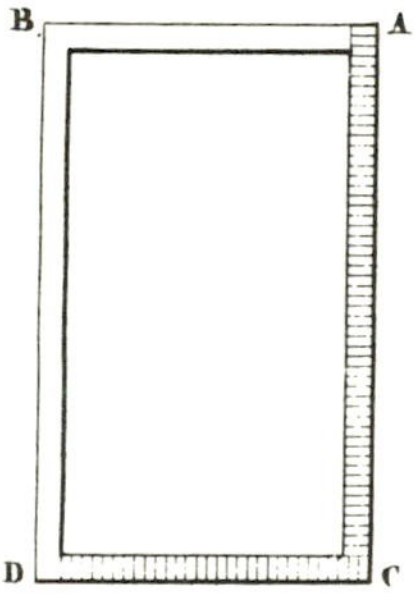

A circuit *ABCD* (fig. 3), the circumference of which was twice that of the circuit in the first experiment, was activated by ice placed on one of its solderings. The deviation was only from 13 to 15 degrees under the same conditions in which the circuit (fig. 1) gave 22 to 23 degrees.

Fig. 3

Fourth Experiment.

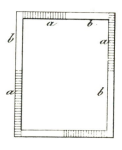

Fig. 4

Another circuit (fig. 4) was formed of the same length as the one in the preceding experiment, but it had four alternations or four thermoelectric elements *ab*, *a* being the antimony, *b* the bismuth. This circuit was activated by ice placed on every other soldering. The deviation of the magnetic needle was then $31^3/_4°$, under the same circumstances in which the simple circuit of the same length, used in the third experiment, produced a deviation of only 13 to 15 degrees, but it must be remembered that the circuit used in the second experiment (fig. 2), which had only half the circumference and half the number of elements, produced nearly the same effect. Thus it is clear, and this result will be confirmed by later experiments, that the deviation of the needle produced by the thermoelectric circuit increases with the number of elements when the length of the circuit remains the same, but that it becomes proportionally weaker as the length is increased. It is also clear, and this will become even more evident in the following, that these two effects balance each other so that the effect of a circuit is not changed when the length of the circumference increases in the same proportion as the number of elements, or, in other words, that elements of equal length form circuits which produce equal deviations, whatever may be the number of these elements. We confirmed these results by comparing the effects of one, two, three, four, six, thirteen, and twenty-two elements.

In order to construct complex circuits capable of producing a very large effect on the magnetic needle, it will be necessary to use very short elements. It is true that this will have the inconvenience that thermal equilibrium will be restored quickly in the circuit unless the solderings are placed alternately in contact with continuous sources of either heat or cold. There is one effect of the complex thermoelectric circuit which is not thus limited by the length of the circumference, but before mentioning it, we shall show the relation which exists between the different elements in the complex apparatus.

Fifth Experiment.

We have examined the effects of a complex circuit by cooling first one, then two, then three, etc., of the solderings which were to be rendered active, and after several experiments, we found the following mean numbers:

In a circuit of two elements, only one soldering cooled	
gave a deviation of	$21°$
the two together	$32^{1}/_{2}°$
In a circuit of three elements, the cooling of only one soldering	
gave a deviation of	$15^{5}/_{8}°$
the first two	$25^{7}/_{8}°$
all three together	$31°$
In a circuit of four elements, the ice placed on one soldering	
gave a deviation of [3]	$13^{1}/_{4}°$
on two solderings	$19°$
on three	$25°$
on four	$31^{3}/_{4}°$
In a circuit of six elements, one soldering gave	
a deviation of	$9°$
the first two [4]	$13^{3}/_{4}°$
the first three	$18^{1}/_{2}°$
the first four	$22°$
the first five	$25^{2}/_{3}°$
all six together	$28^{2}/_{3}°$

[3] [Both the Danish and the English versions of this article have $13^{3}/_{4}°$.]
[4] [Both the Danish and the English versions of this article have $13^{1}/_{4}°$.]

It will be observed that the deviation produced by the first cooled soldering is nearly twice the quotient obtained by dividing the total deviation, produced by the circuit when all its elements are activated, by the number of elements plus one. It is also evident that the other numbers are very close to the value of the simple quotient, but they still appear to form a decreasing series.

Here we still refer to the deviations as measured by the angles and not to the real magnitude of the effects. If it were not necessary to regard the different distances between all the points which act on each other in the different positions of the needle and perhaps even the reciprocal situation more or less oblique to the edges of the conductor and of the needle, the effects might have been represented by the tangents of the deviations. It is, however, remarkable that the experiments which we have made indicate such a constant relation between the deviations. If these experiments were susceptible to greater accuracy, it would no doubt be possible to deduce from them consequences of interest to the theory.

Sixth Experiment.

Thermoelectric action can be made perceptible by means of the eletromagnetic multiplier. In order to produce this effect, one piece a (fig. 5) of one of two kinds of

Fig. 5

metal is combined with two pieces b of the other so that this arrangement constitutes a broken circuit the two ends of which are of the same metal.

After having put some ice on one of the solderings, contact is established between the two pieces b by means of the wire of the multiplier. The effect is noticeable on the needle of the instrument, but it is very weak; weaker, for example, than the effect of a piece of copper and of silver with water as a liquid conductor. The effect is rendered more perceptible by the communication of a fresh impulse to the needle at the end of each return oscillation resulting from the previous impulse.

The extraordinary weakness of this effect is quite remarkable. This result shows that the same thermoelectric elements which produce a great effect on the magnetic needle of the compass when the connection is made by a short and thick conductor have very little effect, even on a needle supported by a wire and therefore very sensitive, when the connection is made by a thin conductor of considerable length. A hydroelectric current produced by one piece of zinc and one of silver, with water as a liquid conductor, produces an effect on the needle of the multiplier which is perhaps a hundred times greater than the thermoelectric current, and still the effect of the former on the needle of the compass is almost imperceptible even when the connection is made between the elements by the best conductors, while the latter produces considerable deviations of the needle. This shows us a very important property of the thermoelectric current, which indeed might have been foreseen by theory, but which is no less worth noticing for that. It is that the thermoelectric circuit contains electrical powers in much greater quantity than any hydroelectric circuit of equal size while, on the other hand, the intensity of the forces in the latter is much stronger than in the former.

Since the first electromagnetic experiments, it has been clear that the deviation of the needle produced by the electric current was regulated by the quantity of electrical power and not by its intensity (electrometric action.) Thus the considerable deviation which the thermoelectric current produces is an indication of the great quantity of power that it contains. On the other hand, it is well known that an electric current penetrates conductors more easily, the more intense it is. The hydroelectric current, which is more easily transmitted through the wire of the multiplier than the thermoelectric current, must therefore also be more intense. The greater quantity of electric power which must be admitted to exist in the thermoelectric current will present no objection to this reasoning, for it is quite evident that in the case in which a current *A*, of an intensity equal to that of another current *B* but greater in quantity, is presented to a conductor which is only sufficient to transmit the quantity of *B*, this conductor must be capable of transmitting a part of current *A* equal to current *B*, and if we suppose *A* to possess a stronger intensity than *B*, the transmission of the former will be greater.

Seventh Experiment.

We tested the effect of the complex circuit on the needle of the multiplier, and we found that it increased considerably with the number of elements in the circuit, even in cases where this multiplication of elements added nothing to the effect on the compass. We obtained this result from experiments with six, thirteen, and twenty-two elements. It appears then that the intensity of the power increases in the circuit with the number of elements, precisely as happens in the voltaic pile. The circuit had no perceptible effect on the compass when the connection was made by the multiplying wire.

Eighth Experiment.

A platinum wire with a diameter of $^1/_{10}$ of a millimetre was not caused to glow by a thermoelectric circuit composed of 13 elements, which was nevertheless intense enough to make the compass deviate 28 degrees. A hydroelectric circuit capable of producing a similar effect on the compass was quite sufficient to cause the same wire to glow.

This difference is due to the extremely weak transmission of the thermoelectric current by the platinum wire. When the connection was established by means of this wire, the needle of the compass indicated a deviation of only 2 or 3 degrees. An iron wire, $^1/_5$ of a millimetre in diameter, did not glow either. The communication effected by this wire produced a greater deviation than the platinum wire, but it only grew to 5 degrees. It must be assumed that a thermoelectric circuit of several hundred elements would produce a current which is intense enough to cause a metallic wire to glow.

Ninth Experiment.

We were unable to produce any perceptible chemical action by means of the thermoelectric circuit. The fluids which have the greatest conducting power resisted its action; for instance, nitric acid, solution of soda, and several metallic solutions. We shall mention only one of those experiments, which, frequently repeated, appeared to produce some chemical effect.

We placed a piece of filter paper moistened with a solution of sulphate of copper between two completely new five-franc pieces from the same year. We took the precaution of putting the two coins in contact with the paper on the sides which had the same impression, and the thermoelectric current was transmitted through the two pieces of metal and the moistened paper. After a quarter of an hour, some parts of the silver had a very weak tinge of copper, but as this trace of metallic precipitation did not resist washing accompanied by slight rubbing, we are inclined to consider this experiment too questionable. During the time when the two pieces of silver and the paper were part of the circuit, there was not the slightest effect on the compass so that this thin piece of moistened paper literally cut off the thermoelectric current completely. In a state of such perfect isolation, no perceptible chemical effect should be expected either. From the weak intensity indicated by the multiplier, there is reason to think that it would require a thermoelectric circuit of several hundred elements to penetrate a fluid as well as a voltaic pile of four or five elements does, but it is likely that such an apparatus would produce effects similar to those which may be expected from hydroelectric piles whose metallic elements are enormously large.

Tenth Experiment.

The effect on the animal body is one of the most remarkable that electric currents exert. The thermoelectric circuit gave no perceptible taste when it was made to act on the tongue, but it produced the effect of two slightly different metals on a prepared frog. This result showed us what excellent conductors the nerves of a frog are.

Eleventh Experiment.

A thermoelectric circuit composed of thirteen elements produced no effect on the most delicate electrometers. Nor did Volta's condenser give us any unequivocal signs of electricity, but we acknowledge that we did not repeat this experiment as often as it deserves.

Twelfth Experiment.

The experiments which we have mentioned are sufficient to show how weak the conducting power of bodies is with regard to the thermoelectric current. The following experiment gives the same result in other forms.

The large circuit, consisting of a rectangle which was almost four times as long as it was wide, was placed in such a way that the two short sides were parallel to the magnetic needle. The compass was placed on one of these sides, and the two adjacent elements were activated. After the deviation of the needle had been observed, a connection was effected between the active parts of the circuit farthest from the compass by means of a copper wire so that all the active parts formed a separate circuit. After this decrease of the circumference of the circuit, the needle indicated a stronger action. This effect, however, would not have been sensible if the transmission of the thermoelectric current had not been so difficult in the metal itself that a difference of distance of two or three feet produced a considerable change. It must be observed that when some part of the complete circuit was opened and the same copper wire used to re-establish the connection, it hardly produced the same effect as the immediate junction of the separated parts.

When the part of the circuit farthest from the compass was rendered active, and a similar communication was effected, the deviation of the needle diminished. However, this difficulty of transmission should not surprise us, for the electricity aroused in a circuit of conductors in consequence of their contact must flow in proportion as it acquires the necessary intensity to break down the resistance which these conductors offer. Therefore this electricity never acquires sufficient intensity to penetrate the conductor with great ease, but it will constitute a current as soon as the circuit no longer opposes the obstacle of a very considerable isolation. It is also easy to see that the quantity of electricity produced by this continuous excitation which takes place in the circuits must be so much the greater as the circuit is a more perfect conductor. Therefore, the thermoelectric circuit supplies an incomparably greater quantity of electricity than any other circuit which has so far been invented. If water, acids, and alkalis have successively been decomposed by means of the old circuits, it is not beyond the limits of probability that we shall even be able to decompose metals by means of the new circuit and thus complete the great change in chemistry which began with Volta's pile.

52

On an Apparent Paradoxical Galvanic Experiment[1]

In a Memoir, published some months ago, by M. Von Moll,[2] at Utrecht,[3] this philosopher (already known from various experimental researches) describes an experiment, which, at first sight, appears to indicate a new class of galvanic phenomena.

I have submitted this experiment to an attentive examination. Fig. 1 is the apparatus of M. Von Moll. *ABCD* is a perpendicular section of a plate of zinc, bent in such a way that its extremities touch, and form a closed circuit. *NS* is a magnetic needle, properly suspended. The part *A* of the circuit is plunged in acidulated water.

If any point of this circuit under the water be touched by a piece of brass, the motion of the needle indicates an electric current. In order to be certain that the metallic continuity was not interrupted by the interposition of a part of the fluid, I substituted for that in Fig. 1, the circuit *ABCDE*, Fig. 2, cut out of a plate of zinc. The effect described by the Dutch philosopher was produced by this circuit likewise; but I soon discovered that it was owing to the ordinary galvanic circuits, like that formed by the copper *GH*, the zinc *GA*, and the fluid between *H* and *A*, or, as in that in Fig. 3 formed by the copper *JK*, the zinc *KC*, and the fluid.

The contact of the copper and zinc above the water, or at the surface of the water, produces no effect. In order to make myself sure that a collateral galvanic circuit

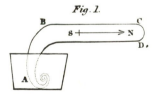

Fig. 1.

[1] Communicated by the author. [*The Edinburgh Philosophical Journal*, ed. by Dr. Brewster and Professor Jameson, Vol. 10, pp. 205–7. Edinburgh 1824. KM II, pp. 282–84. Originally published in English.]

[2] [Moll.]   [3] *Edinburgh Phil. Journal*, Vol. 9, p. 167.

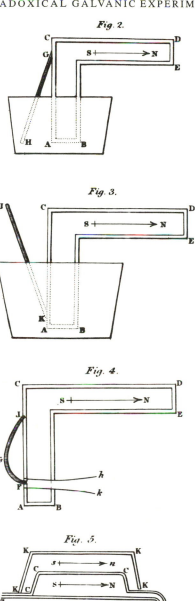

Fig. 2.

Fig. 3.

Fig. 4.

Fig. 5.

was capable of producing such an effect on a homogeneous metallic circuit, I made the construction shewn at Fig. 4, in which *ABCDE* is the same homogeneous circuit as before, but *JGF* is an arch of copper, in contact with the zinc at *J*, and separated from it at *F* by a fold *hk* of paper wetted in acidulated water. In alternately opening and closing this circuit, I found that the needle moved as in the preceding cases. This construction may therefore be considered as composed of a galvanic circuit *JGFJ*, and a conductor *JCDEBAF*, which transmits a current similar to that transmitted by *JGF*.

This explanation is confirmed by an experiment made with the construction represented in Fig. 5 in which *ZZ* is a plate of zinc, *CCCC* a plate of copper, and *AA* a vessel of acidulated water. When a magnetic needle is placed at *NS*, it is deflected according to the known laws; but, if to this circuit there be added the conductor *KKK*, a part of the electricity passes in it, and acts more feebly on the needle, from being at a greater distance from it. The effect of the second conductor becomes more striking when the needle is placed at *ns*, and when (after having noted its deviation) the conductor *KKK* is added; because, in this case, the second conductor being above the needle, tends to give it a contrary deflexion to that given by the first conductor, which is below it. These experiments have the same result when *ZZ* is made of copper, and *CCC* of zinc.

On applying all this to the constructions in Figs. 2 and 3, we observe, that *CDEBA* is the same thing as the second conductor in Fig. 5 and that the current in the part *DE* (Figs. 2 and 3) should have the same direction as in the part *CA* (to which it is parallel), in the same manner that the currents are similar in *CCC* and *KKK*, in Fig. 5. This being granted, we can determine the direction of the current in all the other parts of *CDEBA* (Figs. 2 and 3), and experiments with the needle will confirm the predictions of the theory. In some experiments, M. Von Moll substituted a plate of zinc for the copper, with which he touched the zinc circuit, and produced electro-magnetic effects by these means also. This is likewise reducible to a collateral galvanic circuit; for I have proved, by experiments published two years ago, that a galvanic circuit may be made for a short space, by two plates of zinc and a liquid, provided that one of the plates be brought into contact with the liquid before the other.

53

Experiments Proving That Mariotte's Law Is Applicable to All Kinds of Gases; and at All Degrees of Pressure under Which the Gases Retain Their Aëriform State[1]

THE so-called law of Mariotte,[2] according to which the volumes occupied by a certain quantity of gas or air are found to be in an inverse ratio to the degrees of pressure which they experience, has hitherto been demonstrated by precise experiments for very small degrees of pressure only. Several men of science of the first rank have assumed this law to be in exact conformity with nature for every degree of pressure; others, among them Jacob Bernoulli and Euler, entertained the opinion that the volumes decrease at a smaller rate than that at which the pressure increases; and if, finally, we refer to the small number of experiments made with high pressures, the proportions of volume seem to decrease at a much greater rate than that at which the pressure increases. In the transactions of the Academy of Berlin, Sulzer, a distinguished German scholar, has published experiments with pressures as high as eight atmospheres. Robison, a very respectable English scholar, made similar experiments. The following table furnishes the results obtained by each.

[1] Read before the Royal Society of Copenhagen. [*Journal für Chemie und Physik*, ed. by Dr. J. S. C. Schweigger and Dr. Fr. W. Schweigger-Seidel, Vol. 45, pp. 352–67. Halle 1825. An English translation appeared in Tilloch's *Philosophical Magazine*, Vol. 68, pp. 102–11. London 1826. Also in *The Edinburgh Journal of Science*, Vol. 4, pp. 224–34. 1826. An account is to be found in *Videnskabernes Selskabs Oversigter*, 1824–25, pp. 13–15. Copenhagen. KM II, pp. 285–97. Originally published in German.]

[2] It is well known that this law was first deduced from the experiments of the famous Boyle by his friend Richard Townlay [Towneley]. Nevertheless I have named it after Mariotte, who discovered it at the same time through an experiment of his own, as it is commonly known by this appelation.—Ø

Sulzer's Experiments (complete series).		Robison's Experiments with dry air.	
Density	Compressing forces	Density	Compressing forces
1.000	1.000	1.000	1.000
1.091	1.076	2.000	1.957
1.200	1.183	3.000	2.848
1.333	1.303	4.000	3.737
1.500	1.472	5.500	4.930
1.714	1.659	6.000	5.342
2.000	1.900	7.620	6.490
2.400	2.241		
3.000	2.793		
4.000	3.631		
6.000	5.297		
8.000	6.835		

Captain Schwendsen[3] and myself, in anticipation of making some experiments on the theory of the air-gun, felt the necessity of first testing the extent to which Mariotte's law could serve us as a fundamental principle. The apparatus generally used to prove this law is known to consist of a curved tube $ABCD$ (fig. 1), one part of which DE contains air and the other $ABCE$ mercury, which serves to enclose and compress the air. This apparatus, however, has several inconveniences; it is difficult to subdivide the part DE of the tube into equal volumes; besides, this part is expanded by the pressure within, and there is some risk of its bursting if the pressure becomes too great. To prevent such an accident, tubes of a smaller diameter are used, but this, on the other hand, creates a friction sufficient to disturb the results to a considerable degree. To avoid all these inconveniences, therefore, we resorted to an apparatus which, constructed after the same principle, had formed part of my apparatus for the compression of water. Fig. 2 presents a vertical section of this new apparatus. $ABCD$ is a very strong glass cylinder, equipped with a brass lid. EF is a graduated glass tube supported by an iron frame $lmno$, which terminates at its lower end in an iron cup containing some mercury. This closes the tube EF before it is immersed in the mass of mercury spread at the bottom of the cylinder. IK shows the upper limit of the mercury. GH represents a part of a very strong glass tube, which is cemented into a hollow piece of metal, on whose outer surface are threads for a nut which is placed in the lid of the cylinder. In this lid there is yet another aperture P, which can be closed by means of a screw, which can be found

[3] [Suenssen.]

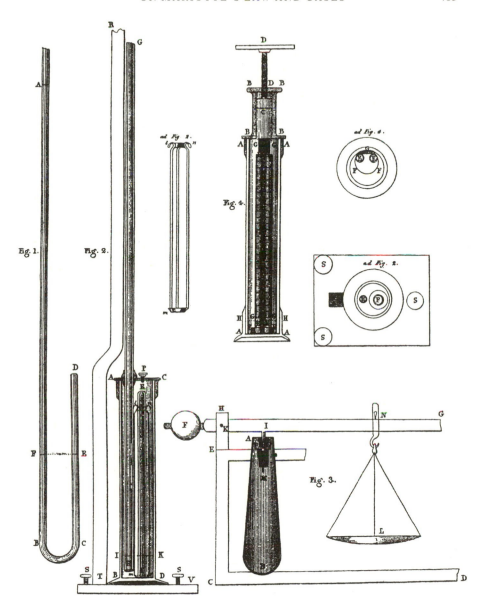

[This figure is reproduced from *Det Kongelige Danske Videnskabernes Selskabs Afhandlinger*, Part 2. Copenhagen 1826. It is identical to the figure in Schweigger's *Journal*. This plate is identical to the first plate in "Contribution to the Determination of the Law of the Compression of Bodies" (Chapter 55, this volume).]

in the place indicated in this drawing. *TV* is a wooden base, from which a rod *RT* rises, the function of which is to support the tube *GH*. The two additional figures, marked *ad* fig. 2, represent the frame *lmno* and the transverse section of the lower part of the apparatus, respectively. When an experiment is to be made with this apparatus, the lid *AC* is screwed off, the tube *EF*, filled with carefully dried air, is lowered into the cylinder, and the lid is screwed on again and carefully secured. Then the tube *GH* is also put in its place, and the cylinder is filled with water by means of a funnel placed in the aperture *P*. The pressure which this produces is measured by the rising of the mercury in the tube *GH*. Finally, the apparatus is closed by means of the screw fitted to the aperture *P*, and the mercury is poured into the tube *GH*. The mercury also rises in the tube *EF* and compresses the air contained in it. The height of the mercury in the tubes *EF* and *GH*, both being graduated alike, show by simple subtraction the magnitude of the compressing force. The tube *EF* is of nearly equal size throughout; nevertheless we have accurately determined the volumes corresponding to the divisions by means of carefully weighed quantities of mercury. The graduation of the tube *GH* extends only a few inches above the cylinder, the other distances having been measured by means of a rule.

The tube *GH* was lengthened for high pressures by our joining together several tubes, each seven feet in length or more, by means of iron screws. The experiment was always made in the stair well of the building which houses the physics collection of the university, as there was no room high enough for the requisite extension of the tube *GH*.

With this apparatus, we made several experiments whose results agreed with Mariotte's law, but not all of them were crowned with the same complete success, for it is very difficult to ensure that all the cemented joints and screws, exposed to such considerable compressing forces, sufficiently resist the penetrating mercury.

In one experiment only, the results of which we shall communicate immediately, were we able to increase the pressure to as much as eight atmospheres. The air contained in the tube *EF* had been well dried by chloride of calcium; the volume of the tube, measured by mercury, amounted to 1054.8 grammes at 20° C; the pressure of the atmosphere on the day of the experiment = 0.7578 of a metre of mercury. The following table shows the ratios which we found between the compression of the air and the pressure of the mercury. The first column of this table contains the quotient of the original volume of air divided by the volume as proportionally reduced by the compressing forces. The second shows these forces numerically, assuming the pressure of the atmosphere on the day of the experiment to be unity; the third shows the differences between the different degrees of density and compressing forces; and, finally, the fourth shows the ratio of those differences to the compressing forces.

In these experiments it is very difficult to determine the volume of the enclosed column of air with accuracy since it is bounded from below by a curved surface, the form of which often varies according to the friction between the mercury and the glass. In all these experiments we endeavoured to divide this curve into two equal portions by eye, but the results show that we underestimated the enclosed air. With-

Density	Compressing forces	Differences	Differences divided by the compr. forces
1.0000	1.0000	0.0000	0.0000
1.1052	1.1051	+0.0001	+0.0001
1.1676	1.1693	−0.0017	−0.0015
1.2736	1.2706	+0.0030	+0.0024
1.4744	1.4694	+0.0050	+0.0035
1.587	1.581	+0.006	+0.004
1.812	1.806	+0.006	+0.003
2.112	2.079	+0.033	+0.016
2.529	2.520	+0.009	+0.004
3.168	3.147	+0.021	+0.007
3.616	3.599	+0.017	+0.005
4.209	4.185	+0.024	+0.006
5.057	5.010	+0.047	+0.009
5.603	5.572	+0.031	+0.005
6.288	6.287	+0.001	0.000
7.175	7.082	+0.093	+0.013
8.030	8.014	+0.016	+0.002

out this error, the differences would have been smaller, and the numbers sometimes higher, sometimes lower. Still, the discrepancies are smaller than could have been expected in experiments which preclude the use of the vernier.

In the last experiments, e.g., the observed height of the column of air amounted to 56.4 millimetres; according to Mariotte's law, it should have been 56.287; the entire difference, therefore, is no more than 0.113 millimetres, an error which is quite inevitable in such observations. In the penultimate experiment, the observed height of the column was = 63.17 millimetres, which, according to Mariotte's law, should have been = 63.99. This discrepancy, the greatest we obtained, amounts to 0.82 millimetres, but being between two observations which offer only a very small deviation, it will not affect the general law.

In order to investigate the compression of air by greater forces, we made use of the air gun. Our King, whose enlightened generosity has already done so much for the advancement of science, placed all the requisite apparatus at our disposal. Everyone knows that the butt of this kind of weapon serves as the reservoir for the compressed air. This part, therefore, must be particularly strong. We first determined the capacity of one by weighing it when empty and, again, when filled with water. By this means, the quantity of air that might be contained in it could easily be ascertained. The one which we used most frequently could contain 0.891 grammes of air at a height of 0.76 metres of mercury. We were also able to determine, by weighing, the degree of density we obtained in our experiments. This method proved to be sufficiently accurate as the scales we generally used were sensitive to one centigramme. We succeeded in forcing 101.2 grammes of air into one butt, a quantity corresponding to a pressure of 110.5 atmospheres. We also took into consideration the expansion of the reservoir, caused by the internal pressure, and determined it by weighing the butt in water both empty and filled with air. In calculating, we assumed that the different degrees of this expansion were in proportion to the quantity of air which had been put into it. If the reservoir had not expanded after 101.2 grammes of air had been forced into it, the density of this air would have equalled that of 113.5 atmospheres, but taking into consideration the expansion of the reservoir, it only amounted to 110.5.

The third figure shows the manner in which we carried out our experiments on the expansive power of the air compressed in such a reservoir. *AB* represents such a reservoir, that is, the butt of an air gun; *CD* is a board with a perpendicular lath *CE*; *EH* is a piece of iron which in its upper part receives the axis, around which turns the lever *FG*, which again is balanced by a counterpoise, *F*. This lever has a tooth *I*, pressing on the valve *M* of the butt *MB*, fastened underneath it. A slide *N* with a scale *L* appended to it serves to determine the force necessary to open the valve. This valve being closed by a spring, we began by investigating the force required to open the valve when the density of the enclosed air corresponded to that of the atmosphere. After this the reservoir was loaded as much as possible, and after having measured the resistance exerted by the enclosed air against the valve, we gradually emptied the reservoir, constantly weighing the remaining air by means of scales and weights and determining its expansive power by means of the apparatus shown in fig. 3. Experiments of this kind, however, are not capable of great accuracy since the valve does not always close uniformly. If the valve is lined with leather, for the purpose of making it close more tightly, this non-uniformity is very great, for which reason we carried out another series of experiments with a precisely ground steel valve, but we could not obtain any very considerable charges in this way. We give the results of both experiments in the following tables. The first column shows the air forced into the reservoir; the second, its density; the third, the force needed to open the valve, that which was necessary before loading being deducted; and, finally, the fourth, the pressure of the atmosphere obtained as the product of the magnitude of this force divided by the degree of density.

TABLE 1: Experiments with a butt having a valve lined with leather.[4]

Weight of enclosed air, in grammes	Density, that of atmosphere = 1.	Pressure on the valve, in grammes.	Pressure, divided by the density.
1	1.122	812	725
2	2.243	1809	806
3	3.364	2552	758
4	4.484	3693	823
5	5.604	3495	784
6	6.732	5750	855
7	7.842	6693	853
8	8.960	6797	758
9	10.077	7711	764
10	11.193	8166	729
10	11.193	8434	753
10	11.193	8480	757
10	11.193	8445	754
10	11.193	8437	753

In the first table the mean number is 797, and we find that the deviations from it are not at all regular. The mean of the second table is 1027 (excluding the first number, the deviation of which is too great), and we also find that most of the numbers do not deviate significantly from it. However imperfect these experiments may be, and must be from their very nature, they contribute to the proof that the compressions produced by very great forces are governed by the same laws as those produced by low compressing forces. However, wishing to ascertain whether the compression of every kind of gas was subject to the same law, we selected such gases as have the property of being changed into a liquid by a pressure of a few atmospheres. Sulphurous acid gas, which (according to Faraday) becomes such a liquid under a pressure of two atmospheres, seemed to us most appropriate for this kind of experiment.

[4][A letter from Ørsted to Hansteen (September 28, 1826) and Ørsted's posthumous papers make it clear that the caption of the table shown is incorrect as it contains the results of a series of experiments with steel valves, and that a completely different table belongs to this caption. This table, found among Ørsted's papers, is the following:

Wgt. of air	101.2	94.3	88.6	81.7	70.3	58.9	47.4	35.9	25.5
In atmosph.	110.5	103.19	97.08	89.69	77.41	65.06	52.52	39.94	28.42
Valve open at press. (gms)	87040	83910	74813	66085	57777	51455	44447	32834	25962
Gives for 1 atm. (gms)	781	806	763	728	736	779	832	804	888

The rigidity of the spring of the valve = 720 grammes is included in the third row.]

TABLE 2: Experiments with a butt with an unlined valve.

Weight of enclosed air, in grammes	Density, that of atmosphere = 1.	Pressure on the valve, in grammes.	Pressure, divided by the density.
1	1.122	1269	1131
2	2.243	2368	1055
3	3.364	3388	1007
4	4.484	4751	1059
5	5.604	5750	1026
5	5.604	5620	1002
5.05	5.657	5790	1023
5.05	5.657	5800	1025
5	5.604	5730	1022
6	6.732	6871	1021
7	7.842	8113	1034
8	8.960	9344	1043
9	10.077	10375	1029
10	11.193	11440	1022
10.2	11.417	11725	1027
15	16.76	16766	1000
15.1	16.87	17243	1022
20	22.326	22988	1029
25.6	28.543	29253	1025
30	33.393	34197	1024
35.2	39.13	40232	1026
40.1	44.52	45633	1025
45	49.894	51641	1035
50	55.362	57467	1038
55	60.816	63102	1037
60	66.254	67798	1023

Two glass tubes of equal sizes, one filled with well-dried sulphurous acid gas, the other with atmospheric air, were placed in a small basin of mercury and put into an apparatus by means of which these gases could be exposed to an appropriate pressure. The result was that both volumes were diminished in a constantly equal ratio until the moment when the sulphurous acid gas began to liquefy.

We add the following details of these experiments.

AAAA (fig. 4) is a very strong glass cylinder, the same which I use for the compression of water. This cylinder has a brass lid; on this is placed another cylinder *BBBB*, in which a piston *C* can be moved up and down by the screw *DD*. *EEEE* are two identical, graduated tubes, the lower extremities of which are placed in a small iron basin *FF*. This basin is fastened to the end of a slip of glass *GGGG*, which serves at the same time to maintain the tubes in a perpendicular position. The cylinder *AAAA* is filled with mercury up to *HH*. The experiment is begun by the two tubes being filled with the gases, put into the small basin, and fastened to the slip of glass *GGGG*. The entire apparatus is then placed in the cylinder *AAAA*, whereby the basin is immersed in the mercury below the line *HH*; the cylinder is then filled with water, the pumping cylinder *BBBB* attached to it and also filled with water, and finally the piston is inserted and made to press on the enclosed water. The water communicates the pressure to the mercury, which again transmits it to the gases in the tubes. A transverse section of the lower part of the apparatus is shown in *ad* Fig. 4.

The following table shows that the differences are very small, and that sometimes one, sometimes the other of the gases, suffers a greater degree of compression up to a pressure of 2.3 atmospheres, where they become greater, and where the sulphurous acid gas shows a continually higher density. At a pressure of 3.2689 moisture begins to be visible, and condensation begins to show itself in a much more powerful and decided manner. Perhaps some small liquefaction takes place before this point where the gases come in contact with the sides of the tubes and with the mercury, for the contact with a heterogeneous body seems to favour the transition from one state to another, as I have shown in an earlier treatise on some experiments of Winterl's.[5]

Experiments with two tubes, one filled with atmospheric air,
the other with sulphurous acid gas. The temperature was $21^1/_4°$ C.[6]

Sulphurous acid gas	Atmos. air	Comp. of sulphurous acid gas	Comp. of air	Differences
131.2	128.5	1	1	
128	125.33	1.0261	1.0259	+0.0002
122.4	120	1.0754	1.0768	−0.0014
117.33	115	1.1229	1.1215	+0.0014
112	110	1.1750	1.1729	+0.0021
106.875	105	1.2302	1.2297	+0.0005
101.5	100	1.2937	1.2942	−0.0005
96.3	95	1.3634	1.3644	−0.0010
91.25	90	1.4396	1.4403	−0.0007
86	85	1.5278	1.5257	+0.0021
80.75	80	1.6228	1.6228	0.0000
75.5	75	1.7329	1.7311	+0.0018
70.6	70	1.8542	1.8539	+0.0003

[5]Gehlen's *Journ. d. Physik und Chemie* 1806. Vol. 1, pp. 276–89. ["Some Passages in Winterl's Writings" (Chapter 23, this volume).]

[6][Misprints at several places in this table, and at a few places in the previous ones, have been corrected according to Ørsted's original tables.]

Experiments with two tubes, one filled with atmospheric air,
the other with sulphurous acid gas, *(cont.)*.

Sulphurous acid gas	Atmos. air	Comp. of sulphurous acid gas	Comp. of air	Differences
65.6	65	1.9971	1.9974	−0.0003
64.5	64	2.0310	2.0307	+0.0003
63.4	63	2.0649	2.0638	+0.0011
62.4	62	2.0976	2.0982	−0.0006
61.3	61	2.1342	2.1336	+0.0006
60.3	60	2.1705	2.1702	+0.0003
59.25	59	2.2101	2.2082	+0.0019
58.2	58	2.2475	2.2474	+0.0001
57.16	57	2.2879	2.2874	+0.0005
56	56	2.3356	2.3289	+0.0067
54.875	55	2.3835	2.3720	+0.0115
53.875	54	2.4279	2.4166	+0.0113
52.8	53	2.4798	2.4629	+0.0169
51.75	52	2.5317	2.5109	+0.0208
50.6	51	2.5831	2.5610	+0.0221
49.6	50	2.6488	2.6171	+0.0317
48.6	49	2.7008	2.6674	+0.0334
47.6	48	2.7595	2.7240	+0.0355
46.6	47	2.8207	2.7819	+0.0388
45.5	46	2.8886	2.8423	+0.0463
44.4	45	2.9556	2.9057	+0.0499
43.33	44	3.0240	2.9717	+0.0523
42.4	43	3.0974	3.0407	+0.0567
41.16	42	3.1733	3.1130	+0.0603
39.33	41	3.3186	3.1889	+0.1297
34.5	40	3.7796	3.2689	+0.5107
20.33	39	6.4890	3.3526	+3.1364

In some experiments we found that the water had penetrated between the sides of the tube and the mercury. This inconvenience was subsequently avoided by our cementing the extremity of each tube into a brass ring, which amalgamates with the mercury and prevents the water from penetrating.

We also compressed cyanogen in the same manner and found that liquefaction of this gas begins when the air has been compressed to $^1/_{3.5}$ of its weight at a temperature of 23° and a barometric pressure of 0.759 metres of mercury.

It would have been easy to multiply these experiments, but those which we have communicated here will, we trust, suffice to prove that the compression of atmospheric air and gases is proportional to the compressing forces, however great these may be, presuming that the gases remain in their aëriform state, and that the caloric liberated by the compression has been carried off. Thence it appears that our investigations have done no more than confirm the opinions of the most distinguished men of science of our time with respect to this subject, but as there still remained some scholars who entertained a contrary opinion, we considered the publication of our experiments to be not altogether useless.

The compression of liquids is subject to the same law as far as our experience goes. Here, too, the compression and the compressing force seem to be in a direct proportion. We may therefore assume that gases converted into liquids begin anew to follow the same law to which they answered as gases. It is also quite likely that liquids converted into solids are subject to the same law. If this should be confirmed by further experiments, it may be said that the compression of a body ceases to conform to this law only at the moment of its transition from one state of aggregation to another.

54

Preliminary Note on the Production of Aluminium, Aluminium Chloride, and Silicon Chloride[1]

I SHALL soon bring you news about some new experiments whereby I have succeeded in producing *aluminium chloride* and, from that, *aluminium*. The aluminium chloride is obtained as a volatile substance when dry chlorine is spread over a glowing mixture of alumina and coal. Silicon chloride is obtained in the same way, but here the volatile substance must be greatly cooled.—More of this shortly.

[1] From a letter from Professor H. C. Ørsted to Professor Schweigger, dated October 9, 1825. [*Journal für Chemie und Physik*, edited by Dr. J. S. C. Schweigger and Dr. Fr. W. Schweigger-Seidel, Vol. 45, p. 368. Halle 1825. The same subject is treated in more detail in *Videnskabernes Selskabs Oversigter,* 1824–25, pp. 15–16. Copenhagen. Notes on the same topic are to be found in *Magazin for Naturvidenskaberne,* Vol. 5, pp. 176–77. Christiania 1825; and in Poggendorff's *Annalen der Physik und Chemie,* Vol. 5, p. 132. Leipzig 1825. KM II, p. 297. Originally published in German.]

55

Contribution to the Determination of the Law of the Compression of Bodies[1]

FOREWORD

The fact that a body can be forced by pressure to occupy a smaller volume, and that its parts can return to their previous position when the pressure has ceased, is a natural effect whose singularity is forgotten because of its daily occurrence, but in which the essence of corporality manifests itself most directly. It is not my intention here to add to the many researches that have been ventured upon regarding the cause or the internal nature of this property, but I shall endeavour to make a contribution to the knowledge of its mode of action. It appears that in all bodies it is subject to one law, which is that *the compression is in proportion to the compressing forces.* For a long time, mathematicians have assumed that this law applies to all weak compressions, but my experiments have shown me that it has an extent which no one seems to have imagined before. These are the experiments I want to present here, not in the chronological order in which they were carried out, but in the order in which they best illuminate each other.

THE COMPRESSION OF AIR

The law, named after Mariotte, which says that the volume of a quantity of air is inversely proportional to the compressing forces, has not yet been proved by accurate experiments, except at low pressures. Whether it also applied at higher pressures could still be open to doubt. Some eminent mathematicians and physicists have apparently assumed, although not proved, its validity for every compression to which a gas could be subjected, while others, among them Jacob Bernoulli and Euler, have assumed that the volume of air decreases at a somewhat lower rate than the

---

[1] [*Det Kongelige Danske Videnskabernes Selskabs Naturvidenskabelige of Mathematiske Afhandlinger*, Part 2, pp. 289–324. Copenhagen 1826. On the half-title page, the year given is 1823, but this cannot be the year of the publication of the paper as Ø., in a letter to Hansteen of May 25, 1826, mentions that this paper, which had been finished for a long time, has finally been printed. Also to be found in *Magazin for Naturvidenskaberne*, Vol. 7, pp. 255–89. Christiania 1826. KM II, pp. 298–325. Originally published in Danish.]

compressing forces increase. Finally, if we resort to the small number of experiments in which air has been subjected to considerable compressing forces, its volume seems to decrease at a higher rate than the compressing forces increase. In the papers of the Berlin Academy, the perspicacious German investigator Sulzer has published experiments which extend to a pressure of 8 atmospheres, and which have this result. He certainly endeavours to explain away the deviations of his experiments from Mariotte's law but only with assumptions which in no way enable us to verify the experiments through calculation. Robison, the astute physicist from Edinburgh, has arrived at the same relation in his much later experiments. The following two tables present the results of some of their most distinguished experiments, from which the others do not differ substantially.

Sulzer's Experiment		Robison's Experiment	
Density	Compressing forces	Density	Compressing forces
1.000	1.000	1.000	1.000
1.091	1.076	2.000	1.957
1.200	1.183	3.000	2.848
1.333	1.303	4.000	3.737
1.500	1.472	5.500	4.930
1.714	1.659	6.000	5.342
2.000	1.900	7.620	6.490
2.400	2.241		
3.00	2.793		
4.000	3.631		
6.000	5.297		
8.000	6.835		

These experiments, which are only examples taken from several different series, have sufficient regularity to compare them with all conclusions advanced so far, according to which it might seem reasonable to assume the validity of Mariotte's law for all degrees of air pressure. These experiments might even invite the belief that the condensation point, at which air becomes liquid, must be rather close and attainable by compressing forces which we could easily produce. When, in the summer of 1824, I joined with Captain v. Suensson of the Royal Engineers to conduct some experiments on the theory of the air gun, the law of the compression of air had to be the first subject that came under investigation. All the experiments to be described in the present section were conducted in the company of this clever officer.

It will be recalled that the primary instrument used so far for experiments on the compression of air consists of a bent tube $ABCD$, fig. 1, whose shorter leg DC is closed at the top, while the longer AB is open. When the air in the shorter and closed part of the tube is blocked and compressed by a quantity of mercury $ABCE$, the decrease of the volume can be determined by comparing the height of the quantity of air DE with its previous height. By measuring the mercury column AF, one also finds the magnitude of the pressure borne by the confined air, in addition to the atmospheric pressure that acts on it both before and after the addition of the mercury. This device has several defects. It is very difficult to divide the short tube into parts of equal volume as such tubes are not easily found so uniformly wide that this object can be achieved by marking equal lengths on them. If a rather wide tube is used, it does not endure any considerable internal pressure. If, perforce, a narrower tube is used, the small quantity of air which always settles between the glass and the mercury is far larger in comparison with the volume of the air than when sufficiently wide tubes are used, and the frictional resistance between the mercury and the glass is proportionately far more appreciable so that a slight shaking can produce unaccountable changes in the position of the mercury. In any event, the glass tube expands due to the pressure it receives from within. In order to avoid these sources of error, we used the instrument shown in fig. 2, where the air is confined in a glass tube, sealed at the top, which is enclosed in a strong glass cylinder, which contains mercury at the bottom to block the air but otherwise is filled with water. A given pressure, when applied to the water, will be transmitted to the mercury, which again acts on the air. This makes it possible to use a wide glass tube which does not have to be made of thick glass as it receives equal pressure from without and from within. I had previously used a device built on the same principle for the compression of water, and it will be described in the following section.

The details of the device used for the experiment will be seen from the vertical section shown in the figure. $ABCD$ is a cylinder of strong glass, fitted at the top with a strong brass ring, to which a lid can be screwed. EF is a graduated glass tube, sealed at the top, approximately 6 lines in diameter and about $1\frac{1}{2}$ foot high. Although the tube we used was almost equally wide everywhere, we still determined the small differences by weighing the mercury which the parts could contain. The tube EF is held in a vertical position by an iron frame $lmno$, whose lower part is a small iron bowl which can hold the mercury that is required to block the mouth of the tube when it is taken in or out. The inside of the cylinder is filled with mercury to a height of approximately two inches; the line IK represents its limit. GH indicates a strong, $\frac{1}{2}$-inch-wide glass tube, open at both ends, which is cemented into a metal ring that has threads on the outside, by means of which it can be screwed into an interior screw placed in the lid for that purpose. In the lid there is another and smaller opening which can be closed by means of a screw P, which is here shown in its place. TV is a wooden base, from which rises a pole, only the lower part RT of which is shown here. The base can be adjusted by three screws, of which only two could be shown in the cross section, designated by S. The iron frame $lmno$ is shown enlarged in an additional portion of fig. 2. Another additional figure shows a cross section of the lower part of the cylinder and its contents along with the base.

Before the start of the experiment, the air in the tube EF must be dried. For this purpose, the tube is first placed in the frame in such a way that the mouth is not

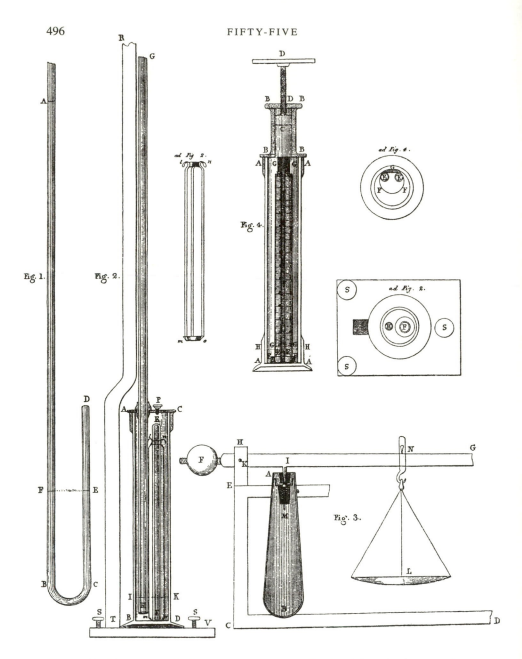

blocked by the mercury in the bowl connected with it, and then the frame with the tube is placed in a cylinder that contains anhydrous chloride of lime. When the drying is completed, the tube is pushed down, by means of a suitable, previously arranged device so that the mouth is blocked. After the lid has been removed from the cylinder, the frame with the tube is placed in it so that now it is also blocked by the mercury at the bottom of the cylinder, and then the lid is screwed back on as tightly as possible. In order that everything may fit all the more closely, a leather ring, saturated in a mixture of wax and turpentine, is placed between the edge of the lid and the ring. Similar leather rings are placed at the other screws. When the lid has been put on, the tube *GH* is also screwed in, whereupon the entire cylinder is filled with water through the opening at *P*. This pressure acts on the air in the tube *EF* through the mercury. At the same time the pressure of the water forces the mercury a small distance up in the tube *GH*. The excess of the height of the mercury in the latter above that which can be observed in the former gives, here and below, the pressure which, in addition to the external air pressure, acts on the confined air. Observation is greatly facilitated by the fact that the bottom part of the tube *GH*, as well as all of *EF*, is graduated in lines. When the cylinder has been filled with water and the first observation made, the opening *P* is closed at its screw. Then mercury is gradually poured into the tube *GH*, and the heights in both tubes are observed. The tube *GH* is graduated for only a few inches above the lid, and the remaining height of the mercury column has to be measured with a rule.

In order to make the tube *GH* long enough for high pressures, we had to join several tubes together. Each was approximately 7 feet long, and they were joined with strong iron screws. No ordinary room was high enough for this experiment, so we had to conduct it in the stair well at the University collection of physical instruments. The cylinder was placed on a solid table, and the stairs served as a ladder during the performance of the experiment.

With this instrument we have conducted several series of experiments, though not all equally satisfactory as it is very difficult to make the cement and the screws tight enough to withstand high pressures. We succeeded only once in making them withstand a pressure of 8 times the height of the mercury column in the barometer, which was 335.93 French lines during the experiment.

The following table shows the relation we have found between the compression of air and the compressing forces. The first column shows the decrease in volume, or the volume of air with which we started divided by the volume to which it was compressed by each pressure. The second column gives the pressure as the height of the mercury column, where the mercury column supported by the atmosphere, 335.93, is regarded as unity. The third presents the differences between the ratios of the compressions and the compressing forces; finally, the fourth shows the ratio of these differences to the entire magnitudes to which they belong. A special difficulty in the observations made during these experiments is the accurate determination of the lower boundary of the confined air because the surface of the blocking mercury is convex, although not always to the same extent. During the observations, an attempt was made to divide this curve by eye into two parts of equal volume, but the results of the experiments show clearly enough that the upper half has been judged too small, or, in other words, too little of it has been attributed to the air. Without this error the differences would have been smaller and would have alternated between + and −. Apart from this, the differences are as small as could be expected in such observations where no vernier can be used. For example, the height of the air column in the last experiment was 25 lines. If it had been assumed to be $^1/_{20}$ of a line longer, the volume would have been $^1/_{500}$ bigger, and the deviation would have disappeared. In the last experiment but one, where the discrepancy was biggest, the height of the air column was 28 lines. Here the error amounts to 0.36 lines, but as this observation is between two others whose deviations are very small, this clearly does not give rise to any objection to the general law.

Decrease in volume	Pressures	Difference of density and pressure	Ratio of difference to primary quantity
1.0000	1.0000	0.0000	0.0000
1.1052	1.1051	+0.0001	+0.0001
1.1676	1.1693	−0.0017	−0.0015
1.2736	1.2706	+0.0030	+0.0024
1.4744	1.4694	+0.0050	+0.0035
1.587	1.581	+0.006	+0.004
1.812	1.806	+0.006	+0.003
2.112	2.079	+0.033	+0.016
2.529	2.520	+0.009	+0.004
3.168	3.147	+0.021	+0.007
3.616	3.599	+0.017	+0.005
4.209	4.185	+0.024	+0.006
5.057	5.010	+0.047	+0.009
5.603	5.572	+0.031	+0.005
6.288	6.287	+0.001	0.000
7.175	7.082	+0.093	+0.013
8.030	8.014	+0.016	+0.002

   In order to investigate the compression of air by greater forces, we used reservoirs from air guns. His Majesty, the King, whose enlightened love of the sciences always encourages and supports its practitioners, graciously granted the use of his own excellent air gun reservoirs and charging machine, to which was added the loan of a large, yet very sensitive scale from Frederiksværk. We began by determining how much water such a reservoir could hold, and then we calculated the quantity of air it contained in its natural equilibrium with the atmosphere. The reservoir which we used most often contained 0.891 French grammes of air at a pressure of 336.9 lines of the mercury column. We were then able to determine, by weighing, the density of the air we put in the reservoir. This method had all the accuracy required as our scale responded to one centigramme. We succeeded in forcing the density to such a degree that the air in the reservoir weighed 101.2 grammes, that is, more than $^1/_5$-pound Danish avoirdupois. On that occasion the air had been compressed more than 110 times. In these experiments the expansion of the reservoir by the internal pressure had been taken into account. This expansion was determined by investigating how much weight the reservoir lost, both in an un-loaded and a loaded condition, when weighed in water. In the calculations it was as-sumed that the expansion is in proportion to the pressure, which again is assumed to be in proportion to the quantity of air pumped in. The result of the calculation was that the air in the experiment just mentioned was only compressed to a $110^1/_2$ times smaller space though the reservoir contained $113^1/_2$ times more air than it could hold in an uncompressed state. Indeed, in this calculation we have assumed the validity of the law whose generality was to be proved, but the whole expansion with which we are dealing here is so small that even if we had completely neglected this correction, the numbers would still not deviate much from the assumed law. The elasticity which we have attributed to the compressed air in our calculations cannot, then, deviate so much from the truth that the correction could be percepti-bly changed by it.

Fig. 3 shows the device we used to measure the elasticity of the pumped-in air. AB is the reservoir of an air gun, CD a board, CE an upright post, EH is an iron fork which supports an axis at K, around which a graduated beam FG turns. When un-stressed, this beam is maintained in equilibrium by means of the mass F. At I, the beam has a pin with which it can push the blocking valve M. The scale L is held by a peg N, which can be moved forwards and backwards in order to discover the mag-nitude of the pressure required to open the blocking valve. As this is not held by the air pressure alone but also by a spring, we investigated how much force was re-quired to open it before the internal air pressure was made bigger than the external. There is hardly any need to mention that the weight of the scale and its fittings was determined before the experiments.

The experiments were now conducted in such a way that a considerable amount of air was first pumped into the reservoir and its weight determined, and then the elasticity of the compressed air was tested by means of the device just described. Then some air is gradually released, and each time it is weighed to determine how much remains, and the above-mentioned apparatus is used to find out how much elasticity it still has. This kind of experiment does not allow the same degree of accuracy as the previous ones because the blocking valve does not always cover the opening completely uniformly, nor does it receive the pressure so completely perpendicularly as the purpose requires. When the blocking valve is fitted with leather, the heterogeneity is very great, and we have therefore conducted a series of experiments with a ground blocking valve of steel, but then we have not been able to obtain such large charges, probably because the violent impetus that the blocking valve receives when the air is pumped in at very high pressure reduces its ability to fit tightly. As the first series of experiments shows considerable devia-tions, we want to describe the second here even though it only goes up to a pressure of 60 atmospheres. In the appended table, the first column shows the weight of the air pumped in, the second the density, the third the force required to open the blocking valve, the fourth the atmospheric pressure calculated from each experi-ment by dividing this force by the density. The average here is 1027, and when the first number is rejected as much too aberrant, it is clear that the others do not differ much from that value.

Weight of Air	Density	Pressure to open blocking valve	Pressure divided by density
1	1.122	1269	1131
2	2.243	2368	1055
3	3.364	3388	1007
4	4.484	4751	1059
5	5.604	5750	1026
5	5.604	5620	1002
5.05	5.657	5790	1023
5.05	5.657	5800	1025
5	5.604	5730	1022
6	6.732	6871	1021
7	7.842	8113	1034
8	8.960	9344	1043
9	10.077	10375	1029
10	11.193	11440	1022
10.2	11.417	11725	1027
15	16.76	16766	1000
15.1	16.87	17243	1022
20	22.326	22988	1029
25.6	28.543	29253	1025
30	33.393	34197	1024
35.2	39.13	40232	1026
40.1	44.52	45633	1025
45	49.894	51641	1035
50	55.362	57467	1038
55	60.816	63102	1037
60	66.254	67798	1023

However imperfect these experiments must be according to their nature, they still help to complete the proof of the law that the compression of air is always in proportion to the compressing forces. However, for the sake of completeness, we had to pursue the compression of air to the point where it becomes a liquid. For this we had to choose gases whose liquefaction does not require too great forces. We first chose sulphuric[2] acid, which according to Faraday's experiment already condenses at a pressure of 2 atmospheres. We filled two equally large, graduated tubes, one with well-dried sulphurous acid, the other with dry atmospheric air. These were then placed in an apparatus in which they could be subjected to the same pressure. The experiments showed that both gases underwent the same compression until the sulphurous acid began to liquefy. The instrument which was used for these experiments is shown in fig. 4. *AAAA* is a very strong glass cylinder, which I have

[2] [sulphurous.]

used in my experiments on the compression of water. The cylinder is carefully fitted with a cemented brass lid, with an opening into which a pumping cylinder *BBBB* can be screwed. This is equipped with a piston *C*, which can be raised and lowered by means of a screw *DD*. The cylinder *AAAA* is filled with water and contains two glass tubes, *EE* and *EE*, which are placed in a small iron bowl *FF* which, together with the tubes, is fastened to a glass slip *GGGG*. The cylinder *AAAA* contains mercury up to *HH*. It is easy to understand the use of this instrument. First the tubes are filled with the two gases, placed with their mouths in the small iron bowl *FF* filled with mercury, and then fastened to the glass slip *GGGG*; the glass cylinder is filled with mercury to *HH*, and the tubes are placed in it. When this has been done, the cylinder is filled with water, the pumping cylinder is screwed on and also filled, and finally the piston is inserted, and the lid of the pumping cylinder, in which its screw moves, is fastened. The piston is now made to exert pressure on the water, which communicates it to the mercury, from where it is further communicated to the confined gases. The table below shows the result of an experiment which was conducted at a temperature of $21^1/_2$ centigrade and an air pressure of 28 inches. The first column shows the volume of the sulphurous acid, the second that of the air, the third shows the ratio of the compressions of the sulphurous acid, the fourth that of the air, the fifth the differences. The differences can be seen to be quite inconsiderable and to alternate between + and − up to a pressure of 2.3 atmospheres, where they become bigger and ever increasing. At 3.2689 the moisture becomes visible, and from this point the decreases occur rapidly. Between the two limits mentioned here, some liquid might have formed on the surface of contact of the sulphurous acid with the glass and the mercury, for contact with a foreign body seems to favour the transition from one form of corporality to another, which I have shown in a treatise in Gehlen's *Journal d. Chemie*, 1806, Vol. 1, pp. 276–89.[3]

We have repeated this experiment with sulphurous acid many times with similar results. We have also repeated it with cyanogen and found that, at a temperature of 23° centigrade and at an air pressure of approximately 28 inches of mercury, it liquefied at a pressure of 3.5 atmospheres. Experiments which confirmed the same law have been conducted with ammonia, but the notes have accidentally been mislaid. As we observed in our experiments that the water forced its way between the glass and the mercury, we fitted the mouths of our tubes with a cemented brass ring, which amalgamated with the mercury in which they were immersed. This formed a coherent metallic mass which the water could not penetrate. We have later learnt that Daniell has used the same principle with the barometer. It appears that he had this idea at the same time as we did.

[3] ["Some Passages in Winterl's Writings" (Chapter 23, this volume).]

Volume		Ratios of compressions		Difference
Sulph. acid gas	Air	Sulph. acid gas	Air	
131.2	128.5	1	1	
128	125.33	1.0261	1.0259	+0.0002
122.4	120	1.0754	1.0768	−0.0014
117.33	115	1.1229	1.1215	+0.0014
112	110	1.1750	1.1729	+0.0021
106.875	105	1.2302	1.2297	+0.0005
101.5	100	1.2937	1.2942	−0.0005
96.3	95	1.3634	1.3644	−0.0010
91.25	90	1.4396	1.4403	−0.0007
86	85	1.5278	1.5257	+0.0021
80.75	80	1.6228	1.6228	0.0000
75.5	75	1.7329	1.7311	+0.0018
70.6	70	1.8542	1.8539	+0.0003
65.6	65	1.9971	1.9974	−0.0003
64.5	64	2.0310	2.0307	+0.0003
63.4	63	2.0649	2.0638	+0.0011
62.4	62	2.0976	2.0982	−0.0006
61.3	61	2.1342	2.1336	+0.0006
60.3	60	2.1705	2.1702	+0.0003
59.25	59	2.2101	2.2082	+0.0019
58.2	58	2.2475	2.2474	+0.0001
57.16	57	2.2879	2.2874	+0.0005
56	56	2.3356	2.3289	+0.0067
54.875	55	2.3835	2.3720	+0.0115
53.875	54	2.4279	2.4166	+0.0113
52.8	53	2.4798	2.4629	+0.0169
51.75	52	2.5317	2.5109	+0.0208
50.6	51	2.5831	2.5610	+0.0221
49.6	50	2.6488	2.6171	+0.0317
48.6	49	2.7008	2.6674	+0.0334
47.6	48	2.7595	2.7240	+0.0355
46.6	47	2.8207	2.7819	+0.0388
45.5	46	2.8886	2.8423	+0.0463
44.4	45	2.9556	2.9057	+0.0499
43.33	44	3.0240	2.9717	+0.0523
42.4	43	3.0974	3.0407	+0.0567
41.16	42	3.1733	3.1130	+0.0603
39.33	41	3.3186	3.1889	+0.1297
34.5	40	3.7796	3.2689	+0.5107
20.33	39	6.4890	3.3526	+3.1364

THE COMPRESSION OF WATER

For almost two centuries the question of whether water can be compressed or not has been the object of scientific experiment. Most of the history of these investigations can be gleaned from Herbert's *Dissertatio de aquæ aliorumque nonnullorum fluidorum elasticitate*, Vienna, 1773, and Zimmermann's *Über die Elasticität des Wassers*, Leipzig, 1779. Here, then, it suffices to point out that the earlier experiments by Baco,[4] by Academia del Cimento,[5] by Boyle, Muschenbrock,[6] and others, mostly confined themselves to filling metal globes with water and then pressing and hammering them so that they changed shape and enclosed less volume than before. According to their nature, these experiments could not provide accurate results, but they showed us the curiosity that water sweated out of the pores of metals, and some physicists observed in these experiments that water continued to sweat after the pressure had ceased, which latter phenomenon, however, could just as well be attributed to the contraction of the metal globe as to the re-expansion of the water but has really been a consequence of both. Nollet and Hamberger tried to compress water in the same way as Mariotte compressed air, but the weak compression they could obtain with this procedure could not but escape their attention. Canton's extremely well devised and accurate experiments consisted mainly in his placing thermometer tubes, filled with water to a given mark and not sealed, under the bell of an air pump, whereby it turned out that the water occupied more space when the air pressure was removed and collapsed when the air pressure was returned. As the water in these experiments was compressed by only forty-odd millionths, and as the effect of heat both on the water and on the glass tube could easily cause substantial deviations, these experiments have not won the general confidence which it is easy to see they deserve now that they can be compared with experiments which have fewer difficulties. What gives Canton's investigative method a considerable advantage, not only over all previous experiments on the compression of water but also over those conducted for a half-century after him, is that he arranged things so that the compressing force acted equally on the outside and the inside of the vessel in which the water was subjected to the experiment, and that this vessel was designed in such a way that very small changes in the volume of the water in it could be observed. His immediate successors departed to some extent from this principle, and the results of their experiments also came to diverge considerably from the truth.

Immediately after Canton, Herbert made experiments on this subject. At first glance, the description of his experiments might easily tempt the reader to consider these as far more complete than those of the British physicist as it was much easier for Herbert to avoid the effect of heat because of his procedure. He confined the water intended for compression in a glass sphere which is connected to a horizontal tube, part of which is also filled with water. The other end of the horizontal

---

[4][Bacon.]   [5][Accademia del Cimento.]
[6][Musschenbroek.]

tube is connected to a vertical tube, into which mercury can be poured, which then exerts pressure on the water. There could be no doubt that the glass sphere had to be expanded by the pressure which the water on this occasion had to exert on its inside, so he placed the sphere in a glass box filled with water and equipped with a tube, in which the water had to rise when the sphere expanded. As the widths of the tubes had been measured with weighed mercury, he believed that he would be able to determine with great accuracy how much volume the water had lost while it was subjected to the effect of the pressure. However, we see here another example of the importance of choosing the simplest experimental methods. The assumption that a volume of water would rise in the tube on the surrounding box equal to that displaced by the glass sphere is not quite correct. As the water rose in the tube of the box, it had to exert a new pressure on its side walls and thus expand it ever so slightly. If it is not too presumptuous to judge the size of the glass box by the drawing of the instrument, its sides were $3^1/_3$ inches each, the area of each side then about 11 square inches, and all sides 66 square inches. According to the same figure, the rise in the tube of the box due to the expansion of the glass sphere must be estimated to be about $^1/_3$ inch. The pressure on the sides of the box would then be as big as the weight of 22 cubic inches of water, which is more than 26 *lod*.[7] Anyone wishing to object that this pressure was too weak to expand the box needs only to be reminded that the bending, expansion, and compression of bodies are all proportional to the forces so that small forces also have an effect, no matter how imperceptible it may be to everyday observations. Here it is only a question of small quantities, for when he made a mercury column of 48 inches exert pressure on a volume of water which filled the same space as 183029 *Richtpfenningstheile* of mercury, the water was only reduced by a volume which could be filled by 88 *Richtpfenningstheile* of mercury. As 17.33 *Richtpfenningstheile* is one grain of apothecaries' weight, this only constitutes 5.08 grains of apothecaries' weight. The space which the rising water filled in the tube of the box constituted 46 *Richtpfenningstheile* = 2.65 grains of apothecaries' weight. The decrease in the volume of the water that he deduced from this, then, became as large as the volume occupied by $88 - 46 = 42$ *Richtpfenningstheile* or 2.43 grains of apothecaries' weight of mercury. If we calculate how much the mercury should have receded, according to experiments which we now regard as certain, that is, 45 millionths for 28 inches or 77 millionths for 48 inches of mercury, this would only have created a volume which could be filled with 14 *Richtpf.* = 0.81 grains of mercury so that a volume is left which could be filled with 28 *Richtpf.* = 1.62 grains of mercury, which should constitute the expansion of the surrounding glass box due to the pressure which the rise of the water caused in the tube of the box. Considering that this space constitutes only 0.717 cubic lines and the cubic content of the box almost 40 cubic inches = 69120 cubic lines, or in other words the former approximately $^1/_{100000}$ of the latter, we can easily understand that such an expansion, which does not constitute as much as one drop of water in a full quart measure, could be produced in the glass box by the pressure that we are dealing with here. It is true that part of this cal-

---

[7] [1 *lod* equals $^1/_{32}$ Danish pound.]

culation is based on measurements of a drawing which may not be accurate in its dimensions, but even if we have erred in this, we could still say with certainty that the glass box was large because, according to Herbert's specification, the glass sphere held almost 3 pounds of mercury. It is a pity that Herbert used different tubes in his various experiments; otherwise it might have been possible to calculate a systematic error in his experiments and thereby make a more accurate comparison between the work of this meticulous and perspicacious natural scientist and that in which the expansion of the vessel has been avoided.

The book in which Zimmermann describes Abich's experiments on the compression of water has gained a reputation which it would hardly have won but perhaps better merited if Abich himself had described his experiments briefly and modestly. What the book owes to Z. is an ostentatious foreword, which leads the reader to believe that the investigation communicated here was a triumph for the name of Germany, a fairly complete historical compilation of older investigations on this subject, in which, however, Herbert is ignored, and finally a number of confusing and partly erroneous calculations related to Abich's experiments. In short, Zimmermann's book seems to owe its excessive reputation to the art of scribbling, to whose products the carefully conceived work of the true scholar has long been considered inferior by public opinion. Abich's apparatus for the compression of water was a very strong brass tube, in which a piston could be driven down with great force. Z. states the measurements of the parts of this instrument and also how many cubic inches of water it can hold, but an easy calculation, in which the proportions of the drawing are also helpful, shows that these two specifications do not agree at all. If a description is made on the basis of the scale which accompanies the drawing, this again gives other sizes. This makes any calculation of these experiments dubious. All the same, it is still possible to deduce from this that the compression of liquids is proportional to the compressing forces. In that respect, the experiments are as correct as experiments of this character could be, not including those experiments in which the piston could not rebound to its initial position as soon as the compressing force ceased, for in these cases the piston itself has probably undergone a change in the size of its cross section. It is true that the metal must also have expanded in these experiments, but as its expansion is proportional to the applied forces, the compression of the water must have the same proportionality so that the combined effect of both should follow the same law. It is true that Zimmermann has not concluded this, but, on the contrary, that no rule for compressional relations can be found, but this is due to a miscalculation which he would hardly have made if he had contented himself with simple arithmetic and not wanted to give his work a mathematical form, which is superfluous here. I would not have made any remarks here about all these errors if Zimmermann's misunderstanding had not gone unchallenged from one writer to another for almost half a century. A similar miscalculation has also led him to assert that boiled water is more compressible than unboiled, which, too, has often been repeated without being challenged. Besides, it is hardly necessary to note that the experiments themselves deserve all the more praise as criticism shows that they are better than their publicizer has described them in his mistaken representation. However, they are not satisfactory because the metal cylinder must undoubtedly have expanded while the water in it was being compressed. Nevertheless, Zimmermann conducted an experiment which was intended to prove that the metal cylinder did not expand during the experiment. This cylinder was surrounded by another which contained water and was completely closed so that the water which would be displaced by the possible expansion of the enclosed cylinder had to move into a narrow glass tube, where it would be quite noticeable. He maintains that the water in this tube rose only during the first moments but then fell to its initial position. He does not determine the cause of this rise and fall, but as he had to have the machine held in a vertical position by 3 people, there is no doubt that the outer metal casing received more heat than the water in it, which could cause both that the water rose a little in the glass tube, and that it fell again. The thickness of the sides cannot be expected to prevent the expansion of the vessel completely, for though it is true that this expansion has a certain dependence on the thickness, it does not disappear. To this must be added that the compression of thick masses of metal themselves is not so insignificant that it can be ignored completely. All experiments on the compression of water in which it is impossible either to prevent the expansion of the vessel completely or to determine it quite accurately must be considered inadequate.

Since 1817, when I presented my first observations and experiments on the compression of water to the Royal Society of Sciences, I have occupied myself with this matter from time to time. Already then I demonstrated that the opinion, propagated by Zimmermann's book, that the compression of water is not in proportion to the compressing forces, is false, and I announced new experiments which confirmed that law, but only in the same way as Abich's experiments when interpreted correctly, that is, in so far as the combined consequence of the compression of the water and the expansion of the vessel conformed to the compressing forces. The device I used then had several advantages over Abich's. The compressing forces in it were measured by the compression of a quantity of air. The pressure was applied in a narrow cylinder which was connected to a much wider one, whereby it was possible to do much with a small force, according to the law that the pressure which is exerted on a small part of the surface of a confined liquid acts on it as if a similar force were applied to every equally large part of its surface. To this was added that a small compression of the entire mass must be indicated by a considerable descent of the piston in the narrow tube. However, it still remained a principal defect that the vessel in which the compression took place could not be prevented from expanding. I therefore returned to the fundamental principle which Canton had followed, and which had too long been abandoned: that the instrument in which the water is compressed ought to receive equal pressure from within and from without. However, as the compression of air is accompanied by so much heating and its rarefaction by so much cooling, I substituted water pressure for air pressure. I then devised the instrument depicted in figs. 5 and 6, which I showed to the Society in 1822. The water which is to be compressed is in a bottle whose neck is a long, narrow tube which ends in a small funnel at the top. Not only the bottle but also most of the tube is filled with water. On top of the water there is mercury, which is first poured into the funnel and then made to descend in the tube by first warming the bottle slightly, whereby a drop of water works its way into the funnel and rises

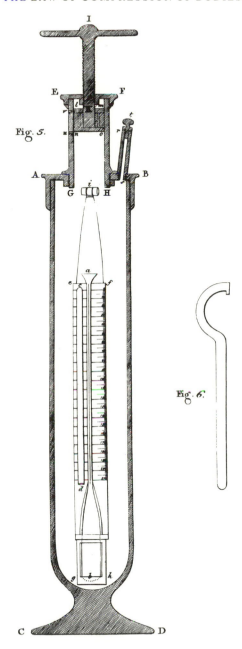

Fig. 5.

Fig. 6.

above the mercury, and then leaving it to cool, whereby the water recedes, and the mercury follows, but it cannot sink to the bottom because such a small drop of mercury, which offers proportionally far greater resistance to a change in its shape than a bigger mass, does not permit the water which it should then displace to move past it. The bottle thus equipped is then placed in a strong glass cylinder which at the top has a brass lid, which supports a pumping cylinder with a piston. Once the entire cylinder and the pumping cylinder have been filled with water, pressure is exerted, by means of the piston, on this water which communicates the pressure both to the outside of the bottle and to the mercury which forms the boundary between the water in the bottle and the cylinder. From the mercury the pressure is communicated to the water which is compressed and again communicates the pressure to the inside of the bottle; therefore, its side walls, under equal pressure from inside and from outside, cannot change their position. In some sense, the mercury in the narrow tube here serves both as piston and as indicator, and the considerable narrowness of the tube relative to the bottle makes any small decrease in the volume of the enclosed water noticeable through a significant displacement of the mercury drop. The details of the above-mentioned device can be seen in figs. 5 and 6. *ABCD* represents the vertical section of a glass cylinder, equipped with a brass cap at the top. At *GH* it has an opening, into which a brass cylinder is screwed, whose upper part has a screwed-in lid *EF*, with an opening in the middle for the screw *IK*, by which the piston *lmno* can be moved up and down. In the glass cylinder is the bottle *ab*, fastened to a brass plate *efgh*, which has been cut out at the bottom in order to give room for the bottle. The tube *cd* is closed at the top but open at the bottom. On the upper part of the plate there are graduations which are used both for the bottle neck and for the tube. At the beginning of an experiment with this device, the bottle, with its equipment, is lowered through the opening *GH* with the aid of a pair of wires which ascend from the upper edge of the brass plate and are joined by a piece of cork *i*. The tube is now filled with water, in which the piece of cork will then rise and serve as a buoy for the immersed apparatus so that it can be lifted up easily. Now the pumping cylinder *EFGH* is screwed on. An iron wrench, depicted in fig. 6, serves to screw it in quite tightly; it is used to grip the tube, letting its tooth take hold of the opening *v*. The empty part of the pumping cylinder is then filled through the tube *rs*, and the air, which would otherwise have resisted the water, finds an outlet at *u*. The tube *rs* is finally closed with the screw *t*. As soon as the piston is moved down, it closes the opening *u* and exerts pressure on the water in the cylinder. From here the pressure is communicated to the water in the funnel *a* and continues through the drop of mercury in the upper part of the bottle neck down to the inside of the bottle and the water contained in it. The same communication of the pressure from the volume of water in the cylinder also affects the air in the tube *cd*. While the mercury drop falls in the bottle neck and indicates the compression of the water, the water rises in the air-filled tube and indicates, through the decrease of the air-filled space, how big the compressing force is.

In order to determine accurately the volume of the bottle as well as that of the neck, I weighed the mercury which could be contained in it. As an example, I choose here one among several bottles which I have used in my experiments. The bottle held 709.48 grammes of mercury, while the tube which constituted the neck held 96 milligrammes of mercury corresponding to a height of 24.6 lines, so that one line constitutes 55 ten millionths of the total volume. The line on the brass plate has been subdivided into quarters, which the naked eye can easily subdivide into halves or even quarters so that it is possible to determine the size of the compression fairly accurately down to millionths, perhaps even half of these. However, this instrument still has many small imperfections which could be removed by future work so that I do not consider it impossible that there could be errors of a few millionths in the results of the experiments conducted so far, but I should say that the experiments that I have conducted after having achieved some proficiency in the use of the instrument have given very nearly the same average, that is, 45 millionths for a pressure equal to that of the atmosphere when it supports 28 inches of mercury at a temperature of 15° (centigrade). A deviation of two or three millionths above or below this average has been rare. I am almost amazed at having obtained such great consistency with an instrument which still leaves something to be desired in several respects. I have repeated these experiments hundreds of times, first to convince myself and later to show them to scholars and other friends of science, both here and abroad. Therefore I do not report any series of the numbers obtained in the experiments, but rather the precautions I have taken as well as a few improvements which I shall add when I have the opportunity to repeat these experiments at greater length.

For all my main experiments the water in the bottle had been boiled. In order to do this, a little distilled water was first brought to the boil in the bottle so that steam rushed out of the narrow neck, like an æolipyle, whereupon the bottle was lowered into a high cylinder with boiling distilled water which had been kept on the boil for two hours. As soon as this water was allowed to cool a few degrees, the vapours in the bottle condensed so much that the water rushed into it and left only a tiny bubble behind, which was easily removed by renewed heating and cooling. However, later when the water had had the opportunity to absorb air, I did not find the compression perceptibly changed, which agrees with Canton's experience.

Naturally, in order to determine the pressure that the water experiences before the piston is set in motion, it is necessary to add the pressure which is produced by the drop of mercury in the bottle neck, the pressure which is produced by the water above it, and finally the atmospheric pressure. Later when contact between the water and the atmosphere is blocked completely, it exerts as much pressure on the blocking parts as it did before on the air and thus receives as much counterpressure. In the same way the air which serves to measure the force is compressed from the beginning, not only by the atmosphere but also by a column of water, whose height is measured from the mouth of the air-filled tube to the surface of the water. The whole instrument must have had time to assume the temperature of the atmosphere before the experiments are begun. The water in the bottle itself provides a fine measure of the temperature. At 15° (centigrade), 1° made the water rise 27 lines. As the measure is divided into quarter lines, $1/108$ of a degree could certainly not escape the attention of the observer, and even $1/216$ and $1/432$ should be perceptible. It is practically impossible to avoid communicating a little heat to the instrument during an experiment. Therefore, as soon as the intended pressure has been exerted on the water and the observation made, the piston should be returned to its original position immediately. Normally the water rises a little, so the average between the point where the water was before the experiment and at the end of it should be taken. When I conducted my experiments with celerity, the above-mentioned rise amounted to only $1/8$ of a line, though sometimes also to $1/4$ of a line.

In the former case the increase of the heat has been only $^1/_{216}$ degree, in the latter $^1/_{108}$ (both centigrade). At the beginning this effect of heat, which cannot be observed with an ordinary thermometer, almost induced me to assume that the compression of water did not grow uniformly with every new increase of the pressure but followed a decreasing progression. I conducted the experiment in such a way that after having compressed the water with a certain force and recorded the observation, I added a new force, and so on, until I had a series of experiments which seemed sufficient to me. During the time when such an experiment took place, the heat rose constantly, and though it was hardly perceptible on an ordinary thermometer, it was still enough, as the experiment progressed, to cause the water to assume a larger volume than it should at the given pressure. In order to examine the effect of a higher pressure, I now produce it as quickly as possible and return the piston to its former position immediately after the observation. I then find that *the compression is proportional to the compressing forces.* My experiments go up to a pressure 5 times that of the atmosphere. However, after the improvements I have recently made on the apparatus, I see that it will be possible to extend experiments with it up to a force of 10 atmospheres. As filling and emptying such a bottle, whose neck is a capillary tube, is very laborious, and as it would be desirable if the compression of several liquids could be examined in the same vessel, it would be an advantage to have a bottle whose neck could easily be taken off and put on. I think this could be accomplished by taking a bottle in whose neck a ground glass stopper could be inserted, to which the glass tube that would serve as a neck was fastened. Of course, it would be a good idea to choose a bottle of as thin glass as possible in order not to allow its compression to have much influence; so far I have always used a bottle which had been blown from a glass tube. In future experiments I would also isolate the air-filled tube from its connection with the bottle, make it longer, seal it with mercury, and equip its opening with a metal ring.

In order to investigate whether heat is generated by the compression of water, I placed a Breguet metal thermometer in the liquid. This consists of a suspended spiral, made of strips of platina and silver soldered together, with a pointer at its lower end. When the spiral is heated, the silver expands more than the platina and changes the curve, whereby the position of the pointer is also changed. It is easy to observe $^1/_{10}$ of a degree on it. This showed no change at a pressure of 5 atmospheres. However, it can hardly be doubted that some heat is generated at this pressure, but then it must be very slight. Some might think that the water in the bottle, which makes such small changes in temperature perceptible, would be the best thermometer for these experiments, but as the water, when re-expanding after the pressure is removed, probably loses as much heat as is gained in its compression, this idea naturally has to be abandoned.

The agreement between the results of the experiments on the compression of water described here and those by Canton is truly extraordinary. The English physicist found a compression of 44 millionths for a pressure of one atmosphere at 64° Fahrenheit = $15^1/_2°$ centigrade,[8] and my experiments gave 45 millionths at the

[8][64° Fahrenheit = $17^7/_9°$ C.]

same temperature. I thought that the experiments reported by the ingenious Jacob Perkins, in which he applied very great forces, also agreed with my experiments. In *Philosophical Transactions* of 1821 he reported some experiments in which water was said to lose one percent at a pressure of 100 atmospheres, but a calculation using the figures provided by him gave only 0.0048, which gives 0.000048 when divided by 100, which is very close to Canton's and my results. Dr. Roget called attention to this miscalculation in Thomson's *Annals of Nat. Phil.*, 1821, p. 131, but in the same periodical, pp. 222 and 223, Perkins declares that the error was in the measurements given for the apparatus, which belonged to another instrument. He then gives new ones which fit his calculation. As it is not likely that a force 100 times larger will produce over 200 times more compression, I must assume that some defect in his apparatus, which is not so easily understood as mine, has caused this. In my previous short reports, I had overlooked the above-mentioned declaration, for which reason I described his experiments as agreeing with mine. It is clear, then, that a new series of experiments with large compressing forces is still demanded by science.

General Remarks on Compression

We have seen that the compression of gases is proportional to the compressing forces no matter how large the pressure is, provided only that the gas does not become a liquid. As vapours are not different from gases, except in that they are at a temperature at which a small pressure or a slight cooling may change them into liquids, there is no doubt that the law of the compression of gases also applies to these, provided that they are sufficiently far from the boiling point of their parent liquid that the applied pressure does not transform them into liquids. If other liquids as well as water may be assumed to follow the same law, the result is a great series of compressions, from the most rarefied air to the most dense liquid, all subject to the same law. However, the compression of solids also obeys this law as long as it does not go so far that the body experiences a change that does not cease with the effect of the compressing force. If we then imagine the same body in all three states, the gaseous, the liquid, and the solid, we see that the compression of all of them is proportional to the compressing forces, and that there is only a break in this regularity at the point of transition from one of these states to another. It is true that we do not yet have any very certain experiments which show that liquids other than water follow this law of compression, but it is so natural that there is no great risk in assuming it for all liquids in so far as the pressure does not cause them to undergo any change in the internal arrangement of their constituents. However, I hope to find an opportunity to fill this gap with some experiments. Everywhere, the things we have found almost always encourage us to look for more. How do liquids behave at different temperatures? Such experiments ought to be conducted in an environment which has the temperature of the liquid; otherwise changes in temperature, which produce greater changes than a pressure of several atmospheres, will cause confusion. Pure ethers as well as anhydrous alcohol would be very suitable for this

because the ease with which they boil permits us to examine them without artificial heating, at temperatures which either coincide with their boiling point, e.g.,
chlorethyl, or are not far from it. A survey of a great number of such experiments
would probably teach us something about the relationship between the compressibility of bodies and the rest of their nature. Concerning vapours which are in contact with a portion of their parent liquid, experiments on the compression of gases
would give us more, and more certain, determinations than the ones we have now.
Furthermore, this kind of experiment has not yet been used as it should be for the
generation of high degrees of artificial cold. Neither have compression experiments been used with any perseverance for chemical investigations.

[Among H. C. Ørsted's posthumous papers (Parcel no. 30 in the Royal Library) is a series of detailed notes on experimental results from the summer of 1824, on which the
previous paper is based; they follow below]:

Made the following experiments with the mercury apparatus on Friday, June 4.
    Barometric height 340‴.66 = 768.3 mm. Temperature 25° C. The mercury in the
comp. tube at 187‴.5, in the apparatus 188‴, and in the pressure tube 188‴. The water
in the apparatus had the same temperature as the atmosphere. The following experiments were made.

Height of mercury			Ratio of	
Compression tube		Pressure tube		
Degrees	mm	mm	Air compression	Mercury pressure
184	42.5	63.65	1.019	1.027
155	104	262	1.207	1.25ʋ
119	181	600	1.567	1.545
97.25	227	898	1.916	1.874
88	247.5	1070	2.115	2.069
78	270	1291	2.383	2.333
67.25	291.2	1582	2.770	2.670
60	307.2	1846	3.102	3.009
52.75	322	2186	3.524	3.433
46.5	335	2535	3.993	3.872
43	342.5	2776	4.320	4.176

The compression tube has been accurately weighed with mercury, and then the ratio
has been calculated, but as some water occasionally got out of the apparatus during
these experiments, they are not considered completely reliable.

    Tuesday, June 15. Barometric height = 335‴.92, the thermometer at 18¼° C. Af-
ter the apparatus had been made as tight as possible, the following experiments were

conducted with atmospheric air, dried by means of chloride of lime, whose height in the compression tube was 204‴.2 Danish measure = 197‴.613 French. Its volume = 1054.8 grammes of mercury.

Dry air compressed to		Difference in heights of mercury	Ratio of	
Dan. lines	Contents		Air comp.	Mercury press.
204.2	1054.8	0	1	1
184.66	954.55	35.‴3 French	1.1052	1.1051
174.05	903.15	56.89 —	1.1676	1.1693
160	828.4	90.93 —	1.2736	1.2706
138	715.36	157.70 —	1.4744	1.4694
128	664.55	195.43 —	1.587	1.581
112	581.95	270.94 —	1.812	1.806
96	499.3	358.73 —	2.112	2.078
80	417.05	510.82 —	2.529	2.520
64	332.95	721.16 —	3.168	3.147
56	291.72	873.09 —	3.616	3.599
48	250.55	1069.68 —	4.209	4.185
40	208.55	1346.94 —	5.057	5.010
36	188.25	1535.73 —	5.603	5.572
32	167.75	1776.86 —	6.288	6.287
28	147	2042.99 —	7.175	7.082
25	131.35	2356.09 —	8.030	8.014

This experiment seems to confirm the validity of Mariotte's Law: that the expansive power of air increases at the same rate as the compression. The following calculation will show that the small differences which are found in these experiments result solely from the impossibility of determining with mathematical precision the real height of the mercury, which retains its full convex surface in spite of a compression of the air of 8 atmospheres.

Air comp.	Diff. of mercury height		Discrepancy	Error in reading	Comp. of air should be
	Found	Calculated			
204.2	0	0	0	0	204.2
184.66	35.3	35.943	+0.643	+0.32	184.94
174.5	56.891	56.39	−0.5	−0.24	174.26
160	90.93	91.79	+0.86	+0.35	160.35
138	157.70	159.37	+1.67	+0.54	138.54
128	195.43	197.23	+1.8	+0.47	128.47
112	270.94	272.91	+1.97	+0.415	112.415
96	358.73	363.54	+4.81	+0.85	96.85
80	510.82	503.65	−7.17	−0.82	79.18
64	721.16	728.23	+7.07	+0.5	64.5
56	873.09	878.75	+5.66	+0.3	56.3
48	1069.68	1078.24	+8.56	+0.34	48.34
40	1346.94	1363	+16	+0.45	40.45
36	1535.73	1546.16	+10.53	+0.23	36.23
32	1776.86	1776.54	−0.32	−0.005	31.995
28	2042.99	2074.37	+31.38	+0.42	28.42
25	2356.09	2361.55	+5.46	+0.056	25.056

In all the experiments the mercury bubble was allowed to rise a little above the line in order to divide the volume equally. At 96′′′ and 80′′′ a major error in observation must surely have occurred.

Therefore, Mariotte's Law has been confirmed by the preceding experiments. In order to investigate whether it also applies to other gases and for how long, experiments were conducted with the water compresssion apparatus to compress sulphurous acid, and the following was found:

Experiments with completely dry and pure sulphurous acid:

Diff. of mer. hgt.	Sulphurous acid		Atmospheric air		Ratio of press.	
	Degs. in tube	Contents	Degs. in tube	Contents	Sulph. acid	Atmos. air
—	101	222.016	101.33	227.334	1	1
0	69.5	153.52	72	160.8	1.4461	1.4137
14.6	60	132.92	61.33	136.834	1.6703	1.6614
29.5	50	110.64	50.66	113.147	2.0066	2.0046
37.5	45	99.1	45.5	100.943	2.2403	2.2521
43.3	40	87.987	40.5	89.784	2.5233	2.5320
48.1	35	77.01	35.33	78.698	2.8829	2.8885[a]
49.7	34	74.815	34.5	76.91	2.9675	2.9558
55.3	30	66.02	32	71.547	3.3628	3.1774
62.7	25	55.02	31.125	69.451	4.0351	3.2733
66.4	22.5	49.52	31	69.163	4.4883	3.2869
70.1	20	44.075	30.775	68.874	5.0372	3.3018
75.5	17.5	38.64	30.666	68.394	5.7457	3.3238
77.7	15	33.21	30.333	67.626	6.6852	3.3616
81.6	12.5	27.779	30.125	67.146	7.9922	3.3856
85.2	10	22.29	29.333	65.322	9.9603	3.4802
88.2	8	17.83	29.125	64.841	12.4518	3.5060
91.1	6	13.37	29	64.553	16.6055	3.5216
94.2	4	8.91	28.775	64.265	24.9176	3.5374
95.7	3	6.68	28.5	63.4	33.2359	3.5857
—	—	—	20	44.075	—	5.1597[b]

[a] At this pressure sulphurous acid begins to condense visibly.
[b] At this pressure there was neither air nor empty space in the tube.

In the following experiments I used sulphurous acid which was not completely pure, and I could also clearly observe that some atmospheric air mixed with the sulphurous acid during compression. The results of these experiments were the following:

Wednesday, July 14, 1824

Diff. of mer. hgt.	Sulphurous acid		Atmospheric air		Ratio of press.	
	Degs. in tube	Contents	Degs. in tube	Contents	Sulph. acid	Atm. air
—	182	411.54	full	229.429	1	1
0	165	373.62	95.33	209.786	1.1014	1.0946
10	145	328.42	82	180.974	1.2530	1.2677
11.7	130	293.42	73.33	161.971	1.4025	1.4164
13.3	115	259.62	65.5	144.741	1.5851	1.5851
18.3	100	224.62	56.33	124.973	1.8322	1.8358
20.4	97	217.62	55.125	121.154	1.8911	1.8936
16.3	91	204.19	51.666	114.496	2.0154	2.0381
19.6	86.5	194.12	49	108.332	2.1200	2.1178
27	76	170.02	43	93.643	2.4205	2.4590
34.2	65.5	145.82	37.125	81.675	2.8222	2.8096
41	54	121.02	30.5	67.120	3.4006	3.4181
—	43.5	96.22	27	59.42	4.2770	3.8611[a]
53.7	37	82.27	26	57.22	5.0023	4.0095
69.8	16.33	36.126	9	20.06	11.3915	11.437

[a] At this pressure sulphurous acid began to condense visibly. —At reduced pressure a strong white smoke was generated which gradually disappeared.

With the mercury apparatus, the following experiments were made with dry sulphurous acid:

Sulphurous acid		Difference of mercury height	Ratio of pressures	
Degrees	Contents	French Lines	Sulph. acid	Mercury
158	768.81	327.807	1	1[a]
155	755.23	334.081	1.0179	1.0206
154	750.74	336.242	1.0240	1.0288
140	683.56	368.435	1.1247	1.1256
125	610.88	412.629	1.2585	1.2606
107	524.38	478.596	1.4661	1.4622
90	443.13	560.371	1.7349	1.7121
80	395.33	618.113	1.9447	1.8885[b]
130.5	637.2	395.613	1.2065	1.1995
107	524.38	482.225	1.4661	1.4621
94	462.248	538.355	1.6632	1.6323
80	395.33	629.322	1.9447	1.9081
62	307.23	827.241	2.5024	2.5082
49.5	245.43	990.774	3.1325	3.0041
40	199.61	1108.355	3.8515	3.3606[c]
8	39.922	1077.742	19.2577	3.2677
1.25	5.09	1348.113	151.0432	4.0876

[a] Barometric height 336'''.5; Temperature in the air and in the apparatus = $19^{1}/_{2}°$ C on 14 July 1824.

[b] Barometric height 339''', Temperature = $20^{1}/_{2}°$ C on 17 July 1824.

[c] The sulphurous acid began to condense visibly and continued to condense at the same mercury height up to 8'''. In the space of $1^{1}/_{4}'''$ there was, besides the liquid sulphurous acid, a small quantity of atmospheric air which in volume equalled 1.4 grammes of mercury, so the liquid sulphurous acid filled as much space as 3.69 grammes of mercury and was thus 208.3496 times smaller than in its expanded fluid state.

56

[Improvements of the Compression
Apparatus 1826.
Measurement of the Compressibility
of Mercury][1]

At the end of the "Contribution to the Determination of the Law of the Compression of Bodies," which is to be found in the last volume of *Det Kongelige Videnskabernes Selskabs Skrifter*,[2] and in which I have presented the finest of the experiments I have conducted on this subject for several years, I called attention to the many points which remained to be elucidated. Later, at the meeting on May 12th of last year, I had the honour of further developing the nature of these needs, which I had only briefly touched on the last time. With gratitude I acknowledge the confidence which the Society showed me in its decision to pay for the proposed experiments with its own funds. I am far from having completed the work which has thus become a duty for me, but as a considerable number of experiments have already been performed, and as they have decided several important issues, I thought that I ought not withhold their announcement, the less so as this would give me an opportunity to receive observations which could help to complete the rest of the work, first from this scholarly Society and later from the learned world in general.

While conducting my numerous experiments on the compression of liquids, I had occasion to make even more improvements of the apparatus which I had previously used. Fig. 1[3] shows this improved apparatus. As in the earlier device, *ABCD* shows the vertical section of a glass cylinder, equipped at the top with a metal cap, the base of which can, of course, be either a continuation of the glass or a metal base. As before, *GH* is an opening in the metal cap to which a pumping cylinder is screwed, with a screwed lid *EF* at the top, through the middle of which is a screw

---

[1][Communication to the Royal Society of Sciences, January 5, 1827. After a manuscript found among H. C. Ørsted's posthumous papers, parcel no. 30 in the Royal Library. KM II, pp. 325–35. Originally published in Danish.]

[2][Chapter 55, this volume.]

[3][This figure does not exist. Ø. describes a slightly changed version of figs. 5 and 6, p. 507 in this edition.]

IK, whereby a piston *lmno* can be moved up and down. As before, the opening *v* still serves to accommodate the tooth of the wrench, fig. 2. Finally, as before, the function of the tube *rs* is to let water in and out of the pumping cylinder, and at *r* it has an opening which is kept shut by a screwed metal stopper during compression. On the other hand, *tuz* represents a subsidiary device which was added later, viz., a siphon fastened to the tube *rs* by means of a pierced stopper, the other end of which is placed in an open jar, and which serves to suck water into the pumping cylinder or expel water from it. When, for our experiments, the pumping cylinder is to be screwed into the glass cylinder, which must first be filled with water, the piston must be moved to the lower part of the pumping cylinder. When it has been fastened to its place, and the siphon to its, and the mouth *z* has finally been placed in a jar of water, the pumping cylinder can be filled with water by raising the piston. When the pumping cylinder is to be unscrewed, the siphon can be attached, the piston driven down, and the water expelled. Before the addition of the siphon, it was not possible to unscrew the pumping cylinder without spilling water.

The bottle *bbb* with the tube *aa*, which is lowered into the cylinder, and in which the compression is measured, consists, now as before, of a wider part and a capillary tube, but instead of the latter being fused to the wider part, as it used to be, I have found it useful in most experiments, and even necessary in some, to allow the capillary tube to be movable. Therefore I have the capillary tube ground to fit the neck of the bottle *bbb*. It is possible to do this when the glass tube has a fairly considerable thickness, and when the neck of the bottle is narrow. Where a wider neck is needed, as in the compression of solid bodies, I have had a capillary tube fused on to a piece of a very thick glass tube and had the resulting stopper ground into shape. The bottle has always been blown for the purpose from a wide glass tube. The upper part of the capillary tube has also been adjusted by grinding to fit a funnel, which is attached when the bottle and the tube have been filled with a liquid whose temperature, when poured, is noticeably different from that of the water in the glass cylinder. When the bottle has been duly filled to the top of the neck, and the tube with its funnel is then attached, the liquid will not only fill the tube but will also rise a little in the funnel. When the bottle and most of the tube are now lowered into the water in the cylinder, and this is colder than the liquid in the bottle, which most often has received heat from being touched, the liquid risen in the funnel serves to fill the space which would otherwise have been empty through the contraction which accompanies cooling. When the experiments are made with ether, a cooling of a few degrees may require the addition of more ether.

In my earlier experiments, the capillary tube had a funnel fused to the top, in which there was mercury whose descent served to show how far down the water had been forced during compression. As this procedure left some doubt about the extent of the effect of the mercury, I have now omitted this and found that it is quite possible to see all that goes on if there is only air above the water in the tube. However, now it became necessary to find a new way of preventing the water in the cylinder from entering the capillary tube. This is very easily done by placing over the tube an inverted funnel *ppp*, whose narrow end is closed, and whose rim is fitted with weights adequate to prevent the water from lifting it. When the funnel is sufficiently narrow at the top, a pressure of many atmospheres can be applied without the water rising to the mouth of the capillary.

At the capillary tube, as in the older apparatus, there is a scale *efgh*, which is preferably divided into quarter lines, but in the older one there was, next to that scale, an air-filled tube which was narrow and short. I have now moved this air-filled tube and made it much longer. The graduations have been ground into the tube, and it has a weight at the bottom. However, in many of the experiments I have used a smaller tube, like the one shown in the figure. Its divisions were very imperfect, but I have calibrated them by weights of mercury. Where this tube has been used, fractions in the number of atmospheres will always be found. In accordance with its intended function, I call this air-filled tube the dynamometer. Both this tube and the bottle are fastened with strings to small pieces of cork *ii*, which serve as buoys for them, and by means of which they can easily be pulled up.

The apparatus thus improved makes the experiments easier and somewhat more accurate, but the results of the experiments on water that I have conducted with it do not differ essentially from those I obtained with the old device, which was as expected. One might fear that the liquid in the bottle would come into contact with the water outside through the interstices which always exist between the irregularities of the ground stopper and the bottle neck, but it was to be expected that these would not permit such contact. The experiments themselves demonstrate beyond doubt that, after the release of the pressure, the liquid always returns to the same place as before the compression in so far as a change in temperature has had no effect on it.

Before every experiment, or when the temperature changes perceptibly during an experiment, the tube *cd* is lifted out of the water so that the air contained in it can be balanced with the surrounding air, but it should be left above the water only for a few moments so that the air in it can keep the temperature of the water, which is not always that of the atmosphere though I always see to it that they are very close to each other. In the calculation, it must be recalled that the air in the dynamometer always experiences some pressure from the water. The premise being that it is full of air which has the same elasticity as the atmosphere, a correction for the pressure of the water is required here. Such a correction is not required for the liquid being tested because it is not observed until the bottle has been lowered into the water of the cylinder. So, when the air in the dynamometer has been compressed to half its original volume, this is not due to the pressure of the piston alone, but to piston pressure + water pressure; a pressure of 2 atmospheres (or, in other words, the pressure of 1 atmosphere added to the one that was there before) has not been applied, rather a pressure of 1 atmosphere—the water pressure has been added. Now this has to be expressed in the height of a mercury column and subtracted from the mercury height of the barometer. It is obvious that the water pressure is relatively less important, the more atmospheres of pressure are exerted. It is also obvious that the mercury in the barometer is always recorded as the value obtained when all the corrections, among others the one for heat, have been made. When I speak about 1 atmosphere, I always assume it to be 336 French lines even though the mean pressure is a little higher but not definitely determined. The thermometer I have used is in degrees centigrade. Of course, the temperature in the bottle must be, as closely as possible, the same as in the glass cylinder.

THE COMPRESSION OF MERCURY

The knowledge of this is important in evaluating many other experiments, for which reason its investigation is of primary importance. It will be seen that it presents several difficulties, and that the compressibility of mercury is so small that any error in its determination will have a large effect on the whole magnitude that is to be determined, but it will also be seen that just because of its smallness, it is sufficient for our purposes to determine it with only moderate accuracy.

For the compression of mercury, it is impossible simply to use a bottle with a ground stopper since mercury, as a consequence of capillarity, is prevented from completely filling the small wedge-shaped space between the lower rim of the stopper and the neck; however, pressure will drive it into this space and destroy all agreement between the results of the experiments. The capillary, then, must be fused to the bottle, but the rest of the device can be like that just described. The bottle used could hold 1043.25 French grammes of mercury at a temperature of 24° C. The tube was carefully calibrated, and a length of 50 lines in it held 0.275 grammes of mercury, thus $\frac{1}{4}$ line, which is one division on the scale, 0.0000013175 of the entire volume. The apparent dilatation of mercury in a glass vessel is $\frac{1}{6400}$ or 0.00015626 for each degree, which gives 118 divisions for each atmosphere.[4] This shows how much caution must be exercised with regard to changes in temperature, which here have the effect of a pressure of more than 100 atmospheres.

Mercury was poured into the bottle under the air pump, and great care was taken that no air bubbles remained in it. Boiling might have been preferable but was found to be accompanied by many difficulties since such a large volume would have to exit through such a narrow tube. In any event, we could be convinced that the possible effect of not boiling would be to show the compression to be larger than it is, but it will be seen below that we can be certain of several conclusions when we know that we have not underestimated it.

When the bottle in question, filled with mercury to about 10 lines from the upper mouth of the capillary, was placed in the cylinder, which had already been filled with water, it was found that the mercury rose and fell almost incessantly, which was to be expected as it is very difficult to obtain a perfect equilibrium of the temperature where not even a change of $\frac{1}{100}$ of a degree per minute takes place. The experiments were conducted on June 29 and 30 at a little over 24° C. They were made when the mercury did not rise or fall more than 1 division per minute, but often such an approximation to equilibrium was obtained that there was only a change of $\frac{1}{2}$ division per minute, indeed, even less. As the experiment normally lasts $1\frac{1}{4}$ minutes, the result was often that the mercury was only $\frac{1}{2}$ division higher or lower when the piston was raised again. It was assumed then that, at the moment

[4][degree.]

when the compression was produced, it was $^1/_4$ above or below the starting point. When the mercury was falling before the experiments, the water in the cylinder and later the mercury in the bottle received heat through the proximity of the observers and their handling of the instrument. Sometimes a point of rest could be achieved for the heat so that, after the experiment, the mercury returned to the very place where it had been before. In these, as well as in most of the following experiments, I observed the position of the liquid substance in the capillary tube of the bottle while someone else observed the compression of the air in the long, wide tube.

The first useful experiments were conducted on June 29 and 30, on which two days the air pressure corresponded to a mercury column which would have measured $338^1/_2$ French lines at the freezing point. Smaller fractions of lines are ignored here as superfluous to the intended accuracy.

With a force such that the air in the dynamometer was reduced to a volume 5.736 times smaller, the addition to the atmospheric pressure thus being 4.736 times itself, which is = 1593 lines of mercury, the mercury was lowered 4 divisions each time. As these constitute $4 \times 0.0000013175 = 0.00000527$ of the total, this makes 0.00000111 for 336 lines. Here it is assumed that the compression is proportional to the compressing forces, and the following experiments confirm this assumption, which is quite likely for several reasons.

At a pressure of 3.55 that of the atmosphere on the day, the result was a compression of 3 divisions, which gives 0.00000112 for each atmosphere. At a pressure of 6.62 that of the atmosphere, the compression was $5^1/_2$ divisions, which gives 0.00000110 for each atmosphere. I must, however, point out that the great agreement in the figures here cannot be regarded as a guarantee of perfect accuracy, for an error in the observation, were it only of $^1/_4$ division, or $^1/_{16}$ of a line, would generate very large discrepancies here, so I must admit that this coincidence of the various experiments, which only became perceptible when all the corrections had been made, surprises even me, without therefore inspiring in me greater confidence in these experiments than I ought to have in the others. During the experiments we noticed a hopping motion of the mercury in the capillary tube, which was probably due to the fact that the pressure which the side walls of the bottle was to receive from within could not act so quickly through the extremely narrow tube as the external pressure could act on their opposite side. With those liquids that have strong adhesion to glass, this motion does not appear.

In order to test these results I conducted several other series of experiments.

At a barometric height of 338 French lines and a temperature of 15° C, we filled a bottle which could hold 963.135 grammes of mercury at $14^1/_2$° C with mercury, which constituted 928.545 grammes, the rest was filled up with water, and the stopper with the capillary tube was put on. The water remaining after the stopper had expelled the superfluous water, then, filled the same volume as 34.59 of mercury. Each division of the tube held 0.000002535 of the total. The air-filled tube was not so high here as indicated in fig. 1,[5] but the capacity of the various divisions had

[5][See note 3, above.]

been determined by weights of mercury. In a series of experiments on the compression of water with the very same instrument, to which I shall return later, we had obtained an average of 17.615 divisions on the scale, or 44.65 millionths of the total, for 336 lines of mercury.

The first column of the table below shows the pressures in French lines of mercury, the second, where the liquid was in the capillary tube before compression, the third shows the same after the release of the pressure, the fourth the average of these two positions, the fifth how far the liquid was driven down by the pressure.

Addition to Atmos. Press., Fr. lines mercury	Liquid height in capillary		Mean of these	Depress. of liquid	Amount of depress.
	before	after			
955.04	59	58.5	58.75	55.5	3.25
same	52.25	52.5	52.375	49.75	2.625[a]
1590.06	56	55	55.5	51	4.50
	53	52.5	52.75	48.5	4.25
	52.5	53	52.75	48.75	4.00
	53	53	53	49	4.00[b]
2227.76	55	53.75	54.375	48	6.375
	52.5	52.25	52.375	46	6.375
	52	52.25	52.125	46.25	5.875[c]

[a] Mean of these is 2.94; 336 lines of pressure corresponds to 1.034 divisions.
[b] Mean of these is 4.19; 336 lines of pressure corresponds to 0.885 divisions.
[c] Mean of these is 6.208; 336 lines of pressure corresponds to 0.936 divisions.

Together they give an average of 0.952 divisions for a pressure of 336 lines, which constitutes 0.000002413 of the total, but after other experiments the water would drop 0.00004465 for a mercury height of 336 lines. However, the water present here was $34.39 / 963.135 = 1 / 27.87$ of the total volume of water which the bottle could hold. If this is divided[6] by 0.00004465, the result is 0.0000016, which subtracted from 0.000002413 leaves 0.00000081, that is, only a little over 8 ten-millionths. The result of these experiments, thus, differs by 3 ten-millionths from that of the others.

As all these experiments still did not satisfy me, and as the latter in particular, in spite of all our meticulousness, had the short-coming that the volume of water in them was far too big so that any error which might lie in the quantities deduced from other experiments in order to calculate these might easily have a great effect, I conducted the same experiments with a more convenient apparatus. The bottle

[6] [multiplied.]

had a very narrow neck and could be almost completely filled with mercury without preventing the ground capillary tube from being in contact with the water. It could hold 1087.637 grammes of mercury at 14°. The tube had been very carefully calibrated, and a volume of mercury which measured 75.5 lines or 302 divisions weighed 0.625 grammes. The weighing of the mercury in the bottle was performed, here as in the other experiments, on a large Fortin scale, which the collection of instruments at the University has on loan from Counsellor Manthei, Knight of the Order of Dannebrog, and the weighing of the mercury in the tube on an extremely sensitive scale from Robertson in London. In order to maintain the calibration of the scale, we used platina weights made by Fortin. The numbers given clearly show that each division corresponded to 0.000001902 of the total capacity of the bottle. For the experiment described here, we filled the bottle with 1085.72 grammes of mercury and let it reach a temperature of 14°; then there was only room left for 1.917 mercury, which constitutes only 0.00176 of the total volume. This space was filled with water. The experiment was conducted on a day when the air pressure was 328.7 lines of mercury at 0° C. The temperature of the room was 14°. Each time the air in the dynamometer was compressed to a 5.736 times smaller volume, thus indicating an additional pressure of 4.736 times that of the atmosphere = 1556.723 lines of mercury, from which 11.956 was subtracted for the pressure of the water on the air in the dynamometer. Thus 1544.77 lines remain. We were careful to equilibrate the temperature in the apparatus as well as possible, but every touch of a hand communicated heat which soon penetrated the cylinder and the water to the bottle and the mercury and made this rise several divisions. As one degree of heat in this instrument would make the mercury rise 80 degrees, it is easy to see that complete stasis would be difficult to attain.

| Hgt. of Water in Capillary | | Mean of | Depress. by | Amount of |
before	after	these	pressure	depress.
208	211	209.5	204	5.50
232	235.5	233.75	230	3.75
236	239	237.5	233	4.50
239.5	242	241.25	237	4.25
247	251	249	244.5	4.50
256	255.5	255.75	251	4.75
255.5	257	256.25	252	4.25
265	267	266	262	4.00
240	242.5	241.25	237	4.25
				Sum 39.75

The average of the 9 experiments, then, gives 4.4166 divisions. If we reject the first two experiments, which seem to have been conducted before the temperature had come to true equilibrium, we get a total of 30.50 or an average of 4.36, which differs only slightly from the previous values. These 4.36 divisions constitute 0.000008293 of the total. At a pressure of 1544.77 the water should have been compressed by 207 millionths, but as the water here constituted only 0.00176 of the total, its compression could only amount to 0.000000364. That leaves only 0.000007929 for the compression of the mercury, which for 336 lines gives 0.00000172. This is 6 ten-millionths more than our first result, and 9 ten-millionths more than the second. If we take the average of all three series, we have 0.00000122 or a little over $1^1/_5$ millionth.

We cannot be surprised by the apparently great discrepancy between these experiments when we consider that we are dealing here with ten-millionths of the total. It is to be hoped that the average does not deviate $^1/_3$ of a millionth from the truth, perhaps not even $^1/_4$ of a millionth. This is indeed considerable in relation to the compressibility of mercury, but at least we now have the satisfaction of knowing that the pressure to which mercury is subjected in a barometer or a thermometer cannot generate such a perceptible change in it that this could have an effect on our observations. As the apparent expansion of mercury in a glass vessel is 0.000156 for one degree, but its compression only 0.00000122, or according to the highest results 0.00000173, for each atmosphere, it is easy to see that a pressure of approximately 100 atmospheres is required to change the volume of mercury by the same amount as one degree of heat.

There could hardly be any doubt that the pressure which was exerted on the mercury was not able to prevent the effect of the expansive power of heat, but I thought that I ought to investigate this matter. I therefore maintained the compressing force for 1 or more minutes and found that the mercury in the capillary tube rose as much during the compression as it did just before or after this pressure. Completely accurate experiments are difficult here as the change in temperature during one minute can easily be somewhat different from the next. In one of the experiments I maintained the compression, as indicated by the dynamometer, at the same point for one minute. Meanwhile the liquid in the capillary tube rose from 223 to 226.5, that is, 3.5 divisions, or $^{3.5}/_{80}$ degrees; as soon as the pressure ceased, it rose to 232.6, or 6 divisions. If we assume that the retraction of the piston, together with the observation of the position of the liquid in the tube after the cessation of the pressure, took $^1/_2$ minute, and that the liquid during this time rose $^1/_2 \times 3.5 = 1.725$ divisions, only 4.375 were left of the 6 divisions, which deviates very little from the average of 4.36. In another experiment I had the pressure maintained unchanged for 3 minutes. The liquid in the capillary tube meanwhile rose from 240.5 to 247, that is, 6.5 divisions. At the cessation of the pressure it rose to 252, that is, 5 divisions higher. If we again assume that re-establishment of equilibrium with the atmosphere took $^1/_2$ minute, and the rise meanwhile constituted $^1/_6 \times 6.5$ or almost 1.1, we are left with 3.9, which is indeed farther from the average than the previous result, but still only by 0.47 divisions, which still does not constitute a full 6 thousandths of a degree.

57

On the Relative Compressibilities of different Fluids at High Temperatures[1]

Copenhagen, December 30th 1826

HAVING in the course of last summer performed a very great number of experiments on the compressibility of different fluids, and particularly on the compressibility of water at high pressures, I am now about to calculate the corrections which must be introduced for the variations of atmospherical pressure, temperature, &c. As soon as the paper is finished I will send you a translation of it. The following results, however, will not be much affected by these corrections.

1. As far as the strength of my apparatus has permitted me to push the compression of water, (viz. seventy times that of the atmosphere) the compressibility is in proportion to the compressing powers.

The compression produced by one atmosphere, as already stated by Canton, is about *forty-five millionth* parts of the volume. Mr. Perkins has obtained by a pressure of one hundred atmospheres, a compression equal to 0.01 (one hundredth of the volume) which is much more than could be expected from my experiments. From calculations founded on the results of experiments made with pressures beneath seventy atmospheres, I have obtained only 0.0045 for 100 atmospheres.

In consequence of this great discrepancy between my results, and those of that highly distinguished inventor, I have repeated them with great care, and, from their simplicity, I believe there is not much room to doubt of their accuracy.

2. In so far as I have tried the temperature of compressed water (to forty-eight atmospheres) *no heat is liberated by its compression.*

3. The compressibility of *mercury* is not much greater than *one-millionth* of its volume by one atmosphere.

[1]Communicated in a letter to Dr. Brewster. [*The Edinburgh Journal of Science*, conducted by David Brewster, LL. D., Vol. 6, pp. 201–2. Edinburgh 1827. Also to be found in Schweigger's *Journal für Chemie und Physik*, Vol. 51, pp. 112–14. Halle 1827.—*Bibliothèque universelle*, Vol. 36, pp. 127–29. Geneva 1827.—*Annales de chimie*, Vol. 37, pp. 104–5. Paris 1828. —Poggendorff's *Annalen der Physik*, Vol. 9, pp. 603–4. Leipzig 1827.—*Videnskabernes Selskabs Oversigter*, 1826–27, pp. 12–13. Copenhagen. KM II, pp. 335–36. Originally published in English.]

4.[2] The compressibility of *sulphuric ether* is nearly *thrice* that of *alcohol*; nearly *twice* that of *sulphuret of carbon*, but only *one and a third* that of *water*.

5. The compressibility of *water* containing *salts, alkalies*, or *acids*, is less than that of *pure water*.

6. The compressibility of glass is exceedingly small, and very greatly beneath that of *mercury*.

[2][This passage must have been misunderstood; it should read as follows: The compressibility of sulphuric ether is nearly thrice, that of alcohol nearly twice, that of sulphuret of carbon only one and a third that of water.]

58

An Electromagnetic Method
for the Assay of Silver and Other Metals
Invented by M. Ørsted[1]

THE preference which has always been accorded to certain metals, regarded as more precious than others, depends in particular on the degree of immutability which they exhibit under the influence of air, of water, of fire, or of other chemical forces of nature, but the resistance to these agents is related to the ability of these metals to combine with oxygen. The less the metal attracts this gas, the better it generally preserves its metallic nature under the influence of foreign bodies. It could also be said that the less combustible a metal is, the more it deserves the designation of precious. It is evident that the word *combustible* is used here in the chemical sense, according to which any change of the same nature as the one which fire causes in bodies is designated by this name. It is in this sense of the word that chemists say that a metal undergoes combustion in an acid the moment before it is dissolved. However, the tendency which metals have to dissolve in that way through combustion is again related to their ability to produce an electric current when two of them are brought in contact with a suitable fluid or with themselves, that is, when together they form a chemical circuit. In such cases, the electric current always goes in the direction from the less combustible metal to the other. The combustion also depends on the nature of the fluid. The metal which is more susceptible to combustion in a certain fluid than another is perhaps less so in a fluid of a different nature, the result of which is that one metal cannot be said to be more or less combustible than another in all cases, but only that it is generally so. The direction of the electric current, then, can serve to indicate to us the principal chemical relations between metals, which must lead us to distinguish different kinds of

---

[1] Article communicated by the author. [*Annales de chimie et de physique*, ed. by Messrs Gay-Lussac and Arago, Vol. 39, pp. 274–87. Paris 1828. Also to be found in Schweigger's *Journal für Chemie und Physik*, Vol. 52, pp. 14–26. Halle 1828. —Erdmann's *Journal für teknische and ökonomische Chemie*, Vol. 2, pp. 89–100. Leipzig 1828. And the same contents in: Ursin's *Magasin for Kunstnere og Haandværkere*, February 1828, pp. 441–50. Copenhagen. A report is to be found in *Videnskabernes Selskabs Oversigter*, 1826–27, pp. 13–14. Copenhagen. KM II, pp. 337–46. Originally published in French.]

alloys. It is true that the presence of a weak electric current is usually imperceptible, but it is possible to make it perceptible in a striking manner through its magnetic effect. The advantage which can be drawn from these phenomena has not entirely escaped the attention of physicists, but before M. Ørsted no one had submitted them to experiments or exact observations.

M. Ørsted explained this theory last year in the public lectures established by the Society for the Propagation of Physics in Denmark, but he only made experiments with two metals. Observations now repeated seem to have given the theory a greater certitude. However, in his report in the journal entitled: *Magazin for Kunstnere og Haandværkere*,[2] February, 1828, M. Ørsted expresses the desire that this new method of assaying metals should only be regarded as one with a potential for greater perfection. What has been sufficiently proved is that this method offers great advantages over the usual one. The assay of silver is only part of this electromagnetic theory. It can be applied to all other metals, but each new metal requires new experiments. We shall give here a detailed description of the manner of assaying silver proposed by M. Ørsted.

The electromagnetic multiplier was invented by M. Schweigger, professor at the University of Halle, some months after the discovery of electromagnetism, and since then several physicists have perfected this instrument. The latest improvements that it has received are due to M. Ørsted. It is used to great advantage to discover quickly which of two metals combines most easily with oxygen.

The theory of the instrument is based on the magnetic effect of the electric current. When a metal wire is placed very close to a mobile magnetic needle in such a way that it is parallel to this needle, and an electric current is sent through it, preferably one which accompanies a chemical process, the needle changes direction. If the current, that is, the transition from positive electricity (+) to negative (−), goes from the right hand of the observer to his left, the upper part of the wire, which is also called the conductor, will move the northern end of the needle away from the observer while the lower side of the wire will turn it towards him. In order to understand the mechanism of this instrument, it is sufficient to know, of all the laws of electromagnetism, only this opposite effect which the two opposite sides of the conductor have on the magnetic needle.

Consequently, if a metal wire was placed above the magnetic needle and another below it, and electricity was sent through both of them in the same direction, the effects of these two wires on the needle would be neutralized. On the other hand, if the electricity moved through the metal wires in opposite directions, their forces would combine to drive the same end of the magnetic needle to the same side. Therefore, if a metal wire is placed around a magnetic needle, as shown in fig. 1, and we let the electricity enter at *a*, the electric current follows opposite directions according as it passes above or below the needle, and the effect on the needle is twice that produced by a single wire.

[2] *Journal for Artists and Craftsmen*, ed. by M. Ursin, professor and doctor of philosophy. Published by Gyldendahl in Copenhagen. [Cf. the previous footnote.]

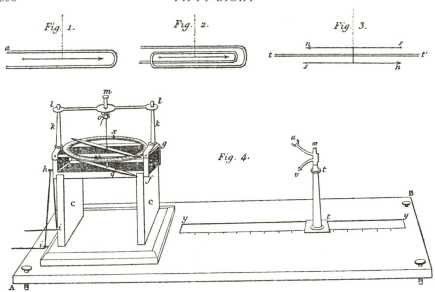

Fig. 1.   *Fig.* 2.   *Fig.* 3.

Fig. 4.

If the metal wire is bent as shown in fig. 2, the effect will be four times greater than that of a single metal wire. The effect keeps increasing in proportion to the number of turns of the wire.

The effect of the electric current could be made even more perceptible if two magnetic needles in opposite directions were connected by a wire of brass or another metal, as shown by *ns* and *sn* in fig. 3, and the metal wire *tt'* was then placed between the needles to serve as conductor for the electricity.

If these needles had the same direction, they would strive to turn to opposite sides, but, having opposite directions, they would both turn to the same side because the conductor is between the two needles, that is, above one and below the other. A further advantage is that the tendency of the needles to point towards the south or the north is more or less neutralized by their reciprocal action. If the needles are equally strong, the neutralization will be perfect, and the weakest external force will be able to make them leave their position. If the needles are unequal, their tendency to maintain their direction is equal to the difference between the respective forces of the needles.

It is evident, then, that the two methods for intensifying the force on the needles, the one which is represented in the second figure and the one shown in the third, can be combined, and this is what has been done in the electromagnetic multiplier which we shall now describe.

AB, fig. 4, is a wooden base with a screw in each corner to orient the instrument horizontally. *C* and *C* are two columns which support the frame *fg*, around which a metal wire, which from now on will be called the *conducting* or *multiplying wire*, is wound several times so that the weakest electric current passing through it can produce a great effect on the magnetic needle on which it must act. Such a conducting wire can easily be 50 to 60 feet long and can be wound more than a hundred times, but the windings must be isolated so that they do not touch each other, which can be achieved by wrapping silk around the metal wire before placing it on the frame.

The height of the frame must be as small as possible so that the turns of the wire are very close to the magnetic needle.

The conducting wire having thus been wound around the frame, each end goes through a small ring. The instrument has two such rings, but the figure only shows one (*h*). The ends of the conductor also go through the rings *i, i*, one of which is hidden by other parts of the instrument. *k, k* are two small columns of ivory or wood which support the cross piece *ll*, through the middle of which a small cylinder *mo*, fitted with a button *m*, can be made to rise or descend. In the middle of the lower part *o*, there is a small opening which is connected with a transverse hole closed by a pin, which can be seen immediately under the ring placed above *o*. One end of a cocoon silk thread passes through the hole *o* and then goes through one of the openings of the transverse hole and is finally fastened to the said pin. It is clear that this arrangement serves to stop the needle when it has been made to oscillate.

The needle is suspended by the cocoon silk thread marked *x*. This needle consists of two connected magnetic needles, as shown in fig. 3. The circle which shows the degrees is made of glass, a material preferable to brass, which is often magnetic. At *q* there is a fork which serves to secure the needle when the instrument has to be moved. Another fork for the same purpose is found on the other side. When the machine is to be used, the needle is first disengaged; though disengaged, it is still at rest until the cylinder *mo* is raised. The ring at *o* stops the cylinder and prevents it from being raised too high. A glass box, which covers the entire frame, protects the needle from draughts; at the top of the glass box there is a hole through which the top of the cylinder *mo* passes; *tt* is a perpendicular column which can slide in a groove *yy*, along the edge of which there is a scale for measuring the distance of the column from the needle. *uv* is a horseshoe magnet with two pins, one of which can be seen at *w*, and the other enters into a hole made in the column. This magnet can be taken out and turned so that it comes to rest on the pin *w*. It serves to augment or to diminish the force with which the needle strives to turn towards the north. This force is augmented when the horseshoe magnet is placed in such a way that each of its poles is opposed to the opposite pole of the needle, and it is weakened when each of the poles of the magnet is placed opposite that of the needle which has the same nature. It is clear that the distance of the horseshoe magnet from the needle has a great influence.

As the needle most often has a tendency to take a certain direction, the instrument is turned until one of its ends can be seen pointing towards zero on the graduated circle or, equivalently, until it makes equal vibrations to both sides of 0. Then the other end of the needle is observed to see whether it is also at rest at 0, or whether it makes equal vibrations to both sides. If that is the case, everything is in order; if not, the middle of the needle, which here means the upper needle, is not exactly above the centre of the circle, and as this is caused by a tilting of the instrument to one side, the defect must be corrected by means of the screws in the base of the instrument.

If the needle has too strong a direction towards one of the poles, the horseshoe magnet must be placed in such a way as to weaken it.

On each side of the circle, there is a peg z at 90°, which prevents the needle from oscillating any farther.

The use of this instrument depends on the fact that the oxidation of metals produces an electric current. Some examples will serve to illustrate this. If a piece of zinc is placed at one end of the multiplying wire and a piece of copper at the other, and these two metals are then brought in contact with water, an electric current will pass through the multiplying wire and cause the needle to turn. The same experiment can be made with nobler metals, for instance, with a piece of silver and another of copper, but then the effect will not be strong unless acid, alkali, or salt is added to the water.

It must be pointed out that in all these experiments the parts of the metals and of the wire which touch each other must be perfectly polished. The end of the needle which has once been seen to turn towards the more precious metal will take the same direction in each new experiment made with other metals.

A greater or smaller difference in the oxidizability of the two metals makes the needle deviate more or less, which follows naturally from the different strengths of the electric current. Thus it is possible to learn the oxidizability of two metals relative to one another.

Silver alloyed with copper must be regarded as less noble compared with pure silver. Consequently, silver can be assayed by means of the electromagnetic multiplier. Instead of touch needles, such an assay requires silver plates of all degrees of purity, from the purest silver to copper. Those that have been used so far have been from three to four inches long and three-quarters of an inch wide.

When a piece of silver is to be assayed, the first step is to examine the electric current which it produces with one of the medium assay plates, for example, with the one made of silver with a fineness of 12 demi-ounces.[3] In order to do this, the plate used for the assay is joined to one end of the multiplying wire and the silver which is to be assayed to the other end of the wire. Then the plate and the silver are brought in contact with a porous body saturated with muriatic acid. The needle of the multiplier will immediately show whether the silver being assayed is more or less pure than that of the plate. If it is more pure, the assay is made with the plate of 14 demi-ounce silver; if it is then inferior to the silver in this plate, an assay is made with the plate of 13 demi-ounce silver; but if it is not yet the same as this plate, it is easy to see, from the deviation of the needle, whether it is between silver of 12 and 13 demi-ounces or between silver of 13 and 14 demi-ounces. In this assay it will also be easy to discover how much the silver differs from that of 13 demi-ounces, for if the needle deviates by 9 degrees to the left or to the right when the plate of 13 demi-ounces is compared with the preceding or the following one, a deviation of 3 degrees will indicate a difference of $^3/_9$ demi-ounces or of 6 grains.

This example easily demonstrates the procedure to follow in all other cases.

[3] In Denmark fineness is divided into 16 parts or demi-ounces, and each ounce [demi-ounce] is subdivided into 18 grains. With 20 assay plates, differing from one another by 5 per cent, it may be easiest to adapt the Danish standard of purity to that of France.

It is not adequate to assay the silver with only one liquid conductor. That sufficed only in the case where it was certain that the silver was only alloyed with copper. It can also be alloyed with brass and sometimes even with white tombac, which is an alloy of copper and arsenic. The use of several liquids as conductors will reveal the nature of the alloy.

If the apparent degree of fineness of a piece of silver containing brass has been found in an assay made with muriatic acid, and the assay is then repeated with a solution of caustic potash as intermediate conductor, the deviation of the needle will put the silver alloyed with brass approximately one ounce lower. Therefore, if a piece of silver whose fineness is not known, shows itself to be an ounce or a demi-ounce lower with a solution of potash than with muriatic acid, it is reasonable to conclude that it is alloyed with brass. Silver alloyed with white tombac will lose even more in the assay made with a solution of potash; it can even be 7, 8, or 10 [demi-ounces] less.

If it turns out that the silver is alloyed with several base metals, it is possible to make the assay with several liquid conductors, always following the rules indicated by chemistry.

We shall now explain what must be observed for the assay to acquire the desired accuracy.

An equal surface of the silver to be assayed and the assay plate must be brought in contact with the liquid conductor. This is done by making the intermediate porous body narrower than the assay plate. The two pieces of silver must, as far as possible, be brought into contact with the liquid conductor at the same time.

The surfaces must be uniform. In order to achieve this, it is good to use powdered pumice to polish the parts of the silver to be assayed and of the assay plate which must be brought in contact with the intermediate liquid. The assay plates, polished once in this way, will later only need a light cleaning, which will not wear them excessively. Cast silver which has not been hammered cannot be assayed with plates of hammered silver. In this case, the silver must either be hammered or assayed with silver of the same nature. Sometimes the surfaces become tarnished during the assay, and then they must be repolished and the assay repeated. This can be avoided by terminating the contact between the metal and the liquid as quickly as possible.

As an intermediate substance, undyed cloth or washed tinder can be used and soaked in the liquid employed. If this is caustic potash, it must not be concentrated in the solution but preferably rather dilute. Muriatic acid can also be diluted a little.

It is important not to forget to connect the two metals and the ends of the multiplying wire with a good metallic contact. The contact must be made on the two inner sides or on the two outer sides of the metals and approximately at equal distances from the fluid.

When everything has been assembled in the prescribed order, and there is a difference in the quality of the silver, the needle, as is well known, will swing to one side, but it will swing back to its initial position, and most of the time it will even go beyond that so that it makes several oscillations. However, these oscillations will be inclined more towards one side than the other. Only four or six oscillations are sufficient to determine the side to which it is more inclined. It is easy to understand the method to follow in order to calculate the absolute deviation of the needle.

If, for instance, the needle still oscillated between 8 degrees to the left of zero and 30 degrees to the right after 6 oscillations, the absolute deviation to the right would be 11 degrees, for if we assume that the force which makes the needle deviate after the sixth vibration remained the same, the needle, coming to rest, would place itself half-way between the two end-points of the arc of these oscillations, which would be the eleventh degree to the right of zero.

If, on the other hand, the needle only oscillated to the right between the 6th and the 30th, the absolute deviation would be 18 degrees.

While the assay is being made, it is necessary to take care that the instrument is not shifted, that the needle does not make a complete rotation, and that the horseshoe magnet is always at the same distance from the needle.

In order to give this method of assaying silver all the perfection that it can obtain, it is of course necessary to acquire much practice. Then experience will teach better than rules the caution to employ and the dexterity which the success of the assay requires. It is to be hoped that this method, which, though still in its infancy, is so superior to the assay made with the touch stone, will acquire with time a very high degree of accuracy through the combined efforts of many philosophers.

This method is very convenient for the daily assays made by goldsmiths or in banks. If, for instance, one silver spoon out of a dozen or more has been assayed, the multiplier will easily indicate whether the fineness of the others is the same. The same method can be used to discover whether both ends of a silver bar are of equal purity.

In the same way as the electromagnetic multiplier is used to assay silver, it can be used to study alloys of other metals in order to find out, for example, whether tin is alloyed with lead or not.

59

Observations Concerning the
Compressibility of Fluids[1]

MESSRS Colladon and Sturm dismiss the procedure earlier suggested by myself to block the water in the narrow tube of the compression bottle with mercury. I agree with them in this, and more than a year ago I showed the Society and many enthusiasts of physics an apparatus in which the narrow tube of the bottle is blocked by air. In addition, I have made another improvement of the compression bottle, which does not yet seem to have been attempted by anyone else. The narrow tube is not fused to the bottle but ground in its neck. This arrangement, which results in great convenience in the use of the apparatus, also makes it possible for me to put pieces of glass into the bottle and thus to conduct direct experiments on the cubic compression of glass caused by pressure. While these experiments have not yet led me to entirely satisfactory results, some remarkable phenomena have occurred, on which I still have to conduct some experiments. In any event, the compression is very small, and I have strong reasons to assume that it is going to be smaller rather than bigger than 3.3 millionths, which Colladon and Sturm have found it to be. Incidentally, for the calculation of my experiments I do not directly use the cubic but only the linear compression of glass, but I wanted to calculate the linear from the cubic compression since I believe that traction makes the glass rods somewhat thinner while pressure makes them somewhat thicker, and therefore the former does not show a purely linear expansion nor the latter a purely linear compression.

I have also performed many experiments on the influence of temperature on compressibility and have found that Canton was also right in saying that water is more compressible at the freezing point than at higher temperatures. The compressibility of water at 0° R. is approximately one tenth bigger than at 10°. At higher temperatures it is even less but not in such a strong proportion.

You know that Perkins's determinations of the compressibility of water deviate considerably from Canton's, which have been confirmed by me and recently supported even further by Colladon and Sturm. This disagreement certainly does not

[1] From a letter from Professor Ørsted to the Editor. [*Annalen der Physik und Chemie*, ed. by J. C. Poggendorff, Vol. 12, pp. 158–59. Leipzig 1828. The same subject is treated in *Videnskabernes Selskabs Oversigter*, 1827–28, pp. 14–17. Copenhagen 1828. KM II, pp. 346–47. Originally published in German.]

come from any observational error by this excellent and intelligent technician, but only from the fact that his apparatus gives a considerable impetus to the water and therefore has a bigger effect, while the other methods, mine in particular, give such a slow impetus that it can be considered to be a mere pressure. In addition, the observation is made while the water column is at rest, while Perkins, with his apparatus, can only observe what has happened at the moment of the largest pressure or impetus.

I have provided my apparatus with several other improvements which make its use much easier, but I cannot give an account of these without exceeding the limits of a letter.

60

On the Compression of Water in Vessels of Varying Compressibility[1]

Among the problems which present themselves in the investigations of the compressibility of liquids, the following has also been mentioned on the occasion of a prize question posed by the Paris Academy of Sciences: What influence would the compressibility of the walls of the vessel containing the object of the experiment have on the results? This influence can be considered from two different points of view. Some physicists have believed that the walls of the vessel are compressed in all directions so that the vessel loses in capacity due to the pressure it experiences from the compressed liquid. Others, on the other hand, have believed that this pressure only has the effect of making the walls of the vessel thinner. In this case, the capacity of the vessel is increased slightly by the pressure but only by a very insignificant amount. I have always been of the latter opinion. The reasons for one or the other of these opinions are too well known to be repeated here. I therefore content myself with a report on experiments with which I have sought to decide the question.

I have performed the compression of water in vessels of very different compressibility. Since the compressibility of lead is more than 18 times larger than that of glass, I have mainly made use of this metal in my new experiments. Mssrs Colladon and Sturm have previously, in their beautiful work on the compression of liquids, measured the extension which a glass rod experiences due to a given traction and, according to this experiment, established the linear contraction of glass to be 11 ten-millionths at a pressure of one atmosphere. Since they believe that the walls of the vessel containing the liquid are compressed in all directions, they draw the conclusion that the glass vessel in which the compression of glass is observed loses 33 ten-millionths of its capacity due to the pressure of one atmosphere, and that this amount should be added to the apparent compression of water in order to determine its true compression.

[1][Annalen der Physik und Chemie, edited in Berlin by J. C. Poggendorff, Vol. 12, pp. 513–16. Leipzig 1828. The same subject can be found in Annales de chimie et de physique, Vol. 38, pp. 326–30. Paris 1828; in Videnskabernes Selskabs Oversigter 1827–28, pp. 14–17. Copenhagen; and in Schweigger's Journal der Physik und Chemie, Vol. 52, pp. 9–10. Halle 1828. KM II, pp. 348–50. Originally published in German.]

According to the experiments of Mr. Tredgold, which he cites in his excellent work on the strength of cast iron and other metals, a lead rod with a cross-section amounting to one square inch in English measure is lengthened by $^1/_{480}$ at a traction of 1500 English pounds. A similar weight would shorten the rod by the same amount. A pressure of 1500 English pounds on an English square inch is equal to the pressure of 101.7 atmospheres. This causes a linear shortening of 0.00002048 at a pressure of one atmosphere. A calculation following the theorem assumed by Mssrs Colladon and Sturm yields a reduction in the capacity of 0.00006144 for a bottle of lead. Since the compression of water due to one atmosphere amounts to only 51 millionths according to these two physicists, and according to my experiments even less, water must show a perceptible expansion when compressed in lead vessels. Even if the skillful English master builder had found too large a magnitude in his experiments, even if he had been wrong by more than one half, which I certainly do not believe, the experiment discussed here would still be decisive.

The lead bottle I used was fastened at the mouth with a brass ring and was completely sealed by a hollow glass stopper, fitted by grinding, and then a well-calibrated glass tube was placed in the stopper. After air had been removed from the water, the stopper with its tube was inserted in such a way that no air remained under the stopper so that the water was forced to rise in the tube. The upper opening of the tube was covered with a small conical bell jar. It goes without saying that the tube was equipped with a scale. Otherwise, the experiment on the compression of water with this bottle was performed in the same way as I have performed it previously with glass bottles.

The same tube with its stopper which I used in the experiment with the lead bottle had very often been used in experiments with a glass bottle, where the stopper was likewise ground to fit its mouth. Therefore it was easy to compare the experiments with the two bottles. After correction for the difference in their capacities, the apparent compression in the lead bottle was found to be a little larger than in the glass bottle. At a pressure of one atmosphere, this difference did not exceed 2 millionths of the volume of the water. This result agrees completely with my opinion and is quite contrary to that which I dispute.

I have performed similar experiments with bottles made of brass and tin and thereby obtained similar results. I shall report all my experiments on the compression of liquids in detail in the 4th volume of the memoirs of the Royal Society of Sciences in Copenhagen. Here, I content myself with observing that in these experiments it is important to be aware of air bubbles which often form during continued contact of the water with the metal. If water has been sitting in the lead bottle for an entire day, little air bubbles can almost always be found. I also believe to have found that water shows a greater compressibility when it has been in contact with a surface, be it of glass or metal, for a short time. I am still busy with experiments on this subject.

61

A New Electromagnetic Experiment
Disproving Ampère's Theory

Remarks on the Relation Between Sound,
Light, Heat, and Electricity[1]

COUNCILLOR of State[2] and Professor Ørsted has informed the Society of a new electromagnetic experiment which he believes to be inconsistent with Ampère's theory.[3] It is a familiar experience in the history of science that opposing theories about a natural phenomenon are able to persist for a long time even though there may be arguments which should decide the issue. In such a case an attempt must be made to devise an experiment which cannot possibly be explained in two ways. If one stopped at a crossroads where one did not know which direction to take, such an *experimentum crucis*, as Baco[4] called it, would show the right way. The controversy between the explanation of the electromagnetic effects given by Ampère and the one given by the discoverer may be said more or less to stand at such a point. Admittedly, Ampère's theory has not retained many defenders outside France, and even there opinions are divided, but the profusion of mathematical expositions which makes it difficult to assess this theory has also prevented many physicists from deciding in favour of one view. As is well known, Ampère assumes that magnetism consists of nothing but an accumulation of small electric currents which move in orbits around the principal elements in planes which are parallel, and which have a very small angle to the axis of the magnet. The ingenuity with which this clever French mathematician has gradually changed and developed his theory

---

[1] [*Videnskabernes Selskabs Oversigter*, 1829–30, pp. 22–26. Reports of the meetings held on April 16, 1830 and April 30, 1830, respectively. KM II, pp. 479–82. Originally published in Danish.]

[2] [Between 1827 and 1840 Ørsted refers to himself in these reports as *Etatsraad*, an honorary title meaning Councillor of State.]

[3] [Summary in Oken's *Isis*, Vol. 22, Col. 260–62, Jena 1829.]

[4] [Bacon.]

in such a way that it is consistent with a variety of contradictory facts is very remarkable. However, Ørsted now believes that he has found a fact which is so obvious that it will be difficult to make it compatible with Ampère's theory. This experiment requires a magnetic needle, approximately 4 inches long, which is bent in such a way that it has a horizontal part, in the middle of which is the centre of suspension, while one end is bent upwards, and the other downwards, like $ABCDE$ in

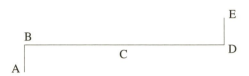

the adjoining figure. When a current-carrying electrical conductor is placed directly opposite and parallel to one end of the needle, e.g., DE, it should not have any effect on it, according to Ampère's theory, which states that the effect of the conductor on the magnet relies on nothing but the law that magnetic currents attract each other when their directions are the same, or when the resolution of the forces leads to the same situation. When their directions are opposite, they repel each other. Therefore, when the conductor is perpendicular to all the currents which the theory assumes to be in the magnet, it cannot have any effect on it. On the other hand, it must produce effects as soon as it is brought away from a parallel position and drive the needle to one side or the other, depending on its inclination, in such a way that the currents either attract or repel each other. However, experiments showed that the conductor drove the end of the needle which was immediately opposite to the same side, whether its direction was perpendicular or tilted to either side provided only that it did not approach the horizontal position too closely. This already seemed conclusive, but the experiment allows even more instructive variations. If the needle is turned around the axis of the horizontal part in such a way that AB points upwards, but DE downwards, the current-carrying conductor drives the needle in the same direction as before, without any change occurring if it is tilted towards one of the sides. If there were such currents in the magnetic needle as Ampère assumes, they would be horizontal in AB and DE and would take the opposite orientation when the needle was reversed, whereby the effect of the perpendicular conductor would also be reversed. The possibility still remained that the perpendicular conductor had no effect at all on the perpendicular part of the needle immediately opposite but only on the horizontal piece. Though such a possibility might be refuted by mathematical arguments, the experimental way seemed to be the shortest. Another needle of the same length was fastened to the horizontal part of the needle but in the opposite direction. Its magnetic force was far greater than that found in the horizontal part, which was easily proved experimentally. In spite of this addition, the current-carrying conductor had the same effect as before. The combined results of these experiments seem incompatible with Ampère's theory.

Councillor of State Ørsted has also made some reflections to the Society on the relationship between sound, light, heat, and electricity. This short summary cannot claim to be completely clear except to those who have already followed the train of his thoughts on this subject. Besides the other important reasons which show that a certain basic similarity between light and sound does not justify the inference of the existence of a music of colour, he draws attention to the fact that the number of oscillations in the light waves with the highest velocity does not increase to twice the number of the ones with the lowest, unless, perhaps, we want to compare the deepest, but faintest violet in the prismatic spectrum with the least visible red, two colours for whose rays we have no proper measurements. However, it would be possible to imagine the relationship between the sensory impression of the red and violet extremes of the spectrum as the consequence of there being twice as many oscillations in the latter as in the former. All the colours of the spectrum would then have the same relationship to each other as the notes which are contained in one octave. Concerning light and heat, he repeats the remark he made a long time ago to the effect that they differ only through the internal frequency of oscillation. Therefore, as light consists of ether waves, so must heat. Hereby he is led to a further confirmation of the opinion which he stated in 1813 that all heat is radiant heat, and that heat which is said to be conducted is only an internal heat radiated backwards and forwards between the basic elements of the body. From this it further followed that when a body is brought into a new state where the internal radiation of heat is increased, the body suddenly emits more heat rays, whereby heat is released, but if it enters a state in which the internal radiation is decreased, or perhaps more correctly, encounters a greater number of obstacles, it does not emit so many heat rays as before and is said to bind heat. Finally he showed that, if it was necessary to accept light and heat as oscillations in the ether, it was impossible to avoid also considering electricity and magnetism as oscillations, but that the difference between the electric and the magnetic effects could not lie in the frequency of oscillations alone, but that an essential difference had to exist in the nature of the oscillations. Previously he has pointed out the necessity of assuming internal motions accompanying chemical effects. Besides he would not have it regarded as definitely settled that light consists in oscillations of the ether. He only wanted to show that, given this opinion which has recently gained so much in probability, the interdependency between electricity, galvanism, and magnetism must be imagined to be just as uninterrupted as in the theory which started from the electrical forces, a truth which he has already pointed out, in another form, in his *Ansichten [sic] der chemischen Naturgesezte*, in 1812.[5]

[5] [Cf. *View of the Chemical Laws of Nature* (Chapter 30, this volume).]

62

Thermo-Electricity[1]

THERMO-ELECTRICITY is a term introduced a few years ago into natural philoso-

Thermo-electricity. phy, to signify the electrical current, excited in a circuit of conductors, when the equilibrium of its heat is disturbed in such a manner as to cause therein a circulation of caloric.

Thermo-electricity being a particular branch of *Electromagnetism*, which has been discovered since the publication of the volume of this work in which it ought to have been treated, it will be necessary to comprehend the whole doctrine of electromagnetism in the present article.

HISTORY

In the earliest period of the history of magnetism and electricity, the minds of philosophers were more struck by the resemblances of these two agencies than by their disparities. The first philosopher who undertook a regular series of comparative

History. experiments upon magnetism and electricity, was the celebrated Dr. William Gilbert, who first published his inquiries in the year 1600. He was aware of so many disparities between them, that he declared their resemblance to be merely accidental. He had indeed strong reasons to think so at that time, for the magnetical polarity was well known to him, and principally by his own experiments, but the discovery of the electrical polarity was reserved for a philosopher of the following century (du Fay). This discovery, and particularly the fundamental law of electrical polarity, brought forward by Franklin, again countenanced the opinion of the resemblance of electrical and magnetical powers; and the sagacity of Æpinus gave great credit to it. But immediately after this acknowledgment of their resemblance, another excellent philosopher, Van Swinden, was struck with the disparities which remained still unexplained, and his ingenious inquiries obtained much approbation. The discoveries of Galvani and Volta, by which the electrical powers were exhibited in forms very different from

---

[1] [*The Edinburgh Encyclopædia*, conducted by David Brewster, LL.D., Vol. 18, pp. 573–89. Edinburgh 1830. According to the editor's statement this article was written by H. C. Ørsted. The original article contains a number of trivial printing errors which have been corrected here. KM II, pp. 351–98. Originally published in English.]

those formerly known, gave the opinions upon this subject a new turn. The German philosopher, Joh. Will. Ritter, was thought during some time to have produced magnetical effects by the Voltaic pile, but his experiment having been repeated without success, the subject remained as it was. Thus the balance inclined alternately sometimes to the one and sometimes to the other side; but at no time have either of these opinions met with general reception. A certain turn of mind has here, as in most other controversial doctrines, exercised a considerable influence. One class of natural philosophers have always a tendency to combine the phenomena and to discover their analogies, another class, on the contrary, employ all their efforts in showing the disparities of things. Both tendencies are necessary for the perfection of science, the one for its progress, the other for its correctness. The philosophers of the first of these classes are guided by the sense of unity throughout nature; the philosophers of the second have their minds more directed towards the certainty of our knowledge. The one are [sic] absorbed in search of principles, and neglect often the peculiarities, and not seldom the strictness of demonstrations; the other considers the science only as the investigation of facts, but in their laudable zeal they often lose sight of the harmony of the whole, which is the character of truth. Those who look for the stamp of divinity on every thing around them, consider the opposite pursuits as ignoble and even as irreligious; while those who are engaged in the search after truth, look upon the others as unphilosophical enthusiasts, and perhaps as phantastical contemners of truth. Happily these two tendencies are in most natural philosophers so well tempered with good sense, that their controversies seldom exhibit any of the exaggerations which have disgraced so many theological and metaphysical controversies; but they always exercise their influence, which is generally a salutary one, in forming an opposition of sentiment in the republic of letters by which stagnation is prevented. This conflict of opinions keeps science alive, and promotes it by an oscillatory progress, though it seems to the common eye a mere fluctuation, without any definite purpose.

The reasons for and against an essential resemblance between magnetism and electricity might, before the discovery of electromagnetism seem to be nearly balanced. The most striking analogies were, that each of them consists of two powers, or directions of powers, of an opposite nature, submitted to the same laws of attraction and repulsion; that the magnetical action on bodies, fit to receive it, has much analogy with the electrical action; that the distribution of the powers in a body, *Analogy between electricity and magnetism.* which has an electrical charge, and still more a series of bodies charged by cascade, differs very little from the distribution of the powers in a magnet; if we imagine a Voltaic pile, and principally the modification denominated after Zamboni, composed of minute and molecular elements, it would have the most perfect analogy with a magnet; and lastly, that the tourmaline differs but little from such an electrical magnet.

We shall not here consider that most of these analogies are overturned by the discovery of electromagnetism; but still confining ourselves to the period before this discovery, it may be objected that the magnetical and electrical powers do not act on each other, which should be the case, if they were of the same nature; that all bodies transmit with ease the magnetical action, but not the electrical; that neither the tourmaline nor any system of charged glass-plates, or of galvanical arrangements, has the effects of the magnet. Although it might be answered that the galvanical circuit, in its first period, seemed no less different from any electrical apparatus than the Voltaic pile from a magnet, these objections did not cease to have considerable weight, but we have hitherto deliberately omitted one of the arguments, viz. the observation of magnetism in bodies struck by lightning, and the experiments made to imitate this effect. It had often been observed, that the magnetical needles in a ship struck by lightning have suffered a change in their polarity.

A very remarkable case of this kind, mentioned in the *Philosophical Transactions*, Vol. 11, No. 127, p. 647, seems to be the earliest on record. It is there related that a vessel, whose mast was struck by lightning, had the poles of the needles in all

Magnetical effects of lightning.

its compasses inverted, yet the compasses themselves were not struck. Some other observations of a similar nature are recorded in Domsdorph's[2] *Treatise upon Electricity, Magnetism, Fire, and Ether*, (*Über Electricität, Magnetismus, Feuer und Ether*, 1783). An accident of this kind, which happened in the year 1751, caused Franklin to try the effect of artificial electricity upon needles of steel. The result was, that when the needles were in a position in which the earth could produce in them some magnetism, this effect was much increased by any electrical stroke; but when the position gave no such advantage, he found that the extremity of the needle, in which the electricity entered (which received the positive electricity) was directed towards the north, when the needle was conveniently suspended. Wilcke, who repeated these experiments, obtained the same results, only with the difference, that in the case when the direction of the electrical stroke seemed to decide the polarity, this was the inverse of that observed by Franklin. (*Transactions of the Royal Academy at Stockholm*, 1766.) The experiments made in the year 1785, upon the same

Experiments of Van Marum and Van Swinden.

subject by van Marum and van Swinden have been considered as decisive against the magnetical effects of electricity, nevertheless the ninth of their experiments was precisely an electromagnetical one, for they led the electrical discharge transversely through a steel needle, and obtained a strong magnetical polarity in a direction perpendicular to the magnetical meridian; but they considered this as a singularity not to be explained, and hence it has been out of the sight of philosophers from the year 1785 until 1820, when electromagnetism was discovered. (See van Marum *description d'une très grande machine électrique*.)

One of the earlier experiments, which probably belongs to electromagnetism, is

Electromagnetic experiment by Cavallo.

that of Cavallo, by which he proved that iron has more efficacy on the magnetical needle, when an acid, particularly diluted sulphuric acid, acts upon it.

Joh. Will. Ritter, already mentioned, pursued a great number of researches upon the analogy of magnetism and electricity. He had in the year 1801 made a series of very delicate experiments upon the galvanical difference between the two magnetical poles of a steel needle. The result deduced from his ex-

---

[2][Donndorph.]

periments was, that the southern extremity of the needle was more oxidable than the northern, and that the galvanical effect of two magnetical needles upon a frog was such, that the south pole acted as the more oxidable, the north pole as the less oxidable metal. It is now acknowledged, that he has been led into error by the difference which a small disparity in the *Experiments of Ritter.* polish of the metal can produce, and which he employed insufficient means to avoid. The same philosopher stated likewise erroneously, that a platina wire, which has been employed to make a liquid communicate with a powerful galvanic circuit, assumes some magnetical direction, and that a needle, of which one half is zinc and the other silver, takes, when conveniently suspended, the same direction as the magnetical needle. The precipitation with which Ritter published these and some other erroneous statements, has thrown a shade over the name of this unhappy but ingenious philosopher, who has enriched science with several discoveries of great importance, and whose profound yet obscure ideas in many cases have anticipated the discoveries of future times. We are far from patronizing a vain exhibition of new ideas, by which it is possible for a very ordinary mind to make pretensions to every new discovery; but when works are marked with the true stamp of genius, it is but justice to acknowledge the merits of their speculations. Some writers have thought that this act of justice would deprive experimental philosophers of a part of the honour due to their exertions; but this honour is quite unimpaired, if the author, who has anticipated their discoveries, has only had a vague and obscure notion of them; while it must be avowed, that when the author has clearly announced the discovery, has derived it from good data and conceived its connections with other truths, the merit of the experimental philosopher is only that of having confirmed it by experiment, which still in many cases can be a work of no smaller claim to glory than the primitive conception itself.

Among the electromagnetical experiments which preceded the discovery of electromagnetism, ought to be mentioned an experiment of Professor Mojon at Genoa, who found that a steel needle having been 22 days in communication with a galvanical apparatus of 100 elements, had become magnet- *Electrical experiment* ical,—an experiment which would have been of no historical *of Professor Mojon* interest, if its author had not founded upon it, 18 years later, a *of Genoa.* pretension to the discovery of electromagnetism. He seems not to have been aware that his pretended discovery, were it true, should be considered as new even now; for the magnetical effect, hitherto proved by experiments, is not in the direction of the electrical current, but perpendicular to it. The experiment of Mojon is described in Aldini's *Essai Théorique et Expérimental sur le Galvanisme*. Paris, 1804, Vol. 1, pp. 339 and 340. Aldini mentions, at the same place, that a certain Mr. Romanesi[3] at Trent had confirmed the experiment of Mojon, and at the same time observed that galvanism makes the magnetical needle deviate. Professor Aldini, whose work upon galvanism comprehends two volumes, does not say a word more upon this subject.

It is, therefore, not surprising, that neither the French institute, nor the other learned societies, nor the numerous natural philosophers, to which the work was

[3][Romagnosi.]

presented in the year 1804, took any notice of this observation, which would have accelerated the discovery of electromagnetism by sixteen years. Romanesi seems likewise to have forgot his observation, until electromagnetism was discovered.

Two or three years before the discovery of electromagnetism, Professor Masch-

Observation of Professor Maschmann.

mann at Christiania, in Norway, observed that the silver tree, formed in a solution of nitrate of silver, when put in contact with mercury, (the *arbor Dianæ*,) takes a direction towards the north; and the celebrated Professor Hansteen found that this direction can likewise be determined by a great magnet. As the metallic precipitation is also of galvanical nature, this observation may be considered as one of the precursors of electromagnetism.

Electromagnetism itself, was discovered in the year 1820, by Professor Hans Christian Ørsted, of the university of Copenhagen. Throughout his literary career,

Electromagnetism discovered by Professor Ørsted.

he adhered to the opinion, that the magnetical effects are produced by the same powers as the electrical. He was not so much led to this, by the reasons commonly alleged for this opinion, as by the philosophical principle, that all phenomena are produced by the same original power. In a treatise upon the chemical law of nature, published in Germany in 1812, under the title *Ansichten [sic] der chemischen Naturgestze*,[4] and translated into French, under the title *Recherches sur l'identité des forces électriques et chymiques*, 1813, he endeavoured to establish a general chemical theory, in harmony with this principle. In this work, he proved that not only chemical affinities, but also heat and light are produced by the same two powers, which probably might be only two different forms of one primordial power. He stated also, that the magnetical effects were produced by the same powers; but he was well aware, that nothing in the whole work was less satisfactory, than the reasons he alleged for this. His researches upon this subject, were still fruitless, until the year 1820. In the winter of 1819–20, he delivered a course of lectures upon electricity, galvanism, and magnetism, before an audience that had been previously acquainted with the principles of natural philosophy. In composing the lecture, in which he was to treat of the analogy between magnetism and electricity, he conjectured, that if it were possible to produce any magnetical effect

Discoveries of Professor Ørsted.

by electricity, this could not be in the direction of the current, since this had been so often tried in vain, but that it must be produced by a lateral action. This was strictly connected with his other ideas; for he did not consider the transmission of electricity through a conductor as an [sic] uniform stream, but as a succession of interruptions and reestablishments of equilibrium, in such a manner, that the electrical powers in the current were not in quiet equilibrium, but in a state of continual conflict. As the luminous and heating effect of the electrical current, goes out in all directions from a conductor, which transmits a great quantity of electricity; so he thought it possible that the magnetical effect could likewise eradiate. The observations above recorded, of magnetical effects produced by lightning, in steel-needles not immedi-

---

[4][Cf. *View of the Chemical Laws of Nature* (Chapter 30, this volume).]

ately struck, confirmed him in his opinion. He was nevertheless far from expecting a great magnetical effect of the galvanical pile; and still he supposed that a power, sufficient to make the conducting wire glowing, might be required. The plan of the first experiment was, to make the current of a little galvanic trough apparatus, commonly used in his lectures, pass through a very thin platina wire, which was placed over a compass covered with glass. The preparations for the experiments were made, but some accident having hindered him from trying it before the lecture, he intended to defer it to another opportunity; yet during the lecture, the probability of its success appeared stronger, so that he made the first experiment in the presence of the audience. The magnetical needle, though included in a box, was disturbed; but as the effect was very feeble, and must, before its law was discovered, seem very irregular, the experiment made no strong impression on the audience. It may appear strange, that the discoverer made no further experiments upon the subject during three months; he himself finds it difficult enough to conceive it; but the extreme feebleness and seeming confusion of the phenomena in the first experiment, the remembrance of the numerous errors committed upon this subject by earlier philosophers, and particularly by his friend Ritter, the claim such a matter has to be treated with earnest attention, may have determined him to delay his researches to a more convenient time. In the month of July 1820, he again resumed the experiment, making use of a much more considerable galvanical apparatus. The success was now evident, yet the effects were still feeble in the first repetitions of the experiment, because he employed only very thin wires, supposing that the magnetical effect would not take place, when heat and light were not produced by the galvanical current; but he soon found that conductors of a greater diameter give much more effect; and he then discovered, by continued experiments during a few days, the fundamental law of electromagnetism, viz., *that the magnetical effect of the electrical current has a circular motion round it.*

When he had discovered this fundamental law, he thought it proper to publish the discovery, in order that it might be as soon as possible perfected by the co-operation of other philosophers. Apprehending that others might lay claim to this discovery, he sent a short Latin description[5] of his experiments to the most distinguished philosophers and learned bodies; and though, by this means, he has not avoided the pretensions which have been made to his discovery by others, still he has rendered them ineffectual. It deserves, perhaps, to be noticed, that the above-mentioned Latin description, consisting of four pages in 4to., of which the first gives the introduction and the description of the apparatus, the last the conclusions, contains upon the two intermediate pages, the results of more than 60 distinct experiments. From this brevity, it has happened, that some philosophers have thought that he had treated his subject in a superficial manner.

As the details of this discovery, and of all those which have originated from it, will be exhibited in this article, we shall in the remainder of this historical sketch, in order to avoid repetitions, confine ourselves to the most striking and leading facts, and insert the other historical notices in the doctrinal part.

[5][Cf. "The Effect of the Electric Conflict on the Magnetic Needle" (Chapter 39, this volume).]

The first discovery to which that of Professor Ørsted gave occasion, was that of

Ampère's discovery of the mutual action of conductors.

Mr. Ampère, member of the French institute. He found that a conductor, conveniently suspended, is attracted by another, when both are transmitting an electrical current in the same direction; but that they repel each other, when the two currents have opposite directions. Professor Schweigger at Halle, invented at the same time, an electromagnetical multiplicator, which is of very extensive use. Mr. Arago found that steel can be magnetized by the electrical current. Mr. Gay Lussac at Paris, and Professor Ermann at Berlin, discovered, that when the current has passed perpendicularly through the plane of a steel ring, or through a steel plate, it shows no magnetical effect, before the circumference was interrupted.

The most remarkable of all of the discoveries, to which that of Ørsted has given occasion, is no doubt the thermo-electricity, discovered in 1822 by Dr. Seebeck, member of the Royal Academy at Berlin.

In the same year, the rotation of a magnetical needle around an electrical current, and of a body, which transmits an electrical current around a magnet, first imagined by Dr. Wollaston, was exhibited in a series of ingenious experiments by Mr. Faraday.

Effect of the Electrical Current upon the Magnetic Needle

The galvanic battery was the first apparatus, by which the magnetic effects of elec-

Effect of the electrical current upon the magnetic needle.

tricity were demonstrated. In order to make it give its magnetic action, its two poles must be joined by a conductor, commonly a metallic wire, which, for brevity's sake, we shall call the *uniting conductor*, or the *uniting wire*.

When not closed, the galvanic circle produces no effect upon the needle of a compass.

When the uniting wire is approached, and placed parallel, or nearly so, to a properly suspended magnetical needle, it is caused to deviate from its ordinary direction.

The magnetical effect of the electrical current is not interrupted by the interposition of other bodies. Already the first experiment showed that it passes like the magnetism of a loadstone through metals, glass, resin, wood, stoneware, water, &c.; even when the magnetical needle was placed in water, it was affected by the electrical current.

When the conducting wire is placed parallel to a conveniently suspended magnetical needle, the direction of the needle is changed.

1) If the needle is above the wire, and the positive electricity passes from the right to the left hand of the observer, the north end of the needle will go from the observer.

2) When the needle is below the wire, the direction of the needle is changed in the opposite way; its north end approaches the observer. It is not necessary, in this and the preceding experiment, that the needle is in the same

perpendicular plane as the conducting wire; it is only required that the needle shall be sufficiently near the wire, and in the first experiment, in a plane above, in the last in a plane below it.

3) When the needle is in the same horizontal plane as the wire, and is placed between the observer and the wire, the north end is elevated.

4) If the needle is upon the opposite side, the north end is forced down. In these two experiments, the needle must be very near to the wire.

From these facts, Professor Ørsted concludes, *that the magnetical action of the electrical current describes circles round the conductor.* It will perhaps not be out of place to quote here his own words, which have been overlooked by several authors, who have written the history of this discovery.

In the original publication he says, "ex observatis colligere licet, hunc conflictum (the electrical current,) gyros peragere; nam hoc esse videtur conditio, sine qua fieri nequit, ut eadem pars fili conjungentis (conducting wire,) quæ infra polum magneticum posita cum orientem versus ferat, supra posita eandem occidentum versus agat." For the sake of brevity we shall, in the following pages, denominate the direction of the current after the system of Franklin; or, to speak according to the system of two electricities, after the direction of the positive electricity in the current. If we now suppose that the electricity of the current enters the conductor at the right hand of the observer, the austral magnetism (the same which predominates in the north-end of the needle,) will, upon the superior surface of the conductor go off from the observer; on the side most distant from the observer, the austral magnetism goes downwards; on the inferior surface it goes towards the observer;

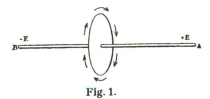

Fig. 1.

on the side nearest the observer it goes upwards. This is represented in Fig. 1, where *BA* is the conductor in which the direction of the current is *AB*, the circle *cdef* [sic] represents a plane perpendicular to the conductor, in which the magnetical circulation takes place. This plane is here and in the other figures represented as if it were material and opaque. The little arrows show the direction of the austral magnetism. We can make the application of this law to experiments, in a very commodious manner. For this purpose take a piece of paper (Fig. 2,) upon which the arrows and

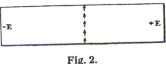

Fig. 2.

letters, there represented, are drawn. This piece of paper is to be wrapt around a cylindrical body, for instance a pencil, in such a way that the arrows lie in a plane perpendicular to the axis of the cylinder. We have thus an electromagnetical index, which, put in the place of any part of the conductor, shows the direction of the magnetical powers in it. The sharp ends of the arrows indicate the direction in which the austral magnetism (and consequently the north-end of the needle,) is repelled, and the contrary attracted; the opposite ends of the arrows indicate also the direction in which the boreal magnetism (and consequently the south-end of the needle) is repelled, and the contrary attracted. The reader may understand without trouble the most complex facts we are here to explain, if he has at hand two such cylinders, during the experiment. The same thing may be expressed in different ways. Mr. Hill, lecturer of mathematics at the University of Lund, in Sweden, has proposed one of the best. Let us imagine, says he, that the observer swims upon the electrical current, with his face turned outwards, (with his back turned towards the axis of the current,) and his head towards the origin of the current, the direction of the austral magnetism of the current will always proceed from his left to his right hand.

This law was confirmed by several other experiments.

When the uniting wire is placed in the same horizontal plane as the needle, but perpendicular to its direction, and near one of its poles, this pole will be elevated, if the current comes from the east, but depressed if it comes from the west. This will

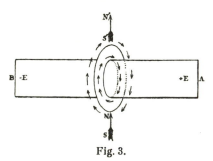

Fig. 3.

easily be understood by the inspection of Fig. 3. *BA* represents here the conductor, *NS* and *N'S'* two needles. All the parts of the drawing have the same signification as in Fig. 1, only that the dotted lines denote the inferior parts of the magnetical circles, but the uninterrupted lines the superior parts. It is evident that *N* (the north-end of one of the needles,) is here driven upwards by the repelling action from below and the attracting one above it. In the same manner, *S'* (the south end of the other needle,) is both drawn and pushed upwards.

The effect is on both sides the same, because not only the magnetical poles, but likewise the opposite sides have contrary effects. If one of the needles were turned by means of a magnet, so that each side of the wire could act upon a pole of the same kind, one of them would be elevated, when the other was depressed.

When the uniting wire is perpendicular, and the current enters its superior part, a needle, of which one of the poles is very near to the wire, will be thrown westwards; but if the wire is placed over against a point of the needle, situated between

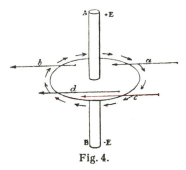

Fig. 4.

one of the poles and the middle, the needle will be turned eastwards. By opposite currents the results are likewise opposite. Fig. 4 will make this easier understood. *AB* is the uniting wire, the notations the same as in the former figures. It is evident, by the inspection of the figure, that the north-end of the needle *a*, having predominant austral magnetism, must be repelled by the similar magnetism of the conductor; and be turned towards the west. The attraction of the opposite magnetism in the conductor tends to give the needle the same direction; but as this coincidence of motions, produced by opposite powers, is constant in electromagnetism, we shall always confine ourselves to mention but one of them. The south-end of *c*, having predominant boreal magnetism, is also repelled by the similar magnetism of the current, which here has the same direction as the austral on the opposite side of the conductor. Thus the north-end of the needle is on one side of the conductor turned the same way as the south-end on the other side. The north-end of *c* receives the strongest impulses from the west, and must, therefore, be pushed eastward; while the south-end of *d* receives the strongest impulses from the east, and must move towards the west, and in consequence of this its north end must also turn eastward like that of *c*. Were the wire placed exactly over against the middle of the needle, this would be solicited equally in opposite directions, and therefore rest at its place.

When the uniting wire is bent in such a manner, that the parts on each side of the flexure are parallel, the exterior surfaces of the two branches are similar, and also

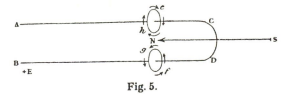

Fig. 5.

the interior ones. In Fig. 5, *ACDB* represents such a wire. As the current enters the superior branch at *C*, and in the inferior at *B*, it is obvious that the directions of the powers in the magnetical circles are the same at *e* and *f*, at *g* and *h*. Suppose that the two branches are in one perpendicular plane, and the north-end of the needle is placed in a plane, below the superior and above the inferior branch, the north end will be repelled, when placed on the west side, and attracted, when placed on the north[6] side of the wire. Above the superior branch, or below the inferior branch the effects are in the opposite direction. All the other cases, belonging to the effects of bent connecting wires upon magnetical needles, may be easily explained in a similar manner.

These are the principal experiments, by which Professor Ørsted endeavoured to establish the fundamental law of electromagnetism. As they all belong to one class, it has been practicable for us here to maintain in our account the historical order, without impairing the systematical one. In order to have a short term, we shall call the magnetical action of the electrical current, the *revolving magnetism*.

The discoverer remarks, in his Latin publication, that the magnetical action of the current being necessarily propagated and not instantaneous, the association of a progressive and revolving motion, must give origin to a spiral motion; still, he adds, this seems not to be required for the explanation of the electromagnetical facts hitherto discovered. His words are, "Præterea motus per gyros cum motu progressivo, juxta longitudinem conductoris conjunctus, cochleam vel lineam spiralem formare videtur, quod tamen, nisi fallor, ad phænomena hucusque observata explicanda nihil confert." Several writers upon the continent have considered it as an essential point in Ørsted's theory, that the magnetical motions in the current, should be of a spiral form; but it is evident that he has well distinguished this theoretical but still necessary consequence from the fundamental law, deduced from the facts. Supposing here spirals in the place of parallel circles, their windings must be so near to parallelism, that the deviations from it must be imperceptible. Thus the question belonging to the spirals may be left for farther research, in which, perhaps, the whole doctrine of vibrations might be considered.

In an appendix[7] published two months later (in Schweigger's *Journal*), Professor Ørsted explained the apparent difference observed between the effect of the galvanical battery, and that of a simple galvanical circuit. In the battery, which is a compound galvanical circuit, as well as in the simple one, the electrical current goes from the more oxidable metal (zinc), through the liquid conductor, to the less oxidable (copper): and when the water is taken away in one of the elements of the battery, and a wire put in its place, the direction of the current remains of course the same; but when we make use of a simple circuit, the water remains at its place, and

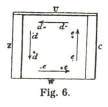

Fig. 6.

the uniting wire connects the two pieces of metal in a place, where the direction of the current is the opposite to that of the water. Fig. 6 will make this more perspicuous; Z represents here the zinc, C the copper, W the water, U the uniting wire; the arrows marked with $+e$ and $-e$ [$+d$ and $-d$] indicate the direction of the electrical current. It is visible that when in W the current goes from zinc through the water to the copper, it must in U go from the copper to the zinc.

In this appendix it is remarked that the magnetical efficacy of the electrical current depends not on its intensity, but on its quantity of electricity, and that the sim-

---

[6] [east?]
[7] [Cf. "New Electro-Magnetic Experiments" (Chapter 40, this volume).]

ple galvanical circuit is preferable for electromagnetical experiments. Some time after the discovery of electromagnetism, the great Swedish chemical philosopher Berzelius was of opinion that all the effects of the uniting wire could be explained in assuming four magnetical poles in its circumference. Fig. 7, where A indicates the austral, B the boreal poles, represents such a distribution. As the appearances in the first electromagnetical experiments may, until a certain degree, be represented by this scheme, it had many adherents, even since Berzelius had abandoned it. In order to decide the question upon this subject, Professor Ørsted made a direct experiment, which will be understood by Fig. 8. AB is a wooden pillar more than twelve feet

Fig. 7.

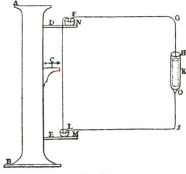

Fig. 8.

high; C is a magnetical needle, protected with glass against motions in the air, DE a wire of brass; K a galvanical apparatus; HGF and OJL brass wire; M and N small cups with mercury. The whole moveable part of this arrangement was supported by a wooden frame, not here represented. It appears that the apparatus K with its conductors, whose extremities are plunged in the mercury, can turn around nearly through the whole circle, without an interruption of the continuity of the conductors; thus the same point of the perpendicular wire, though immoveable itself, changes every moment its relative place in the circuit, when the moveable part FGIL is turned round. The experiment shows that the deviation remains the same, whatever the position of the moveable part may be, and that of consequence the polarity must be the same in all points of the circumference of the conductor. The great distance of the other parts of the circuit is the reason that DE is the only one which can have a sensible effect upon the direction of the needle.

A most useful application of electromagnetism is the *electromagnetic multiplier*, invented by Professor Schweigger at Halle, and im-proved by several other philosophers. We have already seen that when the uniting wire is bent so as to form two parallel branches, each of them acts in the same direction upon one of the poles of a magnetical needle placed between them (Fig. 5.) On proceeding upon this principle it is clearly shown that when the uniting wire is bent several times, as ABCDE, Fig. 9, and a magnetic needle is suspended in the space, inclosed by the windings of the

Schweigger's multiplier.

Fig. 9.

wire, each of its horizontal parts must produce upon the needle an equal effect; thus in the figure the effect is quadrupled. It is to be remarked that the windings should be as near each other as possible, in order to keep them all very near to the needle. At the same time the windings must be isolated from each other, which is effected by covering the wire with silk. As the windings can be repeated a great number of times, the multiplication of the effect may go very far. It should be nearly without limits, were it not that the conducting power decreases when the length of the wire increases. In order to give the instrument the solidity necessary, the wire is wound upon a frame. As it is required that the needle should be as moveable as possible, it is suspended by a fibre of silk, such as is found in the cod of the silk-worm. The instrument may be made much more sensible by means of another magnet placed so as to diminish the directive power of the needle. Mr. Nobili has made

M. Nobili's improved multiplier. a new improvement in this apparatus. In the place of one needle he introduces a compound index, consisting of two needles, *NS* and *S'N'*, Fig. 10, in opposite directions, and joined by a piece of wood or of stout wire, *GH*. When these two needles are of equal strength, the directive power of the index is reduced to nothing; so that the most feeble impulse will move it. But even when one of them has some preponderance, the force required for making the index deviate is still inconsiderable. At the same time this arrangement has the advantage, that both needles receive an impulsion, the needle *NS* from the inferior side of the conductor, and *S'N'* from the superior. The needles being in opposite situations, one will receive the same direction by the superior, as the other by the inferior side of the wire. When the needles approach as much to equality as is required for some nice experiments, the index is too easily moved in some others. In order to make the instrument proper for experiments with various degrees of force, though all of the feebler kind, Professor Ørsted added a bent magnet, *JKL*, which can be placed so as to repel the nearest end of the index,

Fig. 10.

or so as to attract it. The first of these positions is represented in Fig. 10. The magnet can also be approached to the index or removed from it. Fig. 11 represents the whole instrument of half its dimensions. *AB* is a stand of wood, having a screw on

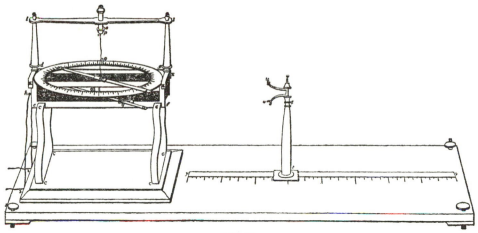

Fig. 11.

each corner for levelling it. *CCC*, *CCC* are two supporters likewise of wood, bearing the frame *defg*, upon which the multiplying wire is wound. This wire may be conveniently 50 to 60 feet long, and make 100 or more windings. From the windings each end of the wire pass [sic] through a little ring *h*, (the other is not to be seen in the figure) at *ii* the ends of the wire passes [sic] also through rings, which are here covered by the other parts of the figure; *KK*, *KKK* are two small pillars of ivory or wood, supporting the transverse piece *ll*, through which passes the cylindrical piece *mp*, having a head at *m*, and being moveable upwards and downwards. At the centre *r* of the inferior extremity, *p* is a little hole, communicating with a transverse hole, which here is represented as shut with a pin, seen immediately under the ring *o*. Through the hole at *r* is introduced one end of the silk *rx*, which is drawn out through one of the openings of the transverse hole, and fastened by means of the pin above-mentioned. By the silk *rx* is suspended the index, consisting of the superior magnetic needle *ns*, and the inferior one, of which the extremity *n* is here visible, the other being covered by other parts of the figure. The boreal pole of one of these needles is turned in the same way as the austral of the other, and both connected with a piece of wire. The circle at whose divisions the index points, is made of glass, preferable to brass, which often is magnetic. At *q* is a slit to receive the needle and keep it, when the instrument shall be transported, a similar one is on the other side of the instrument. The index is cleared from the slits when the instrument is to be employed. Having been thus cleared, it is still at rest until the piece *mp* is drawn upwards, the ring *o* stops it, so that it shall not be elevated too much. The index is sheltered from the air by means of a case of glass which covers the whole frame including the index, and has in the upper part a hole through which the head *m* of the piece *mp* passes; *tt* is a pile moveable in the slit *yy*, which has a scale, showing the distance from a point in the same plane, perpendicular below the centre of the index; *uv* is a bent magnet, which has two points, one of which is visible at *w*, the other is placed in a hole in the pile *tt*. This magnet can be taken out, and the point *w* introduced in the pillar, in order to augment or diminish the directive power of the index, as the purpose may require. When this instrument is to be used, the index must, as already mentioned, be taken out of the slits and the piece *mp* be elevated, so that the index can move freely. When it is made to oscillate too much it may be brought to rest by lowering the piece *mp* a moment. If the two needles of the index have exactly the same power, it will have the highest mobility; but if this is not obtained, the bent magnet *uv* is to be so placed upon the pillar *tt* that the two nearest poles of the index are repelled. By approaching or retiring the pillar, the magnet may be brought into such a position that the directive power of the index is scarcely sensible. When the instrument is in this state it can make sensible the difference between two pieces of metal, of which one differs only from the other by $^1/_{100}$ alloys, when a powerful liquid is applied. When a more considerable effect is to be tried, the bent magnet is put in such a position that it attracts the nearest poles of the index. When the magnet is near the index, and the current makes the index deviate very little, the deviation increases as the magnet is removed. The distance of the magnet being measured by the scale, this arrangement may contribute much to the determination of the powers. As the needles submitted to the effect of the current can never rest at an angle greater than 90°, the needle is prevented from going farther by means of two small pins here marked with the Greek letter φ.

The use of the electromagnetical multiplier is very extensive. Before the invention of this instrument, a prepared frog was considered as the nicest test for galvanism; the multiplier surpasses it by far.

Application of the multiplier.

Mr. Poggendorff has made a very extensive trial upon the galvanic series of metals and other conductors, by means of this instrument. Professor Ørsted has made use of it, for confirming the discovery earlier made by Zamboni, upon electrical currents which two pieces of one metal makes with a liquid.[8] He has also discovered, by means of this instrument, that two equal pieces of metal give galvanical effects, when one of the pieces is earlier introduced in the fluid than the other, a fact which Sir Humphry Davy has confirmed, as it appears, without knowing Ørsted's experiments. Professor Ørsted has also made use of this instrument for trying silver.[9] With a powerful liquid conductor, solution of potash and muriatic acid for instance, silver pieces, whose alloy differs less than a hundredth, give a deviation of several degrees. As silver containing brass gives more effect than silver containing an equal quantity of copper, when muriatic acid is employed, but less when solution of potash is the liquid conductor, the presence of brass in silver is easily discovered by this instrument. It need scarcely be mentioned that gold and other metals may be tried in the same manner. Dr. Seebeck, at Berlin, has investigated, with much care, all the circumstances belonging to the construction of the multiplier. These researches are given in an excellent paper, read at the Royal Academy of Berlin, on the 14th December 1820, and the 8th February 1821, containing a valuable detail of experiments upon several points of electromagnetism. Dr. Seebeck has proved, by experiment, what might be presumed in theory, that the increase of the effects of the multiplier, with the number of the turns, is limited by

[8] [Cf. "An Experiment on Zamboni's Double Galvanic Circuit" (Chapter 44, this volume).]
[9] [Cf. "An Electromagnetic Method for the Assay of Silver . . ." (Chapter 58, this volume).]

the resistance against the transmission increasing with the length. The effects of the multiplier increase also with the breadth of the conductor, which he made of a long and thin lamina, in the place of a wire; still the advantage of broad conductors is only confined to experiments with considerable powers: In feeble currents the effects of broad and narrow conductors are equal.

Several philosophers have given themselves much trouble to produce upon the needle, by means of common electricity, the same effects as those produced by galvanism. A simple electric spark transmitted through a conductor passes too speedily to move the needle. A current produced by the electrical machine does not seem to contain a sufficient quantity of electricity for acting upon the needle without the aid of the multiplier. Even by this instrument it was tried often, without decided success, until of late Mr. Colladon, at Geneva, repeated the experiment with a multiplier, in which the wire was covered with three folds of silk, and thus well isolated. Then he approached the two ends of the wire of this instrument to the two conductors of an electric battery of 4000 square inches, so as to make the discharge go a little distance through the air, before it enters in the wire. In this manner a current sufficiently strong, and of some duration, is produced, whereby a considerable deviation is effected. The current produced by an electric machine caused also a deviation of several degrees in this instrument.

Professor Ørsted proposed, in a paper[10] printed in Schweigger's *Chemical Journal*, 1821, to make use of magnetical needles, suspended in various directions for investigating the electrical currents in the atmosphere; but he has published nothing since that time. Mr. Colladon has, with full success, employed the multiplier, to prove the presence of electromagnetism in a thunder storm.

The idea of magnetical revolutions around the uniting wire experienced much opposition at its first publication. Professor Schweigger objected to it, that when such revolutions did exist, it would be possible to make a magnet circulate round the uniting wire. Dr. Wollaston drew the same conclusion, but with the contrary meaning; finding this result probable, he invented an instrument to prove it. The experiment having been stopped by an accident, Mr. Faraday took it up, and made an extensive series of experiments on the subject, conducted with the same skill which he has displayed in so many other investigations. He found that not only the magnet may be made to turn round the conductor, but that likewise a moveable conductor may be made to turn round the magnet. We shall have an opportunity to return to this subject; here we can only give an account of the experiments by which the motion was communicated to the magnet. Fig. 12, represents an apparatus proper for the experiment, *CCCC* represents a cup of glass, or some other non-conductor, through the bottom of which passes the conductor *EFG*. The cup is filled with mercury, in

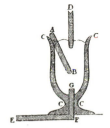

Fig. 12.

which a small magnet *AB* floats, being kept in a vertical position by a piece of platinum, fixed at its inferior extremity. It can also be kept in this position by fixing the inferior extremity to the bottom by means of a short thread of silk. *D* is a conductor

---

[10][Cf. "Observations on Electro-magnetism" (Chapter 42, this volume).]

whose lower end dips in the mercury. When a strong electrical current is now caused to pass through this arrangement, the magnet revolves about the conductor

Fig. 13.

D. The directions of the rotations are in all cases such as the fundamental law of electromagnetism indicates that they should be. A magnet can also be made to turn round its own axis by an electrical current. Let *CCCC*, Fig. 13, be a cup of glass or wood, nearly filled with mercury; *AB* a magnet, having at its lower extremity a steel point, introduced into the agate *H*. *JK* is a slip of brass or ivory, having a hole through which the magnet passes freely, and by means of which it is kept perpendicular at the superior extremity; *A*, is a cavity for receiving mercury; *EF* is a wire, at whose extremity is also a cup for mercury; and at *D* is placed a similar one, from which proceeds a wire amalgamated on its lower extremity, in order to favour the electrical communication. When the electrical current is established by conductors plunged in the mercury at *D* and *F*, the magnet will turn, with great rapidity.

On the power of the Electrical Current in developing Magnetism in other Bodies

In a paper read before the French Institute, the 25th September 1820, Mr. Arago

The electrical current developing magnetism in other bodies.

showed that the electrical current possesses, in a very high degree, the power of developing magnetism in iron and steel. Sir Humphry Davy stated the same facts in a letter to Dr. Wollaston on the 12th November 1820. Dr. Seebeck communicated to the Royal Academy at Berlin, the 14th December, an excellent series of experiments upon the same subject. Thus treated in the space of three months by three so highly distinguished philosophers, the subject was nearly exhausted in the same year that the discovery was made. The uniting wire of a powerful galvanic apparatus attracts iron-filings often with such a power as to form a coating around the wire ten or twelve times bigger than itself. Mr. Arago found that this attraction did not take its origin from any previous magnetism in the iron-filings, which could touch iron without adhering to it; nor was the attraction to be considered as a common electrical one, since brass and copper filings were not attracted. He found also that the iron-filings began to move before they came in contact with the uniting wire. Hence it must be admitted that this attraction is operated by converting each little piece of iron into a temporary magnet. Greater pieces of soft iron were also converted into temporary magnets, and small steel-needles into permanent magnets. Sir Humphry Davy had, in his researches, obtained the same results, before he had got notice of the experiments of the French philosopher. Dr. Seebeck seems to have been in the same case, when he made his experiments; but he had received notice of Arago's experiments when he published his own. The direction of the magnetism produced is always according to the fundamental law. Let the circle in Fig. 14 represent a horizontal section of a perpendicular conductor, in which the current comes from above; let the little arrows indicate the direction of the revolving magnetism, and *BA*, *BA*, *BA*, *BA*, some steel needles; then these needles will obtain austral magnetism at *A*, and boreal magnetism at *B*.

Dr. Seebeck found that a steel needle was strongly magnetized when it was drawn around the conductor. The direction of the magnetism was the same as it should be, if the needle had been laid closely around the conductor, and afterwards removed. He laid also an armour of soft iron on both sides of the conductor, which hereby was made able to bear a considerable weight of iron.

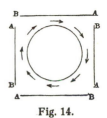

Fig. 14.

Mr. Arago and Mr. Ampère, employed in the development of magnetism the principle of the multiplier, without having notice of the discovery of Schweigger. A steel needle *AB* covered with paper, was surrounded by a winding of the uniting wire *EE*, as represented in Fig. 15.

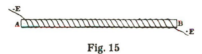

Fig. 15

The steel-needle may also be included in a glass tube. The great galvanic apparatus of the London Institution is now found to develope [sic] magnetism in such an eminently high degree, that a little steel bar, by falling through a glass tube, around which the windings of the uniting wire passed, was magnetized to saturation.

The electricity produced by friction, when employed in sufficient quantity, developes likewise magnetism in steel. The discharge of an electric battery, and even of a single bottle, magnetises a steel needle. All these magnetical effects are submitted to the same law as those of the galvano-electrical current, and hence they are also increased upon the principle of the multiplier. When the discharge passes through the air across the steel-needle, the magnetism developed is feebler than it is when the electricity passes across it through a metallic conductor.

Mr. Savary, at Paris, has of late discovered that steel-needles placed at different, yet small distances from a wire, through which passes an electric discharge, do not all obtain magnetism in the same direction. In one of his experiments he caused to pass the discharge of a battery having twenty-two feet surface through a platina-wire of about three feet in length and one-hundredth of an inch in diameter. The needles in contact with the wire became magnetised in the direction commonly observed, which he calls the positive direction, but a needle placed at a distance of 1.1 millimetre, about $1/24$ inch, becomes magnetic in the opposite direction, which he calls the negative. At the distance of 2 millimetres a needle was not made magnetic by the discharge. At the distance of 3 to 8 millimetres the needles become magnetic in the positive direction, but most at the distance of 5.5 millimetres. From 8.6, to 21.4 millimetres, the magnetic direction was negative, with increasing intensity from 8.6, to 14.6, and with decreasing from this point until 21.4, where it was nearly at zero. From 23 millimetres distance the magnetic direction became again positive.

Mr. Savary's discoveries.

As for different conducting wires, he found, that within certain limits the maximum of effect is the more distant from the wire, and the numbers of alternating directions the greater, in the same degree that the wire is shorter in comparison to its length. In a helix of narrow windings, needles placed parallel to its axis obtain all the same kind of magnetism, but by varying the electrical power, from that of one bottle of Leyden, to that of a battery of twenty-two feet surface, he obtained, in one experiment, six alternations, viz., three positive and three negative. When the needles are included in a metal coating, for instance, wrapt in a lamina of tin, the effect is changed. If the coating is thick, the effect is nothing, but by a coat sufficiently thin the effect may be increased. When the conducting wire is straight, a plate interposed between the wire and the needle, if thin, augments the effect, if thick, diminishes it; a certain thickness may also be found by which the plate is without effect. The needle is in all these experiments in contact with the plate. When the plate is not interposed, but the wire placed upon the plate, the effect of a very feeble discharge is increased by the plate, and still more the thicker it is. At a certain degree of discharge a thin plate diminishes the effect, a thick plate augments it. The effect of very considerable discharges is always reduced to nothing, or inverted by thick plates. By the galvanic arrangement the same effect is not produced, when the current is uninterrupted, but analogous effects to those mentioned may be produced by an apparatus which has intensity enough to give sparks at the moment of closing the circuit. The current must, for this purpose, only be established for a moment; a constant current destroys the alternations.

The analogy of these effects, with those alternations, which may be produced in bad conductors, by common electric experiments is obvious.

Experiments of Mr. Hill. Mr. Hill, at Lund in Sweden, has found that when the discharge passes along a magnetical needle, exactly through its axis, all its magnetism is destroyed. He even considers this as the best means to take away the magnetism of a needle. At the same time he remarks that when the electric charge does not go through the axis, a feeble magnetism is developed on both sides of the line of passage, which probably has led preceding philosophers into an error respecting the magnetical effects of electricity. (Schweigger's *Journal* for the year 1822, No. 3.)

Professor Erman's experiments. Professor Erman at Berlin found that when the electrical discharge passes perpendicularly through the center of a round plate of steel, it reveals no magnetism, but when a split is afterwards made in the plate, or a sector cut out of it, the opposite side of the gap shows the opposite magnetism. The celebrated Gay Lussac and Mr. Welther,[11] without knowing the experiment of the Prussian philosopher, discovered the same fact in a steel ring. This experiment is very illustrative; it shows that the steel disc or steel ring, whose circumference has been in the same state as that of the uniting conductor, preserves after the cessation of the current a latent magnetism, resembling that of a magnetic circle, composed of small magnets, connected by their opposite poles. Such a circle is ineffectual, when the circumference is closed, but be-

[11] [Welter.]

comes a magnet when opened. This magnetism was, however, effectual during the time that the ring or disc was comprehended in the current, wherein its magnetism at every moment received a new impulse. Hence we may conclude that the circumference of the uniting conductor is not to be compared with a magnetic circle, wherein the powers are at rest, which is the theory brought forward by Mr. Prechtel,[12] director of the polytechnic school at Vienna; but our experiment confirms the original idea of the magnetical effect of the current, as produced by a revolving magnetism.

This view of the subject, that the magnetism of the electrical current, is a magnetism in motion, has been overlooked by a great number of authors, who have written upon electromagnetism; while it has been adopted by two highly distinguished philosophers, Dr. Wollaston and Mr. Biot. The difference between magnetism in motion and at rest being until our time unexemplified, this view appeared to many philosophers as a mere postulate, which they tried to avoid, by adopting some other theory, particularly the elaborate theory of Ampère, of which we shall afterwards speak. Now the theory of revolving magnetism has obtained a considerable support by the discovery of Mr. Arago, who, in his researches on the effect of metals upon the oscillations of the magnetic needle, found that it was much affected by a metallic plate, for instance a copperplate, when either the needle or the plate was put in motion. There is [sic] certainly but few philosophers, who have not repeated Arago's remarkable experiment by which a rotatory plate of copper, or some other metal puts a magnetic needle, conveniently suspended, into a revolving motion. We must pass in silence the numerous and skilfully conducted experiments of Mr. Barlow and Dr. Seebeck; and only quote for our purpose those of Messrs. Herschel and Babbage, by which it is proved that a rotating magnet causes a conveniently suspended metallic plate to turn round. Mr. Poisson has read before the French Institute an elaborate mathematical treatise upon the theory of moved magnetism. Thus the theory of revolving magnetism has obtained the only confirmation which could still be desired.

Effects of the Magnet upon the Uniting Wire

Professor Oersted, in the prosecution of his experiments, was well aware that a moveable part of the electrical circuit must be attracted and repelled by a magnet after the same laws by which the uniting wire acts upon the magnet. He published, two months after *Effects of the magnet upon the uniting wire.* his first electromagnetical paper, another paper[13] in which he gives an account of an experiment he made; he found that a little galvanical circuit, suspended by a thin metallic wire was put in motion by a magnet. He complains himself, in this paper, that he had not succeeded hitherto in getting an apparatus sufficiently moveable to be directed by the magnetism of the earth (Schweigger's *Journal*.) Professor Schweigger at Halle, and Professor Erman at Berlin, both invented, without knowing Oersted's experiment, apparatuses fit for the same purpose. It would be te-

[12] [Prechtl.]
[13] [Cf. "New Electro-Magnetic Experiments" (Chapter 40, this volume).]

dious to give an account of all the experiments made upon this subject; a short description of those which are considered as the best, will be sufficient. Fig. 16 represents, with some slight modifications, an apparatus invented by Mr. Ampère.

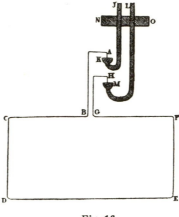

Fig. 16.

ABCDEFGH is a bent wire, of which the two ascending parts at *B* and *G* are isolated from each other by some non-conductor and tied together. At *A*, and also at *H*,

Ampère's apparatus. is soldered a steel point, which reposes on the bottom of a small iron cup filled with mercury, at *K* and *M*. *JK* and *LM* are brass wires, *NO* a piece of wood, in which they are inserted, and by means of which they can be fixed at a convenient place. It appears that when the current enters at the end of one of these wires, for instance at *J*, it is obliged to pass through the whole moveable conductor *ABCDEFGH*, and go out at the other end *L*. This conductor is put in motion with much promptitude by means of the magnet. In comparing this arrangement with Fig. 5, it is obvious that the part *DEFG* of the moveable conductor, in which the current enters at *D*, is quite analogous with *BDCA*, Fig. 5, and that therefore the austral magnetism on the interior side of both, is turned towards a spectator placed over against the place represented by the figure. It is also evident, that the magnetical direction is the same in the part *BCD*, which turns the same side to the space included by the moveable conductor. Thus a magnet whose austral pole is directed against this space, will repel the conductor, but placed near to a point of the exterior side it will attract it. On the opposite side of the plane *BCDEFG*, all the effects are opposite to those here mentioned.

The magnetism of the earth is likewise able to give a direction to the suspended wire. This direction must, in the northern hemisphere, be the same which is produced by a magnet placed below the wire, with its austral pole above, and its magnetical axis put in the direction of the dipping needle; which direction is the same as that which a magnetical needle should tend to give the wire, if it were fixed below it, in the same position which the current gives it. Thus the place *CDEF* must be directed perpendicularly to the magnetical direction; when the current enters at

A, the perpendicular part FE will be placed towards the west, but towards the east, if the current enters at H.

The same reasoning may be employed in all other cases where a moveable uniting wire is exposed to the influence of terrestrial magnetism; for instance, when the wire is suspended in such a way as to permit the particles to move only in vertical planes. Fig. 17 represents an arrangement of this kind. $ABCD$ is a wire, whose two extremities are wrapt round the ends of a thin axis of some non-conductor, and are terminated by two steel points, a and b, destined to be placed in two steel cups filled with mercury, and communicating with a galvanic apparatus. In order to give it the mobility necessary, it is nearly balanced by a counterweight at E. When the axis is placed perpendicularly to the direction of the magnetic needle, and the current enters at a, that is in the west, the plane $ABCD$ will be driven out of its perpendicular position, and deviate towards north: but if the cur-

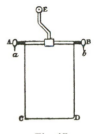

Fig. 17.

rent enters at b, the deviation will be austral. If the axis AB is placed in the direction of the magnetic needle, the deviation will, in the first case, be towards the west, in the last towards the east. The boreal pole of a magnet, placed below DC produces the same phenomena; the deviation goes always to the left of the current.

The principle of the multiplier has also been applied to the moveable uniting wire. Fig. 18 represents one of these contrivances, invented by Mr. Ampère, and somewhat modified by Professor Van de Ross. On the extremity A of the wire is a steel point, resting in a cup with mercury: B is a part of the wire, which forms spirals, fixed on a circular piece of pasteboard, through whose centre it passes at the last, and is prolonged to C, which dips in a cup of mercury. Another apparatus, likewise invented by Mr. Ampère, is represented in Fig. 19. The wire passes through a glass tube, from A to B, it is then wrapt around it, and, being returned to the extremity A, passes also around CD, and being arrived at D is drawn through the tube, and descends finally to the inferior cup.

Fig. 18.

Another apparatus of Mr. Ampère, improved by Mr. Marsh, destined to show the magnetical effect of the earth upon the uniting wire, is represented in Fig. 20; AB is a cup of glass nearly filled with a convenient liquid, containing a galvanical arrangement, and kept swimming upon a liquid by a piece of cork; the uniting wire is like that of Fig. 19.

In the same manner as a magnet can be made to revolve round the uniting wire, so can a moveable uniting wire be made to revolve round a magnet. Fig. 21 shows the principal parts of an apparatus for this experiment, $CCCC$ is a glass cup, having a hole through its foot, into which is inserted a copper tube, soldered to a copper disc, cemented to the foot of the glass. The wire EF is also soldered to another copper disc upon which the glass rests; ns is a magnet inserted in the copper tube. The cup is filled with mercury. At a there is a sort of ball and socket joint, by means

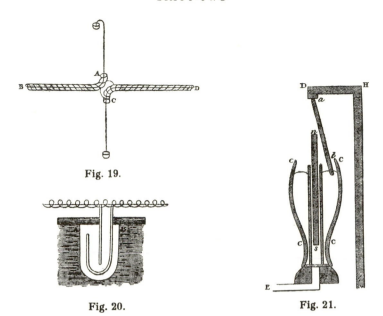

Fig. 19.

Fig. 20.   Fig. 21.

of which a wire *ab* is put in communication with the arm *DH* of a brass pillar: both the socket and the ball are amalgamated, and a piece of silk fixed to the ball or head of the wire, passes through a hole drilled in the arm *DH*, and by which the wire *ab* is suspended, thereby preserving the contact, and leaving to the latter a perfect freedom of motion. When the current is established, the wire *ab* will revolve about the magnet. The directions of the rotations are such as the theory indicates.

We have seen that a magnet can be made to turn round its axis. An apparatus has likewise been contrived for producing the same phenomena in a moveable uniting wire. For shortness sake we shall here omit the description of it; while we give the

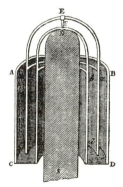

Fig. 22.

description of a very simple turning apparatus invented by Mr. Ampère, and whereof a perpendicular section is exhibited in Fig. 22. *ABCD* and *abcd* are two cylinders of copper, soldered to a bottom of copper, in such a manner that the space between the two cylinders is able to contain a liquid, but the interior cylinder is left open at both its ends. To *a* and *b* is soldered a bent copper wire, having a cavity at *F*. *zz* is a light cylinder of zinc, to which is also soldered a bent wire, in the middle *E* of which is a steel point, resting in the cavity *F*, and consequently the cylinder *zz* will move upon its point of suspension. When the space between the two cylinders is filled with a convenient fluid conductor, an electrical current is established.

Now, if a magnet *NS* is introduced into the cylindrical space of *abcd*, the cylinder *zz* will begin to turn. When the north end (the austral pole) is upwards, the motion is from left to right of the observer, and the contrary with the magnet reversed; all as it could be predicted from the fundamental law of electromagnetism.

Another ingenious contrivance, invented by Mr. Barlow is represented in Fig. 23, where *AB* is a rectangular piece of hard wood, *CD* a wooden pillar, *DEF* a piece of stout brass or copper wire, *ab* a somewhat smaller bent wire, soldered to it at *F*, through the legs of which passes the axis of a wheel *W*, of thin copper, *hf* is a small reservoir for mercury, and *gi* a narrow channel running into it. *H* is a strong horseshoe magnet.

Mr. Barlow's apparatus.

Fig. 23.

Mercury being now poured into the reservoir *hf*, till the tips of the wheel are slightly immersed in it, and the surface covered with weak dilute nitric acid, let the connection with the battery be made at *i* and *D*, and the wheel will immediately begin to rotate. If the current or the magnet be inverted, the motion of the wheel will also be reversed. In order to understand this experiment, it must be remarked, that each radius of the wheel, which touches the mercury, is a part of the uniting conductor, of which one side is repelled by the austral, the other by the boreal pole of the magnet; thus it must either tend to raise or depress each of these radii.

Sir H. Davy has exhibited the rotation of a conductor by means of mercury. When in a shallow non-conducting vessel containing mercury, the conductors of a powerful galvanical arrangement are plunged at some distance from the sides, and one of the poles of a strong magnet is brought from below to the bottom of the vessel, near one of the conductors, the mercury round this conductor will form a vortex about it. The directions of the motions are always according to the poles and conductors in action, such as the fundamental law indicates.

Sir H. Davy's experiments.

When a moveable part of the uniting wire is placed in the direction of the dipping needle, it cannot be put in motion by the magnetism of the earth; but when it is placed in another plane, though under the same inclination, it is put in motion. Professor Pohl at Berlin, has invented an apparatus, represented in Fig. 24 exhibiting this phenomenon. *AB* is a piece of board, supported by screws, by means of which it can be levelled. *CD* is a wooden pillar, whose superior part is moveable, and has on its top an

Professor Pohl's apparatus.

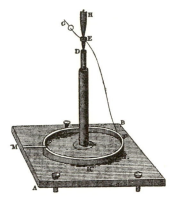

Fig. 24.

agate, which serves to support a steel-point, whereupon rests a wire *EF*, balanced by a counter-weight *G*. At *E* is a cavity containing a drop of mercury, by means of which one of the conductors, whereof only a part *H*, here is represented, may be made to communicate with the moveable wire. *JKL* is a circular channel containing mercury, which can be put in communication with the galvanic apparatus, through a conductor at *M*. When a powerful electric current is transmitted through the apparatus, *EF* can only rest in the position of the dipping needle; in all others, it moves until it arrives at that position, which it nevertheless will leave by the motion already obtained. Hence it must still continue to turn, when it is not stopped, to the position in which it is possible for it to rest.

Mutual Action of Electrical Currents

Mr. Ampère found, soon after the discovery of electro-magnetism, that *two con-*
Mutual action of   *ductors attract each other, when they are transmitting electri-*
electrical currents. *cal currents of the same direction, but that they repel each*
other, when the currents have opposite directions.

The moveable conductor, represented in Fig. 16, and already described, may be employed to prove this by experiment. As the current which passes through the moveable wire *ABCDEFGH*, has in *CD* the opposite direction of that in *FE*, the same uniting wire, which attracts one of these, will repel the other. This experiment may be exhibited in various shapes; but it does not appear that any experiment which could not be made by this simple apparatus, is necessary for confirming the law above-mentioned.

This law may easily be deduced from the fundamental law of electro-magnetism, as may be seen by Fig. 25, which represents two parallel currents of equal direction, and expressed by the same signs of which we have made use in the preceding pages of this article. It is here evident, that the boreal magnetism at *b* meets with the austral at α, and that the austral at *a*, meets with the boreal at β, thus the effect must be attraction. In Fig. 26, two currents of opposite direction are rep-

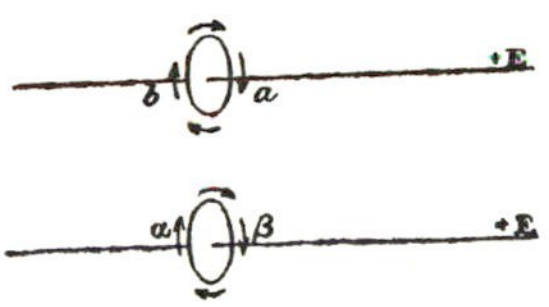

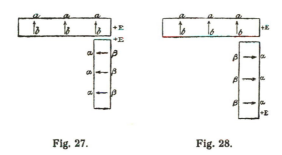

Fig. 25. 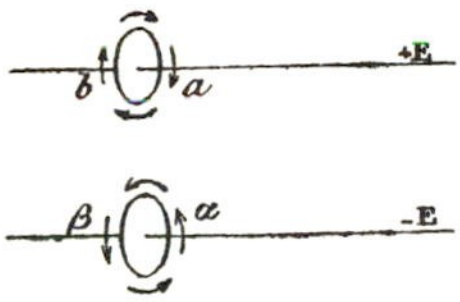   Fig. 26.

resented, where the boreal magnetism at b, meets with that at β, and the austral magnetism at α, with the similar at a: which must produce repulsion.

When the currents are not parallel, but form an angle, they attract each other when both are directed either towards the apex or in the contrary way, but they repel each other when one of the two currents is directed towards the apex, at the

Fig. 27.   **Fig. 28.**

same time that the other goes off from it. Fig. 27 represents two currents which go off from the apex. The boreal magnetism being in one of these directed from b to a; the austral magnetism in the other from β to α, the result must be an attraction by which the conductors, if one of them is moveable, are brought to parallelism. The figure represents only one side of the conductors; but the opposite sides, having both their magnetical directions reversed, will likewise be attractive. It is also easily understood, that the opposite magnetical poles are directed against each other, and produce attraction when the current in both conductors goes towards the apex of the angle. Fig. 28 represents two currents having opposite directions, with respect to the apex of the angle. Here the similar poles in the magnetical rotations, are directed against each other, and therefore produce repulsion, such as to place both conductors in the opposite ends of one straight line, if one of them is moveable.

This may be confirmed by means of the apparatus represented in Fig. 29, consisting of two parts, viz., a moveable conductor, ACB, and a multiplying wire, $DEFG$. The moveable wire is terminated by two steel points at A and B, which are to be placed in two small steel cups, filled with mercury, and communicating with a galvanic apparatus. The multiplying wire is preferred to a straight one, in order to increase the effect. The upper part DE, of the multiplying wire is placed at the same height as the branch BC of the moveable conductor; but in such a position that both conductors prolonged would form an angle. The extremities F and G of the multiplying wire, are to be put in communication with a galvanic apparatus.

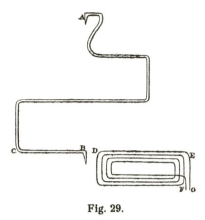

Fig. 29.

Mr. Ampère, to whom we are indebted for the discovery of the mutual attractions and repulsions of the electrical currents, considers the law of this action as a fundamental one, at least so far as our present knowledge extends. He thus admits no rotative action in the electrical current, but he transports it to the magnet, in which he supposes electrical currents, revolving in planes perpendicular or nearly perpendicular to the axis of the magnet. At first he supposed that all the currents had their centres in the axis, and were situated in planes perpendicular to this axis, but as he soon found that this would not represent the phenomena, he supposed that each atom of the magnet was surrounded by electrical currents, still revolving in planes perpendicular to the axis of the magnet. When Mr. Poisson, however, showed, that in consequence of this view the greatest effect of a magnetical bar would be placed in its extremity, contrary to experiment, he changed this supposition, and at present he is of opinion that the currents are situated in a plane somewhat inclined to the axis of the magnet.

By these suppositions, and a considerable exertion of mathematical skill, he is enabled to make this view represent well enough the phenomena, though his theory is very complicated. It is not necessary here to enter into a discussion on all the points of this theory, as simple consideration of the fact upon which it is founded will be sufficient to decide the question.

Let us suppose that electromagnetism had not been discovered before the discovery of the mutual action of electrical currents, the application of the common philosophical rules should enable us to discover therein the rotative character of the action. The fact is, as above mentioned, that parallel currents attract each other, when they have the same direction, and repel each other, when they have opposite directions. Now it is to be remarked, that two parallel things of the same direction have their opposite sides placed against each other: the left of the one is nearest the right of the other; but two parallel things of contrary directions have their similar sides turned against each other: right against right, or left against left. Thus the fact reduced to the simplest philosophical expression is that *two points of electrical currents repel each other by their similar sides, and attract each other by their opposite sides.* The most direct enunciation of the experimental result cannot here be considered as at the same time the expression of the philosophical one; for it is evident that two parallel things cannot act upon each other immediately, but only by some transverse action, which here shows itself as consisting of attractions and repulsions in opposite directions, or in other terms, as having polarity. But such contrary powers forming a circle, should keep themselves in equilibrium, and produce no effect without their limits, were they not in motion. Thus the very experiment of Mr. Ampère should, in the absence of all other evidence, be sufficient to prove, *that the electric current contains a revolving action, exhibiting every appearance of polarity.* We do not mean to ascertain the nature of these attractions and repulsions; but it has been our object only to point out the more immediate consequences of the facts.

ELECTROMAGNETICAL CURRENTS PRODUCED BY HEAT

Dr. Seebeck, in his researches upon electromagnetism, extended at the same timehis investigations to the laws of galvanic action, and among these to the influence of heat in galvanic arrangements. Some phenomena *Electromagnetical* here occurred to him, which led him to think that two metals, *currents produced* forming a circuit, might produce magnetismwhen the equi- *by heat.* librium of heat in it was disturbed. Experiment confirmed this opinion. Fig. 30 represents such a circuit; let *ABC* be a piece of bismuth, and *ADC* apiece of copper, and let one of the junctions, *A* for instance, be heated, an electrical current will be established, which here can only betray its existence by the magnetical needle; this indicates all the magnetical properties of an electrical circuit, and, in the instance here mentioned, the current goes into the heated junction from the bismuth to the copper. Dr. Seebeck is not inclined to consider the effect thus produced as a true electrical current, but an effect *sui generis*; and indeed we have not hitherto been able to discover in this circuit either any chemical effect, nor heat or light; still we can represent all the phenomena of Dr. Seebeck's circuit by the same terms as those of the

Fig. 30.

common electrical current: and in the explanation of all the facts, it will appear highly probable that this current is truly a particular kind of electrical one. Professor Oersted has proposed to call the current discovered by Dr. Seebeck the *thermoelectrical* current, and in consequence of this to distinguish the action hitherto called Galvanism, by the name of the *Hydro-electrical current.* Hence we have now the names *thermo-electricity* and *hydro-electricity,* to which we could add the name *tribo-electricity* for the electricity produced by friction. Dr. Seebeck has made a very considerable number of experiments upon the thermo-electricity produced by the metals and other perfect conductors. In a circuit containing bismuth, together with one of the other metals, he finds that, in the heated junction, the current goes always from the bismuth to the other metal; of course the bismuth loses, at that point, positive electricity. This we shall, for shortness sake, express thus: bismuth becomes negative with all other metals in the thermo-electrical circuit. In the same sense tellurium may be said to become positive with all other metals. It appears already by these two examples, that the thermo-electrical order of the metals is not the same as the hydro-electrical; and indeed the experiments of Dr. Seebeck have proved that these two orders are discrepant throughout.

The order of the metals, beginning with that which becomes negative with all others, is,

1. Bismuth.
2. Nickel.
3. Cobalt.
4. Palladium.
5. Platinum.   Several pieces of this metal gave very different results, even those which came from the same workshop. Three pieces from Jeannetty's platina manufacture were placed in the order of their effects very far from each other. The pieces which kept this place here between palladium and uranium were prepared by Dr. Wollaston, Mr. Bergemann, chemist at Berlin, Mr. Trick, chemist, appointed to the manufacture of china at Berlin, and Mr. Jeannetty at Paris. As one of these pieces was prepared by Dr. Wollaston, and the two Berlin chemists being men of much chemical skill, we may consider this place as that of the pure platinum, if Mr. Becquerel had not found that two parts of the same platinum wire give a considerable thermo-electric action, when one of them was drawn out so as to become much thinner. Hence it appears that the density of the platinum has a considerable influence upon its thermo-electrical effect. this might perhaps also be the case with other metals.

6. Uranium.
7. Copper,   reduced from the oxide, by means of black flux, Comp. No. 12.

8. Manganum.
9. Titanium.
10. Brass,   some specimens. (Comp. No. 13.)
11. Gold,   of Hungarian ducat containing $1/90$ alloy of silver and copper.
12. Copper,   occurring in the trade, and containing no silver, iron, lead or sulphur. (Comp. 21.)
13. Brass,   some specimens. (Comp. No. 10.)
14. Platinum,   a piece of unknown origin. (Comp. No. 5, 18, 29.)
15. Mercury,   the purest occurring in trade.
16. Lead,   specimens occurring in trade, and pure lead.
17. Tin,   English and Bohemian.
18. Platinum,   a bar from Jeannetty's manufacture.
19. Chromium.
20. Molybdænum.
21. Copper,   occurring in trade, and containing neither silver, iron, lead or sulphur. (Comp. 12.)

22. Rhodium.
23. Iridium.
24. Gold,   *a*, purified by antimonium, *b* reduced from the oxide.
25. Silver,   *a*, purified by cupellation, *b* reduced from the chloride
                             of silver.
26. Zinc,   *a*, occurring in trade, *b* pure zinc.
27. Copper,   reduced from sulphate of copper, *a* by iron, *b* by zinc.
                             (Comp. 12 and 21.)
28. Wolfram.
29. Platina,   some specimens, (Comp. 5, 14, 18.)
30. Cadmium.
31. Steel.
32. Iron,   *a*, occurring in trade, *b*, pure iron.
33. Arsenic.
34. Antimony.   *a*, occurring in trade, *b*, pure.
35. Tellurium.

In this series, Dr. Seebeck found that though most of the metals placed here near each other give only a feeble effect, and the more distant a stronger effect, this rule is not constant; tellurium, for instance, gives with bismuth less effect than antimony. With most of the metals in the series tellurium produces a feebler effect than antimony; with silver it produces more effect than with most of the metals placed above it. Antimony produces more effect with cadmium than with mercury. Iron produces only a feeble effect with most of the other metals, and particularly with nickel and cobalt. Of such exceptions Dr. Seebeck has found a great many.

Dr. Seebeck also examined the thermo-electrical powers of several other bodies. Sulphuret of lead becomes negative even in contact with bismuth. Some other sulphurets, as sulphuret of iron, of arsenic, of cobalt and arsenic, of copper, all with a maximum of sulphur, stand in the thermo-electrical series very near to the bismuth. On the contrary, the sulphurets with a minimum of sulphur stand very near to antimony; that of copper stands even under antimony.

Dr. Seebeck found also that concentrated nitric and sulphuric acid are to be placed above the bismuth, but that a concentrated solution of potash or of soda, obtains a place below antimony and tellurium.

Dr. Seebeck constructed also circuits of two pieces of one metal; heating or melting one of the pieces, and putting one extremity of the other piece, which must be bent, in durable contact, while the opposite extremity was in temporary contact with the heated piece. A bent silver wire was, for instance, plunged first with one of its extremities and afterwards with the other in melted silver; the magnetic needle indicated that the current was directed from the melted metal to that extremity which had been the longest time in contact. The same effect, though feebler, was observed when the silver had ceased to be liquid. When a platina wire is tried with a heated piece of platina the direction of the current is opposite. The general result of Seebeck's experiments is, that in the metals of the superior part of the thermo-electric series the direction of the current is as in the platina going from the heated metal to that extremity of the bent piece, which is latest put in contact with it; but in the inferior part of this series the current goes, as in the silver, from the heated metal to that extremity of the other metal, which has been longer in contact with it.

As soon as the thermo-electrical current was discovered, it was obvious that a compound thermo-electrical circuit might be formed, in analogy with Volta's complex hydro-electrical circuit. This consequence did not escape Dr. Seebeck, but discovering some opposing circumstances, which we shall soon mention, he bestowed little labour upon this subject, to which he perhaps proposed to return another time. Baron Fourier and Professor Oersted undertook, without knowing this observation of Dr. Seebeck's, a similar research.[14] Their first complex thermo-electrical circuit was a hexagon formed of three pieces of bismuth and three of antimony soldered together. One of the sides was put in the magnetic direction, and a compass placed below it, when first one of the junctions was heated, then two, not adjacent junctions were heated, at last three, still leaving between two heated junctions one which was not heated. The compass needle changed its direction some degrees by the heating of one of the junctions, still more by the heating of two, and most when all three junctions were heated. By cooling the three junctions by means of ice, and leaving the three others to the temperature of the atmosphere, similar and even more comparable effects were produced. By heating three alternating junctions, and cooling the other with ice, the effect rose to 60° of the compass used in the experiment. In another series of experiments a rectangular circuit of 22 bars of antimony and 22 of bismuth soldered together was employed. Here likewise as in the preceding experiment, the combined effect of heating and cooling was employed. Now the circuit was opened by dissolving one of the junctions, and, in order to establish the circuit, when required, a little cup of brass destined to contain mercury, was soldered to each of the two bars, whose conjunction was interrupted. A copper wire of about 4 inches in length, and $^1/_{25}$th inch in diameter re-established nearly the current; and by two parallel pieces of this wire the current was brought to the full effect. A wire of the same diameter, but a little more than three feet long, was found a tolerably good conductor, while a platina wire of $^1/_{50}$th inch and about 16 inches long scarcely transmitted a fortieth of the effect. Liquid acids and solutions of alkalis or other metallic oxides, which prove excellent conductors in the hydroelectrical current, were found quite isolating in the thermo-electrical circuit. Two discs of silver, separated only by a lamina of the thinnest blotting paper, moistened with sulphate of copper, isolated likewise the whole effect of the thermo-electrical current.

The thermo-electrical current, even the most intense that was tried, produced no visible chemical effect; nor was it capable of producing heat in thin metallic wires, probably because they are too feeble conductors of thermo-electricity.

The thermo-electrical circuit also produces no effect upon the electrical condensation.

It is very remarkable that, notwithstanding all that has been mentioned, the thermo-electric circuit makes a prepared frog palpitate, like the hydro-electrical circuit. The communication between the extremities of the circuit and the nerves

[14][Cf. "On Some New Thermoelectric Experiments" (Chapter 51, this volume).]

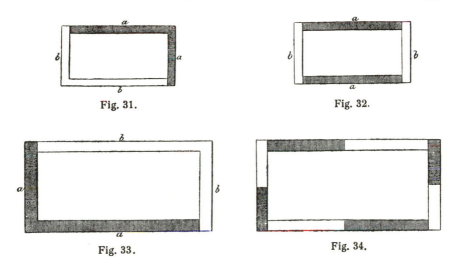

Fig. 31.   Fig. 32.

Fig. 33.   Fig. 34.

of the frog were made by means of platina wire, in order to guard against the influence of unequally oxidated surfaces.

Among circuits differing only by their length, the shortest has the greatest effect. A circuit of double length has not much more than half the effect. Complex circuits do not seem, therefore, at first sight, more efficacious than simple ones; the length being as much increased by the increased number of elements, as the effect should be heightened by the greater number of acting junctions; but comparing circuits of equal length whereof one has only two junctions, the other more, we see the true influence of the increase of acting junctions. Fig. 31 represents a simple circuit of antimony *aa*, and bismuth *bb*, where only one of the junctions is to be heated or cooled. Fig. 32 represents a complex circuit of the same length, formed of two pieces *aa* of antimony, and two pieces *bb* of bismuth. Two of the junctions of the latter arrangement, situated on the extremities of one diagonal are here heated or cooled. Under the same changes of temperature, where the circuit, Fig. 31, made the needle to deviate about 22 degrees, that of Fig. 32 made it to deviate about 30 degrees. Fig. 33 and 34 represent two circuits of double the extent of the former, one simple, one having three[15] alternations. By the same differences of temperature, by which the arrangement, Fig. 33, gave from 13 to 15 degrees, that of Fig. 34 gave nearly 32 degrees.

In several complex circuits, it is found that the heating or cooling of one junction only produces twice the angular deviations of that added by the addition of each active junction more. The effect of one active junction, when the others are at rest, is by experiment found to be twice the effect of all the arrangements, divided by the sum of the elements + one. The effect of each addition of a new active junction is only half this quantity, and seems even to be in a decreasing ratio, when the number of junctions is great.

[15] [four.]

The effect of thermo-electricity upon the multiplier is very instructive. Fig. 35 represents an arrangement formed by two pieces b, b, of bismuth, and one piece a of antimony. When the two free extremities of b, b, are put in communication with

Fig. 35.

the extremities of the wire of the multiplier, and one of the junctions between a and b is heated or cooled, the needle of the multiplier is deviated, but very little; when one of the junctions is only cooled with ice, the effect is not so great as that of a disc of copper with one of silver, having common water as the liquid conductor. But when the extremities of b, b, are put in communication by means of a short piece of metal, the effect on the compass needle is considerable, whereas the effect of the hydro-electrical current of silver and copper, and even of silver and zinc, with common water as the liquid conductor, is scarcely sensible upon the same compass needle. This is a strong additional proof of the difficult transmission of thermo-electricity.

From all these obervations we must conclude that the thermo-electric current produces an enormous quantity of electricity, but in a state of exceedingly small intensity. In order to conceive this well, it is to be remarked that the *intensity* of electricity is measured by the attractions and repulsions, whose force is in the inverse ratio of the squares of the distances, and that the *quantity* of electricity is measured by the number of equal surfaces which can be electrified by it to a certain degree of attraction and repulsion indicated by the electrometer. In the voltaic pile the intensity increases with the number of discs, the quantity with the surface of each of the discs. The greater the intensity the greater is the power of surmounting obstacles, or of penetrating through imperfect conductors; on the contrary, the greater the quantity the more perfect conductor is required to transmit it. The electricity produced by some thousand pairs of discs is able to penetrate a little lamina of air; that of some hundred pairs can at least penetrate through a considerable length of water; that of two pairs cannot easily be transmitted but by the solid conductors and some of the powerful liquid conductors.

The thermo-electrical current has a prodigious quantity of electricity in comparison with the hydro-electrical of silver, zinc, and water, but the intensity of the electricity is much greater in the latter; the electricity of the former is impaired by the resistance of the long multiplying wire, the electricity of the latter surmounting this resistance is on the contrary increased by the multiplying wire.

The complex thermo-electric circuit produces much more effect upon the multiplier, not only when the increased number of elements heightens the effect upon the compass needle, but still also when this increase does not augment the direct effect upon the needle. We must therefore conclude that the intensity increases with the number of the elements in the thermo-electrical as well as in the hydro-electrical current. It must therefore be possible to attain an intensity of the thermo-electrical current great enough for penetrating the liquid conductors, and producing the most considerable chemical effects. Still the construction of a thermo-electrical circuit of a great number of elements is very difficult, because the elements must be as short as possible in order to preserve the conducting faculty; but even the smallness of the distance between the heated and cooled parts must give way to a very speedy re-establishment of equilibrium. The best way seems to be, to produce the heating and cooling of the junctions by some continual current of hot and cold liquids.

A very easy manner of constructing thermo-electric batteries deserves to be

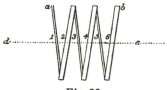

Fig. 36.

mentioned. Fig. 36 represents it. The parts indicated by the odd numbers 1, 3, 5, represent copper slips, and those indicated by the even numbers 2, 4, 6, small bars of bismuth. All the junctions situated on one side of the dotted line *cd*, are to be heated, those on the other side are to be cooled. The extremities *a* and *b* are to be connected by a conductor. The number of elements may here be tolerably great.

Thermo-electric batteries.

That the intensity of the electro-magnetic current must increase with the temperature was to be presumed; but this is not a general law. Dr. Seebeck had already found some exceptions, and also Professor Cumming at Cambridge, who made his experiments without knowing those of Dr. Seebeck upon this subject. We shall not stop here to detail these experiments, as another philosopher, Mr. Becquerel, availing himself of the imposed instruments of research, and making a very ingenious application of them, has given us exact measures of the quantities here occurring.

It was supposed that the declination of the needle, produced by the electrical current is in the ratio of the sine of the angle of deviation. Though this is a consequence of the resolution of powers, he thought that, in a matter so little known as the magnetical effects of the electrical current, it might be advisable to examine the law of this measure by experiment, particularly with regard to the multiplier, where the current makes so many windings round the needle. In order to execute this plan, he formed his multiplier with four parallel and equal wires, covered with silk, and each making an equally great number of windings. Thus he had four multiplying windings about one frame. To the ends of each multiplying wire he soldered the ends of an iron wire, so that four thermo-electrical circuits, consisting of the copper wires of the multiplier and the iron wires were formed. When he wished to put one of these currents in activity, he cooled one of the junctions with ice, and heated the other in mercury. The junction was included in a thin bent glass tube, in order to guard it against the dissolving power of mercury. The mercury was heated by means of a lamp, somewhat above the temperature required, and when heated the lamp was taken away; thus the temperature remains for a short time stationary. In this manner he tried first the effect of one, then of two, three, or four of the multiplying circuits, and noted down the deviations produced, one of the junctions still being kept at the freezing point. Thus he found that one of the circuits gave, by 5° Centigrade

Mr. Becquerel's experiments.

or 9° Fahr. above the freezing point (41° Fahr.) a magnetic deviation of 0.65° French division, or 0.585° of common division of the arc. Two circuits gave by the same temperature twice 0.585°; three gave thrice, and four gave four times this quantity; whence he concluded, that when one circuit produces $4 \times 0°.585$ it has four times the power of that producing only 0°.585. It is easily understood that the greater angles of deviation could not be in the same ratio as the action; but this does not hinder us from drawing analogous conclusions. Thus by a difference of 180° Fah. one circuit gave the deviation 10.71° of the circle; but two circuits gave nearly the same (10°.575) by a difference of 90° Fah. But it is not in all temperatures that this proportion of the effect and temperature takes place; in very high degrees of heat he found that the effect of circuits of copper and iron did not increase so fast as the temperature. From the freezing point (32° Fah.) up to 284° Fah. the magnetical effect increases with the temperature. From this degree to 572° the magnetic power, though increasing with the temperature, still proceeds in a decreasing progression; and exposed to the immediate action of a lamp, the current is inverted. When none of the junctions is at the freezing point, the effect of the circuit is equal to the difference of the effect, which each of the two temperatures applied to one of the junctions, the other being at the freezing point should give; thus, for instance, a circuit of iron and copper, when one junction is heated to 392° F., the other being at 32° F., has an intensity expressed by 37; but when the heat is only at 212°, the intensity is expressed by 22. The difference between these two numbers is 15, which is found by experiment to be the effect of the circuit, in which one junction is heated to 392°, and the other to 212°. He found that a complex circuit of copper and iron produced an effect proportional to the number of elements, which is not the case, when the whole power of the circuit can be exerted, but is only so, when a very small part of the whole effect can be transmitted through a conductor, of such a length or feeble conducting faculty, that it requires much intensity of electricity, for being penetrated. Thus the observation of Mr. Becquerel proves, what had already been shown by less perfect experiments, *that the intensity of thermo-electricity increases as the number of the elements.*

Circuits of iron, with gold or silver, have likewise, as well as those which it forms with copper, a *minimum* of effect, by a certain elevated temperature, and in a still higher one their current changes its direction. In circuits of platina with gold, silver, lead, zinc, copper, and palladium, the differences of the intensities form an increasing arithmetical series.

Mr. Becquerel found that two pieces of platina form an active thermo-electrical current, when they are not of a perfectly equal nature. He cut through a piece of platina wire, and had one of the pieces drawn thinner; these two formed a thermoelectrical circuit. He maintains that the circuit is not efficacious unless a piece of some other metal is soldered to the one end of the wire, upon which statement we cannot but entertain some doubt, though Mr. Becquerel's authority is of no little weight. As Mr. Becquerel had found that the increments of the magnetic effect preserve the more their proportion to the increments of temperature, the more difficult the metal is in being melted [sic]. He considers a circuit of two unequal pieces of platina as a pyrometer. By means of this, he has tried the temperature of the different parts of a spirit-flame, and estimated the temperature of the blue flame bordering the white, at 1350° Centigr., or 2462° Fahr.; in the white part he estimated it to be 1080° Cent. or 1976° Fahr., and in the darker part of the flame to be 780° Cent. or 1436° Fahr. The last he considers as too high, because the other parts of the flame contributed to heat the junction.

TERRESTRIAL ELECTRO-MAGNETISM

We cannot pass by this subject entirely, though we must treat it very briefly. Mr. Ampère, who thinks that magnetism consists only in transverse electrical currents, must, in consequence of his hypothesis, suppose an electrical current round the earth, from east to west. He thinks that the numerous strata, of which our globe is composed, may *Terrestrial Electro-Magnetism.* form considerable galvanic arrangements; still he supposes that the rotation of the earth cannot but have an effect on the electric currents around it. Mr. Ampère, in consequence of his system, admits no other magnetism of the earth than these currents. The opinion, that the earth is surrounded by electrical currents, though not strictly proved, is very probable. As for the galvanic arrangements which the earth is supposed to contain, there can be no doubt that the strata of the earth may form such combinations; but it is not at all proved that they produce a current from east to west. As far as the different currents formed by the strata, do not destroy the effect of each other, it is probable that their resultant effect lies nearly in the perpendicular; for the most general situation of the strata, is that one is placed above the other, generally with some inclination; but as this inclination may have all possible directions, the effects of the galvanic arrangements, (in so far as their action should have a horizontal direction, and thus be founded upon the inclinations,) must destroy each other, even if the inclination towards one side should be somewhat predominant; for galvanic arrangements combined in variable directions of their currents, produce a total effect much feebler than the difference of their positive and negative effects. The most efficacious excitation of electricity upon the earth appears to be produced by the sun. Its light passes round the globe from sunrise to sunrise, and produces evaporation, deoxidation and heat. Evaporation in contact with oxidable matters, produces electricity, as has already been asserted, but first exactly elucidated by the ingenious experiments of Mr. Pouillet. That the deoxidation which the sun produces during the day not only of the surface of plants, but also upon the surface of many other bodies, particularly when moistened, excites electric currents, is a well-known galvanic fact. That the heat produced by the sunbeams, and also circulating from east to west must produce an electrical current can scarcely be doubted; for though the surface of the earth be not composed of perfect conductors, and this resistance should make a common current insensible, the celerity of the circulation may, on the other hand, augment the effect to a degree sufficient for producing some effect upon the magnetic needle. Now, if it be admitted that the sun produces an electric current round the earth, this current must form a zone of considerable breadth, whose most intense part is situated in the plane of the circle, in which the sun seems to make its daily motion. Thus the situation of the most intense part of the zone varies with every day of the year. If we suppose that the earth had no other magnetism than that of this zone, a steel needle made magnetic by an artificial current, and then freely suspended, should take a direction towards the north and the south. Even a steel needle laid across the great natural current should be made magnetic, and suspended, take its direction accordingly. But the great current must also produce magnetism in the body of the earth itself; and as the magnetic effects of the inferior side of the current are opposite to those of the superior one, the magnetic poles of the earth become the opposite to those of the needle directed by the current, and should therefore, if we for a moment suppose the electric zone destroyed, still give it the same direction. Thus the earth seems to have a constant magnetic polarity, produced, in the course of time, by the electrical currents which surround it, and a variable magnetism produced immediately by the same current. As the sun does not produce an equal effect upon water as upon solid bodies, the intensity of the current cannot be equal in all parts of a parallel circle, and therefore the direction of the needle cannot be perpendicular to the equator, nor can it form everywhere the same angle with the equator, for the lines of equal electromagnetic intensity must be twice bent by the influence of the two great masses of continent. The yearly and daily change of the electromagnetic zone, must occasion yearly and daily variations. As to the variations comprehended in greater periods, we might perhaps attribute them to a motion of the coolest points in each continent, which it appears cannot remain the same for ever, because the currents of warmer air must principally be directed towards such points; but we shall leave this research to future times, which may discover causes concealed from us, for explaining the great and secret revolution, which is continually performing in our globe.

It would be to offend against a love of truth, if we proposed these views as ascertained facts. Our researches upon the magnetism of the earth have been, during too short a time, directed by the electromagnetic discoveries, to enable us to give a complete theory of this subject. The great series of profound mathematical and philosophical investigations by which Professor Hansteen, at Christiania, has confirmed and improved the theory established by Dr. Halley, shows how many difficulties are to be surmounted. The accordance of this theory with observation, seems even to exclude the possibility of a new theory; but it must be remarked that this theory is only a mathematical representation of the phenomena, and does not pretend to be a physical one. In the same way as the mathematical laws of the celestial motions were discovered by Kepler, long time before their physical laws were superficially guessed by Hooke, or profoundly recognised and demonstrated by Newton, so the physical laws of the magnetism of the earth may now, perhaps, be fairly conjectured, and in a future age be brought to the requisite degree of perfection. Still we hope that these views will recommend themselves to farther investigation, as they would, if proved, have the great advantage of showing an intimate connection between an extensive series of phenomena upon the earth and those of the universe.

Some Theoretical Considerations

The question has during late years been often proposed, *whether or not magnetism and electricity are identical.* There has been a good deal of misunderstanding in the discussions on this subject. Mr. Ampère pretends that the *Some theoretical* discoverer of electro-magnetism, though he had earlier ad- *considerations.* mitted the identity of these effects has, in his first paper upon electromagnetism, denied it. We must here remark that the words have two acceptations; in one of these Professor Oersted is perhaps the most earnest supporter of this identity, in the other he is a no less decided opponent of it. His opinion is, that all effects are produced by one fundamental power, operating in different forms of action. These different forms constitute all the dissimilarities. Thus, for instance, pressure upon the mercury of the barometer, wind and sound, are only different forms of action of the same powers. It is easy to see that this fundamental identity extends to all mechanical effects. All pressures are produced by the same powers as that of air; all communications of motion, and likewise all vibrations, owe their origin to the same expansive and attractive powers, by which each body fills its space, and has its parts confined within this space. This fundamental and universal identity of mechanical powers has for a long time been more or less clearly acknowledged; but the effects which have hitherto not been reduced to mechanical principles, seemed to be derived from powers so different, that the one could scarcely be deduced from the other. The discoveries which began with galvanism, and which have principally illustrated our century, led us to see the common principles in all these actions. Two or three years before the beginning of the century, Ritter had, by means of the simple galvanic arrangement, pointed out and distinctly stated the principle of the electro-chemical theory; still his ideas were not generally admitted before the discovery of the Voltaic pile had struck the mind of the experimental philosophers with more palpable facts. That heat and light are produced by the union of the opposite electrical powers, had been acknowledged by the Swedish philosopher Wilcke, a cotemporary of Black, but this view was far from being generally admitted. Winterl brought it forward in 1800, and was supported by Ritter and Oersted. The last investigated the subject farther, and developed some of the principal laws of the generation of heat by the electrical and chemical powers.[16] He proved that the electrical powers are present in all cases where heat and light are generated. That magnetical effects can be produced by the same powers need not here be mentioned. As the chemical powers give rise to expansion and contraction, it appears that their nature is not different. Thus acknowledging the fundamental and universal identity of powers, effects must be considered as different, when their form of action differs, and therefore magnetism, in this acceptation of the term, is far from being identical with electricity. It would likewise be erroneous to pretend that all chemical effects are produced by electricity; but the truth seems to be, that the chemical effects are produced by the

[16] *Recherches sur l'identité des forces électriques et chimiques*, pp. 193–233. [Cf. *View of the Chemical Laws of Nature* (Chapter 30, this volume).]

same powers which, in another form of action, produce electricity. The name of *electro-chemical* theory, given to the modern chemical system, seems therefore less admissible than the denomination of *dynamico-chemical* theory, proposed by Oersted so early as 1805. It is still true that the common electro-chemical theory deserves its name, as it does not go out of the limits of an electrical view of the subject. This theory stops throughout in generalities, and gives no account of the disparities of the effects. We will not pretend that a sufficient dynamico-chemical theory has hitherto been pointed out; we must even admit that our knowledge is not ripe enough for this purpose; but we think that some laws, accounting for the disparities, have been pointed out in the work above quoted, upon the identity of electrical and chemical powers (viz., fundamental powers), and that the ideas therein explained deserved attentive examination. The dynamico-chemical theory must still remain very imperfect, until it is decided if the powers acting in magnetism, electricity, heat, light, and chemical affinities are to be ascribed to vibratory, circulating, and other internal motions or not. That these effects do not pass without the most remarkable internal motions, appears from the experiments upon light and upon electro-magnetism. The electrical current is a system of rotative motions, upon whose directions, perhaps, all the disparity of positive and negative electricity depends. It is not improbable that even magnetism involves some rotations, and thus the opinion of Mr. Ampère comes to agree with ours, at least in this point. When the transmission of the electrical current through liquid bodies is accompanied with a chemical decomposition, it seems necessary to admit that the substances styled electro-positives and electro-negatives, must rotate in opposite directions, and we may suppose that their neutralizing powers are connected with the propensities to those opposite rotations. The new discoveries, in short, reveal to us the world of secret motions, whose laws are probably analogous to those of the universe, and which deserve to be the subject of our most earnest meditations.

63

An Explanation of Faraday's
Magneto-Electric Discovery[1]

COUNCILLOR of State Ørsted, Knight of the Order of Dannebrog, has informed the Society of his explanation of Faraday's magneto-electric discovery. From the time when electromagnetism was discovered, it became a natural question whether it would not be equally possible to produce electricity by means of magnetism as magnetism by means of electricity. In spite of many efforts, however, no one had succeeded in doing this until the end of last year, when the English natural philosopher Faraday invented the proper means to do so, which is the following. In the proximity of a good conductor, north and south magnetism are made to combine or separate suddenly. Hereby an electric current arises in a direction which is perpendicular to the line of magnetic action. However, this effect is so weak that it can only be made perceptible when it is made to happen simultaneously at many points on a conductor. This is accomplished most easily if a metal wire is wound many times around a piece of iron, which can serve as the so-called armature of a magnet, and the ends of this wire are then connected to an electromagnetic multiplier. Every time this armature is either placed between the two poles of a powerful magnet or removed, the multiplier indicates an electric current, whose direction in the two opposite cases is also opposite. This is the basic experiment which is to be explained here. Ørsted believes that this naturally, and with some necessity, follows from the fundamental law which he has advanced for electromagnetic effects. According to this law, every electric current is surrounded by a magnetic circuit in such a way that the planes in which these occur are perpendicular to the axis of the electric current. Therefore, when an electric current is produced, a series of magnetic circuits are formed spontaneously. Conversely, these new observations now show us that it is possible to produce an electric current parallel to the axis of the conductor by producing a series of magnetic circuits around it. The direction of this electric current is precisely that which would be obtained as a consequence of the assumption that the magnetic forces which flow through the armature attract the opposite forces in the coiled conductor so that the south magnetism in the latter

[1][*Videnskabernes Selskabs Oversigter*, 1831–32, pp. 19–21. Report of the meeting held on July 6, 1832. KM II, pp. 484–85. Originally published in Danish.]

takes the same direction as the north magnetism in the former, and, conversely, the north magnetism in the latter follows the south magnetism in the former. It is particularly instructive that, according to Ampère's observations, an electric conductor, parallel to another conductor carrying a current in a given direction, remains in a state which shows it to carry a current in the opposite direction. This is easily understood when it is remembered that the magnetic circuit in a current-carrying conductor does not have a dispersive effect, like an ordinary magnet, but produces, in adjacent objects, the same magnetic direction as the one predominant in the active part itself. Hence it follows that the conductor which is placed parallel to the current-carrying one receives the same tangential magnetic direction on the near side, consequently the same direction on its right side as the one found on the left side of the other conductor, or the same direction on its lower side as is found on the upper part of the other, according to the position; in any event, opposite directions of rotation and, consequently, also opposite electric currents.

Therefore, it is obvious that this remarkable new chapter with which Faraday has enriched electromagnetism is most beautifully consistent with the already well known fundamental law.

64

Results of New Experiments on the Compressibility of Water[1]

ØRSTED has continued his experiments on the compressibility of water. Although the agreement of his experiments on this subject with the ones by foreign physicists leaves essentially nothing to be desired, there remain several points in this investigation which merit continued work. One of these is that water is less compressed, the warmer it is. We have some experiments on this subject by Canton from the middle of the last century, and these were confirmed by Ørsted's earlier experiences; it is only the connection between this peculiarity of water and other laws of nature which requires further investigation. This has now been carried out by Ørsted through a series of experiments, whose numerical results are of such a nature that they show no deviation if it is assumed that, for each atmosphere of pressure exerted, there is a temperature increase of $1/40°$ C. It goes without saying that this temperature increase disappears when the pressure ceases. Since the expansion of water for each degree of heat added is very different at different temperatures, it is easy to understand that the heat generated during the compression must give a certain appearance of irregularity to the magnitude of the compression.

At the temperature where water is most dense, it is least expanded by a small increase or decrease of the temperature. Therefore, the magnitude of the compression of water at this temperature is virtually unaffected by the influence of the heat generated. Concerning the temperature at which water is most dense, the determinations of various experimenters do deviate from each other, but they all agree on placing it a little above or below 4° C, and the most accurate investigations seem to fix it at 3.75° C. At this temperature experiments give the compression due to one atmosphere of pressure (equal to 28 French inches of mercury) as being equal to 46.77 millionths of the volume of the compressed water.

[1] A preliminary report sent by the author, from the *Oversigt over det Kongelige Danske Videnskabernes Selskabs Forhandlinger etc*. The results of earlier work by the author on this subject can be found in these annals, Vol. 9, p. 603 and Vol. 12, p. 158 and p. 513. [*Annalen der Physik und Chemie*, edited in Berlin by J. C. Poggendorf, Vol. 31, pp. 361–65. Leipzig 1834. The same subject can be found in *Bibliothèque universelle des sciences*, Vol. 55, pp. 421–25. Geneva 1834; *L'institut; journal des académies et sociétés scientifiques de la France et de l'étranger*, Vol. 2, pp. 219–20. Paris 1834; and *Videnskabernes Selskabs Oversigter* 1832–33, pp. 16–20. Copenhagen. KM II, pp. 399–402. Originally published in German.]

At 10° C, on the other hand, water expands by 84 millionths with an increase in temperature of one degree. Therefore, a warming of $^1/_{40}$° yields an expansion of 2 millionths, and the magnitude of the apparent compression is then only $46.77 - 2 = 44.77$ millionths. At 16° C, a warming of one degree causes an expansion of 160 millionths, so $^1/_{40}$° C will give 4 millionths, and consequently the apparent compression of water amounts to only 42.77. At 20° this decrease amounts to 5 millionths, and at 24° it is 6.

It is well known that water expands when cooled below 3.75° C. At 0°, a warming of $^1/_{40}$° C will result in a compression of 1.5 millionths so that the apparent compression here will amount to $46.77 + 1.5$, that is, more than 48 $^1/_4$ millionths. A long series of experiments, in which integer numbers (which simplify comprehension) rarely occur, consistently yielded numbers very close to the ones calculated.

If it may be assumed that glass experiences an increase in temperature during compression similar to that of water and loses it again when the pressure ceases, and if the linear expansion of glass at 1° C is set equal to 9 millionths so that the bulk expansion is equal to 27 millionths, then a warming of $^1/_{40}$° C produces an increase in the volume of the glass = 0.675 millionths. Therefore, this circumstance would make the apparent compression of water larger than its actual value by this much and by the same amount at all degrees. The true compression of water for one atmosphere of pressure would therefore amount to approximately 46.095 millionths. Incidentally, in accordance with the nature of these experiments, errors can easily occur at the level of 0.1 millionths so that it may be best to retain only the integer number, 46 millionths.

This view of the influence of heat in the above-mentioned experiments is further supported by the fact that the magnitude of the apparent compression of water in bottles or cylinders made of lead or tin is bigger than in bottles made of glass, and in relatively close proportion to the heat expansibility of these metals.

The compression of glass and metals might be thought to have a significant influence here. It has indeed been believed that the cubic compressibility of objects could be derived from the lengthening or shortening which a rod of the same material experiences when it is pulled or pressed by a weight; from such experiments it was concluded that the cubic compression of glass by the weight of one atmosphere amounted to 1.65 millionths. It was concluded from similar experiments that the cubic compression of lead is more than 30 millionths.

Some years ago Ørsted showed that the compression of water in bottles made of various metals leads to results which do not agree with such notions.[2] He has added a new class of experiments to these experiences. For this he uses a glass cylinder which is sealed at the lower end, and which at the open upper end accommodates a ground stopper, drilled through and equipped with a glass tube, like the glass bottle in which the compressibility of water is usually investigated. If the cylinder is to be used in these experiments, it must be very nearly filled with glass or metal whose volume has been accurately determined by being weighed in water, and the

---

[2] *Annalen*, Vol. 12, p. 513. ["On the Compression of Water in Vessels of Varying Compressibility" (Chapter 60, this volume).]

remaining volume must subsequently be filled with water whose weight is known. Then it is possible to measure the compressibility of solid objects by experiments similar to those on the compressibility of water. All of these experiments have given such a small compression for solid objects that it is difficult to distinguish the results from the inevitable errors.

It might well appear that these experiments contradict a mathematical proof since the famous mathematician Poisson has derived a formula for cubic compression from experiments on the linear expansion or compression of objects, from which quantities are obtained which in certain cases are 20 to 30 times larger than the ones resulting from Ørsted's experiments. However, this does not defy mathematics but only shows that the assumptions regarding the internal composition of bodies, from which the honoured French chemist started, cannot be entirely correct.

In these most recent experiments, Ørsted used an improved method for the measurement of the volume of air which is used as a measure of force. The device consists of a glass tube closed at the top, which is drawn out to form a narrower tube at a certain distance from the closed end, and whose open end has a narrow tube with a scale. The narrow part has a mark which the compressed air must reach each time. This provides greater accuracy in the observation as compared with a tube of equal over-all width; the lower tube with the scale indicates any change in temperature and any possible loss of air.

━━━━━━━

65

On the Compressibility of Water[1]

WERE I not withheld by official duties, I should certainly not omit so excellent an opportunity of renewing the very interesting and useful acquaintance I made during my last visit to England and Scotland, and of forming new ones with those distinguished scientific characters that I was not fortunate enough to meet with at that time, or such as have risen to eminence of late years. But though I must now forgo this advantage, I will not let this opportunity pass without giving the illustrious assembly some mark of my high esteem, and of my desire to keep up the friendly intercourse which I have maintained with the British philosophers since my acquaintance with your happy country.

You are, perhaps, aware that I have published several notices upon the compressibility of water, the first as early as 1818, and the first description of the improved method in 1822. Since that time I have still gone on improving my methods, and am now preparing a paper on the subject for the *Transactions of the Royal Society of Sciences at Copenhagen*. I will endeavour to give you a succinct account of my method and its results. It has been found that the apparatus for compressing water, a description of which I published in 1822, can give very accurate results; so that the results it has given in the hands of philosophers in different countries, have agreed more than might have been expected. Next to the accuracy of the measurements, however, one of the most important requisites of such an apparatus is, that the experiments be performed with the greatest celerity possible. When the experiment is protracted, the change of temperature produces great variations in the volume of the water, $^1/_{100}$ of the thermical measure (1° centigrade)[2] causing at high temperatures the volume of the water to vary more than the pressure of 3, 4, or even 5 atmospheres.

The improved apparatus is represented in the diagram fig. 1. Its principal parts

---

[1] From a letter to the Rev. William Whewell, dated Copenhagen, June 18, 1833. [*Report of the Third Meeting of the British Association for the Advancement of Science; held at Cambridge in 1833*, pp. 353–60. London 1834. KM II, pp. 402–11. Originally published in English.]

[2] The unit of thermical measurement is the distance between the freezing and the boiling point. I think that the most natural expression for the temperatures would be this unit and its fractions. Thus, the temperature 0.50 would be the same as 50° centigrade, 19.30 the same as 1930° centigrade. I will mark this metrical measure by Th. If this innovation should not please, I wish that it might be suppressed, and centigrade degrees put in the place, which is an easy change.

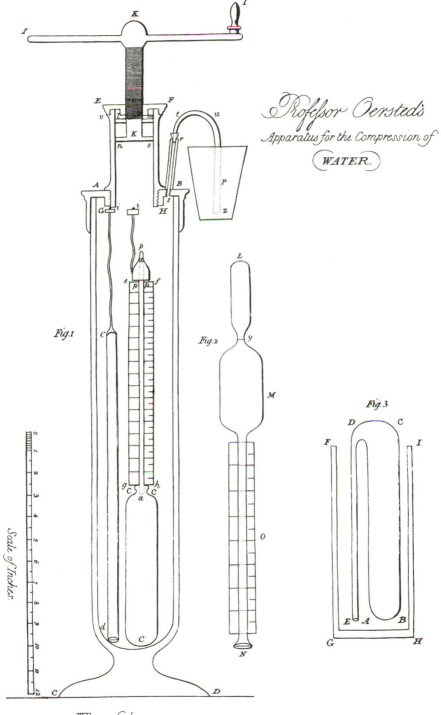

Professor Oersted's
Apparatus for the Compression of
WATER.

Fig.1

Fig.2

Fig.3

Scale of Inches.

W.Brown, sculp.

are the same as in the earlier; in each of them, however, some change is introduced. *ABCD* is a strong glass cylinder, having at the top a cylinder *EFGH*, containing a piston *lmno*, moved by a screw *KK*, as in the first apparatus; but the handle *II* is now arranged in such a manner that the screw can be turned without interruption; by this means the effect is accelerated, and subitaneous strokes avoided. The bottle *ccc*, with its capillary tube *aa*, is different from the earlier only so far, that the tube is not soldered to the bottle, but merely adjusted by grinding. This alteration is not necessary except when solid bodies are to be compressed. The scale *efgh* is divided into parts of $^1/_{40}$ inch. In order to exclude the water with which the large cylinder is filled, from communication with that of the bottle, the top of the tube *aa* is covered with a small diving-bell, or rather diving-cap, *ppp*, whose conical shape has the advantage of preventing the water from reaching the top of the tube *aa*, even when the air is compressed to a tenth or twelfth part of its first volume. Its margin is loaded with a ring of lead or brass. *cd* is a glass tube with proper divisions, containing air, whose compression measures the pressure; its inferior part is loaded with some lead or a ring of brass. *tuz* is a siphon; *P*, a vessel containing water; *ii* are two buoys of cork for lifting up the bottle and the glass tube *cd*; *sr* is a tube of brass, which can be stopped by a screw. In the beginning and at the conclusion of the experiment it serves to introduce water into the space *EFGH*, or to get it out again. Before the experiments the calibre of the two tubes must be exactly ascertained, and the relative capacities of the bottle and its capillary tube determined by the quantities of mercury they can admit. I have had some tubes in which $^1/_{40}$ of an inch (making one division) held only 2 millionths of the capacity of the bottle, in others they have held more, in some even as much as 7 millionths. The capacity of the bottle was not less than $1^1/_2$ pound, often 2 pounds of mercury. It is next filled with water, which must be boiled in the bottle in order to expel the air, which might be suspected of having a great influence in these experiments, though Canton has already observed that this is not the case. When the large cylinder is filled with water, the bottle is to be immersed in it. If the tube *aa* is full of water, a little of it must be expelled, which can be done by heating it gently with the hand, or better, by introducing a wire into it. As the bottle may be considered as a water thermometer, it is easy to ascertain whether it is in thermical equilibrium with the water in the cylinder. The air in the tube *cd* must likewise be brought to the same temperature with the water, before it is ultimately immersed. When the pumping cylinder shall be placed in its box, the piston must be at *GH*. If the large cylinder is full of water, part of it will be expelled through the siphon *tuz*. Now the piston is to be lifted up by means of the screw, whereby the pumping cylinder is filled with water. When this is done, the siphon is taken away, and the tube *rs* is stopped by a screw appertaining to it. The experiment is most conveniently performed by three persons; one turning the screw, the second observing the height of the water in the tube *aa*, and the third observing the volume of the air in the tube *cd*: the last writes down the numbers observed. Now, the point where the water stands in the tube of the bottle is to be noted. The descending piston having reduced the volume of the air in the tube *cd* to the point desired, the observer of it takes hold of the handle of the screw, and keeps the volume unchanged until the other observer has settled the point to which the water is brought down, and writes down the observation. When the piston is lifted up to its first place, the screw at r is to be opened, and the state of the water in the capillary tube again noted down. I commonly make ten or more such observations, one after another, which is performed in less than ten minutes when the operators are accustomed to work together. An example will illustrate the use of the observations.

The height of the mercury in the barometer, reduced to the freezing point, was at the same time 332.36 French lines. The volume of the air was in each experiment reduced to 5.264, or the pressure added to that of the atmosphere was 4.264 atmospheres. The pressure reduced to lines of mercury is thus, $332.36 \times 4.264 = 1417.18$; yet this reduction was not produced by the united pressure of the atmosphere and the piston alone, but was aided by a pressure of 40 lines of water,

Height of the water in the capillary tube

Before the pressure.	Mean.	When the pressure has ceased.	Length of descent in the capillary tube.	Point to which the pressure has driven the water down.
248.9	248.95	249.0	50.35	198.6
249.0	249.2	249.4	50.20	199.0
249.4	249.7	250.0	49.90	199.8
250.0	250.5	251.0	50.10	200.4
251.0	251.5	252.0	49.90	201.6
252.0	252.65	253.3	49.85	202.8
253.3	254.1	254.9	49.90	204.2
254.9	255.5	256.1	49.70	205.8
256.1	256.95	257.8	49.95	207.0
257.8	258.4	259.0	49.80	208.6
259.0	259.5	260.0	49.70	209.8

Mean of descents = 49.96

whose effect is equal to that of 2.94 lines of mercury, which is to be deducted, leaving then a pressure of 1414.24. Now, when a pressure of 1414.24 produces a descent of 49.96 parts, a pressure of 336 must produce a descent of nearly 11.87 parts. Each part makes in the instrument here employed 3.497 millionths of the whole capacity. 11.87×3.497 gives ultimately 41.51. The temperature of the water was at the beginning 0.20 Th., at the end 0.2025 Th., by the thermometer. The water stood 10.1 parts, about 35 millionths, higher at the end of the experiments than at the beginning. This gives 0.202 Th., which is as perfect an agreement as could be desired, the difference being only 0.0005 of the thermical measure, or 0.09 degree of the scale of Fahrenheit. During the last three months I have not made use of the tube cd for measuring the compression of air, but I have employed a glass tube LMN (fig. 2), whose shape is better seen in the diagram than it can be described. The capacity of the part above the line y, and that of the whole, are measured by weights of mercury. When the instrument is sunk in the water, the liquid mounts in the tube which has the scale O, whose parts are likewise measured by mercury. This has the double advantage of giving a more accurate measure, and of showing whether or not the volume of air has changed. In the series of experiments above mentioned this measure has been employed. By a considerable number of experiments, I have found that the compressibility of water is not so great in high temperatures as in lower. Canton had already obtained this result, but some doubts might remain, because his experiments were made by means much more troublesome to make use of, and at a time when all instruments were less perfect. Here, as well as in the whole research into the compressibility of water, the new experiments prove the great skill and acute judgement of this distinguished philosopher. My experiments are much more numerous than his, and have been extended to a greater range of temperatures. Their results[3] may be expressed by supposing that the pressure of one atmosphere equivalent to 336 French lines' height of mercury develops a heat 0.00025 Th. = 0.045° Fahr. In calculating this I have made use of the tables of Professor Stampfer at Vienna, who finds the highest contraction of water at 0.0375 Th., or 38.75° Fahr.[4] At this point the recession of water by the pressure of one atmosphere is 46.77 millionths. At 0.08125 Th. the volume of the water augments 71.75 millionths, having its temperature augmented 0.01 Th. The heat developed by the pressure thus augments its volume $0.00025 \times 71.75 / 0.01 = 1.79$ millionth, or the recession is $46.77 - 1.79 = 44.98$ millionths. Actual experiments have given it 44.89, or 0.09 millionths greater. The coincidence is often less perfect. At 0.1775 Th. the quantity calculated is 42.65, the quantity given by experiment 43.03, a difference of 0.38 millionth. The experiment mentioned above gave a recession = 41.51 at 0.20125 Th. (mean of the temperatures of the beginning and end of the series). The calculation gives 41.63, or a difference of 0.12 millionth. At 0.005 Th. the change of volume produced by one 0.01 is 60.5 millionths, but inversely, as the water at low temperature loses in volume by augmented heat; thus an addition is to be made equal to $60.5 / 0.01 \times 0.00025 = 1.5$. Now $46.77 + 1.5$ gives 48.27, experiment 48.02. At 0.019 the quantity calculated is 47.72, that given by experiment 47.97. I have not yet finished the tedious discussion of all the experiments, but as far as I have proceeded the agreement of the hypothesis with facts is satisfactory. Messrs. Colladon and Sturm have in the calculation of their experiments introduced a correction founded upon the supposition that the glass of the bottle in which the water is compressed should suffer a compression so great as to have an influence upon the results. Their supposition is, that the diminution of volume produced by a pressure on all sides can be calculated by the change of length which takes place in a rod during longitudinal traction or pression. Thus, a rod of glass, lengthened by a traction equal to the weight of the atmosphere as much as 1.1 millionth, should by an equal pression on all sides lose 3.3 millionths, or, according to a calculation by the illustrious Poisson, 1.65 millionth. As the mathematical calculation here is founded upon physical suppositions, it is not only allowable, but necessary, to try its results by experiment. Were the hypothesis of this calculation just,

---

[3] [The following results are briefly summarized in a letter from Ørsted to Chevreul read in the Academie des sciences on July 22, 1833, and printed in *Revue encyclopédique*, Vol. 59, p. 287. Paris 1833.]

[4] [See the table at the end of this article.]

the result would be, that most of the solids were more compressible than mercury. For this purpose I have procured cylinders of glass, of lead, and of tin, which filled the greater part of a cylinder, to which a stopple of glass, perforated by a capillary tube, was adjusted by grinding. I have not yet exactly discussed all the experiments on this subject, but the numbers obtained are such as to show that the results are widely different from those calculated after the supposition above mentioned. The quantity assigned by this calculation to the glass is very small indeed, yet the experiment gives it much less. Lead, which extends, according to Tredgold, 20.45 millionths by a weight equal to that of the atmosphere, and thus much more by the pressure on all sides, does not change one millionth. Tin is not more compressible. The inverse experiment is, perhaps, still more striking. I published it some years ago: however, as I have now repeated these experiments, and as they appear hitherto not to have satisfied philosophers, I shall here mention, that in all my experiments upon the subject, I have invariably found that the recession of the water in the capillary tube is about 1.5 millionth greater in bottles of lead or tin than in those of glass. Supposing the compressibility of the solid bodies to be so small that it cannot be observed in those experiments, yet the heat developed by the compression, feeble as it is, produces a small augmentation of the recession of the water in the capillary tube. If the dilatation of a rod of glass by 1 Th. is 0.0009, its cubical dilatation is 0.0027, and the dilatation by an increase of 0.00025 is 0.000000675, or nearly 7 ten millionths. The dilatation of lead is about 3 times greater, and the bottle containing it must get an increase of 0.00000225, which exceeds the former by more than 1.5 millionth. The dilatation of tin should give only one millionth more than glass, but it seems to give a little more, yet the quantity is not great. After all this, I think that the true compressibility of water is about 46.1 millionths, and that the apparent compressibility depends upon the effect of the heat developed by the compression, by which the liquid and the bottle are dilated.

My continued experiments have confirmed my earlier result, that the differences of volume in the compressed water are proportionate to the compressing power. I do not know if the method I have made use of to try the effects of high compression has been published in England. These experiments cannot be made in a cylinder of glass; one of metal is required. As, in this case, the opacity prevents direct observations being made, an index, nearly like that in Six's register-thermometer, is placed in the capillary tube of the bottle. This tube is dilated a little at the top, so as to form a minute funnel. Some drops of mercury are poured into it, which being pressed, pushes the index forward; thus the recession may be seen when the bottle is taken out of the large cylinder. The compression of the air is measured in another way: a bent tube, of the form shown in fig. 3, is fixed in a glass vessel *FGHI* containing mercury, and exposed to the pressure together with the bottle. The pressure of the piston upon the water in the cylinder is communicated to the mercury, and pushes it into the wide part of the tube, as far as the resistance of the air will permit. The weight of the mercury driven into the wide part *ABCD*, together with that which has filled *DE*, and which may be computed, compared with the weight of mercury which the whole tube can admit, gives the volume of the air compressed. By this kind of experiment I have found that the decrease of volume produced by pressure preserves the same proportion to the pressing power as far as the pressure of 65

atmospheres, and probably much further; but how far, I have not hitherto been able
to try, my apparatus not having resisted a greater pressure.

I have thus given you a short abstract of my researches into the compressibility
of water. They may be considered as a continuation of those of Canton. I should
feel much flattered if they should obtain the approbation of the philosophers of the
country where the first good experiments upon the subject have been made.

Part of Stampfer's Table

Temperatures, centigrade	Volumes of the water	Differences
−3	1.000373	—
2	1.000269	104
1	1.000182	87
0	1.000113	69
+1	1.000061	52
2	1.000025	36
3	1.000005	20
3.75	1.000000	5
4	1.000001	1
5	1.000012	11
6	1.000038	26
7	1.000079	41
8	1.000135	56
9	1.000205	70
10	1.000289	84
11	1.000387	98
12	1.000497	110
13	1.000620	123
14	1.000757	137
15	1.000906	149
16	1.001066	160
17	1.001239	173
18	1.001422	183
19	1.001617	195
20	1.001822	205
21	1.002039	217
22	1.002265	226
23	1.002502	237
24	1.002749	247
25	1.003005	256
26	1.003271	266
27	1.003545	274
28	1.003828	283
29	1.004119	291
30	1.004418	299

66

New Experiments on the Effect of the Electrical Circuit[1]

COUNCILLOR of State Ørsted, Knight Commander of the Order of Dannebrog, has presented to the Society some new experiments on the effect of the electrical circuit. As is well-known, Faraday, in his famous series of papers on electricity and magnetism, has reported a number of experiments in which the magnetic effects of the electrical circuit were very closely related to the oxidations which occurred in it. It cannot be denied that this relation really exists in the experiments which the English philosopher has reported, but there are other experiments which have the completely opposite result. Berzelius had demonstrated earlier that a voltaic apparatus of zinc, nitre solution, nitric acid, copper—zinc, nitre solution, nitric acid, copper, and thus repeated several times, gives the same direction to the electric current as if there had been only one liquid between the zinc and the copper although in one case it is the copper which oxidizes, in the other the zinc. Although experiments made with all the insight and precision customary for Berzelius are sufficient to demonstrate that a general law cannot be deduced from Faraday's experiments, it still seemed useful to extend Berzelius's experiments. Ø. started with a repetition although he had already shown them frequently in his lectures. For this he used a glass tube bent in the shape of a Latin U, whose lower part was filled with sand, while one of the two legs of the tube was filled with nitre solution, the other with nitric acid. The liquids were connected by joined narrow strips of zinc and copper in such a way that the copper was placed in the nitric acid, the zinc in the nitre solution. The experiments took place either with 6 or with 8 tubes. Both the electromagnetic and the chemical effects revealed that the electric current went from the zinc to the liquid and from the liquid to the copper. However, the copper dissolved quite visibly in the nitric acid, but the zinc was changed very little, almost imperceptibly. Although weighing was not necessary, it was done to be certain of this, and it was found that the copper strips had lost many hundred times more weight than the zinc. In one experiment the copper had lost 3.92 grams, the zinc only 0.005 gram.

[1] [*Videnskabernes Selskabs Oversigter*, 1835–36, pp. 26–28. Report of the meeting held on November 20, 1835. KM II, pp. 490–91. Originally published in Danish.]

So far the experiments are merely a repetition of those made by Berzelius. However, it seemed worth while investigating whether the effect of the acid on the metal might produce its own electric current. He therefore connected the liquids with a strap of a single metal, sometimes zinc, sometimes copper. Here, too, the electric current went from the oxidizing acid to the metal and not the other way, as might have been expected if the oxidation was assumed to be the cause of the electric current. The chemical effect, which could only be observed by means of iodized paper, was very weak here but suggested the same direction of the electric current as the multiplier.

In order to make the experiments faster and thereby have the opportunity to give them greater variation, the experiment was performed with single circuits. Each of the multiplier wires was connected to one of the metal strips which were to be used, and each of these strips was dipped into one of the two liquids contained in the bent tube. Here, too, the current proceeded from the oxidizing liquid to the metal when one strip was dipped in the nitric acid, the other in the nitre solution; and this happened both when the one put into the acid was copper and the one in the solution zinc, or the opposite, or when both were copper or zinc or platina. The same result was obtained with two thick iron wires. Neither did it make any difference in the result when, instead of the nitre solution, another saline solution was used, e.g., of sodium chloride or sodium sulphate. *Aqua regia*, chlorine water, or chloric acid substituted for the nitric acid made no difference in the result either. The same experiments were also repeated with dilute sulphuric acid and with hydrochloric acid instead of the nitric acid, and on the whole with the same outcome, but the effects here are much weaker and present more complications.

67

On a New Electrometer[1]

THIS instrument, which was demonstrated to the Royal Society of Sciences in Copenhagen, is shown in fig. 1 at half size.

aa is a thin, annealed brass wire which acts as the indicator, *bbb* a bow made of very thin iron wire which must have the slightest magnetism, *cccccc* is a brass tube which ends in a bow, *ee* a pin around which one end of a cocoon silk thread is wound; this thread supports the indicator.

dddd is a glass tube in which the brass tube with the bowed end has been fastened by means of rubber cement. The cement should not fill the entire length of the tube so that the path through the insulator will be all the longer.

gg is a microscope with vertical threads which can be moved up and down along the rod *ii* by means of a suitable clamp. At *h* there is a peg about which *hi*, and with it also the microscope, can be rotated; *ll* is an indicator which moves across a graduated arc indicated by *kk* during the rotation of the microscope.

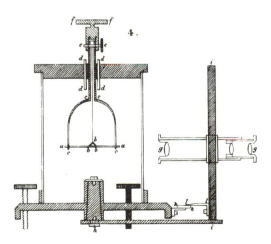

[1] [*Annalen der Physik und Chemie*, ed. by J. C. Poggendorff, Vol. 53, pp. 612–13. Leipzig 1841. The same topic can be found in *Forhandlinger ved de skandinaviske Naturforskeres andet Møde i Kjøbenhavn 1840*, p. 213. Copenhagen 1841.—*Videnskabernes Selskabs Oversigter*, 1840. p. 24. Copenhagen. KM II, pp. 411–13. Originally published in German.]

The microscope, which was taken from another instrument, is more complex than necessary. In most cases it is superfluous, indeed, people with sharp eyesight can do without it altogether. I have not yet used it in actual measurements.

The glass cylinder in which the electrometer is suspended, the wooden lid, and the base with the set screws need no explanation.

I have had the instrument drawn as it was when it was completed. However, it still shows a number of signs of the first experiment where different arrangements were previously made. Work on a new one has already begun. However, the great usefulness of this instrument for delicate measurements has induced me to publish this drawing since anyone can easily make the necessary improvements.

In order to elucidate the use of the instrument, we add the following as reported by the author in the transactions of the Royal Society:[2]   (P).

The balance beam is suspended in a glass cylinder, with a metal bow through its lid, isolated from it by rubber cement and the glass tube, the ends of which come into contact with the ends of the balance beam in such a way that one touches the right and the other the left side. Therefore, when the metal bow receives electricity, this is simultaneously transmitted to the balance beam and produces a rotation. When the magnetic aligning force (of the iron bow) is so weak that it is hardly noticeable, this electrometer shows an extraordinary sensitivity. In order to detect very weak electric effects, one first transmits sufficient electricity to rotate the balance beam by several degrees. When approached, a body which has the same kind of electricity produces a very significant enhancement of the deviation. The electricity which insulated zinc and copper plates show after contact and separation becomes very perceptible in this way, without the help of a condenser.

[2][See the preceding footnote.]

68

A New Device for the Measurement
of Capillarity[1]

THE experimental investigation of the capillary effect has so far been restricted within very narrow limits since tubes or plates of glass had to be used almost exclusively, and yet it would be very important to examine these effects in opaque objects, viz., metals, as well. An apparatus, which is depicted in fig. 1 at $^1/_8$ of its actual size, was constructed in order to remove these restrictions.

aaaa, *bbbb*, *cccc* are glass tubes which are connected with each other. The upper end of *aaaa* carries a copper ring which becomes thicker at the top, and its broad rim is ground flat. Several perforated plates, ground flat at the bottom, like *ll*, fit on it. *LL* (fig. 1a) depicts the cross-section of such a plate in actual size.

If the cover is made of metal like the ring, the tightness of the connection can be achieved with mercury, in other cases with grease.

bbbb is the comparison tube, where the height of the liquid column indicates the magnitude of the capillary effect in the covered tube.

cccc is the piston tube, in which a glass cylinder, fused at the top and at the bottom, can be pushed down or lifted up, whereby the level of the liquid in the two other tubes can be influenced.

gg is a vertical scale, divided in millimeters, from which a slide *ee*, equipped with a vernier, can move up and down.

This slide carries two arms *f* and *f'*, which can be

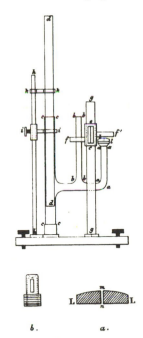

Fig. 1

[1][*Annalen der Physik und Chemie*, ed. by J. C. Poggendorf, Vol. 53, pp. 614–16. Leipzig 1841. The same subject may be found in *Annales de Chimie*, Vol. 4, pp. 379–81. Paris 1842.—Erdmann's *Journal für praktische Chemie*, Vol. 23, pp. 472–75. Leipzig 1841.—*Videnskabernes Selskabs Oversigter 1840*, pp. 22–24. Copenhagen. KM II, pp. 413–15. Originally published in German.]

rotated in the horizontal direction. The upper edge of f and the lower of f' are aligned.

At the beginning of an experiment with a given cover, the height of the uppermost point, where the mouth of a capillary is located, is measured. To this end, the arm f' is drawn across it, and care is taken that it touches the uppermost point. After the measurement it is again turned sideways.

The thickness of the cover is measured with the spherometer in order to determine the exact length of the capillary passing through it.

The width of the capillaries is determined with a fitted metal wire, which is measured afterwards.

The height of the liquid column supported by the capillary force is measured by the distance between the lower opening of the capillary and the surface of the liquid in the comparison tube *bbbb*.

f serves to transmit the level of the liquid in the comparison tube to the scale.

hh is a vertical rod which, through a slide *ii* with a ring and a screw, holds the tube *cccc* and, with it, the entire glass apparatus.

kk is a device with a ring, through which the cylinder which serves as a piston passes and is held in any given position by friction.

The foot with three adjusting screws requires no description.

When the liquid in the tubes *aaaa*, *bbbb* is driven so far upwards that it begins to move above the upper opening of the capillary plate, the pressure can still be increased for some time before the liquid overflows. The magnitude of the pressure at which the overflow begins can be determined from the level of the liquid in the comparison tube, and in this manner the capillary force pressing down is found.

When these experiments are performed with mercury, the cover could easily be lifted by the pressure from below, so a counter-pressure must be applied. The device shown in fig. 1b, which is put over the cover to push it down, has served me well. It is a wooden cylinder, weighted at the bottom with a leaden ring, cut lengthwise on two opposite sides in order to permit the observation of the falling of the liquid in the tube. On the top there is a hemispherical depression, which has an opening of a couple of lines in diameter in the middle.

ADDENDUM: As stated in the *Oversigt* mentioned in the previous article,[2] the author has not yet had the opportunity to perform very many experiments according to this method, but he has performed a sufficiently large number with water and with mercury and there used openings of quite different diameters, as well as covers of various materials, viz., metals and glass. Identical openings in covers of amalgamated copper and of glass raised the water to the same height. Mercury was raised approximately $3/4$ times as high as water by perforated covers of amalgamated copper, from which it follows that the capillary force supports a weight of more than ten times as much mercury as water.—

[2] [Cf. "On a New Electrometer" (Chapter 67, this volume).]

69

An Investigation of Light with Regard to the Physics of Beauty[1]

PROFESSOR[2] Ørsted delivered the first part of a lecture on an investigation of light with regard to the physics of beauty. He first drew attention to the pleasure in light, of which we become especially conscious at the transition from a long darkness to light, but which we also enjoy, although with less pronounced awareness, under many lighting conditions in nature and in art. In order to demonstrate the connection between this feeling and the essence of things, he drew attention to the fundamental laws of light. No matter how much people may disagree about the nature of light, they will agree that it is an activity whose speed exceeds all our everyday ideas of motion. However, its motion is not only external, from one place to another; internally, countless opposite activities alternate incessantly so that millions of these changes take place in each millionth of a second. Where there is light, this hidden activity is to be found, but the activity of light is not entirely excluded from any part of space although at innumerable points it is too weak to be detected by the eye. Consequently, there is no complete darkness. However, to the same extent that the light-creating agencies cease to function in any part of space, the activity taking place within also ceases more and more. Although darkness in the abstract is imagined as a mere absence of light, this absence, in reality, is not an empty, inert state but an internal *lying-down-to-rest*, an internal *dying-away*. In accordance with its innermost nature, then, light is an image of life, darkness of death.

---

[1] [*Videnskabernes Selskabs Oversigter*, 1842, pp. 97–99. Report of the meeting held on December 23, 1842. The same topic is treated extensively and in a popular form in Chapter Two of *To Capitler af det Skjønnes Naturlære*, printed in Copenhagen 1843, published in 1845. German translation by Zeise: *Naturlehre des Schönen*, Hamburg 1845. Chapter One deals with the relationship between tones and a sense of beauty. Under the title *Bidrag til det Skjønnes Naturlære*, these two chapters and two didactic dialogues are included in *Samlede og Efterladte Skrifter* by H. C. Ørsted, Vol. 3, p. 65 ff. Copenhagen 1851. KM II, pp. 506–8. Originally published in Danish.]

[2] [In the transactions of the Royal Society of Sciences from c. 1840 onwards, Ørsted was referred to by the title of *Conferentsraad*, which is a Danish title given to a high functionary. Since this title has no English equivalent, we use his academic title.]

It is true that this internal nature of light does not immediately enter our consciousness, but according to its nature it works everywhere as a way of arousing not only us but all external nature. Through its immense speed, through its ability to connect all, even the most distant, parts of the universe with each other, and by including us in this intercourse and revealing the external world to us to an extent which, without comparison, exceeds what can be conveyed to us through the other senses, it makes us feel like participants in all of existence, whereas we feel more and more excluded from the unity of life and, as it were, isolated, the more darkness prevails.

With regard to colours, Ø. finds himself on the whole in agreement with what Goethe has said about their effects on our feelings, but he partly seeks to find the cause of this in the nature of the colours and partly shows that more can be deduced from this. He points out that the most lively colours, which can, however, also have a disturbing effect in large surfaces, especially red and orange, are produced by the largest but slowest oscillations of the ether. On the other hand, the colours which are furthest from arousing such feelings, and in particular blue and violet, are produced by the least extensive but fastest oscillations of the ether. Yellow and green are produced by oscillations between the two extremes.

He then pointed out that the colours which painters call warm, red, orange, and yellow, are accompanied in sunlight by the greatest heat, and that the ones which are declared to be colder, blue and violet, are also accompanied by the least heat in sunlight.

He likewise pointed out that, among the coloured rays of sunlight, the yellow and orange have the greatest illuminating power, and after that the green and the liveliest red, whereas the blue and especially the violet have the smallest illuminating power. Both in frequency of oscillation, in heat, and in illuminating power green takes the middle road.

Goethe had already stated as a truth based on natural law that the colours which complement each other to form white light are harmonious, which later discoveries have re-confirmed although this was not necessary.

Though white and black cannot be called colours in the same sense as red, green, etc., they can still be called so in a broader sense of the word. The effect and significance of white can easily be understood from its light intensity, its freedom from all real colour contrasts, and the resulting internal harmony. That black, representing darkness, must be the colour of mourning and death needs no further explanation, but it should be noted that the lustre of the surface, whereby a stronger impression of light is produced, diminishes the melancholy impression of black.

70

Continued Reflections on Light with Regard to the Physics of Beauty[1]

PROFESSOR Ørsted continued his reflections on light with regard to the physics of beauty. The object of this continuation was the conditions under which certain figures are produced according to the natural laws of light. Among these is the rainbow. On this occasion, it was not his intention to repeat the well-known theory of the rainbow; it was sufficient to regard this as a settled issue. The shape of the rainbow is a necessary consequence of the mathematical laws of nature. In the same act of nature which forms this arc, the coloured rays contained in white light are also separated, and a harmony of colours is developed as the entire colour content of light stands before us, both in its distinctness, caused by the distribution in space, and in its unity. Therefore it is perceived no less as unity than as plurality. This, however, does not exhaust the contents of the total perception. The contrast between the dark sheet of rain and the bright light also arouses the strange pleasure of light. The rainbow comprises a whole world of thought, in which light, in its struggle against darkness, unfolds its beauty with true, triumphal splendour. Naturally, all of this applies fully only in so far as no other conditions, such as an intervening haze, weakens the purity of the impression.

Through the effect of polarized light on crystals or bodies in which external influence has created a particular distribution of the internal tension, figures are produced which are both satisfying in their own intellectual impression and distinguished by a distribution of often rich colours, which always constitute a harmony. This combination of forms and colours is not accidental either but constitutes a conceptual unity and thus provides us with a most significant multiplication of the examples of beautiful creations as a consequence of the rational laws which reveal themselves in nature. In some of these figures, we find colour contrasts as striking as those in many Pompeian paintings and also a naturally determined harmony of colours.

The interaction (interference) of rays of light also offers some beautiful formations of shape and colour. Among these, special mention was made of the strangely

---

[1][*Videnskabernes Selskabs Oversigter*, 1843, pp. 6–7. Report of the meeting held on February 10, 1843. KM II, pp. 509–10. Originally published in Danish.]

arranged luminous spots which appear in Frauenhofer's experiments when a thin sheet with three holes at the corners of an equilateral triangle is placed in front of a telescope. Even the distribution of the spots, reproduced without colours, satisfies the eye with a rich and curiously arranged variety, but the beauty is further enhanced by the colours which inevitably accompany this phenomenon.

71

Development of the Theory of Lustre[1]

PROFESSOR Ørsted informed the Society of a development in the theory of lustre.

He began with the statement that what he had to communicate was not essentially new but only a compilation of well-known truths, but as this had not been made elsewhere, as far as he had been able to ascertain, he did not consider it inappropriate to offer the insights which he has acquired.

In order better to attract attention to what matters here, he started with the apparent contradiction in the combination of blackness and lustre since as little light as possible is reflected according to the former but as much light as possible according to the latter. In order to resolve this difficulty, it is necessary to distinguish clearly between the two ways in which surfaces reflect the light they receive from a luminous point.

Every such point is the source of a series of ether waves. Each straight line which can be drawn from this point perpendicular to the wave fronts defines a direction of action and is called a ray of light. As the light which departs from a point and falls on a surface occupies a conical space, such a limited portion of a set[2] of light waves, such an externally limited but internally infinite collection of rays of light, is called a *cone of light* or a *cone of rays*. When the cone of rays falls on a polished and plane surface, it is reflected in such a way that all its rays maintain their relative position so that the eye receives this reflected light exactly as if it came from the luminous point with the only difference, which is not relevant here, that the eye, which knows nothing about the change of directions, now imagines the point to be just as far behind the polished surface as it is in front of it in reality. Also, when the polished surface is not plane but has one of certain regular shapes, such as that of the sphere, the hyperbola, the parabola, the cone, or the cylinder, the rays are reflected in such a way that those which reach the eye continue to belong to a common cone of rays although the shape of this is more or less changed. It can be said that the cones of rays are here reflected undistorted though not unchanged. As is

---

[1] [*Videnskabernes Selskabs Oversigter*, 1843, pp. 47–51. Report of the meeting held on May 5, 1843. Also to be found in Poggendorff's *Annalen der Physik und Chemie*, Vol. 60, pp. 49–55. Leipzig 1843. KM II, 510–14. Originally published in Danish.]

[2] In science, the word *set*, in the meaning used here, is not given as wide and as useful an application as it allows. It meets a frequently felt need for an expression to describe a multiplicity of parts which belong together.

well known, surfaces which reflect the cones of rays without distortion show us images of the objects or are mirrors. If a surface consists of many very small but separate polished parts, each thin cone of rays which is reflected from each of these parts will remain undistorted. Some consideration leads to the recognition that, while each of these small, polished parts is a mirror, the surface as a whole can no longer be regarded as one even though its polished nature cannot be denied on these grounds. From each of the polished parts the reflection follows the laws of the mirror, and therefore one can denote this reflection, which is usually called *regular*, by the term *mirror-like*, which is closer to the conception. However, in so far as the rays on the surface are reflected from the receiving parts in all directions, the original cones of rays are *dissolved*. In so far as this happens—it never happens completely—this reflection has rightly been called *dispersive*, but it could more appropriately be called *dissolving*, whereby the unthoughtful could be prevented from mistaking this for the vastly different dispersive reflection caused by convex mirrors.

The light which reaches our eye by mirror reflection gives us no conception of the nature of the reflecting parts but only of the presence of light and, when the parts of the surface have a suitable relative position, of the light-emitting point. From dissolving reflection, however, we receive knowledge of the reflecting parts themselves. It also seems that, with this reflection, some of the received rays of light disappear or are *absorbed*, as it is said. Often more of one kind of ray, that is, light waves with a certain frequency of oscillation, are absorbed than of the others, whereby the reflected rays acquire a certain dominant colour effect.

If a surface existed which only performed this mirror-like reflection, it would not be seen, in the literal sense of the word, although its presence would certainly be noticed by its reflecting effect. As for sight proper, it would behave as if it were black. However, for any surface however perfectly reflecting, light undergoes some dissolving reflection with the result that it also becomes an object of actual sight. On the other hand, there exists no surface at which the received rays of light undergo only dissolving reflection, but we call surfaces shiny or lustreless according as one or the other of the two kinds of reflection produces the most perceptible impression on us.

It is highly worthy of our attention that the same changes which enhance lustre diminish the dissolving reflection and vice versa. This is seen in the polishing of a dull surface or the frosting of a shiny one. In the first instance, the visibility of the individual parts decreases as the lustre reaches greater perfection, and on some surfaces, e.g., steel, the characteristic colour disappears to such an extent that one is inclined to call the surface black. In the second instance, frosting, the particular character of the material recovers its lost influence. A survey of the following old and new experiences will make these matters more familiar. Powdered iron, as obtained through the treatment of ferric oxide with hydrogen gas, is black, but if it is compressed, it acquires the well-known lustre and colour of iron. The same procedure can essentially be used on all metals which can be produced in a powdered state. In their pulverized state, most of them are black or grey, like platina, silver, lead, and arsenic, while others are coloured, like gold or copper, but with compression or an appropriate juxtaposition of their parts, each of them acquires its well-known metallic lustre and colour. It would be incorrect to think that this situation applies only to the metallic state. If a piece of red ferric oxide is polished, the lustre will give it a steel-grey tinge, and its redness will be inversely proportional to the perfection of its polish. The same is true of cinnabar, except that, in the polished state, it has a colour which is more like that of lead or mercury, although with a much less lively lustre. It is well known that indigo, when polished, acquires a coppery lustre. The same treatment gives Prussian blue its own dark blue lustre. Related experiments can be made by scattering some dye on paper, placing it on a hard surface, and rubbing it with a piece of hard polished glass, porcelain, steel, or the like, and the colour will always be seen to decrease at the same rate as the lustre increases. In such experiments similar results can be obtained with the use of painted surfaces whose binder does not produce any significant lustre.

In experiments with all these polished surfaces, reflections from them are found to produce no colour. It is true that a tinge from the reflecting surface is often seen in the image, but this originates from the dissolving reflection, which always mixes with the mirror reflection. The more the mirror is in shade but the object well lit, the less the reflection assumes this additional tinge. The reflection of each coloured object, then, appears almost completely with its own hue even though the body whose surface causes the reflection has a completely different colour due to dissolving reflection.

We know that the light which is reflected from a dull surface into shadow is always coloured. If the same surface is polished, the uncoloured light which falls into the shade acquires a dominance which is in proportion to the polishing. In a darkened room it also appears quite clearly that mirror-reflected light does not have the colour of the object although it readily mixes with some of the light emitted by mirror[3] reflection.

As the surfaces of all liquids are shiny, they should show the same effect, and that is also what is actually found. Even though there are older experiments regarding this, Ø. has confirmed this in experiments with strongly coloured liquids, e.g., dark blue ink or a deep red litmus solution in black vessels. Coloured glass shows the same effect.

The light emitted in dissolving reflection is not polarized, but the light reflected by lustre is. Although this effect has not been in doubt, it is still of some interest to see it confirmed by new experiments which clearly illustrate the issue. This can be done very easily by polishing half of a dull surface and then letting it reflect light at a fairly acute angle on to a mirror which has been placed at the polarizing angle, and which can be duly turned, or into a polariscope. It will then be seen that the light reflected from the shining half is strongly polarized, the other is not. If Savart's polariscope is used, some rather weak lines can be seen on the lustreless surface, while strongly coloured lines are seen on the shiny surface. The colour which

---

[3] [dissolving.]

the surface would have in its dull state is not perceived to have any influence on the colours of the lines in the polarized light, which again confirms the conviction that mirror reflection is achromatic.

Through the polishing of substances which have not been tested before, it will be possible to determine the polarizing angle of many substances and thus deduce their refractive index when other means cannot be used.

All this, then, shows that the light emitted in mirror reflection is not part of the impression of colour which we receive from bodies, but that this impression is due solely to dissolving reflection. It is also seen that whiteness and blackness, which in daily life are called colours, have in common with the proper colours that they are due to dissolving reflection.

72

An Account of Experiments
on the Heat Generated by the
Compression of Water[1]

PROFESSOR H. C. Ørsted gave a preliminary report of a series of experiments on the heat generated by the compression of water. In 1833 he had informed the Society that, on the basis of his experiments on the compression of water at different temperatures, he had to conclude that approximately $1/40°$ C was generated in the water for each atmosphere of pressure applied. As this depended only on conclusions from experiments which, it is true, did not easily permit another interpretation but still did not prove the issue through direct measurement of the heat generated, he decided to make experiments on this. Of all means for this kind of measurement, none was so useful as the thermoelectric circuit in conjunction with the multiplier. A compound thermoelectric circuit, such as the one which is used in Melloni's experiments, was placed in an opening cut in the bottom of the glass cylinder in which the compression takes place. After the considerable difficulties which arise when the insertion has to be made water-tight had been overcome, the operation of the thermoelectric calorimeter was compared with good, calibrated mercury thermometers, whereby it was ascertained that, through the applied thermoelectric effect, each $1/100°$ C produced a deviation of the multiplier of one degree, except for a tiny fraction. In a series of experiments, the heat generated by the pressure of several atmospheres was now tested and the results calculated, using the method of least squares. So far the final result is $1/49.2°$ C. In spite of both the internal consistency of the experiments and the agreement between results which have been obtained in two completely different ways, he still intends to add a few series of experiments as soon as the colder season begins. Only when these have been concluded, will the investigation of the compression of water achieve such a degree of perfection that the magnitude for each degree of temperature can be given. He recounted that the above-mentioned series of experiments had been made according

---

[1] [*Videnskabernes Selskabs Oversigter*, 1845, p. 117. Report of the meeting held on July 11. KM II, pp. 527–29. Originally published in Danish.]

to his plan and with his co-operation by Mr. Colding, Bachelor of Engineering,[2] already favourably known by the Society, who has also made all the calculations and thus shown both his expert knowledge and his conscientious precision.

[The draft of the above report to the Danish Society of Sciences is found among the posthumous papers of H. C. Ørsted (The Royal Library, Parcel 17) and gives somewhat more detailed information about the procedure in the measurement of temperatures. It is printed below:]

As it is a general law that heat is generated by the compression of bodies, it was also imperative to attempt to discover the magnitude of the heat generation which takes place when water is compressed. Already in 1821 I made several experiments on this, using Breguet's thermometer with the application of a pressure of 6 atmospheres, but this gave no indication of any generation of heat. Later Colladon and Sturm made similar experiments with a pressure of 36 atmospheres with the same result. They even obtained an indication which corresponded to a decrease of heat, but which they attributed to a difference in the compression of the metals of which the calorimetric spiral of the instrument was compounded. Through continued experiments, however, I found another means to determine the quantity sought here in vain. I found this means in the difference which is seen in the compression of water at different temperatures, for this compression is apparently smaller, the higher the temperature of the water is. This result of the experiments, incredible at first glance, which had previously been obtained by Canton and fully confirmed by me, offered a means to determine the heat generated. In order to understand this, one must remember that water has its smallest volume at 4°, and that, by being heated above this temperature, it does not expand uniformly but has a greater increase in volume for each new degree that is added; for instance, at the transition from 4° to 5° the increase is only 83 ten-millionths, at the transition from 10° to 11° 896 ten–millionths, and at the transition from 15° to 16° 1515 such parts, etc. Thus, the heat generated at a certain pressure must increase the volume of the water more, the higher the temperature, and consequently decrease the apparent compression of the water. At temperatures below 4° a contraction takes place at each decrease[3] of the temperature, and this increases the apparent magnitude of the compression. I found that an experiment which I could have made with the aids that were at my disposal at that time leads to the conclusion that there is a heat generation of approximately 0.025° for each atmosphere of pressure. I informed the Society of this investigation in 1833. Since that time I have made several improvements in the instruments for my experiments, whereby I shall be able to arrive at a somewhat more reliable determination, but I am already certain that the number arrived at will only be corrected by thousandths of a degree. As it is very difficult to make accurate experiments on this subject when the water has a different temperature from that of the surrounding room, a suitably complete series of such experiments is only possible when made over several seasons, and therefore I must postpone further reports on this line of inquiry. However, even when this work has reached the greatest perfection, a direct measurement would still be desirable. The old instruments were insufficient for this, but after all the improvements which thermoelectric instruments have effected on calorimetry, the undertaking was no longer impossible. However, it presented many difficulties which were only solved by a series of preliminary experiments which it would be too tedious to have explained here. The device which was finally adopted was the following. A

---

[2]Now acting paving inspector.   [3][increase.]

hole was cut in the bottom of the cylinder in which the instruments for measuring the compression were placed, and a thermoelectric apparatus of antimony and zinc, exactly like the one used in Melloni's experiment, was placed in the opening. This thermoelectric apparatus was cemented to a tube to the middle of which was soldered a broad ring, which served both to strengthen the bottom of the glass cylinder and to give the cement which was to fasten the tube to the bottom of the glass cylinder an expansibility that could duly ensure its water-tightness. Naturally, the position of the thermoelectric apparatus was such that the two conductors emerged from its lower part. The glass cylinder was placed on a wooden base, which can best be compared with an inverted box whose up-turned bottom has a hole for the lower part of the thermoelectric apparatus and its conductors, which consist of strong copper wires. These conductors, which first descend vertically, are bent $^{1}/_{2}$ inch below the apparatus so that the greater part of them has a horizontal direction, whereby it is possible to fasten the ends securely in one of the vertical sides of the base. In the same side there is also a hole, which permits the insertion of a horizontal thermometer. In order to keep a very constant temperature around the lower part of the thermoelectric apparatus, the space beneath the box is filled with dry sand. When experiments were made, the two conductors were connected to a very fine electromagnetic multiplier, made by Gourson [?].

The first task was to determine what temperatures corresponded to the effects on the multiplier. Everything must be left undisturbed for several hours in order to equilibrate the temperature. Then a small but carefully determined increase or decrease in the temperature of the upper surface of the thermoelectric apparatus must be produced and the effect on the multiplier observed. The change in temperature is caused by a cylinder of sheet iron, containing several inches of mercury, whose temperature is measured by a very accurate thermometer. This cylinder is equipped with a movable handle, like a bucket, and is lowered into the glass cylinder so that it comes to rest on the top of the thermoelectric apparatus, which is just touched by the bottom of the bucket. The indications on the multiplier thus produced were written down, and the experiment was repeated many times, but this had to be done at long intervals so that the equilibrium of the temperature in the whole apparatus could be restored.

All the various series of experiments prompted by this, both on the meaning of the deviation of the multiplier and on the generation of heat by the compression of water, as well as the calculations, were performed, by pre-arrangement, by Mr. Colding, Bachelor of Engineering, already favourably known to the Society, who has shown in this the most unflagging diligence and the most conscientious precision.

73

Letter, on the Deviation of
Falling Bodies from the Perpendicular,
to Sir John Herschel, Bart.[1]

THE first experiments of merit upon this subject were made last century, I think in 1793, by Professor Guglielmini. He found in a great church an opportunity to make bodies fall from a height of 231 feet. As the earth rotates from west to east, each point in or upon her describes an arc proportional to its distance from the axis, and therefore the falling body has from the beginning of the fall a greater tendency towards east than the point of the surface which is perpendicularly below it; thus it must strike a point lying somewhat easterly from the perpendicular. Still, the difference is so small, that great heights are necessary for giving only a deviation of some tenth-parts of an inch. The experiments of Guglielmini gave indeed such a deviation; but at the same time they gave a deviation to the south, which was not in accordance with the mathematical calculations. De la Place objected to these experiments, that the author had not immediately verified his perpendicular, but only some months afterwards. In the beginning of this century, Dr. Benzenberg undertook new experiments at Hamburg from a height of about 240 feet. The book in which he describes his experiments, contains in an appendix researches and illustrations upon the subject from Gauss and Olbers, to which several abstracts of older researches are added. The paper of Gauss is ill-printed, and therefore difficult to read; but the result is, that the experiments of Benzenberg should give a deviation of 3.95 French lines. The mean of his experiments gave 3.99; but they gave a still greater deviation to the south. Though the experiments here quoted seem to be satisfactory in point of the eastern deviation, I cannot consider them to be so in truth; for it is but right to state that these experiments have considerable discrepancies among themselves, and that their mean therefore cannot be of great value. In some other experiments made afterwards in a deep pit, Dr. Benzenberg obtained only the easterly deviation; but they seem not to deserve more confidence. Greater faith is to

[1] [*Report of the Sixteenth Meeting of the British Association for the Advancement of Science*, 1846. Notices, pp. 2–3. London 1847. The same subject is dealt with in *Amtlicher Bericht über die 24ste Versammlung deutscher Naturforscher und Aerzte in Kiel im September 1846*, pp. 192–93. Kiel 1847. KM II, pp. 416–18. Originally published in English.]

be placed in the experiments tried by Professor Reich in a pit of 540 feet at Freiberg. Here the easterly deviation was also found in good agreement with the calculated result; but a considerable southern deviation was observed. I am not sure that I remember the numbers obtained; but I must state that they were means of experiments which differed much among themselves, though not in the same degree as those of Dr. Benzenberg. Professor Reich has published his researches, an abstract of which is to be found in Poggendorff's *Annalen der Physik*. After all this there can be no doubt that our knowledge upon this subject is imperfect, and that new experiments are to be desired; but these are so expensive, that it is not probable that they would be performed with all means necessary to their perfection without the concurrence of the British Association. I will here state the reasons which seem to recommend such an undertaking. 1. The art of measurement has made great progress in these later times, and is here exercised in great perfection. 2. All kinds of workmanship can be obtained here in the highest perfection. I think it would not be impossible to have an air-tight cylinder of some hundred feet high made for the purpose. This would indeed be expensive, but it would present the advantage that the experiments could be made in the vacuum and in different gases. 3. With these experiments others could be connected upon the celerity of the fall and the resistance opposed to it by the air and by gases. Professor Wheatstone's method for measuring the time would here be of great use. 4. If the southern deviation should be confirmed, experiments could be undertaken in order to discover in how far this could be effected by magnetism in motion. For this purpose balls of different metals might be tried. Very moveable magnetical needles, well-sheltered, but placed sufficiently near to the path of the falling bodies, would indicate magnetical effects induced in them.

[The same is contained in a letter from Ørsted to Sir John Herschel sent from Portswood House September 14, 1846 (The Royal Library, Copenhagen. Parcel No. 17), which has in addition the following unprinted page:] "I am far from thinking that the experiments in question give a certain result, or even a highly probable one, but I think that its probability is sufficient to call forth new experiments, and to utter some opinion upon the cause of the southern deviation. I think that the southern deviation originates from the action, which a magnetic pole exercices [sic] upon a body approaching it. Now in the experiments here mentioned the falling body approaches with increasing celerity to the northern pole and is therefore repelled from it. Though this could be considered as a consequence of well known experiments, I have tried some ones [once?] more, particularly related to the subject.

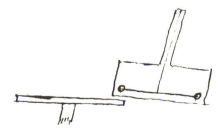

The diagramma here joined represents the very simple apparatus. It consists of two parts: a glasscase, containing a balance of torsion, in which two leaden bulbs are fastened at the ends of a wooden rod, and of a magnetic bar fixed on a perpendicular axis, connected with a machinery by which it can be turned round, so that the magnet moves in a plane perpendicular to that represented here by the paper, and very near to the bottom of the glass case. The motion of the magnet produces a considerable deviation of the horizontal rod from its original situation, towards which it returns through oscillations, when the magnet ceases to be moved."

On the Changes which Mercury sometimes suffers in Glass Vessels hermetically Sealed[1]

IT HAS been frequently noticed that mercury inclosed in glass tubes, even when those tubes were hermetically sealed, undergoes a remarkable change. It first becomes covered by a thin film of a yellow colour, which adheres to the glass, and becomes eventually nearly black. This has been attributed to oxidation, but the oxidation which would arise from the exceedingly small quantity of atmospheric air which could be contained within the bulbs exhibited by Professor Ørsted was too small to account for the formation of such a quantity of dark and yellow powder as many of them exhibited. Professor Ørsted referred the change on the mercury to the action of that metal on the glass of which the bulb was formed. It appears that sulphate of soda is frequently employed in the manufacture of glass, and it is thought that a sulphuret of mercury is formed by the decomposition of the glass itself. This is not however satisfactorily proved, and the subject has only been brought forward that attention might be directed to a subject which appeared to involve some remarkable conditions.

[1] [*Report of the Sixteenth Meeting of the British Association for the Advancement of Science*, 1846. Notices, p. 37. London 1847. The same subject is dealt with in *Videnskabernes Selskabs Oversigter*, 1845, pp. 11–12. Copenhagen. KM II, p. 418. Originally published in English.]

———

On Faraday's Diamagnetic Experiments[1]

———

PROFESSOR Ørsted demonstrated the highly extraordinary experiments whereby Faraday has discovered the effect which he calls diamagnetism and the change which a magnetic field can produce in certain transparent bodies, according to which they are made to rotate the plane of polarization of transmitted, previously polarized light. It is now well known that Faraday uses the term *diamagnetic* to describe those bodies which are repelled by both magnetic poles, whereas the ones which we call *magnetic* are attracted by both if they have not previously been given a fixed polarization. These two different classes of bodies are also given characteristic directions by the action of a magnet when they have decidedly greater length than width, and when they are duly suspended in such a manner that they can easily rotate horizontally. Magnetic bodies then turn in such a way that they point towards the two magnetic poles, which we call assuming the magnetic direction. Diamagnetic bodies, on the other hand, orient themselves perpendicular to the magnetic direction, which can be explained by the fact that both extremities are repelled by the magnetic poles.

Among all the bodies investigated so far, none is so highly suited to undergo the change by the magnet whereby the polarization plane of the transmitted light is changed as a special lead borate glass, which is certainly not available commercially, but of which Faraday possessed some pieces from an earlier chemical investigation finished many years ago. After this, flint glass is the most convenient to show this effect. A cylinder with polished, parallel bases is made of clear flint glass. Previously polarized light is sent through this parallel to the axis and then on to a polarizing crystal which, when rotated to a certain position, does not allow light of a given polarization free passage.

Now, the observer first places this crystal in such a way that the polarized light reaches his eye but then turns it until he perceives virtually no light. The magnetic force is now allowed to influence the flint glass cylinder, and light reaches the eye. While the effect lasts, the crystal can be turned further so that the light disappears. If this is done, light will be received as soon as the magnetic effect ceases.

This is a brief summary of these important discoveries. However, it was not Ø's

---

[1] [*Videnskabernes Selskabs Oversigter*, 1847, pp. 47–49. Report of the meeting held on April 23, 1847. KM II, pp. 550–52. Originally published in Danish.]

primary purpose to report these, which have been explained both in Faraday's papers and in many journals, but he thought that it would be of interest to the members of the Society to see the relevant experiments. These require considerable magnetic forces. These effects can be produced to an extremely small degree by very powerful steel magnets, but in order to see them clearly and unequivocally, and especially the effects on the polarization of light, the most powerful electromagnets are necessary. A quite powerful electromagnet has recently been added to the collection of physical instruments at the university, and there may be more to say about this on another occasion, but here it suffices to say that it is made of excellent soft iron, which easily changes the magnetic poles which the earth produces in it when its position is changed appropriately. It weighs 220 pounds and is shaped like a large Latin U, whose height is $2^1/_2$ feet, and whose arms are 8 inches apart and measure $3^1/_2$ inches in diameter. It is covered by varnished shirting and surrounded by a spiral made of slightly alloyed copper cast in two hollow cylinders. These cylinders have been cut into spirals, and in order to prevent the windings from touching each other, oiled pieces of cardboard have been inserted between them. The length of the copper strip which forms the spiral is 900 inches, when imagined to be straightened out, and its average area is $^1/_6$ square inch. 30 Bunsen cells, joined into one, were used to produce the electromagnetic effect applied. Its effects are extremely powerful, but, as yet, it has not been calibrated although arrangements have been made to do so.

Faraday's experiments can be made with sufficient certitude by means of this electromagnet. Some new experiments have also been initiated for the further investigation of this matter. To the extent that they lead to additional enlightenment, they will be published in due course.

76

Experiments on the Carrying Capacity of the
Large Electromagnet of
the Polytechnic School[1]

AT THE meeting of the Society on April 23, Professor Ørsted had demonstrated the large electromagnet of the Polytechnic School, by means of which he had performed Faraday's famous experiments on diamagnetism and on the change which many bodies show with regard to their ability to polarize light. At this meeting he reported some new experiments on the carrying capacity of this electromagnet. The yoke, which is fastened to a base on wheels, was duly secured to the floor so that the forces used to attempt to tear the armature away could not lift it. The first experiments on this were made in some good-sized workshops where large weighing devices were available, and where every obliging assistance was given, but where all the means for such experiments were not to be found in one place. Therefore the Royal Artillery Corps was approached, which most obligingly granted access to a weighing device in the armoury. This is both very advantageously placed for such experiments, and it will weigh up to 12,000 pounds. Moreover, every possible assistance was given during the experiments. The very first experiments revealed that the size of the armature has to be considerable if the greatest possible attraction between this and the yoke is to be obtained. The first armature used was a thick iron plate weighing $18^1/_2$ pounds. With the electrical current from a Bunsen cell, it was attracted by the yoke with a force of 475 pounds. The armature was connected to two other iron plates of half the length but the same thickness. With 2 galvanic cells, combined into one, this composite armature gave an attraction of 1000 pounds, which (according to other experiments) would give over 700 pounds for 1 cell. Direct experiments on the effect of 1 cell on this composite armature were not recorded. Later, a 62-pound iron bow, which would have measured approximately $22^1/_2$ inches in a straight line, was chosen as the armature. By means of this bow, a Bunsen cell had an average effect of 1425 pounds. This size

[1] [*Videnskabernes Selskabs Oversigter*, 1847, pp. 99–102. Report of the meeting held on December 17, 1847. Also to be found in *Forhandlinger ved de skandinaviske Naturforskeres femte Møde*, 1847, pp. 82–87. Copenhagen 1849. KM II, pp. 552–55. Originally published in Danish.]

of the armature seemed satisfactory, but one dare not consider this issue settled. The investigation of this must await future experiments.

For the moment it should be mentioned that all the experiments described here are not only of such a nature that they do not allow accurate determinations, but that they are very far from this goal. The uneven exactitude of contact in the various experiments, the ease with which the armature glides when the pull is slightly oblique, the irregularities which often occur in a galvanic apparatus even when it seems relatively steady, produce differences in the results which would require many more experiments to eliminate than it has so far been possible to perform.

In a series of experiments in which 16 apparently [similar] Bunsen cells were tested, two yielded an attraction of 1860 pounds, one 1680, another 1580, and so on downwards to the one which had the smallest effect, 1120 pounds. The sum of the individual effects of all these cells totalled 22,800 pounds, and thus the average effect of one cell was 1425 pounds. Of course, continued experiments could easily have provided greater similarity between these cells, but this goal had to be set aside for the moment.

In experiments on the combined effect of several cells it was discovered, as could have been expected, that the greatest effect was not obtained by combining them into a multi-element electric generator but by having several form one bigger cell. This is the kind of arrangement, then, which is mentioned in the description of the following experiments. The effect of two or more combined cells was now compared with the sum of their individual effects. According to the average of the experiments, it was found that

2	cells gave	0.72	of the sum of their effects
3	—	0.48	—
4	—	0.44	—
8	—	0.26	—
16	—	0.125	—

No more than the first two decimals have been given as the deviations between the various experiments were so considerable that even the second decimal cannot be regarded as quite accurate. Only in the last instance has an exception been made because the third decimal was so close to $^1/_2$ of the one immediately preceding. All 16 cells had an attraction of 2860 pounds, which is only a little more than twice the average 1425 of the effects of a single one.

It would, however, give a completely false impression of the matter if it were assumed that the magnetic attraction was increased so little with the multiplication of the galvanic effect under all conditions. When there is no direct contact, the situation is entirely different. In one series of experiments two sheets of cardboard were placed between the yoke and the armature, and after having been compressed by the effects of the applied attractions, they formed an intermediate layer of 0.6 lines; in a second series, a 1–line thick piece of wood was used. The following table shows the results of these experiments after averaging:

Total effect of single elements in armature contact	Effect with armature distance of 0.6 line	Fraction of the sum of their effects
1387	247	0.178
2600	420	0.161
3840	593	0.154
5208	727	0.140
10300	1000	0.097
22860	1340	0.059

Total effect of single elements in armature contact	Effect with armature distance of 1 line	Fraction of the sum of their effects
1373	70	0.051
2600	127	0.049
3840	167	0.043*
5208	257	0.049
10300	400	0.039
22800	660	0.029

It is quite likely that the entry marked * in the last column owes its large deviation from the rest of the column to some error which will only be discovered in future experiments. It is likely to be much too small, whereas the one below is presumably a little too large, but regardless of how imperfect and incomplete these experiments are, it was still possible to draw several important conclusions.

1) No complete knowledge of the carrying power of the electromagnet can be obtained without knowledge of the size of the armature used.

2) The entire attractive force of the electromagnet cannot be determined by testing the force which occurs during contact between the armature and the yoke. In this kind of experiment it is tempting to assume that little is gained by increasing the magnitude of the galvanic effect, but

3) when the armature is placed at a perceptible distance from the yoke, it turns out that the effect increases very considerably with the increase of the quantity of electricity, however in decreasing order, which incidentally decreases less, the bigger the distance. At a distance of 0.6 line, the last entry is only 0.33 of the first, but it is 0.57 at a distance of 1 line.

It is obvious that the investigation reported here cannot be regarded as concluded, but so far its complicated character has not permitted continuation.

Professor Ørsted concluded by adding that Candidate Holten had not only assisted him in the execution of the experiments mentioned here but should be regarded as a scientific collaborator.

77

Abstract of a Series of Experiments
on Diamagnetism[1]

AT THE meeting of the Royal Society of Science in Copenhagen on June 30,[2] I presented the results of researches which I had made on diamagnetism, and I gave a report of the same in the transactions of the Society. During the recess of this society, I have continued my researches and obtained several new results. As the memoir which describes these will not appear for several months, I have decided to give an abstract of my results which can be communicated to my foreign friends.

My researches relate to the celebrated diamagnetic discoveries of M. Faraday and to the developments which these have received from some German philosophers.

In the experiments with his large electromagnet, M. Faraday encountered a class of bodies[3] which are repelled by both poles of the magnet. One or two examples of this repulsion had long been well known, but the researches of the illustrious English philosopher have given to this fact a generality and an importance which have made it the object of attention for all physicists. As early as 1778, M. A. Brugmanns had noticed that bismuth is repelled by both poles of the magnet. M. Becquerel the Elder again observed this repulsion as regards both bismuth and antimony. M. Faraday found that his large electromagnet produced this repulsion in virtually all bodies which it did not attract. At the same time he discovered that bodies which were longer than they were wide, when thus repelled under the influence of the electromagnet, assumed a position perpendicular to that which an attracted body would take under the same circumstances. This is the property which he called *diamagnetism*.

M. Reich of Freiberg, well known for his beautiful experiments on the deviation

---

[1] [*Annales de chimie et de physique*, 3rd series, Vol. 24, pp. 424–35. Paris 1848. Printed for the first time in Copenhagen on September 9, 1848.—The same contents are to be found in *The Philosophical Magazine*, Vol. 34, pp. 81–88. London 1849.—Silliman's *American Journal of Sciences and Arts*, Vol. 7, pp. 233–39. New Haven 1849.—Poggendorff's *Annalen der Physik*, Vol. 75, pp. 445–55. Leipzig 1848.—*Videnskabernes Selskabs Oversigter*, 1848, pp. 49–56, and 1849, pp. 2–9. Copenhagen. Cf. "Investigations on Diamagnetism" (Chapter 78, this volume) and "Further Investigations on Diamagnetism" (Chapter 79, this volume). KM II, pp. 419–27. Originally published in French.]

[2] [1848.]

[3] [Throughout this article, Ørsted uses *corps*, which has consistently been translated as "body" or "bodies" even though "substance[s]" was probably intended in some cases.]

in the descent of bodies which fall from a great height, applied to the discovery of diamagnetism the observation, neglected by other physicists, that the two poles of the magnet used together did not produce a repulsion in these bodies equal to the sum of the repulsions effected by each but equal to their difference. Consequently, their combined effect would be nil when their forces were equal. At the same time he made some experiments which seem to indicate that the pole repelling a diamagnetic body produces a magnetic force similar to its own in the adjoining parts of the diamagnetic body, and not the opposite force as is the case in attracted bodies. M. Wilhelm Weber confirmed M. Reich's idea through his learned researches and showed that diamagnetic bodies, through the influence of the electromagnet, acquire a transverse magnetism having two poles but distributed in such a way that each of them has the same sort of magnetism as the nearest pole of the electromagnet.

M. Poggendorff devised some very decisive experiments, which have the advantage of proving the new idea in an easy way, and M. Plucker[4] added to this another new experiment, which increased if not the certainty of the idea at least the facility of convincing oneself of it.

These are the labours which served as the starting point of my researches.

For my experiments I used the large electromagnet of the Polytechnic School in Copenhagen, in the form of a U and capable of supporting 1400 kilogrammes.[5]

However, it must be noted that it was not necessary to put all its power in action for these experiments, but we rarely used less than half of this power although the greater part of them could be performed with a much weaker power, even with one element. Each extremity of the electromagnet carries a horizontal piece of iron, which we shall call a pole piece. These pole pieces serve to give a horizontal direction to the effect of the electromagnet. The diamagnetic body is made to oscillate between the two perpendicular faces placed opposite each another. These faces we shall call pole faces. In all cases where I have not indicated exceptions, I have used rectangular pieces. At the beginning of my experiments I used cylindrical pieces, but this shape is less convenient for discovering all the circumstances which ought to be taken into consideration in these researches.

---

[4] [Plücker.]

[5] I here conform to the usual manner of indicating the force of a magnet although it is quite uncertain as has been proved by some experiments with this electromagnet, which I reported to the Royal Society at their session of December 17, 1847. In these experiments the weight which the electromagnet was capable of supporting was tested when its poles were armed with different masses of iron. Within certain limits, the carrying power increased nearly in proportion to the mass of the armature, but what merits our attention much more is that the force of the electromagnet, expressed in weight, does not have the same relation to the electromotive force of the galvanic apparatus when the armature is in contact with the electromagnet as when it is at a certain distance from it. In contact, the mean power of each galvanic element was 712.5 kilogrammes. However, two elements combined gave only 0.72 of the sum of the individual effects of the elements; three elements combined gave only 0.48, eight 0.26, sixteen 0.125, of the sum of the individual effects so that the effect of sixteen elements was only twice that of one element. At a distance of 1.33 millimetres, the effect of one element was only 0.178 of that of the same element when in contact, but the effect increased quite differently with the number of elements. Sixteen elements here gave four times the effect of one. At a distance of 2.225 millimetres, the effect of one element was only 0.051 of that produced in case of contact, but sixteen elements gave 9.4 times the effect of one. These researches, which demand much time, will be continued as soon as my other occupations permit.

As is well known, a diamagnetic needle suspended horizontally between the pole faces takes the so-called *equatorial* position, which is parallel to the pole faces, but if it is raised a little above the borders of the pole faces, it takes the direction perpendicular to the extension of the pole faces. This position is also axial, but we shall see from what follows that it is the perpendicularity to the pole faces which is relevant here. This phenomenon appears with remarkable promptness, which makes this experiment very convenient for many diamagnetic researches. When the needle is turned away from its position perpendicular to the faces, it oscillates back again. Its directive force gradually diminishes as it is raised farther above the pole pieces. This experiment has been made with many diamagnetic bodies, with bismuth, amber, mother-of-pearl, tortoise-shell, alabaster, quills, sulphur, pit coal, etc.

The change in direction observed in these experiments diminishes in proportion to the distance between the pole faces. At a distance of 17 millimetres, the effect was still marked, but it is much stronger at short distances. When the distance was reduced to the point where the diamagnetic body could not enter between the pole faces, the part of the effect which takes place above the pole faces, that is, the position perpendicular to the pole faces, showed itself with considerable force. When the diamagnetic needle is suspended above the upper border of one of the pole faces, it also takes the so-called *axial* position, perpendicular to this border, but with less force than when under the influence of both faces. When the position of the needle above the other borders of the pole piece is studied, it is found everywhere to take a position perpendicular to the border to whose influence it is exposed. In the cases where it is exposed to the action of both borders at the same time, it assumes an intermediate position. Above the border of a wedge of iron, placed with its base on one of the poles of the electromagnet, the needle also assumes a position perpendicular to this border. On a cylindrical pole piece, the needle located with its centre above the border of the pole face places itself perpendicular to this face, but when it is at some distance from the border, it turns and takes a position perpendicular to the line which can be drawn parallel to the axis in the highest part of the cylindrical surface. When a perforated cylinder is used as a pole piece, and the diamagnetic needle is made alternately to descend and rise parallel to the pole face, this needle is found to leave its position parallel to the pole faces and assume the so-called axial position as soon as it is placed opposite the perforations. For this experiment I used a needle of bismuth which was only 16 millimetres long. When two identical pole pieces are used, the same effect is obtained only much stronger.

When the diamagnetic needle is suspended between the pole faces, it has, in accordance with the experiments of the already-quoted German philosophers, magnetic poles in a transverse direction, distributed in such a manner that the magnetism of each side is of the same nature as that of the nearest pole of the electromagnet. The easiest method to confirm this is that used by Plucker, who introduces, between the pole faces and parallel to them, a small rod of iron separated from the faces by some non-magnetic body. Since the sides of this rod acquire an induced magnetism opposite that of the nearest face, but each side of the needle has the same magnetism as the face nearest to it, the needle, which is now held by two forces, oscillates with much greater speed than when under the influence of the pole faces alone. When the diamagnetic needle is raised above one of the pole pieces so as to change direction, its magnetic poles change places at the same time. In the beginning I was misled by several phenomena which seemed very complicated given the novelty of the research, but which now appear very simple since the law for them has been found. In the beginning I believed that the diamagnetic needle above the pole pieces had the magnetism opposite that of the neighbouring pole piece in each extremity, for the lower part of an iron rod, influenced by the piece, repelled the extremity of the needle which was above this piece. I found this effect not only when placing the repelling iron pole close to each side but also above and below. However, later experiments refuted the conclusion which I had drawn from the first experiments. I found that a piece of iron which is not particularly small acquires from the pole piece which acts on it a magnetic force of sufficient strength to repel the diamagnetic matter of the needle, in spite of the poles which it received through the influence exerted on it by the electromagnet. In order to discover the diamagnetic poles in this case, it is necessary to use very small pieces or plates of iron; generally they should weigh only two or three grammes. In order to manage them more easily, I attached them to plates of zinc or pieces of wood. By this means I finally succeeded in convincing myself that the lower part of the diamagnetic needle, suspended above a pole piece, has the same magnetism as that, and that its upper part has the opposite. In my experiments on this subject, I finally used a thin plate of iron, shaped like (and attached to a piece of wood. When this plate is placed on the pole piece, it has the same magnetism as the pole piece in its upper part and the opposite in the lower part. When the opening of this curve is opposite the needle, it attracts it, but when its upper part is below or its lower part above the needle, it repels it.

When the needle is suspended above one of the pole pieces in such a way that the extension of one of the perpendicular faces of this piece divides the needle in two, we find that the diamagnetic poles produced by the electromagnet extend beyond the part which projects over the upper surface of the piece. In experiments made with a needle of bismuth of 56 millimetres, this effect extends almost 14 millimetres.

When the needle was divided into two equal parts by the extension of the perpendicular faces, we found that the extremity of the needle farthest from the pole piece was without polarity.

When the electromagnet was equipped with two pole pieces placed at a distance of 48 millimetres, I found that the same needle had diamagnetic poles in all its parts. The half of the needle which was turned towards the north pole received boreal magnetism on its lower border and austral on its upper border. The other half of the needle had, through the influence of the south pole, the magnetism of this pole on the lower border and the boreal on the upper. There is, then, an opposition of magnetism in the two halves of each border taken separately and in each half between the two borders, the upper and the lower.

When the diamagnetic body is made to oscillate between the pole faces, we find that it makes its oscillations much more quickly the closer it is to one of the borders of this face. In an experiment in which the electromagnet was acted on by 16 of Bunsen's galvanic cells, and where the distance of the pole faces was 6 millimetres, a bismuth needle at an equal distance from the upper and lower borders of these faces made 25 oscillations in 30 seconds, but at the level of the borders, it made 100 oscillations in the same time. Above the pole pieces in an axial position, the needle made only 19 oscillations in the same time. These experiments were sufficiently repeated and varied to give the most perfect certainty to what has been stated here, but the research has not yet been carried far enough to deduce an exact numerical law.

When a horizontal needle of bismuth is suspended by a thread of cocoon silk at one end of a balance in such a way that the balance can be made to descend or rise, it turns out that the needle is much more strongly repelled when it is closer to one of the borders of the pole faces. Naturally, this repulsion causes the needle to rise when it is close to the upper borders and to descend when it is close to the lower borders. In the intermediate position it neither rises nor descends. When the needle is suspended above the pole pieces, and consequently in the direction perpendicular to the pole faces, it is still repelled, though much more feebly than when in the so-called equatorial position.

Hitherto we have only recognized diamagnetic effects in bodies which are repelled by both poles of the magnet. My experiments have shown that a similar effect can be produced in most bodies which are attracted by the two magnetic poles so that these bodies constitute a new species of diamagnetic bodies. These two classes of diamagnetic bodies can be distinguished by the terms *repulsive* and *attractive*.

It is well known that when a needle, made of a body which can be attracted by the magnet, but whose magnetism is not of the same nature as that of iron and nickel, is suspended between the two pole faces of the electromagnet, it takes the position which M. Faraday called axial, but if made to rise above the upper borders or descend beneath the lower borders of the pole face, it takes the so-called equatorial position. The bodies in which I have so far found this property are platina, palladium, iridium, titanium, an alloy of 0.825 tin, 0.0024 bismuth, and 0.108 iron, brass, argentan, charcoal, coke (raw pit coal belongs to the repulsive diamagnetic bodies), obsidian, natural carbonate of iron, attractable glass, Prussian blue, and solutions of iron.

In most of these bodies the magnetic poles which they acquire under the influence of the electromagnet disappear almost as soon as its influence ceases. However, they betray themselves when the poles of the electromagnet are suddenly changed, for then several of these bodies turn a half circle as a magnetic needle would. Others do not turn completely but oscillate, which indicates their tendency to change positions. However, there are some attractive diamagnetic bodies, such as a piece of iridium in my possession, charcoal, and coke, which retain the poles that they have acquired through induction for so much longer that they can be demonstrated through experiments with a compass. The experimental researches into the phenomena of these bodies are complicated by this duration of the polarity, but they will probably lead to the discovery of the relation which exists between magnetism and diamagnetism.

When a needle made of an attractive diamagnetic body is suspended above the upper or beneath the lower border of the pole piece, it takes a position parallel to this border. In this parallel position, which may be either perpendicular to the magnetic axis of the pole piece, or parallel, or have any other position which the shape of the pole piece allows, the distribution of the magnetic forces in the needle is transverse as in a repulsive diamagnetic body, but with the difference that its lower part has the magnetism opposite that of the pole piece and the upper part the same.

I have not succeeded in putting iron itself into a diamagnetic state. An iron wire of only $^1/_{10}$ of a millimetre in diameter still takes the axial direction both above the pole faces and between them, and that with a force which seems close to breaking the thread of cocoon silk. To vary this experiment, I placed a fragment, only 2 millimetres long, of the same iron wire in a quill, which is repulsive, but this arrangement showed the same effects as the iron alone. The same effect was also obtained when the fragment of iron wire was replaced by a very tiny iron filing, but when a piece of straw which had been moistened in an iron solution was used instead of the iron, this resulted in the diamagnetic effects of attractive bodies. Nickel gives the same results as iron. Thus iron and nickel must be called magnetic in a strict sense. Some other bodies may be of the same kind; I presume cobalt is.

Consequently, there is a decreasing magnetic progression which includes proper magnetic bodies, attractive diamagnetic bodies, and repulsive diamagnetic bodies. The magnetism of the last may be considered negative if the magnetism of iron and attractive diamagnetic bodies is considered positive.

As is the case with regard to repulsive magnetic bodies, the effect which the pole faces exert on the attractive ones is stronger when the body is placed closer to the upper or lower borders than to their intermediate parts. A 27-millimetre-long piece of attractable glass, suspended between pole faces which are 29 millimetres apart in such a way that the ends of this needle were not farther from the pole faces than one millimetre, was made to oscillate each time for 30 seconds. At an equal distance from the upper and lower borders it made only 4.5 oscillations in the 30 seconds, but level with the borders of the pole faces it made 19 oscillations.

When the pole faces are at this distance, the needle does not take the so-called equatorial position when it is suspended above their borders. At a distance of 4.5 millimetres it made 5.5 oscillations; at a distance of 13.5 millimetres it made only 2.5 oscillations. The pole faces were brought to a distance of 3 millimetres from each other. The needle, which now could not assume the axial position between the faces, still showed all its tendency to take this position, but when raised to a distance of 2 millimetres above the borders, it took the equatorial position and made 18 oscillations in 30 seconds. At a distance of $^3/_{10}$ of a millimetre it made 35 oscillations. At the shortest distance at which contact with the pole pieces could be avoided, it made 45 oscillations.

We see that diamagnetic bodies, repulsive as well as attractive, make more numerous oscillations in a position parallel to the pole faces than in a perpendicular position. However, it must be remarked, as has already been done in connection with another series of experiments reported here, that the numerical results do not yet have the accuracy necessary for calculating their laws.

I have lately made some experiments on the influence of heat on diamagnetic bodies. These experiments are not yet sufficiently numerous, but they show that some attractive diamagnetic bodies become repulsive at an elevated temperature. The only body which has shown this effect to a high degree is brass. Similar experiments on other bodies are not yet sufficiently decisive to be reported here.

78

Investigations on Diamagnetism[1]

PROFESSOR Ørsted gave a report of his investigations on diamagnetism and demonstrated the relevant principal experiments.

As is well known, diamagnetism was discovered by Faraday. During experiments with his large electromagnet, this famous physicist chanced upon a fact, several examples of which had admittedly been known before, but which his investigations have given a new and much broader significance. In 1778 Anton Brügmanns[2] had discovered that bismuth was repelled by both poles of a magnet, and Becquerel the Elder had rediscovered the same effect in 1827 and added that antimony also shows it. However, Faraday now discovered, by using his powerful electromagnet, that virtually all bodies[3] which are not attracted by the magnet are repelled. Furthermore, he found that the same bodies, when used in pieces which have greater length than width, orient themselves perpendicular to the line joining the magnet poles and not parallel to it, as attractable bodies usually do. He called this property diamagnetism.

To this, Reich in Freiberg added the observation, which had not occurred to his predecessors, that the two magnetic poles, which separately repel the diamagnetic body, far from producing increased repulsion when used together, produce either none at all when they are equally strong or only a repulsion which corresponds to the difference between the forces if they are unequal. In addition, he made it likely that a magnetic force had to be induced in the repelled body similar to that of the repulsive pole, contrary to what happens in attracted bodies. Wilh. Weber confirmed this idea with new and perspicacious investigations, which further revealed that diamagnetic bodies have a distribution of forces such that the magnetic force which dominates is the same as that of the nearest influencing magnetic pole. Poggendorff further confirmed this in well-devised experiments, to which Plücker added another instructive and useful, though not necessary, confirmation.

---

[1] [*Videnskabernes Selskabs Oversigter*, 1848, pp. 49–56. Report of the meeting held on June 30, 1848. KM II, pp. 568–74. Originally published in Danish.]

[2] [Brugmanns.]

[3] [Here and below, Ørsted uses *Legeme[r]*, which has consistently been translated as "body" or "bodies" even though "substance[s]" was probably intended in some places.]

This preliminary work had occasioned Ørsted to undertake new investigations. He found everything which had been said by the above-mentioned physicists regarding the transverse position and the distribution of magnetic forces in diamagnetic bodies to be completely valid when these bodies are suspended between the magnetic pole faces, but not when they are outside them.

In order to provide a simple survey of the various observations, it will be necessary to say a few words about the arrangement of the electromagnet. The large electromagnet belonging to the Polytechnic School (see *Oversigt*, 1847, p. 48 and pp. 100–102)[4] was used though not with its full potential power. Usually it was activated by from 8 to 16 Bunsen cells. The magnet is shaped like a large Latin U; its 0.785-metre-tall legs are 0.209 metre apart. For most of the experiments, pieces of iron were placed on the two terminal surfaces so that these could be brought farther or closer to each other or in contact. Thus the terminal surfaces of these pole pieces became pole faces during the experiments. In some experiments these pole pieces were flat pieces of iron, 0.09 metre wide and 0.026 metre thick; in many others they were cylinders which were connected with the magnet's own pole faces by means of a suitable intermediate link and were 0.047 metre in diameter.

Now imagine a bismuth rod, suspended between two pole faces by cocoon silk in such a way that it can turn horizontally. It will then adjust itself perpendicular to the magnetic direction, and if it is removed from this position, it will be made to oscillate, and then it will return to the same position, as already indicated. However, if the pole faces are relatively close to each other, the bismuth rod will assume the longitudinal position as soon as it is raised a little above the border of the pole faces by shortening the supporting cocoon silk and lose all orienting force only when it is approximately one decimetre above the pole pieces. The experiments had the same results when amber, mother-of-pearl, tortoise-shell, alabaster, etc. were used instead of bismuth.

In these experiments, the greatest distance between the pole faces was 17 millimetres, but this was not the greatest distance at which the experiment still had approximately the same result. At a smaller separation of, e.g., 3 millimetres the effect was very strong. Even smaller distances could be used to show the longitudinal position if there was no need first to see the transverse position between the pole faces. In that case, even the thickness of a map sheet, $^3/_{10}$ millimetre, was sufficient distance.

While the diamagnetic body was suspended between the pole faces, the magnetic poles of the opposite sides were investigated. This cannot be done so conveniently with ordinary magnets, whose poles are easily changed by the powerful electromagnet, as with pieces of soft iron in the parts of which closest to a magnetic pole the opposite activity is invariably induced. For the sake of brevity, iron which has thus been forced into opposition is called pole-induced, and as we know with certainty which kind of magnetic force it has, we can use it as a means to discover

---

[4][See this volume, "On Faraday's Diamagnetic Experiments" (Chapter 75) and "Experiments on the Carrying Capacity of the Large Electromagnet . . ." (Chapter 76).]

the nature of the magnetic force in another body. Through the attraction which the closest pole-induced iron parts had on the sides of the diamagnetic body, Ørsted saw a confirmation of the German physicists' assertion that the magnetic force in the parts closest to a magnetic pole is the same as that of the magnetic pole itself. When, on the other hand, the diamagnetic body was suspended above the border of the pole face and oriented itself parallel to the direction of the magnetic field, the repulsive effect of the closest pole-induced iron parts showed that the magnetic force was now distributed in the longitudinal direction of the rod, and that each end had a magnetic force which was the opposite of the closest magnetic pole.

When the diamagnetic rod is placed beneath the lower border of the pole faces, it assumes the same orientation and distribution of forces as above the border.

If the small diamagnetic rod was suspended above one pole piece while the other was completely removed, it still assumed the longitudinal position when its centre was suspended above or almost above the border of the pole face, but if it was moved farther in above the pole piece, it again adjusted itself in the transverse position.

From this we learn that the reason why there is no longitudinal position above the pole pieces when the pole faces are placed at a considerable distance from each other is simply that the two active places are too far from the centre of the suspended rod.

When the diamagnetic body suspended between two pole faces is caused to oscillate, the speed of the oscillations is found to be lowest between the central part of the pole faces but increasingly greater, the closer the body is to either the upper or the lower border of the pole face. The numerical results have not yet been determined with sufficient accuracy. With the use of 16 Bunsen cells and with a distance of 6 millimetres between the pole faces, a bismuth rod placed halfway between the upper and the lower borders made 25 oscillations in 30 seconds, but 100 at the upper border. The same bismuth rod, suspended as closely above the border of the pole faces as possible so that it arranged itself in the longitudinal direction, made only 19 oscillations in 30 seconds.

The experiments reported here show us a hitherto unknown property of diamagnetic bodies, which is that, at the transition from one relationship with the poles of the affecting magnet to another, the magnetic lines of force cross each other. This crossing of the lines of force is altogether a most remarkable phenomenon, and below it will prove to be the noblest characteristic of diamagnetism.

The reported experiments also show that this crossing of the lines of force takes place at the border of the pole faces, that is, at the line of intersection of the two planes, where one is parallel to the axis, the other perpendicular to it.

If one of the diamagnetic bodies is suspended horizontally by cocoon silk at one end of a lever which can be raised and lowered, it is possible to measure the repulsion of the magnet under different conditions. It is then found that the diamagnetic body is strongly repelled when it is suspended between the pole faces, and also when it is close to their borders. It is pushed upwards the closer it is to the upper border and downwards the closer it is to the lower. When it is equally far from both, the repulsions are in equilibrium. However, when the body is suspended above or below the pole borders in the longitudinal position, it is repelled though much more weakly.

Through these investigations, Ørsted was led to the discovery of a new class of diamagnetic bodies. It had so far been assumed that only those bodies which are repelled by the magnet are diamagnetic while those which are attracted are all magnetic in the same manner as iron, but it now appears that the bodies which are attracted extremely weakly by the magnet have such a fundamental similarity to the previously known diamagnetic bodies that they must be grouped with these, although in a new subdivision.

Naturally, all the diamagnetic bodies mentioned above belong to the older subdivision, and the effects reported are attributed only to them.

Between the pole faces, a suspended, weakly attractive body which is longer than it is wide positions itself parallel to the magnetic axis, but if it is raised above the border of the pole faces, it orients itself in the transverse position. These experiments have been made with various kinds of glass, with brass, and with platina. Since the meeting, he has found that the following bodies belong to the same series: Several iron solutions of very different strengths, bismuth electroplated with an almost immeasurably thin layer of iron, obsidian, blackband iron ore, palladium, iridium, German silver, charcoal (at least those tested), and coke. Burned coke belongs to the older class of diamagnetic bodies. Coke and iridium acquired magnetic poles of some duration and therefore often create complications.

Of course, the distribution of magnetic forces has some, although often inexpressibly short, duration in all diamagnetic bodies. Some turn whenever the electric current which provides the yoke with its magnetic power is reversed. However, others are thus made to oscillate, although often very weakly. The effect seems to last longer, the more attractable they are.

In the attracted diamagnetic bodies, we see the same crossing effect as the one which appeared in the repelled bodies, only in the opposite order, in that the attracted bodies orient themselves in the longitudinal position under the same conditions under which the repelled bodies orient themselves in the transverse position, and in the transverse position under the same conditions under which the latter assume the longitudinal position.

When attractive diamagnetic bodies are in the transverse position, the line of their magnetic force is also in the tranverse direction and not in the longitudinal direction. Initially, Ørsted believed that here, too, each side had the same kind of magnetic force as that of the nearest influencing magnetic pole. He had actually discovered that the nearest pole-induced iron parts attracted the side turned towards them, but this only occurs in so far as these, through their magnetic force, produce a new magnetic distribution which cancels the one produced by the electromagnet. The pole-induced iron parts were able to produce such an effect in the parts of the suspended rod which were very close to them. In continued experiments he found that each side of the transverse rod has the kind of magnetic power which is opposite that of the nearest influencing magnetic pole. When these attractive diamagnetic bodies are suspended between the pole faces and there have the longitudinal position, the distribution of their forces is also in the longitudinal direction. We then have a crossing of the lines of force in attracted as well as in repelled bodies. Thus, this property seems to be the fundamental nature of diamagnetism, but it is also obvious that the repulsive and the attractive diamagnetic bodies constitute two subdivisions of this class. It is possible to distinguish them either by calling such bodies, as has already been done, the repulsive and the attractive diamagnetic bodies, or by calling the ones in the former subdivision the *positive* diamagnetic and the ones in the latter the *negative* diamagnetic. For the negative diamagnetic bodies to orient themselves in the transverse position, then, the pole faces must be somewhat closer to each other than for the positive diamagnetic.

So far it has not been possible to place iron itself in a diamagnetic state. Extremely thin iron wires of $1/10$ millimetre in diameter are still attracted so strongly that the cocoon silk is greatly tightened thereby. For another experiment the barrel of a quill was used, which in itself is positively diamagnetic. In this was placed a piece of iron wire which was scarcely two millimetres long, but it still behaved like iron. Even when a tiny iron filing was placed in it, it still exhibited the property of iron. On the other hand, when a straw, which had been moistened in an iron solution, was placed in the quill, negative diamagnetic effects were clearly obtained.

Just as the positive diamagnetic bodies between the pole faces perform their transverse oscillations most quickly when they are suspended close to the edges, the negative diamagnetic bodies also perform their longitudinal oscillations most quickly near the edges of the pole faces. A 27-millimetre-long attractable glass rod, suspended between pole faces which were 29 millimetres apart so that each end of the rod was only one millimetre from the pole face, made $4^1/2$ oscillations in 30 seconds when its ends were just opposite the centre of the pole faces. At various distances above and below, it oscillated more quickly, and at the border of the pole faces it made 19 oscillations in 30 seconds. When the pole faces have this distance, the rod does not assume a transverse position above the border of the pole faces, but with an unchanged longitudinal position, it made $5^1/2$ oscillations at a distance of $4^1/2$ milimetres, 4 oscillations at a distance of 9 millimetres, $2^1/2$ oscillations at a distance of $13^1/2$ millimetres, all in 30 seconds. The pole faces were now placed at a distance of 3 millimetres. Of course, the glass rod could no longer adjust itself in the longitudinal position between them, but it still exhibited the most lively efforts to assume this position. On the other hand, above the border of the pole faces it made rapid transverse oscillations. At a distance of 2 millimetres above, it made 18 oscillations in 30 seconds, at the distance of a map sheet, 35, at the shortest distance possible without contact, 45 oscillations.

The oscillations of both the positive and the negative diamagnetic bodies are seen to be more rapid when they have the transverse position than when they have the longitudinal position. Incidentally, it must be noted that the numbers given here have not been obtained by means of such perfect standards of comparison that accurate calculations can be based on them.

The above-mentioned experiment with a positive diamagnetic body, suspended by cocoon silk at one end of a lever, was repeated with a negative diamagnetic one.

In the transverse position above the border of the pole faces, it was pulled downwards vigorously, below them upwards. What happens in other positions has not yet been investigated.

Ørsted concluded by observing that though he regards the experiments reported here only as the beginning of a far more extensive investigation, he did not want to postpone the announcement of the results so that others might contribute to the elucidation of this matter.

79

Further Investigations on Diamagnetism
and Their Results[1]

AT THE meetings of January 5th and 19th, Professor H. C. Ørsted, who, at the meeting of June 30th, had informed the Society of the investigations on diamagnetism which he had made up to that time but had since continued, now reported the results which he had later obtained. As his continued work had taken place during a period of time when the Society did not hold meetings, he had had a brief notice[2] about it printed in French for the benefit of scientists in foreign countries. However, it is not a Danish translation of this notice that he ventured to submit to the Society, but a new account, in which the diamagnetic effects discovered by him are seen in a clearer context, which is so much easier to do now that the most important relevant basic experiments are well known. Several new results, which he had obtained since the said announcement, were added to this. It is assumed known from his earlier experiments that there are two kinds of diamagnetic bodies, the *repulsive* and the *attractive*, and that the latter, although they orient themselves in the same position as an iron or nickel needle between two magnetic pole pieces, still distinguish themselves sufficiently from these proper magnetic bodies in that they turn and take up a position which is perpendicular to their previous position when raised above the pole pieces, whereas the proper magnetic bodies maintain the same direction in both cases. To designate the two classes of diamagnetic bodies, he had previously, in addition to *repulsive* and *attractive*, used the terms *positive* and *negative*. Now he uses only the former terms and excludes the latter, even though they are suitable in themselves, because it now appears that the proper magnetic and the diamagnetic bodies can be regarded as one long series which goes almost continuously from the magnetic bodies to the more attractive diamagnetic ones and on to the least attractive and finally from these to the more and more repulsive ones so that the series ends with the most repulsive ones. In such a series, all diamagnetism represents a contrast to magnetism, and the expressions *positive* and

---

[1] [*Videnskabernes Selskabs Skrifter*, 1849, pp. 2–9. Report of the meetings held on January 5 and 19, 1849. KM II, pp. 574–81. Originally published in Danish.]

[2] [Cf. "Abstract of a Series of Experiments on Diamagnetism" (Chapter 77, this volume).]

negative might then be applied to this relationship. Therefore, he now prefers to avoid them altogether.

According to the more complete investigations which are now available, the positions which diamagnetic bodies assume in the vicinity of a magnetic pole can be regarded as characteristically dependent on the borders of the terminal surfaces or pole pieces of the magnet, in such a way that the attractive diamagnetic bodies assume a position which is parallel to the border under the same conditions under which the repulsive ones position themselves perpendicular to it, and the former, on the other hand, orient themselves perpendicular to the border when the former[3] orient themselves in a position parallel to it. If the pole piece is bordered above and below by two parallel horizontal surfaces, and on the sides by vertical ones, an attractive diamagnetic body, when suspended close to one of the vertical surfaces, will assume a position which points towards this, that is, it crosses a line which is parallel to the horizontal borders. However, if it is suspended just above or below one of these borders, it will orient itself parallel to it. A repulsive body acts in the opposite way. It assumes a position parallel to the border, which, for the sake of brevity, is called the *border position*, when it is suspended close to one of the vertical surfaces and the transverse position when it is suspended above or below one of the horizontal borders. As the transverse position relative to the border may often be parallel to the magnetic axis, it is designated the border-crossing position in order to prevent any misunderstanding.

This law is valid regardless of the direction of the magnetic force. Diamagnetic bodies follow it whether the borders are parallel to the so-called axial direction or form any possible angle with it. When the diamagnetic body is suspended above the horizontal surface of a pole piece, and this surface is bounded by lines which form various angles, the body, when its centre is above a vertex, takes the position which it should take according to the well-known laws of composite motive forces. At a rather considerable distance from all the borders, it will show only a weak or no tendency to assume a definite position.

This also applies close to the magnet's own pole ends. The attractive diamagnetic body, suspended just above the border of one of the round pole faces of the magnet, assumes a position which is tangential to the round terminal surfaces but, when suspended below the border, a position which points towards the axis. Conversely, in the former case the repulsive body points towards the axis, in the latter it takes a tangential position.

Above a pole piece which ends in a horizontal wedge-shaped border, the attractive body assumes a position parallel to the border, that is, the border position, but a repulsive one takes the transverse position.

Above a horizontal cylindrical pole piece, with vertical terminal surfaces, it is true that the diamagnetic body assumes the previously described positions close to the vertical terminal surfaces and just above and below its borders, but if it is suspended just above the cylinder at some distance from the border of the terminal

---

[3] [latter.]

surface, it assumes the position that it would have taken if the highest line of the surface had been the border of a wedge-shaped pole piece. When a hollowed cylinder was used as a pole piece, the border of the bore hole behaved in the same way as other borders.

All this shows that the expressions *axial* and *equatorial* ought not to be used in the future to designate diamagnetic directions while the terms *border position* and *border-crossing position* will probably be found convenient to designate these phenomena.

Of course, two pole pieces with opposite forces which are placed immediately opposite one another produce in a heightened degree all the phenomena in which they mutually support each other's effects.

Through this intensified effect, it is also seen that a bismuth needle which is suspended between two such opposing pole faces, at a rather considerable distance from the borders, endeavours to assume the longitudinal position. The pole faces in the experiment were 25 millimetres high. The bismuth rod assumed the border position as long as its distance from either the upper or the lower pair of borders did not exceed 5 millimetres, whereas it turned to the transverse position in the intervening band of 15 millimetres. Attractive diamagnetic bodies show no corresponding change in similar experiments.

As regards the pole distribution which the magnet induces in diamagnetic bodies, Ø. had pointed out in his previous report that the pieces of iron used in the investigation ought to be small because, under the influence of the electromagnet, bigger pieces become such powerful magnets that, through their own power, they overwhelm the pole distribution they were intended to indicate. In the investigation of repulsive diamagnetic bodies, however, he had not exercised sufficient caution and had therefore not quite found the correct pole distribution in the cases where the diamagnetic needle is suspended above a magnetic surface. This has later been done with the use of very small pieces of iron of various shapes. From all these investigations, a rule now emerges which could very easily have been predicted if new and unknown effects could be understood with the same clarity as well-tried ones. The rule is simply that a diamagnetic needle which is suspended above a magnetic surface acquires the same pole distribution (working from bottom to top) as that which the perpendicular magnetic surface immediately opposite has in the horizontal direction. Consequently, an attractive diamagnetic needle which is suspended above a pole piece and therefore has the border position acquires, on its lower side, a magnetic force which is opposite that of the pole piece

but on its upper side the same as in the pole piece. If it is suspended above the border in such a way that part of its width is outside, or if it is suspended between the borders of opposite pole pieces, it also acquires a pole distribution across its width according to the same laws. Thus, a cross section, which is here shown enlarged, suspended above two pole pieces N and S, has the magnetic forces distributed in the manner indicated by the letters n and s. According to the same rule, a repulsive diamagnetic needle which has the transverse position has the same magnetic force as the pole piece at its bottom but the opposite on top in that portion of the needle which is suspended above the pole piece and a little beyond. If it is suspended above two opposite polepieces, the distribution is also determined by this so that a lengthwise vertical section must have the pole distribution indicated in the adjoining figure. Faraday had already discovered that bodies which are weakly attracted by the magnet are repelled when they are suspended in a more attractive fluid, but because of the state of knowledge at that time, and with so many new objects of investigation on his mind, he could not concern himself with finding the true facts of the case. Now we can, if not explain them in an exhaustive way, at least contribute something to the knowledge of the true facts. With this in mind, Ørsted has conducted a number of experiments which show, as could have been expected, that the least attractive diamagnetic bodies take the same position as the repulsive ones when they are suspended in an iron solution. He also found that the pole distribution in them became the same. In order to examine the pole distribution when the glass vessel containing the iron solution was placed above a pole piece, he used an iron wire in the shape of

whose lower horizontal part could easily be placed above or below a portion of the diamagnetic needle. As this horizontal part acquires a magnetic force opposite the one dominant in the pole piece, it reveals the pole distribution of the repulsive bodies by attracting the lower and repelling the upper side of the needle. When, on the other hand, the glass vessel containing the suspended needle was opposite the vertical surface of a pole piece, the iron wire probe was in the shape

whose shorter, vertical part could easily be placed as desired on either side of the needle, which was parallel to the vertical surface. Since this short vertical piece is always placed closer to the pole piece than the other parts of the iron wire, it acquires a magnetic force opposite that of the pole piece and attracts the side of the diamagnetic needle nearer to it but repels the farther.

These experiments show that the internal distribution of forces in diamagnetic bodies has the same dependence on the surroundings as the one Faraday has shown with regard to the external effects. Now it seems evident from the experiments that the pole distribution induced in the diamagnetic liquid plays a significant role in the pole distribution induced in an attractive diamagnetic body which has become repulsive; e.g., a pole piece in which the north magnetism is dominant attracts the south magnetism in the nearest parts of the iron solution and repels the north magnetism, which seems to settle at the boundary between the solution and the body suspended in it.

This leads to the thought that the state of repulsive diamagnetic bodies might in general depend on the state of its surface relative to the surroundings. The normal repulsiveness of these bodies, then, would depend on the air being more attractive than the bodies investigated. In order to examine this idea, Ø. conducted some experiments whose results admittedly were not favourable, but which should not yet be considered absolutely decisive. He suspended a bismuth rod by cocoon silk in a device made of glass tubes with brass couplings from which the air could be evacuated. Under the influence of the magnet the bismuth needle oscillated almost equally rapidly in ordinary air, in highly rarefied air, in hydrogen gas, and in carbonic acid gas. In this connection we may well regard the fact that the needle does not oscillate more rapidly in the space filled with rarefied air as already proving some influence from the surroundings, but the effect of air on the frequency of oscillation, especially in such a confined space, is such a complicated matter that it can hardly be used in the discussion of the present question. Ø. has decided to conduct other experiments with a view to its solution. If the idea is correct, the most weakly repulsive bodies should pass into the class of attractive ones in rarefied air.

However, another much more important objection can be raised against the idea put forward here. If this notion were correct, it must be expected that diamagnetic repulsion was determined by the surface. In order to investigate this, Ø. conducted some experiments whose results show that the diamagnetic effect permeates the entire mass, a result which is not favourable to this otherwise so reasonable idea either. For one of these experiments he chose two bismuth rods of equal size, 56.6 millimetres long, 1.9 millimetres wide, and 5 millimetres high. First he suspended each of them on edge between the pole pieces, whose separation was 12 millimetres, and counted the oscillations. One gave 64 oscillations in one minute, the other, which was made of pure bismuth, gave 68 oscillations in the same period. After this preliminary test, the two needles were to be placed in close contact with each other, but as this would create a source of error because the needle would come considerably closer to the pole faces than had been the case in the previous experiments, the pole faces were placed at a separation of 38 millimetres. Now one of the bismuth rods made only 13 complete oscillations in 123 seconds. The other, which should have made $12\,^4/_{17}$ oscillations in the same period, was not examined in this respect, but when they were joined, 122 seconds were required to perform 13 oscillations. A few seconds more should have been expected, but since a great difference would have appeared if the repulsion of the mass of twice the size had been a surface effect, no greater accuracy was sought. However, similar experiments were made with quills. The barrel of a quill, suspended between the pole pieces, whose separation was 8 millimetres, made $11^1/_2$ oscillations in 20 seconds. A piece of quill, from the part which bears the vane and slightly longer than the barrel, made 11 oscillations in the same time. The quill was now inserted into the barrel in such a way that it protruded a little at each end. Now the number of oscillations was $11^1/_4$. The counting had to be limited to 20 seconds because air resistance diminished the oscillations very quickly.

It was now possible to describe the positions which the diamagnetic needles assume relative to the pole pieces of the magnet with a somewhat closer connection to the laws of repulsion and attraction. The border position of attractive diamagnetic needles, when suspended above the borders of a pole piece, seems to arise from the fact that their lower side has a magnetic force opposite the one which prevails in the pole piece, and which is pre-eminently concentrated at the border. The border-crossing position of the repulsive ones seems to arise from the fact that their lower side has the same magnetic force as the pole piece and, above all, as its powerful border and is therefore repelled from all sides. When diamagnetic needles are under the influence of the vertical pole faces, attractive diamagnetic bodies orient themselves in the same direction as iron wire and according to the same force laws even though the pole distribution in them experiences far greater resistance than in iron wire. The repulsive ones, on the other hand, in which the pole distribution is established only over a very short distance each time, can only realize this distribution transversely. It is obvious that there are many questions about the true relationship between diamagnetic bodies which we could wish answered here, but which must be left for the future.

Among experiments which are still rather exceptional, it must be mentioned that a brass needle, which has the diamagnetism of attractive bodies, appears to have all the properties of repulsive ones when strongly heated.

SUBJECT INDEX

ELECTROMAGNETISM

HEAT

LIGHT

NAME INDEX